GW00702987

NUMERICAL METHODS IN THERMAL PROBLEMS

Vol. VIII, Part 1

NUMERICAL METHODS IN THERMAL PROBLEMS

Vol. VIII, Part 1

Edited by:
R. W. LEWIS
Institute of Numerical Methods in Engineering,
University College of Swansea, Wales.

Proceedings of the Eighth International Conference
held in Seansea, Wales on
July 12-16th, 1993.

PINERIDGE PRESS
Swansea, UK

First Published 1993 by
Pineridge Press Limited
54, Newton Road, Mumbles, Swansea, U.K.

ISBN 0-906674-80-8

British Library Cataloguing in Publication Data

Numerical Methods in Thermal Problems:
Proceedings of the Eighth International Conference

1. Heat -Transmission - Mathematics - Numerical Methods

I. LEWIS, R. W. (Roland, 1940-)
536.2001511

Printed and bound in Great Britain by
The Cromwell Press Ltd., Melksham, Wiltshire

PREFACE

This proceedings contains the papers presented at the Eighth International Conference on Numerical Methods in Thermal Problems held in Swansea during the period July 12th-16th, 1993. Over two hundred and eighty abstracts were submitted to the Organizing Committee and their high standard made it difficult to leave out so many but constraints of space and time demanded that almost a half be rejected. The organizers would, however, like to thank the authors concerned for sending in their manuscripts for consideration.

The ever continuing interest in the application of numerical methods for the solution of thermal based problems is aptly demonstrated in these proceedings. The papers, contributed by authors from many countries, have been separated into nine sections which cover the main areas of interest, eg, 'Heat Conduction', 'Free and Forced Convection', 'Turbulent Flow', and a large interest in Casting Simulation necessitated a section being given to this topic alone. In general, the manuscripts have been placed into the appropriate field if interest apart from the more diversified groups of 'Novel Computational Techniques' and 'Industrial and Scientific Applications'.

The proceedings are printed from direct lithographs of the authors' manuscripts and the editors do not accept responsibility for any erroneous comments or opinions expressed herein. As in previous conferences of the highly successful series, a planned 'state-of-the-art' archival text will be produced after the conference.

The Conference Organizing Committee wishes to acknowledge the collaboration of the *International Journal for Numerical Methods in Engineering*, *Communications in Numerical Methods in Engineering* and the *International Journal of Numerical Methods for Heat and Fluid Flow.*

R W LEWIS
May 1993

CONTENTS

PREFACE

PART 1

SECTION 1

PHASE CHANGE PROBLEMS

SECTION 2

HEAT CONDUCTION

SECTION 3

SOLIDIFICATION AND CASTING PROCESSES

SECTION 4

NATURAL AND/OR FORCED CONVECTION

SECTION 5

TURBULENT FLOW

SECTION 6

RADIATIVE HEAT TRANSFER

PART 2

SECTION 7

THERMOMECHANICAL ANALYSIS

SECTION 8

NOVEL COMPUTATIONAL TECHNIQUES

SECTION 9

INDUSTRIAL AND SCIENTIFIC APPLICATIONS

PART 1

SECTION 1

PHASE CHANGE PROBLEMS

SELF ADAPTIVE TIME INTEGRATION AND THE KINETICS OF PHASE TRANSFORMATION

M A Keavey

Nuclear Electric plc, Berkeley, Gloucestershire GL13 9PB, England

SUMMARY

When the enthalpy method is applied to phase change problems involving only two states, it is usually sufficient simply to solve the thermal diffusion equation together with a source term derived from the enthalpy-temperature relation. When multiple phases exist concurrently, however, as in the case of solid state transformations in carbon steels, it is necessary to represent the kinetics of each transformation explicitly.

This paper examines a novel approach to problems of melting and solidification in which the algebraic enthalpy-temperature relation is replaced by a differential equation describing the evolution of the new state. The resulting equations may, however, be stiff. Typical time constants for steels on heating can be measured in fractions of a second, whilst for cooling, times may be measured in hours.

Although standard methods for stiff equations may be applied, adaptive time stepping is essential. Experience with an implicit single step integration scheme is described in which iterations are required to be contractive and where time steps are predicted on the basis of accurate a posteriori estimates of local truncation and convergence errors.

Preliminary results are presented for both solid state metallurgical transformations and simple non-dimensional liquid-solid benchmarks.

1. INTRODUCTION

When the finite element method is used to solve phase change problems, the thermal energy equation is usually expressed in terms of enthalpy [1,2]. Earlier methods based on specific heat alone were too prone to inaccuracy arising from inadequate discretisation in time. Even so, some enthalpy approaches [3] can only be applied to second order transformations, that is, when melting or freezing takes place over a non-zero temperature range. Such transformations are usually associated with solutions (eg. salt in water or carbon in iron). First order transformations, which take place at a single temperature, can sometimes cause difficulty.

The reason for this is, of course, that whilst temperature is always a unique function of enthalpy, enthalpy as a function of temperature can be multivalued. A formulation solving for enthalpy as the independent variable and in which temperature is treated as a dependent variable is given in Section 2.

Much published work on the subject of phase change is concerned with water-ice systems. However, leaving aside for the moment problems of vaporisation and boiling, there is still a very significant area of interest associated with the welding and casting of steels and other alloys. The metallurgy of the iron-carbon system is of particular importance in engineering. Here, below the melting point, a number of complex solid state transformations can take place as body centred cubic ferrite transforms into face centred cubic austenite on heating, and as carbon diffuses, reacts to form carbide, and crystalises on cooling, to give bainite, pearlite and martensite.

Although the latent heat effects arising from these transformations are small, volumetric changes, particularly in low alloy steels such as 2¼CrMo, are significant and can give rise to strains of the order of 5%. These in turn result in high residual stresses and subsequent cracking. It follows that a great deal of work on modelling these effects has been carried out, particularly in the nuclear industry [4-6].

Since multiple materials are involved, it is no longer possible to define a single enthalpy-temperature relation. Instead, each component has its own material properties, weighted by the proportion in existence at any one time. The evolution of each phase proportion is governed by an ordinary differential equation. These equations are solved simultaneously with those derived from the finite element discretisation defining the temperature.

The aim of this paper is to compare the evolution approach with the enthalpy method and to evaluate their relative merits.

2. THE ENTHALPY METHOD

This is presented here in a form suitable for first order transformations to illustrate the solution scheme and to provide a comparison with the evolution method. Spatial discretisation of the thermal energy equation

$$\nabla.(k\nabla\phi) + Q - \frac{\partial H}{\partial t} = 0 \tag{1}$$

where ϕ is the temperature and k the conductivity, yields a system of ordinary differential equations for the nodal enthalpies $\mathbf{H}$ of the form

$$\mathbf{G}\frac{\partial \mathbf{H}}{\partial t} = \mathbf{f}(t, \mathbf{H}) \tag{2}$$

where **G** is the Gramian derived from the shape function **N**,

$$\mathbf{G} = \int_V \mathbf{N}^T \mathbf{N}\, dV \tag{3}$$

and where the vector valued function **f** of the vector variable **H** is given by

$$\mathbf{f} = \int_V \mathbf{N}^T Q\, dV + \int_A \mathbf{N}^T q\, dA - \int_V \nabla^T \mathbf{N} k \frac{\partial \phi}{\partial H} \nabla H\, dV \tag{4}$$

where Q represents possible source terms and q represents the applied flux boundary conditions. This follows from the fact that

$$\phi = \phi(H) \tag{5}$$

and

$$\nabla \phi = \frac{\partial \phi}{\partial H} \nabla H \tag{6}$$

The function (5) is single valued and, in particular, at the point at which a first order transformation takes place, the value of $\frac{\partial \phi}{\partial H}$ is zero rather than infinity. For an implicit integration scheme it is then only necessary to supply the Jacobian

$$\frac{\partial \mathbf{f}}{\partial \mathbf{H}} = \int_V \mathbf{N}^T \frac{\partial Q}{\partial \phi} \frac{\partial \phi}{\partial H} \mathbf{N}\, dV + \int_A \mathbf{N}^T \frac{\partial q}{\partial \phi} \frac{\partial \phi}{\partial H} \mathbf{N}\, dA - \int_V \nabla^T \mathbf{N} k \frac{\partial \phi}{\partial H} \nabla \mathbf{N}\, dV \tag{7}$$

3. THE EVOLUTION APPROACH

Following Leblond and Devaux [4], the proportion of each phase is governed by an evolution equation, typically

$$\frac{\partial p}{\partial t} = \frac{p_{eq} - p}{\tau} \qquad \phi \le \phi_0 \tag{8}$$

where p is the proportion of the phase concerned, $p_{eq}(\phi)$ is the equilibrium proportion as a function of temperature, and $\tau(\phi)$ is an experimentally determined time constant. Conductivity, specific heat and enthalpy are then described by a decomposition rule, eg.

$$H = \sum p_i H_i \tag{9}$$

from which it follows that

$$\begin{aligned} \frac{\partial H}{\partial t} &= \sum p_i \frac{\partial H_i}{\partial \phi} \frac{\partial \phi}{\partial t} + \sum \frac{\partial p_i}{\partial t} H_i \\ &= \sum p_i \rho_i c_i \frac{\partial \phi}{\partial t} + \sum \frac{\partial p_i}{\partial t} H_i \end{aligned} \tag{10}$$

In particular, for a two phase system

$$p_1 + p_2 = 1 \tag{11}$$

and

$$\sum_{i=1}^{2} \frac{\partial p_i}{\partial t} H_i = \frac{\partial p_1}{\partial t}(H_1 - H_2) \tag{12}$$

$$= -\frac{\partial p_1}{\partial t} L$$

where L is the latent heat of transformation. This then appears as a source term in the thermal energy equation

$$\nabla.(k\nabla\phi) + \frac{\partial p_1}{\partial t} L - \rho c \frac{\partial \phi}{\partial t} = 0 \tag{13}$$

where k, and ρc are now understood to take their proportion weighted values and the proportion itself is governed by equation (8). Finite element discretisation of (13) again yields a system of ordinary differential equations in the form

$$\mathbf{G}\frac{\partial \phi}{\partial t} = \mathbf{f}_\phi(t, \phi, \mathbf{p}, \frac{\partial \mathbf{p}}{\partial t}) \tag{14}$$

where, dropping the subscript on p,

$$\mathbf{f}_\phi = \int_V \mathbf{N}^T \frac{L}{\rho c} \frac{\partial p}{\partial t} dV - \int_V \nabla^T \mathbf{N} \frac{k}{\rho c} \nabla\phi \, dV \tag{15}$$

The evolution equation (8) is similarly expressed in the form

$$\frac{\partial \mathbf{p}}{\partial t} = \mathbf{f}_p(t, \phi, \mathbf{p}) \tag{16}$$

This represents values of p evaluated at either nodes or integration points (the latter being necessary if spatial discontinuities exist). Finite element discretisation is not required. The evolution equations are, indeed, just another form of constitutive law expressed in differential form.

4. TIME INTEGRATION

Since equations (2, 14 and 16) are cast in virtually standard form, any method appropriate to a system of ordinary differential equations can be used, given due regard for accuracy, stability and convergence. A predictor-corrector is chosen, iterated to convergence for stiff stability. Complete flexibility requires that the method be self starting, and therefore, single step. The only second order single step theta method is the trapezoidal rule, so allowing for the Gramian, the corrector becomes

$$\mathbf{G}\mathbf{y}_{t+\Delta t} = \mathbf{G}\mathbf{y}_t + \frac{1}{2}\Delta t\,[\mathbf{f}_t + \mathbf{f}_{t+\Delta t}] \tag{17}$$

Forward Euler could be used for the predictor, but this is only first order and therefore precludes the use of Milne's device in the normal way. A block Runge-Kutta method is therefore employed, such methods having been devised by Milne [7] for just this purpose. Thus

$$\mathbf{G}\mathbf{y}_{t+\frac{1}{2}\Delta t} = \mathbf{G}\mathbf{y}_t + \frac{1}{2}\Delta t\, \mathbf{f}_t \tag{18}$$

$$\mathbf{G}\mathbf{y}_{t+\frac{1}{2}\Delta t} = \mathbf{G}\mathbf{y}_t + \frac{1}{4}\Delta t\, [\mathbf{f}_t + \mathbf{f}_{t+\frac{1}{2}\Delta t}]$$

$$\mathbf{G}\mathbf{y}_{t+\Delta t} = \mathbf{G}\mathbf{y}_t + \Delta t\, \mathbf{f}_{t+\frac{1}{2}\Delta t}$$

Only the corrector requires inversion of the full system Jacobian, and even then approximations may suffice, although they are not used here. The Gramian remains unchanged throughout the analysis and therefore, in the absence of adaptive meshing, need be factored only once.

Following Lambert [8] and expressing the corrector (17) in the form $\mathbf{F}(\mathbf{y}) = 0$ for solution by Newton's method gives

$$\mathbf{F}(\mathbf{y}_{t+\Delta t}) = \mathbf{G}\mathbf{y}_{t+\Delta t} - \frac{1}{2}\Delta t\, \mathbf{f}_{t+\Delta t} - \mathbf{g} \tag{19}$$

$$\mathbf{F}'(\mathbf{y}_{t+\Delta t}) = \mathbf{G} - \frac{1}{2}\Delta t\, \mathbf{f}'_{t+\Delta t} \tag{20}$$

and

$$\mathbf{y}^{i+1}_{t+\Delta t} - \mathbf{y}^{i}_{t+\Delta t} = -[\mathbf{G} - \frac{1}{2}\Delta t\, \mathbf{f}'_{t+\Delta t}]^{-1}\, [\mathbf{G}\mathbf{y}^{i}_{t+\Delta t} - \frac{1}{2}\Delta t\, \mathbf{f}_{t+\Delta t} - \mathbf{g}] \tag{21}$$

where $\mathbf{g}$ remains constant throughout the iteration process and is given by

$$\mathbf{g} = \mathbf{G}\mathbf{y}_t + \frac{1}{2}\Delta t\, \mathbf{f}_t \tag{22}$$

It will be noted that the zero value of $\frac{\partial \phi}{\partial H}$ during a first order transformation and the consequent zero value of $\mathbf{f}'_{t+\Delta t}$ is of no consequence since the final system Jacobian also contains the Gramian $\mathbf{G}$.

Adaptive time step control is based on estimates of the truncation error ε_T and the convergence properties of the iterations (21). Although the Newton-Raphson process can converge without being globally contractive, requiring it to be so allows a corollary of the contraction mapping theorem [9] to be invoked, ie.

$$||\mathbf{y}^* - \mathbf{y}^{i+1}|| \leq \frac{\alpha}{1-\alpha}\, ||\mathbf{y}^{i+1} - \mathbf{y}^{i}|| \tag{23}$$

where $\mathbf{y}^*$ represents the exact solution. An estimate of the contraction constant α may be found [10] from,

$$\alpha = \max_i \frac{||\mathbf{y}^{i+1} - \mathbf{y}^{i}||}{||\mathbf{y}^{i} - \mathbf{y}^{i-1}||} \tag{24}$$

When convergence is slow, $\frac{1}{2} < \alpha < 1$ and an error estimate based on $||y^{i+1} - y^i||$ alone is over optimistic. For $0 < \alpha < \frac{1}{2}$, the simple error estimate is too conservative. If $\alpha \geq 1$, the step is discarded.

The truncation error is estimated from the standard Milne approximation,

$$\varepsilon_T = \frac{C_c}{C_p - C_c} ||y^{i+1} - y^0|| \tag{25}$$

where for the midpoint rule, $C_p = 1/3$, or allowing for the half interval, $C_p = 1/24$. For the trapezoidal rule, $C_c = -1/12$, yielding

$$\varepsilon_T = \frac{1}{3} ||y^{i+1} - y^0|| \tag{26}$$

5. COUPLING

The predictor equations for the evolution problem take the general form

$$\begin{bmatrix} \mathbf{G} & 0 \\ 0 & \mathbf{I} \end{bmatrix} \begin{Bmatrix} \Delta\phi \\ \Delta\mathbf{p} \end{Bmatrix} = \begin{Bmatrix} \mathbf{F}_\phi \\ \mathbf{F}_p \end{Bmatrix} \tag{27}$$

where the vectors $\mathbf{F}_\phi$ and $\mathbf{F}_p$ successively take the three forms given in equation (18) with $\mathbf{f}_\phi$ and $\mathbf{f}_p$ defined by equations (14) and (16) respectively.

The final system of corrector equations (21) for the evolution problem may be written

$$\begin{bmatrix} \mathbf{F}'_{\phi\phi} & \mathbf{F}'_{\phi p} \\ \mathbf{F}'_{p\phi} & \mathbf{F}'_{pp} \end{bmatrix} \begin{Bmatrix} \Delta\phi \\ \Delta\mathbf{p} \end{Bmatrix} = - \begin{Bmatrix} \mathbf{F}_\phi \\ \mathbf{F}_p \end{Bmatrix} \tag{28}$$

This raises the general question of how the solution of such systems should be approached. For fully coupled stiff systems, the equations must be solved as they stand. Remembering that the left hand side is merely the Jacobian and that the problem is actually defined by the requirement that the right hand side vector valued function be equal to zero, the off-diagonal terms $\mathbf{F}'_{\phi p}$ and $\mathbf{F}'_{p\phi}$, in the right circumstances, may simply be ignored, ie.

$$\begin{bmatrix} \mathbf{F}'_{\phi\phi} & 0 \\ 0 & \mathbf{F}'_{pp} \end{bmatrix} \begin{Bmatrix} \Delta\phi \\ \Delta\mathbf{p} \end{Bmatrix} = - \begin{Bmatrix} \mathbf{F}_\phi \\ \mathbf{F}_p \end{Bmatrix} \tag{29}$$

The equations are now decoupled and may be solved separately in sequence. This in no way compromises nonlinear solutions as the right hand side terms are, in any case, based on the previous iteration.

In practice, the evolution approach may be decoupled in this way, which greatly reduces the resources required for solution. Although the finite element equations are coupled in space, the evolution equations are not, being coupled only between phases at each sampling point, allowing further simplification.

As a final point, it should be noted that, unlike conventional finite element formulations, expressions (4) and (15) contain a term, equivalent to the conventional 'stiffness' matrix, which appears in the right hand sides of (28) and (29), this term being recomposed as a vector.

$$\int_V \nabla \mathbf{N}^T \frac{k}{\rho c} \nabla \mathbf{N} \, dV \, \phi = \int_V \nabla \mathbf{N}^T \frac{k}{\rho c} \nabla \phi \, dV \tag{30}$$

It is this fact that allows a full range of Newton and quasi Newton methods to be used at will without changing the fundamental physical problem that is being solved. The appearance of the conventional 'stiffness' matrix on the left hand side can now be seen merely as a condition for convergence.

6. EXAMPLES

The examples given are of a preliminary nature and illustrate the behaviour of a single point in specimens subject to uniform heating or cooling.

Using data for A 508 steel taken from [4], Figure 1 is an example of the solution of equations of type (8) showing transformations between ferrite+pearlite and austenite when a sample is heated from 600°C to 800°C in 10 seconds, cooled back to 620°C in 1000 seconds and then heated again to 800°C, this time also in 1000 seconds. The linearly ramped temperature history is prescribed; indeed, no data is given for latent heat release. The transformations are asymmetric, giving rise to a characteristic hysteresis loop. For the austenitic transformation, the time constant is less than 1 second. For the reverse transformation, the time constant varies between 500 and 10,000 seconds. Both time constants are temperature dependent. The points marked on the curve represent calculated values and illustrate the behaviour of the adaptive control algorithm.

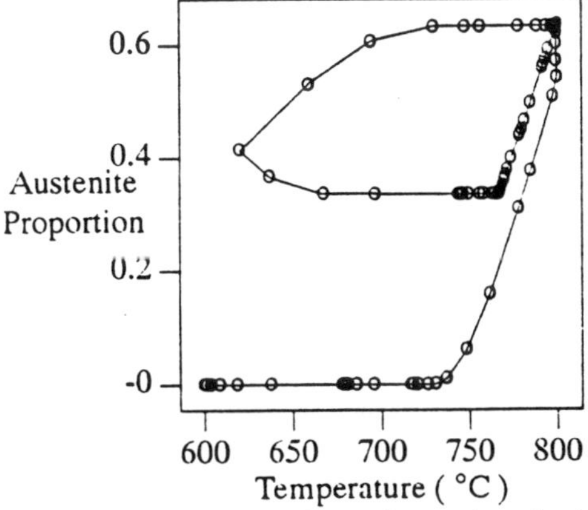

Fig. 1 Asymmetric Transformations in A 508 Steel

Leblond and Devaux [4] suggest that their evolution equations may be used for liquid-solid transformations, though this is not demonstrated. Purely to test the theory, an artificial problem is posed here, identical to one used previously [3]. This is a freezing problem driven by a temperature dependent heat sink which has a known analytical solution in terms of falling exponentials.

$$Q - \frac{\partial H}{\partial t} = 0 \qquad \phi = 1 \; at \; t = 0 \qquad Q = -\phi$$

$$\rho c = \frac{\partial H}{\partial \phi} = \begin{cases} 1 & 0.6 < \phi \le 1.0 \\ 10 & 0.5 < \phi \le 0.6 \\ 1 & 0.0 < \phi \le 0.5 \end{cases} \tag{31}$$

Figure 2a shows enthalpy and temperature against time using the enthalpy method. Figure 2b shows corresponding results for the evolution method, together with the evolving solid proportion.

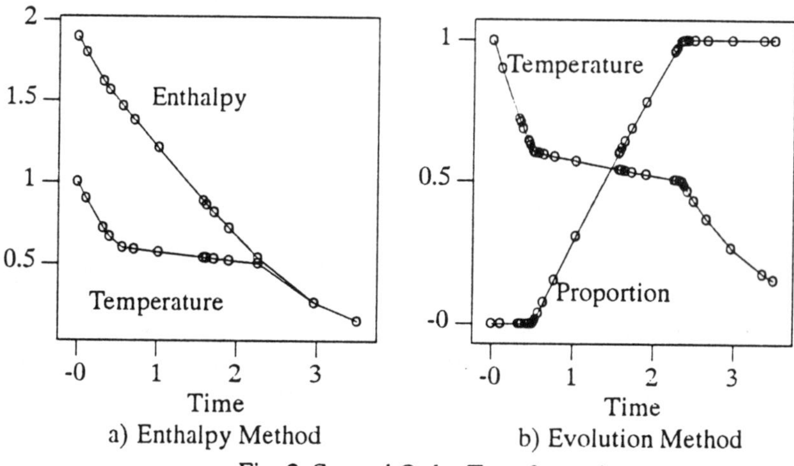

a) Enthalpy Method b) Evolution Method

Fig. 2 Second Order Transformation

In order to simulate the specified enthalpy-temperature relationship (31), the evolution equation must take a specific form.

$$\frac{\partial p}{\partial t} = \phi_0 - \Delta\phi p \qquad 0.5 < \phi \le 0.6 \tag{32}$$

where ϕ_0 is the temperature at which the phase change begins, and $\Delta\phi$ is the phase change interval. This corresponds to values of p_{eq} and τ given by

$$p_{eq} = \frac{\phi_0}{\Delta\phi} \qquad \tau = \frac{1}{\Delta\phi} \tag{33}$$

Both methods also yield accurate results for a first order transformation at $\phi = 0.5$. The latent heat release is chosen to be unity.

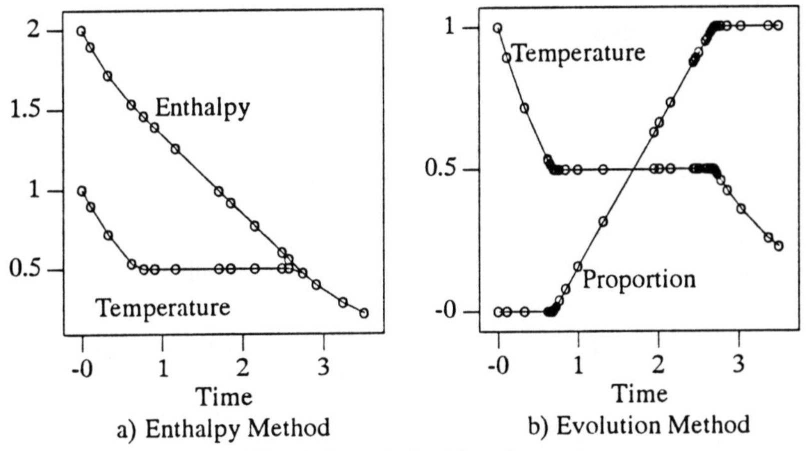

Fig. 3 First Order Transformation

Perhaps contrary to expectation, the transient behaviour is not rapid, which is reasonable when it is remembered that cooling is arrested at the freezing point, not accelerated. The evolution equation corresponding to Figure 3 is simply

$$\frac{\partial p}{\partial t} = \phi_0 \qquad \phi = 0.5 \tag{34}$$

Integration of the evolution equations begins when the appropriate temperature is reached and ceases when the proportion reaches unity. Note that it is perfectly possible to specify an enthalpy release rate high enough to cause the temperature to rise, a phenomenon encountered physically in supercooled liquids.

7. DISCUSSION

Although some of the results presented here may be viewed as purely academic, there is a serious relevance to the evolution approach. The importance of cooling rates and dwell times in the heat treatment of steels has long been recognised. Viewed in terms of an appropriate time scale, no physical process takes place instantaneously, and the freezing of water is no exception; crystals take time to form [11]. Slow cooling rates result in a small number of nucleation sites and large crystals. Rapid rates result in supercooling followed by an explosion of a large number of nucleation sites and small crystals. The analogy with the metallurgy of steels is obvious.

In biological materials, cooling rates also affect the time available for cell membrane osmosis, chemical reactions, and the diffusion of solutes, the latter process having a direct effect on the depression of the freezing point [12]. All these processes are relevant both to medical cryobiology, particularly with regard to organ preservation for transplant, and to food technology, where the economy and microbiological safety of freezing tunnels and the consumer's perception of taste and texture are of paramount importance. In the process of cryofixation, where biological materials are prepared for microscopic examination, cooling rates of up to $10^6 K/s$ [13] are used to achieve almost total glassification, necessary to prevent ice crystals from obscuring features of interest.

8. CONCLUSION

Given a sufficiently robust adaptive scheme for integration in time, both the enthalpy method and the evolution approach are capable of dealing with the simple examples presented in this paper. It is apparent, however, that for conventional problems, the enthalpy method is both easier to implement and more efficient.

In situations where detailed modelling of the transformation kinetics is important, however, the evolution approach is perfectly practical. First order transformations may be represented by a particularly simple form of transient. With physically realistic data, coupling with other phenomena becomes possible and, in particular, grain size predictions based on the Arrhenius rate law, such as those put forward for austenitic steels [14], should in theory also be able to model the growth of crystals in ice.

ACKNOWLEDGEMENT

This work was carried out at Berkeley Technology Centre and is published by permission of Nuclear Electric plc. The author would like to thank the many colleagues who contributed to the ideas contained herein.

REFERENCES

1. ROLPH, W.D. and BATHE, K.J. - 'An Efficient Algorithm for Analysis of Non-Linear Heat Transfer with Phase Changes'. Int. J. Num. Meths. Engng., 1982, 18, p119.

2. HARDER, R.L. - 'A Method to Model Latent Heat for Transient Analysis using Nastran'. Computational Aspects of Heat Transfer, NASA CP-2216, 1981, p83.

3. KEAVEY, M.A. - 'A Simple Enthalpy Technique for Heat Conduction with Phase Change'. 4th Int. Conf. Num. Meths. Thermal Problems, Swansea, 1985, p270.

4. LEBLOND, J.B. and DEVAUX, J. - 'A New Kinetic Model for Anisothermal Metallurgical Transformations in Steels including the Effect of Grain Size'. Acta Metallurgica, 1984, 32, p137.

5. LEBLOND, J.B. - 'A Theoretical and Numerical Approach to the Plastic Behaviour of Steels During Phase Transformations - I. Derivation of General Relations. J. Mech. Phys Solids, 1986, 34, 4, p395.

6. LEBLOND, J.B. - 'A Theoretical and Numerical Approach to the Plastic Behaviour of Steels During Phase Transformations - II. Study of Classical Plasticity for Ideal Plastic Phases. J. Mech. Phys Solids, 1986, 34 , 4, p411.

7. MILNE, W.E. - Numerical Solution of Differential Equations , Wiley, 1953.

8. LAMBERT, J.D. - Computational Methods in Ordinary Differential Equations , Wiley, 1973.

9. ODEN, J.T. - Applied Functional Analysis , Prentice-Hall, 1979.

10. SCHMIDT, W.F. - 'Adaptive Step Size Selection for Use with the Continuation Method'. Int. J. Num. Meths. Engng., 1978, 12 , p677.

11. BALD, W.B. - 'Ice Crystal Growth in Idealised Freezing Systems', Food Freezing Today and Tomorrow , Ed. Bald, W.B., Springer, 1991.

12. BLOND, G. and COLAS, B. - 'Freezing in Polymer-Water Systems', Food Freezing Today and Tomorrow , Ed. Bald, W.B., Springer, 1991.

13. BALD, W.B. - 'Optimising the Cooling Block for the Quick Freeze Method'. J. Microscopy, 1983, 131 , Pt 1, p11.

14. ALBERRY, P.J. and JONES, W.K.C. - 'A Computer Model for the Prediction of Heat Affected Zone Microstructures in Multi-Pass Weldments'. C.E.G.B. Report R/M/R282, 1979.

A Finite Element Enthalpy Technique for Solving Coupled Nonlinear Heat Conduction/Mass Diffusion Problems with Phase Change

R.L. McAdie
Senior Research Assistant
Dept. of Civil Engineering, University College of Swansea
Formerly: Centre for Research in Computational and Applied Mechanics
University of Cape Town
J.B. Martin
Dean of Engineering and Professor of Applied Mechanics
University of Cape Town
R.W. Lewis
Professor of Civil Engineering
Dept. of Civil Engineering, University College of Swansea

April 1993

Abstract

A rigorous Finite Element (FE) formulation based on an enthalpy technique is developed for solving coupled nonlinear heat conduction/mass diffusion problems with phase change. The FE formulation consists of a fully coupled heat conduction and solute diffusion formulation, with solid-liquid phase change, where the effects of pressure and convection are neglected. A full enthalpy method is employed eliminating singularities resulting from abrupt changes in heat capacity at the phase interfaces. The FE formulation is based on the fixed grid technique where the elements are two dimensional, four noded quadrilaterals with the primary variables being enthalpy and average solute concentration. Temperature and solid mass fraction are calculated on a local level at each integration point of an element.

A fully consistent Newton-Raphson method is used to solve the global coupled equations and an Euler backward difference scheme is used for the temporal discretization. The solution of the enthalpy-temperature relationship is carried out at the integration points using a Newton-Raphson method. A secant method employing the *regula falsi* technique takes into account sudden jumps or sharp changes in the enthalpy-temperature behaviour which occur at the phase zone interfaces. The Euler backward difference integration rule is used to calculate the solid mass fraction and its derivatives.

A phase change problem is analysed and results are presented. The accuracy and efficiency of the FE model is checked for different phase change situations.

1 Introduction

Many phase change problems encountered in the material processing industry are not only heat conduction dependent but are also influenced by the mass diffusion of the material constituents, where the two processes are coupled by the phase change mechanism. This phase change mechanism is, in turn, dependent on the local temperature and constituent concentration fields. Depending on the process this mechanism can be governed by the equations of state (*i.e.* the phase diagram) where local thermodynamic equilibrium with respect to the material constituents is assumed – this is the case for most numerical models [1], [2] and [3]

Different models representing the phase change process on the microscopic level have been developed; the lever rule model, where local thermodynamic equilibrium with respect to the material constituents is assumed in both the liquid and solid phases (" ***equilibrium model*** ") or the Scheil model, where solid constituent diffusion is assumed to be negligible and only the constituent concentration in the liquid phase is at thermodynamic equilibrium (" ***non-equilibrium model*** ") . The type of model chosen not only affects the local phase change description but also affects the global variations in constituent concentration and temperature. The solid mass fraction, which is used as a measure of phase change, is therefore a function of temperature and average constituent concentration. If the global conservation equations are cast in a continuum form then the material constitutive and thermodynamic coefficients will, in turn, be functions of solid mass fraction. Therefore, what happens on a global level, governed by the continuum conservation equations, affects what happens on a local level, governed either fully or partialy by the equations of state, and *vice-versa*. This results in a tighly coupled system of equations which describe the process.

Typical industrial processes that would be described by this set of equations are, among others, the alloy solidification process, the freezing or melting of solutions (i.e. the desalination of water) and the industrial diamond synthesis process. It is appreciated that in some of these process fluid convection plays a dominant role but the model described in this work is considered as a first step in modelling the particular process.

Depending on the material and process the phase change can be either discrete or take place over a region (i.e. mushy phase change) or both. Different numerical techniques such as the front tracking and fixed grid techniques have been used to tackle these phase change problems. The front tracking technique is used for discrete phase change where an accurate description of the phase interface is required [4] and [5]. Conversely, the fixed grid method is preferred for diffuse mushy type or mushy/discrete phase change. Some of the advantages of using the fixed grid method are, that it lends itself to easy implementation of continuum based formulations and it is computationaly inexpensive when compared to the front tracking scheme. Much finite volume work in the field of alloy solidification has been conducted using the fixed grid continuum approach [2], [3] and [6]. In contrast, little work in the finite element field has been published, besides that of Porier and Heinrich [7]. This paper therefore presents a mathematically rigious fixed grid finite element formulation using an enthalpy technique to model the coupled phase change problem.

2 Problem Definition

The global conservation equations are that of energy (Fourier's law) and the conservation of solute (Fick's law) which are respectively expressed as

$$\rho C_p \frac{\partial T}{\partial t} = \nabla \cdot \ \boldsymbol{K} \nabla T + Q_H \tag{1}$$

and

$$\rho \frac{\partial \bar{C}}{\partial t} = \nabla \cdot \rho \boldsymbol{D} \nabla C^L + Q_C \ . \tag{2}$$

Both essential and natural boundary conditions are defined on different parts of the boundary of the domain. The essential boundary condition prescribes a given temperature g_H and/or a given solute concentration g_C which may depend on both position and time. The natural boundary condition can be described as a heat flux h_H and/or solute flux h_C. The heat flux h_H may depend on temperature, position and time, and can constitute an applied heat flux, radiative flux or convective flux. The solute flux h_C may depend on solute concentration, position and time, and constitutes an applied solute flux.

The internal heat generation Q_H is expressed as $Q_H = Q_H(T, \boldsymbol{x}, t)$ and the solute source Q_C is expressed as $Q_C = Q_C(\bar{C}, \boldsymbol{x}, t)$.

The material properties vary with respect to the phase of the material. The material is considered to have a maximum of three phases (i.e. solid, liquid and mush) at any one time . The thermodynamic coefficients (i.e. effective specific heat C_p), the constitutive coefficients (i.e. conductivity $\boldsymbol{K}$, and diffusivity $\boldsymbol{D}$) and density ρ of the phase mixture are expressed in terms of a lever rule with respect to the solid mass fraction ϕ, with the exception of conductivity, which is expressed in terms of the phase volume fraction g^α where α represents solid (S) or liquid (L) phase. These properties admit a nonlinear dependence on temperature T and average solute concentration $\bar{C}$.

A summary of the material properties defining the phase mixture follows.

a) The solid mass fraction is expressed as

$$\phi = \phi(T, \bar{C}, \boldsymbol{x}, t) = \begin{cases} 1 & \text{solid region} \\ 0 < \phi < 1 & \text{mushy region} \ . \\ 0 & \text{liquid region} \end{cases}$$

b) The effective specific heat of the phase mixture C_p is expressed as

$$C_p = C_{p_0} - L \partial \phi / \partial T \ , \tag{3}$$

where C_{p_0} is the lever rule specific heat of the phase mixture and is expressed as

$$C_{p_0} = \phi \, C_p^S(T, \boldsymbol{x}, t) + (1 - \phi) \, C_p^L(T, \boldsymbol{x}, t) \tag{4}$$

and L is the latent heat (enthalpy of fusion), which is expressed as

$$L = \int_{Tref}^{T} (C_p^L(\tau, \boldsymbol{x}, t) - C_p^S(\tau, \boldsymbol{x}, t)) d\tau + \Delta H_{ref}^f \ . \tag{5}$$

where $C_p^S = C_p^S(T, \boldsymbol{x}, t)$ is the specific heat of the solid, $C_p^L = C_p^L(T, \boldsymbol{x}, t)$ is the specific heat of the liquid and ΔH_{ref}^f is the heat of formation of the liquid alloy which is constant. Dependency on the solute concentration of C_p^α in both phases are ignored.

c) The average solute concentration of the phase mixture is expressed as

$$\bar{C} = \phi\bar{C}^S + (1-\phi)\bar{C}^L, \tag{6}$$

where $\bar{C}^S$ is the average solute concentration of the solute in the solid and is expressed for the case of zero local solid diffusion by $\bar{C}^S = \frac{1}{\phi}\int_0^\phi C^S(f^S)\,dV^S$, and for the local solid diffusion case as $\bar{C}^S = C^S$ where $C^S = C^S(T,\boldsymbol{x},t)$ in the mushy region and $C^S = \bar{C}(\boldsymbol{x},t)$ in the solid. The term $\bar{C}^L$ is the average solute concentration of the solute in the liquid, where $\bar{C}^L = C^L$, which is expressed in the mushy region as $C^L = C^L(T,\boldsymbol{x},t)$ and in the liquid region as $C^L = \bar{C}(\boldsymbol{x},t)$. Complete solute mixing is assumed within the liquid phase and undercooling is neglected. As a result, position of the dendrite tips are located at the equilibrium liquidus temperature.

d) The phase mixture density ρ is expressed as

$$\rho = g^S\rho_*^S + g^L\rho_*^L, \tag{7}$$

where $g^S = (\rho/\rho_*^S)\,\phi$ is the solid volume fraction, $g^L = (\rho/\rho_*^L)\,(1-\phi)$ is the liquid volume fraction, ρ_*^S is the actual solid density and ρ_*^L is the actual liquid density. Note that if the Oberbeck-Boussinesq approximations are used where it is assumed that ρ = constant (except in the bouyancy terms for the fluid flow case). To maintain phase mixture saturation, solid and liquid densities are assumed to be equal, therefore $\rho_*^S = \rho_*^L$ = constant. From these assumptions therefore, $g^S = \phi$ and $g^L = (1-\phi)$.

e) The conductivity of the phase mixture is expressed as

$$\boldsymbol{K} = g^S\boldsymbol{K}^S(T,\bar{C},\boldsymbol{x},t) + g^L\boldsymbol{K}^L(T,\bar{C},\boldsymbol{x},t), \tag{8}$$

where $\boldsymbol{K}^S$ is the conductivity tensor for the solid and $\boldsymbol{K}^L$ is the conductivity tensor for the liquid. Note that $\boldsymbol{K}^\alpha$ is allowed to be anisotropic although symmetry is assumed [8].

f) The diffusivity of the phase is expressed as

$$\boldsymbol{D} = \phi\boldsymbol{D}^S(T,\bar{C},\boldsymbol{x},t) + (1-\phi)\boldsymbol{D}^L(T,\bar{C},\boldsymbol{x},t), \tag{9}$$

where $\boldsymbol{D}^S$ is the diffusivity of the solid and $\boldsymbol{D}^L$ is the diffusivity of the liquid. It is assumed that the solid diffusion is zero, therefore $\boldsymbol{D}^S = 0$.

2.1 The Enthalpy Approach

In the phase change region the specific heat C_p experiences a dirac-delta behaviour, Figure 1, which leads to significant numerical difficulties when trying to solve the conservation of energy equation. To circumvent this difficulty, enthalpy is introduced into the conservation equation (1) to give

$$\rho\frac{\partial H}{\partial t} = \nabla\cdot\boldsymbol{K}\nabla T + Q_H. \tag{10}$$

By rewriting the rate term in (10), in terms of enthalpy, numerical difficulties are only circumvented when a non-discrete phase change takes place where the $T-H$ curve is a smooth function throughout the domain as shown in Figure 1. However, as illustrated in Figure 1, the $T-H$ curve exhibits a discontinuity at the eutectic point of an alloy or at the melting temperature of a pure substance. Therefore the enthalpy becomes multivalued, making the accurate solution of equation (10) very difficult and energy conservation cannot be ensured. If enthalpy is chosen as the field variable, rather than

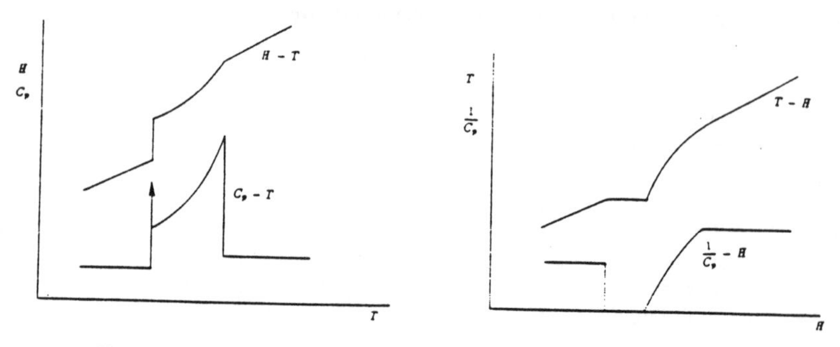

Figure 1: Enthalpy, specific heat, temperature relationships.

temperature, the function becomes single valued and this problem can be overcome with the energy being totally conserved. This is illustrated in Figure 1, it is clear that the $H-T$ curve experiences no discontinuity and also $\frac{\partial T}{\partial H}$, which is $\frac{1}{C_p}$, is zero in the discrete phase change region. Equation (10) expressed in terms of enthalpy is therefore

$$\rho\frac{\partial H}{\partial t} = \nabla\cdot\boldsymbol{\mathcal{K}}\nabla T(H) + Q_H = \nabla\cdot\frac{\boldsymbol{\mathcal{K}}}{C_p}\nabla H + Q_H, \tag{11}$$

where the enthalpy for the phase mixture is expressed as

$$H = \phi H^S + (1-\phi)H^L\,, \tag{12}$$

where

$$H^S = H^S(T(H),\boldsymbol{x},t) = \int_{Tref}^{T(H)} C_p^S(\tau,\boldsymbol{x},t)d\tau \tag{13}$$

and

$$H^L = H^L(T(H),\boldsymbol{x},t) = \int_{Tref}^{T(H)} C_p^L(\tau,\boldsymbol{x},t)d\tau + \Delta H_{ref}^f. \tag{14}$$

The advantage in using this enthalpy formulation is that it is characterized by a strictly decreasing or increasing enthalpy for solidification or melting respectively, where energy conservation is ensured. It must be noted that even when using the full enthalpy approach the discontinuity which is implicit in the formulation cannot be correctly modelled using the fixed grid finite element formulation. The standard finite elements cannot handle a discontinuity within the elements and the jump in enthalpy will be smeared over the element or elements depending on the solution scheme used. It is also clear that to model this jump condition reasonably a fine mesh would have to be used.

3 Finite element formulation

In this section, the solution of the global conservation equations of energy and solute, using a fixed grid finite element approach, is discussed.

First, the strong form of the initial boundary-value problem (IBVP) is defined. This leads to the weighted residual or weak form of the problem. Next, the Galerkin approximation is introduced which, along with an assumed spatial discretization, leads to the finite element matrix form of the problem. The finite element matrix equations are then temporally discretized. The solution of the resulting nonlinear matrix equations motivates the development of iterative algorithms. Finally, the linear problem generated by the iterative algorithms is discussed.

4 Strong form of the initial boundary-value problem

4.1 Problem Domain

The problem is posed for a body occupying a spatial domain Ω, a finite region of $\Re^{N_{sd}}$ where $\Re$ is the set of real numbers and N_{sd} is the number of space dimensions. A general point in $\bar{\Omega}$ will be denoted as $\boldsymbol{x} = \{x_i\}$, $i = 1, 2, ..., N_{sd}$ where $\bar{\Omega}$ denotes a closed domain, i.e. the total domain including boundaries. The closed domain $\bar{\Omega}$ is divided up into different subregions, with $\bar{\Omega} = \overline{\Omega^S \cup \Omega^L \cup \Omega^M}$ where $\Omega^S \subset \Re^{N_{sd}} \times \tau$ (*the solid region of the domain*), $\Omega^L \subset \Re^{N_{sd}} \times \tau$ (*the liquid region of the domain*), $\Omega^M = \overline{\Omega^S \cap \Omega^L}$ (*the phase change region of the domain*) and τ denotes time. The three subdomains Ω^S, Ω^L and Ω^M vary with time. When the width of Ω^M tends to zero a discrete phase change results between Ω^S and Ω^L. This is described as $\Omega^S \cap \Omega^L = \emptyset$, where $\emptyset$ denotes the empty set.

The boundary of Ω, denoted Γ, is assumed to be piecewise smooth. At almost every point on Γ there is a unique outward normal unit vector $\boldsymbol{n} = (n_i)$, $i = 1, 2, \ldots, N_{sd}$. In addition, Γ can be subdivided into two disjoint sets, Γ_g and Γ_h. Thus Γ admits the following decomposition $\Gamma = \overline{\Gamma_g \cup \Gamma_h}$ and $\emptyset = \Gamma_g \cap \Gamma_h$, where $\Gamma_g = \overline{\Gamma_{g^L} \cup \Gamma_{g^M} \cup \Gamma_{g^S}}$ and $\Gamma_h = \overline{\Gamma_{h^L} \cup \Gamma_{h^M} \cup \Gamma_{h^S}}$, where the superposed bar represents set closure.

4.2 The Strong Form

The IBVP describing the coupled heat conduction/mass diffusion phase change process is defined. Therefore the strong form (**S**) of the IBVP can thus be stated as follows.

Given ρ, C_P, $\boldsymbol{K}$, $\boldsymbol{D}$, g_H, g_C, h_H, h_C, T_0 and $\bar{C}_0$, as in § 2,

find

$$H \;:\; \bar{\Omega} \times [0, \tau] \longrightarrow R^+, \text{ expressed in § 2,}$$

and

$$\bar{C} \;:\; \bar{\Omega} \times [0, \tau] \longrightarrow R^+, \text{ expressed in § 2,}$$

such that the following coupled equations hold:

$$\rho \frac{\partial H}{\partial t} = \nabla \cdot \frac{\boldsymbol{K}}{C_p} \nabla H + Q_H \text{ on } \Omega \times]0, \tau[\tag{15}$$

and

$$\rho \frac{\partial \bar{C}}{\partial t} = \nabla \cdot \rho \boldsymbol{D} \nabla C^L + Q_C \text{ on } \Omega \times]0, \tau[\,; \tag{16}$$

where the essential boundary conditions are

$$H = g_H \text{ on } \Gamma_g \times]0, \tau[\tag{17}$$

and

$$\bar{C} = g_C \text{ on } \Gamma_g \times]0,\tau[\,, \tag{18}$$

the natural boundary conditions are

$$\boldsymbol{n} \bullet (\frac{\boldsymbol{\mathcal{K}}}{C_p} \nabla H) = h_H \;=\; h_{f_1} + h_{conv} + h_{rad} \text{ on } \Gamma_h \times]0,\tau[\tag{19}$$

and

$$\boldsymbol{n} \bullet (\rho \boldsymbol{D} \nabla C^L) = h_C \;=\; h_{f_2} \text{ on } \Gamma_h \times]0,\tau[\,, \tag{20}$$

and the initial conditions are

$$H(\boldsymbol{x},0) = h_0(\boldsymbol{x}) \;\forall\; \boldsymbol{x} \in \Omega \tag{21}$$

and

$$\bar{C}(\boldsymbol{x},0) = \bar{C}_0(\boldsymbol{x}) \;\forall\; \boldsymbol{x} \in \Omega\,. \tag{22}$$

5 Weighted residual form of the initial boundary-value problem

The 'weighted residual' or 'weak' form of (**S**) is generated by a suitable choice of solution and variational spaces and the application of the divergence theorem, Mitchell [9] and Strang [10].

In order to develop a weak formulation for the IBVP, a total solution and a variational space are defined. Let H and w_1 denote the enthalpy fields and $\bar{C}$ and w_2 denote the solute concentration fields. The solution space $\mathcal{S}$ is defined as $\mathcal{S} = \mathcal{S}_H \cup \mathcal{S}_C$ where $\mathcal{S}_H = \{H \mid H = g_H \text{ on } \Gamma_g \text{ with } H = H^L \text{ in } \Omega^L, H = \phi H^S + (1-\phi)H^L \text{ in } \Omega_M \text{ and } H = H^S \text{ in } \Omega^S\}$, and where $\mathcal{S}_C = \{\bar{C} \mid \bar{C} = g_C \text{ on } \Gamma_g \text{ with } \bar{C} = C^L \text{ in } \Omega^L,$ $\bar{C} = \phi \bar{C}^S + (1-\phi)C^L \text{ in } \Omega_M, \text{ and } \bar{C} = \bar{C}^S \text{ in } \Omega^S\}$.

The variational space $\mathcal{V}$ is defined as $\mathcal{V} = \mathcal{V}_H \cup \mathcal{V}_C$ where $\mathcal{V}_H = \{w_1 \mid w_1 = 0 \text{ on } \Gamma_g\}$ and $\mathcal{V}_C = \{w_2 \mid w_2 = 0 \text{ on } \Gamma_g\}$.

Note that $\mathcal{S}$ is time dependent due to its use of the g-type condition, while $\mathcal{V}$ is time independent.

5.1 The Weak Form

The weak form of the problem (**W**) is obtained by multiplying (15) and (21) by $w_1 \in \mathcal{V}_H$ and (16) and (22) by $w_2 \in \mathcal{V}_C$, integrating over Ω, applying the divergence theorem, and making use of the boundary conditions (17) – (20) to simplify the result. This yields weak form for the IBVP as follows.

Given $\rho, C_p, \boldsymbol{\mathcal{K}}, \boldsymbol{D}, g_H, g_C, h_H, h_C, H_0$ and $\bar{C}_0$, as in § 2,

find

$$H : [\,0,\tau\,] \longrightarrow \mathcal{S}_H$$

and

$$\bar{C} : [\,0,\tau\,] \longrightarrow \mathcal{S}_C\,,$$

such that for every $w_H \in \mathcal{V}_H$ and $w_C \in \mathcal{V}_C$

$$\mathcal{M}_H(\dot{H}, w_H) + \mathcal{K}_H(H, w_H) = \mathcal{F}_H(Q_H, w_H) + \mathcal{H}_H(h_H, w_H) \text{ on }]\,0, \tau\,[\,, \tag{23}$$

$$\mathcal{M}_C(\dot{\bar{C}}, w_C) + \mathcal{K}_C(C^L, w_C) = \mathcal{F}_C(Q_C, w_C) + \mathcal{H}_C(h_C, w_C) \text{ on }]\,0, \tau\,[\,, \tag{24}$$

$$(H(\boldsymbol{x}, 0) - H_0, w_H) = 0 \text{ on } \Omega, \tag{25}$$

and

$$(\bar{C}(\boldsymbol{x}, 0) - \bar{C}_0, w_C) = 0 \text{ on } \Omega. \tag{26}$$

The operators $\mathcal{M}_H$, $\mathcal{K}_H$, $\mathcal{F}_H$, $\mathcal{H}_H$, $\mathcal{M}_C$, $\mathcal{K}_C$, $\mathcal{F}_C$, $\mathcal{H}_C$ and $(.,\ )$ are defined respectively as:

$$\mathcal{M}_H(\dot{H}, w_H) = \int_\Omega \dot{H}(\boldsymbol{x}, t)\, w_H(\boldsymbol{x})\; d\Omega\,, \tag{27}$$

$$\mathcal{K}_H(H, w_H) = \int_\Omega \nabla H(\boldsymbol{x}, t) \bullet \frac{\boldsymbol{\mathcal{K}}(T(H(\boldsymbol{x}, t)), \bar{C}(\boldsymbol{x}, t), \boldsymbol{x}, t)}{C_p(T(H(\boldsymbol{x}, t)), \bar{C}(\boldsymbol{x}, t), \boldsymbol{x}, t)}\, \nabla w_H(\boldsymbol{x})\; d\Omega\,, \tag{28}$$

$$\mathcal{F}_H(H, w_H) = \int_\Omega Q_H(T(H(\boldsymbol{x}, t)), \boldsymbol{x}, t)\, w_H(\boldsymbol{x})\; d\Omega\,, \tag{29}$$

$$\mathcal{H}_H(h_H, w_H) = \int_{\Gamma_h} h_H(T(H(\boldsymbol{x}, t)), \boldsymbol{x}, t)\, w_H(\boldsymbol{x})\; d\Gamma\,, \tag{30}$$

$$\mathcal{M}_C(\dot{\bar{C}}, w_c) = \int_\Omega \dot{\bar{C}}(\boldsymbol{x}, t)\, \rho(\boldsymbol{x})\, w_C(\boldsymbol{x})\; d\Omega\,, \tag{31}$$

$$\mathcal{K}_C(C^L, w_C) = \int_\Omega \nabla C^L \bullet \rho(\boldsymbol{x}) \boldsymbol{D}(T(H(\boldsymbol{x}, t)), \bar{C}(\boldsymbol{x}, t), \boldsymbol{x}, t) \nabla w_C(\boldsymbol{x})\; d\Omega\,, \tag{32}$$

$$\mathcal{F}_C(Q_C, w_C) = \int_\Omega Q_C(\bar{C}(\boldsymbol{x}, t), \boldsymbol{x}, t)\, w_C(\boldsymbol{x})\; d\Omega\,, \tag{33}$$

$$\mathcal{H}_C(h_C, w_C) = \int_{\Gamma_C} h_C(\bar{C}(\boldsymbol{x}, t), \boldsymbol{x}, t)\, w_C(\boldsymbol{x})\; d\Gamma\,, \tag{34}$$

$$(H, w_H) = \int_\Omega H(\boldsymbol{x}, t)\, w_H(\boldsymbol{x})\; d\Omega \tag{35}$$

and

$$(\bar{C}, w_C) = \int_\Omega \bar{C}(\boldsymbol{x}, t)\, w_C(\boldsymbol{x})\; d\Omega. \tag{36}$$

Note that:

a) given suitable smoothness conditions a solution of **(S)** is a solution **(W)**, **(S)** $\Longleftrightarrow$ **(W)**,

b) $\mathcal{M}_H(\dot{H}, w_H)$, $\mathcal{K}_H(H, w_H)$, $\mathcal{M}_C(\dot{\bar{C}}, w_C)$, $\mathcal{K}_C(C^L, w_C)$, $(\bar{C}, w_C)$ and (H, w_H) are symmetric bilinear forms.

6 Galerkin approximation of the initial boundary-value problem

The Garlerkin form is derived from the weak form by approximating the variational and solution spaces with finite-dimensional subspaces.

The Garlerkin approximation uses a finite number of linear independent functions to span a subspace $\mathcal{V}^h$ and $\mathcal{S}^h$ where $\mathcal{V}^h \subset \mathcal{V}$ and $\mathcal{S}^h \subset \mathcal{S}$. We represent $\mathcal{V}^h$ as $\mathcal{V}^h = \mathcal{V}_H^h \cup \mathcal{V}_C^h$, where $\mathcal{V}_H^h = \{w_H^h \mid w_H^h = \sum_{A=1} N_A(\boldsymbol{x})d_A,\ w_H^h = 0 \text{ on } \Gamma_g\}$ and $\mathcal{V}_C^h = \left\{w_C^h \mid w_C^h = \sum_{A=1} N_A(\boldsymbol{x})\hat{d}_A,\ w_C^h = 0 \text{ on } \Gamma_g\right\}$, where N_A, $A = 1, 2 \cdots n$, are linearly independent functions in $\mathcal{V}$ and d_A and $\hat{d}_A$ are constants.

Similarly, the approximation to the trial solution space $\mathcal{S}^h$ is defined as $\mathcal{S}^h = \mathcal{S}_H^h \cup \mathcal{S}_C^h$, where $\mathcal{S}_H^h = \{H^h \mid H^h = v_H^h + g_H^h,\ v_H^h \in \mathcal{V}_H^h,\ g_H^h \in \mathcal{S}_H^h\}$ and $\mathcal{S}_C^h = \{\bar{C}^h \mid \bar{C}^h = v_C^h + g_C^h,\ v_C^h \in \mathcal{V}_C^h,\ g_C^h \in \mathcal{S}_C^h\}$.

6.1 The Galerkin Form

The Galerkin approximation (G) of the IBVP may therefore be stated as follows.

Given $\rho, C_p, \boldsymbol{\mathcal{K}}, \boldsymbol{D}, g_H, g_C, h_H, h_C, H_0$ and $\bar{C}_0$, as in § 2, find

$$H^h = v_H^h + g_H^h \ : \ [\,0,\tau\,] \longrightarrow \mathcal{S}_H^h$$

and

$$\bar{C}^h = v_C^h + g_C^h \ : \ [\,0,\tau\,] \longrightarrow \mathcal{S}_C^h\ ,$$

such that for every $w_H^h \in \mathcal{S}_H^h$ and $w_C^h \in \mathcal{S}_C^h$

$$\mathcal{M}_H(\dot{v}_H^h, w_H^h) + \mathcal{K}_H(v_H^h, w_H^h) = \mathcal{F}_H(Q_H, w_H^h) + \tag{37}$$

$$\mathcal{H}_H(h_H^h, w_H^h) - \mathcal{M}_H(\dot{g}_H^h, w_H^h) - \mathcal{K}_H(g_H^h, w_H^h) \quad \text{on }]\,0,\tau\,[, \tag{38}$$

$$\mathcal{M}_C(\dot{v}_C, w_C^h) + \mathcal{K}_C(v_C^h, w_C^h) = \mathcal{F}_H(Q_C, w_C^h) + \tag{39}$$

$$\mathcal{H}_H(h_C^h, w_C^h) - \mathcal{M}_C(\dot{g}_C^h, w_C^h) - \mathcal{K}_C(g_C^h, w_C^h) \text{ on }]\,0,\tau\,[, \tag{40}$$

$$(v_H^h(\boldsymbol{x},0), w_H^h) = (H_0 - g_H^h(\boldsymbol{x},0)) \text{ on } \Omega \tag{41}$$

and

$$(v_C^h(\boldsymbol{x},0), w_C^h) = (\bar{C}_0 - g_C^h(\boldsymbol{x},0)) \text{ on } \Omega. \tag{42}$$

7 Finite element matrix approximation of the initial boundary-value problem

The finite element matrix equations are derived from the Galerkin form by defining the approximation of the variational and solution spaces based on a given spatial discretization.

A finite element basis for $\mathcal{S}^h$ and $\mathcal{V}^h$ is defined by using a finite number of linear independent functions $N_a(\boldsymbol{x})$ which span $\mathcal{S}^h$ and $\mathcal{V}^h$,

thus we can write

$$v_H^h(\boldsymbol{x},t) = \sum_{A \epsilon \eta - \eta_{gH}} N_a(\boldsymbol{x}) h_a(t) : [\,0,t\,] \longrightarrow \mathcal{V}_H^h, \tag{43}$$

$$v_C^h(\boldsymbol{x},t) = \sum_{A \epsilon \eta - \eta_{gC}} N_a(\boldsymbol{x}) c_a(t) : [\,0,\tau] \longrightarrow \mathcal{V}_C^h \tag{44}$$

and

$$g_H^h(\boldsymbol{x},t) = \sum_{A \epsilon \eta_{gH}} N_a(\boldsymbol{x}) g_{H_a}(t) : [\,0,\tau] \longrightarrow \mathcal{S}_H^h, \tag{45}$$

$$g_C^h(\boldsymbol{x},t) = \sum_{A \epsilon \eta_{gC}} N_a(\boldsymbol{x}) g_{C_a}(t) : [\,0,\tau] \longrightarrow \mathcal{S}_C^h. \tag{46}$$

From (43), we see that a function in $\mathcal{V}_H^h$ may be represented in terms of a time-varying vector $\boldsymbol{h}$ of N_{EQ}^H components that are coefficients associated with shape functions. Similarly a function in $\mathcal{V}_C^h$ may be represented in terms of a time-varying vector $\boldsymbol{c}$ of N_{EQ}^C components that are coefficients associated with shape functions. Note that the time-dependent coefficients g_{H_a} and g_{C_a} are chosen so that g_H^h is a 'good' approximation of g_H and g_C^h is a 'good' approximation of g_C.

7.1 The Finite Element Matrix Form

The finite element matrix form (M) of the IBVP is:

Given ρ, C_p, $\boldsymbol{\mathcal{K}}$, $\boldsymbol{D}$, g_H, g_C, h_H, h_C, H_0 and $\bar{C}_0$, as in § 2,

find

$$\boldsymbol{h} \; : \; [\,0,\tau\,] \longrightarrow \Re^{N_{eq}^H}$$

and

$$\boldsymbol{c} \; : \; [\,0,\tau\,] \longrightarrow \Re^{N_{eq}^C}$$

such that

$$\boldsymbol{M}\,\dot{\boldsymbol{h}} + \boldsymbol{K}_H(\boldsymbol{h},\boldsymbol{c},t)\boldsymbol{h} = \boldsymbol{F}_H(\boldsymbol{h},t), \tag{47}$$

$$\boldsymbol{M}\,\dot{\boldsymbol{c}} + \boldsymbol{N}_C(\boldsymbol{h},\boldsymbol{c},t) = \boldsymbol{F}_C(\boldsymbol{c},t), \tag{48}$$

$$\boldsymbol{h}(0) = \boldsymbol{h}^0 \tag{49}$$

and

$$\boldsymbol{c}(0) = \boldsymbol{c}^0, \tag{50}$$

where $\boldsymbol{h}(t)$ and $\boldsymbol{c}(t)$ are respectively vectors of nodal enthalpy and solute concentration at time t, and $\boldsymbol{h}^0$ and $\boldsymbol{c}^0$ are a 'good' approximation to the exact initial enthalpy H_0 and $\bar{C}_0$ respectively.

8 Temporal Algorithm

The semi-discrete matrix form (**M**) of the coupled nonlinear ordinary differential equations is discretized in terms of time where the real enthalpy $\boldsymbol{h}(t_n)$, the real solute concentration $c(t_n)$ and their rate terms are approximated by discrete values $\boldsymbol{h}_n$ and $\boldsymbol{c}_n$, and, $\dot{\boldsymbol{c}}_n$ and $\dot{\boldsymbol{h}}_n$, respectively. The discrete solution times are given by $t_n = n\Delta t$, where Δt can be a constant or varying time step depending on the degree of nonlinearity of the problem.

8.1 Generalized Trapezoidal rule

The time-integration method chosen is the generalized trapezoidal rule (**T**). Applying it to the matrix form of the problem (**M**) leads to the following time-integration scheme:
Given $\boldsymbol{M}, \boldsymbol{K}_H, \boldsymbol{F}_H, \boldsymbol{N}_C$ and $\boldsymbol{F}_C$, as in equations (47) - (48),
find

$$\boldsymbol{h}_n \ , \ n \in \{0, 1 \cdots N_{steps}\}$$

and

$$\boldsymbol{c}_n \ , \ n \in \{0, 1 \cdots N_{steps}\} ,$$

such that

$$\boldsymbol{M}\ \dot{\boldsymbol{h}}_{n+1} + \boldsymbol{K}_H(\boldsymbol{h}_{n+1}, \boldsymbol{c}_{n+1}, t_{n+1})\boldsymbol{h}_{n+1} = \boldsymbol{F}_H(\boldsymbol{h}_{n+1}, t_{n+1}), \tag{51}$$

$$\boldsymbol{M}\ \dot{\boldsymbol{c}}_{n+1} + \boldsymbol{N}_C(\boldsymbol{h}_{n+1}, \boldsymbol{c}_{n+1}, t_{n+1}) = \boldsymbol{F}_C(\boldsymbol{c}_{n+1}, t_{n+1}), \tag{52}$$

$$\boldsymbol{h}_0 = \boldsymbol{h}^{\circ}, \tag{53}$$

$$\boldsymbol{c}_0 = \boldsymbol{c}^{\circ}, \tag{54}$$

$$\boldsymbol{h}_{n+1} = \boldsymbol{h}_n + \Delta t \left\{ (1-\alpha)\ \dot{\boldsymbol{h}}_n + \alpha\ \dot{\boldsymbol{h}}_{n+1} \right\} \tag{55}$$

and

$$\boldsymbol{c}_{n+1} = \boldsymbol{c}_n + \Delta t \left\{ (1-\alpha)\ \dot{\boldsymbol{c}}_n + \alpha\ \dot{\boldsymbol{c}}_{n+1} \right\}, \tag{56}$$

where $\alpha = [0, 1]$.
In this algorithm α is chosen so that the solution will be unconditionally stable, as in most solidification problems the solution is sought over very long time periods compared to the stability limit for the explicit form of the operator (i.e. when $\alpha = 0$), Abaqus [11]. Of these algorithms, the central difference method (i.e. $\alpha = \frac{1}{2}$) has the highest accuracy. However, this method tends to produce oscillations in the early time solution. These oscillations are not present in the backward difference method (i.e. $\alpha = 1$). Thus the backward difference method is used.

9 Nonlinear Solution Scheme

The iterative scheme proposed for solving the nonlinear algebraic problem is a variant of Newton-Raphson iteration, which avails of the continuity of the temporal discretization. This *predictor - corrector* method makes use of a fully 'consistent' *linearized* operator to compute solution increments, which results in a non-symmetric linear equation system where quadratic convergence is ensured, Hughes [12].

9.1 Newton-Raphson Iteration Scheme

Obtain values h_{n+1} and c_{n+1} such that the residual $r(h_{n+1}, r_{n+1}) = 0$ where

$$r(h_{n+1}, c_{n+1}) = \begin{bmatrix} r_H(h_{n+1}, c_{n+1}) \\ r_C(h_{n+1}, c_{n+1}) \end{bmatrix} = \begin{bmatrix} 0 \\ 0 \end{bmatrix} = 0, \tag{57}$$

where

$$\begin{aligned} r_H(h_{n+1}, c_{n+1}) &= M(\frac{h_{n+1} - h_n}{\Delta t}) + K_H(h_{n+1}, c_{n+1}, t_{n+1})h_{n+1} \\ &- F_H(h_{n+1}, t_{n+1}) \end{aligned} \tag{58}$$

and

$$\begin{aligned} r_C(h_{n+1}, c_{n+1}) &= M(\frac{c_{n+1} - c_n}{\Delta t}) + N_C(h_{n+1}, c_{n+1}, t_{n+1}) \\ &- F_C(c_{n+1}, t_{n+1}). \end{aligned} \tag{59}$$

Using a Taylor series expansion about the exact solutions c_{n+1} and h_{n+1}, we may approximate the residual r at the values h^i_{n+1} and c^i_{n+1} and by ignoring higher order terms in the Taylor expansion, we may write

$$\begin{bmatrix} \frac{\partial r_H}{\partial h} & \frac{\partial r_H}{\partial c} \\ \frac{\partial r_C}{\partial h} & \frac{\partial r_C}{\partial c} \end{bmatrix}_{\substack{h = h^i_{n+1} \\ c = c^i_{n+1}}} \begin{bmatrix} \Delta h^i_{n+1} \\ \Delta c^i_{n+1} \end{bmatrix} = \begin{bmatrix} -r_H(c^i_{n+1}, h^i_{n+1}) \\ -r_C(c^i_{n+1}, h^i_{n+1}) \end{bmatrix}. \tag{60}$$

The solution of this equation allows a better approximation to the exact solution, thus

$$c^{i+1}_{n+1} = c^i_{n+1} + \Delta c^i_{n+1} \tag{61}$$

and

$$h^{i+1}_{n+1} = h^i_{n+1} + \Delta h^i_{n+1}. \tag{62}$$

In order to evaluate the residual and the Jacobian operator at the specific sampling point $\bar{\xi}(x)$, the mass fraction $\phi(\bar{\xi})$ and the temperature $T(\bar{\xi})$ have to be calculated from the given enthalpy and average solute concentration fields as follows: for time $n+1$ and for the ith global iteration, $\bar{H}^i_{n+1}(\bar{\xi}) = \sum_{i=1}^{nnode} N_i(\bar{\xi})\,(h_i)^i_{n+1}$ and $\bar{C}^i_{n+1}(\bar{\xi}) = \sum_{i=1}^{nnode} N_i(\bar{\xi})\,(c_i)^i_{n+1}$.

These unknown variables, temperature $T(\bar{\xi})$ and solid mass fraction $\phi(\bar{\xi})$ at $\bar{\xi}(x)$, are termed the state variables of the problem. The state variables are based on the micro-mechanics of the phase change problem. Therefore, the state variables, will describe the phase change kinetics of the problem depending on the type of microscopic phase evolution model chosen.

The temperature $T(\bar{\xi})$ and solid mass fraction $\phi(\bar{\xi})$ are calculated using the temperature (T) - enthalpy (H) relationship (shown in Figure 1) coupled with the microscopic phase evolution model. The microscopic phase evolution model is based on the expression for the average solute concentration, equation (6). As the (T - H) relationship is unknown and the phase evolution model is not explicit, a predictor-corrector Newton Raphson type algorithm is used. This is coupled with a trapezoidal integration scheme of the mass fraction evolution, McAdie [13]. The (T - H) curve experiences kinks or sharp change in slope at phase interfaces. At these points, the Newton-Raphson method will not converge, so the secant method employing the *reguli falsi* technique is then used to obtain convergence, McAdie [13] and Minkowycz et al [14].

10 Numerical Considerations

To solve the equations which make up the residual vector and Jacobian matrix numerically, a Newton Cotes numerical integration scheme is chosen. The reason being that the sampling points are at the nodes of the element resulting in a diagonal the mass matrix contribution to the Jacobian. This is equivalent to using a lumped mass method. In contrast, the Gaussian quadrature scheme, with sampling points positioned inside the element, would cause the mass matrix contribution to the Jacobian to be fully consistent. It has been found that oscillations, which appear in the solution when using a fully consistent mass matrix Jacobian contribution, do not occur when this matrix is diagonal (or lumped), as in the Newton Cotes integration scheme. It must be noted that a combination of the two schemes can also be used (ie. Newton Cotes for the mass terms and Gaussian quadrature for the conductivity and diffusivity terms), Abaqus [11].

As the enthalpy solution is not smooth at the phase front low ordered linear elements are used since better behaviour from higher-order elements cannot be expected in this case.

11 Phase Change Problem

The problem under consideration is the solidification of a rectangular cavity as shown in Figure 2. The material ($NH_4Cl - H_20$) is initially all liquid at a temperature T_0 which is greater than the liquidus temperature corresponding to the initial uniform average solute concentration $\bar{C}_0$, see Figure 2 . At time $t = 0$, the temperature of the left wall is lowered to value T_b which is less than the eutectic temperature T_e. All other walls of the cavity are insulated. Due to cooling, the material near the left wall begins to freeze, so that the cavity consists of a solid phase near the left wall, liquid near the right wall, and a two phase mushy region in between. Thermophysical data used in this problem is based on that for $NH_4Cl - H_20$, which is given in McAdie [13]. A linear phase diagram is used for the problem, see Figure 2. This problem is similar to that considered by Bennon and Incropera, [2] and McAdie *et al* [15].

The initial temperature T_0 is 570K, an initial average solute concentration $\bar{C}_0$ of H_20 is 15% and a chill wall temperature T_b is 100K as shown in Figure 2.

Due to the nature of the problem there will be no vertical gradients in temperature and solute concentration in these examples. Therefore the whole domain needs not be analyzed. Instead, a single strip of elements extending through the width of the domain will be adequate to obtain results for the whole domain, as shown in Figure 2. The boundary conditions on the two sides of the mesh in contact with the solidifying liquid are treated as being insulated as there will be no flow of heat or solute across these boundaries.

12 Results and Discussion

In this example two different local models ("local solute equilibrium" and local solute non-equilibrium") were used to observe the degree of the local/global coupling and whether significant differences arise when comparing the global influence of the two local models. Accuracy of the finite element solution was check by comparing the results for a reasonably coarse mesh against that of a refined mesh and the results were compared to that in the literature [2].

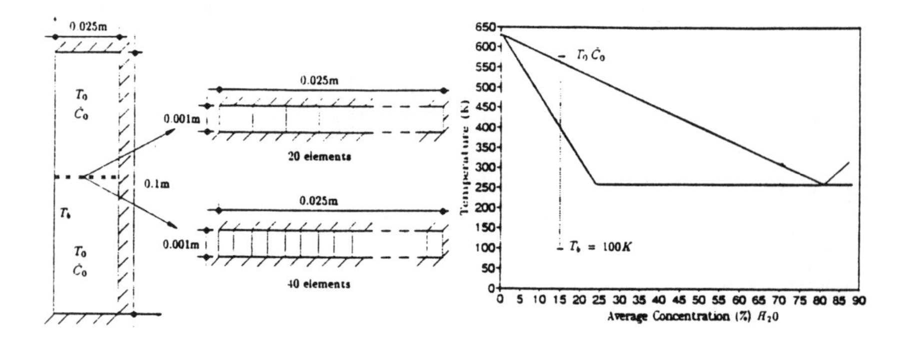

Figure 2: Phase diagram of a $NH_4Cl - H_20$ showing the initial and boundary values of T and $\bar{C}$, and the domain of the problem with the representative finite element meshes that are used.

The solid mass fraction plot in Figure 3 shows a clear difference in the evolution in mass fraction between the local "equilibrium" and local "non-equilibrium" based models. The "equilibrium" based model solidifies at a higher temperature than the "non-equilibrium" based model as more solute is rejected into the liquid for the "non-equilibrium" case lowering the freezing temperature. This results in all the solidification being dendritic for the "equilibrium" case and approximately 90 % being dendritic, with the remainder being eutectic, for the "non-equilibrium case". The concentration of the liquid solute ahead of the advancing front, which is a direct function of temperature, is approx 30 % greater for the "non-equilibrium" case than for the equlibrium case. Even though the maximum change in average solute concentration from $\bar{C_0}$ is of the order of 1%, the variations in $\bar{C}$ for the local "non-equilibrium" based model were greater than for the local "equilibrium" based model. Therefore, greater segregation occurs when using the local "non-equilibrium" based model. There is little difference between results obtained using the 40 element mesh as opposed to the 20 element mesh thus showing a convergence of the results to a consistent solution similar to that in the literature [2]. For the "non-equilibrium" case the final solidification, approx 10 %, occured at the eutectic. This discrete eutectic phase change did not result in any numerical difficulties as the jump in enthalpy was within the tolerance of the Newton Raphson method.

13 Conclusion

A coupled nonlinear heat conduction/mass diffusion phase change formulation has been developed using an enthalpy technique for the fixed grid finite element method. The results clearly demonstrate that the local/global coupling, and consequently the coupling between the solute and temperature/enthalpy fields, have a significant effect on the solution of the phase change problem. The enthalpy method is demonstrated to be proficient and accurate when dealing with this type of mushy/discrete phase change problem where energy conservation is ensured.

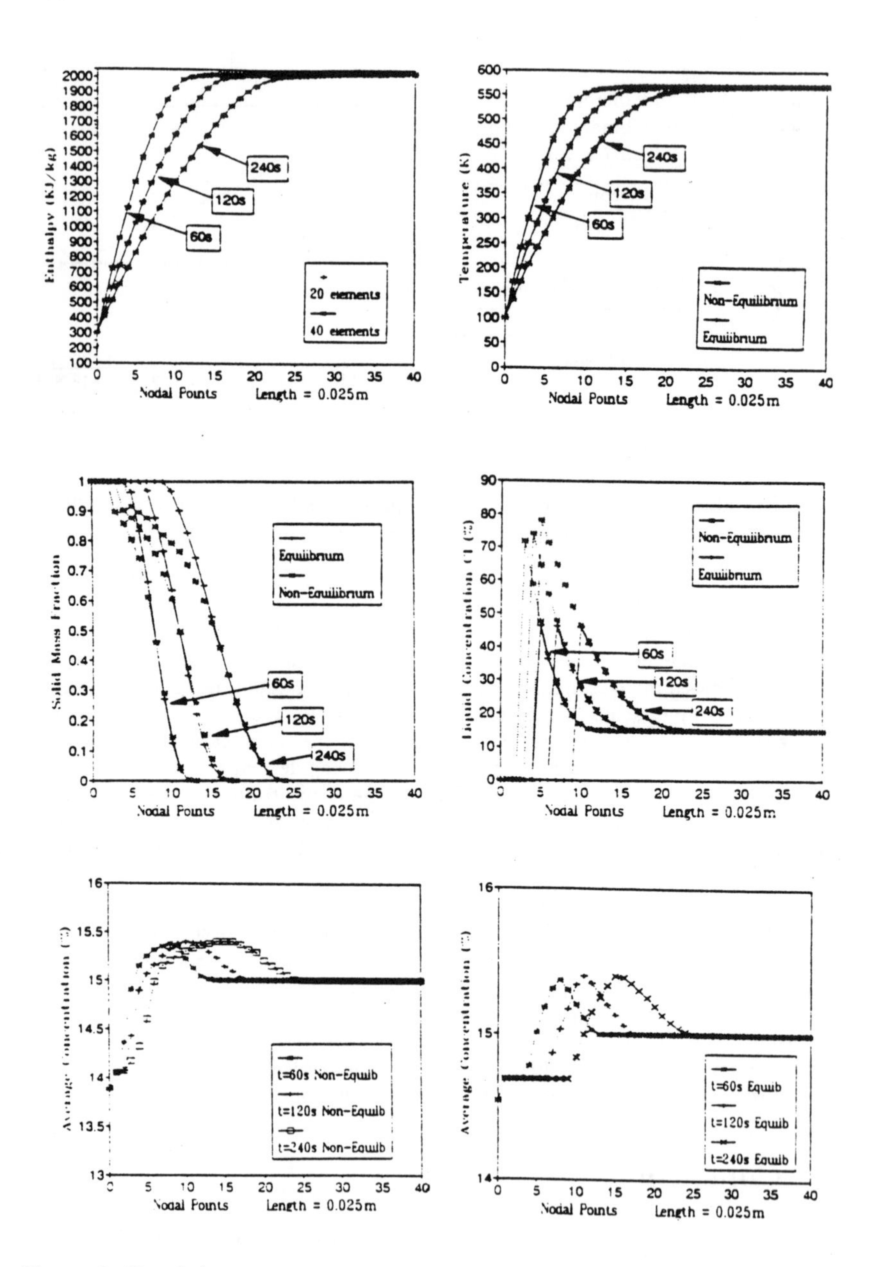

Figure 3: Entahalpy, temperature, Solid mass fraction, liquid concentration and average solute concentration

References

[1] Bennon, W.D. and Incropera, F.P., A continuum model for momentum, heat and species transport in binary solid-liquid phase change systems - 1. model formulation. *Numerical Heat Transfer*, **30**, 2161–2170, 1987.

[2] Bennon, W.D. & Incropera, F.P., Numerical analysis of binary solid-liquid phase change using a continuum model, *Numerical Heat Transfer*, **13**, 277–296, 1988.

[3] Voller, V.R., Brent, A.D. & Prakash, C., The modelling of heat, mass and solute transport in solidification systems, *International Journal of Heat Mass Transfer*, **32**, No. 9, 1719–1731, 1989.

[4] Kececioglu, I. & Rubinsky, B., A Continuum model for the propagation of discrete phase-change fronts in porous media in the presence of coupled heat flow, fluid flow and species transport processes, *International Journal of Heat and Mass Transfer*, **32**, 1111-1130, 1989.

[5] Bonnerot, R. & Jamet, P., Numerical computation of the free boundary for the two dimension Stefan problem by space time finite elements, *Journal of Computational Physics*, **25**, 163-181, 1977.

[6] Prakash, C. & Voller, V., On the numerical solution of continuum mixture model equations describing binary solid-liquid phase change, *Numerical Heat Transfer*, **15**, Part B, 171-189, 1989.

[7] Poirier, D.R. & Heinrich, J.C., Simulations of Thermosololutal Convection in Direction Solidification, *Modelling of Casting and Welding and Advanced Solidification Processes VI*, Edited by Piwonka, T.S., Voller, V. & Katgerman, L., TMS, 227–234, 1993.

[8] Carslaw, H.S. & Jaeger, J.C., Conduction of Heat in Solids, Oxford University Press, Oxford, 1959.

[9] Mitchell, A.R. & Wait, R., The Finite Element Method in Partial Differential Equations, Wiley, New York, 1977.

[10] Strang, G. & Fix, G.J., Analysis of the Finite Element Method, Prentice-Hall, New Jersey, 1973.

[11] Hibbit, Karlson, and Inc. Sorenson., *ABAQUS Users Manual, Version 4.9.* Providence, Rhode Island, USA.

[12] Hughes, T.J.R. & Pister, K.S., Consistent Linearization in mechanics of Solids, *Computers and Structures*, **8**, 391-397, 1978.

[13] McAdie, R.L., *PhD Thesis: Modelling of the Binary Alloy Solidification Process.* University of Cape Town, Cape Town, R.S.A., 1992.

[14] Minkowycz, W.J., Sparrow, E.M., Schneider, G.E. and Pletcher, R.H., *Handbook of Numerical Heat Transfer*. John Wiley & Sons, INC., New York, New York, U.S.A., 1988.

[15] McAdie, R.L., Martin, J.B. & Lewis, R.W.L., Finite Element Modelling of the Binary Alloy Solidification Process, *Modelling of Casting and Welding and Advanced Solidification Processes VI*, Edited by Piwonka, T.S., Voller, V. & Katgerman, L., TMS, 169–177 1993.

HEAT FLOW IN MATERIALS DURING MELTING AND RE–SOLIDIFICATION INDUCED BY A SCANNING LASER BEAM

M. Zerroukat and C.R. Chatwin

Department of Mechanical Engineering, University of Glasgow, Glasgow G12 8QQ, Scotland, U.K.

ABSTRACT

A new explicit finite difference equation for solution of the non–linear heat conduction equation is developed and validated by solving a normalised problem which has an analytical solution. A virtual sub–interval elimination technique is incorporated into the algorithm to insure the stability of the scheme irrespective of the mesh size; hence, avoiding the necessity for large array sizes. The finite difference scheme is then utilised to simulate the melting and re–crystallisation of silicon when subjected to radiation by a scanning cw–laser beam. The model incorporates temperature dependence of material properties and the surface reflectivity; the variation of the incident power density with time is also taken into account. The numerical scheme is also applied to the laser heating and melting of nodular cast iron; the results are compared with published experimental results.

1. INTRODUCTION

Laser processing of surfaces, which involves rapid localised heating of materials to modify its physical properties by altering the surface structure or distribution of impurities, has found many applications in manufacturing and particularly in semi–conductor technology [1–3]. Typically a high–power laser beam is utilised to rapidly heat a thin surface region, with the underlying bulk substrate providing self quenching.

To achieve a desired effect on a specific material it is necessary to make the optimum choice of process parameters such as: power, scan velocity and the beam diameter. To avoid a costly time–consuming experimental trial–and–error search to obtain the optimal combination; and in order to assess the influence of the laser parameters on the process, it is necessary to construct a mathematical model describing the laser–material interaction process.

Laser materials processing technology allows beam–workpiece interaction times ranging from a few picoseconds duration to several seconds; the concomitant incident energy densities, which induce observably reproducible effects, range from several mJ/cm^2 to some thousands of J/cm^2. For processes with high power and picoseconds interaction times, the heat diffusion length becomes comparable both to the radiation absorption length – which is of the order of 20 nm in most metals – and to the mean free path l of the heat carriers (i.e. the conduction electrons for which l=20 nm at room temperature). Under these extreme conditions, the linearity between the heat flux and temperature gradient breaks down; for these circumstances the energy transport into the material can be described using the kinetic theory rather than the Fourier conduction theory [4]. However, for interaction times greater than some picoseconds, the heat conduction process is correctly described by the Fourier conduction equation [5].

A new explicit finite difference equation for the heat conduction equation – where thermal properties are a function of temperature – is developed, and validated by solving a normalised problem which has an analytical solution. In order to eliminate the constraint of using a small mesh size, a virtual sub interval elimination technique is incorporated to insure the stability of the scheme irrespective of the mesh size without loss of accuracy. The finite difference scheme is then utilised to simulate the melting and re–crystallisation of silicon when subjected to radiation by a scanning cw–laser beam. The process is physically described as follows: a laser beam moving with a uniform velocity scans a silicon substrate of infinite length and width but finite depth. The incident power is partly reflected and partly absorbed, depending on the reflectivity of the surface. Some of the absorbed energy is lost by re–radiation and convection from both the upper and lower surfaces into the surrounding environment, whilst the rest is conducted throughout the substrate. When the beam has completely passed, the melted zone is self quenched by the solid substrate and the heat lost from the surfaces. The numerical scheme is also applied to the laser heating and melting of nodular cast iron; the results are compared with published experimental results.

2. FINITE DIFFERENCE EQUATION FOR NON-LINEAR HEAT CONDUCTION EQUATION

The one-dimensional heat conduction equation in homogenous material with variable thermal properties is given by:

$$C(T)\frac{\partial T}{\partial t}=\frac{\partial}{\partial x}\left(K(T)\frac{\partial T}{\partial x}\right) \tag{1}$$

where $T(x,t)$, $C(T)$ and $K(T)$ are temperature at the coordinates (x,t), heat capacity/unit volume and thermal conductivity respectively. Equation (1) can also be written as:

$$C(T)\frac{\partial T}{\partial t}=K(T)\frac{\partial^2 T}{\partial x^2}+\frac{\partial K}{\partial x}\left(\frac{\partial T}{\partial x}\right) \tag{2}$$

Writing (2) for the grid line i, situated at $x_i = i\Delta x$ relative to x=0, gives:

$$\left(\frac{\partial T}{\partial t}\right)_i=\frac{K(T_i)}{C(T_i)}\left(\frac{\partial^2 T}{\partial x^2}\right)_i+\frac{1}{C(T_i)}\left(\frac{\partial K}{\partial x}\right)_i\left(\frac{\partial T}{\partial x}\right)_i \tag{3}$$

Using a three point Lagrange formula, for the grid lines $(i-1)$, i and $(i+1)$; $(\partial K / \partial x)_i$ can be approximated by:

$$\left(\frac{\partial K}{\partial x}\right)_i = \frac{K_{i+1} - K_{i-1}}{2\Delta x} \tag{4}$$

Making use of the following notations:

$$\frac{1}{C(T_i)}\left(\frac{\partial K}{\partial x}\right)_i = -\frac{K_{i-1} - K_{i+1}}{2\Delta x C_i} = -\gamma_i \quad , \quad \frac{K(T_i)}{C(T_i)} = \frac{K_i}{C_i} = \alpha_i \tag{5}$$

equation (3) is written as:

$$\left(\frac{\partial T}{\partial t}\right)_i = \alpha_i \left(\frac{\partial^2 T}{\partial x^2}\right)_i - \gamma_i \left(\frac{\partial T}{\partial x}\right)_i \tag{6}$$

Following the procedure of Bhattacharya [6]; the finite difference solution of (6) – expressing the temperature at the node $(x_i = i\Delta x, t_{j+1} = t_j + \Delta t)$, as a function of those at nodes (x_{i-1}, t_j), (x_i, t_j) and (x_{i+1}, t_j)– is given by:

$$T_{i,j+1} = T_{i,j} \exp\left(-r_{i,j}\, \psi_{i,j}\right) \tag{7}$$

where

$$\psi_{i,j} = \left(\frac{1}{T_{i,j}}\right)\left\{2T_{i,j} - \left(1 + \beta_{i,j}\right)T_{i-1,j} - \left(1 - \beta_{i,j}\right)T_{i+1,j}\right\} \tag{8}$$

where

$$\beta_{i,j} = \frac{K_{i-1,j} - K_{i+1,j}}{4K_{i,j}} \quad , \quad \alpha_{i,j} = \frac{K_{i,j}}{C_{i,j}} \quad , \quad r_{i,j} = \alpha_{i,j}\frac{\Delta t}{\Delta x^2} \tag{9}$$

$\psi_{i,j}$ can also be evaluated using the approach in [7] yielding to a more lengthy equation:

$$\psi_{i,j} = \beta_{i,j}^2 + \left(\frac{1}{T_{i,j}}\right)\left\{\begin{aligned}&2T_{i,j} - \left(1 + \beta_{i,j}\right)T_{i-1,j} - \\ &\left(1 - \beta_{i,j}\right)T_{i+1,j} - \frac{\beta_{i,j}^2}{2}\left(T_{i-1,j} + T_{i+1,j}\right)\end{aligned}\right\} \tag{10}$$

When $T_{i,j} \to 0$, $T_{i,j}\exp(-r_{i,j}\psi_{i,j}) \to T_{i,j} - r_{i,j}\psi_{i,j}T_{i,j}$; this eliminates $T_{i,j}$ from denominator of $\psi_{i,j}$, given by either (8) or (10); replacing $T_{i,j} = 0$ in what remains of the equation yields the finite difference equation for the special case of $T_{i,j} = 0$.

The classical finite difference equation of (1) is given by [8]:

$$C_{i,j}\frac{T_{i,j+1} - T_{i,j}}{\Delta t} = \frac{1}{\Delta x}\left(K_{i-1/2,j}\frac{T_{i-1,j} - T_{i,j}}{\Delta x} - K_{i+1/2,j}\frac{T_{i,j} - T_{i+1,j}}{\Delta x}\right) \tag{11}$$

where

$$K_{i-1/2,j} = \frac{K_{i-1,j} + K_{i,j}}{2} \quad , \quad K_{i+1/2,j} = \frac{K_{i,j} + K_{i+1,j}}{2} \tag{12}$$

In equation (12), a linear interpolation is used for the thermal conductivity between each grid point giving K at the midpoint of the space interval separating the grid points. $K_{i-1/2,j}$ and $K_{i+1/2,j}$ can also be evaluated using the approach of summing the thermal resistance as suggested by Patankar [9]. This gives:

$$K_{i-1/2,j}=\frac{2K_{i-1,j}K_{i,j}}{K_{i-1,j}+K_{i,j}} \quad , \quad K_{i+1/2,j}=\frac{2K_{i,j}K_{i+1,j}}{K_{i,j}+K_{i+1,j}} \tag{13}$$

2.1 Virtual Sub-Interval Elimination Technique (VSIET)

This technique permits the use of large time steps for any of the explicit equations (Eq.(7) or (11)) without loss of accuracy or stability. It consists of automatic generation of Virtual Sub-time-Steps (VSSs), which are progressively eliminated as the computation moves forward [10]. For illustration, let us consider that the temperature distribution $T_{i,j}, i=0,N$ is known and we wish to compute the temperature distribution at the time $t_{j+1}=t_j+\Delta t$, using (7) and (8). The temperature distribution $-T^n_{i,j}, i=0,N$ – at any VSS of index n=1,k is calculated from that of previous VSS using:

$$T^n_{i,j}=T^{n-1}_{i,j}\exp\left\{\frac{-r^{n-1}_{i,j}}{T^{n-1}_{i,j}}\left[2T^{n-1}_{i,j}-\left(1+\beta^{n-1}_{i,j}\right)T^{n-1}_{i-1,j}-\left(1-\beta^{n-1}_{i,j}\right)T^{n-1}_{i+1,j}\right]\right\} \tag{14}$$

where

$$r^n_{i,j}=\frac{\alpha^n_{i,j}}{\max\left(\alpha^n_{i,j}, i=0,N\right)}r_f \quad , \quad \alpha^n_{i,j}=\frac{K^n_{i,j}}{C^n_{i,j}} \quad , \quad \beta^n_{i,j}=\frac{K^n_{i-1,j}-K^n_{i+1,j}}{4K^n_{i,j}} \tag{15}$$

where $r_f \le 0.5$ is a fixed Fourier number throughout the computation process. At some intermediate VSS $n=k$, so that $t^k_j \ge t_{j+1}$, the temperature distribution at $t=t_{j+1}$, is calculated from that of $t=t^{k-1}_j$, using equation (14) (i.e. $T_{i,j+1}=T^k_{i,j}, i=0,N$) with $r^{k-1}_{i,j}=\alpha^{k-1}_{i,j}(t_{j+1}-t^{k-1}_j)/\Delta x^2$.

In order to validate the finite difference equation and compare its performance with the classical equations, let us consider the following problem:

$$C(T)\frac{\partial T}{\partial t}=\frac{\partial}{\partial x}\left(K(T)\frac{\partial T}{\partial x}\right) \quad , \quad 0\le x\le 1 \quad , \quad t>0 \tag{16}$$

$$T(x,t)=\phi(t) \quad , \quad x=0 \quad , \quad t\ge 0 \tag{17}$$

$$T(x,t)=\varphi(t) \quad , \quad x=1 \quad , \quad t\ge 0 \tag{18}$$

$$T(x,t)=\eta(x) \quad , \quad 0\le x\le 1 \quad , \quad t=0 \tag{19}$$

Let us assume the following dependencies:

$$K(T)=d\,\mathrm{Log}_e(T) \quad , \quad C(T)=\frac{dc^2}{b}\left(1+\mathrm{Log}_e(T)\right) \tag{20}$$

If we choose the following boundary conditions:

$$\eta(x)=\exp(-cx) \quad , \quad \phi(t)=a\exp(bt) \quad , \quad \varphi(t)=a\exp(bt-c) \tag{21}$$

The following is an exact solution for equations (16) to (19):

$$T(x,t)=a\exp(bt-cx) \tag{22}$$

where a, b, c and d are constants.

The problem defined by (16) to (19) is solved using: the classical equation (i.e. (11)&(12)), the equation of Patankar (i.e. (11)&(13)) and the new finite difference equations (i.e. (7)&(8) and (7)&(10)). The VSIET is incorporated with all the equations. For accuracy assessment, the numerical results are compared to the analytical solution and expressed in terms of percentage deviations defined as:

$$MPD_j = \max\left(\left|PD_{i,j}\right| i = 0, N\right) \quad , \quad PD_{i,j} = 100 \times \frac{\left(T_{i,j}\right)_{Nu} - \left(T_{i,j}\right)_{An}}{\left(T_{i,j}\right)_{An}} \tag{23}$$

where An and Nu refer to temperature calculated Analytically and Numerically respectively. Numerical results, omitting the units of variables, (when a=20, b=4, c=1/2, d=2.0, Δx=0.1, Δt=0.01 and $r_f = 0.35$), are shown in tables 1 and 2. These are discussed in section 5.

Table 1: *Comparison of percentage deviations (x10²) at the centre of the slab.*

time step		Classical (11)&(12)	Patankar (11)&(13)	Present equations (7)&(10)	Present equations (7)&(8)	Analytic
j	t_j	$PD_{N/2,j}$	$PD_{N/2,j}$	$PD_{N/2,j}$	$PD_{N/2,j}$	$T\left(\frac{1}{2}, t_j\right)$
1	0.01	−0.1694	−0.1788	0.05647	+0.03765	16.2117
2	0.02	−0.2080	−0.2261	0.05426	+0.03617	16.8733
3	0.03	−0.2346	−0.2520	0.01738	+0.01738	17.5619
4	0.04	−0.2421	−0.2588	0.00000	−0.00835	18.2786
5	0.05	−0.2567	−0.2647	0.00802	−0.00802	19.0246
6	0.06	−0.2620	−0.2697	0.00000	−0.03082	19.8010
7	0.07	−0.2369	−0.2517	0.00740	+0.00000	20.6091
8	0.08	−0.2276	−0.2419	0.00711	+0.00711	21.4501
9	0.09	−0.2119	−0.2255	0.02050	+0.02050	22.3255
10	0.10	−0.2036	−0.2101	0.02627	+0.02627	23.2367

Table 2: *Comparison of maximum percentage deviation (x10²).*

time step	Classical (11)&(12)	Patankar (11)&(13)	Present equations (7)&(10)	Present equations (7)&(8)
j	MPD_j	MPD_j	MPD_j	MPD_j
1	0.1694	0.1788	0.07297	0.07021
2	0.2080	0.2261	0.07683	0.05844
3	0.2346	0.2520	0.04801	0.05581
4	0.2457	0.2633	0.02667	0.02728
5	0.2614	0.2698	0.01864	0.02071
6	0.2620	0.2697	0.00852	0.03240
7	0.2369	0.2517	0.01557	0.00818
8	0.2276	0.2468	0.00786	0.00869
9	0.2227	0.2299	0.02601	0.02050
10	0.2071	0.2140	0.03123	0.02761

3. MATHEMATICAL FORMULATION OF LASER BEAM–MATERIAL INTERACTION

As the laser beam diameter is very small compared to the lateral dimensions of the material, and the power distribution is symmetric with respect to the beam centre (i.e. Gaussian); the approximation of the heat flow by a one-dimensional model is not expected to introduce significant errors [11], since the maximum heat effect occurs when the surface is irradiated by the centre part of the laser beam. The process can be adequately modelled as a 1–D moving boundary problem with variable thermal properties. The governing equations, describing the heat conduction into a material of thickness X with change of phase, are written as:

$$C_l(T_l)\frac{\partial T_l}{\partial t} = \frac{\partial}{\partial x}\left(K_l(T_l)\frac{\partial T_l}{\partial x}\right) \quad , \quad 0 \le x \le M(t) \quad , \quad t_m \le t \le t_e \tag{24}$$

$$C_s(T_s)\frac{\partial T_s}{\partial t} = \frac{\partial}{\partial x}\left(K_s(T_s)\frac{\partial T_s}{\partial x}\right) \quad , \quad M(t) \le x \le X \quad , \quad t > 0 \tag{25}$$

$$\left.\begin{aligned} \lambda\frac{dM}{dt} &= K_s(T_m)\left(\frac{\partial T_s}{\partial x}\right) - K_l(T_m)\left(\frac{\partial T_l}{\partial x}\right) \\ T_s(x,t) &= T_l(x,t) = T_m \end{aligned}\right\} \quad , \quad x = M(t) \quad , \quad t_m \le t \le t_e \tag{26}$$

where: M, λ, T_m, t_m and t_e refer to the liquid/solid interface position, latent heat of melting/solidification per unit volume, melting temperature, time at which melting starts and time at which solidification ends, respectively. The subscripts s and l refer to solid and liquid regions respectively. Equations (24)–(26) may also be written with a heat generation term, due to penetration of the radiation. However, calculations show that energy absorption is predominantly a surface phenomenon [12]; hence, the heat generation term is neglected in (24)–(26). The boundary conditions are written as:

$$-K_l(T_l)\left(\frac{\partial T_l}{\partial x}\right) = F(T_l,t) - h(T_l,T_0)(T_l - T_0) \quad , \quad x = 0 \quad , \quad t_m \le t \le t_e \tag{27}$$

$$-K_s(T_s)\left(\frac{\partial T_s}{\partial x}\right) = h(T_s,T_0)(T_s - T_0) \quad , \quad x = X \quad , \quad 0 \le t \le t_e \tag{28}$$

where h is the heat transfer coefficient and $F(T,t)$ is the absorbed power density – at the centre of the beam – by the surface material, which is at temperature T, at time t. Assuming that the laser beam has a Gaussian distribution of radius ω at the $1/e$ point; if the laser beam has a total incident power P and is moving with a constant velocity v, then $F(T,t)$ is given by:

$$F(T,t) = \begin{cases} [1 - R(T)]\dfrac{P}{\pi\omega^2}\exp\left\{-\left(\dfrac{vt-\omega}{\omega}\right)^2\right\} & , \quad 0 \le t \le \tau \\ 0 & , \quad t > \tau \end{cases} \tag{29}$$

where $\tau = \omega / v$ and R are the beam–workpiece interaction time and the reflectivity of the surface, respectively. From both surfaces (x=0 and x=X) the heat is lost by a combination of free–convection and radiation. The function $h(T_1,T_2)$– expressing the heat transfer coefficient, due to both convection and radiation, of a plate of thickness X at temperature T_1 sub-merged into a fluid at temperature T_2 – is given by:

$$h(T_1,T_2) = \sigma[1-R(T_1)](T_1^2+T_2^2)(T_1+T_2)+ z\frac{\kappa(T_f)}{X}\left(\frac{c_p(T_f)}{\kappa(T_f)}\frac{2g\rho^2(T_f)(T_1-T_2)X^3}{(T_1+T_2)\mu(T_f)}\right)^y \tag{30}$$

where $T_f = \frac{1}{2}(T_1+T_2)$, σ and g: are temperature of the air film, Stefan-Boltzman constant and gravity respectively; c_p, μ, ρ, and κ denotes heat capacity, dynamic viscosity, density and thermal conductivity of air respectively; these variables are temperature dependent. The correlation factors z and y are equal to: z=0.14 and y=1/3, for the surface facing upwards; z=0.58 and y=1/5 for the surface facing downwards [13].

3.1. Application to silicon

The variation of the thermal conductivity and the heat capacity/unit volume of silicon with temperature is given by [14]:

$$K(T) = 299\times 10^2\left(\frac{1}{T-99}\right),\quad C(T) = 2.336\times 10^6\left(\frac{T-159}{T-99}\right) \tag{31}$$

The dimensions in (31) are: $T\,[^\circ\mathrm{K}]$, $K[\mathrm{W/m^\circ K}]$ and $C[\mathrm{J/m^{3\circ}K}]$. The melting temperature of silicon is $T_m = 1690\ ^\circ\mathrm{K}$ and the latent heat of fusion per unit volume $\lambda = 3282\times 10^6\ \mathrm{J/m^3}$. For liquid silicon the thermal conductivity increases to 0.64 [W/mK], while $C(T)$ is taken to be constant [11]. Experimental evidence [15,16] shows that the reflectivity of silicon varies linearly with temperature for temperatures less than the melting point. However, a more accurate function, for a large range of temperatures including melting, is given by [14]:

$$R(T) = 0.372 + 2.693\times 10^{-5}T + 2.691\times 10^{-15}T^4,\quad T \le 3130\ ^\circ\mathrm{K} \tag{32}$$

3.2. Application to nodular cast–iron

The dependence of thermal conductivity and the heat capacity/unit volume of nodular cast iron with temperature are given in [17]. The reflectivity as a function of temperature is approximated with a linear function [18], which is given by:

$$R(T) = 0.72 - 25\times 10^{-5}(T-293)\quad,\qquad T\,[^\circ\mathrm{K}] \tag{33}$$

For both processes (silicon and nodular cast-iron), the functions describing the variation of thermal properties of air with temperature are obtained by approximating the experimental data in [19] by a polynomial of degree 5.

4. NUMERICAL COMPUTATION SCHEME

Prior to melting ($T(0,t) < T_m$) when only one phase is present, the numerical computation is similar to that illustrated in section 2. When the temperature of the surface, exposed to laser radiation, exceeds the melting point, the heat in the superheated region $0 \le x \le x_1$ (i.e. $T(x,t_m) > T_m$) is converted to heat of fusion, giving an initial melt depth x_2 at a uniform melting temperature [20]. This can be expressed mathematically as:

$$x_2 = \frac{C(T_m)}{\lambda} \int_0^{x_1} (T(x,t_m) - T_m)dx \tag{34}$$

Generally the superheated region extends less than a space increment ($x_1 < \Delta x$); therefore, $T(x,t_m)$ in (34) is approximated by a linear function of x; hence, x_2 is directly obtained without recourse to numerical integration.

During the melting and solidification stages, the problem is to compute the temperature distribution $T_{i,j+1}, i = 0,N$ as well as the melting front position M_{j+1} from those at the previous time t_j. The computing procedure is the same as that in section 2, except for the nodes near the moving interface. Consider that the moving boundary is located between i_m and $i_m + 1$. For grid lines $i \in \{0 \le i \le i_{m-1}\} \cup \{i_{m+2} \le i \le N\}$ equation (14) is used. For the grid lines i_m and $i_m + 1$, instead of (14) the following is used:

$$T_{i,j}^n = T_{i,j}^{n-1} \exp\left(-\frac{r_{i,j}^{n-1}}{T_{i,j}^{n-1}} \Omega_{i,j}^{n-1} \right) \quad , \quad i = i_m, i_m + 1 \tag{35}$$

where

$$\Omega_{i,j}^{n-1} = \begin{cases} \left[1 + \xi + (\xi - 1)\beta_{i,j}^{n-1}\right]T_{i,j}^{n-1} - \left(\xi + \xi\beta_{i,j}^{n-1}\right)T_{i-1,j}^{n-1} - \left(1 - \beta_{i,j}^{n-1}\right)T_m, i = i_m \\ \left[2 - \xi\left(1 - \beta_{i,j}^{n-1}\right)\right]T_{i,j}^{n-1} - \left[(1-\xi)\left(1 - \beta_{i,j}^{n-1}\right)\right]T_{i+1,j}^{n-1} - \left(1 + \beta_{i,j}^{n-1}\right)T_m, i = i_m + 1 \end{cases} \tag{36}$$

where

$$\beta_{i,j}^{n-1} = \begin{cases} \left[(1-\xi)K_{i,j}^{n-1} + \xi K_{i-1,j}^{n-1} - K(T_m)\right] / 4K_{i,j}^{n-1} & , \quad i = i_m \\ \left[-\xi K_{i,j}^{n-1} + (\xi - 1)K_{i+1,j}^{n-1} + K(T_m)\right] / 4K_{i,j}^{n-1} & , \quad i = i_m + 1 \end{cases} \tag{37}$$

where $\xi = \mathrm{mod}\left(M_j^{n-1}, \Delta x\right)$. Equations (36) and (37) are derived using the same approximations, for the nodes near the moving boundary, used in [10].

The temperature distribution at the time t_{j+1} is calculated from that of the last VSS (i.e. $T_{i,j+1} = T_{i,j}^k, i = 0,N$). Equation (14) is used for grid lines $i \in \{0, i_m - 1\} \cup \{i_m + 2, N\}$ and (35) for $i = i_m, i_m + 1$. The position of the melting front is given by:

$$M_{j+1} = M_j + \sum_{n=1}^{k} \chi_j^n \tag{38}$$

where χ_j^n, which is the variation of the melting front during each virtual time step $\delta t_j^n = t_j^n - t_j^{n-1}$, is given in the appendix.

5. NUMERICAL RESULTS AND DISCUSSION

Table 1 shows that the relative errors using the classical equations are greater than that of the new equations. It also shows that both the classical finite difference equations under-estimate the solution; whereas, (7)&(10) give an over-estimate; however, the numerical results due to (7)&(8) oscillate around the analytical solution. Table 2 shows the maximum percentage deviation at each time step; overall the new finite difference equations are 4 times more accurate than the classical ones. Equations (8) and (10) achieve similar accuracy; however, equation (8) requires less operations.

Figure 1 and 2 shows the variation of the melt depth with the scanning velocity and the incident power, for a silicon substrate of thickness X=0.5mm irradiated with a laser beam of ω=1mm. Figure 1 shows that as the velocity of the beam increases the melt depth decreases, whereas it increases when increasing the total incident power, as shown in figure 2. Figure 2 also shows that the melt depth is sensitive to the laser power at low scanning speeds, this sensitivity is less apparent at high scan speeds. Numerical results show that the melt depth is generally sensitive to the change in either the total incident power or the beam diameter; therefore, the scanning speed is the most useful laser parameter for high precison control of the process.

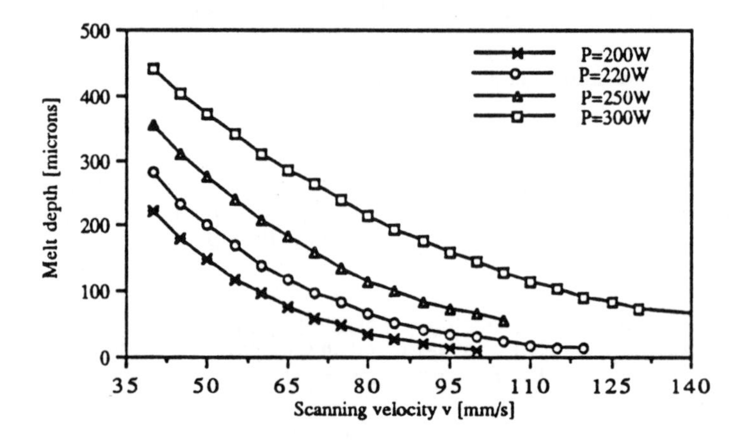

Figure 1: Variation of melt depth with the scanning speed at different incident powers – Silicon.

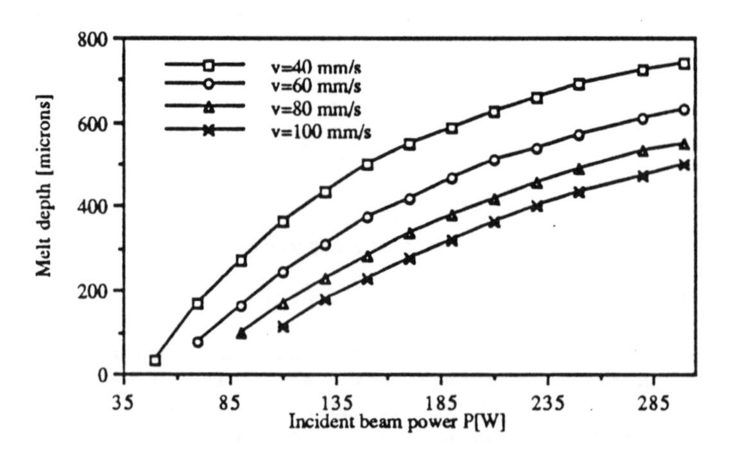

Figure 2: Variation of melt depth with the incident power at different scanning velocities – Silicon.

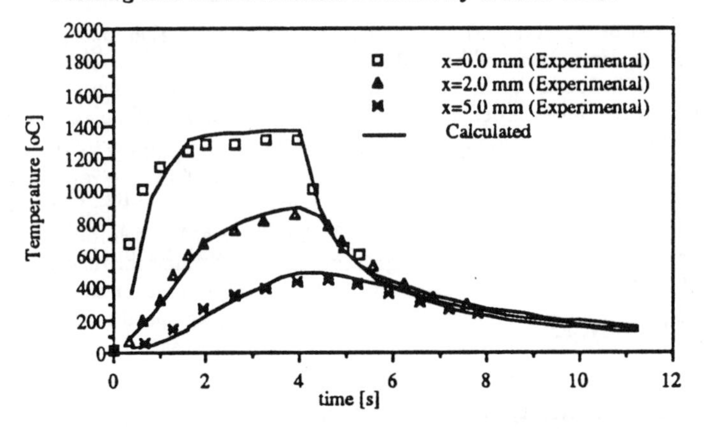

Figure 3: Comparison of measured and calculated temperature histories at different depths for nodular cast-iron. P=3.3KW, ω=16 mm

In order to measure the temperature evolution in a substrate material when irradiated by a laser beam, Gay [17] used large interaction times (2-4s) and large beam diameters (14-16mm) to achieve lower heating rates. The same experimental environment was used in the calculation of temperature profile during the radiation of nodular-cast-iron substrate. Figure 3 shows a comparison between the measured [17] and calculated temperature evolution at different depths, when a 15 mm thick substrate is irradiated by a 16 mm diameter, 3.3 KW laser beam, for 4 seconds. Figure 4 shows the comparison between the measured and predicted melt depth for different interaction times when the substrate is irradiated by a 3KW laser beam, 10 mm diameter. The agreement is satisfactory.

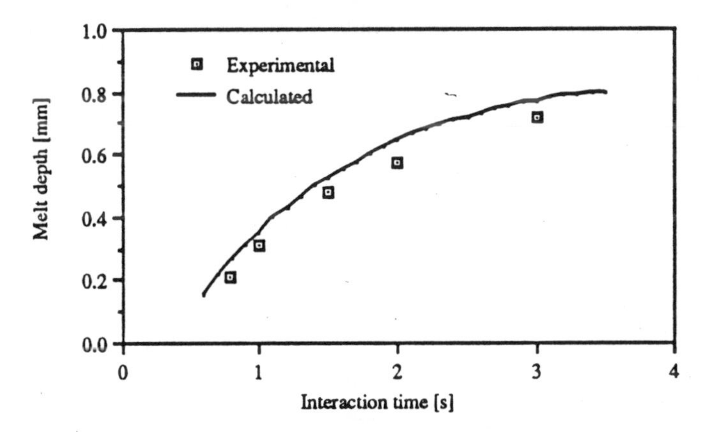

Figure 4: Comparison of measured and calculated melt depths for different interaction times for nodular cast-iron. P=3KW, ω=10 mm.

6. CONCLUSION

A finite difference equation for non-linear heat conduction problems has been presented and proved to be superior in terms of accuracy, when compared to the classical equations. The equation is also incorporated into a model for simulation of heat flow during melting and re–crystallisation of material induced by a scanning laser beam. The calculation includes temperature dependence of thermal properties and surface reflectivity. The numerical results compare very favourably with experimental results. Silicon was chosen due to its importance to the electronics industry, and nodular cast iron for comparison with experimental results. The large array sizes and consequently large computer memory, generally required for such calculations, are avoided by using the virtual time step elimination technique.

7. REFERENCES

1. MIAOULIS, I.N. and MIKIC, B.B. – Heat source power requirements for high quality re-crystallisation of thin silicon films for electronic devices. J. Appl. Phys., Vol. 59, pp. 1658-1662, 1986.
2. MIAOULIS, I.N. and MIKIC, B.B. – Temperature distribution of silicon-on-insulator systems during re-crystallisation processing. J. Appl. Phys., Vol. 59, pp. 1663-1666, 1986.
3. WAECHTER, D., SCHVAN, P., THOMAS, R.E. and TARR, N.G. – Modelling of heat flow in multilayer cw-laser annealed structures. J. Appl. Phys., Vol. 59, pp. 3371-3374, 1986.
4. ALWAN, A.A. – Surface Temperature Transients in Laser Machining Metals, M.Sc. Thesis, University of Glasgow, 1989.
5. ROSE, L.G.D.D. – Modelling of heat and mass transfer, Laser Surface Treatment of Metals, DRAPER, C.W. and MAZZOLDI, P., Martinus Nijhoff, Dordrecht, 1986.
6. BHATTACHARYA, M.C. – An explicit conditionally stable finite difference equation for heat conduction problems. Int. J. Numer. Meth. Eng., Vol. 21, pp. 239-265, 1985.
7. BHATTACHARYA, M.C. – Finite difference solutions of partial equations. Commu. Appl. Numer. Meth., Vol. 6, pp. 173-184, 1990.
8. JALURIA, Y. and TORRANCE, K.E. – Computational Heat Transfer, Hemisphere P.C., Washington D.C., 1986.
9. PATANKAR, S.V. – Numerical Heat Transfer and Fluid Flow, Hemisphere P.C., Washington D.C., 1980.
10. ZERROUKAT, M. and CHATWIN, C.R. – An explicit unconditionally stable variable time step for 1-D Stefan problems. Int. J. Numer. Meth. Eng., Vol. 35, pp. 1503-1520, 1992.
11. SCHVAN, P. and THOMAS, R.E. – Time dependent heat flow calculation of cw-laser-induced melting of silicon. J. Appl. Phys., Vol. 57, pp. 4738-4741, 1985.
12. KUBOTA, K., HUNT, C.E. and FREY, J. – Thermal profiles during recrystallisation of silicon on insulator with scanning incoherent light line source. Appl. Phys. Lett., Vol. 46, pp. 1153-1155, 1985.
13. FUJII, J. and IMURA, H. – Natural convection heat transfer from a plate with arbitrary inclination. Int. J. Heat Mass Transfer, Vol.15, pp. 755-767, 1973.
14. MOODY, J.E. and HENDEL, R.H. – Temperature profiles induced by scanning cw-laser beam. J. Appl. Phys., Vol. 53, pp. 4364–4371, 1982.
15. JELLISON Jr., G.E. and MODINE, F.A. – Optical functions of silicon between 1.7 and 4.7 eV at elevated temperatures. Phys. Review, Vol. B27, pp. 7466-7472, 1983.

16. JELLISON Jr., G.E. and BURKE, H.H. – The temperature dependence of the refractive index of silicon at elevated temperatures at several laser wavelength. J. Appl. Phys., Vol. 60, pp. 841–843, 1986.
17. GAY, P. – Application of mathematical heat transfer analysis to high power CO2 laser material processing, Laser Surface Treatment of Metals, DRAPER, C.W. and MAZZOLDI, P., Martinus Nijhoff, Dordrecht, 1986.
18. DULEY, W.W. – Laser material interactions of relevance to metal surface treatment, Laser Surface Treatment of Metals, DRAPER, C.W. and MAZZOLDI, P., Martinus Nijhoff, Dordrecht, 1986.
19. ROGERS, G.F.C. and MAYHEW, Y.R. – Thermodynamic and Transport Properties of Fluids, Basil Balackwell, Oxford, 1980.
20. HSU, S.C., CHAKRAVORTY, S. and MEHRABIAN, R. – Rapid melting and solidification of a surface layer. Metall. Trans., Vol. B9, pp. 221-229, 1978.

Appendix: Computation of liquid/solid motion

Integrating (24) with respect to x, from x=0 to x=M, and (25) from x=M to x=X, we obtain:

$$\int_0^M C_l \frac{\partial T_l}{\partial t} dx + \int_M^X C_s \frac{\partial T_s}{\partial t} dx = -\lambda \frac{dM}{dt} + Q_1(t) - Q_2(t) \tag{39}$$

where $Q_1(t)$ and $Q_2(t)$ are the heat flux absorbed by the top surface (i.e. right side of (27)) and the heat flux lost by the bottom surface (i.e. right side of (28)), respectively. Further integration of (39) with respect of t, from $t = t_j^{n-1}$ to $t = t_j^n = t_j^{n-1} + \delta t_j^n$, gives:

$$\int_{t_j^{n-1}}^{t_j^n} \int_0^M C_l \frac{\partial T_l}{\partial t} dx dt + \int_{t_j^{n-1}}^{t_j^n} \int_M^X C_s \frac{\partial T_s}{\partial t} dx dt = -\lambda \chi_j^n + \int_{t_j^{n-1}}^{t_j^n} Q_1(t) dt - \int_{t_j^{n-1}}^{t_j^n} Q_2(t) dt \tag{40}$$

where χ_j^n is the variation of the position of the liquid/solid interface during the time step $\delta t_j^n = t_j^n - t_j^{n-1}$. Using the trapezoidal rule to approximate the integrals in (40), χ_j^n is given by:

$$\chi_j^n = \frac{1}{\lambda} \left\{ \left(\frac{Q_{1,j}^n + Q_{1,j}^{n-1}}{2} \right) \delta t_j^n - \left(\frac{Q_{2,j}^n + Q_{2,j}^{n-1}}{2} \right) \delta t_j^n - \theta_j^n \right\} \tag{41}$$

where $Q_{1,j}^n = Q_1(t_j^n)$, $Q_{2,j}^n = Q_2(t_j^n)$ and θ_j^n is given by:

$$\theta_j^n = \frac{\Delta x}{2} \sum_{i=0}^{N-1} C \left(\frac{T_{i,j}^n + T_{i+1,j}^n + T_{i,j}^{n-1} + T_{i+1,j}^{n-1}}{4} \right) \left(T_{i,j}^n + T_{i+1,j}^n - T_{i,j}^{n-1} - T_{i+1,j}^{n-1} \right) \tag{42}$$

Numerical simulation of thermocapillary convection in a melted pool

D. Morvan and Ph. Bournot
Institut Méditerranéen de Technologie CNRS
Technopôle de Château Gombert
13451 Marseille cedex 20 FRANCE

April 1, 1993

Abstract

A two-dimensional laser surface remelting problem is numerically simulated. The mathematical formulation of this multiphase problem is obtained using a continuum model, constructed from classical mixture theory. This formulation permits to construct a set of continuum conservation equations for pure or binary, solid–liquid phase change systems. The numerical resolution of this set of coupled partial differential equations is performed using a finite volume method associated with a PISO algorithm. The numerical results show the modifications caused by an increase of the free surface shear stress (represented by the Reynolds number R_e) upon the stability of the thermocapillary flow in the melting pool. The solutions exhibit a symmetry – breaking flow transition, oscillatory behaviour at higher values of R_e.

1 Introduction

The temperature gradient induced at the surface of a melted pool during laser material processing (such as surface treatment or welding), produces high convective motions by capillary effects which affect very significantly the thermal coupling between the working piece and the laser beam [1, 2]. The main advantage of lasers is the possibility to deposite the energy with high precision and therefore to realize an accurate and very localized heat treatment. The flow of melted material could create a deformation of the free surface which is frozen in as the beam passes to other part of the piece [3]. This surface rippling and roughning that may result, is the main problem for the development of the industrial applications of lasers.

These defaults could result from the oscillations in the melted pool which appear for high values of the Reynolds number which measures the ratio between the inertial and the viscous forces. The aim of this study is to predict numericaly the behaviour (steady or oscillatory) of flow in a two - dimensional melted pool for various free surface shear stress conditions represented by differents Reynolds numbers.

2 Mathematical formulation

The mathematical model is developped from the following physical problem (see also Fig.1).

- A laser beam having a constant power distribution strikes the surface of the material. All the incident radiation is assumed to be absorbed by the material.
- The heat absorbed develops a molten pool. The surface tension gradient, produced by the surface temperature distribution, induces a convective motion in the melting pool.

The following assumptions are made for the present model:

- To limit the computational time, the study is limited to a two–dimensional geometry, with a general unsteady formulation.
- The surface of the melt is flat and adiabatic outside the beam.
- The specific heat is constant inside a single phase (these value can be different in the liquid and solid phase).

The mathematical model is constructed from a continuum formulation [4, 5], based on the integration of semiempirical laws and microscopic descriptions of transport behavior with principles of classical mixture theory. This formulation permits to treat various solidification problems of pure and alloy materials, without the need of imposing boundary conditions at the melting front. Therefore the problem can be solved with a fixed cartesian mesh , reducing the computation time [6, 7]. Using these assumptions, the present problem can be mathematically defined by the following governing equations and boundary conditions:

- Continuity equation

$$\frac{\partial}{\partial t}(\rho) + \frac{\partial}{\partial x_j}(\rho v_j) = 0 \tag{1}$$

- Momentum equation

$$\frac{\partial}{\partial t}(\rho v_i) + \frac{\partial}{\partial x_j}(\rho v_j v_i) = \frac{\partial}{\partial x_j}\left(\mu \frac{\rho}{\rho_l}\frac{\partial v_i}{\partial x_j}\right) +$$
$$\rho_l g_i \beta (T - T_f) - \frac{\mu}{K}\frac{\rho}{\rho_l}(v_i - v_i^s) - \frac{\partial p}{\partial x_i} \quad (2)$$

- Energy equation

$$\frac{\partial}{\partial t}(\rho h) + \frac{\partial}{\partial x_j}(\rho v_j h) = \frac{\partial}{\partial x_j}\left(\frac{k}{c_s}\frac{\partial h}{\partial x_j}\right) +$$
$$\frac{\partial}{\partial x_j}\left(\frac{k}{c_s}\frac{\partial}{\partial x_j}(h_s - h)\right) - \frac{\partial}{\partial x_j}\left(\rho (h_l - h)\left(v_j - v_j^s\right)\right) \quad (3)$$

Boundary conditions:

- West, south and east boundaries:

$$v_j = 0 \quad T = T_0 \quad (4)$$

- North boundary (free surface)

$$\frac{\partial T}{\partial n} = -\frac{q}{k} \quad (5)$$

$$\frac{\partial (v_j.\tau_j)}{\partial n} = \frac{1}{\mu}\frac{\partial \sigma}{\partial \tau} \quad \text{(liquid phase)}$$
$$v_j = 0 \quad \text{(solid phase)} \quad (6)$$

The different variables introduced in the previous equations are:

- ρ_ϕ, c_ϕ, k_ϕ density, specific heat and conductivity in the solid ($\phi = s$) and in the liquid phase ($\phi = l$)
- g_i, β gravitational acceleration and thermal coefficient expansion.
- h_ϕ, v_i^ϕ enthalpy and velocity components in the solid and in the liquid phase.
- p, K.μ presure, permeability and kinematical viscosity.
- g_s, f_s volume and mass fraction of the solid phase.
- σ, h_f surface tension and latent heat of fusion.
- T_0 and T_f ambiant and melted temperature.

- τ and n tangential and normal direction at the free surface.

The mixture density, velocity, enthalpy and thermal conductivity are, recpectively:

$$\rho = g_s \rho_s + (1 - g_s)\,\rho_l \tag{7}$$

$$v_j = (1 - f_s)\,v_j^l \tag{8}$$

$$h = f_s h_s + (1 - f_s)\,h_l \tag{9}$$

$$k = g_s k_s + (1 - g_s)\,k_l \tag{10}$$

Phases enthalpies are:

$$h_s = c_s T \tag{11}$$

$$h_l = c_l T + \left[(c_s - c_l)\,T_f + h_f\right] \tag{12}$$

The permeability appearing in the momentum equations is assumed to vary with the liquid volume fraction according to the Carman – Kozeny equation :

$$K = K_0 \left[\frac{g_l^3}{(1 - g_l)^2} \right] \tag{13}$$

where the constant K_0 depends upon the specific structure of the multiphase region. The governing equations (1-3) are nondimensionalized with the following reference scales:

- length : beam diameter d
- temperature : $\Delta T = T_f - T_0$
- density, specific heat and conductivity : ρ_l, c_s and k_l
- velocity : $U = \gamma \Delta T / \mu$ avec $\gamma = |d\sigma / dT|$
- enthalpy : $c_s\,(T_f - T_0)$

The dimensionless variables are now:

$$x_j^* = \frac{x_j}{d} \quad , \quad t^* = \frac{tU}{d} \quad , \quad v_j^* = \frac{v_j}{U} \quad , \quad T^* = \frac{T}{\Delta T} \tag{14}$$

$$h^* = c_l^* T^* + (1 - c_l^*)\,T_f^* + \frac{1}{S_{te}} \tag{15}$$

$$\rho^* = \frac{\rho}{\rho_l} \quad , \quad k^* = \frac{k}{k_l} \quad , \quad c_\phi^* = \frac{c_\phi}{c_s} \tag{16}$$

Introducing these dimensionless variables, the complete mathematical description of the problem is as follows:

$$\frac{\partial}{\partial t^*}(\rho^*) + \frac{\partial}{\partial x_j^*}\left(\rho^* v_j^*\right) = 0 \tag{17}$$

$$\frac{\partial}{\partial t^*}(\rho^* v_i^*) + \frac{\partial}{\partial x_j^*}\left(\rho^* v_j^* v_i^*\right) = \frac{\partial}{\partial x_j^*}\left(\frac{\rho^*}{R_e}\frac{\partial v_j^*}{\partial x_j^*}\right) -$$
$$\frac{G_r}{R_e^2}\left(T^* - T_f^*\right)\delta_{i2} - \frac{1}{R_e D_a}\frac{\rho^*}{K^*}(v_i^* - v_i^{*s}) - \frac{\partial p^*}{\partial x_i^*} \quad (18)$$

$$\frac{\partial}{\partial t^*}(\rho^* h^*) + \frac{\partial}{\partial x_j^*}\left(\rho^* v_j^* h^*\right) = \frac{\partial}{\partial x_j^*}\left(\frac{k^*}{R_e P_r}\frac{\partial h^*}{\partial x_j^*}\right) +$$
$$\frac{\partial}{\partial x_j^*}\left(\frac{k^*}{R_e P_r}\frac{\partial}{\partial x_j^*}(h_s^* - h^*)\right) - \frac{\partial}{\partial x_j^*}\left(\rho^*(h_l^* - h^*)\left(v_j^* - v_j^{*s}\right)\right) \quad (19)$$

The corresponding boundary conditions are:

- West, south and east boundary:

$$T^* = T_0^* \quad \text{and} \quad v_j^* = 0 \quad (20)$$

- North boundary:

$$\frac{\partial T^*}{\partial n^*} = -\frac{q^*}{k^*} \quad (21)$$

$$\frac{\partial v_j^* . \tau^*}{\partial n^*} = sign\left(\frac{d\sigma}{dT}\right)\frac{\partial T^*}{\partial \tau^*} \quad \text{for the liquid phase}$$
$$v_j^* = 0 \quad \text{for the solid phase} \quad (22)$$

From the previous equations we see that there are five non – dimensional parameters which govern the behaviour of this system :

- the Reynolds, Prandtl and Grashof numbers:

$$R_e = \frac{\rho_l U d}{\mu} \quad G_r = \frac{g\beta\Delta T d^3 \rho_l^2}{\mu^2} \quad P_r = \frac{\mu c_s}{k_l} \quad (23)$$

- the Darcy and Stefan numbers

$$D_a = \frac{K}{d^2} \quad S_{te} = \frac{c_s \Delta T}{h_f} \quad (24)$$

The numerical resolution is performed using a finite volume method [8, 9] with a non-itterative PISO algorithm [10, 11].

3 Results and discussion.

This study shows the effects of an increase of the Reynolds number upon the pattern and the behaviour of the liquid flow in the melted pool. The dimensions (HxL) of the target must be sufficiently large to assume that the heat affected zone is totaly included inside it (for the present study H=5xd and L=10xd, d designs the beam diameter). All the results are obtained for the following set of physical parameter (see Table 1) (the effects of natural convection and latent heat are neglected):

For relatively small Reynolds number (< 200) the flow field is charac-

Reynolds number R_e	from 10 to 1000
Prandtl number P_r	0.01
liquid to solid specific heat ratio c_l/c_s	1.0
solid to liquid conductivity ratio k_s/k_l	2.4
solid to liquid density ratio ρ_s/ρ_l	1.0
Non – dimesionnal heat flux q^*	5.0

Table 1: Typical values of non – dimensionnal physical parameters.

terized by the formation of two convective cells which fill all the melted zone (Figure 2). When the Reynolds number increases ($R_e = 200$) these convective cells rise near the free surface,and by viscous effects two small cells appear at the bottom of the melted pool (Figure 3). In these two cases the flow remains stationnary. The interaction between these four vortices produces the symmetry - breaking and oscillations in the melted pool, then the flow becomes instationnary. (Figure 4, 5 and 6). For $R_e = 400$ the flow is mono – periodic (Figure 7) then for higher values we observe the introduction of various modes ($R_e = 500$) (Figure 8) and ($R_e = 1000$) (Figure 9) which rapidly increase the band width of the corresponding power spectrum.

References

[1] A.B. Vannes, editor. *Laser de puissance et traitements des matériaux*, chapter 24, pages 556–567. Presses Polytechniques et Universitaires Romandes, 1991. Ecole de printemps CNRS-EPFL, Sireuil.

[2] C. Chan, J. Mazumder, and M.M. Chen. A two-dimensional transient model for convection in laser melted pool. *Metallurgical Transactions*, 15(12):2175–2184, 1984.

[3] T.R. Anthony and H.E. Cline. Surface rippling induced by surface-tension gradients during laser surface melting and alloying. *J. Appl. Phys.*, 48(9):3888–3894, 1977.

[4] W.D. Bennon and F.P. Incropera. A continum model for momentum, heat and species transport in binary solid-liquid phase change systems: 1 model formulation. *Int. J. Heat Mass Transfer*, 30(10):2161–2170, 1987.

[5] W.D. Bennon and F.P. Incropera. A continum model for momentum, heat and species transport in binary solid-liquid phase change systems: 2 application to solidification in a rectangular cavity. *Int. J. Heat Mass Transfer*, 30(10):2171–2187, 1987.

[6] W.D. Bennon and F.P. Incropera. Numerical analysis of binary solid–liquid phase change using a continuum model. *Numerical Heat Transfer*, 13:277–296, 1988.

[7] V.R. Voller and C. Prakash. A fixed grid numerical modelling methodology for convection-diffusion mushy region phase change problems. *Int. J. Heat Mass Transfer*, 30(8):1709–1719, 1987.

[8] S.V. Patankar. A calculation procedure for two-dimensional elliptic situations. *Numerical Heat Transfer*, 4:409–425, 1981.

[9] J.P. Van Doormaal and G.D. Raithby. Enhancements of the simple method for predicting incompressible fluid flow. *Numerical Heat Transfer*, 7:147–163, 1984.

[10] R.I. Issa, A.D. Gosman, and A.P. Watkins. The computation of compressible and incompressible recirculating flows by a non-iterative implicit scheme. *Journal of Computatinal Physics*, 62:66–82, 1986.

[11] D.S. Jang, R. Jetli, and S. Acharya. Comparison of the piso,simpler and simplec algoritms for the treatment of the pressure-velocity coupling in steady flow problems. *Numerical Heat Transfer*, 10:209–228, 1986.

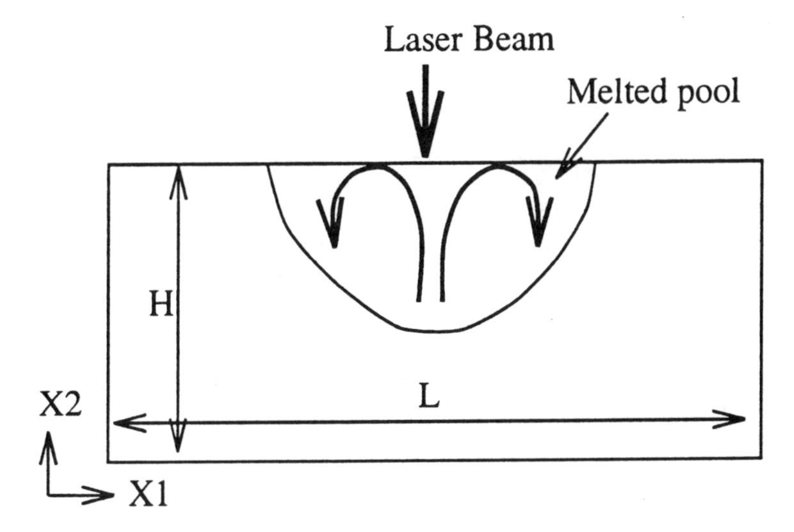

Figure 1: Schematic illustration of the melted pool and coordinates system.

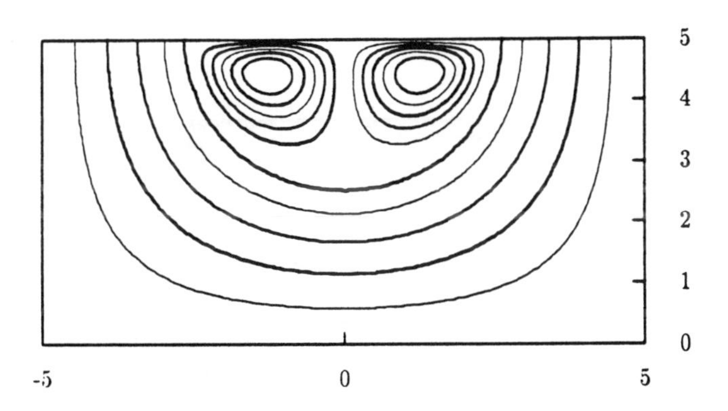

Figure 2: Streamlines (in the liquid phase) and temperature field (in the solid phase) for $R_e = 10$.

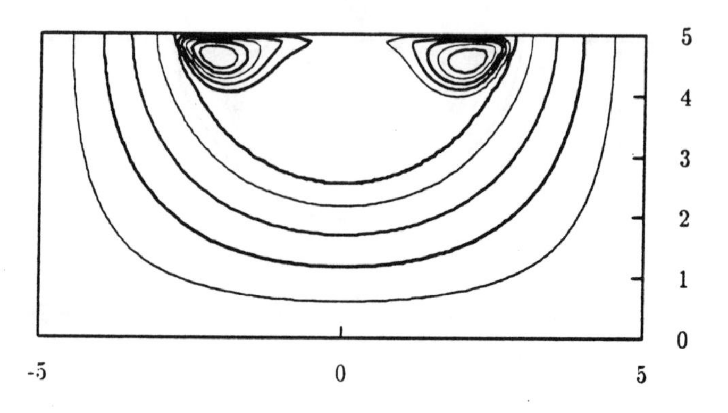

Figure 3: Streamlines (in the liquid phase) and temperature field (in the solid phase) for $R_e = 200$.

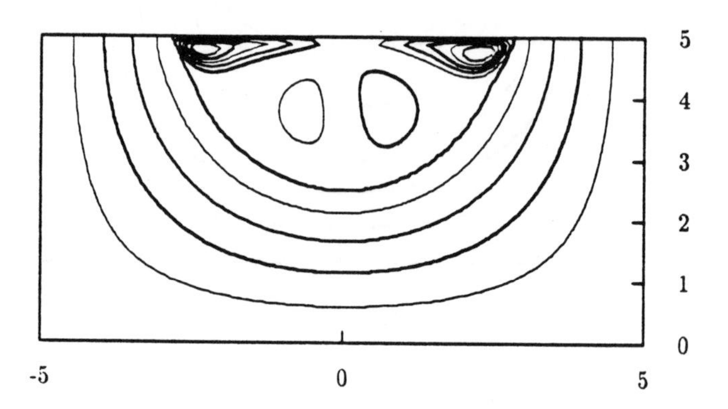

Figure 4: Streamlines (in the liquid phase) and temperature field (in the solid phase) for $R_e = 400$.

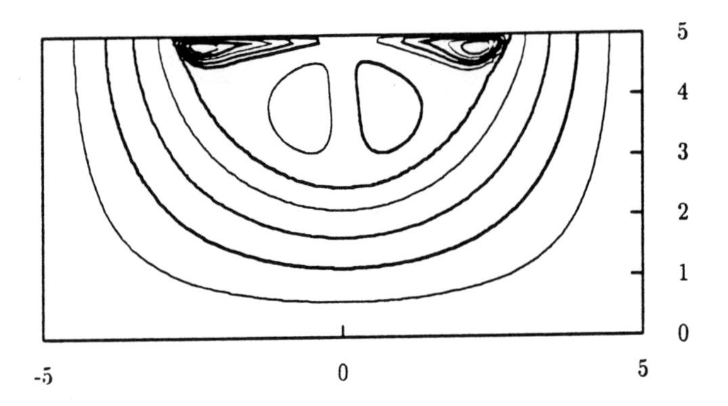

Figure 5: Streamlines (in the liquid phase) and temperature field (in the solid phase) for $R_e = 500$.

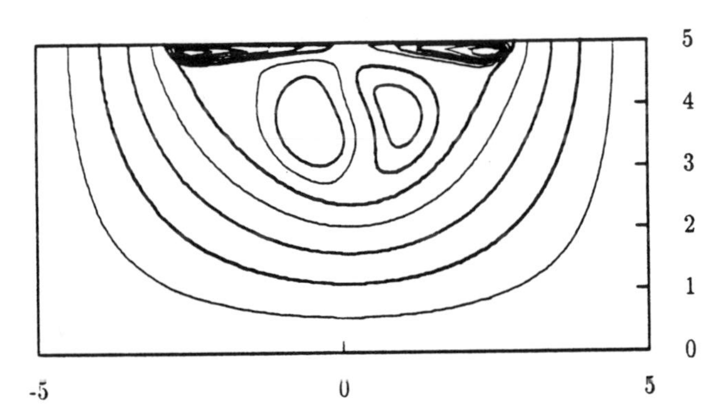

Figure 6: Streamlines (in the liquid phase) and temperature field (in the solid phase) for $R_e = 1000$.

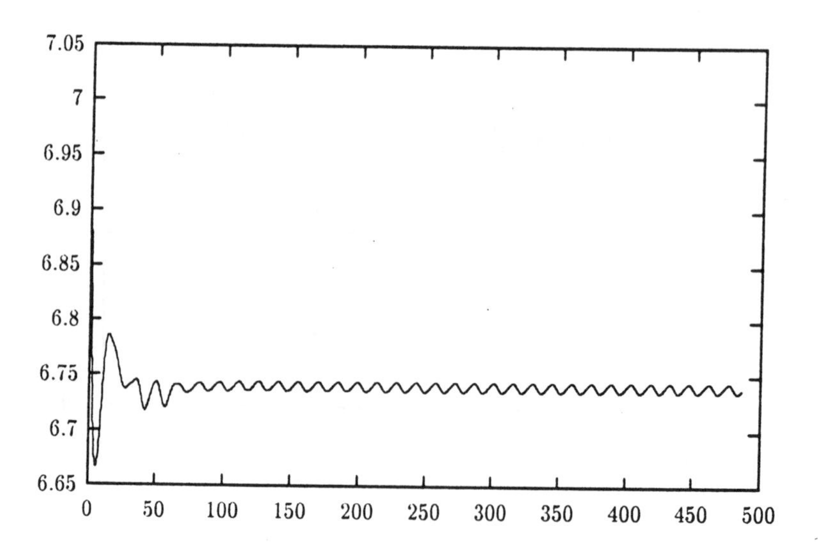

Figure 7: Temperature signal at a given point for $R_e = 400$.

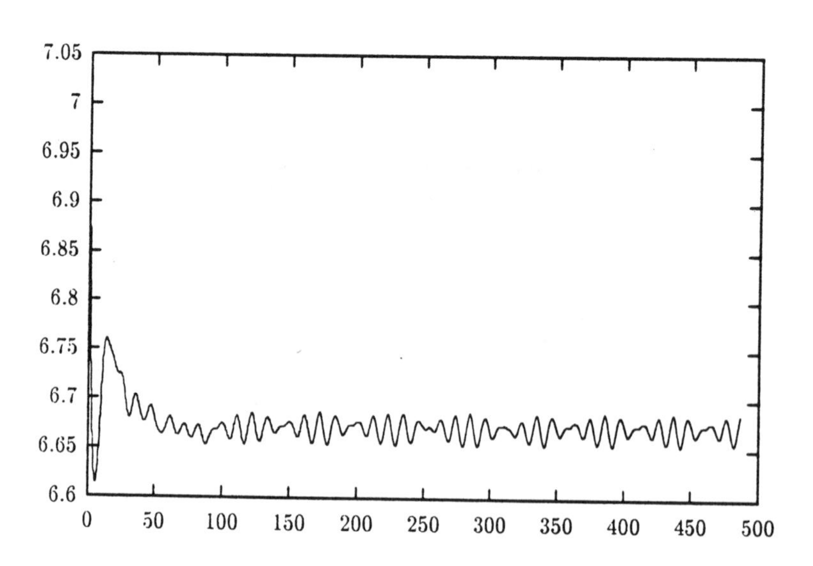

Figure 8: Temperature signal at a given point for $R_e = 500$.

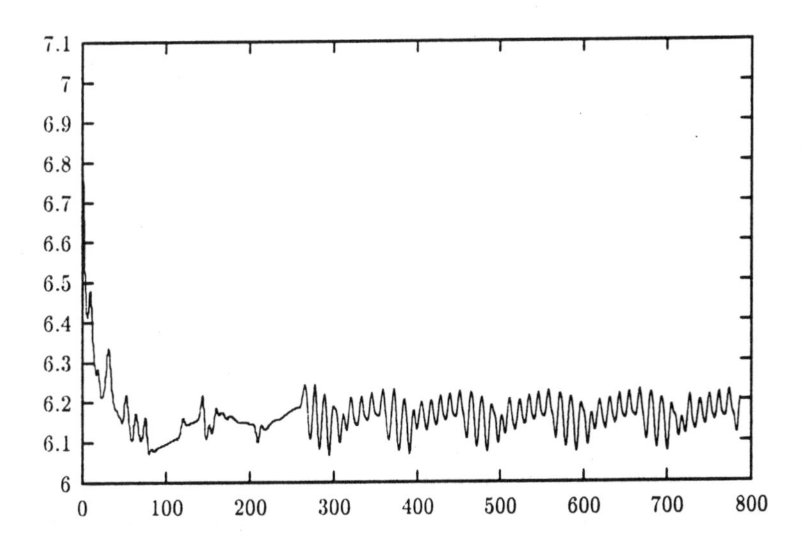

Figure 9: Temperature signal at a given point for $R_e = 1000$.

MELTING OF A PHASE CHANGE MATERIAL INSIDE A VERTICAL CYLINDRICAL CAPSULE

YONGKE WU[1], MARCEL LACROIX[2]

ABSTRACT

A numerical study is reported of natural convection melting of a PCM within a vertical cylindrical enclosure with isothermal boundaries. A stream-function-vorticity-temperature formulation is employed in conjunction with body-fitted coordinates for tracking the irregular shape of the timewise changing solid-liquid phase front. Results show that the convective flow patterns and time evolution of the phase front, resulting from simultaneous bottom, side and top heating are far more complicated than those for the melting from a single isothermal boundary. The heat transfer rate at the top surface was found to decrease monotonically to zero as convection is fully developed in the melt. The highest heat transfer rates are observed at the bottom surface as Bénard convective cells develop. Due to the convective motion of the melt along the vertical heated wall, the onset of Bénard convection occurs at a much earlier time as for the case of melting within a cylinder heated from below.

1. INTRODUCTION

A fundamental understanding of heat transfer during melting and solidification is required for efficient design of latent heat-of-fusion energy storage systems. The effect of natural convection in the melt region is of particular importance in such phase-change problems, especially in the melting stage. Indeed, it has long been recognized that natural convection heat transfer in the melt strongly perturbs the position and the shape of the solid-liquid interface[1].

Over the past decade, the problem of natural convection dominated melting in rectangular enclosures has received considerable attention[2–5].

[1] Research Professor, Industrial Chair: Département des sciences appliquées
Université du Québec à Chicoutimi, Québec, Canada.
[2] Professor: Département de génie mécanique
Université de Sherbrooke, Québec, Canada.

Another important arrangement for latent heat-of-fusion energy storage systems, *i.e.*, the phase-change process around a single or an array of cylinders or inside cylindrical capsules, has also been the subject of many experimental and theoretical investigations [6–8]. Ramsey *et al.*[6] reported an experimental study of melting about a horizontal row of heating cylinders. Bathelt and co-workers conducted experiments around a single horizontal heated cylinder[7] and around an array of three staggered cylinders[8]. Sasaguchi and Viskanta[9] investigated experimentally the phase change heat transfer during melting and resolidification of melt around two cylindrical heat exchangers spaced vertically. Katayama *et al.*[10] conducted experiments to study the heat transfer characteristics of a latent heat thermal energy storage cylindrical capsule.

The problem of melting for cylindrical geometries has also been analyzed by means of numerical methods. Sparrow *et al.*[11] analyzed the melting of a solid in a cylindrical enclosure considering natural convection in the liquid. They applied an implicit finite difference scheme to the transformed domain in which the interface has tractable shape. Prusa and Yao[12],[13] developed a perturbation solution for the melting around a heated cylinder. Rieger *et al.*[14] tackled the same problem by formulating the conservation equations in terms of a stream-function, vorticity and temperature. Difficulties associated with the complex structure of the timewise changing physical domain were successfully overcome by applying a numerical mapping technique. Inward melting in a horizontal cylindrical capsule was studied experimentally and analytically by Ho and Viskanta[15]. A Landau coordinate transformation was adopted for immobilizing the moving solid-liquid interface. More recently, Wu and co-workers[16] carried out an analysis of melting around a vertical circular cylinder imbedded in a solid matrix. The governing equations for heat and fluid flow in the melted region were solved using computer-generated, body-fitted coordinates. The problem was analyzed for two types of wall boundary conditions, *i.e.*, constant temperature and constant heat flux.

In the present paper, natural convection dominated melting of a PCM within a vertical isothermal cylinder is investigated numerically. The objective is not to perform detailed parametric calculations but rather to analyze the timewise complex flow patterns and thermal behavior of the melt resulting from simultaneous bottom, side and top heating of the cylinder. The effect of the Rayleigh number is also examined.

2. PHYSICAL MODEL AND BASIC EQUATIONS

The PCM is contained in a cylindrical enclosure of height H and radius r_0 shown in Fig. 1 (a). The PCM is assumed to be initially at its fusion

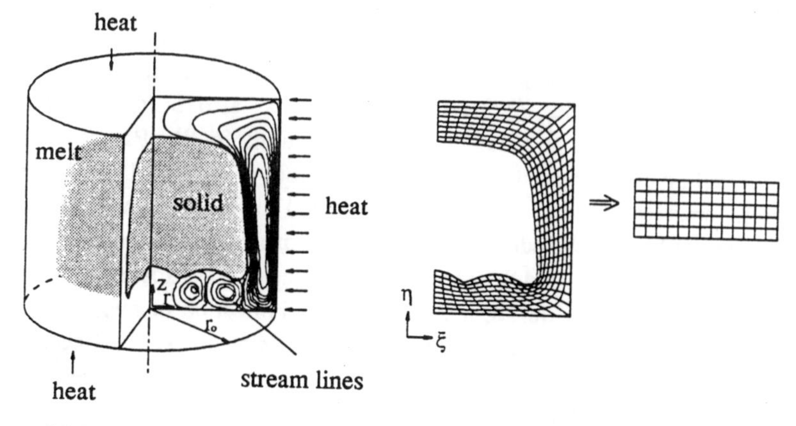

(a) Physical description (b) Grid Transformation

Figure 1. Schematic representation of the cylindrical enclosure (a) and grid transformation (b).

temperature T_f, eliminating the need for solution of the energy equation in the solid. At time $t = 0$, the surface temperature of the entire capsule is raised impulsively to a prescribed temperature above the fusion point $T_w > T_f$. As a result, inward melting is triggered. It is assumed that the thermophysical properties of the PCM are independent of temperature. The Boussinesq approximation is valid *i.e.*, liquid density variations arise only in the buoyancy source term, but are otherwise neglected. The molten liquid is Newtonian. Fluid motion in the melt is laminar and symmetrical about the vertical centered axis. Volume changes and viscous dissipation are neglected.

Under the foregoing assumptions, the partial differential equations governing the conservation of mass, momentum and energy are formulated in terms of stream function-vorticity-temperature in a curvilinear coordinate system (ξ, η):

$$Ste\frac{\partial \omega}{\partial \tau} + \tilde{U}\frac{\partial \omega}{\partial \xi} + \tilde{V}\frac{\partial \omega}{\partial \eta} = Pr\tilde{\nabla}^2\omega + S_\omega \tag{1}$$

$$\tilde{\nabla}^2\psi = S_\psi \tag{2}$$

$$Ste\frac{\partial \theta}{\partial \tau} + \tilde{U}\frac{\partial \theta}{\partial \xi} + \tilde{V}\frac{\partial \theta}{\partial \eta} = \tilde{\nabla}^2\theta + S_\theta \tag{3}$$

where $\tilde{U}$ and $\tilde{V}$ are contravariant velocities expressed by:

$$\tilde{U} = \xi_r U + \xi_z V \quad ; \qquad \tilde{V} = \eta_r U + \eta_z V \tag{4}$$

and

$$U = \frac{1}{R}\left(\xi_z\frac{\partial\psi}{\partial\xi} + \eta_z\frac{\partial\psi}{\partial\eta}\right) ; \qquad V = -\frac{1}{R}\left(\xi_r\frac{\partial\psi}{\partial\xi} + \eta_r\frac{\partial\psi}{\partial\eta}\right)$$

$$S_\psi = \frac{2}{R}\left(\xi_r\frac{\partial\psi}{\partial\xi} + \eta_r\frac{\partial\psi}{\partial\eta}\right) - R\omega ; \qquad S_\theta = Ste\left(\xi_r\frac{\partial\theta}{\partial\xi} + \eta_r\frac{\partial\theta}{\partial\eta}\right)$$

$$S_\omega = Ste\left(\xi_r\frac{\partial\omega}{\partial\xi} + \eta_r\frac{\partial\omega}{\partial\eta}\right) + \left(\frac{U}{R} - \frac{Pr}{R^2}\right)\omega + Pr\,Ra\left(\xi_r\frac{\partial\theta}{\partial\xi} + \eta_r\frac{\partial\theta}{\partial\eta}\right)$$

$$\tilde{\nabla}^2 = g^{11}\frac{\partial^2}{\partial\xi^2} + 2g^{12}\frac{\partial^2}{\partial\xi\eta} + g^{22}\frac{\partial^2}{\partial\eta^2} + P\frac{\partial}{\partial\xi} + Q\frac{\partial}{\partial\eta} + \frac{1}{R}\left(\xi_r\frac{\partial}{\partial\xi} + \eta_r\frac{\partial}{\partial\eta}\right) \tag{5}$$

$\tilde{\nabla}^2$ is the transformed Laplacian operator in cylindrical coordinates.

The energy balance equation for the moving interface becomes:

$$\frac{\partial R}{\partial\tau} = -\frac{\partial\theta}{\partial\xi}\xi_r \quad ; \qquad \frac{\partial Z}{\partial\tau} = -\frac{\partial\theta}{\partial\xi}\xi_z \tag{6}$$

The boundary conditions are summarized in Table 1:

Table 1. Boundary conditions for the problem

Boundary	Stream function	Vorticity	Temperature
heated wall	$\psi = 0$	$\omega = \eta_r V_\eta - \eta_z U_\eta$	$\theta = 1$
axis of symmetry	$\psi = 0$	$\omega = 0$	$\partial\theta/\partial\xi = 0$
melting front	$\psi = 0$	$\omega = \eta_r V_\eta - \eta_z U_\eta$	$\theta = 0$

3. NUMERICAL PROCEDURE

The governing equations (1)-(3) and (6) with the corresponding boundary conditions are solved numerically with a finite-difference method. A first order forward difference approximation is used for the time derivatives. The diffusion terms are replaced by second-order central difference approximations. Special attention is paid, however, to the convection terms. It is well known that the use of second-order central difference approximations for these terms may produce unstable and divergent solutions for high Peclet cell numbers (or high Rayleigh numbers)[17]. Although the use of a first-order upwind scheme may eliminate these wiggly solutions, it introduces truncation errors and produces significant artificial diffusion. In the present study, this problem is overcome by adopting a second-order upwind scheme.

The proposed scheme has the following form

$$u\frac{\partial f}{\partial\xi} = A^u f_{i-2} + B^u f_{i-1} + C^u f_i + D^u f_{i+1} + E^u f_{i+2} \tag{7}$$

where A^u, B^u, C^u, D^u and E^u are functions of u. These coefficients are defined in the Appendix. The resulting finite-difference scheme for the vorticity(1) and temperature equations(3) has the form

$$\begin{aligned} a_1 f_{i-2,j} + a_2 f_{i-1,j} + a_3 f_{i+1,j} + a_4 f_{i+2,j} \\ + a_5 f_{i,j-2} + a_6 f_{i,j-1} + a_7 f_{i,j+1} + a_8 f_{i,j+2} \\ + a_9 f_{i+1,j+1} + a_{10} f_{i+1,j-1} + a_{11} f_{i-1,j-1} \\ + a_{12} f_{i-1,j+1} + a_{13} f_{i,j} = S \end{aligned} \tag{8}$$

Expressions for the coefficients in Eq. (8) may be found in reference[16]. This finite-difference equation is solved by means of an alternating Penta-diagonal matrix algorithm[17]. For the stream function (Eq. 2), only second-order finite differences are used and the resulting discretized equation is solved with a standard alternating tri-diagonal matrix solver.

According to the energy balance equation (6), the local velocity of the interface should be locally orthogonal to the interface. Generally, the melting is nonuniform along the interface because of natural convection. Therefore, the interface can become curved as the boundary is moving. If the interface becomes locally convex, the moving interface grid points have a tendency to move towards their reflex center. As melting proceeds, the generated grids can be distorted and eventually the grid nodes may overlap. To overcome these difficulties, an implicit rezoning procedure is employed. Once the interface is determined at time level, $\tau + \Delta\tau$, a spline interpolation procedure is used to redistribute the boundary grid points at equal arc length intervals along the interface. Thereby, a proper grid network system is available for carrying out the calculations at the next time step $\tau + \Delta\tau$.

4. RESULTS AND DISCUSSION

The foregoing computational methodology has been thoroughly tested for natural convection dominated melting around a vertical heated cylinder[16] and within a vertical cylinder heated from below[17]. The numerical predictions were testified against other numerical solutions and experimental data. These validation analyses are reported in references[16] and [17] and need not to be repeated here.

To avoid computational difficulties at time $\tau = 0$, a very thin uniform thickness melt layer parallel to the heated bottom, top and side walls was assumed to exist initially. The layer thickness was chosen such that the Rayleigh number based on this initial gap width was small enough so that pure conduction could be considered as the prevailing mechanism of heat transfer.

Following a grid refinement study and as a compromise between cost and accuracy, the calculations presented here were done with a grid size

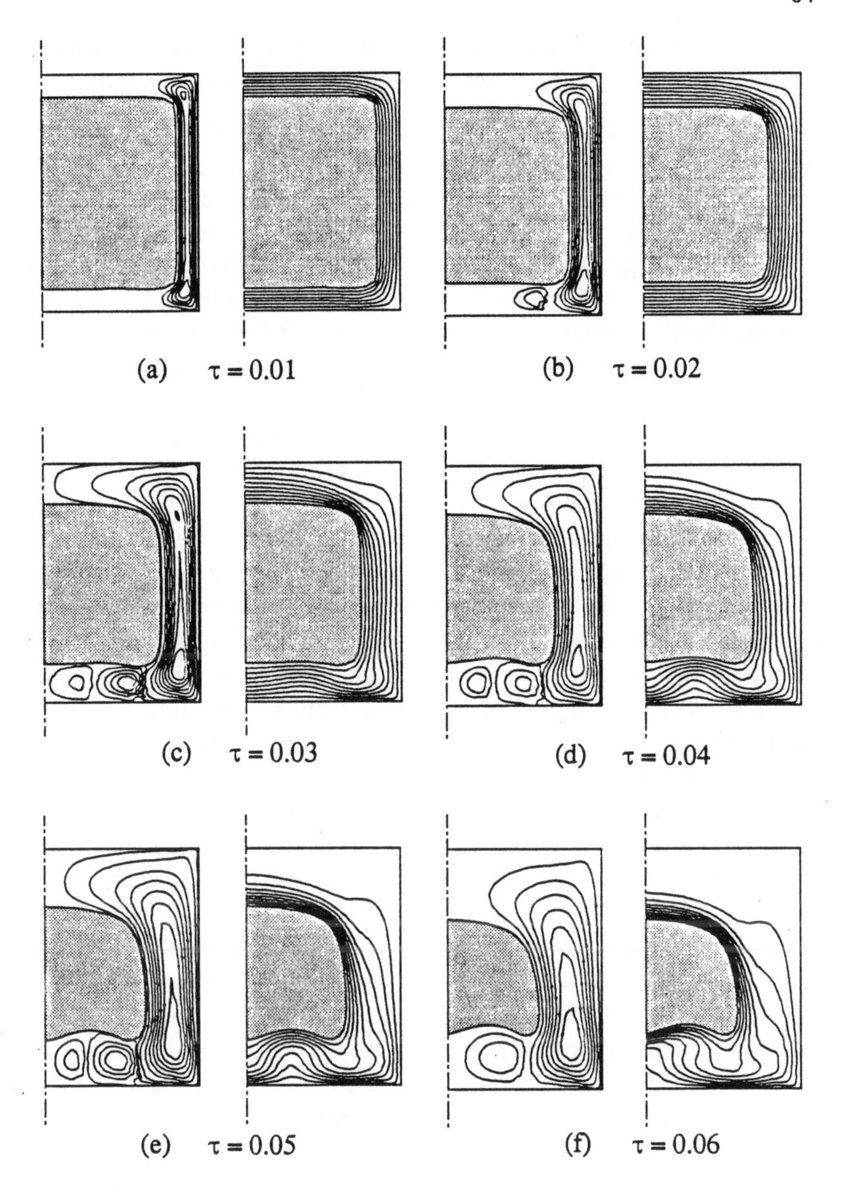

Figure 2. Time evolution of the streamlines (left) and isotherms (right) for $Ra = 10^5$.

of 11×31 nonuniformly distributed nodes. This makes it possible to concentrate several grid points in the critical regions near the heated surfaces and near the solid-liquid interface where large temperature and vorticity gradients prevail. A constant time step of 10^{-3} was utilized in order to assure small interface motion from one time step to the next. No attempts were made to optimize (increase) the time step as melting proceeds.

As a typical example, Fig. 2 shows the time evolution of the streamlines and isotherms for a case with $Pr = 7.0$, $Ste = 0.15$, $A = 2.0$, and $Ra = 10^5$. The increments between the streamlines and the isotherms are constant between their minimum and maximum values. At early times $\tau \leq 0.01$ (Fig. 2a), heat transfer in the melt zone is predominated by conduction *i.e.*, the isotherms remain parallel to the heated walls, and the solid-liquid interface moves uniformly inward from the surface of the enclosure. The isotherms at the top and bottom of the cavity are horizontal and no convective flow exists. In the upper part of the capsule, heat is transferred through the melt from the top heated surface to the bottom cold melt front. As a result, layers of lighter fluid rest on layers of heavier fluid and the flow is stagnant and in a stable condition. On the other hand, at the bottom of the enclosure heat is transferred through the melt from the bottom heated surface to the top cold interface. In this case, however, the situation is potentially unstable as layers of cold and denser fluid adjacent to the solid-liquid interface lie above layers of hot and lighter fluid near the bottom heated wall. For as long as the temperature gradients remain perfectly vertical, the source term for the vorticity equation (Eq. 5, last term on the right hand side of S_ω) is null and the flow is stagnant. In the melt layer near the vertical heated wall, a weak convective recirculating flow has already established itself. Along the vertical cylinder surface, heat is transferred to the melt and the fluid moves upward. Along the vertical phase front, heat is transferred to the interface and the fluid descends. At $\tau = 0.02$ (Fig. 2b), the melt layer around the solid phase is thicker and the unstable thermal situation at the bottom of the enclosure leads to a Bénard clockwise recirculating cell. Due to the long counterclockwise recirculation bubble along the side of the cylinder, it is observed that the onset of Bénard convection occurs at a much earlier time as for the case of bottom heating only[17]. At $\tau = 0.03$ (Fig. 2c), a second Bénard counterrotating cell has appeared in the bottom layer. As melting proceeds, the Bénard cell at the right, entrained by the lateral counterclockwise recirculation bubble, grows faster and stronger than the left one (Fig. 2d-e). As a result, the left cell is pushed leftward and shrinks until it vanishes completely (Fig. 2f).

The timewise variations of the average Nusselt number at the bottom (Nu_B), at the side (Nu_S) and at the top (Nu_T) of the cylinder are depicted in Fig. 3. These Nusselt numbers were calculated from the converged temperature field at each time step. The results display a rapid decrease in the heat transfer rate at the early stages of melting which is indicative of transient heat conduction. As soon as natural convection sets in the lower

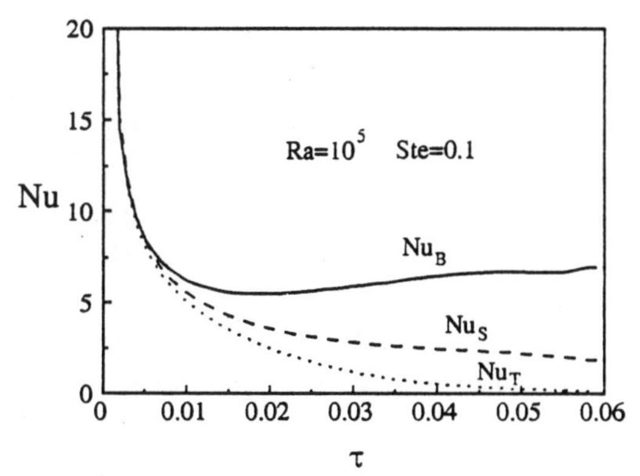

Figure 3. Time variation of the average Nusselt numbers at the bottom Nu_B, at the vertical Nu_S and at the top Nu_T surface for $Ra = 10^5$.

part of the enclosure with the appearance of Bénard cells ($\tau \lesssim 0.02$), the heat transfer rate, *i.e.* Nu_B, starts increasing. As time passes, the melt layer between the vertical cylinder surface and the vertical phase front expands and the thermal resistance across this layer increases. This results in a constant decrease in the magnitude of Nu_s. On the other hand, it is seen that the Nusselt number at the top, Nu_T, decreases monotonically to zero. This is the result of the gradual decrease of the temperature gradients in the upper region of the capsule generated by the continuous upward flow along the vertical heated surface. As the phase front moves away from the top heated surface, the melt region at the top of the cylinder becomes an isothermal zone as shown by the isotherms in Fig. 2f.

Figs. 4 show the time evolution of the streamlines and isotherms for $Ra = 10^6$ with the other parameters remaining unchanged. Due to the stronger convective motion inside the melt layer along the vertical heated wall, Bénard convection appears much earlier ($\tau < 0.004$). The momentum of the clockwise recirculating Bénard cell at the right is large enough to create and entrain two additional cells (Fig. 4c). As a result, the isotherms are considerably perturbed and so are the heat transfer rates and the time evolution of the phase front at the bottom of the cylinder. Once again, as melting proceeds, the melt layer at the bottom expands and the Bénard cell at the right grows faster and stronger than the left ones. The left cells are pushed leftward and shrink until they vanish (Fig. 4d). The remaining clockwise recirculating cell at the bottom then starts decreasing as the lateral counterclockwise recirculating cell grows in size (Fig. 4e). Eventually the latter engulfs the former (Fig. 4f).

Fig. 5 illustrates the corresponding timewise variation of the average Nusselt numbers. Due to the more intense convective motion, the monotonic decrease of Nu_T is faster than that for the previous case. The onset and the

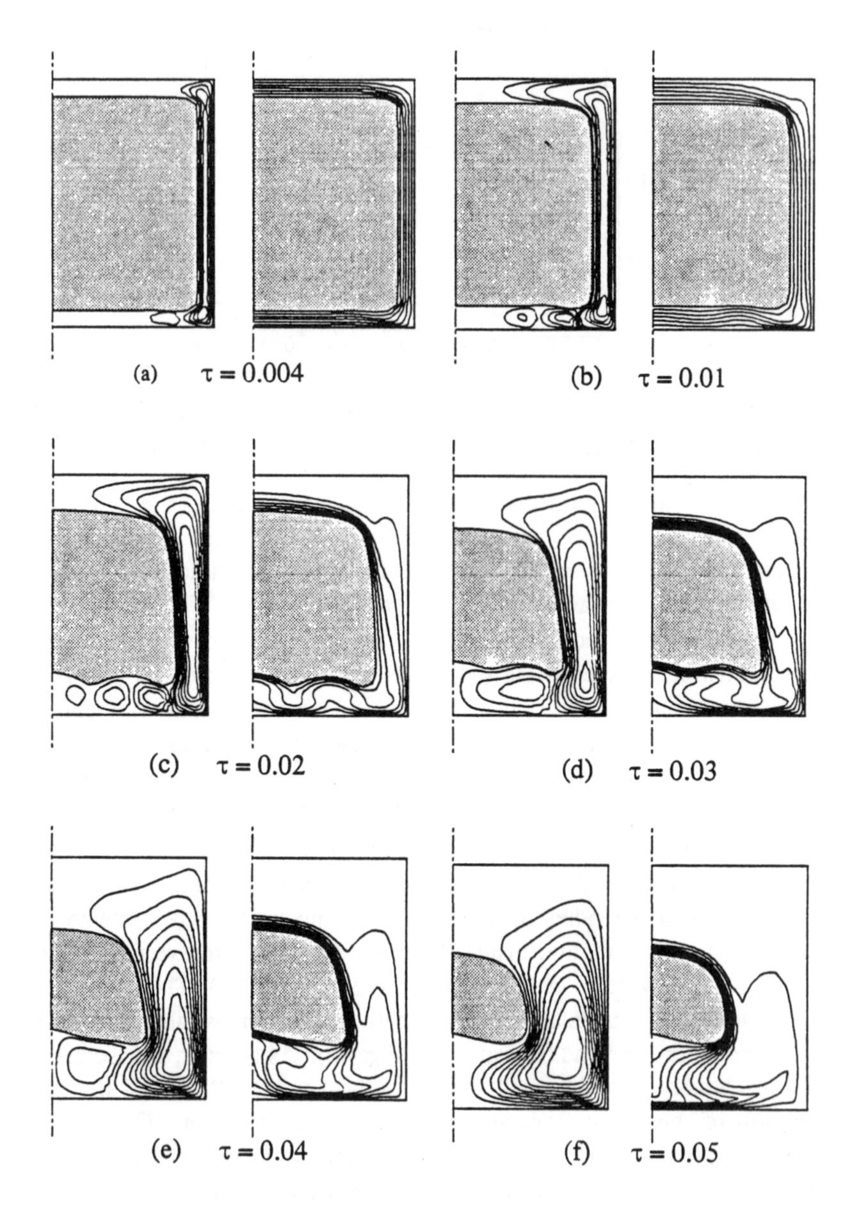

Figure 4. Time evolution of the streamlines (left) and isotherms (right) for $Ra = 10^6$.

development of Bénard convective cells is clearly seen as Nu_B increases from $\tau = 0.01$ to $\tau = 0.02$. For $\tau \gtrsim 0.02$, these cells merge to become one and Nu_B starts decreasing again. The qualitative behavior of Nu_S is the same as for the previous case.

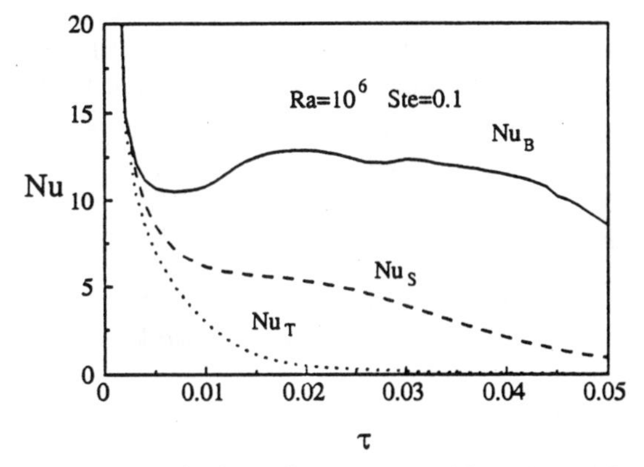

Figure 5. Timewise variation of the average Nusselt numbers at the bottom Nu_B, at the vertical Nu_S and at the top Nu_T surface for $Ra = 10^6$.

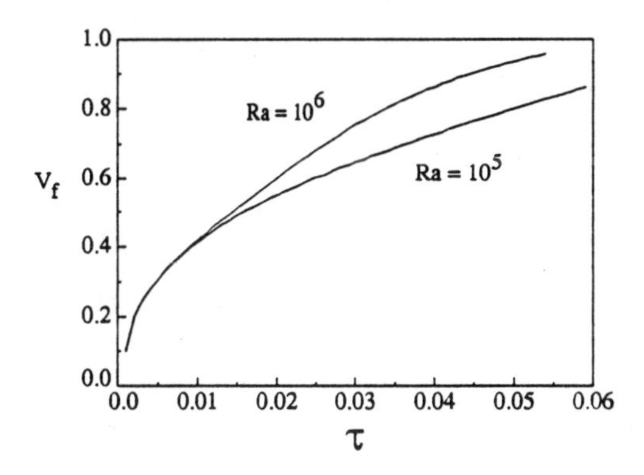

Figure 6. Temporal variation of the molten volume fraction.

The temporal variation of the molten volume fraction was also determined from a numerical integration of the melt cavity volume and is shown in Fig. 6. It is seen that this fraction increases almost linearly with time once the convective motion is well established throughout the melt.

4. CONCLUDING REMARKS

A numerical study of natural convection dominated melting within an isothermal vertical cylinder has been conducted. A robust computational methodology based on body-fitted coordinates was adopted for handling the complex motion and irregular shape of the time-varying solid-liquid interface. Results have shown that heat transfer form the top heated surface is predominated by conduction. The Nusselt number at the top surface decreases monotonically to zero as melting progresses which shows that no heat is transferred across the top layer once natural convection is fully developed in the melt. The highest heat transfer rates are observed for the bottom surface with the appearance and development of Bénard convective cells. Due to the convective motion of the flow along the vertical heated wall, the onset of Bénard convection occurs at a much earlier time as for the case of melting within a cylinder heated from below.

5. ACKNOWLEDGMENTS

The authors are very grateful to the National Sciences and Engineering Research Council of Canada for supporting this work.

6. NOMENCLATURE

A	aspect ratio H/r_0
c_p	specific heat
g	acceleration of gravity
g^{11}	$\xi_r^2 + \xi_z^2$
g^{12}	$\xi_r \eta_r + \xi_z \eta_z$
g^{22}	$\eta_r^2 + \eta_z^2$
h	latent heat of fusion
H	height of cylinder
J^{-1}	$\xi_r \eta_z - \xi_r \eta_z$
n	unit vector
Nu	Nusselt number $q/(Hk\Delta T)$
p	pressure
PCM	phase change material
Pr	Prandtl number ν/α
r	radial coordinate
r_o	radius of the cylinder
R	Radius in dimensionless form r/r_o
Ra	Rayleigh number $g\beta r_0^3 (T_w - T_f)/\alpha\nu$
Ste	Stefan number $c_p(T_w - T_f)/h$
t	time
T	temperature
u, v	radial and axial velocities
U, V	dimensionless velocities
$\tilde{U}, \tilde{V}$	contravariant velocities
V_f	molten fraction
z	axial coordinate
Z	dimensionles axial coordinate z/r_o

Greek letters

α	thermal diffusivity $k/\rho c_p$
β	thermal expansion coefficeint
ξ, η	coordinate in transformed plane
ξ_r	r_η/J
ξ_z	$-r_\eta/J$
η_r	$-z_\xi/J$
η_z	r_ξ/J
ξ_τ	$\xi_r r_\tau + \xi_z z_\tau$
η_τ	$\eta_r r_\tau + \eta_z z_\tau$
θ	dimensionless temperature $(T - T_f)/(T_w - T_f)$
ν	kinematic viscosity
ρ	density of water
τ	dimensionless time $Ste \cdot Fo$
ψ	stream function
ω	vorticity $\frac{\partial v}{\partial r} - \frac{\partial u}{\partial z}$
∇^2	Laplacian in cylindrical coordinate
$\tilde{\nabla}^2$	transformed Laplacian

Subscripts

f	fusion
ω	cylinder wall

7. REFERENCES

1. HALE, N. W. and VISKANTA, R. - Photographic Observation of the Liquid-Solid Interface Motion during Melting of a Solid Heated from an Isothermal Vertical Wall, Lett. Heat Transfer, 5, 329–337, 1978.
2. RAMACHANDRA, N. and GUPTA, J. P. - Thermal and Fluid Flow Effects during Solidification in a Rectangular Enclosure, Int. J. Heat Mass Transfer, 25, 187–194, 1982.

3. OKADA, M. - Analysis of Heat Transfer during Melting from a Vertical Wall, Int. J. Heat Mass Transfer, 27, 2057-2066, 1984.
4. YOO, J. and RUBINSKY, B. - A Finite Element Method for the Study of Solidification Processes in the Presence of Natural Convection, Int. J. Num. Meth. in Engineering, 23, 1785-1805, 1986.
5. BENARD, C., GOBIN, D. and MARTINEZ, F. - Melting in Rectangular Enclosure: Experiments and Numerical Simulations, J. Heat Transfer, 107, 794-803, 1985,
6. RAMSEY, J. W. and SPARROW, E. M. and VAREJAO, L. M. C. - Melting about a Horizontal Row of Heating Cylinder, J. Heat Transfer, 101, 732-733, 1979.
7. BATHELT, A. G., VISKANTA and LEIDENFROST, R. W. - An Experimental Investigation of Natural Convection in the Melted Region around a Heated Horizontal Cylinder, J. Fluid Mech., 90, 227-239, 1979.
8. BATHELT, A. G., VISKANTA and LEIDENFROST, R. W. - Latent Heat-of-Fusion Energy Storage: Experiments on Heat Transfer from Cylinders during Melting, J. Heat Transfer, 101, 453-458, 1979.
9. SASAGUCHI, K. and VISKANTA, R. - Phase Change Heat Transfer during Melting and Resolidification of Melt around Cylindrical Heat Source(s)/Sink(s), J. Energy Resources Technology, 111, 43-49, 1989.
10. KATAYMA, K. *et al.* - A. Saifullah, Heat Transfer Characteristics of the Latent Heat Thermal Energy Storage Capsule, Solar Energy, 27, 91-97, 1981.
11. SPARROW, E. M., PATANKA, S V. and RAMADHYANI, S. - Analysis of Melting in the Presence of Natural Convection in the Melt Region, J. Heat Transfer, 99, 521-526, 1977.
12. PRUSA, J. and YAO, L. S. - Melting around a Horizontal Heated Cylinder: Part I - Perturbation and Numerical Solution for Constant Heat Flux Boundary Condition, J. Heat Transfer, 106, 376-384, 1984.
13. PRUSA, J. and YAO, L. S. - Melting around a Horizontal Heated Cylinder: Part II - Numerical Solution of Isothermal Boundary Condition, J. Heat Transfer, 106, 469-472, 1984.
14. RIEGER, H., PROJAHN, U. and BEER, H. - Analysis of the Heat Transport Mechanisms during Melting around a Horizontal Circular Cylinder, Int. J. Heat Mass Transfer, 25, 137-147, 1982.
15. HO, C. J. and VISKANTA, R. - Heat Transfer during Inward Melting in a Horizontal Tube, Int. J. Heat Mass Transfer, 27, 705-716, 1984.
16. WU, Y. K., PRUD'HOMME, M. and NGUYEN, T. H. - Étude numérique de la Fusion autour d'un Cylindre Vertical Soumis à Deux Types de Conditions Limites, Int. J. Heat Mass Transfer, 32, 1927-1938, 1989.
17. PRUD'HOMME, M. and NGUYEN, T. H. and WU, Y. K. - Simulation numérique de la fusion à l'intérieur d'un cylindre adiabatique chauffé par le bas, Int. J. Heat Mass Transfer, 32, 2275-2286 1991.

8. APPENDIX

Coefficients in equation (7) are:

$$A^u = \frac{|u| + u}{4\Delta\xi} \qquad B^u = -\frac{|u| + u}{\Delta\xi}$$

$$C^u = \frac{3|u|}{2\Delta\xi} \qquad D^u = -\frac{|u| - u}{\Delta\xi} \qquad E^u = \frac{|u| - u}{4\Delta\xi}$$

THERMAL AND FLUID-DYNAMIC BEHAVIOUR OF DOUBLE-PIPE CONDENSERS AND EVAPORATORS. A NUMERICAL STUDY

F. Escanes, C. D. Pérez-Segarra, A. Oliva

Laboratori de Termotècnia i Energètica
Dept. Màquines i Motors Tèrmics, Universitat Politècnica de Catalunya
Colom 9, 08222 Terrassa, Barcelona (Spain)

SUMMARY

This paper deals with a numerical simulation of the thermal and fluid-dynamic behaviour of double-pipe condensers and evaporators. The governing equations of the fluid flow (continuity, momentum and energy) in both the tube and the annulus, together with the energy equation in the tube wall, are solved iteratively in a segregated manner using a one-dimensional, transient formulation. An analysis of the different parameters used in the discretization is made. Some illustrative results are presented together with results of a refrigerating compression unit simulation, where double-pipe condenser and evaporator are used.

1. INTRODUCTION

Accurate methods for the prediction of the heat exchangers' behaviour are required to minimize the energy consumption. This is particularly applicable to situations involving two-phase flow inside tubes like in the cases of double-pipe and shell and tube heat exchangers among others. The inherent difficulty of the heat exchanger design in aspects such as complex geometries and fluid flow patterns, makes the possibilities of analytical solutions very limited without strong restrictive assumptions (e.g. analytical approaches such as F-factor, ε-NTU, etc.). On the other hand, the use of numerical methods allows the governing equations to be solved with fewer restrictions.

1.1 Change of phase liquid-vapor inside tubes

Both the evaporating and condensing flows inside tubes generally present three different regions: a single-phase vapor region, a two-phase region, and a single phase liquid region.

In the case of the evaporator, two sub-regions where the flow is in metastable equilibrium also exist: subcooled boiling (in the liquid region) and post dry-out flow (in the two-phase region) [1]. In these situations, the flow patterns are strongly affected, and significant changes in the friction, the convective heat transfer and the void fraction are produced. However, the one-dimensional governing equations can be written taking into consideration thermodynamic equilibrium, because the amount of both the non-equilibrium latent energy due to evaporation (subcooled boiling), and the non-equilibrium sensible energy due to the vapor superheating (post dry-out), represent small fractions in a global energy balance.

1.2 Objectives and methodology

The objective is to implement and develop numerical criteria which allow a simulation of the thermal and fluid-dynamic behaviour of double-pipe condensers and evaporators, in both transient and steady state. The characteristic parameters that define a specific situation to be analyzed are:

- Geometry: length; roughness; inside diameter of the tube; diameters of the annulus; flow arrangement (cocurrent flow or counter flow).
- Temporal distribution of inlet temperature or vapor mass fraction, pressure and velocity, in both the tube and the annulus; heat flux or temperature at the ends of the tube.
- Initial conditions in transient situations i.e. the value of all dependent variables at each point at t=0.
- Thermophysical properties of both fluids and the tube material.

2. MATHEMATICAL FORMULATION

2.1 Flow inside ducts

The governing equations have been integrated assuming the following hypotheses:

- One-dimensional flow: $T(z,t)$, $p(z,t)$, $v_g(z,t)$, $v_l(z,t)$, ...
- Fluid: pure substance; newtonian behaviour.
- Non-participant radiation medium and negligible radiant heat exchange between surfaces.
- Negligible axial heat conduction inside the fluid.
- Null heat generated by internal focus.
- Constant cross section.

The integration of the governing equations for two-phase flow over finite control volumes have the form (for more details see ref. [2]):

- Continuity:
$$\dot{m}_i = \dot{m}_o + \frac{\partial m}{\partial t} \tag{1}$$

- Momentum:
$$(p_i - p_o)\, S - \bar{\tau}_w\, P\, \Delta z - m\, g \sin\theta = \\ = \dot{m}_{go}\, v_{go} - \dot{m}_{gi}\, v_{gi} + \dot{m}_{lo}\, v_{lo} - \dot{m}_{li}\, v_{li} + \Delta z \frac{\partial \bar{\dot{m}}}{\partial t} \tag{2}$$

where the evaluation of the shear stress is done by means of the two-phase friction factor f_{tp}, which is usually calculated using empirical correlations. This factor is defined from the expression: $\tau_w = (f_{tp}/4)(\dot{m}^2/2S^2\rho_{tp})$.

- Energy:
$$\bar{\dot{q}}_w\, P\, \Delta z = \dot{m}_p\,(e_{lo} - e_{li}) + \dot{m}_{go}\,(e_{go} - e_{lo}) - \dot{m}_{gi}\,(e_{gi} - e_{li}) + \\ + (\bar{e}_g - \bar{e}_l)\frac{\partial m_g}{\partial t} + m_g \frac{\partial \bar{e}_g}{\partial t} + m_l \frac{\partial \bar{e}_l}{\partial t} - S\,\Delta z \frac{\partial \bar{p}}{\partial t} + (\bar{e}_l - e_{lp})\frac{\partial m}{\partial t} \tag{3}$$

where the last term can be neglected. The specific energy is defined as $e = h + v^2/2 + gz\cdot\sin\theta$. In order to relate the convective heat transfer and the wall temperature, the convective two-phase heat transfer coefficient α_{tp} is introduced, which is defined from the equation: $\dot{q}_w = \alpha_{tp} (T_w - T)$.

In these equations the subscript p means arithmetic average between the inlet and the outlet sections of the control volume, while the terms $\bar{\tau}_w$, $\bar{\dot{q}}_w$, $\bar{e}_g$, $\bar{e}_l$, $\bar{p}$ and $\bar{\dot{m}}$ represent integral averages over the control volume.

This mathematical model requires information about f_{tp} and α_{tp} together with the vapor and liquid velocities. In order to evaluate them, the void fraction ε_g is a widely used parameter (generally obtained from empirical sources), defined as the time-averaged volume fraction of the vapor phase in the mixture (or the time-averaged area fraction of the vapor phase in a given cross-section) [3]. The relation between the gas and liquid velocities and the void fraction is: $\varepsilon_g=[1+(1-x_g)\rho_g v_g/x_g \rho_l v_l)]^{-1}$

This formulation lets as particular cases the case of single-phase flow.

2.2 Heat conduction in the tube wall

The conduction equation has been written assuming the following hypothesis:

- One-dimensional temperature distribution.
- Negligible heat exchanged by radiation.
- Null heat generated by internal focus.

Integrating the energy equation over the control volume shown in fig. 1, the following equation is obtained:

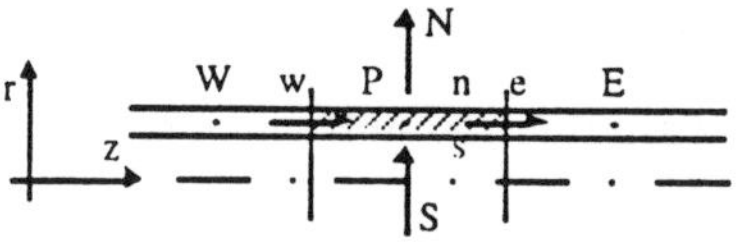

Fig. 1 Heat flux distribution in a control volume of a tubular geometry

$$(\bar{\dot{q}}_s P_s - \bar{\dot{q}}_n P_n) \Delta z + (\dot{q}_w - \dot{q}_e) S = m \frac{\partial \bar{h}}{\partial t} \qquad (4)$$

In fig. 1, "E" and "W" represent the neighboring points with conduction heat flux, and "N" and "S" the points corresponding to the annulus and tube flows respectively, with convective heat flux; "e", "w", "n" and "s" are the respective surfaces of the control volume.

The convective heat fluxes $\bar{\dot{q}}_n$ and $\bar{\dot{q}}_s$ are evaluated using the proper convective heat transfer coefficients and fluid temperatures. The conduction heat fluxes are evaluated from the Fourier law, that is: $\dot{q}_e=-\lambda_e(\partial T/\partial z)_e$, $\dot{q}_w=-\lambda_w(\partial T/\partial z)_w$.

3. NUMERICAL SOLUTION

The domain is divided into control volumes. For each control volume a set of algebraic equations is obtained by a discretization of the governing equations (1) to (4) using a fully implicit scheme. The fluid flow zones of the domain (inside the tube and the annulus) are solved using a step by step numerical scheme: while the tube wall is solved using a TDMA numerical scheme [4]. The three zones are solved iteratively in a segregated manner.

The friction, the convective heat transfer and the void fraction are evaluated, in each control volume, by means of empirical correlations obtained from the available bibliography. Null heat exchanged by the annulus flow with the environment has been considered.

In the following sub-sections a brief description of the numerical procedure is presented. For more details see ref. [2].

3.1 Spatial and temporal discretization procedure

Fig. 2 shows the spatial discretization. The discretization nodes are located at the inlet and outlet sections of the control volumes in the fluid flow zones while the discretization nodes are centered in the control volumes in the tube wall. Each zone contains n control volumes of length Δz.

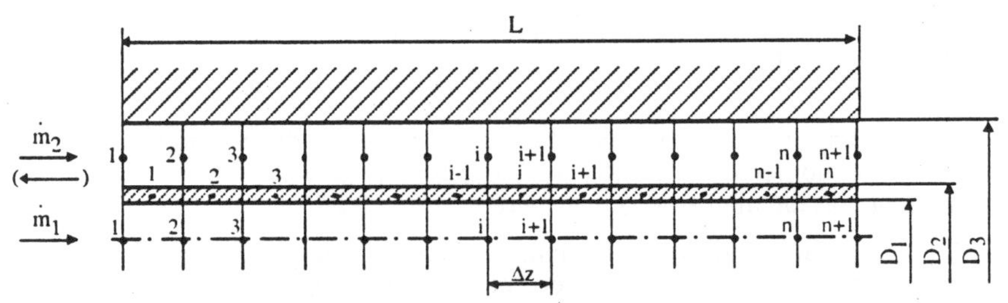

Fig. 2 Node distribution along a double-pipe heat exchanger

The transitory solution is made every time step Δt. Depending on the time evolution of the boundary conditions, a constant or variable value of Δt can be selected.

3.2 Discretization equations

In section 2, the governing equations have been directly presented on the basis of their spatial integration over finite control volumes. Thus, only their temporal integration is required. A fully implicit scheme has been used. The transient terms of the governing equations are discretized using the following approximation: $\partial\phi/\partial t \cong (\phi-\phi^o)/\Delta t$, where ϕ represents any dependent variable. In the same way, the axial heat fluxes in the tube wall are numerically approximated to $(\partial T/\partial z)_e \cong (T_P-T_E)/\Delta z$ and $(\partial T/\partial z)_w \cong (T_W-T_P)/\Delta z$.

The mean values of the different variables have been estimated by the arithmetic average between the inlet and outlet sections as: $\bar{\phi}=(\phi_i+\phi_{i+1})/2$. The mean thermophysical properties have been evaluated at the correspondent mean variables. The enthalpy variations have been evaluated neglecting their dependence on pressure variations, that is: $\Delta h=c_p(T)\cdot\Delta T$.

- *Single-phase flow:* the discretized equations are applicable inside the tube where liquid or vapor flow exists, and also inside the annulus, using the hydraulic diameter to evaluate the friction term of the momentum equation. Applying the numerical approach explained above, the outlet temperature, pressure and mass flow rate are obtained from the discretized energy, momentum, and continuity equations respectively.

- *Two-phase flow:* In this region the discretized energy equation is solved for the outlet vapor mass fraction; the discretized momentum and continuity equations yield the outlet pressure and mass flow rate respectively. The outlet temperature is calculated from the saturation condition.

- *Tube wall:* The balance of energy, for each node of the grid, yields the following discretized equation: $aT_i=bT_{i+1}+cT_{i-1}+d$, where a, b, c and d are the discretization coefficients of the node i. Two special balances are required for the extreme control volumes to account for the axial heat flux or temperature boundary conditions.

3.3 Differentiation between regions

Neglecting metastable phenomena in the evaporating flow, the differentiation between the three main zones existing in both the condensation and the evaporation processes is given by the temperature and the vapor mass fraction.These conditions are:

- Liquid region: $T<T_{sat}$, $p>p_{sat}$, $x_g=0$
- Two-phase region: $T=T_{sat}$, $p=p_{sat}$, $0<x_g<1$
- Vapor region: $T>T_{sat}$, $p<p_{sat}$, $x_g=1$

T_{sat} varies along the duct due to the pressure drop produced by the friction, the momentum variation and the mass forces (when the heat exchanger is not in a horizontal position).

Using the conditions of change of region shown above, the control volume where transition occurs is detected. In order to evaluate the position of the transition point, two criteria have been tested:

- *Transition criterion 1:* the transition point is assigned to the outlet section of the control volume.
- *Transition criterion 2:* the control volume is divided in two. The length of the first control volume is calculated from the energy equation, imposing saturated conditions with $x_g=0$ or $x_g=1$ at the outlet section. The length of the second one is calculated by simple difference.

A numerical analysis shows that using the second criterion, more accurate results are obtained for a given grid, and fewer grid points are needed for a given accuracy.

3.4 Numerical algorithm

At each time step the solution process is carried out on the basis of a global algorithm that solves iteratively, in a segregated manner, the flow inside both the tube and the annulus, and the heat conduction in the tube wall. The flow zones are solved on the basis of a numerical implicit scheme, moving forward step by step in the flow direction. The energy discretized equation for the tube wall has been solved using a TDMA [4]. The convergence in the different iterative loops is controlled using adequate rates of convergence. For example, the convergence of the fluid flow equations is considered to have been reached when $|T^{n+1}-T^{n}|<\varepsilon$ for single-phase flow, and $|x_g^{n+1}-x_g^{n}|/\max(x_g^{n+1}, 1-x_g^{n+1})<\varepsilon$ for two-phase flow, where ε is the required precision, and n and n+1 denote two consecutive iterations.

The governing equations corresponding to a steady state situation are the same equations shown above without considering the temporal derivative terms. In order to use the transient algorithm solving steady situations, an infinite time step has been considered; in this manner, all temporal derivative terms become null in the discretized governing equations.

The initial conditions have usually been obtained from the solution in steady state with boundary conditions corresponding to t=0.

The coupling between the three main subroutines has been performed iteratively following the three next steps:

- Inside the tube, the equations are solved considering the tube wall temperature distribution as boundary condition, evaluating the convective heat transfer coefficient in each control volume.
- Inside the annulus the same process is carried out.
- In the tube wall, the temperature distribution is calculated using the convective heat transfer coefficients evaluated in the preceding step.

The process starts from a given tube wall temperature distribution, which is arbitrary at t=0 and the correspondent to t at $t=t+\Delta t$.

3.5 Evaluation of the empirical coefficients

- *Single-phase flow:* the convective heat transfer coefficient has been calculated according to Gnielinski [5]; the friction factor from the well known expressions of Haegen-Poiseuille and Colebrook-Nikuradze, for laminar and turbulent regimes respectively, using the hydraulic diameter for the annulus.

In the case of subcooled boiling, the convective heat transfer coefficient and the friction factor are treated separately. For the convective heat transfer, the beginning of the subcooled boiling has been estimated according to Frost and Dzakowic [6]; the method proposed by Bergles and Rohsenow [7] has been used to consider the transition between pure liquid convection heat transfer and boiling heat transfer, using the correlation proposed by Forster and Zuber [8] for nucleate boiling. For the friction factor, the point of net vapor generation has been estimated according to Saha and Zuber [9]; the friction factor has been estimated assuming the homogeneous flow model described by Hewitt [3], using the two-phase viscosity proposed by McAdams (cited by Hewitt [3]), and the real vapor mass fraction proposed by Levy [10].

- *Two-phase flow:* the void fraction has been estimated from the semi-empiric equation by Zivi [11]. The friction factor has been calculated according to Lockhart and Martinelli [12], using a simple analytical representation of the Lockhart and Martinelly curve suggested by Chisholm (cited by Hewitt [3]).

In the case of condensation, the convective heat transfer coefficient has been calculated from the Nusselt equation (stratified flow, cited by Butterworth [13]) with the correction of Jaster and Kosky [14], and the expression proposed by Boyko and Kruzhlin [15] (in the case of annular flow); the equation of Wallis [16] has been taken as a transition criterion.

In the case of evaporation, the convective heat transfer coefficient has been evaluated according to Chen [17]. The point of dry-out is considered at $x_g=0.9$ for refrigerating purposes [18]. The convective heat transfer, the friction factor and the void fraction have been calculated from the homogeneous flow model and the single-phase expressions explained above.

4. RESULTS

4.1 Numerical aspects analysis

Using the following numerical parameters: n=600, $\varepsilon=10^{-6}$, a typical computational time consumption is 10 and 20 seconds per time step for the condenser and the evaporator respectively in a workstation of 75 MIPS.

In order to analyze the influence of the two transition criteria considered (see subsection 3.3), and the numerical parameters used in the modelization, results of a condensing flow inside a tube at constant wall temperature are shown.

(a) Case 1: comparison between the results obtained using the transition criteria 1 and 2. Table 1 shows the results corresponding to the following situation:

- Condensation inside a tube at constant wall temperature in steady state.
- Geometry: $L=3$ m, $D_1=5$ mm, $\theta=0$, $\zeta\equiv0$
- Fluid: R-12
- Boundary conditions:
  Tube: $T_w=20$ C
  Fluid (z=0): $T_i=40$ C, $p_i=8$ bar
  $v_i=10$ m/s, superheated vapor.
- Reference numerical results: $z_{bc}=0{,}283$ m, $z_{ec}=2{,}758$ m, $T_o=28{,}08$ C, obtained using the transition criterion 2 with $\varepsilon=10^{-7}$ and n=500.

	Transition criterion 1			Transition criterion 2		
n	z_{bc} (m)	z_{ec} (m)	T_o (C)	z_{bc} (m)	z_{ec} (m)	T_o (C)
10	0,300 (6,0 %)	3,000 (8,8 %)	31,56 (12,4 %)	0,278 (1,8 %)	2,754 (0,1 %)	27,96 (0,4 %)
20	0,300 (6,0 %)	2,850 (3,3 %)	29,23 (4,1 %)	0,281 (0,7 %)	2,756 (0,1 %)	28,01 (0,2 %)
50	0,300 (6,0 %)	2,820 (2,2 %)	28,87 (2,8 %)	0,282 (0,4 %)	2,758 (0,0 %)	28,06 (0,0 %)
100	0,300 (6,0 %)	2,790 (1,2 %)	28,46 (1,4 %)	0,282 (0,4 %)	2,759 (0,0 %)	28,07 (0,0 %)
200	0,285 (0,7 %)	2,775 (0,6 %)	28,28 (0,7 %)	0,283 (0,0 %)	2,759 (0,0 %)	28,08 (0,0 %)
500	0,288 (1,8 %)	2,766 (0,3 %)	28,17 (0,3 %)	0,283 -	2,759 -	28,08 -
1000	0,285 (0,7 %)	2,763 (0,2 %)	28,13 (0,2 %)	- -	- -	- -
2000	0,284 (0,4 %)	2,760 (0,1 %)	28,10 (0,1 %)	- -	- -	- -

Table 1 Results and relative difference respect the reference case (between brackets) offered by the modelization using the transition criteria 1 and 2. z_{bc}, z_{ec}: position of the points of begin and end of condensation respectively.

(b) Case 2: analysis of the influence of the numerical parameters used in the modelization, in a transient situation. Table 2 and 3 show the results obtained using different n, Δt and ε. The analyzed situation is:

- Condensation inside a tube at constant wall temperature in transient state.
- Geometry: $L=3$ m, $D_1=5$ mm, $\theta=0$, $\zeta\equiv0$
- Fluid: R-12
- Boundary conditions:
  Tube: $T_w=20$ C
  Fluid (z=0): $T_i(t)=T_\infty+(T_0-T_\infty)\cdot\exp(-t/t_0)$,
  $p_i=p_\infty+(p_0-p_\infty)\cdot\exp(-t/t_0)$, $v_i=10$ m/s, superheated vapor
  $T_0=40$ C, $T_\infty=46$ C, $p_0=8$ bar, $p_\infty=10$ bar and $t_0=200$ s

n	t (s)				
	0	50	100	150	200
10	27,97	28,24	28,59	28,88	29,06
20	28,01	28,50	28,90	29,19	29,38
50	28,07	28,56	28,95	29,27	29,49
100	28,07	28,56	28,96	29,27	29,51
200	28,08	28,56	28,96	29,27	29,51
500	28,08	28,57	28,96	29,28	29,51
1000	28,10	28,58	28,99	29,29	29,53
2000	28,11	28,60	29,00	29,31	29,55

$\varepsilon=10^{-5}$

n	t (s)				
	0	50	100	150	200
10	27,97	28,24	28,59	28,88	29,06
20	28,02	28,50	28,90	29,19	29,38
50	28,07	28,56	28,95	29,26	29,49
100	28,07	28,56	28,96	29,27	29,51
200	28,07	28,56	28,96	29,27	29,51
500	28,08	28,56	28,96	29,27	29,51
1000	28,08	28,56	28,96	29,27	29,51
2000	28,08	28,56	28,96	29,27	29,51

$\varepsilon=10^{-7}$

Table 2 Outlet temperature T_0(C) obtained for different number of control volumes n, and different time instants.

Δt (s)	t (s)				
	0	50	100	150	200
100	28,08	-	28,96	-	29,51
50	28,08	28,56	28,96	29,27	29,51
25	28,08	28,56	28,96	29,27	29,51
12,5	28,08	28,56	28,96	29,27	29,51
6,25	28,08	28,56	28,96	29,28	29,51
3,125	28,08	28,56	28,96	29,28	29,51
1,5625	28,08	28,56	28,96	29,27	29,51
0,78125	28,08	28,56	28,96	29,28	29,51
0,39063	28,08	28,55	28,94	29,26	29,49
0,19531	28,08	28,55	28,95	29,26	29,50
0,09766	28,08	28,61	29,01	29,32	29,55
0,04883	28,08	28,53	28,92	29,26	29,49

$\varepsilon=10^{-5}$

Δt (s)	t (s)				
	0	50	100	150	200
100	28,07	-	28,95	-	29,51
50	28,07	28,56	28,96	29,27	29,51
25	28,07	28,56	28,95	29,27	29,51
12,5	28,07	28,56	28,96	29,27	29,51
6,25	28,07	28,56	28,96	29,27	29,51
3,125	28,07	28,56	28,96	29,27	29,51
1,5625	28,07	28,56	28,96	29,27	29,51
0,78125	28,07	28,56	28,96	29,27	29,51
0,39063	28,07	28,56	28,96	29,27	29,51
0,19531	28,07	28,56	28,96	29,27	29,51
0,09766	28,07	28,55	28,95	29,27	29,51
0,04883	28,07	28,54	28,94	29,27	29,50

$\varepsilon=10^{-7}$

Table 3 Outlet temperature T_0(C) obtained for different time steps Δt, and different time instants.

4.2 Illustrative results

(a) Case 1: solution of a double-pipe heat exchanger working as a condenser and an evaporator, in transient state. Fig. 3 shows the temperature of the flow inside both the tube and the annulus. The analyzed situations corresponds to:

- Geometry: condenser: L=5.5 m, D_1, D_2, D_3=5, 6, 10 mm, $\zeta\equiv0$, $\theta=0$
  evaporator: L=3 m, D_1, D_2, D_3=5, 6, 10 mm, $\zeta\equiv0$, $\theta=\pi/2$, up flow inside the tube.
- Flow arrangement: counter flow
- Tube:

Fluid: R-12

Boundary conditions (z=0):

condenser:

$$T_i(t) = T_\infty+(T_0-T_\infty)\cdot\exp(-t/t_0),$$
$$p_i(t) = p_\infty+(p_0-p_\infty)\cdot\exp(-t/t_0),$$
$(T_0=40$ C, $T_\infty=46$ C,
$p_0=8$ bar, $p_\infty=10$ bar, $t_0=200$ s)
$v_i= 10$ m/s, superheated vapor

evaporator:

$$p_i(t) = p_\infty+(p_0-p_\infty)\cdot\exp(-t/t_0),$$
$(p_0= 8$ bar, $p_\infty= 10$ bar, $t_0= 200$ s)
$v_{li}= 0{,}5$ m/s, $v_{gi}= 2{,}2$ m/s, liquid+vapor $(x_g=0.1)$

- Annulus:
  - Fluid: Water
  - Boundary conditions (z=L):
    $T_i(t)=T_\infty+(T_0 - T_\infty)\cdot\exp(-t/t_0)$,
    (T_0=20 C, T_∞=15 C, t_0=200 s)
    p_i=2 bar, v_i=1 m/s.
- Tube wall:
  - material: cooper
  - Boundary conditions: adiabatic ends.
- Numerical parameters: $\varepsilon=10^{-6}$, n=200, Δt=50 s.

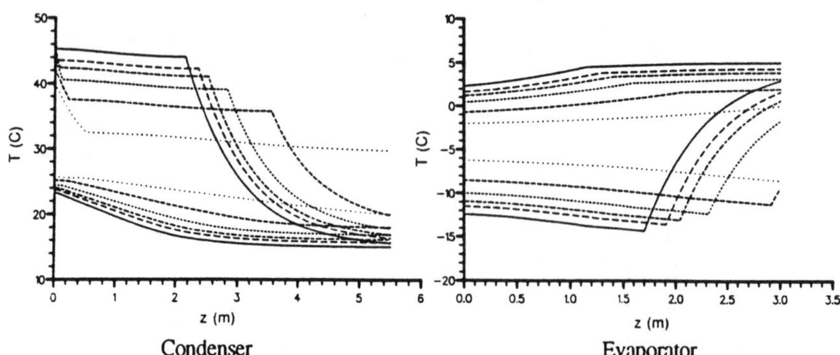

Condenser Evaporator

Fig 3 Tube and annulus temperature distribution in a double-pipe heat exchanger.
···· t=0 s, - - - t=100 s, · - · - t=200 s, - · - t=300 s, — — t=400 s, — t=∞

(b) Case 2: Simulation of a refrigerating unit [19][20], in transient state, using both R-12 and R-134a as working fluid. Fig. 4 shows the Molière diagram and table 5 the working parameters of the refrigerating unit. The analyzed case is:

- Double-pipe condenser and evaporator, capillary tube expansion device and reciprocating compressor refrigerating unit.
- Refrigerant: R-12 , R-134a
- Compressor:
  - Geometry: displacement: 5.5 cm^3; dead volume: 0,33 cm^3;
  - Frequency: 50 Hz; polytropic compression index: 1,09
  - Adiabatic boundary
- Condenser:
  - Geometry: L=5 m, D_1=4 mm, D_2=5 mm, D_3=10 mm, $\zeta\equiv0$, $\theta=0$ counter flow
  - Annulus: water at T_i=32 C, p_i=2 bar, v_i=0,5 m/s
  - Adiabatic boundary
- Evaporator:
  - Geometry: L=2 m, D_1=5 mm, D_2=6 mm, D_3=10 mm, $\zeta\equiv0$, $\theta=\pi/2$, up flow inside the tube, counter flow
  - Annulus: water at $T_i(C)=32-21,05\,(1-\exp(-t(s)/1400))$ for $0\leq t\leq4200$,
    $T_i(C)=12$ for $t>4200$,
    p_i=2 bar, v_i=0,5 m/s
  - Adiabatic boundary
- Capillary tube:
  - Geometry: length: 1,5 m for R-12; 2,4 m for R-134a; inside diameter: 0,6 mm; smooth tube
  - Adiabatic boundary

- Initial conditions: corresponding to the steady state solution with boundary conditions at t=0 and condensation temperature approximately equal to 55 C (compressor outlet pressure: 13,63 bar for R-12; 14,84 bar for R-134a)
- Numerical parameters: n = 400 (condenser), 300 (evaporator), 600 (capillary tube); Δt has been taken variable from 10 to 80 min; $\varepsilon=10^{-6}$

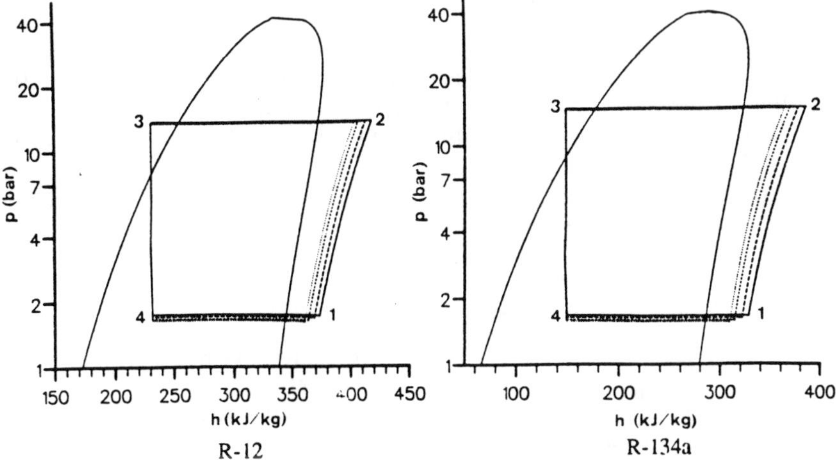

Fig. 4 Molière diagram obtained from the simulation of a refrigerating unit.
— t=0 min, – – t=10 min, - - - t=10 min, ···· t=70 min, · · · t=150 min

	t (min)	p_1 (bar)	p_2	p_3	p_4	T_1 (C)	T_2	T_3	T_4	$\dot{m}_1$ (kg/h)	$\dot{W}_{cp}$ (w)	$\dot{Q}_c$	$\dot{Q}_e$
R-12	0	1,72	13,6	13,6	1,75	30,0	114,3	32,1	-16,1	5,04	59,5	262,4	198,7
	10	1,68	13,6	13,5	1,71	22,6	106,9	32,1	-16,7	5,00	90,0	253,1	191,1
	30	1,64	13,5	13,5	1,67	14,3	98,8	32,1	-17,4	4,97	62,3	243,3	182,7
	70	1,61	13,4	13,4	1,64	9,2	93,8	32,1	-17,8	4,95	60,8	237,2	177,4
	150	1,61	13,4	13,4	1,64	9,2	93,8	32,1	-17,8	4,95	59,5	237,3	177,4
R-134a	0	1,65	14,8	14,8	1,67	30,0	105,1	35,5	-14,6	3,76	60,1	245,3	186,1
	10	1,61	14,7	14,7	1,64	22,6	98,0	35,3	-15,2	3,73	58,4	236,0	178,5
	30	1,57	14,6	14,6	1,60	14,4	90,2	35,2	-15,8	3,71	56,7	226,3	170,1
	70	1,55	14,5	14,5	1,57	9,3	85,4	35,1	-16,1	3,70	55,6	220,3	164,8
	150	1,55	14,5	14,5	1,57	9,3	85,3	35,1	-16,1	3,70	55,6	220,3	164,7

Table 4 Working parameters obtained from the simulation of a refrigerating unit.

5. CONCLUDING REMARKS

A numerical method to analyze the behaviour of double-pipe condensers and evaporators has been developed by means of a transient one-dimensional analysis of the fluid flow governing equations (continuity, momentum and energy) and the heat conduction in the tube wall. Empirical information is needed in order to evaluate shear stress, heat flux and two-phase flow structure.

The simulation has been implemented on the basis of an implicit step by step numerical scheme inside the tube and the annulus, and a TDMA scheme in the tube wall. The different zones have been solved iteratively in a segregated manner. In order to minimize computational time consumption, a special treatment has been implemented solving the control volume that contains the transition between the single and the two-phase flow.

A study of the influence of the numerical parameters used in the modelization is made. Illustrative results of the behaviour of double-pipe condensers and evaporators have been presented, together with their implementation in order to simulate the behaviour of a complete refrigerating unit.

ACKNOWLEDGEMENTS

This study has been supported by the company Unidad Hermética S.A., the Comisión Interministerial de Ciencia y Tecnología, Spain (ref. no. PTR90-0060), the Dirección General de Investigación Científica y Técnica, Spain (ref. no. PB90-0606), and by the Direcció General d'Universitats, Catalonia, Spain (Program AD1).

The authors thank J. M. Serra, J. Pons, A. Castillo, of Unidad Hermética S.A. for the technical support given.

NOMENCLATURE

c_p specific heat at constant pressure
D diameter
f friction factor
g gravity acceleration
h specific enthalpy
e specific energy defined as: $e=h+v^2/2+gz\cdot\sin\theta$
i index grid node
L length
m mass
$\dot{m}$ mass flow rate
n number of control volumes
p pressure
P perimeter
$\dot{q}$ heat flux
$\dot{Q}$ heat power
r radial coordinate
S Cross section
t time
Δt time discretization step
T temperature
v velocity
x vapor mass fraction
$\dot{W}$ mechanical power
z axial coordinate
Δz spatial discretization step

Greek

α convective heat transfer coefficient
ε void fraction; precision
θ inclination angle
λ thermal conductivity
ρ density
τ shear stress
ζ roughness

Subscript

b boundary
c condenser
cp compressor
e evaporator
g gas, vapor
i inlet; index grid node
l liquid
o outlet
sat saturation
tp two-phase
w wall

REFERENCES

1. COLLIER, J.G. - Convective Boiling and Condensation, McGraw-Hill International Book Company, New York, 1981.
2. ESCANES, F., PEREZ-SEGARRA, C.D., OLIVA, A. - Simulación Numérica del Comportamiento Térmico e Hidráulico de Condensadores y Evaporadores con Disposición Tubular Concéntrica, Servei de Publicacions de la U.P.C., Barcelona, 1991.

3. HEWITT, G.F. - Gas-Liquid flow, Heat Exchanger Design Handbook, sec. 2.7.3, Ed. Schlünder, E.U. et al., Hemisphere Publishing Corporation, Washington, D.C., 1983.
4. PATANKAR, S.V. - Numerical Heat Transfer and Fluid Flow, Hemisphere Publishing Corporation, Washington, D.C., 1980.
5. GNIELINSKI, V. - Forced Convection, Heat Exchanger Design Handbook, sec 2.5.1, Ed. Schlünder, E.U. et al., Hemisphere Publishing Corporation, Washington, D.C., 1983.
6. FROST, W., DZANOWIC, G.S. - An Extension of the Method of Predicting Incipient Boiling on Commercially Finished Surfaces, ASME/AIChE Heat Transfer Conf., paper 67-ht-61, pp. 1-8, Seattle, 1967.
7. BERGLES, A.E., ROHSENOW, W.M. - The Determination of Forced-Convection Surface-Boiling Heat Transfer. J. Heat Transfer, Vol. 86C, pp. 365-372, 1964.
8. FORSTER, H.K., ZUBER, N. Dynamics of Vapour Bubble Growth and Boiling Heat Transfer. AIChE J., Vol. 1 No. 4, pp. 531-535, 1955.
9. SAHA, P., ZUBER, N. Point of Net Vapour Generation and Vapour Volumetric Void Fraction, Proc. 15th Int. Heat Transfer Conf., paper B4.7, pp. 175-179, 1974.
10. LEVY, S. Forced Convection Subcooled Boiling Prediction of Vapour Volumetric Fraction. Int. J. Heat Mass Transfer, Vol. 10, pp. 951-965, 1967.
11. ZIVI, S.M. - Estimation of Steady-State Steam Void Fraction by Means of the Principle of Minimum Entropy Production. J. Heat Transfer, Vol. 86C, pp. 247-252, 1964.
12. LOCKHART, R.W., MARTINELLI, R.C. - Proposed Correlation of Data for Isothermal Two-Phase, Two-Component Flow in Pipes. Ch. Eng. Progress, Vol. 45 No. 1, pp. 39-48, 1949.
13. BUTTERWORTH, D. - Film Condensation of Pure Vapor, Heat Exchanger Design Handbook, sec 2.6.2, Ed. Schlünder, E.U. et al., Hemisphere Publishing Corporation, Washington, D.C., 1983.
14. JASTER, H., KOSKY, P.G. - Condensation Heat Transfer in a Mixed Flow Regime. Int. J. Heat Mass Transfer, Vol. 19 pp. 95-99, 1976.
15. BOYKO, L.D., KRUZHILIN G.N. - Heat Transfer and Hydraulic Resistance During Condensation of Steam in a Horizontal Tube and in a Bundle of Tubes. Int. J. Heat Mass transfer, Vol. 10, pp. 361-373, 1969.
16. WALLIS, G.B. - One-Dimensional Two-Phase Flow, McGraw-Hill Book Company, New York, 1969.
17. CHEN, J.C. Correlation for Boiling Heat Transfer to Saturated Fluids in Convective Flow. Ind. Eng. Chem. Process Design and Development, Vol. 5 No. 3, pp. 322-327, 1966.
18. ASHRAE Handbook, 1981 Fundamentals, ASHRAE Inc., Atlanta, 1982.
19. ESCANES, F., PEREZ-SEGARRA, C.D., OLIVA, A. - Simulación Numérica de un Ciclo Frigorífico por Compresión, Internal Report, Barcelona , 1991.
20. ESCANES, F., PEREZ-SEGARRA, C.D., OLIVA, A. - Numerical Simulation of Capillary Tube Expansion Devices. In phase of reviewing.

THE APPLICATION OF THE ENTHALPY METHOD TO ANALYSIS OF A LATENT HEAT STORAGE UNIT

J.GOSCIK and J.LACH

Department of Mechanics, Technical University of Bialystok, Wiejska 45C, 15-351 Bialystok, Poland

ABSTRACT

In this article, we have attempted to show the most essential quantitative aspects of solutions obtainable by making use of the finite-difference based enthalpy formulation of the natural convection controlled phase change processes. Our main goal in this context is a direct confrontation of the numerical results with those provided by the latest reports on known closed form correlations for temporal variations of the average Nusselt number and the molten volume fractions. The most important problems are highlighted via the test problem, namely melting in a quadratic latent heat storage unit. The parametric domain covered by the discussion is $10^4 \le Ra \le 10^6$, $Pr = 50$, $Ste = 0.1$.

1. INTRODUCTION

It is known [1,2] that among the numerical methods useful for solution of conduction controlled solid - liquid phase change problems, one of the most effective is the finite - difference based enthalpy method. Quite recently [3,4] the enthalpy method has also appeared to be promising for predicting more complicated problems in which melted material is in motion due to the action of buoyancy forces. In this way another very important criterion for choosing the enthalpy method as the best for handling heat storage processes with phase change is fulfilled (the importance of this mechanism in the description of storage processes was emphasized by Viskanta [5]).

In general, in the one-domain approach to the modelling phase change processes when coupled with fluid flow in the melted phase, two crucial aspects have to be taken into account:

(i) the formulation of an accumulation energy term appropriate

to the discretization method used in the approach;
(ii) the treatment of the interphase momentum transport (and conjugate to them the interphase energy transport).

The first of these is quite effectively accomplished by representing phase change thermal effects in the form of the source term [3]. Unfortunately this approach seems to be only justified in the case of metals for which it is appropriate to carry out the momentum analysis as for porous media (the liquid phase can flow through the porous two phase solid-liquid mushy region). In the case of amorphous substances (e.g. waxes being the most interesting in latent heat storage application), the transition from liquid to solid (or reverse) is continuous and there is no relative motion between the melt and the solid [6]. Thus for these systems the enhanced viscosity formulation is far more appropriate.

The treatment of the interphase momentum transport by increasing viscosity from small values (where the material behaves like a fluid) to very large values (where the material behaves like a solid) have been undertaken in [7÷10]. However, up till now, in the presentation of their new approaches researchers have restricted their comments mainly to the computational aspects of the calculation schemes. To our knowledge, the correctness of the solutions is mostly indicated by qualitative agreement with the results of the appropriate experiment. Thus, the problem of reliability of the results obtained on the basis of the one-domain formulation of the natural convection controlled phase change processes has not so far been sufficiently elucidated.

In view of the above, we have attempted to verify the most essential quantitative aspects of solutions obtainable by making use of fixed grid methodology with

- the finite - difference based enthalpy formulation, where the sensible and latent enthalpies are separated in the transient term of the energy equation and the latent heat is included in the source term [11], and
- the formulation of the momentum balance in which the description of the interphase transport is achieved by means of the growing viscosity.

To this aim in view, this approach has been confronted with the available closed form correlations based on a test problem corresponding to the physical situations. The numerical results comprise the most interesting quantities from the engineering point of view, namely:

(i) identification and verification of time scales for a melting process with natural convection,
(ii) evolution of the average value of the Nusselt number,
(iii) temporal variations of the molten volume fractions.

2. TEST PROBLEM STATEMENT

As a test problem, we consider the melting process of

paraffin wax (e.g. n-octadecan with a Prandtl number exceeding 50) packed tightly in a storage tank which is an infinitely long, rectangular storage unit with a square cross-section D (l x l). Such an assumption enables one to perform a two-dimensional analysis (end-wall effects are negligible) for which many numerical results are available.

It is assumed that the phase change material (PCM) is a pure substance, changing phase at a well defined temperature θ^* with a known enthalpy **h** - temperature θ relation. The liquid PCM is a Newtonian, incompressible fluid (the density of the pure substance is constant in the entire physical domain except for the buoyancy term in the momentum equation), and satisfies the Boussinesque approximation.

Furthermore, the heat transfer is supposed to fulfil the following hypotheses:

A1. heat transfer at the solid - liquid interface is not strongly perturbed by the expansion of the PCM on melting.

A2. the convection flow in the melt is in the laminar regime.

A3. the flows driven by density differences between the phases are negligibly small as compared to the bouyancy - driven flow.

A4. viscous heat dissipation is neglected.

The general balance equations for phase change processes governed by above given assumptions are well established, eg. [10]. Here we concentrate on their dimensionless form. In order to be able to directly confront theoretical considerations with numerical results, balance equations are restated in a dimensionless form using the scales employed in [12,13]. Consequently,

$$\mathbf{x}=\mathbf{x}/l_c,\quad \mathbf{v}=\mathbf{v}/v_c,\quad \tau = t/t_c,\quad \mu_\eta(s) = \eta(s)/\eta_c,\quad p = p/p_c$$

with characteristic quantities defined as

$$l_c = l,\quad v_c = \eta^*/(\rho\cdot l),\quad t_c = (l^2/a)\cdot Ste^{-1},\quad \eta_c = \eta^*,\quad p = \rho\cdot v_c^2$$

where $\mathbf{v}(v_1,v_2)$ is the velocity vector with components in the 0_1 and 0_2 directions (gravity is assumed to be parallel to the 0_2 direction), l is a geometrical dimension, p is the pressure and η, ρ, a are dynamic viscosity, density and thermal diffusitivity respectively.

The balance equations of, respectively, the mass, the momentum and the energy take the following forms:

$$0 = \partial_1 v_1 + \partial_2 v_2 \quad , \tag{1}$$

$$\begin{aligned}(Ste/Pr)\cdot\partial_\tau(v_1) = &- (\partial_1 v_1\cdot v_1 + \partial_2 v_2\cdot v_1) - \partial_1 p_d \\ &+ \partial_1(2\cdot\mu_\eta(s)\cdot\partial_1 v_1) + \partial_2(\mu_\eta(s)\cdot(\partial_2 v_1+\partial_1 v_2)) \ ,\end{aligned} \tag{2}$$

$$(Ste/Pr)\cdot\partial_\tau(v_2) = -(\partial_1 v_1\cdot v_2 + \partial_2 v_2\cdot v_2) - \partial_2 p_d + Ra\cdot Pr^{-1}\cdot s \tag{3}$$
$$+ \partial_1(\mu_\eta(s)\cdot(\partial_2 v_1 + \partial_1 v_2)) + \partial_2(2\cdot\mu_\eta(s)\cdot\partial_2 v_2) ,$$

where the viscosity in a nondimensional form is given by

$$\mu_\eta(s) = \begin{cases} 1 , & s > 0 , \\ \exp(\text{const} = 20) , & s \le 0 , \end{cases}$$

$$Ste\cdot\partial_\tau s(\theta) = - Pr\cdot(\partial_1 v_1\cdot s(\theta) + \partial_2 v_2\cdot s(\theta)) \tag{4}$$
$$+ (\partial^2_{11} s + \partial^2_{22} s) - \partial_\tau \Gamma(\theta) .$$

According to Voller's concept [11], sensible and latent heat in the energy equation are given by

$$s(\theta) = \int_0^\theta d\theta \quad \text{and} \quad \Gamma(\theta) = \mu_\Gamma(\theta) \text{ with } \mu_\Gamma(s) = \begin{cases} 0, & s \le 0 , \\ 1, & s > 0 . \end{cases}$$

We pose the test problem analogous to the physical situations considered by Okada [12] and Jany and Bejan [13]. Therefore, we assume that the heat transfer conditions on the external storage cavity walls are homogeneous with the one which permits an energy exchange (vertical left wall Γ_1), the other walls being perfectly thermally insulated. Accordingly, the transport processes are governed by a driving force perpendicular to the direction of the gravitational field interaction. The melting process is initiated by a step temperature θ increment of the wall Γ_1 with the assumption that the initial temperature θ_o equals the theoretical temperature θ^* of the phase change (equivalently, the initial subcooling of the solid phase equals zero). Therefore the following initial-boundary conditions are associated with the equations (1)÷(4)

$$\left.\begin{array}{l} s(\mathbf{x}) = s_o , \\ \mathbf{v}(\mathbf{x}) = 0 , \end{array}\right\} \forall \mathbf{x} \in D,\ \tau = 0 , \tag{5}$$

for s function ,

$$\left.\begin{array}{l} s(\mathbf{x}) = s_w , \quad \forall \mathbf{x} \in \Gamma_1 , \\ \partial_\nu s\Big|_{\mathbf{x}} = 0 , \quad \forall \mathbf{x} \in \bigcup_{I=4} \Gamma_I \setminus \Gamma_1 , \\ \text{for velocity ,} \\ \mathbf{v} = 0 , \quad \forall \mathbf{x} \in \bigcup_{I=4} \Gamma_I , \end{array}\right\} \forall \tau \in (0,T]. \tag{6}$$

In the above equations, the parameters Pr, Ra and Ste are

defined by

$$Pr = \eta^{*}/(\rho\cdot a),$$

$$Ra = g\cdot\rho\cdot\beta\cdot(l_c)^3\cdot\Delta\theta/(\eta\cdot a),$$

$$Ste = c\cdot(\theta_w - \theta_o)/\Delta h^{*}.$$

Since the range of numerical calculations is of a strictly test form, the following parameters were assumed to be: (i) Pr = 50, while the actual value of Pr is not essential if it exceeds 7.0; (ii) A = 1.0, i.e. l /l (the cross-section of the storage unit is a square); (iii) Ste = 0.1, (iv) Ra = 10^4, 10^5, 10^6 , which permits the comparison of our results with those presented in [12,13].

3. NUMERICAL SOLUTION

The equations (1)÷(6) were solved iteratively using the control volume - based finite - difference procedure described by Patankar [14]. In detail the code was generated with the following options

- for the solution of momentum and continuity equations the SIMPLE algorithm [14] was applied, with
  - a fully implicit formulation for time - dependent terms (including modifications according to Van Doormal and Raithby's [15] suggestions),
  - convection-diffusion coefficients were evaluated using the Patankar-Spalding power law scheme,
  - source terms in momentum equations (from viscosity) were linearized,
  - for the solution of algebraic discretization equations a line - by - line solver based on the TDMA was used,
- the iteratively updated enthalpy procedure in the energy equation is solved by means of the procedure given by Brent et al. [4].

All the programs generated by us have been tested in detail in the convective part, including the De Vahl Davis [16] benchmark data as well as the results of Patterson and Imberger [17] given for natural convection in nonstationary conditions. After conducting a grid-refinement study, a uniform 40 x 40 grid was chosen for this work. A time step in a range of $\{1.0\cdot10^{-4} \div 1.0\cdot10^{-3}\}$ was applied.

Underrelaxation factors for the solution of the two momentum equations (α_v) the pressure correction equation (α_p), and the energy equations (for sensible enthalpy (α_s) and for latent enthalpy (α_Γ) updating) were chosen as follows

α_v	α_p	α_s	α_Γ
0.5	0.5	0.8	0.5

The convergence criteria for the iterative solution of a

full set of four nonlinear, coupled partial differential equations were based on the maximum local mass imbalance and overall energy balance. Convergence was declared after the above mentioned mass imbalance and the absolute value of energy imbalance became less than

$$\varepsilon_m = \varepsilon_e = 1.0\cdot 10^{-6} \quad .$$

3.1 Analysis of Nusselt's number on the heating wall

It is known, see Viskanta [5], that the melting process under consideration consists of four successive regimes: pure conduction <c>; mixed <c_cv> (here the convection process occurs in the upper fluid layers, while the conduction takes place in the lower ones); fully developed natural convection <cv>; and shrinkage <sh>.

These four stages reflect a way of changing the Nusselt number on the heating wall, the first three of which are now quite well known. In this context, the Okada [12] and Jany and Bejan [13] correlations enable one to verify the correctness of the Nusselt number numerically found.

Taking the average Nusselt number to be defined by

$$\overline{Nu}^{(x)} = \int_{\Gamma_1} (\, \partial_1 \theta \,)\, d\Gamma \;\longrightarrow\; \int_{\Gamma_1} (\, \partial_1 s \,)\, d\Gamma$$

Okada [12] found the formula

$$\overline{Nu}^{(x)} = \begin{cases} (\, 2\cdot\tau \,)^{-0.5} & , \quad \tau \le \tau_{\langle c\rangle} \;, \\ \overline{Nu}^{(x)}_{\langle c\_cv\rangle} \cdot \left\{ 1 + c\cdot(\tau - \tau_{\langle c\rangle}) \right\} , & \quad \tau > \tau_{\langle c\rangle} \;. \end{cases} \qquad (7)$$

The above correlation has been verified in the range of $10 \le Ra \le 5\cdot 10^{6}$, in which the formula

$$Nu_{\langle c\_cv\rangle} = 0.234\cdot Ra^{0.266} \quad ,$$

serves to find the corresponding Nusselt number and the time point $\tau_{\langle c\rangle}$ (at which the third regime is initiated)

$$\tau_{\langle c\rangle} = 0.5\cdot(\; \overline{Nu}^{(x)}_{\langle c\_cv\rangle} \;)^{-2.0} \quad .$$

On the other hand, Jany and Bejan [13] established another relationship

$$\overline{Nu}^{(x)} = (\; 2\cdot\tau \;)^{-0.5} + \left[\, c_1\cdot Ra^{0.25} - (\; 2\cdot\tau \;)^{-0.5} \right]\cdot\left[1 + (\; c_2\cdot Ra^{0.75}\cdot\tau^{1.5})^{n} \right]^{1/n} , \qquad (8)$$

where for the range $0 \le Ra \le 10^{8}$ the constants $c_1 = 0.35$, $c_2 = 0.0175$, and $n=-2$. Figure 1 presents the data obtained by means of the two correlations above as well as our numerical results for three different Rayleigh numbers. It is worth noting that all the curves adequately feature a monotonic decrease of the Nusselt number of the order $\tau^{-0.5}$, which are characteristic of the pure conduction regime <c>. The step-like behaviour of

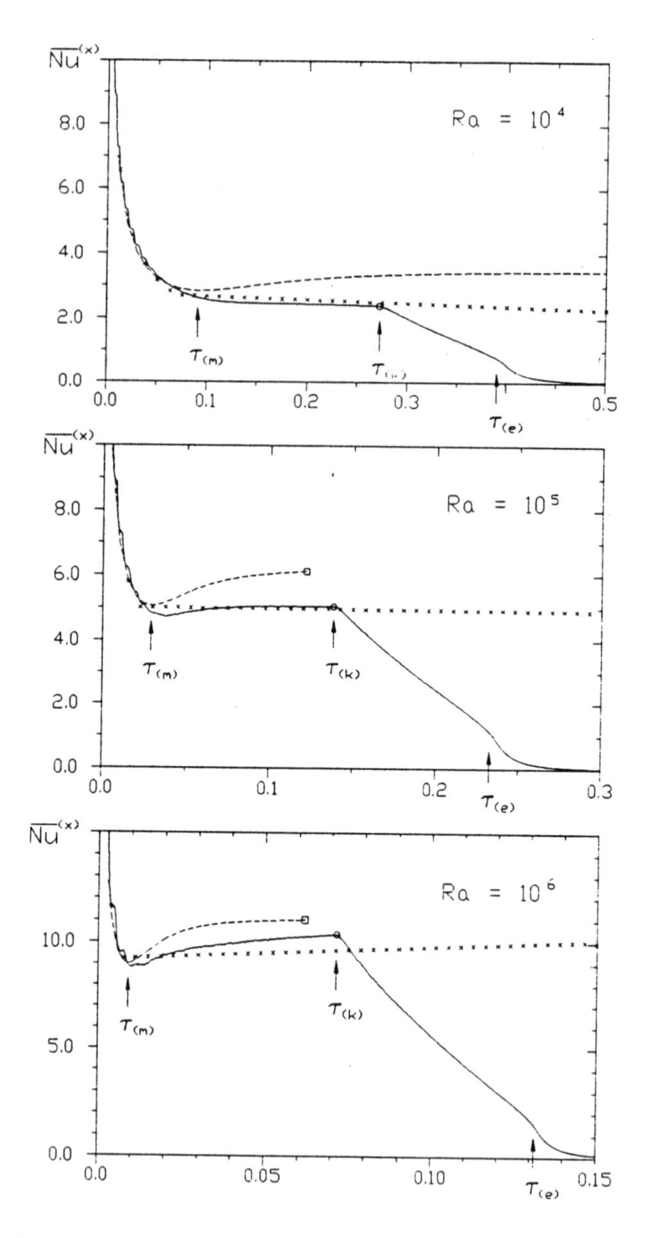

Fig. 1 Average Nusselt number evolution at the heated wall - comparison of the present numerical results (solid lines) and the correlation (8) (dashed lines) and with correlation (7) (crosses), established by [13] and [12], respectively.

the numerical curve, typical for solutions obtained by means of the enthalpy method, is also easily observable. The pure conduction regime terminates according to the scale analysis [13] at the point $\tau_{<c>}$. Its value was determined by Benard et al. [18]

$$\tau_{<c>} = 4.59 \cdot Ra^{-0.5} \quad ,$$

but it is difficult to distinguish this point in the figure.

The mixed regime <c_cv> indicates its characteristic minimum, with

$$\tau_{(m)} \cong 9 \cdot Ra^{-0.5} \quad (\cong 2 \cdot \tau_{<c>}) \quad ,$$

as shown in [13]. From this time point, one can observe a marked discrepancy between the data available from the various formulae. The differences seen between (7) and (8) have their origin in the assumptions concerning the nature of transfer phenomena in the <c_cv> regime. The Okada [12] relation (7) is based on the hypothesis that the transition from pure conduction to fully developed natural convection is very short and in the extreme variant may be reduced to just one point $\tau_{<c>}$ (which is characteristic of the two-regime models for Nu). In consequence, with the development of natural convection, the Nusselt number changes its value linearly with the Rayleigh dependent slope. In contrast, Jany and Bejan [13] on the basis of a three-regime model assumed that when the third regime is reached, the Nusselt number resembles a constant function with values of the order of $Ra^{0.25}$, the latter being typical for fully developed pure convection in the rectangular enclosure.

The numerical solutions, obtained without any simplified assumptions indicate, with particular clarity for increasing values of Ra, that these values depend significantly on the current geometry of a region filled with a liquid phase.

Behaviour of the Nusselt number in the shrinkage regime

Physically, the shrinkage regime starts at the moment $\tau_{(k)}$ when the highest point of the phase boundary for the first time hits the right vertical adiabatic wall. These moments are indicated by small open squares for those given by Jany and Bejan and by the open circles for those obtained in this study (points corresponding to the Okada analysis are not presented because in his study, the storage tank had the form of one heated vertical wall between insulated top and bottom horizontal plates). Numerically,

$$\tau_{(k)} = \arg \chi_{UR}(\tau) \quad \text{when} \quad \chi_{UR}(\tau) = 1.0 \quad ,$$

where χ_{UR} stands for the liquid phase fraction in the balance cell located at the top right-hand corner.

One can easily see that starting from $\tau_{(k)}$ the Nusselt number exhibits a rapid, almost linear, decrease of its values. Similar observations were made in [13], where the following formula was suggested for estimating the Nusselt number in the <sh> regime:

$$\overline{Nu}^{(x)}_{<sh>} \cong [\ 1 - Ra^{0.25}\cdot(\ \tau - \tau_{(k)})\ ]^{0.6}\quad .$$

From this formula it can be concluded that the $Nu(\tau)$ curvature is negative. On the other hand, our numerical solution gives rise to a positive curvature function. This discrepancy may be explained by the simplified assumptions made in [13] and, possibly, by the kind of approximation of convective and diffusion fluxes.

3.2 Analysis of the temporal variations of the molten volume fractions

A verification of the correctness of the melting intensity is based on comparison of the temporal variations of the molten volume fractions,

$$\chi(\tau) = \int_{\vartheta} \Gamma(x,\tau)\ dx\quad ,$$

obtained numerically, with the averaged values of the phase boundary location

$$\bar{\psi}^{(x)}(\tau) = \int_0^1 \psi(x,\tau)\ dx\quad ,$$

obtained from the Okada [12]

$$\bar{\psi}^{(x)}(\tau) = \begin{cases} (\ 2\cdot\tau\ )^{0.5}\quad , & \tau \le \tau_{<c>}, \\ (\ 2\cdot\tau\ )^{0.5} + & \\ Nu_{<c\_cv>}\cdot(\ \tau-\tau_{<c>})\cdot\{1 + c\cdot(\tau-\tau_{<c>})/2\} & \tau > \tau_{<c>}, \end{cases} \tag{9}$$

and the Jany and Bejan [13] correlations

$$\bar{\psi}^{(x)}(\tau) = \{\ [\ (\ 2\cdot\tau\ )^{0.5}]^{m} + [\ c_1\cdot Ra^{0.25}\cdot\tau\ ]^{m}\ \}^{1/m}\ , \tag{10}$$

where for the range $0 \le Ra \le 10^8$, the constants are $c_1 = 0.35$, $m = 5$.

Figures 2a) and 2b) show the results of this comparison over a time interval long enough to cover the first three regimes. Our results are located considerably below (to 18%) the ones predicted by [13]. At this moment it is not clear if this discrepancy results from the simplifying assumptions of the scale analysis carried out in [13]. Generally speaking, our results seem to support the trends already observed by Jany and Bejan on the basis of comparisons with the experimental works carried out by other authors. The trends confirm the good agreement of our numerical results with those given by the Okada correlation (9). As is shown in the figure they are slightly below those for $Ra = 10^4$, 10^5 and almost the same as for $Ra = 10^6$. This is significant because the constant c (see Eq.(7,9)), depends on Ra, and unfortunately was presented only in graphic form. Here we took the following three values for c: $c(Ra = 1.0\cdot10^4) = -\ 0.36$, $c(Ra = 1.0\cdot10^5) = -\ 0.07$, $c(Ra = 1.0\cdot10^6) = 0.64$.

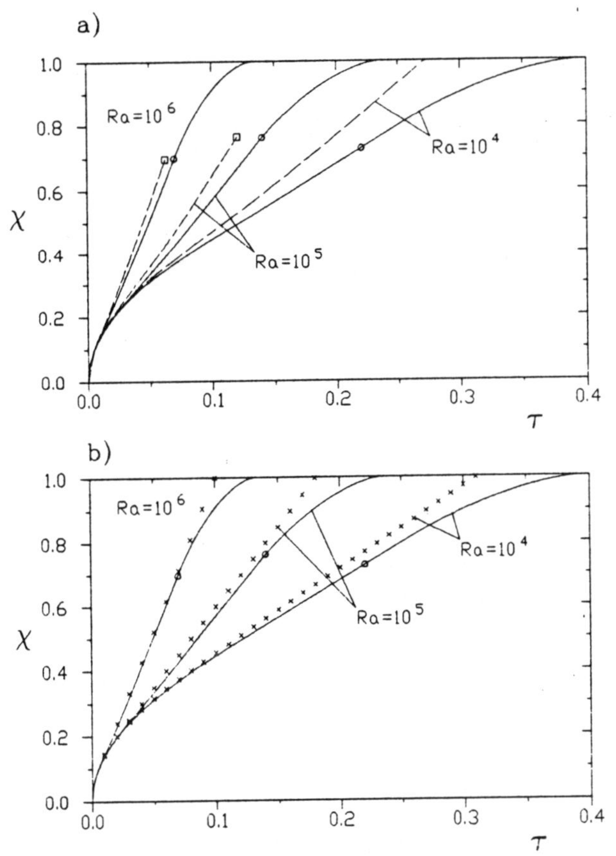

Fig. 2 Time evolution of the melted fraction - comparison of the present numerical results with: a) the correlation (10) established in [13] and b) with correlation (9) given by [12].

Temporal variations of the molten volume fractions in the shrinkage regime

As in the Nusselt number case, Figures 2.a and 2.b clearly show some differences (although less drastic) in the intensity of time evolution of the melted fraction after contact of the melted front with vertical right wall (here adiabatic). In [13] the authors estimated that the material in the solid state will disappear in the time interval

$$\Delta\tau^{T}_{<sh>} \cong Ra^{-0.2} \quad . \tag{11}$$

The table below presents our data calculated according to

$$\Delta\tau^{N}_{<sh>} = \tau_{(e)} - \tau_{(k)} \quad ,$$

where $\tau_{(e)} = \arg \chi_{BR}(\tau)$ when $\chi_{BR}(\tau) = 1.0$,

and χ_{BR} stands for the liquid phase fraction in the balance cell located at the bottom right-hand corner and those obtained via formula (11)

Ra	$\tau_{(k)}$	$\tau_{(e)}$	$\Delta\tau^{N}_{<sh>}$	$\Delta\tau^{T}_{<sh>}$
$1.00\cdot10^{4}$	0.269	0.392	0.123	0.159
$1.00\cdot10^{5}$	0.138	0.234	0.096	0.100
$3.27\cdot10^{5}$	0.102	0.176	0.074	0.079
$6.95\cdot10^{5}$	0.080	0.146	0.066	0.068
$1.00\cdot10^{6}$	0.071	0.131	0.060	0.063

It can be seen that the discrepancies between the above two groups of data do not exceed 5%, except for the case of Ra= 10^4 corresponding to the not fully developed natural convection.

4. CONCLUSIONS

All our results were obtained on a PC (25-MHz, 486i microcomputer running a Fortran code). This fact confirms the generally known advantages of the enthalpy method.

It is worth noting that if one supplements a numerical code with convective transport some of the disadvantages of the enthalpy solutions mentioned above originate from the essential problems for fixed grid methodology previously discussed by other authors. In our opinion, the most important problems to be addressed, are

- nonphysical oscillations occurring in the average Nusselt number evolution (unfortunately, direct application of remedies characteristic of conduction results in a drastic increase in the computational time), and
- the reduction in the melting process intensity (with the unclear quantitative specification as to how far this reduction depends on the models used for description of the interphase momentum and energy transport).

REFERENCES

1. SHAMSUNDAR, N. - Comparison of numerical methods for diffusion problems with moving boundary boundaries, Moving Boundary Problems, Eds. Wilson, D.G. et al., Academic Press Inc., 1978, pp. 165-185.

2. VOLLER, V.R. and SWAMINATHAN, C. and THOMAS, B. - Fixed grid techniques for phase change problems: A review, Int. J. Numer. Methods Eng., Vol. 30, No. 4, 1990, 875-898.

3. VOLLER, V. and BRENT, A. - Investigation of fixed grid techniques for convection\diffusion controlled phase change problems, Numerical Methods in Thermal Problems VI, Eds. Lewis, R. and Morgan, K., Pineridge, 1989, pp. 24-34.

4. BRENT, A., VOLLER, V. and REID, K. - Enthalpy-porosity for

modeling convection-diffusion phase change: application to the melting of a pure metal, Num. Heat Transfer, Vol. 13, No. 2, 1988, 297-318.

5. VISKANTA,R. - Natural convection in melting and solidification, Natural Convection Fundamentals and Applications, Eds. Kakac,S. et al., Hemisphere Publishing Corp., 1985.

6. OLDENBURG,C.M. and SPERA,F.J. - Hybrid model for solidification and convection, Num. Heat Transfer, Part B, Vol. 21, 217-229, 1992.

7. GARTLING,D.K.-Finite element analysis of convective heat transfer problems with change of phase, Computer Methods in Fluids, Eds. Morgan,K. et al., Pentech, London, 1980.

8. DANTZIG,J.A. - Modelling liquid - solid phase changes with melt convection, Int. J. Num. Meth. in Engng., Vol. 28, No. 11, 1989, 1769-1785.

9. USMANI,A.S. and LEWIS,R.W. - Solidification in a square cavity in the presence of natural convection, NUMETA 90 - Numerical Methods in Engineering: Theory and Applications, Eds. Pande,G.N. and Middleton,J., Elsevier Applied Science, 1990, Vol. I, pp. 374-380.

10. VOLLER,V.R., BRENT,A.D. and PRAKASH,C. - The modelling of heat, mass and solute transport in solidification systems, Int. J. Heat Mass Transfer, V. 32, No. 9, 1989, 1719-1731.

11. VOLLER,V.R. - Implicit finite-difference solutions of the enthalpy formulation of Stefan problems, IMA J. Numer. Anal., Vol. 5, No. 2, 1985, 201-214.

12. OKADA,M. - Analysis of heat transfer during melting from a vertical wall, Int. J. Heat Mass Transfer,Vol. 27, No. 11, 1984, 2057-2066.

13. JANY,P. and BEJAN,A. -Scaling theory of melting with natural convection in an enclosure, Int. J. Heat Mass Transfer, Vol. 31, No. 6, 1988, 1221-1235.

14. PATANKAR,S.V. - Numerical Heat Transfer and Fluid Flow, McGraw-Hill Book Co., 1980, New York.

15. VAN DOORMALL,J.P. and RAITHBY,G.D. - Enhancements of the simple method for predicting incompressible fluid flows, Num. Heat Transfer, Vol. 7, No. 1, 1984, 147-163.

16. DE VAHL DAVIS,G. - Natural convection of air in a square cavity: a benchmark numerical solution, Int. J. Numer. Meth. Fluids, Vol. 3, 1983, 249-264.

17. PATTERSON,J. and IMBERGER,J. - Unsteady natural convection in a rectangular cavity, J. Fluid Mech., Vol. 10, Part 1, 1980, 65 - 86.

18. BENARD,C., GOBIN,D. and MARTINEZ,F. - Melting in rectangular enclosures: experiments and numerical simulations, J. of Heat Transfer, Vol. 107, 1985, 794-803.

A NUMERICAL STUDY OF THE STEADY STATE FREEZING OF WATER IN A RECTANGULAR ENCLOSURE

P. H. Oosthuizen
HEAT TRANSFER LABORATORY, Dept. Mechanical Engineering
Queen's University, Kingston, Ontario, Canada K7L 3N6

ABSTRACT

The heat transfer across a vertical enclosure containing pure water when one vertical wall of the enclosure is kept at a temperature that is below the freezing point of water while the opposite wall is kept at a temperature that is above this freezing temperature has been numerically studied. The conditions dealt with in the study are such that there is significant natural convection in the water. Only the steady state has been here considered. The flow situation considered has been the subject of a number of previous studies but almost all of these have been concerned with the evolution of the flow with time and have not basically been concerned with a detailed study of the effects of the various governing parameters on the final steady state situation. This is the aim of the present work. The flow has been assumed to be laminar and two-dimensional. It has also been assumed that fluid properties are constant except for the density change with temperature that gives rise to the buoyancy force, this being treated by assuming a quadratic type relationship. The governing equations have been expressed in terms of the stream function and vorticity and written in dimensionless form. These governing equations, subject to the boundary conditions, have been solved using a finite element method in which the position of the solid-liquid interface is obtained using an iterative approach. Solutions have been obtained for modified Rayleigh numbers of between 10^3 and 10^8 for a Prandtl number of 11 for various degrees of under-cooling and for enclosure aspect ratios of between 0.5 and 2. The effects of these parameters on the heat transfer rate across the enclosure and on the shape and position of the interface has been studied.

1. INTRODUCTION

The present study is concerned with the heat transfer across a vertical enclosure containing pure water. One vertical wall of the enclosure is kept at a temperature that is below the freezing point of water while the opposite wall is kept at a temperature that is above this freezing temperature. Ice therefore forms in part of the enclosure, the conditions being such there is significant natural convection in the water. The upper and lower surfaces of the cavity are assumed to be adiabatic. The flow situation is thus as shown in Fig. 1. Only the steady state has been here considered i.e. the evolution of the flow with time from some prescribed initial state has not been considered.

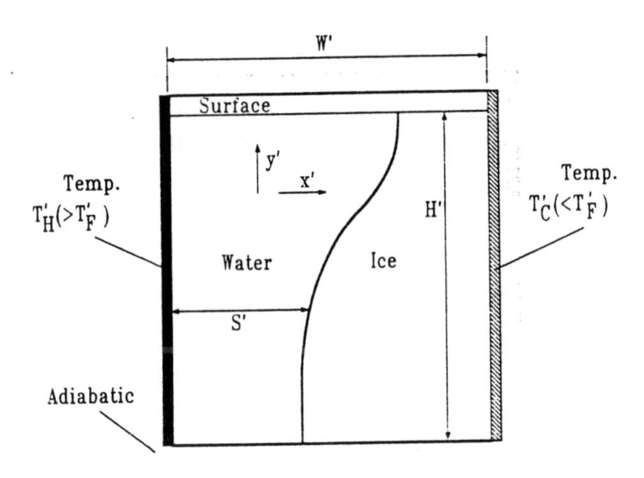

Figure 1. Flow configurations considered and coordinate system used.

There have been number of previous studies of solidification and melting of liquids in an enclosure. Almost all of these studies have, however, been concerned with the evolution of the flow with time and have not been concerned with a detailed study of the effects of the various governing parameters on the final steady state for the case where there is under-cooling. A review of much of this work is given by Yao and Prusa [1] and Fukusako and Yamada [2]. Experimental and numerical studies of the particular case of the freezing of pure water in a rectangular enclosure are given by Braga and Viskanta [3] and de Vahl Davis et al [4] respectively. These studies also provide reviews of past work on the subject. The density maximum that occurs near the freezing point with water plays an important role in the freezing of water in a rectangular enclosure. There have been a number of previous studies of natural convection in an enclosure under such conditions that this density maximum is important but without freezing - for example, see [5], [6], [7] and [8] which also contain reviews of past work in this field.

2. GOVERNING EQUATIONS AND SOLUTION PROCEDURE

It has been assumed that the flow is steady, laminar and two-dimensional and that liquid and solid properties are constant except for the water density change with temperature which gives rise to the buoyancy forces, this being treated by assuming a quadratic type relationship, i.e. by assuming a relation of the form:

$$(\rho_M - \rho)/\rho = a(T - T_M)^2 \tag{1}$$

the subscript M referring to conditions at the temperature of maximum density.

The solution for the liquid has been obtained in terms of the stream function and vorticity defined, as usual, by:

$$u' = \frac{\partial \psi'}{\partial y'} , \qquad v' = -\frac{\partial \psi'}{\partial x'}$$

$$\omega' = \frac{\partial v'}{\partial x'} - \frac{\partial u'}{\partial y'} \tag{2}$$

The prime (′) denotes a dimensional quantity.

The following dimensionless variables have then been defined:

$$\psi = \psi'/\alpha , \qquad \omega = \omega' W'^2/\alpha$$

$$T = (T - T'_F)/(T'_H - T'_C) \tag{3}$$

where $\alpha = k_l/\rho c$ is the thermal diffusivity of the water. The coordinate system used is shown in Figure 2.

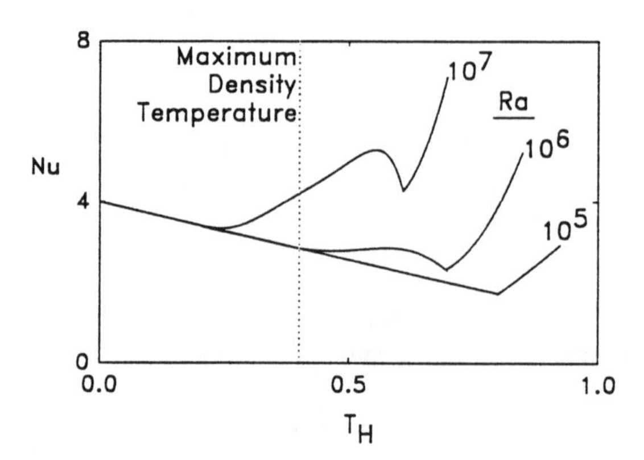

Figure 2. Variation of mean Nusselt Number with :r with dimensionless hot wall temperature for various modified Rayleigh Numbers for a square enclosure with $T_M = 0.4$.

In terms of these dimensionless variables, the governing equations for the liquid flow are:

$$\frac{\partial^2 \psi}{\partial x^2} + \frac{\partial^2 \psi}{\partial y^2} = -\omega \tag{4}$$

$$\left(\frac{\partial^2 \omega}{\partial x^2} + \frac{\partial^2 \omega}{\partial y^2}\right) - \frac{1}{Pr}\left(\frac{\partial \psi}{\partial y}\frac{\partial \omega}{\partial x} - \frac{\partial \psi}{\partial x}\frac{\partial \omega}{\partial y}\right)$$

$$= -2\, Ra\, T \frac{\partial T}{\partial x} \tag{5}$$

$$\frac{\partial^2 T}{\partial x^2} + \frac{\partial^2 T}{\partial y^2} - \left(\frac{\partial \psi}{\partial y}\frac{\partial T}{\partial x} - \frac{\partial \psi}{\partial x}\frac{\partial T}{\partial y}\right) = 0 \tag{6}$$

where Ra is the modified Rayleigh number defined by:

$$Ra = a g W'^{3} (T_H' - T_C')^{2} / \nu \alpha \tag{7}$$

The equation governing the temperature distribution in the solid phase is:

$$\frac{\partial^2 T}{\partial x^2} + \frac{\partial^2 T}{\partial y^2} = 0 \tag{8}$$

The boundary conditions on the solution are:

On all walls:

$$\psi = 0, \quad \frac{\partial \psi}{\partial n} = 0$$

At $x = 0$:

$$T = T_H$$

At $x = 1$:

$$T = T_C \; (= T_H - 1)$$

On remaining wall segments:

$$\frac{\partial T}{\partial n} = 0$$

where n is the coordinate measured normal to the wall surface considered.

On surface, which is assumed to remain flat:

$$\psi = 0, \quad \omega = 0, \quad \frac{\partial T}{\partial y} = 0$$

On the interface between the liquid and solid phases, the following conditions apply:

$$\psi = 0, \quad \frac{\partial \psi}{\partial n} = 0$$

$$\left.\frac{\partial T}{\partial n}\right|_l = \left.\frac{\partial T}{\partial n}\right|_s \left(\frac{k_s}{k_l}\right)$$

where the subscripts l and s refer to conditions on the liquid and solid sides of the interface respectively.

The above dimensionless equations, subject to the boundary conditions, have been solved using a finite element procedure. The solution is iterative in nature. The position of the solid-liquid interface is first guessed and the element distributions in the solid and liquid regions are selected, nodal points being selected to lie along solid-liquid interface. The solution is then started, the interface position being modified according to the difference between the calculated rates of heat transfer at the interface on the solid and liquid sides, the element shapes being adaptively modified to follow the changing interface shape. The solution is continued until a converged solution is obtained.

The main results that will be presented here are the mean Nusselt number, Nu, based on the overall temperature difference and on the full width of the enclosure and the dimensionless liquid volume V. These are defined by:

$$Nu = \frac{qW'}{k_l (T_H' - T_C')} \tag{9}$$

and

$$V = \frac{1}{A} \int_0^A S \, dy \tag{10}$$

3. RESULTS

The solution has the following parameters:

- the dimensionless temperature of the hot wall, T_H
- the dimensionless temperature at which the maximum density occurs, T_M
- The modified Rayleigh number, Ra
- the Prandtl number, Pr
- the aspect ratio of the rectangular enclosure, A
- the ratio of the thermal conductivity of the frozen material to that of the unfrozen material k_r

Solutions have been obtained for modified Rayleigh numbers of between 10^3 and 10^8 for a Prandtl number of 11 for various degrees of under-cooling and for enclosure aspect ratios of between 0.5 and 2. These results have all been obtained for a conductivity ratio k_r of 4.

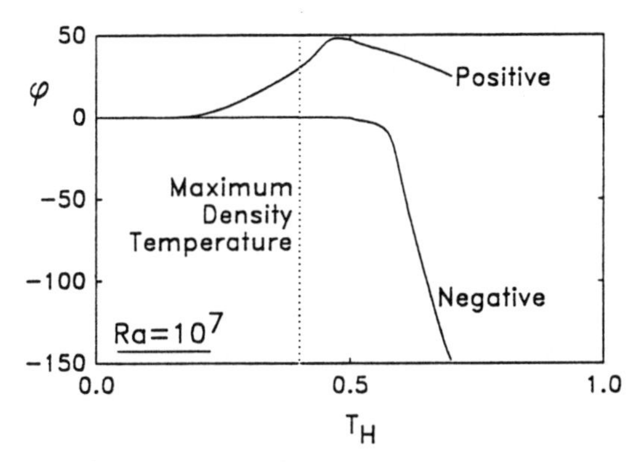

Figure 3. **Variation of maximum and minimum values of the dimensionless stream function with dimensionless hot wall temperature for $Ra = 10^7$ for a square enclosure with $T_M = 0.4$.**

Figure 2 shows the variation of mean Nusselt number with dimensionless hot wall temperature for a fixed value of the dimensionless maximum density temperature for various values of the modified Rayleigh number. Physically, this would involve a situation in which, at a given value Ra, the temperatures of the hot and cold walls are varied in such a way that the difference between these two temperatures remains constant. It will be seen from Figure 2 that at low values of the dimensionless hot wall temperature, the Nusselt number is equal to the pure conduction value, this being given:

$$Nu = T_H + k_r(1 - T_H) \tag{11}$$

the conductivity ratio, k_r, here, it will be recalled, being taken as 4. As the dimensionless hot wall temperature increases, a point is reached at which the convective motion starts to become strong enough to influence the heat transfer rate and the Nusselt number starts to rise above the pure conduction value. At the higher Rayleigh numbers, this occurs at hot wall temperatures that are well below the maximum density temperature. Under these circumstances, then, the flow is predominantly down the hot wall and up the cold wall because of the density inversion. When the hot wall temperature rises above the maximum density temperature an opposite flow starts to develop which tends to first decrease the heat transfer rate. However, with further increase in hot wall temperature, this opposite flow, which involves flow up the hot wall and down the cold wall, rapidly increases in intensity and becomes dominate bringing about an almost discontinuous rise in Nusselt number with increasing hot wall temperature. These changes in flow pattern are illustrated by the results given in Figure 3. This shows the variations of the maximum and minimum values of the dimensionless stream function with dimensionless hot wall temperature for a fixed value of the dimensionless maximum density temperature for

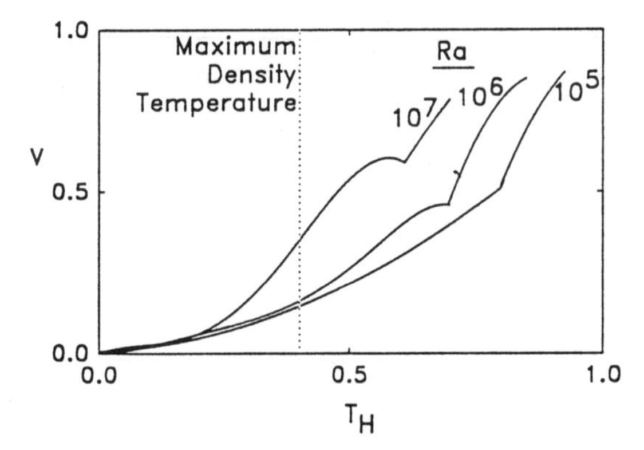

Figure 4. **Variation of dimensionless liquid volume with dimensionless hot wall temperature for various modified Rayleigh Numbers for a square enclosure with $T_M = 0.4$.**

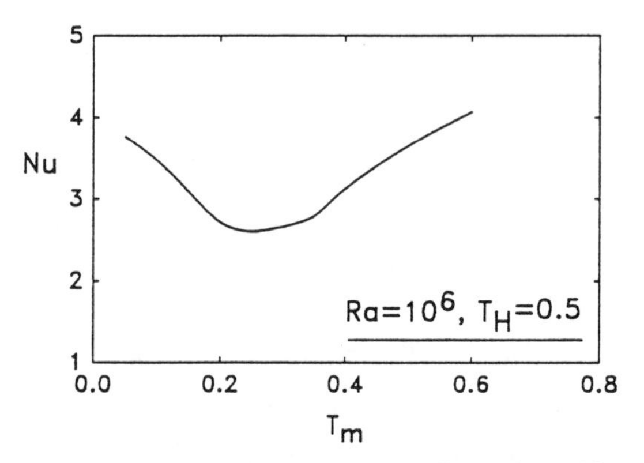

Figure 5. **Variation of mean Nusselt Number with dimensionless maximum density temperature for $Ra = 10^6$ for a square enclosure with $T_H = 0.5$.**

a fixed value of the modified Rayleigh number. Positive values of the stream function are associated with a a downward flow along the hot wall and an upward flow along the cold wall while negative values are associated with a flow up the hot wall and down the cold wall. It will be seen from Figure 3, as mentioned above, that at low wall temperatures the flow is entirely down the hot wall and up the cold wall. When the hot wall temperature increases above the maximum density temperature, however, this motion starts to decrease in intensity and then there is a sudden sharp rise in the motion associated with flow up the hot wall and down the cold wall. Figure 4 shows the variation of dimensionless liquid layer thickness with dimensionless hot wall temperature for the same conditions as those considered in Figure 2. If there is no convective motion, the dimensionless liquid layer thickness

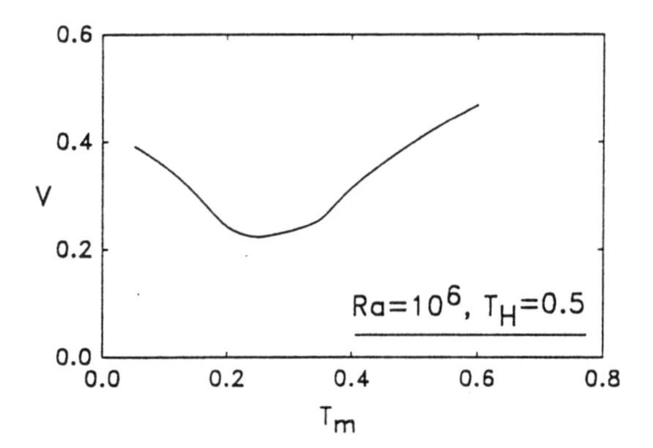

Figure 6. **Variation of dimensionless liquid volume with dimensionless maximum density temperature for $Ra = 10^6$ for a square enclosure with $T_H = 0.5$.**

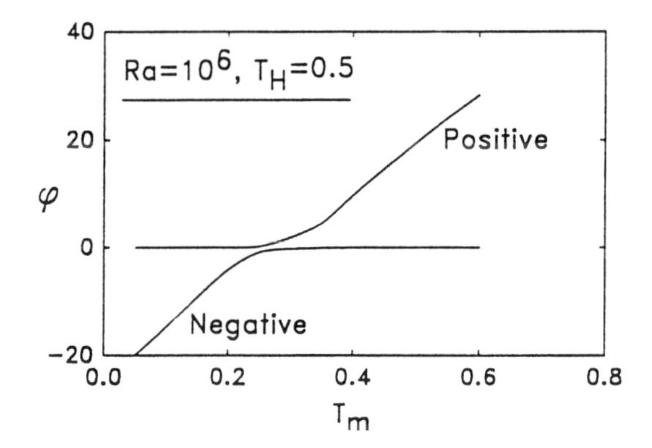

Figure 7. **Variation of maximum and minimum values of the dimensionless stream function with dimensionless maximum density temperature for $Ra = 10^6$ for a square enclosure with $T_H = 0.5$.**

is given by:

$$V = \frac{1}{1 + k_r(1 - T_H)/T_H} \tag{12}$$

and at low values of the dimensionless wall temperature, the results for all Rayleigh numbers tend to this variation.

The effect of the dimensionless maximum density temperature on the mean Nusselt number, the dimensionless liquid volume and the maximum and minimum values of the stream function is illustrated by the results given in Figures 5, 6 and 7. At low values of T_M, the heat transfer rate and dimensionless liquid volume are high, the flow consisting of an upward flow along the hot wall and a downward flow along the cold wall. As T_M increases, the heat transfer rate and

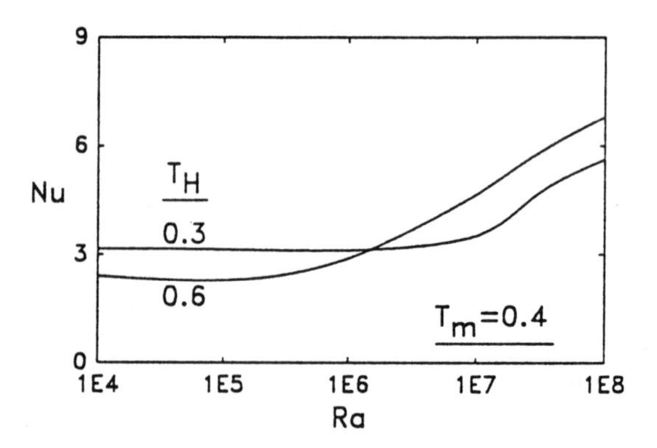

Figure 8. **Variation of mean Nusselt Number with modified Rayleigh number for a square enclosure with T_M = 0.4 for two values of T_H.**

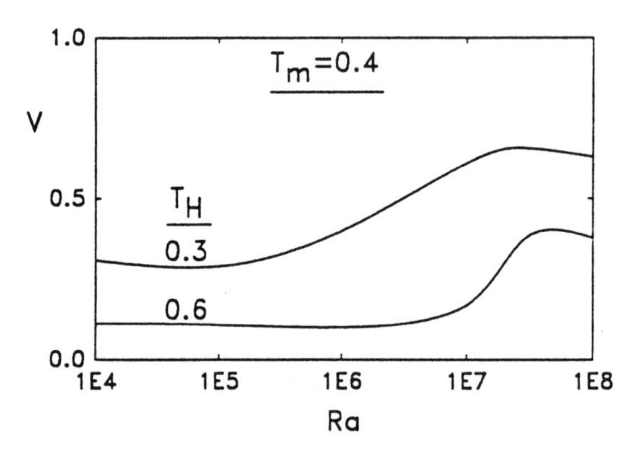

Figure 9. **Variation of dimensionless liquid volume with modified Rayleigh number for a square enclosure with T_M = 0.4 for two values of T_H.**

dimensionless liquid volume fall and pass through a minimum before rising again. When T_M is above that of minimum heat transfer the flow consists of a downward flow along the hot wall and an upward flow along the cold wall.

Figures 8, 9 and 10 illustrate the effect of modified Rayleigh number on the results for two values of the dimensionless hot wall temperature, one below and one above the dimensionless maximum density temperature.

4. CONCLUSIONS

The density inversion has been shown to have a large effect on the steady state freezing of water in an enclosure. Sharp changes in the

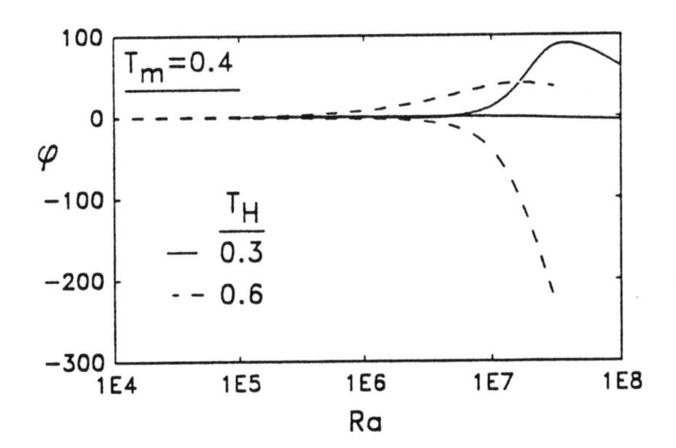

Figure 10. **Variation of maximum and minimum values of the dimensionless stream function with modified Rayleigh number for a square enclosure with T_M = 0.4 for two values of T_H.**

variation of mean heat transfer rate with hot wall temperature have been shown to occur as a result of this density inversion.

5. NOMENCLATURE

A	=	aspect ratio, H' / W'
a	=	coefficient in density-temperature relation
c	=	specific heat
g	=	gravitational acceleration
H'	=	height of cavity
k	=	thermal conductivity
k_r	=	ratio of solid to liquid thermal conductivities
Nu	=	Mean Nusselt number based on W'
n	=	n' / W'
n'	=	coordinate measured normal to surface
Pr	=	Prandtl number
q	=	mean heat transfer rate
Ra	=	modified Rayleigh number based on W'
S	=	S' / W'
S'	=	local distance from hot wall to interface
T	=	dimensionless temperature
T'	=	temperature
T'_C	=	temperature of cold surface
T'_F	=	solidification temperature
T'_H	=	temperature of hot wall
T'_M	=	temperature of maximum density
u	=	dimensionless velocity component in x' direction
u'	=	velocity component in x' direction
v	=	dimensionless velocity component in y' direction

v'	=	velocity component in y' direction
W'	=	width of cavity
x	=	dimensionless x' coordinate
x'	=	horizontal coordinate position
y	=	dimensionless y' coordinate
y'	=	vertical coordinate position
α	=	thermal diffusivity
ν	=	kinematic viscosity
ρ	=	density
ψ	=	dimensionless stream function
ψ'	=	stream function
ω	=	dimensionless vorticity
ω'	=	vorticity

Subscripts

l	=	liquid
s	=	solid

6. ACKNOWLEDGEMENTS

This work was supported by the Natural Sciences and Engineering Research Council of Canada.

7. REFERENCES

1. YAO, L. S. and PRUSA, J. - Melting and Freezing, Advances in Heat Transfer , Vol. 19, pp. 1-95. 1989.
2. FUKUSAKO S. and YAMADA, M. - Recent Advances in Research on Freezing and Melting Heat-Transfer Phenomena, Experimental Heat Transfer, Fluid Mechanics and Thermodynamics 1991 , J. F. Keffer, R. K. Shah and E. N. Ganic, eds., Elsevier, pp. 1157-1170, 1991.
3. BRAGA, S. L. and VISKANTA, R. - Effect of the Water Density Extremum on the Solidification Process, Experimental Heat Transfer, Fluid Mechanics and Thermodynamics 1991 , J. F. Keffer, R. K. Shah and E. N. Ganic, eds., Elsevier, pp. 1185-1192, 1991.
4. DE VAHL DAVIS, G., LEONARDI, E., WONG, P. H. and YEOH, G. H. - Natural Convection in a Solidifying Liquid, Numerical Methods in Thermal Problems , R. W. Lewis and K. Morgan, eds., Pineridge Press , Vol. 6, Part 1, pp. 410-420. 1989.
5. INABA, H. and FUKADA T. - Natural Convection in an Inclined Square Cavity in Regions of Density Inversion of Water, Journal of Fluid Mechanics Vol. 142, pp. 363-381. 1984.
6. LANKFORD, K. E. and BEJAN A. - Natural Convection in a Vertical Enclosure Filled With Water Near 4 °. C, Journal of Heat Transfer Vol. 108, pp. 755-763. 1986.

7. IVEY, G. N. and HAMBLIN P. F. - Convection Near the Temperature of Maximum Density for High Rayleigh Number, Low Aspect Ratio, Rectangular Cavities, Journal of Heat Transfer Vol. 111, pp. 100-104. 1989.

8. OOSTHUIZEN, P. H. and PAUL. J. T. - Unsteady Free Convective Flow in an Enclosure Containing Water Near its Maximum Density, Proceedings 1990 AIAA/ASME Thermophysics and Heat Transfer Conference ASME HTD - Vol. 140, pp. 83-91, 1990.

SIMULATION ROUTINES FOR PHASE CHANGE OF WATER IN GREENHOUSES

Guri Krigsvoll[1]
Department of Agricultural Engineering, Agricultural University of Norway

SUMMARY

The phase change routines were made to be implemented in a simulation program for climate in greenhouses. This paper describes the equations and routines used to describe the phase changes, condensation and evaporation, and the simultaneous heat transfer. The routines calculates the phase transition at cold surfaces and leaves, and in the greenhouse air.

1. PREFACE

Since 1988 Department of Agricultural Engineering has worked with the project "Simulation of greenhouse climate, temperature and humidity". The aim for the project was to develop a numerical mass and heat transfer model to describe the greenhouse climate. The program code used for calculation of fluid flow, mass and heat transfer, Kameleon II, is developed by SINTEF, Division of Thermodynamics . Kameleon II is a computer program system using the finite difference technique for prediction of fluid flow, mass and heat transfer[2,4]. The program also contains a model for prediction of chemical reactions due to turbulent combustion. Kameleon II is expressed in general orthogonal co-ordinates and has been applied to simulate a variety of parabolic and fully elliptic fluid flow phenomena.

This model is made for calculations of enthalpy, mass transfer and fluid flows in greenhouses, and to find how these results are affected by changes in outside temperature, solar radiation or other weather parameters. Also the influences of the greenhouse crop are taken into account.

---

[1] Researcher. Department of Agricultural Engineering. Agricultural University of Norway.

2. PROPERTIES OF GREENHOUSE AIR

Kameleon II has a gas fraction model, that calculates the properties of the total gas from the concentrations of the gases and the equations of state of an ideal gas. The gas fractions used for greenhouse air are, CO_2, H2O (water vapour) and dry air. Concentration of CO_2 is 500-1000 ppm and relative humidity 75-90%. In this study the CO_2 fraction is a part of the dry air component.

Kameleon II works with enthalpy equations, and the enthalpy is defined as: $h=T*CP+H2O*h_{fg}$ J/kg

Changes in H2O and enthalpy for a control volume are given as source term on the form: $S = S_c + S_p * \Phi_p$

Rules for discretization and linearization of source term are described in Patankar[3].

First step in the simulation routines is searching expressions for the source term. Source term for the gas fraction equation, H2O, is the amount of water vapour that condenses/evaporate and the total amount of condensate is a restriction for evaporation. Energy from the evaporation/condensation process represent a source term for the enthalpy equation.

3. PHASE CHANGES IN GREENHOUSE AIR

The high humidity in greenhouses, usually limited to 85-90%, increases the danger of local condensation even with small temperature reductions. Phase transition in the greenhouse air occur either when the vapour pressure exceeds the saturation vapour pressure at the air temperature, or when the vapour pressure is lower than the saturation pressure and there are condensate that can evaporate.

Since the phase transition occur in a control volume, and the volume of the condensed or vaporized water is negligible, volume of the gas is constant. The amount of energy in the control volume is constant, but since the enthalpy equation takes no account of the liquid water, the enthalpy will change. The enthalpy difference is the energy of condensed water vapour, and allow only p and T to change:

$$h = T*CP + H2O*h_{fg}$$

$$dh = dT*CP + dH2O*h_{fg} = dH2O*T*CP_{wat}$$

$$dT = -dH2O*\frac{h_{fg} - T*CP_{wat}}{CP} \qquad (1)$$

Using equations for an ideal gas and for greenhouse air, $p = p_a + e$. Assume that both gas fractions has the same temperature.

$$(p_a + e)*dV + Vd(p_a + e) = \frac{R}{M_t}*dT$$

$$de = e_s(T+dT) - e$$

$$e_s(T+dT) = e_s(T) + \gamma * dT = e_s(T) - \gamma * dH2O * \frac{h_{fg} - T*CP_{wat}}{CP} \quad (2)$$

$$de = e_s(T) - e - \gamma * dH2O * \frac{h_{fg} - T*CP_{wat}}{CP} \quad (3)$$

Assumptions for the routines:

$CP(I,J,K)_{t1} = CP(I,J,K)_{t2}$ $\rho(I,J,K)_{t1} = \rho(I,J,K)_{t2}$

$dV = 0$ $dp_a = 0$

There is no limitation of the velocity of the phase change process. All available water is allowed to evaporate until saturation during one time step, and when the water vapour pressure is higher than the saturation pressure, the routines allow immediate condensation.

3.1. Source term for H2O

$$H2O = \frac{e*M_w}{T*R*\rho} \quad \text{kg/kg}$$

$$dH2O = \frac{de*M_w}{T*R*\rho} - \frac{e*dT*M_w}{T^2*R*\rho} \quad \text{kg/kg}$$

Rearranging the equation by use of eq.1 and 3:

$$dH2O = \frac{e_s(T) - e}{\frac{e}{H2O} + (\gamma - \frac{e}{T}) * \frac{h_{fg} - T*CP_{wat}}{CP}} \quad \text{kg/kg} \quad (4)$$

When the saturation pressure exceeds the vaour pressure evaporation will occur limited by available water, m_{wat} (kg/kg), in the air.

$$S_{H2O} = \min(dH2O, m_{wat}) * \frac{\rho*V}{dt}$$

$$S_{H2O} = \min\left(\frac{e_s(T) - \frac{H2O*T*R*\rho}{M_w}}{\frac{e}{H2O} + (\gamma - \frac{e}{T}) * \frac{h_{fg} - T*CP_{wat}}{CP}}, m_{wat}\right) * \frac{\rho*V}{dt} \quad (5)$$

3.2. Source term for mass of enthalpy

$$dh = dH2O*CP_{wat}*T$$

$$S_h = \min\left(\frac{e_s(T) - e}{\frac{e}{H2O} + (\gamma - \frac{e}{T}) * \frac{h_{fg} - T*CP_{wat}}{CP}}, m_{wat}\right) * \frac{\rho*V}{dt} * CP_{wat}*T \quad (6)$$

$$S_h = \min\left(m_{wat}, \left(\frac{e_s(T)*CP + H2O*h_{fg}*\frac{e}{T} - h*\frac{e}{T}}{\frac{CP*e}{H2O} + (\gamma - \frac{e}{T})*(h_{fg} - T*CP_{wat})}\right)\right) * \frac{\rho*V}{dt} * CP_{wat}*T \quad (7)$$

3.3. Changes in mass of water

Changes in liquid water is the opposite of changes in water vapour.

$$m_{wat} = \max(0, m_{wat} - dH2O) \quad (8)$$

3.4. Linearization of source terms

The source term are linearized after the rules in Patankar[3]. The H2O source or sink terms depend on whether there is evaporation or condensation, and the limitation of available water. For evaporation limited by water liquid the source terms are:

$$S_{H2O,c} = m_{wat} * \frac{\rho * V}{dt} \qquad S_{H2O,p} = 0$$

$$S_{h,c} = m_{wat} * \frac{\rho * V}{dt} * CP_{wat} * T \qquad S_{h,p} = 0$$

For condensation and evaporation not limited by available water the H2O source terms are:

$$S_{H2O,c} = \frac{e_s(T)}{\frac{e}{H2O} + (\gamma - \frac{e}{T}) * \frac{T * CP_{wat} - h_{fg}}{CP}} * \frac{\rho * V}{dt}$$

$$S_{H2O,p} = -\frac{CP * e}{CP * e + H2O * (\gamma - \frac{e}{T}) * (h_{fg} - T * CP_{wat})} * \frac{\rho * V}{dt}$$

The source terms for enthalpy are linearized the same way.

4. PHASE CHANGES AT THE BOUNDARIES

At the boundaries we in addition to the earlier described phase changes have condensation or evaporation processes at the greenhouse cover. Condensation will occur when the vapour pressure in the neighbour control volume exceeds the saturation vapour pressure for the surface temperature, or the dew point temperature exceeds the surface temperature. Evaporation will occur if the saturation vapour pressure at the surface exceeds the vapour pressure of air, and available water will evaporate.

An expression used to define the velocity of the condensation process are given by Al-Attar et al[1]. Mass transfer coefficient, kg/m2*sec:

$$K_w = \exp(-6.0447 - 0.22176 * T_{min}) \tag{9}$$

where T_{min} = minimum outside temperature.

Mass of water condensed per unit area of condensing surface area is then:

$$Cond(t) = K_w * \frac{h_a(t) - h_{gl}(t)}{h_{fg}} \tag{10}$$

where $h_a(t)$= enthalpy of inside air

$h_{gl}(t)$= enthalpy at inside glass conditions

The validation of the expression is limited to night time condition. Maximum evaporation is limited to available condensate, $m_{wat,gl}$.

From eq.10 source term for H2O is:

$$S_{H2O} = \min(-Cond * A_{gl}, \frac{m_{wat,gl}}{dt}) \tag{11}$$

When the phase transition occurs at the boundaries, the temperature of the condensate is no longer the same as the air. If Cond is positive a condensation process will occur. Water vapour that condense has the temperature as the air, and is chilled to the temperature of the boundaries. The latent heat and the heat from the chilling process are transferred to the boundaries. Source term for enthalpy in air is:

$$S_h = \min(-Cond*A_{gl}, \frac{m_{wat,gl}}{dt})*(h_{fg} + T'*CP') \qquad (12)$$

where $T' = T_a$ for condensation and T_{gl} for evaporation, $CP' = CP_{gl}$ for evaporation and $CP' = CP_a$ for condensation.

4.1. Mass of water on boundaries

New amount of condense at boundaries:

$$m_{wat,gl} = m_{wat,gl} - S_{H2O}*dt \qquad (13)$$

In the routines in Kameleon II the S_{H2O} terms due to phase transition for all element next to a boundary are found before the source term routines. This procedure tries to minimize the fault due to the direct calculation of water and temperature, instead of source terms.

4.2. Temperatures on the boundaries

The energy transferred to or from the boundary due to phase changes is used as a term in the heat transfer coefficient. Effect on the surface = $-S_h$. The source term is found directly from the mass of condensation. Heat transfer caused by phase transition is a term in the heat transfer coefficient for the inner surface of the greenhouse.

$$\alpha = \alpha_{sw} + \alpha_{lw} + \alpha_c + \alpha_{pc} \qquad (14)$$

$$\alpha_{pc} = -\frac{S_h}{A_{gl}*(T_{gl} - T_a)} \qquad (15)$$

New temperature on the inner surface is found by backward difference method, using outside and inside air temperature as boundaries with constant temperature.

4.3. Linearization of source terms

Linearization of the source terms depend of whether there water vapour evaporate or condense. For condensation the source terms are:

$$S_{H2O,p} = K_m * \frac{h_{gl}}{h_{fg}} * A_{gl} - K_m * \frac{T_a*CP_a}{h_{fg}} * A_{gl}$$

$$S_{H2O,c} = -K_m*A_{gl}$$

$$S_{h,p} = K_m * h_{gl} (1 + \frac{T_a*CP_a}{h_{fg}})*A_{gl}$$

$$S_{h,c} = -K_m * (1 + \frac{T_a*CP_a}{h_{fg}})*A_{gl}$$

Evaporation, limited by the available water gives:

$$S_{H2O,p} = \frac{m_{wat,gl}}{dt} \qquad S_{H2O,c} = 0$$

$S_{h,p} = \frac{m_{wat\,gl}}{dt} * (h_{fg} + T_{gl} * CP)$ $\qquad S_{h,c} = 0$

Evaporation without limitation gives the same source term for H2O as condensation, but the temperature in the enthalpy terms are changed to the surface temperature.

$$S_{h,p} = K_m * h_{gl} * (1 + \frac{T_{gl} * CP}{h_{fg}}) * A_{gl}$$

$$S_{h,c} = -K_m * (1 + \frac{T_{gl} * CP}{h_{fg}}) * A_{gl}$$

5. CHANGES OF AIR PROPERTIES CAUSED BY PHASE CHANGES

The air properties CP(I,J,K), RHO(I,J,K), H2O(I,J,K) for the time t+dt are calculated in the end of the procedure, when all source terms are found. New temperature is calculated from the new enthalpy, water vapour fraction and specific heat.

6. SIMULATION RESULTS

The greenhouse model was tried out with different humidities and inside and outside temperatures. The test cases were boxes with 3 walls with constant temperature as initial air temperature, 290 K and H2O = 0.009. No phase changes were simulated at these surfaces. The 4th wall was a cold surface. In case 1 the temperature was changing due to phase changes and heat transfer by convection and radiation. In case 2 and 3 the temperature was constant. Case 1 had incoming air with T=320 K and H2O=0.015. For case 2 and 3 T=290 K and H2O=0.009. Time steps during calculation were about 0.02 seconds.

6.1. Case 1

The model has no restriction on humidity for incoming air. When warmer saturated air entered the box, temperature increases in the air, and an upward air flow appeared. This is shown in figure 2 and 3. Air reaching the cold surface was already dehumidified, and the condensation process mainly occurred at the surface. The heat loss lead to temperature decreases in neighbour volumes, where also condensation occurred.

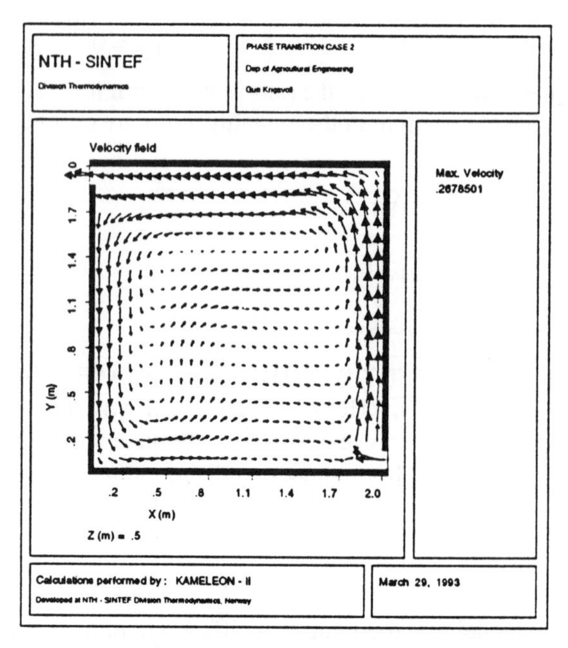

Figure 1. Air flows after 3200 iterations, case 1.

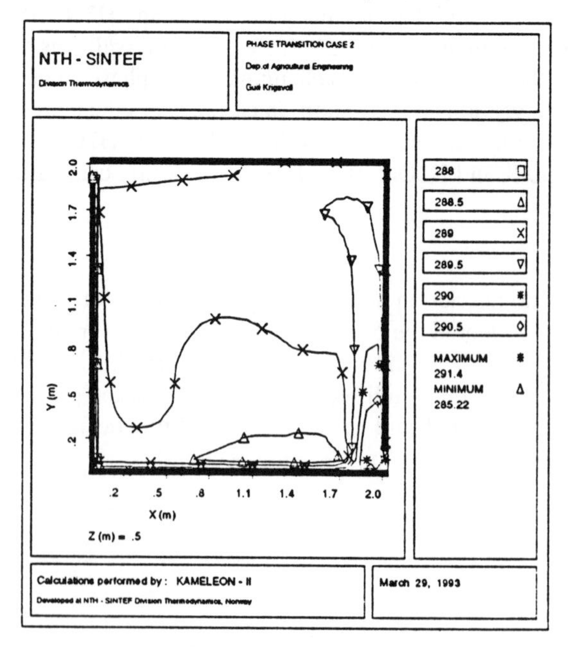

Figure 2. Temperature after 3200 iterations, case 1.

6.2. Case 2

As case 1 the second case get a downwards air flow near the cold surface, and the water vapour in air will condense at the end of the wall. Temperature field is shown in figure 3.

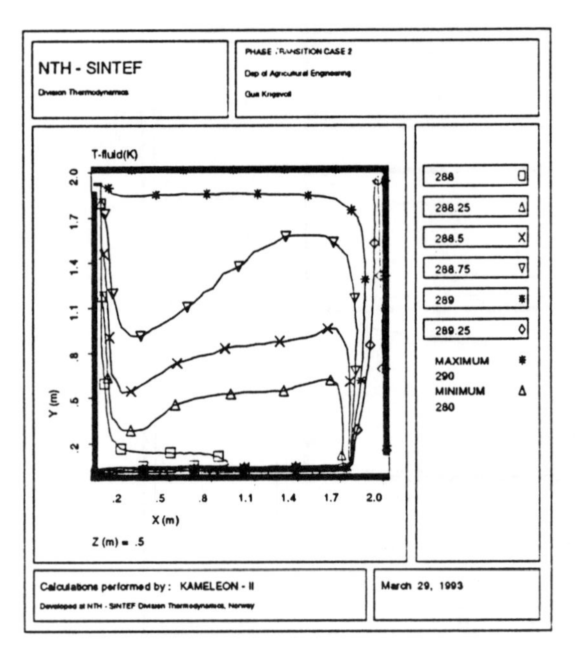

Figure 3. Temperature case 2

6.3. Case 3

For case 3 the inlet was moved to the top of the wall, while the exhaust were on the bottom of the cold surface. In this case all phase transition were on the cold surface.

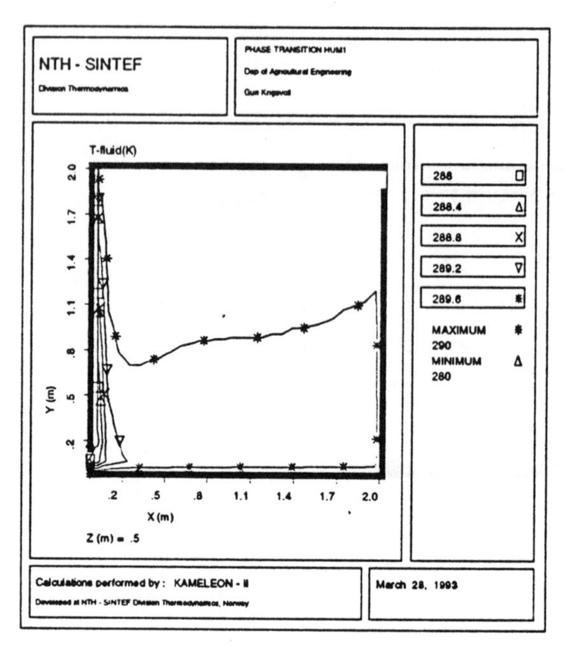

Figure 4. Temperatures after 6400 iterations. Case 3.

7. CONCLUSION

The effect of radiation, both solar and long wave, are not directly influencing the phase change processes at the surfaces. Indirectly radiation influence the temperature, and then the enthalpy in the next time step. Since the time steps are short, this method is considered to be accurate enough for the purpose of simulate the temperature and humidity in greenhouses.

The water liquid does not follow the flows of the greenhouse air. Routines to simulate the flow of water liquid are not put into the model.

LIST OF SYMBOLS

A	area	m2
CP	specific heat at constant pressure	J/kg*K
CV	specific heat at constant volume	
dt	time step	sec
e	vapour pressure	Pa
e_s	saturation vapour pressure	Pa
h	enthalpy	J/kg
h_{fg}	latent heat of vaporization of water	J/kg
M	mole weight	
m	mass	kg
p	pressure	
R	gas constant	
S	source term	
T	temperature	K
t	time	sec
V	volume	m3
ε	ratio of molecular weights of water vapour and air	
Φ	dependent variable	
γ	psychrometer constant = $\frac{CP*p}{hfg*e}$	
ρ	density of air	kg/m3

subscript

a	greenhouse air
c	constant part of source term
con	convection
gl	condensation surface
lw	long wave radiation
sw	short wave radiation
p	coefficient of Φ in source term
pc	phase change
t	total
w	water vapour
wat	liquid water

REFERENCES

1. AL ATTAR, F. et al. - An Experimental evaluation of mass transfer coefficient inside a model greenhouse. Energy and the environment. Into the 90s. Proceedings of the 1st world renewable energy congress. 1990.

2. LAKSÅ, B. and B.E.VEMBE - KAMELEON II a general purpose program for simulation of fluid flow, heat and mass transfer. SINTEF Report STF15F91048. 1991.

3. PATANKAR, S.V. - Numerical Heat Transfer and Fluid Flow. Hemisphere Publishing Corporation, 1980.

4. VEMBE, B.E. - How to run KAMELEON II under LIZARD. SINTEF Division of Thermodynamics.

NUMERICAL SIMULATION OF SOLID-LIQUID PHASE CHANGE STORAGE UNITS

M. Costa, A. Oliva, C.D. Pérez-Segarra, N.R. Reyes

Laboratori de Termotècnia i Energètica
Dept. de Màquines i Motors Tèrmics. Universitat Politècnica de Catalunya
Colom 9, 08222 Terrassa, Barcelona (Spain)

SUMMARY

A two-dimensional numerical simulation for the prediction of the thermal behaviour of solid-liquid phase change units has been developed. The units can be composed of a Phase Change Material (PCM), embedded heat sources, extended surfaces, and a thermal working fluid for discharging and/or for charging heat. All the regions have been solved altogether with the use of a fixed grid. The resulting governing equations have been discretized by the Power-Law technique and the nonlinearities have been accounted for by the segregated method SIMPLEX. The model developed can be easily used to treat different geometries. Difficulties have been encountered when trying to analyse embedded cylinder heat sources due to the three-dimensional behaviour of the natural convection developed.

1. INTRODUCTION

Energy storage is becoming increasingly important in several fields: in the field of alternative energies, in which its discontinuity makes the energy storage inevitable; in the field of traditional systems of heating and air conditioning to take advantage, for instance, of the reduced electrical fee during off-peak hours; and in industrial processes in general whenever there is a delay between the production and utilization periods. Among the different ways of energy storage stand out the latent energy in the form of solid-liquid phase change. It provides a high thermal storage capacity at relatively low temperatures reducing considerably the thermal losses [1].

Despite the attractive features above mentioned, both the lack of confidence in Phase Change Materials (PCM) and difficulties in making an accurate design, have limited the spread of the solid-liquid phase change storage units. Recently, new phase change materials of low cost that claim to stand very large cycles of melting-solidification without suffering degradation have appeared on the market, giving a boost to the use of latent heat storage units. In a parallel way the development and enhancement of numerical techniques in heat transfer and the availability of powerful digital computers makes it possible to carry out a numerical prediction of the phenomena involved and eventually to obtain efficient and cost-effective thermal energy storage units.

Phase change phenomena are classified mathematically into the group of problems with moving boundaries and, even for the simpler cases with conduction in both phases, analytical solutions in closed form only exist for bodies of infinite length [2]. Therefore, numerical methods have been recoursed in order to predict real situations. There are commonly two approaches to carry out the numerical discretization: transformed grid and fixed grid [3]. With the transformed grid method a curvilinear body-fitted coordinate system is used to track dynamically the moving boundary, whereas with the fixed grid method the boundary is accounted for formulating the energy equation with the enthalpy as the dependent variable. The main advantage of the transformed grid is that no loss of accuracy in the discretization of the solid-liquid interface is produced; nevertheless, this accuracy comes at expense of more complex governing equations. This paper has recourse to the fixed grid method due to its flexibility in treating complex boundaries and different interfaces; moreover, the predictions are of the same order of accuracy as those obtained with the transformed grid [3].

The latent heat storage units usually are made of a bundle of PCM encapsulated tubes or of a rectangular encapsulation system. Most of the experimental and numerical available results correspond to the phenomena of solid-liquid heat transfer inside rectangular or cylinder domains subjected to isothermal wall conditions [4][5]. More recently in [6] correlations of the melt time and Nusselt number has been numerically obtained and the downward motion of the solid core is considered by means of an analytical equation.

The purpose of this paper is to report on a numerical study of thermal energy storage units solving all the domains (PCM, container and working fluid) altogether without needing to specify interface equations neither within the PCM nor between the solid elements and the PCM or the working fluid. The comparison of the numerical results here presented with the experimental data available in the literature reveal the need, in certain situations, of a three-dimensional analysis for a successful prediction.

2. GOVERNING EQUATIONS AND NUMERICAL PROCEDURE

The governing equations (mass, momentum and energy) have been formulated using appropriate source terms to make them valid in every region. The energy equation has been formulated using the enthalpy method [7][8], so that both the boundary conditions between the solid regions and the boundary conditions between the solid and liquid regions are not explicitly required. Also, this formulation allows the use of a cartesian fixed grid to track the moving solid-liquid interface boundary. The governing equations, assuming an incompressible newtonian fluid, laminar flow, negligible viscous dissipation, isothermal phase change, and the Boussinesq approximation, are the following:

$$\mathrm{div}\,(\rho \mathbf{u}) = 0 \tag{1}$$

$$\frac{\partial(\rho u)}{\partial t} + \mathrm{div}\,(\rho \mathbf{u} u) = \mathrm{div}\,(\mu\ \mathrm{grad}\ u) - \frac{\partial P}{\partial x} + Au \tag{2}$$

$$\frac{\partial(\rho v)}{\partial t} + \mathrm{div}\,(\rho \mathbf{u} v) = \mathrm{div}\,(\mu\ \mathrm{grad}\ v) - \frac{\partial P}{\partial y} + \rho g \beta (T-T_o) + Av \tag{3}$$

$$\frac{\partial(\rho h)}{\partial t} + \mathrm{div}\,(\rho \mathbf{u} h) = \mathrm{div}\left(\frac{k}{c}\ \mathrm{grad}\ h\right) - \rho L \frac{\partial f_l}{\partial t} + Bq_v \tag{4}$$

In the momentum equations in the "x" and "y" directions, the source terms "A.u" and "A.v" have been added. "A" being zero in the liquid region, while it takes a large value in solid regions forcing the velocities, u and v, to be practically zero. The source terms in the energy equation account for the latent heat and the heat generated by Joule (resistive) heating; the former has been treated according to the scheme proposed by Voller [7], for the latter a correction coefficient "B" has been used, being equal to one in the region where an internal heat is generated and zero elsewhere.

The discretization of the conservation equations has been made applying the first order scheme, so-called Power Law Scheme [9]. A segregated method, SIMPLEX [10], has been applied for the solution of the coupled non-linear Navier-Stokes equations. The MSIP (Modified Strongly Implicit Method) [11] method has been used to solve the linearized discretization equations. The software developed permits to consider any number and arrangement of the solid heat elements and, also, any geometry in general. The spatial grid allows concentration in the regions of large gradients.

3. RESULTS AND DISCUSSION

3.1 Experimental Validation

In order to check the validity of the numerical simulation two situations with available experimental results have been chosen. One corresponds to the melting within a rectangular enclosure with isothermal vertical sidewalls and adiabatic horizontal walls, hereafter called the benchmark case because it has been used as a reference in many investigations, and the other corresponds to the melting of n-octadecane around a horizontal cylinder with constant heat flux.

(a) *The Benchmark Case*

In fig. 1 the melting of gallium (T_m=29.78 C) inside a container of dimensions W=9 cm. and H=4.5 cm. is studied. The initial temperature in all the domain has been taken as 27 C. At time t=0, the temperature of the left-hand sidewall is raised at T_w=38 C. The values of the governing dimensionless numbers are the following:

A	[H/W]	= 0.5
Ra	[$g\beta(T_w-T_m)H^3/\nu\alpha$]	= $2.2*10^5$
Pr	[ν/α]	= 0.021
Ste	[$c(T_w-T_m)/L$]	= 0.042

Two spatial grids have been used (40*20 and 80*40), with a constant time step of 1 s in both cases. The melt front has been compared with the experimental results provided by Gau and Viskanta [12]. There is a reasonable agreement between the experimental and numerical front-phase position.

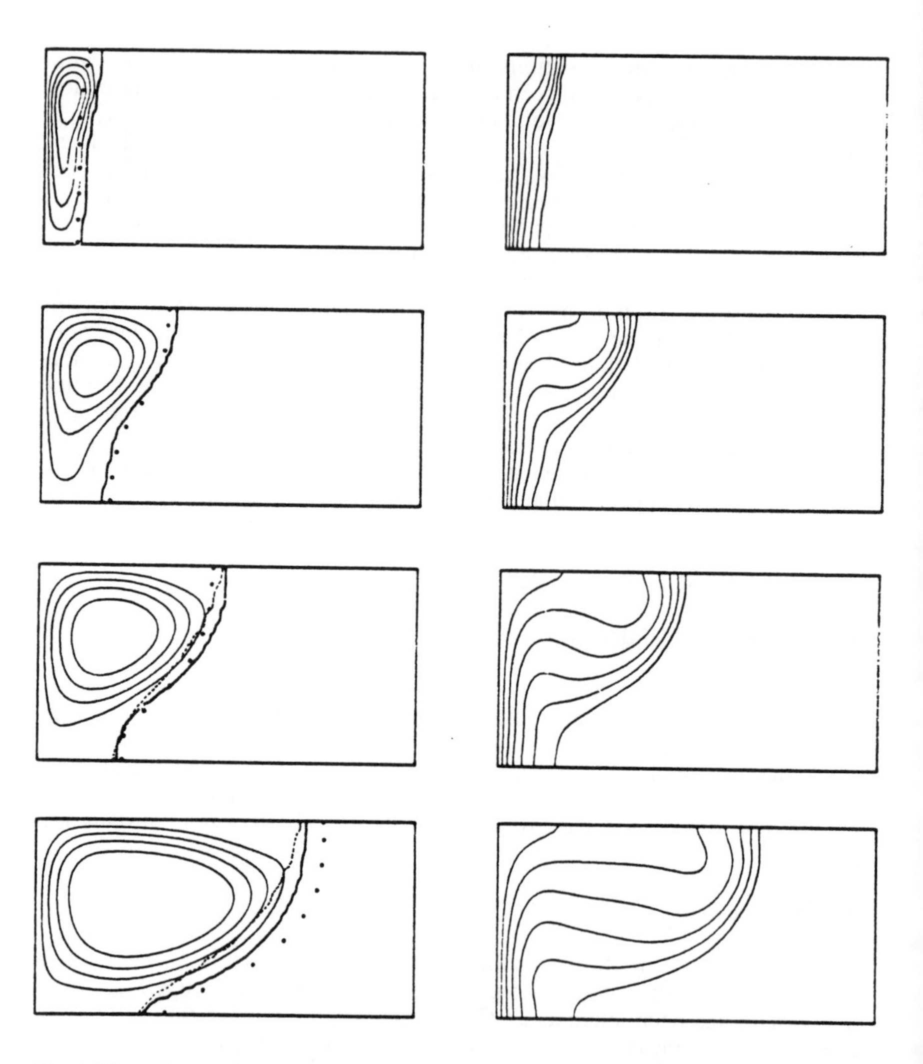

Fig. 1 Phase-front and streamlines on the left and isotherms on the right at four times (2, 6, 10 and 17 minutes) obtained with a grid of 80*40. The dashed line correspond to a grid of 40*20 and the dots denote the experimental results of Gau and Viskanta [12].

(b) *Melting Around an Embedded Horizontal Cylinder*

The phenomenon of melting of a PCM around a horizontal cylinder has been chosen both because it can be a situation quite common in thermal units charged by means of electrical resistances or by a working heating fluid flowing inside a tube, and because there are several experimental results reported on the literature [13][14]. This situation has been solved numerically with the cartesian coordinate system and the method mentioned before, approximating the cylinder with the rectangular grid.

The situation analysed is the melting of n-octadecane (T_m=28.2 C) around a carborundum cylinder of radius $R_i=9.5.10^{-3}$ m. and with constant volumetric heat q_v= 331480 W/m^3, starting from an initial temperature of 26.3 C. It corresponds to a Stefan Number ($Ste=cq_vR^2{}_i/2kL$) of 0.881, and a Prandtl Number of 55.9. In fig. 2 the predicted solid-liquid interface at different times has been compared with the experimental interface obtained by Bathelt and Viskanta[14]. As it is notorious, there is a severe deviation of the numerical and experimental interface shape particularly strong at the top. Different grids, Ste numbers and geometries have been tested and the mentioned discrepancies have always emerged [15].

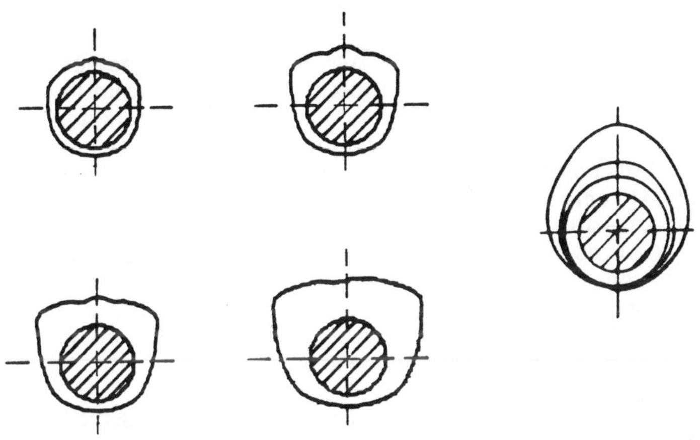

Fig. 2 Comparison at different times of the solid-liquid interface. On the two left columns our present numerical results at times t=600, 900, 1184 and 1800 s, and on the right the experimental results from Bathelt and Viskanta[14] at times t= 1184, 2369 and 4739 s.

In order to rule out the approximation of the cylinder surface by means of a rectangular grid as the cause of the disagreement above mentioned, the situation of natural convection around an isothermal horizontal cylinder is presented. A cylinder of radius R=2 cm. has been located at the middle of two vertical plates 80 cm high separated 30 cm, and at a height of 15 cm. In fig. 3 the isotherms and streamlines corresponding to a Rayleigh Number of $Ra=g\beta(T_w-T_\infty)R_i^3/\nu\alpha=10^5$ and a Prandtl Number of Pr=0.7 are depicted. The flow pattern compares favorably with the numerical results obtained with cylinder coordinates presented by Goldstein and Kuehn[16]. Moreover, the mean Nusselt number has been of 8.9 which is in the range of the mean average of the available experimental results summarized by Morgan [17].

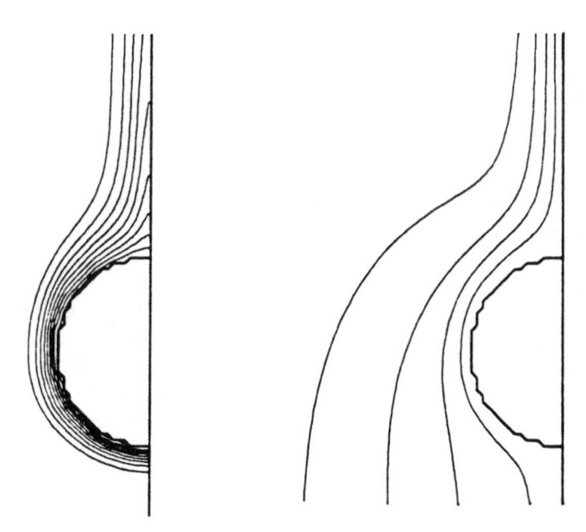

Fig. 3 Numerical isotherms (left) and streamlines (right) corresponding to the case of natural convection without phase change around an isothermal cylinder. ($Ra=10^5$, $Pr=0.7$)

The most probable reason of disagreement between the numerical and experimental results shown in fig. 2 seems to be, apart from the temporal deviation of the melt fraction probably due to different working conditions and/or physical properties of the PCM, the numerical assumption of a two-dimensional phenomenon. Thereby, the experimental results of Bathelt and Viskanta [14] confirm that after an initial period of time, in which the heat transfer is basically by conduction (the mean Nusselt number decreases monotonically) and the solid-liquid interface is concentric around the cylinder, three-dimensional vortices of an unsteady kind appear. After this transition period the interface eventually acquires a two-dimensional "pear-shape" like growing basically at the top (the mean Nusselt number is constant and very similar to the one of natural convection without phase change).

This phenomenon of instability has also been clearly observed in the case of a fluid confined in a horizontal annulus. The natural convection inside of two concentric isothermal cylinders, being higher the temperature of the inner one, has received considerable attention, and the different flow patterns that develop have been classified [18]. The non-dimensional governing parameters in this case are the radii ratio, $R=R_o/R_i$, and the Rayleigh number based on the gap width, $Ra=g\beta(T_w-T_m)(R_o-R_i)^3/\nu\alpha$. Under the assumption that the transient terms do not affect the flow pattern, the solution obtained with an horizontal annulus can be applied to the liquid phase of the problem of melting around a cylinder when the radii ratio of the annulus coincides with the ratio between the solid-liquid interface and the radius of the cylinder.

Based on the last assumption and for the experiment reported by Bathelt and Viskanta [14], at time t=1184 s (time of the first occurrence of instabilities) the solid-liquid interface is concentric with the following non-dimensional parameters: $R = 1.415$ and $Ra = 14356$. For this situation, and acording to the flow pattern map by Powe et al [18], fig. 4, the flow pattern is clearly within the three-dimensional spiral region. This three-dimensional behaviour is more intensive at the top of the annulus as it has been visualized by Rao et al. [19], that will explain the slender form at the top that exhibits the experimental solid-liquid interface. Even though the three-dimensional effects seems to disappear as the melting region increases, the shape predicted by the two-dimensional model drives the melt region to another pattern.

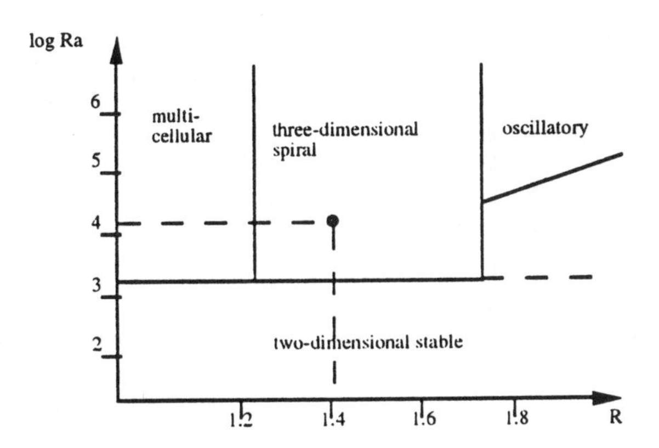

Fig. 4 Flow pattern chart characteristic of horizontal annuls by Powe et al.[18] . The location at the melting time t= 1184 s has been drawn: ●

3.2 Modelization of a Phase Change Solid-Liquid Unit

In this section two illustrative situations in which different domains are present are studied by means of the modelization developed. They are two possible models of latent heat sorage units. The situations illustrated corresponds to the period of charging, that is the period of melting of the PCM (n-octadecane). Initially all the domain is at a uniform temperature (25 C), the PCM is in the solid phase and the working fluid (air at 1 atm.) is at rest. At time zero a working fluid at a temperature above the PCM melting point (T_m=28.2 C) enters in the duct provoking the melting of the PCM. The boundary inflow and outflow condition for the velocity have been that of $\partial v/\partial y=0$ along with a constant flow rate of 0.0157 m^3/s. The inflow condition for the temperature has been of constant value, 40 C, and the outflow condition has been that of $\partial T/\partial y=0$

Two different units have been considered, shown in fig. 5. One, unit (a), corresponds to a PCM material confined in a rectangular adiabatic enclosure with a channel at the middle made of a conducting material (aluminium) which detaches the PCM container from the working fluid (air). One of the possible ways to enhance heat transfer in latent energy storage devices is by means of extended surfaces as it has been proposed by different investigators [20][21]. Therefore, the second case analysed, unit (b), corresponds to the same traits as the first but with the inclusion of an extended surface consisting of an horizontal fin at the half of the height.

Due to symmetry only half of the units has been modelized. The spatial grid used has been of 48(40+2+6, 40 grid points in the PCM, 2 in the plate and 6 in the duct)∗60 in the unit depicted in fig. 5 (a) and of 48(40+2+6)∗60(29+2+29) in the other unit; and the time step used has been of 1 s in both cases.

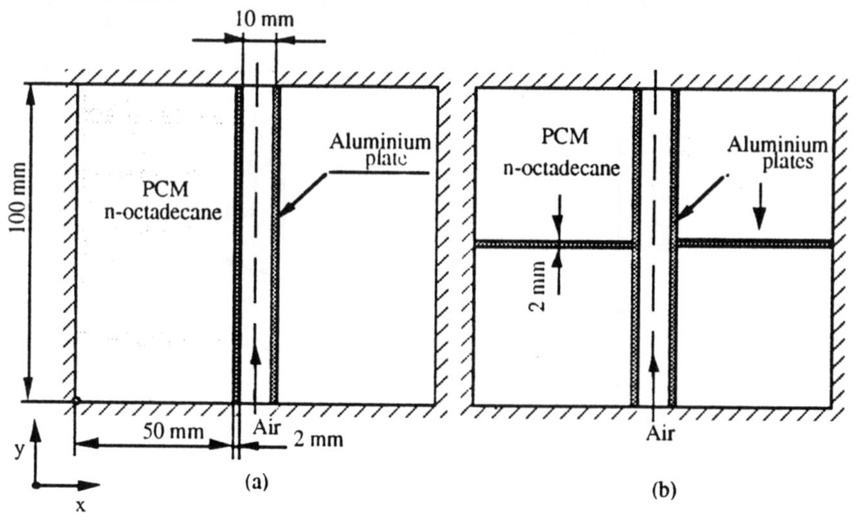

Fig. 5 Scheme of the thermal units under investigation.

The thermocapillary (Marangoni) convection has not been considered; neither the phenomenon of volume expansion of the liquid as a consequence of the variation of density with respect to the solid phase. The consequence of this last phenomenon is an acceleration of the process of melting at the top of the PCM at the early stages [20]. The boundary conditions at this zone have been taken as if the PCM were in close contact with the wall and, thus the boundary conditions of non-slip have been considered.

In figs. 6 and 7, the time evolution of the front phase and the velocities is represented. The importance of the natural convection and the effect of the fin on it is clearly illustrated. In tables 1 and 2 the instantaneous heats at the different regions of the latent storage units are indicated. The capacity of unit (b) to extract more heat from the working fluid is quantified.

Finally, in fig. 8 a comparison of the melt fraction of the two units analysed has been made. At the beginning, the melt fraction is lower in the unit depicted in fig. 5 (b) than that of fig. 5 (a) due to the fin effect of distribution of the energy through the unit; after this period the melt fraction in both cases increases linearly but with different slopes, being the slope of the unit of fig. 5 (b) more important.

The computational cost for analyzing 6 hours of the process has been of 131 CPU-hours for the unit depicted in fig. 5. (a) and 184 CPU-hours for the unit depicted in fig. 5 (b) in a work-station of 75 MIPS.

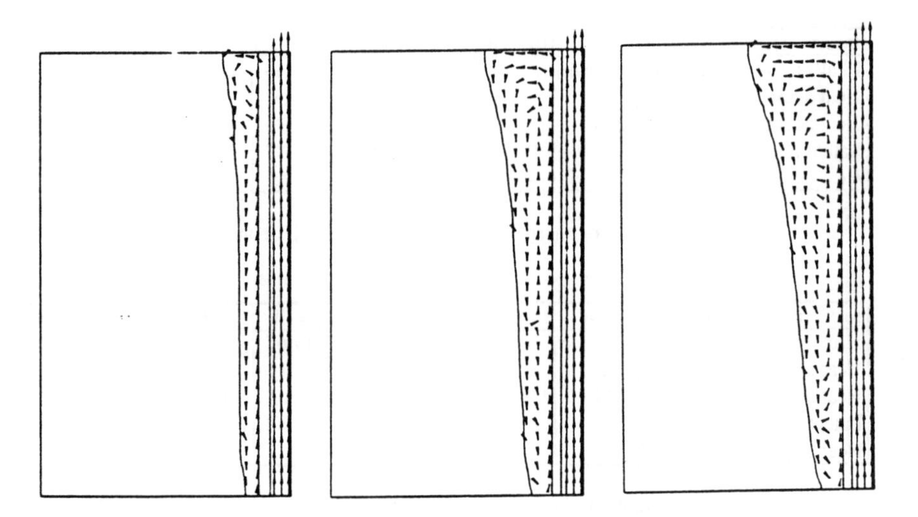

Fig. 6 Thermal evolution of the interface of the unit depicted in fig. 5 (a), at times t=2, 3 and 4 h. Note: the arrow scale is different for forced and natural convection.

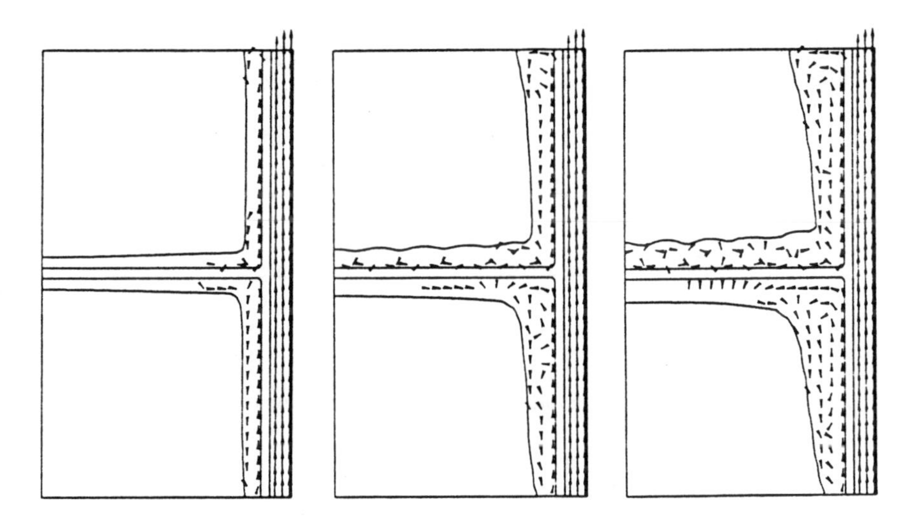

Fig. 7 Thermal evolution of the interface of the unit depicted in fig. 5 (b), at times t=2, 3 and 4 h. Note: the arrow scale is different for forced and natural convection

Instantaneous heats (W)	t= 0.5h	t= 1h.	t= 2h.	t= 3h.	t=4h.	t=5h.	t=6h
Heat transfer from air to container	21.31	18.10	17.30	17.69	17.87	17.97	18.03
Heat acum. in the alum. plate	2.25	0.09	0.00	0.05-	0.02	0.04	0.01-
Heat acum. in the liquid PCM	1.75	0.08	0.22	0.28	0.32	0.50	0.33
Latent heat	12.91	15.59	15.93	16.97	17.36	17.38	17.71
Heat acum. in the solid PCM	4.39	2.43	1.14	0.48	0.17	0.04	0.01

Table 1 Distribution of energy corresponding to the unit depicted in fig.5 (a)

Instantaneous heats (W)	t= 0.5h	t= 1h.	t= 2h.	t= 3h.	t=4h.	t=5h.	t=6h
Heat transfer from air to container	23.57	22.34	20.24	19.67	19.68	19.60	19.49
Heat acum. in the alum. plate	0.21	0.311	0.01	0.01	0.01	0.01	0.01-
Heat acum. in the liquid PCM	0.03	0.41	0.45	0.19	0.38	0.56	0.17
Latent heat	19.06	19.73	19.41	19.40	19.21	19.02	19.33
Heat acum. in the solid PCM	4.25	1.87	0.38	0.07	0.07	0.00	0.00

Table 2 Distribution of energy corresponding to the unit depicted in fig.5 (b)

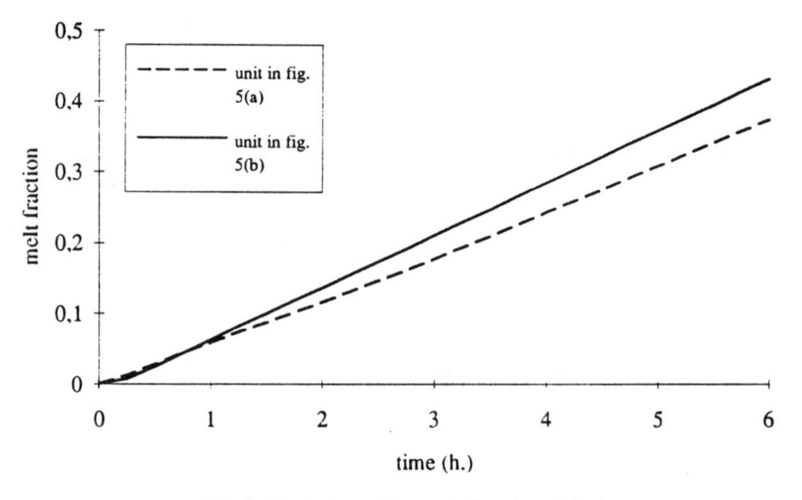

Fig. 8 Evolution of the melt fraction with time

4. CONCLUSIONS

Based on this paper, the following conclusions can be reached:

The source term method and the fixed cartesian grid have proved to be very useful to account for the different governing equations and treat irregular geometries typical of latent-heat storage units. Heat conduction in the solid regions, natural and forced heat convection in the fluid regions can be both considered and solved altogether without needing to explicitly specify interface heat equations.

The two-dimensional assumption is not valid in the period of transition from a conductive regime to a fully developed one in melting around an horizontal cylinder. Three-dimensional vortices are generated and need to be taken into account because they strongly affect the subsequent advance of the melting shape, and also of the flow pattern.

ACKNOWLEDGEMENTS

This study has been supported by the Comisión Interministerial de Ciencia y Tecnología, Spain (ref. no. PTR89-0188) and the Dirección General de Investigación Científica y Técnica, Spain (ref. no. PB90-0606).

NOMENCLATURE

A	source term coeff., aspect ratio
B	source term coefficient
c	specific heat
f_l	liquid fraction
g	gravity acceleration
h	sensible specific enthalpy
H	height of cavity
k	thermal conductivity
L	latent heat of fusion
P	effective pressure
Pr	Prandtl number
q_v	volumetric heat source
R_i	radius of cylinder
R_o	external radius of annulus
Ra	Rayleigh number
Ste	Stefan number
t	time
T	temperature
T_m	melting temperature
T_o	reference temperature
T_w	temperature at hot wall
T_∞	temperature of ambient fluid
u	velocity vector
u,v	cartesian velocity comp.
x,y	cartesian spatial coord
W	width of cavity

Greek symbols

α	thermal diffusivity
β	thermal exp. volum. coef
μ	dynamic viscosity
ν	kinematic viscosity
ρ	density

REFERENCES

1. LANE, G.A. and VISKANTA, R. - Solar Heat Storage: Latent Heat Materials, Volume I: Background and Scientific Principles. CRC Press, Inc. Florida, 1983.
2. CRANK, J. - Free and Moving Boundary Problems, Clarendon Press, Oxford, 1984.
3. LACROIX, M. and VOLLER, V.R. - Finite Difference Solutions of Solidification Phase Change Problems: Transformed Versus Fixed Grids. Numerical Heat Transfer, Part B, Vol. 17, pp. 25-41, 1990.
4. Ho, C.J. - Solid-Liquid Phase Change Heat Transfer in Enclosures, Purdue University, thesis, 1982.5
5. SAITOH, T. and HIROSE, K. - High Rayleigh Number Solutions to Problems of Latent Heat-Thermal Energy Storage in a Horizontal Cylinder Capsule. J. Heat Transfer, Vol. 104, pp. 545-553, 1982.

6. PRASAD. A. and SENGUPTA. S. - Nusselt Number and Melt Time Correlations for Melting Inside a Horizontal Cylinder Subjected to an Isothermal Wall Temperature Conditon. J. Solar Energy Engineering, Vol. 110, pp. 340-345, 1988.
7. VOLLER, V.R. - Fast Implicit Finite-Difference Method for the Analysis of Phase Change Problems. Numerical Heat Transfer, Part B, Vol. 17, pp. 155-169, 1990.
8. COSTA, M., OLIVA, A., PÉREZ SEGARRA C.D. and ALBA, R. - Numerical Simulation of Solid-Liquid Phase Change Phenomena. Computer Methods in Applied Mechanics and Engineering, Vol. 91, pp. 1123-1134, 1991.
9. PATANKAR, S.V. - Numerical Heat Transfer and Fluid Flow. Hemisphere Publishing Corporation, Washington, D.C., 1980.
10. VAN DOORMAL, J.P. and RAITHBY, G.D. - An Evaluation of the Segregated Approach for Predicting Incompressible Fluid Flows, National Heat Transfer Conference, 85-HT-9 ,1985.
11. SCHNEIDER, G.E. and ZEDAN, M. - A Modified Strongly Implicit Procedure for the Numerical Solution of Field Problems. Numerical Heat Transfer, Vol. 4, pp. 1-19, 1981.
12. GAU, C. and VISKANTA, R. - Melting and Solidification of a Pure Metal on a Vertical Wall. J. Heat Transfer, Vol. 108, pp. 174-181, 1986.
13. SPARROW, E.M., SCHMIDT, R.R. and RAMSEY, J.W. -Experiments on the Role of Natural Convection in the Melting of Solids, J. Heat Transfer, Vol. 100, pp. 11-16, 1978.
14. BATHELT, A.G. and VISKANTA, R. - Heat Transfer at the Solid-Liquid Interface during Melting from a Horizontal Cylinder. Int. J. of Heat and Mass Transfer, Vol. 23, pp. 1493-1503, 1980.
15. COSTA, M. - Simulació Numèrica de Fenòmens de Conducció, Convecció i Canvi de Fase Sòlid-Líquid. Contrastació Experimental. Universitat Politècnica de Catalunya, Thesis, 1993.
16. GOLDSTEIN, R.J. and KUEHN, T.H. - Numerical Solution to the Navier-Stokes Equations for Laminar Natural Convection about a Horizontal Isothermal Circular Cylinder, Int. J. of Heat and Mass Transfer, Vol. 23, pp. 971-979, 1980.
17. MORGAN, V.T. - Thermal Behaviour of Electrical Conductors, Research Studies Press Ltd., John Wiley & Sons Inc., New York, (1991).
18. POWE, R.E., CARLEY, C.T. and CARRUTH, S.L. -Free convective flow patterns in cylindrical annuli, J. Heat Transfer, Vol.91, 310-314, 1969.
19. RAO, Y., MIKI, Y., FUKUDA, K., TAKATA, Y. and HASEGAWA, S. - Flow Patterns of Natural convection in Horizontal Cylindrical Annuli, Int. J. of Heat and Mass Transfer, Vol. 28, pp. 705-714,1985.
20. BATHELT, A.G. - Experimental Study of Heat Transfer During Solid-Liquid Phase Change around a Horizontal heat Source/Sink, Purdue University, thesis, 1979.
21. HENZE, R.H. and HUMPHREY, J.A.C. - Enhanced Heat Conduction in Phase-Change Thermal Energy Storage devices, Int. J. Heat Mass Transfer, Vol. 24, pp. 459-474, 1980.

FREEZING / THAWING AROUND A FROZEN BODY IMMERSED IN A FINITE BATH OF LIQUID OF THE SAME MATERIAL.

Graham E Bell and Ian D Wedgwood.

Department of Mathematical & Computational Sciences,
University of St Andrews, St Andrews,
Fife, SCOTLAND.
KY16 9SS

ABSTRACT

The transient solidification layer generated from immersing a cold body into a finite liquid of the same material is investigated using a boundary immobilisation technique. The problem presents considerable computational difficulties due to the substantially different time scales involved in the freezing and thawing processes and the nature of the solutions in the solid and liquid regions. Representation of the temperature profiles becomes increasingly difficult since regions may become very small as freezing or thawing occurs and changes in the method of solution are necessary as the freeze front approaches material boundaries. Starting profiles are generated from a small time approximation and the solution is advanced using explicit finite difference formulae. The validity of the immobilisation procedure is assessed in terms of large time solutions and by comparison with a previously developed, and less efficient, fixed grid technique, which is adapted to fit the problem. Different parameter regimes induce possible scenarios of rapid initial thawing or freezing and the freeze front may then grow or decay to a limiting value as a steady state develops. The physical and mathematical consequences of altering certain parameters is investigated and the results for these are discussed in detail.

INTRODUCTION

A frozen body at temperature T_c is immersed in a finite bath of warm liquid of the same material at temperature T_0 where $T_c < T_f < T_0$, T_f representing the freezing temperature of the liquid, see figure 1.

Unlike the single characteristic motion described in Bell and Wedgwood [1] the resulting heat transfer process may progress in many different ways dictated by the parameters involved in the system. The initial velocity of the freeze front may have negative or positive values, so giving possible initial thawing or freezing. Since the warm liquid is finite, large time steady state solutions are possible which are dependent on the thermal parameters of the system, but independent of the initial motion of the freeze front. If the liquid region is relatively large the process may be terminated when the cold body thaws completely leaving purely liquid (figure 2), or conversely if the liquid region is

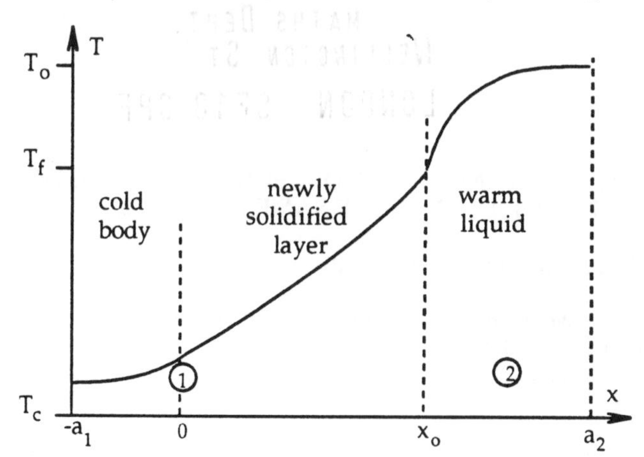

Figure 1: Typical freezing temperature profile.

small enough it may freeze completely leaving purely solid.

This problem is tackled by Tadjbakhsh and Liniger [3] by deriving the leading term of an asymptotic expansion for the displacement of the freeze front from its initial position in terms of the Stefan number. The solution obtained is valid for small Stefan number but less realistic in other situations.

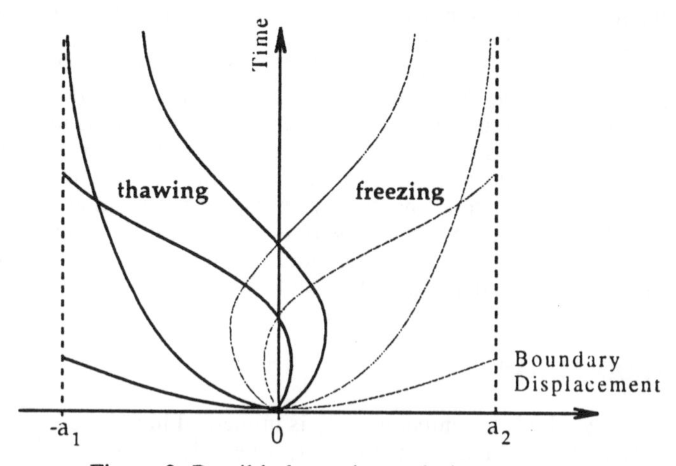

Figure 2: Possible large time solutions.

In computational terms, this system is a slight simplification of that described by Bell and Wedgwood in [1] since there are essentially only two finite regions of solid and liquid materials, however the problems are compounded by the fact that the moving front may approach either boundary a_1 or a_2 and may cross the initial freeze interface at the origin. This causes difficulties in choosing transformations as will become apparent.

This paper is concerned with modelling all possible stages of the problem as illustrated in figure 2. For simplicity a one-dimensional model is assumed and thus the system may be described by the diffusion equations

$$\frac{\partial T}{\partial \tau} = \kappa_1 \frac{\partial^2 T}{\partial x^2} , \qquad -a_1 < x < x_0(\tau) \ ;$$

$$\frac{\partial T}{\partial \tau} = \kappa_2 \frac{\partial^2 T}{\partial x^2} , \qquad x_0(\tau) < x < a_2 \ ;$$

together with the boundary conditions:

$$\left.\frac{\partial T}{\partial x}\right|_{-a_1} = 0 \qquad \text{on } x = -a_1 \qquad (\text{ a line of symmetry }) \ ;$$

$$T(x_0(\tau),\tau) = T_f \qquad (\text{ solid / liquid interface })$$

and

$$K_1 \left.\frac{\partial T}{\partial x}\right|_{x_0-} - K_2 \left.\frac{\partial T}{\partial x}\right|_{x_0+} = L \rho_2 \frac{dx_0}{dt}$$

(heat balance at phase change).

There are various possibilities for a boundary condition at the boundary a_2 involving the heat flux across it, but the simplest case is to assume the vessel being insulated, which is equivalent to the condition imposed at a_1, thus

$$\left.\frac{\partial T}{\partial x}\right|_{a_2} = 0 \qquad (\text{ no heat transfer }).$$

The temperatures and displacements are represented by T and x respectively. The constants κ, K, L and ρ denote thermal diffusivity, conductivity, latent heat of fusion and density with subscripts 1 and 2 identifying the solid and warm liquid respectively.
The initial conditions for the process are

$$T(x,0) = T_c \qquad -a_1 < x < 0 \qquad (\text{ cold body temperature });$$

$$x_0(0) = 0 \qquad (\text{ freeze front at origin })$$

and

$$T(x,0) = T_0 \qquad 0 < x < a_2 \qquad (\text{ ambient temperature of liquid }).$$

In order to simplify the equations they are non-dimensionalised using the transformations

$$U = \frac{T - T_f}{T_0 - T_f} \quad \text{and} \quad t = \frac{\kappa_2 \tau}{a_1^2} \; .$$

It is also appropriate to immobilise the moving boundary using a further transformation, however terms involving $\frac{1}{x_0(t)}$ must be avoided since this creates a division by zero in the event of the freeze front passing through the origin, which will occur in certain situations. Thence the following transformations are used

$$z = \frac{x - x_0}{a_1 + x_0} \quad \text{in the solid region;}$$

$$z = \frac{x - x_0}{a_2 - x_0} \quad \text{in the liquid region.}$$

These transformations map the regions 1 and 2 i.e. $(-a_1,x_0)$ and (x_0,a_2), onto (-1,0) and (0,1) respectively. The diffusion equations governing the process become

$$\frac{\partial U}{\partial t} = \frac{\eta}{(1 + X_0)^2} \frac{\partial^2 U}{\partial z^2} + \frac{(1 + z)}{(1 + X_0)} \frac{dX_0}{dt} \frac{\partial U}{\partial z} \, , \qquad -1 < z < 0 \; ;$$

$$\frac{\partial U}{\partial t} = \frac{1}{(X_1 - X_0)^2} \frac{\partial^2 U}{\partial z^2} + \frac{(1 - z)}{(X_1 - X_0)} \frac{dX_0}{dt} \frac{\partial U}{\partial z} \, , \qquad 0 < z < 1 \; ;$$

where $X_0(t) = \frac{x_0(t)}{a_1}$, $X_1 = \frac{a_2}{a_1}$ and $\eta = \frac{\kappa_1}{\kappa_2}$.

Both the transformations satisfy the condition imposed concerning terms $\frac{1}{x_0(t)}$, however as the freeze front approaches $x = a_2$ or $x = -a_1$ problems will undoubtedly occur, so further considerations will have to be made.

The boundary conditions transform in a similar manner to give

$$\left. \frac{\partial U}{\partial z} \right|_{-1} = 0 \, , \quad \text{at the centre of the cold body ;}$$

$$\left. \frac{\partial U}{\partial z} \right|_{1} = 0 \, , \quad \text{at the extremity of liquid region ;}$$

with

$$\frac{\beta}{(1 + X_0)} \left. \frac{\partial U}{\partial z_1} \right|_{0-} - \frac{1}{(X_1 - X_0)} \left. \frac{\partial U}{\partial z_2} \right|_{0+} = \alpha \frac{dX_0}{dt}$$

and $\qquad U = 0 \qquad$ at $z = 0$, $\qquad$ at the solid / liquid interface.

The dimensionless thermal parameters α and β are described as

$$\alpha = \frac{L}{c_2\,(T_0 - T_f)} \quad \text{and} \quad \beta = \frac{K_1}{K_2}$$

where α is the reciprocal of the Stefan number for the process. The initial conditions become

$$X_0(0) = 0 \ ,$$

$$U = 1 \ , \ \text{for} \quad 1 < z \le 2 \ ,$$

and $$U = U_B = \frac{T_c - T_f}{T_0 - T_f} \ , \ \text{for} \ -1 \le z_1 < 0 \ .$$

The mathematical model contains a discontinuity at X=0 when t=0 and thus small time starting solution must be used to overcome this. If t is very small, effects due to the sizes of the cold body and liquid region may be ignored, since heat transfer is confined to a small region close to the interface between the two regions. A suitable starting process is thus to assume error function profiles for the dimensionless temperature U, as described by Carslaw and Jaeger [4]. The initial velocity of the freeze front is thus

$$X_0(t) = \lambda \sqrt{t} \quad \text{for} \ 0 < t \le t_0 \ ,$$

where λ is the solution of the transcendental equation

$$\frac{\alpha\,\lambda\,\sqrt{\pi}}{2} + \frac{\beta\,U_B\,e^{-\lambda^2/4\eta}}{\sqrt{\eta}\left\{\operatorname{erf}\left[\frac{\lambda}{2\sqrt{\eta}}\right] + 1\right\}} + \frac{e^{-\lambda^2/4}}{\operatorname{erfc}\left[\frac{\lambda}{2}\right]} = 0 \ ,$$

and the dimensionless temperature profiles in the two regions are:

$$U = U_B \left\{ \frac{\operatorname{erf}\left[\frac{\lambda}{2\sqrt{\eta}}\right] - \operatorname{erf}\left[\frac{x}{2\sqrt{\eta t}}\right]}{\operatorname{erf}\left[\frac{\lambda}{2\sqrt{\eta}}\right] + 1} \right\} \qquad \text{in the solid;}$$

and $$U = \frac{\operatorname{erf}\left[\frac{x}{2\sqrt{t}}\right] - \operatorname{erf}\left[\frac{\lambda}{2}\right]}{\operatorname{erfc}\left[\frac{\lambda}{2}\right]} \qquad \text{in the liquid respectively.}$$

Given the starting profiles, the solution is advanced using explicit finite difference formulae in much the same way as described in Bell and Wedgwood [1]. Due to the sensitivity of these equations, it is necessary to choose the size of the time increment Δt with some care so as to avoid cancellation errors when updating the temperature fields. At each cycle of the algorithm Δt is thus required to satisfy the smaller of the two bounds

$$\frac{h^2(1+X_0)^2}{2\eta} \quad , \quad \frac{h^2(X_1-X_0)^2}{2} \ .$$

Both of these bounds become appropriate at some stage during the freeze front lifetime since, neither remains smaller than the other throughout the process.

For the results that follow the common mesh size h in the two regions is taken to be 0.05 throughout, that is both regions are divided into 20 sub-intervals. The start time t_0 is chosen for each parameter regime to ensure that any heat transfer in the time $t<t_0$ is confined to a region close to the interface between the two regions and the outer extremities are not affected. For most cases a start time of $t_0=0.001$ is used.

The initial velocity of the front is fundamental to the solution, however the behaviour of the system is complex so the effects of altering the parameters α, β, η and U_B are examined. A basic parameter regime is taken as follows:

$$\eta = 12 \qquad \beta = 2 \qquad \alpha = 2 \qquad U_B = -2\, , \qquad (1)$$

which corresponds to $\lambda = 0.061184$.

Given this regime, the effect on λ of altering each of these parameters in turn whilst holding the others fixed is examined (figures 3a-d). Clearly, the effects of each parameter on λ are relatively simple, however it is only β and U_B that truly dictate the sign of λ and it is these that are most easily varied to examine different starting scenarios. As expected, the more negative U_B becomes the greater the initial ratio of freezing as the solid becomes much colder in relation to the liquid.

The freeze front motion is also dependent on the existence of any limiting large time solutions. These are examined by considering the net loss or gain of heat from the two regions once a steady state has been established:

$$\text{Gain in heat of solid} = c_1 \rho_1 a_1 (T_f - T_B)$$
$$\text{Loss of heat of liquid} = c_2 \rho_2 a_2 (T_0 - T_f) + L \rho_2 X_0.$$

Equating these and non-dimensionalising gives

$$X_1 + \alpha X_0 + \frac{U_B \beta}{\eta} = 0.$$

It is obvious that by selecting the value of X_1, the limiting front position may be predicted and is independent of initial motion of the front, since the value of X_1 has no bearing on the value of λ. For the parameter regime (1) this becomes

$$\lim_{t\to\infty} \left[\frac{x_0(t)}{a_1} \right] = \frac{1}{6} (1 - 3 X_1) .$$

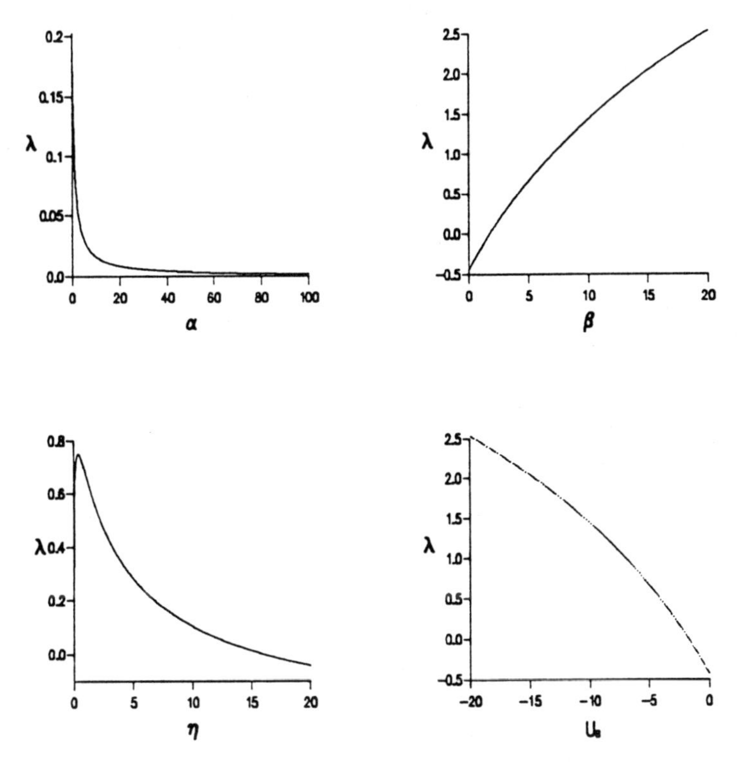

Figure 3: The effect of varying α, β, η, U_B on λ.

This gives a means of checking results obtained, especially if the limiting position of the front remains within the spatial bounds of the system i.e. $-1 < X_0 < X_1$. For the case when the bodies are the same size, $X_1 = 1$ and the limiting front position should be at $X_0 = -1/3$. The front history calculated using the immobilising technique is shown in Figure 4 and clearly shows a good accuracy in the limiting front position (X_0 calculated = -0.339919 which represents an error of only about 2.0%).

Essentially the best way to check the validity of the earlier stages of the front lifetime calculated by this technique, is to match it against an intrinsically differing method. The method adopted is that of using a fixed grid technique first described by Crank [2]. The two regions are again divided into a number of smaller sub-intervals and the front is tracked across these using a Lagrangian interpolation approximation of the boundary partial differential equations. The temperature profile away from the boundary is updated using explicit finite difference approximations of the diffusion equations as before.

Here the initial solid and liquid regions are each sub-divided into 100 intervals. The front is allowed to progress from its starting position at a time $t_0 > 0$ across the mesh. A typical front position is shown in Figure 5 where the front is in the interval between the rth and $r+1^{th}$ mesh points. It is obvious that the main difficulty involved in this method of solution is the existence of unequal grid spacings at the boundary if the front is considered to be a grid point. Thus, when running the algorithm, an interpolation technique is adopted to generate equations at the freeze front. The temperature profile is updated across the mesh points ...r-2, r-1, x_0, r+1, r+2..., the rth mesh point being omitted from the equations for the time being.

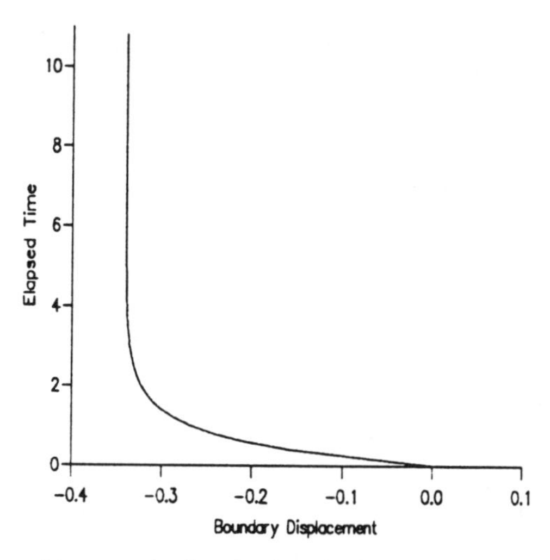

Figure 4: Front history calculated using the immobilising technique

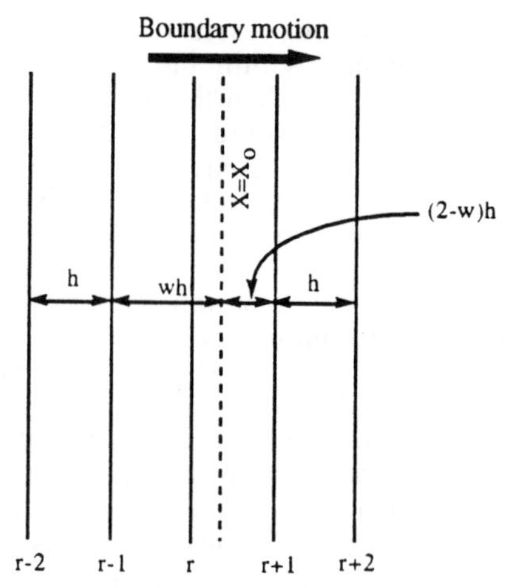

Figure 5: Fixed grid boundary motion.

The solid temperature profile is calculated up to the r-2th mesh point using the usual finite difference approximation to the diffusion equation. To calculate the new temperature value at the r-1th mesh point, a Lagrangian formulated diffusion equation is developed by interpolating over the old temperatures at the points r-2, r-1 and x_0 (ignoring the rth point) giving

$$U^{p+1}_{r-1} = U^{p}_{r-1} + \frac{2\,\kappa\,k}{h^2}\left\{ \frac{U^{p}_{r-2}}{w^{p}+1} - \frac{U^{p}_{r-1}}{w^{p}} \right\},$$

h being the common mesh spacing in both the solid and liquid regions and (r+w)h being the front position at any time. Similarly, the new temperature value of the r+1th mesh point is found by interpolating over the old temperature values at the points x_0, r+1, r+2 giving

$$U^{p+1}_{r+1} = U^{p}_{r+1} + \frac{2\,k}{h^2}\left\{ \frac{U^{p}_{r+2}}{3 - w^{p}} - \frac{U^{p}_{r+1}}{2 - w^{p}} \right\}.$$

The remaining liquid temperature profile is calculated in a similar way to the solid temperature profile by finite difference approximation of the relevant diffusion equation in the liquid region.

In order to track the freeze front it is still necessary to make a Lagrangian interpolation approximation to the usual moving boundary equation, giving

$$\frac{\beta}{h}\left\{ \frac{w^{p}U^{p}_{r-2}}{w^{p}+1} - \frac{(w^{p}+1)U^{p}_{r-1}}{w^{p}} \right\} - \frac{1}{h}\left\{ \frac{(w^{p}-2)U^{p}_{r+2}}{3-w^{p}} + \frac{(w^{p}-3)U^{p}_{r+1}}{w^{p}-2} \right\} =$$

$$= \alpha\frac{dx_0}{dt} = \frac{\alpha\, h}{k}\left\{ w^{p+1} - w^{p} \right\}$$

which is used to calculate the updated value w^{p+1} at each cycle.

When the freeze front passes from one grid space to the adjacent one, so w^p increases by unity then the temperature value U_r is found by Lagrangian interpolation over the points r-2, r-1 and x_0. The interpolation is taken over these three points since the rth grid line lies within the region (r-1,x_0) and so all these points lie within the solidified layer:

$$U^{p+1}_{r} = \frac{(1-w^{p})\,U^{p+1}_{r-2}}{1+w^{p}} + \frac{2\,(w^{p}-1)\,U^{p+1}_{r-1}}{w^{p}}.$$

The procedure presented by Crank [2] only covers the case when the solidified layer is growing, but for the system here, the algorithm must allow for thawing. When w^p decreases by unity, the temperature value U_r is found by interpolating over the points x_0, r+1 and r+2, since the rth grid line lies within the region (x_0,r+1) and thus all these points lie within the liquid region:

$$U^{p+1}_{r} = \frac{2\,(2-w^{p})\,U^{p+1}_{r+1}}{3-w^{p}} - \frac{(2-w^{p})\,U^{p+1}_{r+2}}{4-w^{p}}.$$

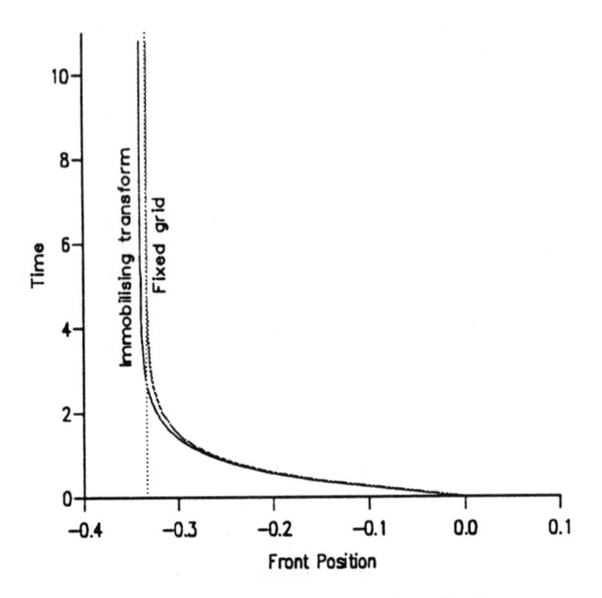

Figure 6: Graphical comparison of methods.

The starting profile used is the same as that used earlier and the parameters are those in (1). To check the validity of the immobilising technique, X_1 is chosen as unity, corresponding to a limiting front position of $X_0 = -1/3$. Figure 6 compares the results for the immobilising transform technique ,the fixed grid technique and the theoretical front position for large time. Clearly the match is very good, but since more sub-intervals are used in the fixed grid method a better representation of the temperature profile is gained and its large time limit is slightly better (X_0 calculated = –0.33153 which represents an error of only 0.05% compared to the immobilising error of 2.0%).

Returning to the use of the immobilising technique, by altering the value of X_1 whilst holding parameters (1) fixed, the initial motion of the boundary remains the same as one of freezing, but the steady state solution can be varied to literally any position across the spatial regions. Figure 7 shows results involving initial freezing, but varying steady state solutions for small values of X_1.

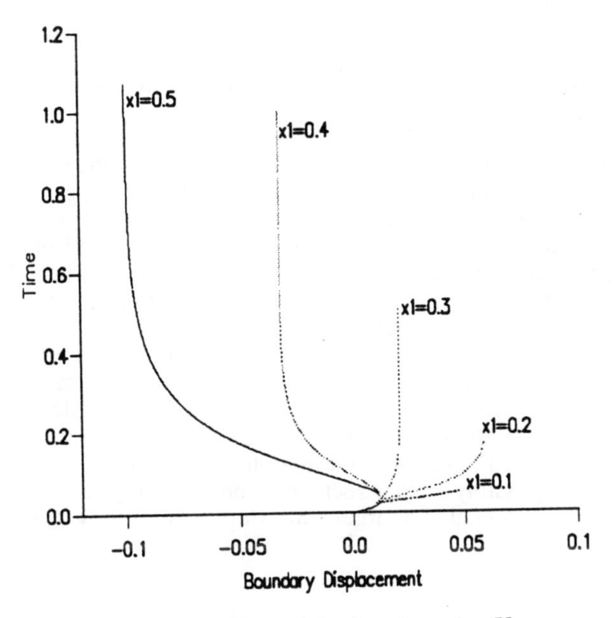

Figure 7: The effect of altering the value X_1 on the steady state solution.

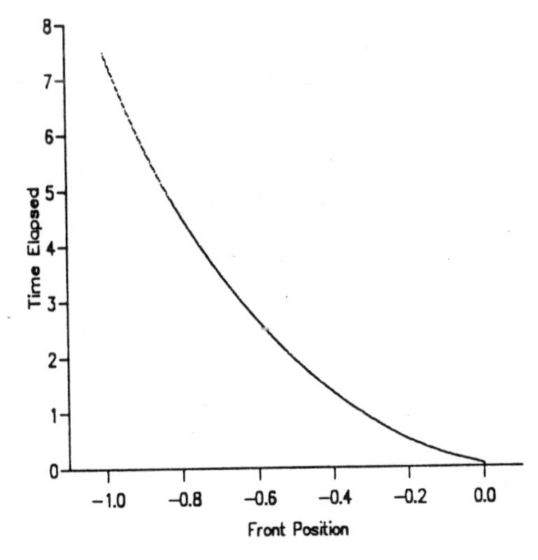

Figure 8: Results for X_1=3 yielding total thawing.

The final problem remaining is that of what happens as the front approaches either boundary $X=-1, X_1$ causing a division by zero in the appropriate diffusion equation. This is overcome by examining the relevant temperature profiles in the two regions as the front approaches a boundary. It is clear that the temperatures in the vanishing region are so close to the freezing temperature T_f $(U=0)$ as to be deemed negligible. This simplification allows the smaller region to be omitted from considerations and the moving boundary equation is simplified accordingly.

Figure 8 shows the solution for the case when $X_1=3$ and thus gives a theoretical limiting value of $X_0=-4/3$, which is outside the realms of the system, hence corresponding to a solution of complete thawing. By examining the temperature profiles in the regions as the front approaches $X_0=-1$, an appropriate time to omit the solid temperature profile is when $X_0 \approx -0.83$ at time $t \approx 4.8$. The transition is smooth and the solution continues until $X_0=-1$ when the system becomes invalid. After this time any heat profiles remaining flatten to a uniform temperature distribution.

Obviously, the greatest restriction on the methods used is the size of the possible time step available, due to the explicit nature of the computational procedures. Nevertheless, the methods developed are simple and relatively easy to apply to the problem and give accurate results. Advancement of this work may be to utilise implicit approximations to the partial differential equations generated, or possibly to construct a method involving variable grid sizes in the Crank method to enable a larger time step to be used with the infinite liquid problem.

REFERENCES

1. BELL, G.E. and WEDGWOOD, I.D. - 'Predicting the Extent and Lifetime of a Transient Solidified Layer.' Proceedings of the 2nd Int. Conf. Heat Transfer, Vol 1: Conduction, Radiation and Phase Change, pp351-363. 1992.
2. CRANK, J. - 'Two methods for the numerical solution of moving-boundary problems in diffusion and heat flow.' Q. J. Appl. Math., Vol 10 pp220-231 (1957).
3. TADJBAKHSH, I. and LINIGER, W. - 'Free boundary problems with regions of growth and decay - Analysis of heat transfer in dip soldering process.' Q.J. Mech. Appl. Math., vol 17(2), pp. 141-155, 1964
4. CARSLAW, H.S. and JAEGER, J.C. - Conduction of Heat in Solids. 2nd ed., Chapter 11, Oxford University Press, 1959.

SOLUTION OF MELTING AND SOLIDIFICATION PROBLEMS BY THE DUAL RECIPROCITY BOUNDARY ELEMENT METHOD

B.Šarler, A.Košir

J.Stefan Institute, University of Ljubljana, Jamova 39, Ljubljana, Slovenia

SUMMARY

This paper describes the application of the dual reciprocity boundary element method for the solution of energy transport in melting and solidification systems composed of incompressible distinct or continuous phase-change materials. The physical model is based on the enthalpy mixture continuum formulation, Fourier constitutive heat flux relation, and includes temperature dependent material properties. The discretization of the Kirchhoff's transformed and heat source term reformulated governing equation is structured by the Green's function of the Laplace equation, the $1+r$ space spline dual reciprocity boundary-only representation of the domain integrals, and straight line boundary elements with constant space and linear time splines. The timestep iterations follow the new Voller and Swaminathan scheme. The sensitivity of the results with respect to space-time discretization, Stefan number, and melting interval was investigated on a two-phase analytical solution for the solidification of an infinite rectangular corner and on the classical one-phase Stefan problem.

1. INTRODUCTION

The research of the solid-liquid phase change involves interdisciplinary theoretical, experimental, and computational modelling of phase transformation kinetics, solid mechanics, and transport phenomena. It has an important impact on many basic-science, engineering, and medical developments. A comprehensive data base of relevant references, including several key words is listed in [1]. Due to the demonstrated [2] suitability of the Boundary Element Method (BEM) for discrete approximative solution of nonlinear transport phenomena models, a great interest exists to enable this method also for coping with melting and solidification.

A comprehensive survey of the related BEM applications is published in [3]. The complete boundary-domain integral equation framework, structured by the fundamental solution of the Laplace equation, for solving the mixture continuum formulation [4] of the coupled transport of mass, energy, momentum, and species during melting and solidification was already developed in [5]. The principal incitement for the present work was an advanced numerical solution of the relevant energy equation.

2. FORMULATION

Consider a connected fixed domain Ω with boundary Γ occupied by a phase change material described with the density ρ_0 the temperature dependent specific heat $c_{\mathcal{P}}$ and the thermal conductivity $k_{\mathcal{P}}$ of the solid $\mathcal{P} = \mathcal{S}$ and the liquid $\mathcal{P} = \mathcal{L}$ phase, and the specific latent heat of solid-liquid phase change $H^0_{\mathcal{M}}$. The mixture continuum formulation of the energy transport for the assumed system is

$$\frac{\partial}{\partial t}(f_{\mathcal{S}}\, \rho_0\, H_{\mathcal{S}} + f_{\mathcal{L}}\, \rho_0\, H_{\mathcal{L}}) + \nabla \cdot (f_{\mathcal{S}}\, \rho_0\, \mathbf{V}_{\mathcal{S}}\, H_{\mathcal{S}} + f_{\mathcal{L}}\, \rho_0\, \mathbf{V}_{\mathcal{L}}\, H_{\mathcal{L}})$$

$$= -\nabla \cdot (f_{\mathcal{S}}\, \mathbf{F}_{\mathcal{S}} + f_{\mathcal{L}}\, \mathbf{F}_{\mathcal{L}}) + f_{\mathcal{S}}\, q_{\mathcal{S}} + f_{\mathcal{L}}\, q_{\mathcal{L}}, \quad f_{\mathcal{S}} + f_{\mathcal{L}} = 1. \tag{1}$$

Function $f_{\mathcal{P}}$ presents the temperature dependent volume fraction, $H_{\mathcal{P}}$ the specific enthalpy, $\mathbf{V}_{\mathcal{P}}$ the known solenoidal velocity, $\mathbf{F}_{\mathcal{P}}$ the heat flux, and $q_{\mathcal{P}}$ the heat source of the phase $\mathcal{P}$. Heat sources could depend arbitrary on temperature and independent time and position variables. Due to the local thermal equilibrium between the phases, are the mixture temperature T and the phase temperatures $T_{\mathcal{S}}$ and $T_{\mathcal{L}}$ equal. The pure equiaxed type of solidification is modelled by $\mathbf{V}_{\mathcal{S}} = \mathbf{V}_{\mathcal{L}} = \mathbf{V}$ and the pure columnar one with $\mathbf{V}_{\mathcal{L}} = \mathbf{V}$, $\mathbf{V}_{\mathcal{S}} = \mathbf{V}_{\mathbf{sys}}$, where $\mathbf{V}_{\mathbf{sys}}$ presents the system velocity. Constitutive equations for the two heat fluxes are based on the Fourier relation

$$\mathbf{F}_{\mathcal{S}} = -\, k_{\mathcal{S}}\, \nabla T, \quad \mathbf{F}_{\mathcal{L}} = -\, k_{\mathcal{L}}\, \nabla T, \tag{2}$$

and the enthalpy-temperature relationship is defined as

$$H_{\mathcal{S}} = c_{\mathcal{S}}(T_H)\, T_H + \int_{T_H}^{T} c_{\mathcal{S}}(\theta)\, d\theta, \quad H_{\mathcal{L}} = c_{\mathcal{L}}(T_H)\, T_H + \int_{T_H}^{T} c_{\mathcal{L}}(\theta)\, d\theta + H^0_{\mathcal{M}}, \tag{3}$$

with T_H representing the enthalpy reference temperature. The governing equation could be rewritten in the following latent heat source term form

$$\rho_0\, c\, \frac{\partial T}{\partial t} + \rho_0 \Big([\, (1 - f_{\mathcal{L}})\, c_{\mathcal{S}} - \frac{df_{\mathcal{L}}}{dT}\, H_{\mathcal{S}}\,]\, \mathbf{V}_{\mathcal{S}} + [\, f_{\mathcal{L}}\, c_{\mathcal{L}} + \frac{df_{\mathcal{L}}}{dT}\, H_{\mathcal{L}}\,]\, \mathbf{V}_{\mathcal{L}} \Big) \cdot \nabla T$$

$$= \nabla \cdot (\, k\, \nabla T\,) + q - \rho_0\, H_{\mathcal{LS}}\, \frac{df_{\mathcal{L}}}{dT}\, \frac{\partial T}{\partial t}, \tag{4}$$

with

$$q = f_S\, q_S + f_{\mathcal{L}}\, q_{\mathcal{L}}, \quad H_{\mathcal{L}S} = H_{\mathcal{L}} - H_S. \tag{5}$$

The thermal conductivity k and the specific heat c of the mixture are defined as

$$k = k_0 + k_T = f_S\, k_S + f_{\mathcal{L}}\, k_{\mathcal{L}}, \quad c = c_0 + c_T = f_S\, c_S + f_{\mathcal{L}}\, c_{\mathcal{L}}. \tag{6}$$

Constants k_0, c_0 present mean values, and functions k_T, c_T temperature behaviour of the respective mixture quantities. We seek the solution of the governing equation for mixture temperature at final time $t = t_0 + \Delta t$, where t_0 presents initial time and Δt positive time increment. The solution is constructed by the initial and boundary conditions that follow.

The initial temperature $T(\mathbf{p}, t_0)$ at point with position vector $\mathbf{p}$ and time t_0 is defined through known function T_0

$$T(\mathbf{p}, t_0) = T_0; \quad \mathbf{p} \in \Omega \oplus \Gamma. \tag{7}$$

The boundary Γ is divided into not necessary connected parts Γ^D, Γ^N, and Γ^R

$$\Gamma = \Gamma^D \oplus \Gamma^N \oplus \Gamma^R, \tag{8}$$

with Dirichlet, Neumann, and Robin type of boundary conditions respectively. These boundary conditions are at point $\mathbf{p}$ and time $t_0 \leq t \leq t_0 + \Delta t$ defined through known functions T_Γ, F_Γ, and h

$$T(\mathbf{p}, t) = T_\Gamma; \quad \mathbf{p} \in \Gamma^D, \tag{9}$$

$$-\, k\, \nabla T(\mathbf{p}, t) \cdot \mathbf{n}_\Gamma(\mathbf{p}) = F_\Gamma; \quad \mathbf{p} \in \Gamma^N, \tag{10}$$

$$-k\, \nabla T(\mathbf{p}, t) \cdot \mathbf{n}_\Gamma(\mathbf{p}) = h\left[T(\mathbf{p}, t) - T_\Gamma\right]; \quad \mathbf{p} \in \Gamma^R, \tag{11}$$

where the heat transfer coefficient h and other known functions are allowed to depend arbitrary on thermal field. The outward pointing normal on Γ is denoted by $\mathbf{n}_\Gamma(\mathbf{p})$. Constraint

$$T_0 = T_\Gamma; \quad \mathbf{p} \in \Gamma^D, \quad t = t_0, \tag{12}$$

is required for the problem to be well-posed.

The equation (4) is rewritten into boundary-domain integral shape by the introduction of the Kirchhoff variable

$$\mathcal{T} = T_{\mathcal{T}} + \int_{T_{\mathcal{T}}}^{T} \frac{k(\theta)}{k_0}\, d\theta = T + \int_{T_{\mathcal{T}}}^{T} \frac{k_T(\theta)}{k_0}\, d\theta, \tag{13}$$

with $T_{\mathcal{T}}$ denoting the Kirchhoff variable reference temperature, and by weighting it over space-time $[\Omega] \times [t_0, t_0 + \Delta t]$ with the fundamental solution of the Laplace equation $T^*(\mathbf{p}; \mathbf{s})$. After lengthy procedure, detailed in [6], the following boundary-domain integral expression is obtained

$$\int_\Omega \rho_0\, c_0\, \mathcal{T}(\mathbf{p}, t_0 + \Delta t)\, T^*(\mathbf{p}; \mathbf{s})\, d\Omega - \int_\Omega \rho_0\, c_0\, \mathcal{T}(\mathbf{p}, t_0)\, T^*(\mathbf{p}; \mathbf{s})\, d\Omega$$

$$+ \int_{t_0}^{t_0+\Delta t} \int_\Omega \mathbf{\Lambda} \cdot \nabla \mathcal{T}\, T^*\, d\Omega\, dt$$

$$= \int_{t_0}^{t_0+\Delta t} \int_\Gamma k_0\, T^*\, \nabla \mathcal{T} \cdot d\mathbf{\Gamma}\, dt - \int_{t_0}^{t_0+\Delta t} \int_\Gamma k_0\, \mathcal{T}\, \nabla T^* \cdot d\mathbf{\Gamma}\, dt$$

$$+ \int_{t_0}^{t_0+\Delta t} c^*(\Omega, \mathbf{s})\, k_0\, \mathcal{T}(\mathbf{s}, t)\, dt + \int_{t_0}^{t_0+\Delta t} \int_\Omega \left[q + \Upsilon\, \frac{\partial \mathcal{T}}{\partial t} \right] T^*\, d\Omega\, dt;$$

$$\mathbf{\Lambda} = \rho_0\, \frac{k_0}{k} \Big(\big[(1 - f_\mathcal{L})\, c_\mathcal{S} - \frac{df_\mathcal{L}}{dT}\, H_\mathcal{S} \big]\, \mathbf{v}_\mathcal{S} + \big[f_\mathcal{L}\, c_\mathcal{L} + \frac{df_\mathcal{L}}{dT}\, H_\mathcal{L} \big]\, \mathbf{v}_\mathcal{L} \Big),$$

$$\Upsilon = \rho_0 \left[c_0 - \frac{k_0}{k} \left(c + H_{\mathcal{LS}}\, \frac{df_\mathcal{L}}{dT}\, \frac{\partial \mathcal{T}}{\partial t} \right) \right],$$

$$c^*(\Omega, \mathbf{s}) = \int_\Omega \nabla^2 T(\mathbf{p}; \mathbf{s})\, d\Omega, \quad T_2^*(\mathbf{p}; \mathbf{s}) = \frac{1}{2\pi} \log \frac{r_0}{|\mathbf{p} - \mathbf{s}|}. \qquad (14)$$

Function T_2^* stands for the two dimensional planar symmetry form of the fundamental solution T^*. Equation (14) is solved by the related Kirchhoff transformed initial and boundary conditions

$$\mathcal{T}(\mathbf{p}, t_0) = \int_{T_\mathcal{T}}^{T_0} \frac{k}{k_0}\, d\theta; \quad \mathbf{p} \in \Omega \oplus \Gamma, \qquad (15)$$

$$\mathcal{T}(\mathbf{p}, t) = \int_{T_\mathcal{T}}^{T_\Gamma} \frac{k}{k_0}\, d\theta; \quad \mathbf{p} \in \Gamma^D, \qquad (16)$$

$$- k_0\, \nabla \mathcal{T}(\mathbf{p}, t) \cdot \mathbf{n}_\Gamma(\mathbf{p}) = F_\Gamma; \quad \mathbf{p} \in \Gamma^N, \qquad (17)$$

$$- k_0\, \nabla \mathcal{T}(\mathbf{p}, t) \cdot \mathbf{n}_\Gamma(\mathbf{p}) = h \left[\mathcal{T}(\mathbf{p}, t) - T_\Gamma - \int_{T_\mathcal{T}}^{T[\mathcal{T}(\mathbf{p}, t)]} \frac{k_T}{k_0}\, d\theta \right]; \; \mathbf{p} \in \Gamma^R. \qquad (18)$$

3. SOLUTION PROCEDURE

The solution procedure is based on the Dual Reciprocity Method (DRM) [7] that handles the boundary-domain integral equation of the type (14) through the calculation of the boundary integrals only. This advantageous property is founded on the approximation of an arbitrary scalar valued function $\mathcal{F}(\mathbf{p}, t)$ over the domain Ω with $n = 1, 2, \ldots, N$ space $\Psi_n^p(\mathbf{p})$ and time splines $\Psi_n^t(t)$

$$\mathcal{F}(\mathbf{p}, t) \approx \Psi_n^p(\mathbf{p})\, \Psi_n^t(t),\; \mathcal{F}(\mathbf{p}_m, t) = \Psi_{nm}^p\, \Psi_n^t(t),\; \Psi_n^t(t) = [\Psi_{nm}^p]^{-1} \mathcal{F}(\mathbf{p}_m, t). \qquad (19)$$

The Einstein summation is used in this text wherever possible. Using the Green's second identity and the definition for the functions $\hat{\Psi}_n^p$

$$\nabla^2 \hat{\Psi}_n^p(\mathbf{p}) = \Psi_n^p(\mathbf{p}), \tag{20}$$

the domain integrals of functions $\mathcal{F}(\mathbf{p},t)$ and $\mathcal{G}(\mathbf{p},t)\cdot\nabla\mathcal{F}(\mathbf{p},t)$ ($\mathcal{G}$ presents an arbitrary vector valued function) weighted with the Green function T^* over Ω approximately transform into a series of N integrals over its boundary

$$\int_\Omega \mathcal{F}\, T^*\, d\Omega \approx \Psi_n(\mathbf{s})\, [\Psi_{nm}^p]^{-1}\, \mathcal{F}(\mathbf{p}_m, t), \tag{21}$$

$$\int_\Omega \mathcal{G}\cdot\nabla\mathcal{F}\, T^*\, d\Omega \approx \Psi_n(\mathbf{s})\, [\Psi_{nu}^p]^{-1}\, \mathcal{G}(\mathbf{p}_{\underline{u}}, t)\cdot\nabla\Psi_l^p(\mathbf{p}_u)\, [\Psi_{lm}^p]^{-1}\, \mathcal{F}(\mathbf{p}_m, t); \tag{22}$$

$$\Psi_n(\mathbf{s}) = \Big[\int_\Gamma T^*\, \nabla\hat{\Psi}_n^p \cdot d\mathbf{\Gamma} - \int_\Gamma \hat{\Psi}_n^p\, \nabla T^* \cdot d\mathbf{\Gamma} + c^*(\Omega, \mathbf{s})\, \hat{\Psi}_n^p(\mathbf{s})\Big]. \tag{23}$$

The efficiency of the transformations (21,22) depends strongly on the choice of the splines Ψ_n^p which is not unique. We select the form

$$\Psi_n^p(\mathbf{p}) = \sum_{i_\Psi=0}^{I_\Psi} |\mathbf{p}-\mathbf{p}_n|^{i_\Psi}, \quad \hat{\Psi}_n^p(\mathbf{p}) = \sum_{i_\Psi=0}^{I_\Psi} \frac{|\mathbf{p}-\mathbf{p}_n|^{i_\Psi+2}}{(i_\Psi+2)^2}, \tag{24}$$

with $I_\Psi = 1$, as suggested by Partridge et al. [7].

The numerical solution of the nonlinear integral equation (14) inherently requires timestep iterations. The recently developed robust and accurate Voller and Swaminathan scheme [8] has been used for the iterative updating of the nonlinear terms with $\mathbf{\Lambda}$ and Υ. The DRM adapted essentials of this scheme could be perceived from the equation (30).

The governing discretization equation (14) is discretized by the introduction of the linear time splines over the time interval $[t_0, t_0+\Delta t]$. The boundary is discretized by N_Γ boundary elements Γ_k with piecewise straight-line geometry and piecewise constant space splines. The first N_Γ points $\mathbf{p}_n$ in the splines (24) coincide with the nodes (geometric centers) of the boundary elements, and the last N_Ω points are arbitrary distributed in Ω. All subsequently involved boundary integrals have been evaluated analytically.

The equation (14) is solved by constructing an algebraic equation system of $j = 1, 2, \ldots, N$ equations. These equations are obtained by writing the discretized form of equation (14) for source point $\mathbf{s}$ to coincide with the nodal points $\mathbf{p}_n$. The deduced system of algebraic equations could be cast in a symbolic form

$$\mathrm{F}_{jm}^{t_0+\Delta t\, i}\, \mathcal{T}^i(\mathbf{p}_m, t_0+\Delta t) + \mathbf{T}_{jm}^{t_0+\Delta t} \cdot \nabla \mathcal{T}^i(\mathbf{p}_m, t_0+\Delta t)$$

$$= \mathrm{F}_{jm}^{t_0} \mathcal{T}(\mathbf{p}_m, t_0) + \mathbf{T}_{jm}^{t_0} \cdot \nabla \mathcal{T}(\mathbf{p}_m, t_0) + \mathrm{q}_{jm}^{t_0+\Delta t}\, q(\mathbf{p}_m, t_0+\Delta t) + \mathrm{q}_{jm}^{t_0}\, q(\mathbf{p}_m, t_0), \tag{25}$$

which has to be rearranged according to the boundary condition types before the solution. Superscript i denotes the value of quantity at i-th iteration. The matrix elements are defined by

$$\mathrm{F}_{jm}^{t_0+\Delta t\, i} = \frac{\Delta t}{2} \Psi_n(\mathbf{s}_j)\, [\Psi_{nu}]^{-1}\, \mathbf{\Lambda}^i_{t_0+\Delta t\, \underline{u}} \cdot \nabla \Psi_l^p(\mathbf{p}_u)\, [\Psi_{lm}]^{-1}$$

$$+\Psi_n(\mathbf{s}_j)\, [\Psi^p_{n\underline{m}}]^{-1} \left[\rho_0\, c_0 - \Upsilon^i_{t_0+\Delta t\, \underline{m}} + \frac{1}{2} \left(\frac{d\Upsilon}{d\mathcal{T}} \mathcal{T} \right)^i_{t_0+\Delta t\, \underline{m}} \right]$$

$$+ \delta_{km} \frac{\Delta t\, k_0}{2} \int_{\Gamma_k} \nabla T^*(\mathbf{p}; \mathbf{s}_j) \cdot d\mathbf{\Gamma} - \delta_{jm} \frac{\Delta t\, c^*(\Omega, \mathbf{s}_j)\, k_0}{2}, \tag{26}$$

$$\mathrm{F}_{jm}^{t_0} = -\frac{\Delta t}{2} \Psi_n(\mathbf{s}_j)\, [\Psi_{nu}]^{-1}\, \mathbf{\Lambda}[\mathcal{T}(\mathbf{p}_{\underline{u}}, t_0)] \cdot \nabla \Psi_l^p(\mathbf{p}_u)\, [\Psi_{lm}]^{-1}$$

$$+\Psi_n(\mathbf{s}_j)\, [\Psi^p_{n\underline{m}}]^{-1} \left[\rho_0\, c_0 - \Upsilon[\mathcal{T}(\mathbf{p}_{\underline{m}}, t_0)] + \frac{1}{2} \frac{d\Upsilon}{d\mathcal{T}}[\mathcal{T}(\mathbf{p}_{\underline{m}}, t_0)]\, \mathcal{T}(\mathbf{p}_{\underline{m}}, t_0) \right]$$

$$- \delta_{km} \frac{\Delta t\, k_0}{2} \int_{\Gamma_k} \nabla T^*(\mathbf{p}; \mathbf{s}_j) \cdot d\mathbf{\Gamma} + \delta_{jm} \frac{\Delta t\, c^*(\Omega, \mathbf{s}_j)\, k_0}{2}, \tag{27}$$

$$\mathbf{T}_{jm}^{t_0+\Delta t} = -\mathbf{T}_{jm}^{t_0} = -\delta_{km} \frac{\Delta t\, k_0}{2} \int_{\Gamma_k} T^*(\mathbf{p}; \mathbf{s}_j)\, d\mathbf{\Gamma}, \tag{28}$$

$$\mathrm{q}_{jm}^{t_0+\Delta t} = \mathrm{q}_{jm}^{t_0} = \Psi_n(\mathbf{s}_j)\, [\Psi^p_{nm}]^{-1} \frac{\Delta t}{2}, \tag{29}$$

The value of $\Xi_{t_0+\Delta t\, m}$ (Ξ denotes $\mathbf{\Lambda}$ or Υ) at $(i+1)$-th iteration level is obtained through the values at i-th and $(i-1)$-th iteration level

$$\Xi^{i+1}_{t_0+\Delta t\, m} = \Xi[\mathcal{T}^i(\mathbf{p}_m, t_0+\Delta t)]$$

$$+\frac{d\Xi}{d\mathcal{T}}[\mathcal{T}^i(\mathbf{p}_m, t_0+\Delta t)]\, [\mathcal{T}^i(\mathbf{p}_m, t_0+\Delta t) - \mathcal{T}^{i-1}(\mathbf{p}_m, t_0+\Delta t)],$$

$$\left(\frac{d\Xi}{d\mathcal{T}} \mathcal{T} \right)^{i+1}_{t_0+\Delta t\, m} = \frac{d\Xi}{d\mathcal{T}}[\mathcal{T}^i(\mathbf{p}_m, t_0+\Delta t)]$$

$$\times [2\mathcal{T}^i(\mathbf{p}_m, t_0+\Delta t) - \mathcal{T}^{i-1}(\mathbf{p}_m, t_0+\Delta t)]. \tag{30}$$

The timestep iterations are stopped when the absolute Kirchhoff variable difference of the two successive iterations does not exceed some predetermined positive margin $\mathcal{T}_\delta$ in any of the meshpoints $\mathbf{p}_m$.

4. NUMERICAL EXAMPLES

Two physical situations have been treated in order to check the proper performance of the solution procedure. The first one is the standard two-dimensional planar symmetry two-phase benchmark representing the solidification of a semi-infinite rectangular corner and the second one is the classical one-dimensional one-phase Stefan problem. The reference solution of the first problem is constructed by the quasi-analytical technique [9] and the second one analyticaly [10]. The computational domain is a square region $0\,[\mathrm{m}] \leq X \leq 1.5\,[\mathrm{m}]$ and $0\,[\mathrm{m}] \leq Y \leq 1.5\,[\mathrm{m}]$ and the material properties are $\rho_0 = 1\,[\mathrm{kg/m^3}]$, $c_{\mathcal{S}} = c_{\mathcal{L}} = 1\,[\mathrm{J/(kg\,K)}]$, $k_{\mathcal{S}} = k_{\mathcal{S}} = 1\,[\mathrm{W/(m\,K)}]$ with melting temperature $T_{\mathcal{M}}^0 = 1\,[\mathrm{K}]$ in both cases.

In the Test Case I is the square initially filled with material at constant temperature $T_0 = 1.3\,[\mathrm{K}]$, the Dirichlet boundary condition with $T_\Gamma = 0\,[\mathrm{K}]$ are imposed on boundaries $X = 0\,[\mathrm{m}]$ and $Y = 0\,[\mathrm{m}]$, and the Neumann boundary conditions at $X = 1.5\,[\mathrm{m}]$ and $Y = 1.5\,[\mathrm{m}]$ are $F_\Gamma = 0\,[\mathrm{W/(m\,K)}]$.

In the Test Case II is the square initially filled with material at constant temperature $T_0 = 1.0\,[\mathrm{K}]$, the Dirichlet boundary condition with $T_\Gamma = 0\,[\mathrm{K}]$ is imposed on the boundary $Y = 0\,[\mathrm{m}]$ and the Neumann boundary conditions at $X = 1.5\,[\mathrm{m}]$, $Y = 0\,[\mathrm{m}]$, and $Y = 1.5\,[\mathrm{m}]$ are $F_\Gamma = 0\,[\mathrm{W/(m\,K)}]$.

Distinct phase change is for computational purposes approximated by a narrow continuous range from $T_{\mathcal{S}}^0 = T_{\mathcal{M}}^0 - \Delta T_{\mathcal{M}}^0$ to $T_{\mathcal{L}}^0 = T_{\mathcal{M}}^0 + \Delta T_{\mathcal{M}}^0$.

$$f_{\mathcal{L}} = \begin{cases} 0; & T < T_{\mathcal{S}}^0, \\ (T - T_{\mathcal{S}}^0)/(T_{\mathcal{L}}^0 - T_{\mathcal{S}}^0); & T_{\mathcal{S}}^0 \leq T \leq T_{\mathcal{L}}^0, \\ 1; & T > T_{\mathcal{L}}^0. \end{cases} \tag{31}$$

The numerically obtained temperatures $T_{\mathrm{cal}} = \Psi_n^p(\mathbf{p})\,\Psi_n^p(t)$ have been compared with the adjacent solutions from [9, 10] in $N_{\mathrm{com}} = 10201$ uniform meshpoints $\mathbf{p}_{\mathrm{com}}$ (these points coincide with the crossections of the lines $Y = const.$ and $X = const.$ as seen in Figures 2 and 3). The maximum T_{max} absolute error and the average T_{ave} absolute error of the numerical solution are

$$T_{\mathrm{max}} = \max |T_{\mathrm{cal}}(\mathbf{p}_{\mathrm{com}\,n}, t) - T_{\mathrm{ana}}(\mathbf{p}_{\mathrm{com}\,n}, t)|;\; n = 1, 2, \ldots, N_{\mathrm{com}}, \tag{32}$$

$$T_{\mathrm{ave}} = \frac{1}{N_{\mathrm{com}}} \sum_{n=1}^{N_{\mathrm{com}}} |T_{\mathrm{cal}}(\mathbf{p}_{\mathrm{com}\,n}, t) - T_{\mathrm{ana}}(\mathbf{p}_{\mathrm{com}\,n}, t)|. \tag{33}$$

The timestep iteration margin T_δ used in all present calculations is $0.001\,[\mathrm{K}]$.

Mesh	N_Γ	N_Ω	N	
^1M	12	16	28	
^2M	32	64	96	
^3M	48	144	192	
Timestep				Δt[s]
$^1\Delta t$				0.01
$^2\Delta t$				0.005
$^3\Delta t$				0.001

Table 1: Discretization parameters used in calculations.

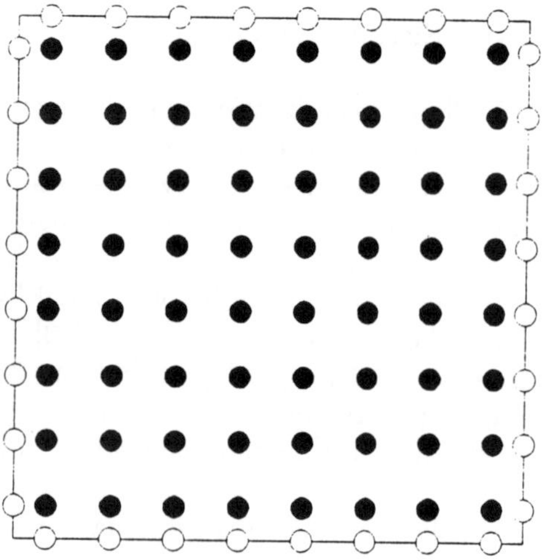

Figure 1: Arrangement of the mesh ^2M. Boundary nodes are denoted with $\circ$ and the domain nodes with $\bullet$. Meshes ^1M and ^3M have similar displaced configuration.

5. CONCLUSIONS

This paper presents the first attempts to computationally solve a multidimensional moving boundary energy transport problem through the calculations that reduce to the integration of the fixed boundary quantities only. The results preliminary confirm the suitability of the described method for coping with such Stefan type of problems.

The principal advantages of the method are the ease of the implementation of the different boundary condition types, straightforward mesh generation, sequel elimination of large data handling, and the potential ability to easily cope with geometrically moving boundaries as well. The main disadvantage of the method is the resulting large algebraic system of equations, since the domain meshpoints have to be present. This unfavourable property could be set out efficiently by substructuring technique in combination with the adaptive strategy, that are both under development.

In addition, complementary benchmarking of the developed method is needed. It will concentrate on the situations with convection such as the Siegel's problem [11], and inclusion of boundary elements upgraded to piecewise linear space splines which will allow to comparably treat all aspects of the method by the standards set in the excellent Dalhuijsen and Segal's study [12].

errors	$^1H^0_{\mathcal{M}}$	$^2H^0_{\mathcal{M}}$	$^3M^0_{\mathcal{M}}$
T_{max} [K]	0.276	0.513	0.649
T_{ave} [K]	0.093	0.167	0.234
T_{max} [K]	0.175	0.370	0.453
T_{ave} [K]	0.062	0.113	0.150

Table 2: Test Case II. Absolute error in the numerically calculated temperatures with mesh $^2M\,^2\Delta t$ at time $t = 0.1$ [s] for different latent heats and melting intervals. The melting enthalpies $^1H^0_{\mathcal{M}}$, $^2H^0_{\mathcal{M}}$, and $^3H^0_{\mathcal{M}}$ are 0.25 [J/kg], 0.50 [J/kg], and 1.00 [J/kg] respectively. The first two result rows were calculated with $\Delta T_{\mathcal{M}} = 0.01$ [K], and the last two with $\Delta T_{\mathcal{M}} = 0.001$ [K].

6. NOMENCLATURE

ROMAN LETTERS

c, c_0, c_T	specific heat, constant-, variable part
f	volume fraction
$\mathbf{F}$	heat flux
h	heat transfer coefficient
H	enthalpy
$H_{\mathcal{LS}}$	enthalpy difference
$H^0_{\mathcal{M}}$	latent heat
k, k_0, k_T	thermal conductivity, constant-, variable part
N_Γ, N_Ω	number of boundary-, domain meshpoints
$\mathbf{p}, \mathbf{s}$	field-, source point
t, t_0	time, initial time
Δt	time increment
T	temperature
T^*	Green's function
T_0	initial condition
T_Γ	boundary condition
$T^0_{\mathcal{S}}$	solidus temperature
$T^0_{\mathcal{M}}$	melting temperature
$T^0_{\mathcal{L}}$	liquidus temperature
$\Delta T_{\mathcal{M}}$	melting interval
$\mathbf{V}$	velocity
q	heat source
$\mathcal{F}, \boldsymbol{\mathcal{G}}$	scalar-, vector function
$F, \mathbf{T}, q$	matrix elements

GREEK LETTERS

ρ_0	constant density
Γ	boundary
Γ_k	boundary element k
Ω	domain
$\mathbf{\Lambda}$	convective term
$\mathcal{T}$	Kirchhoff variable
Υ	source term
Ξ	$\mathbf{\Lambda}$ or Υ
Ψ^p_n	space spline
Ψ^t_n	time spline
Ψ	integral expression

SUBSCRIPTS

$\mathcal{P}$	phase
$\mathcal{S}, \mathcal{L}$	solid-, liquid phase
i	iteration index
$jklmnu$	summation indexes
ave	average
cal	calculated
com	compared
max	maximum
sys	system

SUPERSCRIPTS

D	Dirichlet
N	Neumann
R	Robin

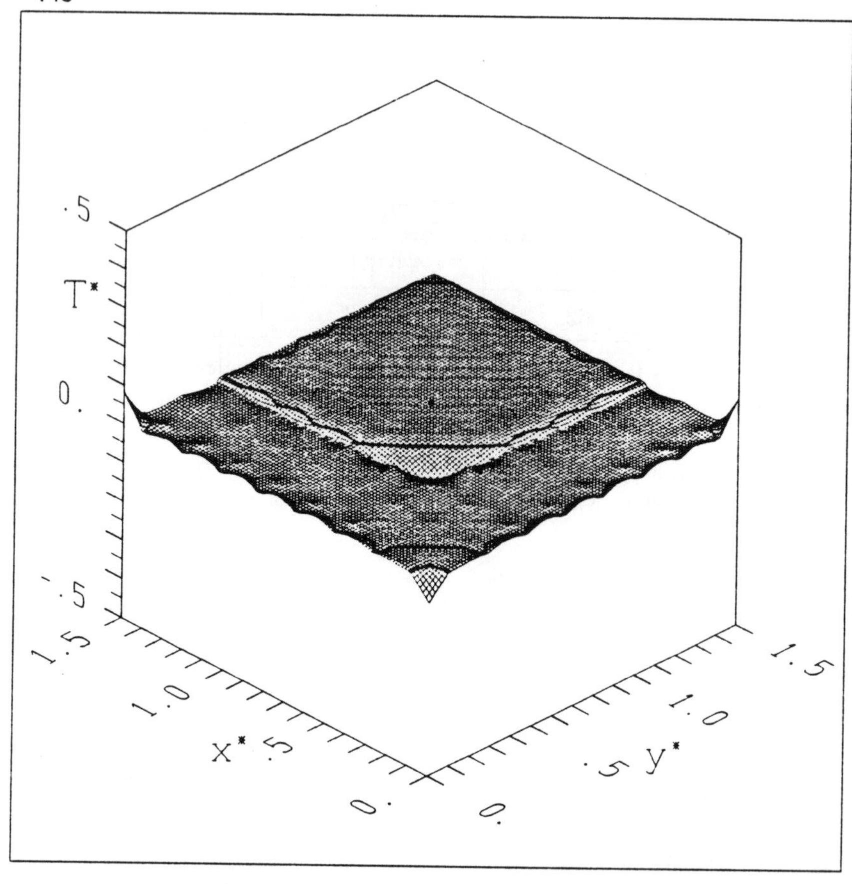

Figure 2: Test Case I. Axonometric view of the error of the interpolated solution for the discretization ${}^2M\,{}^2\Delta t$, $\Delta H^0_{\mathcal{M}} = 0.25$ [J/kg], $\Delta T_{\mathcal{M}} = 0.05$ [K] at time $t = 0.1$ [s]. The dimensionless scales are simply $T^* = (T_{\text{cal}} - T_{\text{ana}})/(1\,[\text{K}])$, $x^* = X/(1\,[\text{m}])$, and $y^* = (Y/1\,[\text{m}])$. The bold curve is isotherm at $T^* = 0$. The positions of the maximum errors are in the corners of the square due to extrapolation and, characteristically for all one-domain schemes, near the solid-liquid interphase boundary.

errors	${}^1M\,{}^1\Delta t$	${}^1M\,{}^2\Delta t$	${}^1M\,{}^3\Delta t$	${}^2M\,{}^2\Delta t$	${}^3M\,{}^2\Delta t$
T_{max} [K]	0.437	0.372	0.293	0.141	0.088
T_{ave} [K]	0.106	0.089	0.068	0.055	0.014

Table 3: Test Case I. Absolute error in the numerically calculated temperatures with $H^0_{\mathcal{M}} = 0.25$ [J/kg] and $\Delta T_{\mathcal{M}} = 0.01$ [K] at time $t = 0.1$ [s] for different space-time discretizations. The Table shows convergence of the method with shorter timesteps and finer meshes. The computer time for solving one iteration on mesh ${}^1M$, ${}^2M$, and ${}^3M$ is approximately 45, 200, and 1000 CPU seconds respectively on 25 MHz PC i486 based compatible with NDP Fortran 77 compiler.

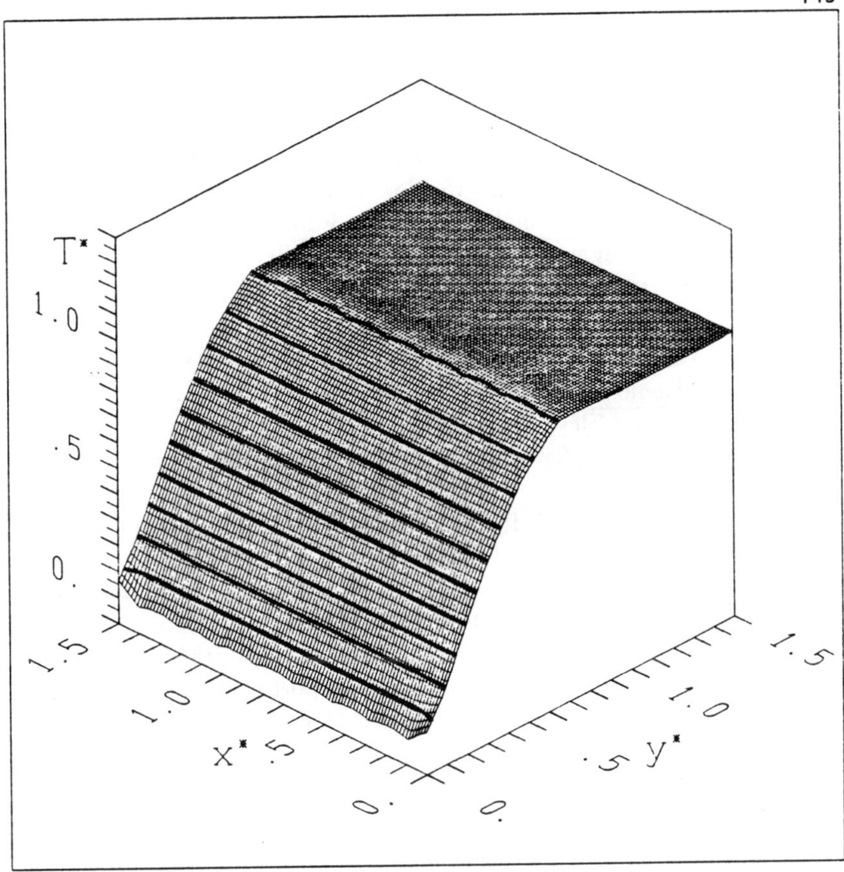

Figure 3: Test Case II. Axonometric view of the interpolated solution for the discretization ${}^2M\,{}^2\Delta t$, $\Delta H^0_{\mathcal{M}} = 0.5\,[\mathrm{J/kg}]$, $\Delta T_{\mathcal{M}} = 0.01\,[\mathrm{K}]$ at time $t = 0.1\,[\mathrm{s}]$. The dimensionless scales are simply $\mathrm{T}^* = T/(1\,[\mathrm{K}])$, $\mathrm{x}^* = X/(1\,[\mathrm{m}])$, and $\mathrm{y}^* = (Y/1\,[\mathrm{m}])$. The bold curves are isotherms at $\mathrm{T}^* = 0.0, 0.1, 0.2, \cdots 1.0$.

7. ACKNOWLEDGEMENT

The work described in this paper is a part of the project Computational Mechanics of Melting and Solidification. The authors would like to acknowledge *Ministry for Science and Technology, Republic of Slovenia* for financial support. Additionally, the authors wish to thank *International Bureau, Research Centre Jülich, FRG* for partial funding through the project Computational Modelling of Solid-Liquid Phase Change Systems. The described method found application in the continuous casting process simulator of the following Slovenian companies *IMPOL*, *Mariborska Livarna Maribor*, *Steelworks Jesenice* and *TALUM*.

8. REFERENCES

[1] ŠARLER, B. - Bibliography on Stefan problem 1992, Technical report IJS-DP-6561, "Jožef Stefan" Institute, Ljubljana, 1992.

[2] WROBEL, L.C. and BREBBIA, C.A. - An Overview of Boundary Element Applications to Nonlinear Heat Transfer Problems, Nonlinear Computational Mechanics - State of the Art, Ed. Wriggers, P. and Wagner, W., Springer-Verlag, Berlin, pp.226-239, 1991.

[3] ŠARLER, B., MAVKO, B. and KUHN, G. - Chapter 16: A Survey of the Attempts for the Solution of Solid-Liquid Phase Change Problems by the Boundary Element Method, Computational Methods for Free and Moving Boundary Problems in Heat and Fluid Flow, Ed. Wrobel, L.C. and Brebbia, C.A. Computational Engineering Series, Elsevier Applied Science, London, pp.373-400, 1993.

[4] BENNON, W.D. and INCROPERA, F.P. - A Continuum Model for Momentum, Heat and Species Transport in Binary Solid-Liquid Phase Change Systems - I.Model Formulation, Int.J.Heat Mass Transfer, Vol.30, pp.2161-2170, 1987.

[5] ŠARLER, B., MAVKO, B. and KUHN, G. - A BEM Formulation for Momentum, Energy and Species Transport in Binary Solid-Liquid Phase Change Systems, Z.angew.Math.Mech., Vol.73, pp.T868-T873, 1993.

[6] ŠARLER, B., MAVKO, B. and KUHN, G. - Mixture Continuum Formulation of Convection-Conduction Energy Transport in Multiconstituent Solid-Liquid Phase Change Systems for BEM Solution Techniques, Eng.Anal., Vol.9, (in print), 1993.

[7] PARTRIDGE, P.W., BREBBIA, C.A. and WROBEL, L.C. - The Dual Reciprocity Boundary Element Method, Elsevier Applied Science, London, 1992.

[8] VOLLER, V.R. and SWAMINATHAN, C.R. - General Source-Based Method for Solidification Phase Change, Num.Heat Transfer, Vol.19B, pp.175-189, 1991.

[9] RATHJEN, K.A. and JIJI, L.M. - Heat Conduction With Melting or Freezing in a Corner, J.Heat Transfer, Vol.93, pp.101-109, 1971.

[10] STEFAN, J. - Über Einige Probleme der Theorie der Wärmeleitung, Aus den Sitzungsberichten d.kais.Akademie d.Wissenschaften in Wien, Mathem.-naturw.Classe, Bd.XCVIII.Abth.II.a.März, 1889.

[11] SIEGEL, R. - Boundary Perturbation Method for Free Boundary Problem in Convectively Cooled Continuous Casting, J.Heat Transfer, Vol.230, pp.230-235, 1986.

[12] DALHUIJSEN, A.L. and SEGAL, A. - Comparison of Finite Element Techniques for Solidification Problems, Int.J.Numer.Methods Eng., Vol.23, pp.1807-1829, 1986.

Simulation of welding with phase transformation

J. Ronda, G.J. Oliver and N. Meinert

Senior Lecturer (D.Sc.), M.Sc. student, Honours student: Applied Mathematics Department and FRD/UCT Centre for Research in Computational and Applied Mechanics, University of Cape Town, Rondebosch 7700, South Africa.

Summary

The applicability of three constitutive models with internal parameters, proposed by Estrin & Mecking, Robinson and Leblond, for modelling of welding has been tested, using a simple bench-mark problem formulated for butt welded plates. The sensitivity of the von Mises norm of residual stresses on the material model is presented and described.

1 Introduction

The effects of heat transfer, microstructure evolution and thermal stress in welding is the focus of the current work, but our interest is concentrated on the study of constitutive equations. We would like to effectively model the welding processes to analyze the sensitivity of residual stresses on constitutive modelling of welded materials, welding parameters controlling the heat influx and cooling intensity, and even the effects of the type of jig or clamping used during welding. Notably, we would like to take into account the austenite-martensite transformation, which is of apparent importance in welding, and which welding designers are keenly aware and wish to limit due to its detrimental effect on fatigue life. Therefore the three following constitutive models with internal parameters have been used: the Estrin-Mecking [4] and the Robinson [2] unified viscoplasticity as well as transformation plasticity proposed by Leblond [6]. The model implementation has been done by specifically writing subroutines for the finite element package ABAQUS through the interfaces such as the UMAT (user material) subroutine and the UEL (user element) subroutine. The UMAT routine allows one to specify the constitutive relation for stress as a function of strain, that is suitable for the Estrin-Mecking and Robinson models. The user element allows one to define an element in a general way and is used for implementation of the Leblond model.

2 Finite Element Formulation

The Lagrangian formulation is used in a description of a thermo- mechanical problem of welding, where large displacements and large rotations but small strains, less than 2%, occur. The stress measure, used in the Lagrangian formulation, is the second Piola- Kirchhoff stress $\tilde{\mathbf{S}}$, which is energetically conjugate to the Green-Lagrange strain $\tilde{\mathbf{L}}$. We assume an incremental decomposition for strains and stresses, i.e. $\tilde{\mathbf{L}}^{n+1} = \tilde{\mathbf{L}}^{n} + \tilde{\mathbf{L}}^{\Delta}$ and $\tilde{\mathbf{S}}^{n+1} = \tilde{\mathbf{S}}^{n} + \tilde{\mathbf{S}}^{\Delta}$, where the superscripts $n, n+1, \Delta$ refer respectively to the previous, current and incremental values. The Green-Lagrange strain is further decomposed into its linear and nonlinear components $\tilde{\mathbf{L}}^{n+1} = \mathbf{L} + \mathbf{L}_{\nu}$. The total Lagrangian formulation is the formulation built into ABAQUS for the standard element and is thus the method of analysis used in the UMAT routines, which will be presented for the viscoplastic models. We have used also the Lagrangian formulation for the UEL routines, which implement transformation plasticity models. It is necessary to use a UEL routine due to the explicit incorporation of the finite element, as a representative volume, for considering phase transformations. The combined global stiffness equation, assembling the mechanical and thermal problems, is then given by:

$$\begin{bmatrix} \mathbf{K}_u \\ \frac{1}{\Delta t}\mathbf{C} + \mathbf{K}_\theta \end{bmatrix} \begin{bmatrix} \Delta \mathbf{u} \\ \Delta \theta \end{bmatrix} = - \begin{bmatrix} \mathbf{F}_u \\ \mathbf{F}_\theta \end{bmatrix} \tag{1}$$

where $\Delta \mathbf{u}$, $\Delta\theta$ are increments of displacement and temperature $\mathbf{K}_u, \mathbf{K}_\theta$ are the mechanical and thermal parts of the static contribution to the global stiffness matrix. The vector of heat flux $\mathbf{F}_\theta$, which contains the dissipation of plastic work, constitutes a coupling between mechanical and thermal effects of welding and the externally applied heat sources. The vector $\mathbf{F}_u$ contains equivalent element forces and externally applied loads. The contribution of $\mathbf{K}_\theta$ stiffness matrix due to thermal conduction $\mathbf{K}_k$ is evaluated to simulate a droplet penetration [11], p.944 during welding, when the matrix of conductivities k_{ij}; $i, j = 1, 2, 3$ is diagonal and contains the two following anisotropic components:

$$\begin{aligned} k_{33} &= k_{mean}\left(1 + 10\exp\left(\frac{-r^2}{2\delta_r^2}\right)\right), \\ k_{22} &= k_{mean}\left(1 + 3\exp\left(\frac{-r^2}{2\delta_r^2}\right)\right), \end{aligned} \tag{2}$$

Here k_{mean} is the mean conductivity, r is the horizontal distance from the weld centre, δ_r is the standard deviation for heat, which this distribution

1. *Begin*
2. *Form Deviatoric Strain Increment :*
   $\Delta \mathbf{E} = \Delta \mathbf{L} - \frac{1}{3}\mathrm{tr}\Delta \mathbf{L}$
3. *Form Equivalent Strain Increment:*
   $\Delta e^{n+1} = (\frac{2}{3}\ \Delta \mathbf{E} : \Delta \mathbf{E})^{\frac{1}{2}}$
4. *Calculate Elastic Predictor:*
   ${}^{*}s^{n+1} = s^{n} + 3\mu \Delta e^{n+1}$
5. *Initialize:*
   $s^{n+1} = {}^{*}s^{n+1},\ z_i^{n+1} = z_i^{n}\ ,\ (i = 1..k)$
6. *Form Residual Equations:*
   $F(s, z_1, ..., z_k) = s^{n+1} - {}^{*}s^{n+1} + 3\mu\Delta t f(s, z_1, ..., z_k)$
   $G_i(s, z_1, ..., z_k) = z_i^{n+1} - z_i^{n} - \Delta t g(s, z_1, ..., z_k)\ ,\ (i = 1..k)$
7. *Solve residual equations in 7 by N-R*
8. *Update solution*
9. *Check For Convergence:*
   IF $F(s, z_1, ..., z_k) > \mathrm{TOL}$
   OR $G_i(s, z_1, ..., z_k)\ ,\ (i = 1..k) > \mathrm{TOL}$
   THEN goto 5
10. *Radial de-magnification of Deviator*
    $\mathbf{S}^{n+1} = {}^{*}s^{n+1}/s^{n+1}\ {}^{*}\mathbf{S}^{n+1}$
11. *Update total stress tensor*
    $\mathbf{T}^{n+1} = \mathbf{S}^{n+1} + \frac{1}{3}\mathbf{1}\ \mathrm{tr}\ {}^{*}\mathbf{T}^{n+1}$
12. *Update Jacobian for contribution to global stiffness matrix*
13. *End*

Figure 1: Unified Viscoplastic Model With Scalar Strain Rate

function models. Subscripts $3 = z$ and $2 = y$ are related to the directions taken for vertically perpendicular to the weld and parallel to motion of electric arc respectively.

3 Viscoplasticity with Scalar Parameter

We consider unified viscoplastic models without an explicitly defined yield surface. These unified models assume plasticity at all levels of strain, as opposed to the elastic-plastic and elasto-viscoplastic models, which assume that plastic flow is initiated at some level of combined stress. Such models as proposed by Anand [1] and Estrin [4] do not include the explicitly defined unloading criteria. The case of unloading results in the measure of the increment of plastic strain becoming negligible rather than being set to zero. This class of viscoplastic models is based on concept of a constitutive equation for the plastic strain rate

$$\dot{\epsilon}^p = f(s, z_1, ..., z_k) \tag{3}$$

which is related to the von Mises stress s and k scalar internal parameters $z_1, ..., z_k$, determined by the following set of evolution equations:

$$\dot{z}_i = g_i(s, z_1, ...z_k) \ , \ i = 1, ..., k \tag{4}$$

Values of stress $s(z_1, ..., z_k)$ define the solution of the stress-strain relationship rather than stress states, which lie on a yield surface. Values $s, z_1, ..., z_k$ are evaluated by formulating residual equations, satisfied at any time step of the integration process. Suppose that Euler's method is used to approximate the solutions to the evolution equations [3] and [4] we have the following $(i+1)$ residual equations [1]:

$$\begin{aligned} F(s, z_1, ..., z_k) &= s^{n+1} - {}^*s^{n+1} + 3\mu\Delta t f(s, z_1, ..., z_k) && (5) \\ G_i(s, z_1, ..., z_k) &= z_i^{n+1} - z_i^n - \Delta t g(s, z_1, ..., z_k) && (6) \end{aligned}$$

where s^* is the equivalent elastic stress predicted by ${}^*s^{n+1} = s^n + 3\mu\Delta e$, in which Δe is the increment of the equivalent total strain and μ is the shear modulus. To approximate the solution of the equations [5] and [6] at the $(n+1)th$ step the Newton-Raphson's scheme is used, where values are initially evaluated for the purely elastic strain increment. The complete procedure is presented in figure [1].
The Estrin-Mecking theory [4] with two internal parameters is the example of the viscoplasticity theory with scalar internal parameters. The constitutive law for this case has the form

$$\dot{\epsilon}^{in} \equiv f(s, z_1, z_2) = C_0 {z_2}^2 \left(\frac{|s|}{z_1}\right)^{\frac{A}{\theta}} \text{sign}(s) \tag{7}$$

The evolution equations for the internal parameters are

$$\begin{aligned} \dot{z}_1 &\equiv g_1(s, z_1, z_2) = \left[C_2 \frac{{z_2}^2}{z_1} + \left(C_3 \frac{1}{z_1} + C_4\right) - C_5 z_1\right] |\dot{\epsilon}^{in}| && (8) \\ \dot{z}_2 &\equiv g_2(s, z_1, z_2) = \left[C_1 \frac{{z_1}^2}{{z_2}^3} - C_2 z_2 - \left(C_3 \frac{1}{z_2} + C_4 \frac{z_1}{z_2}\right)\right] |\dot{\epsilon}^{in}| && (9) \end{aligned}$$

where θ is the absolute temperature, C_i, $i = 1, ..., 5$, A and B are parameters for the Estrin model.
In this case the residual equations are

$$\begin{aligned} F &\equiv s^{n+1} - {}^*s^{n+1} + 3\mu\Delta t f(s, z_1, z_2) = 0 && (10) \\ G_1 &\equiv z_1^{n+1} - z_1^n - \Delta t g_1(s, z_1, z_2) = 0 && (11) \\ G_2 &\equiv z_2^{n+1} - z_2^n - \Delta t g_2(s, z_1, z_2) = 0 && (12) \end{aligned}$$

1. *Begin*
2. *Form Deviatoric Strain Increment :*
   $\Delta\mathbf{E} = \Delta\mathbf{L} - \frac{1}{3}\mathrm{tr}\Delta\mathbf{L}$
3. *Calculate Elastic Predictor:*
   ${}^{*}\mathbf{S}^{n+1} = \mathbf{S}^{n} + 2\mu\Delta\mathbf{E}^{n+1}$

4 *Initialize:*
   $\mathbf{S}^{n+1} = {}^{*}\mathbf{S}^{n+1}, \; \mathbf{Z}^{n+1} = \mathbf{Z}^{n}$

5. *Form Residual Equations:*
   $\mathbf{F}(\mathbf{S},\mathbf{Z}) = \mathbf{S}^{n+1} - {}^{*}\mathbf{S}^{n+1} + 2\mu\Delta t\mathbf{f}(\mathbf{S},\mathbf{Z})$
   $\mathbf{G}(\mathbf{S},\mathbf{Z}) = \mathbf{Z}^{n+1} - \mathbf{Z}^{n} - \Delta t\mathbf{g}(\mathbf{S},\mathbf{Z})$
6. *Solve residual equations in 5 by N-R*
7. *Update solution*
8. *Check For Convergence:*
   IF $\mathbf{F}(\mathbf{S},\mathbf{Z}) > \mathrm{TOL}$
   OR $\mathbf{G}(\mathbf{S},\mathbf{Z})) > \mathrm{TOL}$
   THEN goto 5
9. *Update total stress tensor*
   $\mathbf{T}^{n+1} = \mathbf{S}^{n+1} + \frac{1}{3}\mathbf{1}\ \mathrm{tr}\ {}^{*}\mathbf{T}^{n+1}$
10. *Update Jacobian for contribution to global stiffness matrix*
11. *End*

Figure 2: Unified Viscoplastic Robinson's Model

where ${}^{*}s$ is the elastic predictor of the von Mises stress norm. The Newton-Raphson method coupled with Gauss elimination and partial pivoting technique is used to solve the set of non-linear residual equations [10], [11], [12].
The contribution of the Estrin-Mecking model to the Finite Element stiffness matrix is defined by the Jacobian $\frac{\partial\Delta\mathbf{T}}{\partial\Delta\mathbf{L}}$, which has the form

$$\frac{\partial}{\partial\Delta\mathrm{L}}\left(\mathrm{T}^{*} - \sqrt{6}\mu\Delta t\left(\mathrm{N}\otimes\frac{\partial\dot{\epsilon}^{in}}{\partial\Delta\mathrm{L}} + \dot{\epsilon}^{in}\frac{\partial\mathrm{N}}{\partial\Delta\mathrm{L}}\right)\right) \tag{13}$$

where S is the deviatoric stress tensor, $\|\mathsf{S}\| = \sqrt{\mathsf{S}:\mathsf{S}}$, the normal vector is $\mathrm{N} = \mathsf{S}/\|\mathsf{S}\|$.

4 Robinson's Model

The constitutive equation for the inelastic strain rate for Robinson's model [2] is governed by a flow law of the following form :

$$2\bar{\mu}\dot{\epsilon}_{ij}^{in} = \begin{cases} f(F)\Sigma_{ij}; & F > 0 \quad\quad\quad\quad\;\; \text{and } S_{ij}\Sigma_{ij} > 0 \\ 0; & F \leq 0 \text{ or } F > 0 \;\; \text{and } S_{ij}\Sigma_{ij} \leq 0 \end{cases} \tag{14}$$

where the flow potential takes the form $f(F) = (J_2/\kappa^2 - 1)^n / J_2{}^{1/2}$, $F = J_2/\kappa^2 - 1$ and the effective stress is $\Sigma_{ij} = S_{ij} - Z_{ij}$ in which S_{ij} is the deviatoric stress.

The evolution equation for the internal parameter tensor Z_{ij} has the form

$$\dot{Z}_{ij} = h(Z_{kl})\dot{\epsilon}_{ij} - r(Z_{kl})Z_{ij} \tag{15}$$

Particular forms for the functions $h(Z_{ij})$ and $r(Z_{ij})$ are taken as:

$$h(Z_{ij}) = \begin{cases} \frac{2\mu H}{G^\beta}; & G > G_0 \text{ and } S_{ij}Z_{ij} > 0 \\ \frac{2\mu H}{G_0^\beta}; & G \leq G_0 \text{ or } S_{ij}Z_{ij} \leq 0 \end{cases} \tag{16}$$

$$r(Z_{ij}) = \begin{cases} \frac{RG^{m-\beta}}{\sqrt{I_2}}; & G > G_0 \text{ and } S_{ij}Z_{ij} > 0 \\ \frac{RG_0^{m-\beta}}{\sqrt{I_2}}; & G \leq G_0 \text{ or } S_{ij}Z_{ij} \leq 0 \end{cases} \tag{17}$$

where J_2 is the second invariant of the effective stress, I_2 is the second invariant of the tensor of internal parameters and $G = \sqrt{I_2}\kappa^{-1}$. The parameters μ, H, β, m and G_0 are temperature independent constants, and R and $\bar{\mu}$ are temperature dependent constants defined in [2]. Here κ is the scalar threshold stress, which accounts for the isotropic hardening or softening and reflects the 'radius of the yield surface'. As soon as J_2 goes below κ, the inelastic strain rate $\dot{\epsilon}_{ij}^{in}$ vanishes. The inequalities in equations [14], [16] and [17] define boundaries across which the growth and flow laws change form discontinuously. The residual equations used for evaluation of the stress and internal variable tensors are given by:

$$\mathbf{F} = \mathbf{S}^{n+1} - \mathbf{S}^* + 2\mu\Delta t\bar{\mu}^{-1}\mathbf{f}(\mathbf{S},\mathbf{Z}) \tag{18}$$

$$\mathbf{G} = \mathbf{Z}^{n+1} - Z^n - \Delta t\mathbf{g}(\mathbf{S},\mathbf{Z}) \tag{19}$$

where $\mathbf{g}(\mathbf{S},\mathbf{Z}) = h(\mathbf{Z})\mathbf{f}(\mathbf{S},\mathbf{Z}) - r(\mathbf{Z})\mathbf{Z}$; $\mathbf{f}(\mathbf{S},\mathbf{Z}) = 0.5\bar{\mu}^{-1}\bar{F}\ \mathbf{n}$, and $\mathbf{S}^*$ is the elastic predictor of deviatoric stress $\mathbf{S}$, $\mathbf{Z}$ is the internal variable tensor, $\bar{F} = (J_2/\kappa^2 - 1)^n 2^{1/2}$.
The contribution $\hat{\mathbf{D}} = \frac{\partial \Delta \mathbf{T}}{\partial \Delta \mathbf{L}}$ of Robinson's constitutive equation to the Finite Element stiffness matrix is given by

$$\hat{\mathbf{D}} = \hat{\mathbf{C}} - \Delta t \frac{2\mu^2}{\bar{\mu}} \frac{\hat{\mathbf{I}}_{dev}}{\|\Sigma\|} \otimes \left[f(F) J_2{}^{\frac{1}{2}} \left(\hat{\mathbf{I}} - \frac{\Sigma \otimes \Sigma}{\|\Sigma\|^2} \right) + F' \Sigma \otimes \Sigma \right] \tag{20}$$

where $F' = \frac{n}{\kappa^2}\left(\frac{J_2}{\kappa^2} - 1\right)^{n-1}$, $\hat{\mathbf{C}} = \frac{\partial \mathbf{T}^*}{\partial \mathbf{L}}$ is the usual elastic tangent modulus, $\Sigma = \mathbf{S} - \mathbf{Z}$ is the effective stress, and the deviatoric "unit" tensor is $\hat{\mathbf{I}}_{dev} = \hat{\mathbf{I}} - \frac{1}{3}\mathbf{1} \otimes \mathbf{1}$, in which $\mathbf{1}$ is the 2nd rank unit tensor.
The complete procedure for updating of the stress and internal variable tensors and Jacobian, which is a contribution to the global Finite Element stiffness matrix, is presented in fig.2.

1. *Begin*
2. *Form entire element strain vector*
   2.1 *For each material point, k, form:*
   $\mathbf{B}_L \leftarrow (H_{,x} \leftarrow (J^{-1}, H_{,r}))$
   2.2 *Form:*
   $\mathbf{L}_k = {}_k\mathbf{B}_L\mathbf{u}_k$
   2.3 *Form:*
   $\mathsf{L}_e = [\mathbf{L}_1 ... \mathbf{L}_k]^T$
3 *Form element elastic predicted stress*
   $\mathsf{T}_e = [\mathbf{T}_1 ... \mathbf{T}_k]^T \,,\; \mathbf{T}_k = \mathbf{T}_k^{n-1} + \hat{\mathbf{C}}\Delta\mathbf{L}_k$
4 *Form volume averaged deviatoric stress*
   $\langle \mathbf{S}^* \rangle = 1/(\sum_{i=1..k} w_i^g)\,[w_1^g\mathbf{S}_1 + ... + w_k^g\mathbf{S}_k]$
5. *Goto figure 4*

Figure 3: Elastic predictor for Leblond's model

5 Transformation Plasticity Models

Models of transformation plasticity incorporate material phase transformations, which is measured by the volume fraction of a phase, denoted by z; $0 \leq z \leq 1$, into calculation of the stress state. In this model, proposed by Leblond's, strain and stress measures are averaged for a volume of an entire finite element. The total strain-rate is expressed as the sum of the elastic, thermal and plastic strain rates

$$\dot{\mathsf{E}}^t = \dot{\mathsf{E}}^e + \dot{\mathsf{E}}^{thm} + \dot{\mathsf{E}}^p \tag{21}$$

The macro-plastic strain rate $\dot{\mathsf{E}}^p$ is composed of the transformation strain rate $\dot{\mathsf{E}}^{tp}$ and the classical plastic strain rate $\dot{\mathsf{E}}^{cp}$. The second rate is expressed as the sum of thermally induced, $\dot{\mathsf{E}}^{cp}_T$, and stress induced $\dot{\mathsf{E}}^{cp}_\Sigma$, therefore the plastic strain rate is given by:

$$\dot{\mathsf{E}}^p = \dot{\mathsf{E}}^{tp} + \dot{\mathsf{E}}^{cp}_\Sigma + \dot{\mathsf{E}}^{cp}_T \tag{22}$$

with components defined by

$$\dot{\mathsf{E}}^{tp} = \frac{-\Delta\epsilon^{th}_{1\rightarrow 2}}{\sigma_1^y \mathsf{E}_1^{eff}} \mathsf{S} h\left(\frac{\Sigma^{eq}}{\Sigma^y}\right) \ln(z)\dot{z} \tag{23}$$

$$\dot{\mathsf{E}}^{cp}_\Sigma = \frac{3(1-z)}{2\sigma_1^y \mathsf{E}_1^{eff}} \frac{g(z)}{E} \mathsf{S}\dot{\Sigma}^{eq} \quad \text{and} \quad \dot{\mathsf{E}}^{cp}_T = \frac{3(\alpha_1 - \alpha_2)}{\sigma_1^y \mathsf{E}_1^{eff}} z\ln(z)\mathsf{S}\dot{T} \tag{24}$$

where $\Delta\epsilon^{th}_{1\rightarrow 2} = \mathsf{E}^{thm}_2 - \mathsf{E}^{thm}_1$, and

$$\Sigma^y = [1 - f(z)]\sigma_1^y \mathsf{E}_1^{eff} + f(z)\sigma_2^y \mathsf{E}_2^{eff} \quad \text{and} \quad \Sigma^{eq} = (3/2\mathsf{S} : \mathsf{S})^{1/2} \tag{25}$$

The ultimate stress Σ^y is the maximal possible value of Σ^{eq} and S is the macro-deviatoric stress. The functions $f(z)$, $g(z)$ are defined discreetly by Leblond [6] and in order to incorporate them into the finite element code is necessary to fit curves to these functions. Yield limits of the two phases, at the particular time step, are denoted by σ_1^y, σ_2^y.
Expressions for the evaluation of E_1^{eff} and E_2^{eff} contribute to the relations for Σ^{eq} and Σ^y. When Σ^{eq} is less than Σ^y, the effective strain rates, used in calculating the macro-yield, are defined by the following equations:

$$\begin{aligned} \dot{\mathsf{E}}_1^{eff} &= -\frac{z\Delta\epsilon_{1\to 2}^{th}}{1-z} h\left(\frac{\Sigma^{eq}}{\Sigma^y}\right)(\ln z)\dot{z} + \frac{g(z)}{E}\dot{\Sigma}^{eq} + \frac{2(\alpha_1-\alpha_2)z\ln z}{1-z}\dot{T} & (26) \\ \dot{\mathsf{E}}_2^{eff} &= -\frac{\dot{z}}{z}\mathsf{E}_2^{eff} + \vartheta\frac{\dot{z}}{z}\mathsf{E}_1^{eff} & (27) \end{aligned}$$

where $\Delta\epsilon_{1\to 2}^{th} = K\sigma_1^y/2$ and K is a constant, determined experimentally for various phases. Here the thermal expansion coefficients are denoted by α_1, α_2. The correction function h is given by [6]. In the second case, when Σ^{eq} is equal to Σ^y, the macro-plastic strain rate is expressed by

$$\begin{aligned} \dot{\mathsf{E}}^p &= \dot{\mathsf{E}}^{tp} + \dot{\mathsf{E}}_\Sigma^{cp} + \dot{\mathsf{E}}_T^{cp} = (3\dot{\mathsf{E}}^{eq}/2\Sigma^{eq})\mathsf{S} & (28) \\ \dot{\mathsf{E}}_1^{eff} &= \dot{\mathsf{E}}^{eq} \quad \text{and} \quad \dot{\mathsf{E}}_2^{eff} = \dot{\mathsf{E}}^{eq} - \frac{\dot{z}}{z}\mathsf{E}_2^{eff} + \vartheta\frac{\dot{z}}{z}\mathsf{E}_1^{eff} & (29) \end{aligned}$$

where $\dot{\mathsf{E}}^{eq} = (2/3\dot{\mathsf{E}} : \dot{\mathsf{E}})^{1/2}$. The ϑ factor reflects the material recovery phenomena. The deviatoric stress at each material point is given by

$$s = (\sigma^y/\Sigma^{eq})\,\mathsf{S} \qquad (30)$$

where the yield of a two-phase mixture is $\sigma^y = z\sigma_1^y + (1-z)\sigma_2^y$, and σ_2^y is taken for the current material phase as the yield limit for pearlite, bainite or martensite. The eight residual equations, used for evaluation of the macro-deviatoric stress S, have the following form:

$$\begin{aligned} F(\mathsf{S}) &= \mathsf{S} - \langle \mathbf{S}^* \rangle_v + 2\mu\Delta t\dot{\mathsf{E}}^p \\ G &= (\mathsf{E}_1^{eff})^{n+1} - (\mathsf{E}_1^{eff})^n - \Delta t\dot{\mathsf{E}}_1^{eff} \\ H &= (\mathsf{E}_2^{eff})^{n+1} - (\mathsf{E}_2^{eff})^n - \Delta t\dot{\mathsf{E}}_2^{eff} & (31) \end{aligned}$$

The contribution $\partial\Delta\mathbf{T}/\partial\Delta\mathbf{L}$ arising from Leblond's model to the Finite Element stiffness matrix is given by

$$\frac{\partial \mathbf{T}}{\partial \mathbf{L}} = \begin{cases} \frac{\partial}{\partial \mathbf{L}}\left(\mathbf{T}^* - 2\mu\Delta t\left\{\dot{\mathsf{E}}^{tp} + \dot{\mathsf{E}}_\Sigma^{cp} + \dot{\mathsf{E}}_T^{cp}\right\}\right) & \text{if } \Sigma^{eq} < \Sigma^y \\ \frac{\partial}{\partial \mathbf{L}}\left(\mathbf{T}^* - 2\mu\Delta t\left\{(3\dot{\mathsf{E}}^{eq}/2\Sigma^{eq})\mathsf{S}\right\}\right) & \text{if } \Sigma^{eq} = \Sigma^y \end{cases} \qquad (32)$$

The phase fraction is determined by the relation: $z = 1 - \sum_p y_p$, where y_p are defined by the Johnson-Mehl-Avrami and the Koisten- Marburger models [3].

1. *Initialize Macroscopic Deviator* $S = \langle S^* \rangle_v$
2. *Form Equivalent Macro Stress and Equivalent Strain*
   $\Sigma^{eq} = (3/2 S : S)^{0.5}$, $E^{eq} = (2/3 E : E)^{0.5}$
3. *Recover internal variables and initialize effective strains*
   $(E_1^{eff})^{n+1} = (E_1^{eff})^n$, $(E_2^{eff})^{n+1} = (E_2^{eff})^n$
4. *Form:* Σ^y
5. *Form:* E^p
   $\Delta E^p = \Delta t(\dot{E}^p + \dot{E}_\Sigma^{cp} + \dot{E}_T^{cp})$ if $\Sigma^{eq} < \Sigma^y$
   $\Delta E^p = \Delta t (3/2)\dot{E}^{eq}/\Sigma^{eq} S$ if $\Sigma^{eq} = \Sigma^y$
6. *Form eight Residual Equations*
   $F = S - \langle S^* \rangle_v + 2\mu E^p$
   $G = (E_1^{eff})^{n+1} - (E_1^{eff})^n - \Delta t \dot{E}_1^{eff}$
   $H = (E_2^{eff})^{n+1} - (E_2^{eff})^n - \Delta t \dot{E}_2^{eff}$
7. *Perform N-R step on the eight equations in 6.*
8. *Update solution of macroscopic stress tensor and internal variables*
9. *Check for convergence*
   IF $F(S, E_1^{eff}, E_2^{eff})$ or $G(S, E_1^{eff}, E_2^{eff})$
   or $H(S, E_1^{eff}, E_2^{eff}) > TOL$ GOTO 2
10. *Calculate deviatoric stress by* [30]
11. *Form k contributions to global stiffness*

Figure 4: Plastic solution for Leblond's model

6 Welding Problem and Conditions

The welding bench-mark problems for the Estrin-Mecking, Robinson and Leblond models are formulated for the butt welded plates with a lengthwise single V-groove. Due to symmetry one plate of 75×25×15 mm is considered. The shape of the V-groove root is standard. The finite element model of the plate is composed of 800 elements and 1134 nodes. The thermal boundary conditions for heat input, considered here, simulate fusion welding welding in air such as gas-metal-arc welding i.e. GMA or MIG process.

The welding arc is modelled by a travelling Gaussian distribution:

$$f_\theta^{arc} = \frac{\phi \eta V I}{2\pi \delta_{arc}^2} \exp(-r^2/2\delta_{arc}^2) \tag{33}$$

where V is the voltage and I is the current of the arc, δ_{arc} is the standard deviation of this distribution, which represents the thermal "impression" of the weld electrode, η is the heating efficiency of an electric arc, ϕ is the net fraction of heat input, and r is the horizontal radial distance from the weld centre. The molten pool is roughly assumed to be a moving ellipsoid, whose projection onto welding surface is 5.32 mm long and 4.28 mm wide. The power of the welding arc transmitted to the weld is 560 W, from a welding current of I=20 A, voltage V=28 V with an efficiency, η, of 0.68. The electric arc covers the distance l_v=43 mm with the speed v=8.5 mm/s.

The non-dimensional convection coefficient, defined by an empirical relation [5], which arises from consideration of the boundary layer motion of a welded plate is given by

$$h_c = \bar{N}_u \, k_{air} \tag{34}$$

where the Nusselt number is defined by $\bar{Nu} = C_c Gr_f^m Pr_f^m$, and Gr_f,Pr_f are the Grashof and Prandtl numbers evaluated at the film temperature $\theta_f = \theta_\infty + \theta_w/2$, and k_{air} is the conductivity of the boundary layer. Values of these quantities, for the different surfaces of the plate, are as follows:
a) upper surface: $Gr_f = 2 \times 10^4$, $Pr_f = 8 \times 10^6$, $C_c = 0.54$, $m = 0.25$,
b) lower surface: $Gr_f = 10^5$, $Pr_f = 10^{11}$, $C_c = 0.27$, $m = 0.25$,
c) vertical surfaces: $Gr_f = 10^9$, $Pr_f = 10^{13}$, $C_c = 0.1$, $m = 0.333$.
The heat flux due to convection is $q_c = h_c(\theta_f - \theta_\infty)$, where θ_f, θ_∞ are the film and environmental temperatures respectively. The radiation coefficient is given by $h_r = \beta_s \epsilon_r (\theta^4 - \theta^4_{sink})$, where β_s is the Boltzmann constant, ϵ_r is the emissivity and θ_{sink} is the sink temperature.

7 Residual stresses in butt welded plates

The contours of the Von Mises norm of residual stresses,for the three constitutive models with internal parameters proposed by: a) Leblond, b) Estrin-Mecking and c) Robinson, have been shown in figure [5]. The contours are taken for the material data appropriate for high-strength low alloy steel HSLA, after cooling of the weld bath of butt welded plates, from the solidification temperature. The Leblond model [6] produces the highest residual stress $max||\mathsf{S}||$=335 MPa and contour imperfection inside the weldment. The Estrin-Mecking model generates the residual stresses closest to the J_2 elastic-plastic theory and $max||\mathsf{S}||$=316 MPa. The shapes of the contours are smooth and slightly different than for the case of the Leblond model. Robinson's model produces the lowest residual stresses with $max||\mathsf{S}||$=268 MPa. All models produces residual stresses less than the static yield limit (360 MPa at $\theta < 100^oC$). On the basis of a few numerical tests and without experimental verification of the numerical results it is impossible to evaluate which model predicts the more realistic situation. However, more can be said about the given results, after analysis of the sensitivity of residual stresses on constitutive modelling.

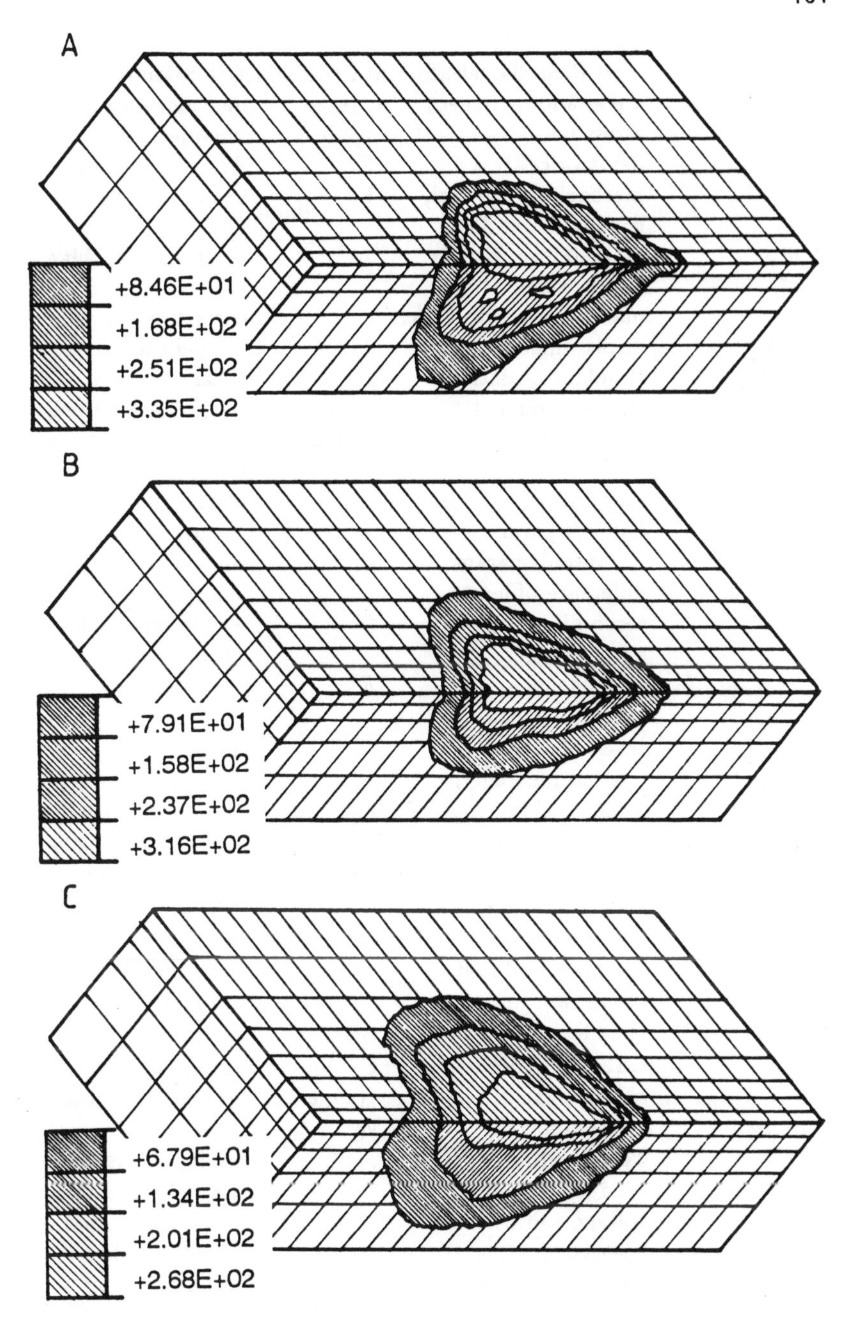

Figure 5: The Mises invariant of stress tensor in the butt welded plate for the constitutive models of material proposed by a) Leblond, b) Estrin & Mecking and c) Robinson.

$\tilde{\mathbf{S}}$, $\tilde{\mathbf{S}}^{\Delta}$	2-nd Piola-Kirchhoff stress and increment	$\tilde{\mathbf{L}}$, $\tilde{\mathbf{L}}^{\Delta}$	Green-Lagrange strain and increment
$\mathbf{T}$	stress tensor	$\mathbf{L}_{\nu}$, $\mathbf{L}$	nonlinear, linear strain
$\mathbf{S}$, S_{ij}	deviatoric stress	$\mathbf{E}$	deviatoric strain
$\mathbf{K}_u$, $\mathbf{K}_{\theta}$	mechanical, thermal,	$\mathbf{F}_u$, $\mathbf{F}_{\theta}$	external loads, heat
$\mathbf{K}_k$	conduction stiffness		flux
$H_{,x}$	grad. interp. matrix	J^{-1}	inverse of Jacobian
w_i^g	Gauss weight		
S	macro-deviatoric stress	h	correction function
u,$\Delta\mathbf{u}$	displacement	θ, $\Delta\theta$	temperature
δ_r, δ_{arc}	standard deviation	μ	shear modulus
Δe^{n+1}	strain increment	s, ${}^{*}s^{n+1}$	stress, elastic predictor
z_i	internal param.	$f(.)$, $g_i(.)$	evolution functions
t	time	$F(.)$, $G(.)$	residual functions
$\dot{\epsilon}^p$, $\dot{\epsilon}^{in}$	strain rates	e, Δe	total strain
C_i, A, B	Estrin's parm.	$\mathbf{N}$	normal vector
$\bar{\mu}$	Robinson's param.	Σ_{ij}	effective stress
$\Delta\epsilon_{1\to 2}^{th}$	difference of length	Z, Z_{ij}	param. tensor
J_2	second invariant of effective stress	I_2	second invariant of param. tensor
H, β, m, G_0	θ independent Robinson's const.	R	θ dependent Robinson's const.
κ	threshold stress	$\hat{\mathbf{C}}$	elastic modulus
$h(.)$, $r(.)$	int. param. funct.	$\bar{F}$	$= (J_2/\kappa^2 - 1)^n 2^{1/2}$
$f(F)$	$= (J_2/\kappa^2 - 1)^n / {J_2}^{1/2}$	$\hat{\mathbf{I}}_{dev}$	deviatoric tensor
z	phase vol. fraction	$\mathbf{B}_L$	linear strain-displa -cement matrix
L_e	$= [\mathbf{L}_1 ... \mathbf{L}_k]^T$		
$\dot{\mathsf{E}}^t$, $\dot{\mathsf{E}}^e$	total and elastic,	$\dot{\mathsf{E}}^{thm}$, $\dot{\mathsf{E}}^p$	thermal and plastic,
$\dot{\mathsf{E}}^{tp}$	transformation,	$\dot{\mathsf{E}}^{cp}$	classical plastic,
$\dot{\mathsf{E}}_T^{cp}$	thermally induced strain rate	$\dot{\mathsf{E}}_{\Sigma}^{cp}$	stress induced strain rate
σ_i^y	yield limit	Σ^y	mixture yield
Σ^{eq}	$= (3/2\mathsf{S} : \mathsf{S})^{1/2}$	E_i^{eff}	hardening param.
ϑ	recovery factor	$f(z)$, $g(z)$	Leblond's functions
α_1, α_2	expansion coeff.	y_k	phase fraction
f_{θ}^{arc}	welding arc	ϕ, η	heating efficiency
k_{ij}, k_{mean}	conductivity	r	distance from a weld
V, I	voltage and current	h_c, h_r	convection, radiation
k_{air}	layer conduct.	$\bar{N}_u$	Nusselt number
Gr_f^m	Grashof number	Pr_f^m	Prandtl number

References

[1] Lush, A.M., Weber, G., Anand, L., An implicit time- integration pro-

cedure for a set of internal variable constitutive equations for isotropic elasto-viscoplasticity, International Journal of Plasticity , Vol. 5, pp. 521-549, 1989

[2] ARYA V.K., KAUFMAN, A. - Finite element implementation of Robinson's unified viscoplastic model and its application to some uniaxial and multiaxial problem, Eng. Computations, 1989, Vol. 6

[3] DENIS S., FARIAS D., SIMON A. - Mathematical Model Coupling Phase Transformations and Temperature Evolutions in Steels. ISIJ International Vol 32 (1992) No 3 p316-325

[4] ESTRIN, Y. A unified constitutive model based on dislocation density evolution. High Temperature Constitutive Modelling Theory and Application Ed. A.D. Freed and K.P. Walker. ASME 1991 p. 65.

[5] HOLMAN, J.P., - Heat Transfer. 1989

[6] LEBLOND J.B., MOTTET G., DEVAUX J. C. - A Theoretical and Numerical Approach to the Plastic Behaviour of Steels during Phase Transformations of General Relations–I Derivation of General Relations. Journal of the Mechanics and Physics of Solids Vol. 34 No 4., pp. 395-409, 1986

[7] LEBLOND J.B., MOTTET G., DEVAUX J.C. A Theoretical and Numerical Approach to the Plastic Behaviour of Steels during Phase Transformations of General Relations–II Study of Classical Plasticity for Ideal-Plastic Phases. Journal of the Mechanics and Physics of Solids Vol. 34 No 4., pp. 411-432, 1986.

[8] LEBLOND J. B., DEVAUX J., DEVAUX J.C., Mathematical Modelling of Transformation Plasticity in Steels I: Case of Ideal-Plastic Phases. International Journal of Plasticity Vol 5 pp-573-591, 1989.

[9] LEBLOND J.B., DEVAUX J., DEVAUX J.C., Mathematical Modelling of Transformation Plasticity in Steels II: Coupling with Strain Hardening Phenomena. International Journal of Plasticity, Vol 5 pp-573-591, 1989.

[10] OLIVER G.J. Mathematical modelling of Welding in Phase Transforming Metals with Finite Elements M.Sc. Thesis University of Cape Town 1993.

[11] PARDO E. ,WECKMAN D.C Prediction of Weld Pool and Reinforcement Dimensions of GMA Welds Using a Finite-Element Model, Metallurgical Transactions B Vol: 20B,pp. 937-946 December 1989

SECTION 2

HEAT CONDUCTION

ADVANCES IN SOLVING INVERSE HEAT CONDUCTION PROBLEMS BY BEM

Kazimierz Kurpisz[(1)] and Andrzej J. Nowak[(1)]

SUMMARY

Analysis of inverse heat conduction problems generally requires two steps. First, assuming certain boundary heat fluxes, temperature field is determined. In the second step calculated temperatures are compared with experimental measurements and boundary fluxes are corrected according to observed errors. In this paper the Boundary Element Method is applied in the first step of the analysis and a combination of function specification method (FSM) with the regularization method (RM) in the second one. Such approach combines the simplicity of FSM and efficiency of RM.

Unknowns are determined sequentially step by step. Required sensitivity coefficients, usually calculated separately, are proposed to be determined directly from Boundary Element influence matrices built for each time step. Thus, computational effort has been reduced considerably.

1. INTRODUCTION

The problem considered in the paper is the estimation of the temperature and boundary heat fluxes as a functions of time from transient temperature measurements at interior locations of a heat conducting body. Such a problem belongs to inverse heat conduction problems that are referred to as ill-posed. It means that they are normally sensitive to measurement errors. In order to obtain stable and accurate results special numerical techniques have to be employed.

In general, any solution method for inverse problems requires two procedures. The first one is to solve a direct boundary problem, or more precisely, it should allow to derive the relationship between known values

[1]Institute of Thermal Technology, Technical University of Silesia, Konarskiego 22, 44-101 Gliwice, Poland

of the temperature and unknown values of the heat flux. Such a relationship is in fact the essence of the Boundary Element Method (BEM). All unknowns in an inverse problem are associated with the boundary only and using BEM one can take advantage of very simple numerical meshes required by this technique. Contrary to other numerical methods, BEM does not need any internal cells and location of internal points where temperature is measured can be chosen in a quite arbitrary way. For the sake of simplicity, in the formulation applied in this paper, constant boundary elements are used when integral equation is discretized.

The second procedure employed when solving ill-posed inverse problem is to stabilize results. There exist numerous approaches but regarding heat conduction the following two have become the most popular:

- function specification technique, [1]
- regularization method, [2]

The function specification method (FSM) provides stable and accurate results for a wide range of one-dimensional and for some multidimensional cases. Application of the method for a two-dimensional body was shown in [3]. It has been however found that there is a number of cases the method is not efficient. On the other hand, the regularization method (RM) seems to be much more efficient, in particular, when it comes to multidimensional cases. This approach, based on BEM procedure was described in [4]. It should however be stressed that the regularization method needs much more CPU time than the function specification method. Thus, it seems reasonable to combine the simplicity of FSM and efficiency of RM. Such an approach was shown for the zeroth-order regularization in a semi-infinite body, [5]. For numerically solved multidimensional cases the procedure was investigated in [6, 8]. It has been found that the combination of FSM and RM in addition to BEM is much more efficient than the original FSM and offers stable results for much wider range of inverse problems still preserving the simplicity of FSM.

Function Specification Method requires that the direct relationship (*e.g.* in matrix form) between temperatures and heat fluxes is known. Thus, the concept of sensitivity coefficients has been widely utilized [1]. Sensitivity coefficients depend on the geometry of the body and usually are calculated separately and stored for future use. In this paper we investigate the possibility of calculating sensitivity coefficients very cheaply making use the boundary elements influence matrices. Such approach reduces computational effort considerably.

2. PROBLEM FORMULATION

The temperature field inside the heat conducting body can be described by the following boundary problem:

- differential equation

$$\frac{\partial T(\mathbf{r},t)}{\partial t} = a\nabla^2 T(\mathbf{r},t) \qquad (1)$$

where T is the temperature at point $\mathbf{r}$ at time t and a denotes the diffusivity coefficient

- boundary conditions
External surface S of the body is divided into two parts S_A and S_B. It is assumed that along part S_A boundary condition is known, whereas along part S_B boundary condition is unknown. It is chosen in the form of the second kind because the heat flux is more difficult to be calculated accurately than the surface temperature. In addition, when knowing the heat flux it is easy to determine the temperature distribution and finally a convective heat transfer coefficient, if needed (provided the ambient temperature is known). The known boundary condition can be chosen in any form and this does not introduce any new difficulty.

$$-\lambda\frac{\partial T}{\partial n} = q_A(\mathbf{r},t) \qquad \mathbf{r}\rightarrow\mathbf{r}_A\epsilon S_A \qquad (2)$$

$$-\lambda\frac{\partial T}{\partial n} = q(\mathbf{r},t) \qquad \mathbf{r}\rightarrow\mathbf{r}_B\epsilon S_B \qquad (3)$$

where λ stands for thermal conductivity, n is outward normal to the boundary, $q_A(\mathbf{r},t)$ and $q(\mathbf{r},t)$ represent known and unknown heat fluxes on the surface S_A and S_B, respectively.

- initial condition

$$T(\mathbf{r},t) = T_o(\mathbf{r}) \qquad t\rightarrow 0 \qquad (4)$$

The thermal diffusivity and conductivity are assumed to be known and constant. Since for inverse problem the boundary condition is not specified for a part or, in particular, along the whole external surface of the body, the mathematical description of the boundary value problem has to be completed by some measured temperature histories at interior locations at discrete times

$$T(\mathbf{r}_i,t_k) = U_{i,k} \qquad i=1,2,\ldots,I \text{ and } k=1,2,\ldots,M \qquad (5)$$

where number I is the total number of interior locations and M stands for the total number of discrete times. The values $U_{i,k}$ are referred to as internal temperature responses. The objective is to estimate the unknown heat flux at locations $\mathbf{r}_j$ along the surface S_B and at discrete times t_k

$$q_{j,k} = q(\mathbf{r}_j,t_k), \qquad j=1,2,\ldots,J \text{ and } k=1,2,\ldots,M. \qquad (6)$$

Because of the discrete nature of equation(6) the problem can be solved in a sequential manner. It is assumed that the temperature distribution and heat flux components are known at times $t_{k-1}, t_{k-2}, \ldots$ and it is desired to determine the heat flux components at time t_k.

3. REGULARIZATION PROCEDURE

The regularization method was developed by Tikhonov and Arsenin [2] to improve the conditioning of ill-posed problems. The method modifies the least squares approach by adding terms that are intended to reduce excursions in the unknown functions such as the boundary heat flux. The method has a number of forms that are referred to as, *e.g.* the zeroth-order, the first-order and the second-order [2]. This results from a kind of bounds that is prescribed for the solution. If the problem is solved at $t = t_{k-1}$ and it is desired to solve the problem at $t = t_k, t_{k+1}, \ldots, t_{k+R-1}$, *i.e.* for *R future steps* the function Δ (representing a measurement error) to be minimized can be written as

$$
\begin{aligned}
\Delta &= \sum_{i=1}^{I} \sum_{r=1}^{R} \left(U_{i,k+r-1} - T_{i,k+r-1} \right)^2 + \\
&+ \gamma \sum_{j=1}^{J} \left[\sum_{r=1}^{R} q_{j,k+r-1}^2 + \sum_{r=1}^{R-1} \left(q_{j,k+r} - q_{j,k+r-1} \right)^2 + \right. \\
&+ \left. \sum_{r=1}^{R-2} \left(q_{j,k+r+1} - 2\, q_{j,k+r} + q_{j,k+r-1} \right)^2 \right] \qquad (7)
\end{aligned}
$$

The last three terms refer to the adequate regularization order and the γ number is called the regularization parameter. A number of approaches can be applied to select the parameter γ. The method used herein is the quasioptimum method that consists in minimizing the following Euclidian norm [2]

$$
\left\| \gamma \frac{d\,\mathbf{Q}^{\gamma}}{d\gamma} \right\| \qquad (8)
$$

where $\mathbf{Q}^{\gamma}$ is a vector of heat flux components for the specified γ value.

4. FUNCTION SPECIFICATION PROCEDURE

It can be seen from equation (7) that the regularization method requires specification of heat flux components at least at t_k for the regularization of zeroth-order, t_k, t_{k+1} for the first order and t_k, t_{k+1}, t_{k+2} for the second order. As regards the classical regularization method the heat flux components $q_{j,k+r-1}$ where $r = 1, 2, \ldots, R$ are to be determined simultaneously. Thus, the number of unknowns is equal to $J \cdot R$. For multidimensional cases such an approach implies a large set of algebraic equations to be solved. Moreover, in order to obtain overdetermined problem, which improves the stability of calculations, the number of nodal points I, at which temperature measurements are taken should be greater than the number of boundary elements J. It is worth stressing that introducing too many temperature sensors may strongly interfere the temperature distribution within the body. These difficulties can be avoided by employing the function specification method.

The method consists in assuming the functional form of heat flux for future time steps. For the simplest formulation of the method a temporary assumption that the heat flux is constant over R future steps is made, cf. [1, 3]. As a result, the number of the unknowns is reduced back to J. However, this assumption has to be modified for the higher order regularization method. Therefore, the linear or square form of heat flux should be specified for future time steps. All these possibilities have been investigated in [8].
In general, the heat flux components are described by the function

$$\begin{aligned} q_{j,k+r-1} &= \left[a_o + a_1(r-1) + a_2 \frac{r(r-1)}{2} \right] q_{j,k} + \\ &+ a_2 \frac{q_{j,k-2} - 2\,q_{j,k-1}}{2} r(r-1) - a_1\,(r-1)\,q_{j,k-1} \end{aligned} \qquad (9)$$

where $a_o = 1$ if only constant term is taken into account, $a_1 = 1$ for a linear term and $a_2 = 1$ for a square term. Otherwise, the a_p number is equal to zero.

Equation (9) is then introduced into equation (7). As a consequence, the only unknowns are $q_{j,k}$ components which means that the total number of unknowns remains equal to J for each time step. Employing $U_{i,k+r-1}$ for $r > 1$ makes the problem overdetermined. For $k = 1$ and $k = 2$ the values of $q_{j,k-1}$ and $q_{j,k-2}$ are taken zero, respectively.

The temperature field depends continuously on the unknown heat flux components. Therefore, in Function Specification Method temperature $T_{i,k}$ at location $\mathbf{r}_i$ and at time t_k is expanded into Taylor series about arbitrary but known value of heat flux $q^*_{i,k}$ [1]

$$T_{i,k} = T^*_{i,k} + \sum_{j=1}^{N} \left. \frac{\partial T_{i,k}}{\partial q_{j,k}} \right|_{q_{j,k}=q^*_{j,k}} \left(q_{j,k} - q^*_{j,k} \right) \qquad (10)$$

where $T^*_{i,k}$ is the temperature at location $\mathbf{r}_i$ and at time t_k with $q_{j,k} = q^*_{j,k}$ over the time interval $t_{k-1} < t \le t_k$. The higher derivatives in equation(10) are equal to zero due to the linearity of the considered problem. Writing equation(10) in matrix form and introducing sensitivity coefficient matrix $\mathbf{Z}$ one obtains

$$\mathbf{T}_k = \mathbf{T}^*_k + \mathbf{Z}_k \left(\mathbf{Q}_k - \mathbf{Q}^*_k \right) \qquad (11)$$

where vector $\mathbf{T}_k$ contains temperatures at time t_k at all nodal points whereas unknown boundary fluxes form vector $\mathbf{Q}_k$. Vector $\mathbf{T}_k^*$ is associated with arbitrary chosen vector $\mathbf{Q}_k^*$. The common choice [1] of vector $\mathbf{Q}_k^*$ is

$$\mathbf{Q}_k^* = \mathbf{Q}_{k-1}, \text{ and } \mathbf{Q}_1^* = \mathbf{0}. \tag{12}$$

According to the equation(11), solution of the problem at any time t_k is obtained in two steps. In the first step, temperature field is calculated for assumed boundary heat fluxes $\mathbf{Q}_k^*$. In other words, direct boundary value problem has to be solved. The Boundary Element Method is applied in this paper. The second step consists of correcting boundary heat fluxes in order to minimize objective function(7) simultaneously satisfying equation(9).

5. SENSITIVITY COEFFICIENTS

Function Specification Method uses the sensitivity coefficient matrix $\mathbf{Z}$ components of which are defined as follows

$$Z_{i,k}^{j,l} = \frac{\partial T_{i,k}}{\partial q_{j,l}} \tag{13}$$

The sensitivity coefficients are usually determined by solving a very similar boundary problem with some changes of the right hand side of conditions (2)-(4), (cf. [1, 3, 6]). Although this step of the analysis has to be performed for linear cases only once, it is generally cumbersome and time consuming. Boundary Element Method applied to solve the direct problem offers the unique opportunity to calculate sensitivity coefficients very cheaply. The proposed approach utilizes the BEM influence matrices $\mathbf{H}$ and $\mathbf{G}$ [5].

As it was indicated previously, for any time t_k BEM provides a direct relationship between boundary heat fluxes and boundary temperatures

$$\mathbf{H}_k \mathbf{T}_k = \mathbf{G}_k \mathbf{Q}_k + \mathbf{R}_k \tag{14}$$

where subscript k indicates the actual time and vector $\mathbf{R}$ represents the influence of initial condition.

Solution at all selected internal points is expressed in terms of boundary values

$$\mathbf{T}^i = -\mathbf{H}_k^i \mathbf{T}_k + \mathbf{G}_k^i \mathbf{Q}_k + \mathbf{R}_k^i \tag{15}$$

where superscript i indicates appropriate matrices for internal points.

Solving equation(14) for vector $\mathbf{T}_k$ one obtains for boundary nodes

$$\mathbf{T}_k = \mathbf{A}_k \mathbf{Q}_k + \mathbf{C}_k \tag{16}$$

Introducing now (16) into (15) the following can be obtained

$$\mathbf{T}^i = (-\mathbf{H}_k^i \mathbf{A}_k + \mathbf{G}_k^i)\mathbf{Q}_k + \mathbf{R}_k^i - \mathbf{H}_k^i \mathbf{C}_k \tag{17}$$

The equations (16) i (17) constitute two sets of relationships which can be directly used to determine sensitivity coefficients. One simply needs to collect appropriate rows and columns of $\mathbf{A}_k$ and $(-\mathbf{H}_k^i \mathbf{A}_k + \mathbf{G}_k^i)$ matrices, according to numbers of unknown boundary fluxes. Since influence matrices are already calculated this step of the analysis is rather not expensive in computing time.

6. TEST CASES AND CONCLUSIONS

The proposed technique was examined for the same test case as the simple function specification method [3] and combined function specification and regularization method [6, 8]. The numerical example refer to 2-D transient heat conduction within the region shown in Fig.1.

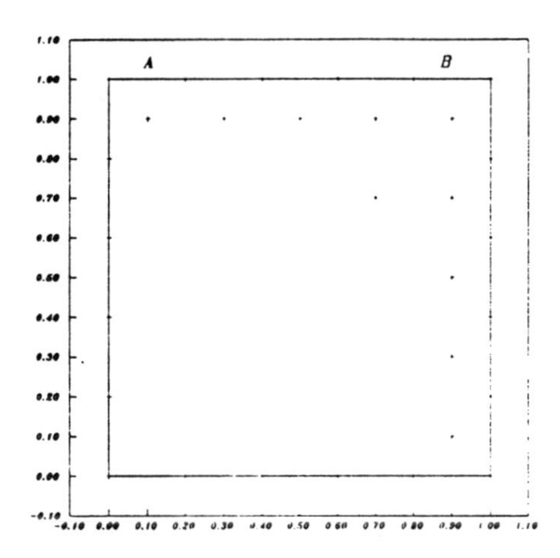

Figure 1. Model for 2-D heat conducting body showing the location of temperature sensors

The temperature histories which were necessary to solve the test case were obtained from the direct boundary problems of heat conduction. To make the situation more realistic stochastic errors were introduced into temperatures according to the equation

$$U_{i,k} = \overline{U}_{i,k} + (1 - 2\varrho)\,\delta \tag{18}$$

where $\overline{U}_{i,k}$ is the exact temperature at node $\mathbf{r}_i$ at time t_k, ϱ stands for a random value produced by a random value generator of a uniform distribution within the range [0, 1] and δ represents the maximum absolute error. The problem is formulated in a dimensionless form. Surfaces $x = 0$ and $y = 0$ are assumed to be insulated whereas the boundary conditions of the third kind are prescribed on the other surfaces. The respective Biot numbers are defined as follows

$$Bi_x = \frac{h_x \, d_x}{\lambda} \qquad Bi_y = \frac{h_y \, d_y}{\lambda} \tag{19}$$

where h_x and h_y are convective heat transfer coefficients along x and y axis, respectively, and d_x and d_y are dimensions of the body in x and y directions. Temperature sensors were located at internal points marked by + signs in Fig.1.

The obtained results are practically the same as those published in [8]. Estimated heat fluxes for the boundary element B are displayed in Fig.2 and 3. Fig.2 refers to the Biot number $Bi_x = Bi_y = 0.5$ whereas Fig.3 refers to the same Biot numbers equal to 2.5. In both cases dimensionless time step is equal to 0.05. The solid lines correspond to the exact solution derived from the direct boundary problem. Symbols are associated with different regularization orders: circles with the zeroth one, squares with the zeroth and the first one, triangles with the zeroth, the first and the second one. It should be stressed that the displayed results correspond to the substantial absolute errors δ: in Fig.2 $\delta = 0.025$, in Fig.3 $\delta = 0.01$.

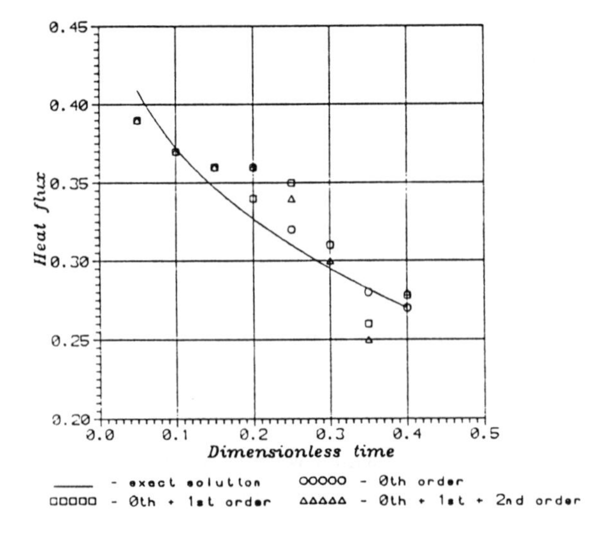

Figure 2. Estimation of the heat flux at boundary element B. Biot numbers are equal to 0.5, dimensionless time step is equal to 0.05 and absolute error is equal to 0.025

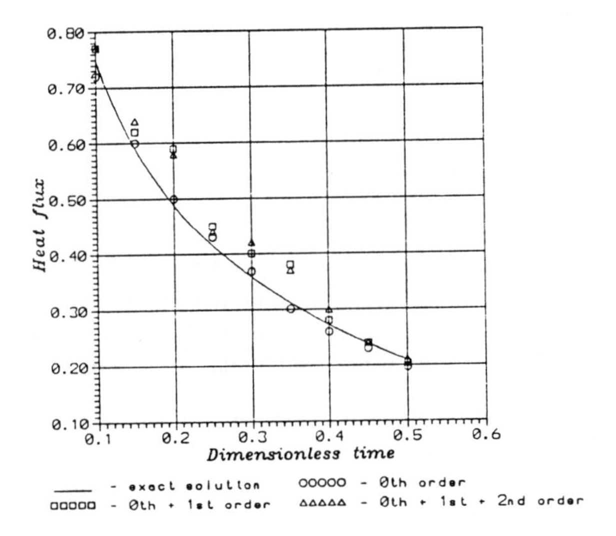

Figure 3. Estimation of the heat flux at boundary element B. Biot numbers are equal to 2.5, dimensionless time step is equal to 0.05 and absolute error is equal to 0.01

The calculations that have been carried out proved that proposed technique allows to calculate sensitivity coefficient from boundary element influence matrices. Since these matrices are necessary to solve direct problem no extra effort to create them is required. Matrix operations in equations(16) and (17) are similar (in terms of CPU time) to those operations when sensitivity coefficients are determined separately. Thus, one can say that in the proposed approach, the time for calculating influence matrices in the stage of determining sensitivity coefficients is saved. This is of great importance when nonlinear cases are solved. Then, sensitivity coefficients have to calculated for each time step and the multiplication effect in CPU time can easily be observed.

Some additional calculations are being currently carried out and their results will be presented during the conference.

7. ACKNOWLEDGEMENTS

The financial assistance of the Polish Committee of Scientific Research within the grant 3 1137 91 01 is gratefully acknowledged herewith.

8. REFERENCES

[1] Beck, J.V.,et al. - Inverse Heat Conduction, J.Wiley Intersc. Publ., New York, 1985.

[2] Tikhonov, A. and Arsenin, V. - Solutions of Ill-Posed Problems, Winston & Sons, Washington, D.C., 1977.

[3] Kurpisz, K. and Nowak A.J. - Applying BEM and the Sensitivity Coefficient Concept to Inverse Heat Conduction Problems, Proc. of the First Intern. Conf. on Advanced Computational Methods in Heat Transfer, Wrobel, L.C., et al (eds.), Springer-Verlag, 1990, pp 309-321.

[4] Pasquetti, R. and Le Niliot, G. - Boundary Element Approach for Inverse Heat Conduction Problems: Application to Transient Bidimensional Numerical Experiment, Numerical Heat Transfer, Part B, Vol. 20, 1991, pp 169-189.

[5] Beck, J.V. and Murio, D.A. - Combined Function Specification-Regularization Procedure for Solution of Inverse Heat Conduction Problems, AIAA Journal, Vol. 24, 101 (1986), pp 180-185.

[6] Kurpisz, K. and Nowak A.J. - BEM Approach to Inverse Heat Conduction Problems, Engineering Analysis with Boundary Elements, Vol.10, No.3, pp.259–266, 1992.

[7] Brebbia, C.A., et. al - Boundary Element Techniques: Theory and Applications in Engineering, Springer-Verlag, 1984.

[8] Kurpisz, K. and Nowak, A.J. - Boundary Elements and combined techniques for the analysis of the inverse heat conduction problems, Vol.1 - Heat Conduction, Radiation and Phase Change, L.C. Wrobel, C.A. Brebbia, A.J. Nowak, editors, pp 399–408, Comp. Mech. Publications and Elsevier Applied Science, 1992.

EXPLICIT AND IMPLICIT FINITE ELEMENT METHODS FOR INVERSE HEAT CONDUCTION PROBLEM

T.V. Radhakrishnan* and R.C. Mehta**
Vikram Sarabhai Space Centre, Trivandrum, India
K.N. Seetharamu+ and P.A. Aswathanarayana++
Indian Institute of Technology, Madras, India

Abstract

Estimation of unknown surface condition using known temperature history inside solid material is carried out in the present paper inconjunction with Beck's sequential method. One-dimensional transient heat conduction equation with different heating conditions is analyzed in order to get simulated temperature-time history. The behaviour of the sensitivity coefficient is investigated for two-different types of heating conditions. Explicit and implicit finite element methods were used to solve heat conduction equation. A comparative study of explicit and implicit finite element methods to the solution of inverse heat conduction problem is studied.

Introduction

The inverse heat conduction problem involves estimation of transient surface thermal characteristics by utilising known temperature history at a given location inside a solid material. The importance of the inverse heat conduction problem arises where the time-dependent heating condition is difficult to measure in actual situation.

Different mathematical techniques have been developed to analyze inverse heat conduction problem. These include graphical [1,2], polynomial [3], Laplace transform [4], finite difference [5,6], finite element methods [7,8] and series solution [9].

Of these finite difference method (FDM) and finite element method (FEM) were found to be most straight forward and offers great deal of flexibility in the analysis of the complex problem. In the explicit numerical method, the stability restriction placed on the allowable time step size makes it less efficient for prediction of large time duration. The computational time step in the explicit scheme is usually much smaller than the measured time-step. In this situation, the computational time step has to be multiplied by an integer and at each measurement time interval the computations has to be repeated so that both time intervals will coincide.

The implicit scheme is unconditionally stable for large time-step, therefore, the multiplication of computational time step with integer is not required to match the measured and calculated time step intervals.

An analytical study of sensitivity coefficient is done in order to understand its behaviour in the prediction of unknown wall condition.

The Governing Equation

The governing one-dimensional transient heat conduction equation is,

$$\frac{1}{\alpha}\frac{\partial T(x,t)}{\partial t} = \frac{\partial^2 T(x,t)}{\partial x^2} \tag{1}$$

Where, T, t and α are temperature, time and thermal diffusivity of the material respectively.

The initial and boundary conditions are,

$$T(x,0) = To \text{ for all } x \tag{2a}$$

$$-K\frac{\partial T(0,t)}{\partial x} = \text{unknown}, \quad t > 0 \tag{2b}$$

$$\frac{\partial T(L,t)}{\partial x} = 0, \quad t > 0 \tag{2c}$$

Let us now consider that thermocouple is embedded inside the region. We are interested in using the information provided by sensor. Mathematically, this means that the following information is available,

$$T(E,t) \simeq Y(t) \tag{3}$$

Where the symbol $\simeq$ indicates the approximate nature of equation (3), E is the position of the sensor, and Y(t) is the measured temperature history at E.

Now the parameter estimation problem can be written as follows,

Knowing the thermocouple measurement Y(t) at x = E. Solve the above transient heat conduction equation to determine the unknown boundary condition (2b).

Weak Formulation

We need to define a weak solution of the problem because $(\partial^2 T/\partial x^2)$ may not exist in a classic sense. In order to use Galerkin's formulation of the finite element method, we introduce the semidiscretised form of T for each element as,

$$T^e(x,t) = \sum_{i=1}^{M} N_i^e(x)\, T_i^e = [N]^T\{T\}^e \qquad (4)$$

where M is the node number of the unidimensional element, while N_i^e are the shape function as well as the test functions. For all the test functions N_i^e the weak equation for a element of length is given as,

$$\frac{1}{\alpha}\int [N] \frac{\partial\, [N]^T\{T\}}{\partial\, t}\, dx = \int [N] \frac{\partial^2 [N]^T\{T\}}{\partial\, x^2}\, dx \qquad (5)$$

Integrating by part yields the following,

$$\frac{1}{\alpha}\int [N][N]^T\, dx\, \frac{d\, T(x,t)}{d\, t}$$

$$= \{f\} - \int \frac{\partial\, [N]}{\partial\, x} \frac{\partial\, [N]^T}{\partial\, x} \{T\}\, dx \qquad (6)$$

where {f} is the load factor which can be derived according to the boundary condition required to be estimated. The simplified form at the element level can be expressed as,

$$\sum_{J=1}^{M} [C]_{I,J} \frac{d\, T}{d\, t} = [r]_I \qquad (7)$$

where $[r]_I$ is the right hand side of Eq.(6). [C] represents the standard finite element mass matrix. The subscripts I,J represent element and node numbers respectively. Equation (7) may be approximated by the use of a standard two-level θ method, so leading to a time-stepping scheme.

$$\sum_{J=1}^{M} [C]_{I,J}\left[\left\{ T_J \right\}^{n+1} - \left\{ T_J \right\}^{n}\right] = (1-\theta)\Delta t[r]_I^n + \theta\, \Delta t[r]_I^{n+1} \qquad (8)$$

where superscript n is time level.

(a) Explicit Method

If fthe value θ = 0 is substituted into Eq.(8), the time-stepping scheme reduces to

$$\sum_{J=1}^{M}[C]_{I,J}\Delta T_J = \Delta t[r]_I^n \quad (9)$$

where ΔT_J represents the increment in the temperature at node J over the time step. A fully explicit procedure can be produced by lumping mass matrix [10].

Eq.(9) is a relatively inefficient solution method for heat conduction problem because of the stability restrictions placed on the allowable time step size as

$$\left[\frac{\alpha\Delta t}{l^2}\right]_e \leq \frac{1}{2}$$

where l is element length.

(b) Implicit Method

A non-zero value of θ is used in Eq.(9), results an implicit time stepping scheme. In the present numerical algorithm, θ = 1 is used in Eq.(9).

Algorithm for the solution

The Beck procedure is to determine at each time step the unknown boundary condition that minimizes the least-square error between the calculated and measured temperatures. The least-square function is

$$F_r = \sum_{j=1}^{r}\left[T_E^{M+j} - Y^{M+j}\right]^2 \quad (10)$$

where T_E^{M+j} and Y^{M+J} are the calculated and measured temperatures at time (M+j)Δt. The number of future temperatures used in the minimization is 'r'. A solution for the unknown boundary condition is given by Beck et al. [11] is

$$q_E^{M+1} = \sum_{j=1}^{r} K_E^j \left(Y^{M+j} - \psi_E^{M,j}\right) \quad (11)$$

with $$K_E^j = \phi_E^j / \sum_{j=1}^{r}(\phi_E^j)^2$$

The symbol $\psi_E^{M,j}$ represents the decay of temperature at x=E when heat flux is set to zero for the 'r' future time steps. The ϕ s are called sensitivity coefficients and K s are called gain coefficient [11].

Calculation of Sensitivity Coefficients

The sensitivity coefficient ϕ is an important quantity because it indicates the step response T at the location E and at time $(M+j)\Delta t$. To calculate the sensitivity coefficient ϕ, let us now perturb q to a nearby value $(1+\varepsilon)q$, with ε being 10% of q, and solve the direct problem again. Then the new value of the temperature can be obtained. Then using a finite difference method, the sensitivity coefficient can be approximated.

Constant Heat Flux

Consider the transient heat conduction Eq.(1), with one surface is exposed to constant heat flux and other surfacce is insulated. The analytical solution of this problem is given as [12]

$$\frac{(T-T_o)k}{q\,L} = \frac{\alpha t}{L^2} + \frac{3(1-x/L)-1}{6} - \frac{2}{\pi^2}\sum_{m=1}^{\infty}\frac{(-1)^n}{m^2}\cos\left[m\pi\left(1-\frac{x}{L}\right)\right]\exp\left(\frac{-n^2\pi^2\alpha t}{L^2}\right) \tag{12}$$

The sensitivity coefficient can be written using Eq.(12) after simplification as

$$\phi = \frac{T(x,t) - T_o(x,0)}{q} \tag{13}$$

It can be seen from eq.(13) that the sensitivity coefficient is directly proportional to temperature, $T(x,t)$.

Convective heat-transfer Coefficient

Now the surface of the finite slab is exposed to the Newtonian type boundary conditions. The exact solution of this problem can be written as [12].

$$\left[\frac{T(x,t)-T_o}{T_\infty - T_o}\right] = 1 - 2\sum_{m=1}^{\infty}\frac{B_i}{(B_i^2+\lambda_m^2+B_i)}\,\frac{\cos[\lambda_m(1-x/L)]}{\cos\lambda_m}\exp(-\alpha t\lambda_m^2/L^2) \tag{14 a}$$

$$\lambda\tan\lambda = B_i \tag{14 b}$$

where B_i is Biot number (hL/K).

The sensitivity coefficient for higher B_i can be expressed as

$$\phi = \frac{[\cos\lambda_m]\exp(-\alpha t\lambda_m{}^2/L^2)}{2\cos[\lambda_m^2\ (1-x/L)]}\ [1-T(x,t)]^2 \tag{15}$$

It can be seen that the ϕ is a non-linear function of temperature.

Numerical Examples

The procedure to predict surface condition is by Beck's sequential procedure. Explicit and implicit finite element methods are employed to reconstruct the boundary conditions. The time step in the explicit should be carefully chosen in order to avoide numerical instability. The stability criteria for explicit scheme is $(\alpha\Delta t/l^2)_e \leq 0.5$. If time interval of measurements is larger than the computational time step, then the computational time step has to be multiplied by an integer in the explicit method. The computation has to be repeated inorder to satisfy the time interval of the measurement.

The implicit scheme is unconditionally stable, therefore larger time step is permissible in the computation of temperature.

A numerical example is worked out with following material properties.
k=35W/m K, Cp=486J/Kg K, ρ=7900Kg/m^3, To=300K and T∞=500K. The thickness of the slab is 10mm, and divided in 10 equal intervals. Tempretures are simulated at third node from the heated surface. The computation was done on IBM PC 386. Table1 shows the estimated heating values employing explicit and implicit finite element methods (FEM) at r = 2.

Table 1 Estimated Heating Condition

Fig.	q(t)	h(t)	FEM	Observations
-	linear	-	explicit	solution diverged
1	linear	-	implicit	oscillations
2	square	-	explicit	good agreement
3	square	-	implicit	good agreement
4	-	linear	explicit	oscillations
5	-	linear	implicit	oscillations
6	-	square	explicit	good agreement
7	-	square	implicit	good agreement

The results show very good agreement between

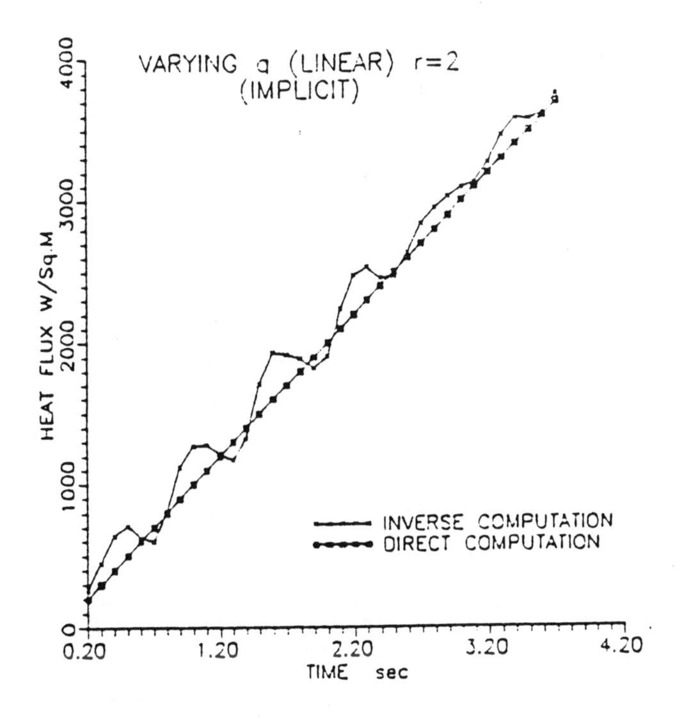

Figure 1 Estimated heat flux, r = 2

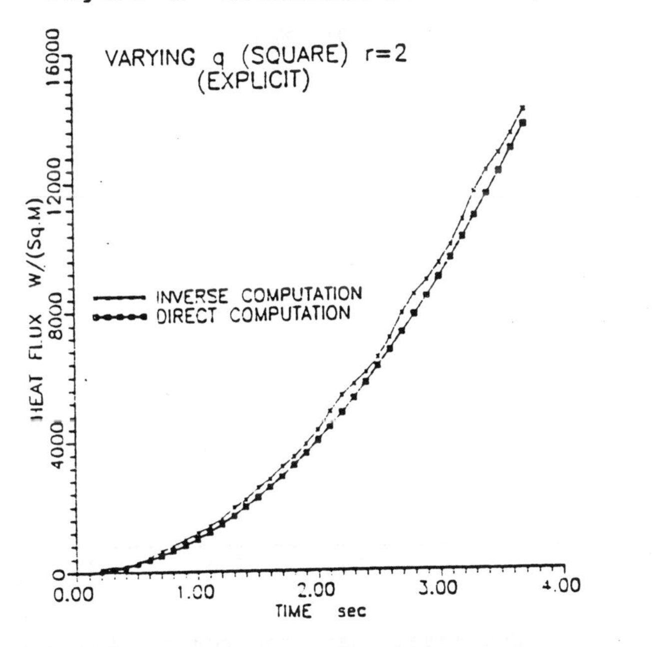

Figure 2 Estimated heat flux, r = 2

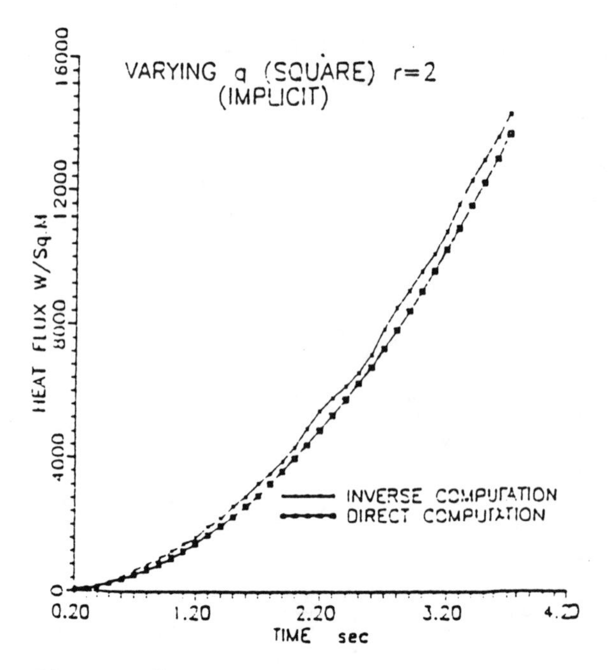

Figure 3 Estimated heat flux, r = 2

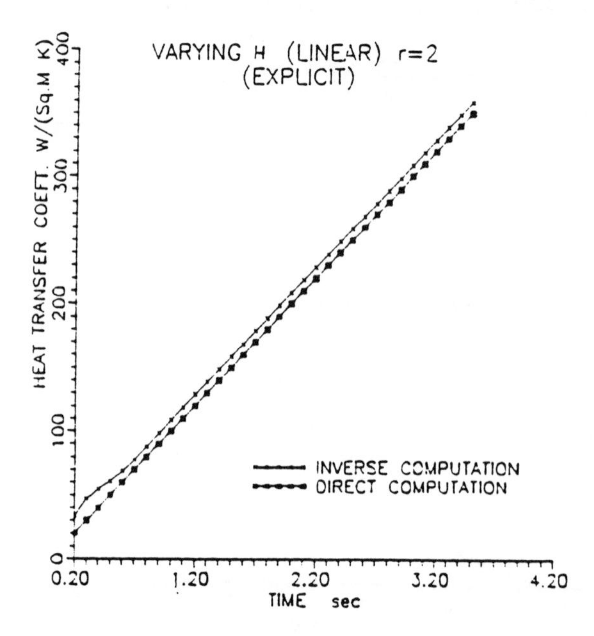

Figure 4 Estimated heat transfer coefficient, r = 2

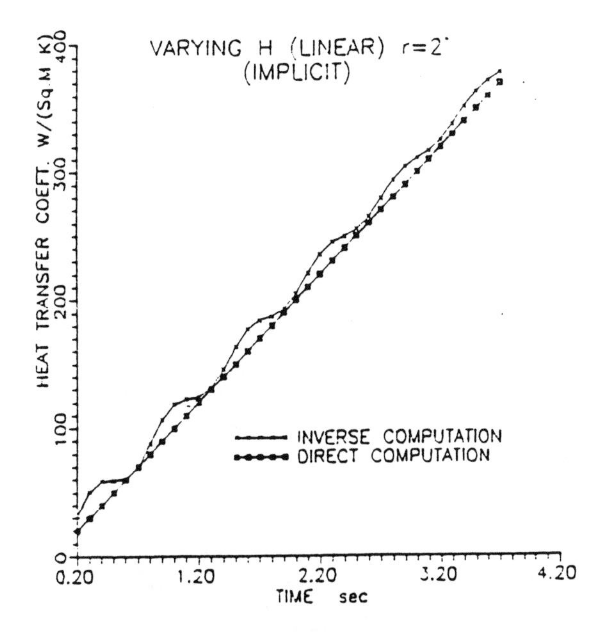

Figure 5 Estimated heat transfer coefficient, r = 2

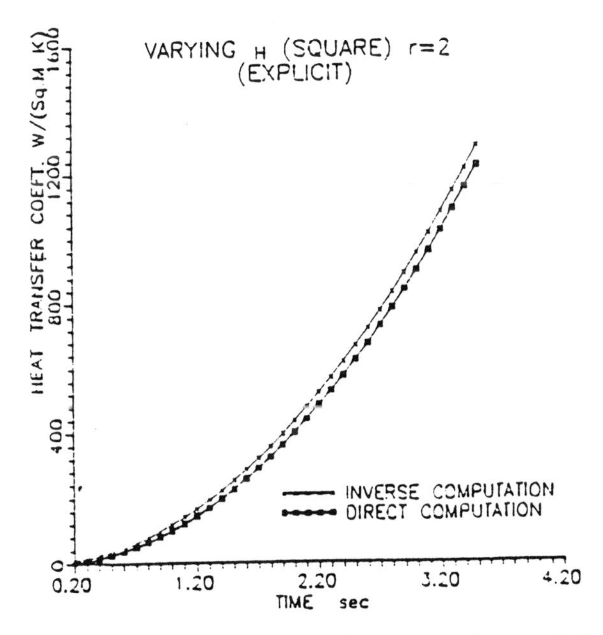

Figure 6 Estimated heat transfer coefficient

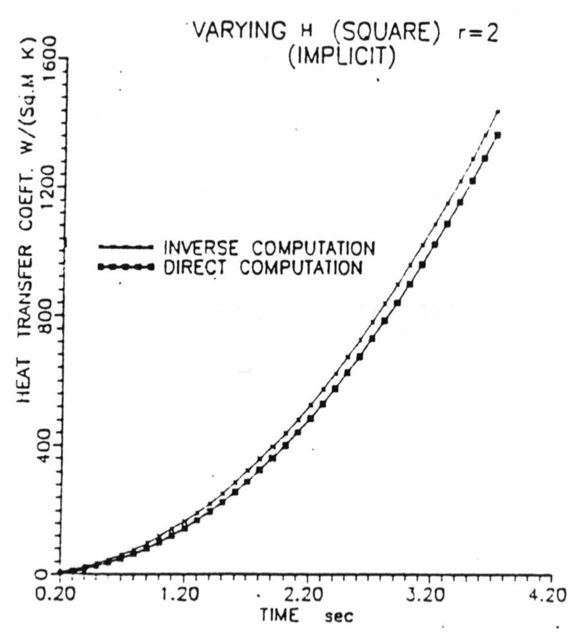

Figure 7 Estimated heat transfer coefficient

estimated and direct solution (simulated values) when the heating rate is very high or at the square heating condition. The computations are also done at large value of r and found excellent agreement. The disagreement between estimated and simulated values is more when the sensor position away from the heated surface. This is attributed to the small magnitude of the gain coefficient. Further work in progress to extend the present analysis in two-dimensional inverse heat conduction problem.

Conclusions

The explicit and implicit finite element methods are used to predict numerically the unknown surface conditions. The following observations are noted in the solution of inverse heat conduction problem:

1. The behaviour of the sensitive coefficient is studies. It is observed the sensitive coefficient is directly proportional to tempreture at the constant heat flux condition. It varies quadratric in the case of constant convective heat transfer condition. Therefore, a knowledge of sensitive coefficient could be used to fixed the sensor location.

2. The explicit and implicit finite elements methods give identical estimated values when the sensor is located near to the heated sureface or at very high heating conditions.

References

1. Hartree, D.R, Numerical Analysis, Oxford University Press, London, (1952).
2. Stolz, G, Journal of Heat Transfer, Transactions of ASME, 82c, pp. 20-26, (1960).
3 Frank, I, Journal of Heat Transfer, Transactions of ASME, 85c, pp. 378-379, (1963).
4 Krzystztof, G., Cialkowski, M.J.,and Kaminski, H., Nuclear Engineering Design, 64, pp. 169-184,(1981).
5 Beck, J.V., International Journal of Heat and Mass Transfer, 13, pp. 703-716, (1970).
6 Mehta, R.C., AIAA J, Vol. 19, No. 8, pp. 1085, (1981).
7 Bass, B.R., J. Engg. For Industry, Trans ASME, 102, pp. 168-176, (1980).
8. Mehta, R.C., Jayachandran, T., Warme-und Stoffubertragung., Vol. 26, pp 1-5, (1991).
9 Mehta, R.C., AIAA J., Vol. 15, pp. 1355-1356,

(1977).

10 Zienkiewicz, O.C.,Morgan, K, Finite Elements and Approximations., Wiley, NY., (1983).

11 Beck, J.V., Blockwell, B., St.Clair, C.R., Inverse Heat Conduction Problem, Wiely, NY., (1985).

12 Carslaw, H.D and Jaeger, J.C., Condition of Heat in Solids, Oxford Univ. Press, London, (1959).

APPLYING THE LEAST SQUARES ADJUSTMENT TECHNIQUE FOR SOLVING INVERSE HEAT CONDUCTION PROBLEMS

Janusz SKOREK(1)

SUMMARY

The problem of identification of temperature field within the body and heat flux on the boundary of the body involves the class of inverse heat conduction problems. A general algórithm for solving the inverse boundary heat conduction problem basing on the statistically optimal estimation method (least squares adjustment technique) is presented in the paper. Proposed method provides the statistical estimation of obtained results of calculations, and can be applied for solving both transient and steady-state problems. Sample numerical results of calculations are presented in the paper.

1. INTRODUCTION

The wide range of inverse heat conduction problems involves also the problem of identification of temperature field within the body and heat fluxes on the surface of the body (so called *boundary inverse problem*). The sough for quantities can be evaluated basing on the mathematical model of the process and any additional information concerning the considered problem e.g. measurements of temperature (so called thermal response) in certain points of the body. One assumes usually that both boundary conditions and initial condition can remain unknowns. There are a great deal of methods which are applied for solving boundary inverse problems of heat conduction [1,2]. Among them we can distinguish statistical estimation methods which base upon *the least squares* principle.

In contrary to the most methods, the least squares adjustment technique gives directly the statistical estimation of the obtained results. It is a very important information when solving inverse problems because in reality usually does not exists any other possibility to check the accuracy of the results of calculations.

Statistical estimation methods are generally used for solving boundary inverse problems of heat conduction formulated in discrete form.

(1) Institute of Thermal Technology, Silesian Technical University, Konarskiego 22, 44-101 Gliwice, Poland

Discrete model of the analyzed process can be here evaluated applying of the well known numerical methods e.g.:Finite Volume Method FEM, Control Volume Method CVM etc. Due to that complexity of the algorithms of estimation methods remains nearly the same when considering one or multidimensional problems.

The other important feature of estimation methods is the lack of any special requirements concerning the location and the number of thermal response sensors.

We can distinguish two essential groups of estimation methods are to be applied for solving inverse boundary problems:

- statistically optimal dynamic filtering methods (Kalman's filtering),
- least square adjustment technique (Bayes' statistical estimation, maximum likelihood method).

Dynamic filtering methods are multistep methods and can be applied only for solving transient problems which can be described by the so called state equation:

$$T_{k+1} = \Phi T_k + b \qquad (1)$$

where Φ is the square transition matrix, T_{k+1} and T_k denote the vectors of state quantities. In this case the thermal response is gathered and processed step by step along the time. Dynamic filtering method has been used for solving transient heat conduction identification problems [3,4]. On the other hand when solving strongly ill posed problems such a procedure can be sometimes insufficient to keep convergency of the solution (especially at the initial steps of time).

We can avoid of certain limitations of the dynamic filtering methods when applying least squares adjustment technique. Here we process thermal response gathered from more than two adjacent points of time, what can significantly improve the convergence of calculations. To effectively solve such an ill posed problems as boundary inverse problems, all the "a priori" information about the quantities which describe the analyzed process should be considered [5],[6]. Such an approach to the solution of inverse problems of heat transfer as well as to the other ill posed problems is a subject of presented paper.

Presented method can be applied for solving discrete inverse boundary problems for the both transient and steady-state problems. When analyzing transient problems we can simultaneously consider several points of time what improve convergency of the methods.

Accuracy of the method has been tested basing on the results of the several numerical tests. As the sample of practical usage of the least square adjustment technique, the transient heat flux on the surface of the wall of the experimental combustion chamber has been calculated basing on the measurements of transient temperature within the wall. Sample result of calculations are presented in the paper.

2. GENERAL ASSUMPTIONS OF THE METHOD

Let us assume that the model of the analyzed physical process, e.g. discrete model of the temperature field within the body, can be described in the form of the linear matrix equation (so called constraint equation)

$$Ax + By + c = 0 \tag{2}$$

where:

A,B - rectangular matrix of coefficients,
x - vector of the measured quantities,
y - vector of the unknown quantities.

The aim of estimation is to evaluate the unknown quantities y_i (i=1,2,...,n) basing on the equation (2) and on the values of the measured quantities (thermal response) x_j (j=1,2...s). Equation (2) is fulfilled for the exact values of **x** and **y**. When we put into equation (2) results of measurements $\mathbf{x}_0$ and initially estimated values of unknowns y_0 we get

$$Ax_o + By_o + c = w \tag{3}$$

where **w** denotes the vector of discrepancies. The aim of the least square adjustment technique is to find such the corrections δx_j and δy_j which fulfil the constraint equation

$$A(x_o + \delta x) + B(y_o + \delta y) + c = 0 \tag{4}$$

and simultaneously minimize the quadratic form

$$\delta x^T V^{-1} \delta x \rightarrow \min \tag{5}$$

where **V** is the covariance matrix of the results of measurements.

The statistic weights of the quadratic form (5) are the element of the matrix $\mathbf{V}^{-1}$.

When solving the problem (4) and (5) one obtains the relationships defining the vectors of corrections δ**x**, δ**y** and covariance matrices $\mathbf{G}_x$, $\mathbf{G}_y$:

$$\delta y = -G_y B^T F^{-1} w \tag{6}$$

$$G_y = (B^T F^{-1} B)^{-1} \tag{7}$$

where: $\mathbf{F}=\mathbf{AVA}^T$.

Relationships (6) and (7) are usually used for adjusting the wide range of problems including the thermal processes [7]. On the other hand the above procedure leads to the wrong results when solving extremely ill posed problems [6], for example the most of inverse heat conduction problems. The modified approach based on the utilization of the all attainable "a priori" data concerning the unknown quantities should be applied [5,6] in such a case. When analyzing any physical process we are able to estimate the real values of the sough for quantities $\mathbf{y}_0$ and their covariance matrix $\mathbf{G}_{y0}$. Diagonal elements of covariance matrix $\mathbf{G}_{y0}$ express the range of possible (from physical point of view) change of given quantities $\mathbf{y}_0$.

Solving now the modified problem (4),(5) one obtains the relationships describing the vector of corrections $\delta\mathbf{y}$ and covariance matrix $\mathbf{G}_y$ in the form

$$\delta y = -G_y B^T F^{-1} w \tag{8}$$

$$G_y = (G_{y0}^{-1} + B^T F^{-1} B)^{-1} \tag{9}$$

From the mathematical point of view the elements of the matrix $\mathbf{G}_{y0}$ are the regularization factors which stabilize the solution of the problem [2].

From the equation (9) results that in the case the elements of covariance matrix $\mathbf{G}_{y0}$ tend to infinity (what means we have no any "a priori" information about unknown quantities) the relationship (7) and (9) become identical.

The great number of tests have proved that the proposed method of the least squares adjustment technique is an effective tool for solving ill posed problems including inverse heat conduction problems.

3. SOLUTION OF THE INVERSE BOUNDARY HEAT CONDUCTION PROBLEM

The problem we consider herein is the identification of the transient temperature field within the body and transient heat flux on the part or whole the boundary of the body. One assumes that the thermal response at the certain points located within the body or on the boundary of the body is known. We can also assume that the initial condition (initial temperature distribution) can remain unknown and the problem we consider is formulated in the discrete form. The inverse problem stated under above assumptions requires the formulation of the model of temperature field at the discrete net of nodes distinguished within the body.

One assumes that transient temperature field is forced by the heat flux $q_s(\tau)$ on the boundary of the body (boundary condition of the 2-nd kind). The heat flux $q_s(\tau)$ is the function of location and the time τ and is the main subject of the identification.

Using one of the discrete method of the formulation of the direct boundary problem e.g. Finite Element Method FEM or Finite Volume Method FVM we formulate the discrete model on the temperature field in the form of the matrix equation

$$T_{k+1} = \Phi\ T_k + \Psi\ q_{k+1} + \Phi\ c_{k+1} \tag{10}$$

where Φ and Ψ are the matrices of coefficients and **c** is the vector of quantities which are not identified. Indexes k+1 and k denote two adjacent points of time.

We assume that the sensors which measure the temperature of the body are located at the certain nodes resulting from the discrete division of the body. Hence we can write

$$x_k = H\ T_k \tag{11}$$

where:

x - vector of the measurement results (thermal response),
k - k-th moment of time,
H - index matrix (consists of elements 0 or 1).

When multiplying equation (10) by matrix **H** and taking into account (11) we obtain

$$H\ T_{k+1} = x_{k+1} = H\ \Phi\ T_k + H\ \Psi\ q_{k+1} + H\ \Phi\ c_{k+1} \tag{12}$$

Equation of (12) type is now written for each of k=1,2,...,l time steps. The task is to eliminate from the equations (12) all the unknown quantities except of initial temperature distribution T_0 and boundary heat flux q_1, q_2,...q_l. Elimination of these unknowns yields

$$\begin{aligned} x_1 &= HT_1 = H\Phi T_o + H\Psi q_1 + H\Phi c_1 \\ x_2 &= HT_2 = H(\Phi T_1 + \Psi q_2 + \Phi c_2) \\ &\cdots \\ &\cdots \\ x_l &= HT_{l-1} = H\Phi^{(l)} T_o + H\sum_{k=o}^{l-1} (\Psi^{(k)} q_{l-k-1} + \Phi^{(k)} c_{l-k-1}) \end{aligned} \tag{13}$$

where $\Phi^{(k)}$ denotes the k-th power of matrix Φ.

Gathering all the equations of type (13) written for the each of l considered points of time we finally obtain the equation of the form (2) where we put

$$x = [x_1, x_2, \dots, x_l]^T \tag{14}$$

$$y = [T_o, q_1, q_2, \dots, q_l]^T \tag{15}$$

At last we have to estimate the elements of vector $\mathbf{T}_0$ (initial temperature distribution) and elements of the covariance matrix $\mathbf{G}_{y0}$.
Presented formulation of the identification problem leads to the relatively complicated form of matrices **A** and **B** (see equation (2)). On the other hand the formulas for calculation of matrices **A** and **B** have to be derived only once as a functions of matrices $\mathbf{\Phi}$, $\mathbf{\Psi}$ **and** **H** and then can be applied for solving other problems of identification.

4. NUMERICAL EXAMPLES

4.1. Numerical test

Let us consider the problem proposed by Beck [1] to test the accuracy of different methods of solution of the certain class of inverse problem of heat conduction. We consider the one-dimensional transient temperature field within the plate. One surface of the plate is insulated, the second one is heated by the transient heat flux q(Fo):

$$q(Fo) = \begin{cases} Fo & , \textit{for } 0.0<Fo<0.6 \\ 1.2-Fo & , \textit{for } 0.6<Fo<1.2 \end{cases} \tag{16}$$

The task is to evaluate the heat flux q(Fo) basing on the knowledge temperature distribution T_1 only at the one point located on the insulated surface of the plate. Temperature T_1 is here simulated from the direct problem and then disturbed by the stochastic errors to simulate measurement errors. The initial temperature is uniform and equal to 0. Time step ΔFo=0.1.
Sample results of identification are shown in the figure 1. For the exact data one obtains the entire accordance between the results of estimation and real distribution of heat flux q(F). For the disturbed data accuracy of the estimation is still relatively high, especially taking into consideration that the initial estimation of vector $\mathbf{y}_0$ is very rough.

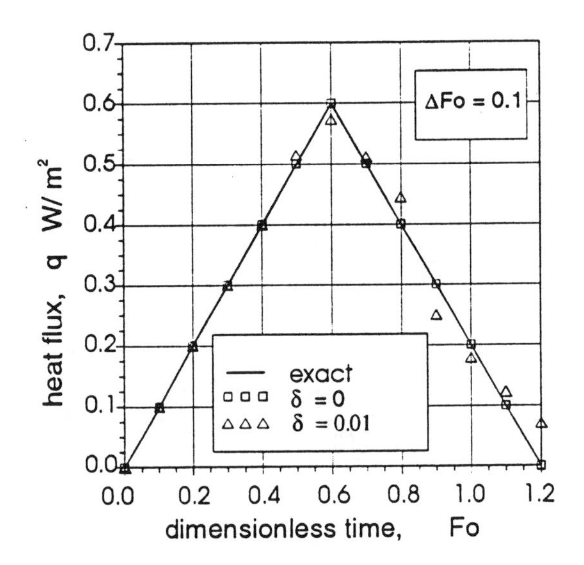

Fig.1. Estimation of the heat flux on the surface of the 1D plate

4.2. Estimation of the heat flux on the surface of combustion chamber

Presented method was applied for estimation of the transient heat flux $q_s(x,\tau)$ on the internal surface of walls of the experimental combustion chamber. The cross section of the chamber wall is shown in the figure 2.

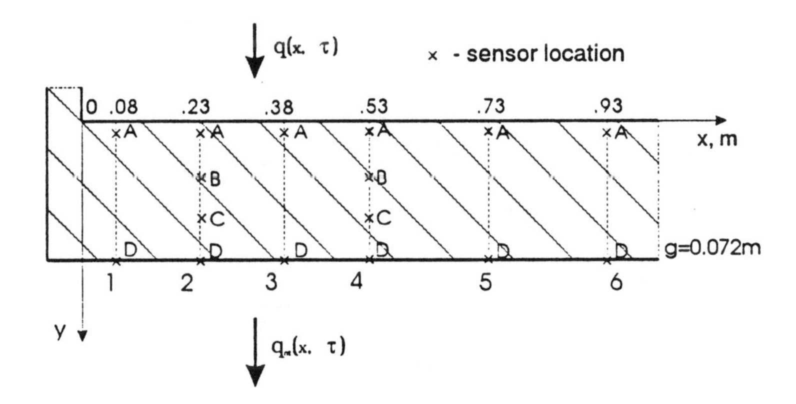

Fig.2. Cross section of the wall of the combustion chamber

Results of calculations of the heat flux $q_s(x,\tau)$ are presented in the figures 3-4. The distribution of $q_s(x,\tau)$ along the chamber length and with respect to the time are shown. The statistical estimation of the $q_s(\tau)$ is presented in the figure 5. The shadowed band presents the area of $q_{ls}(\tau) \pm S(q_{ls})$ where $S(q_{ls})$ denotes the standard deviation of the $q_{ls}(\tau)$.

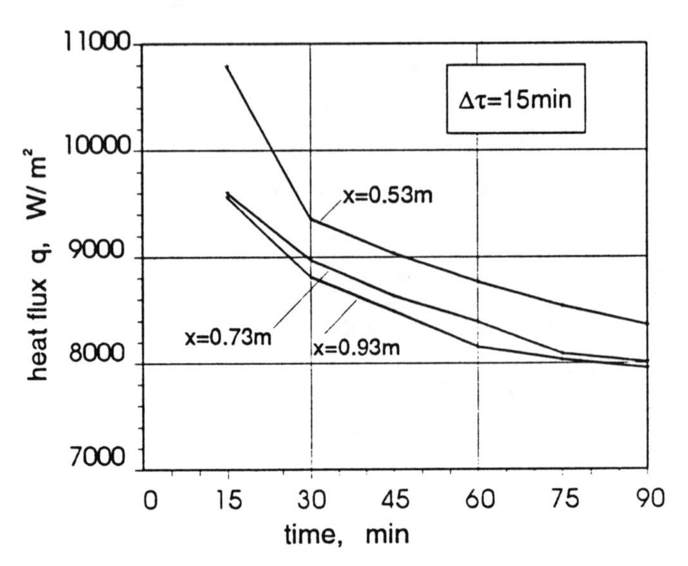

Fig.3. Heat flux on the surface of the combustion chamber

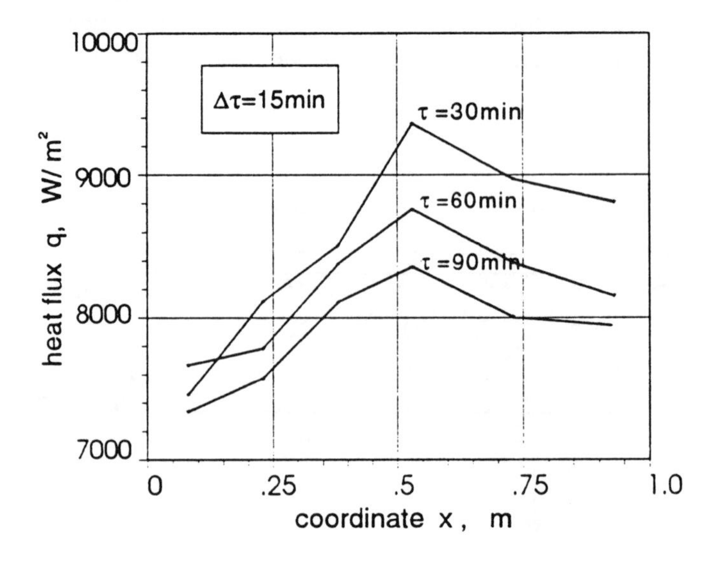

Fig.4. Heat flux on the surface of the combustion chamber

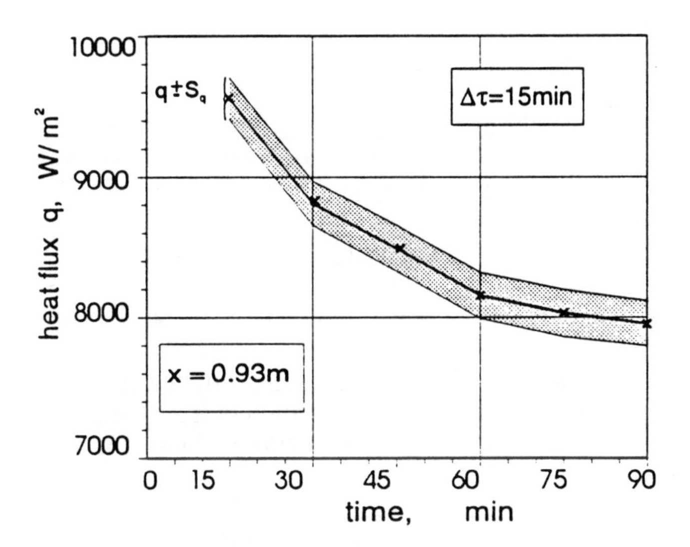

Fig.5. Statistical estimation of the heat flux $q(\tau)$

5. CONCLUSIONS

The carried on researches and numerical experiments have proved that the group of statistically optimal estimation methods (dynamic filtering and the least square adjustment technique) can be effectively used for solving certain class of inverse heat conduction problems. The important advantage of statistical methods is they give directly the statistical estimation of the results of calculations. These methods do not also forced any limitations on the location of temperature measurement sensors.

Presented formulation of the identification problem is of general form what allows to solve the wide class of inverse problems including transient and steady-state heat conduction problems.

7. REFERENCES

[1] BECK J.V.: Inverse Heat Conduction. A Villey Intersc. Publ.,New York 1985.

[2] TIKHONOV A.N., ARSENIN W.: Solution of Ill Posed Problems. Winston & Sons, Washington D.C. 1977.

[3] MATSEVITY J.M., MULTANOVSKI A.V.: Identification in Heat Conduction Problems, (*in russian*). Naukova Dumka, Kiev 1982.

[4] SKOREK J.: Identification of the Transient Temperature of the Charge in the Soaking-Pit Furnace, (*in Polish*). Proceeding of the XXV-th Symposium "Modelling in Mechanics", vol. II, Kudowa, Poland, 1986, p. 113.

[5] SKOREK J.: Applying of Statistical Estimation Methods for Solving Inverse Problems of Heat Conduction. (*in Polish*). Proceedings of the 8th Symposium of Heat and Mass Transfer, Białowieża, Poland, 1992. p. 439.

[6] SZARGUT J., SKOREK J.: Influence of the Preliminary Estimation of Unknowns on the Results of Coordination of Material and Energy Balances. Bulletin of the Polish Academy of Sciences, Technical Sciences, Vol. 39, No. 2, 1991, p. 335.

[7] SZARGUT at. all: The least square adjustment technique in the thermal technology, (*in Polish*), PAN, Wrocław 1984.

NOMENCLATURE

$\mathbf{A},\mathbf{B}$ - matrix of coefficients,
Fo - dimensionless time,
$\mathbf{G}_x,\mathbf{G}_y$ - covariance matrix of unknowns and measured quantities,
$\mathbf{G}_{y0}$ - covariance matrix of initial estimate of unknown quantities,
$\mathbf{\Phi}$ - transition matrix,
$\mathbf{T}_{k+1},\mathbf{T}_k$ - vector of nodal temperatures at the k+1 and k points of time,
$\mathbf{x},\mathbf{y}$ - vector of measured quantities (thermal response) and unknowns,
$\mathbf{x}_0$ - vector of the results of measurements,
$\mathbf{y}_0$ - initial estimate of unknown quantities,
$\mathbf{w}$ - vector of discrepancies of the constraint equations,
$\mathbf{V}$ - covariance matrix of the results of measurements.

ACKNOWLEDGMENTS

The work bas been supported by the Polish Committee of Scientific Research under the grant 1249/3/91.

MODELING OF SINGLE SHOT INDUCTION HEATING BY INVERSE FINITE ELEMENT METHOD

G. Maizza(*) and M. Calì(**)

ABSTRACT

The mathematical model for an induction heating process generally requires integration, in space and in time, of several partial differential equations.

The model herein couples the standard heat conduction equation with an electro-magnetic model based on the proximity-skin effect theory. This requires the introduction into the equations, as unknown parameter, of the current flowing into the coil.

The aim of this paper is to show how a parameter estimation problem can be solved with an Inverse Finite Element procedure on the base of prior *deterministic* and *probabilistic* concepts. For this purpose simulated experiments with different noises in the input data have been carried out in order to determine their influence on the parameter estimation. Actual measurements of the unknown parameter and temperatures in some points of the workpiece surface are performed during on-line process to validate the parameter estimation procedure.

1. INTRODUCTION

Induction heating of materials is extensively used because its great flexibility in industrial application. The main feature of induction heating is to keep under control the size and the shape of the heat affected zone; this is especially important in the heat treatment of workpieces with complex geometry. Induction heating is also applied due to the repeatibility properties of the process. Furthermore, it requires relatively low energy consumption and keeps the pollution of the indoor environment relatively low.

The optimization of the process usually implies the analysis of several working parameters and the control of the most critical ones. This analysis may be accomplished experimentally or more conveniently by the use of specifically

(*) Researcher, Dipartimento di Scienza dei Materiali, Politecnico di Torino

(**) Professor, Dipartimento di Energetica, Politecnico di Torino

designed computational tools. Finite elements method seems to be one of the most versatile techniques for modeling complex problems. In fact, high irregular geometries and strong non linearity in the partial differential equations may be handled without considerable supplemental efforts.

A rigorous mathematical model for an induction heating process generally requires the analysis of fields, say thermal, magneto-electric and micro-structural, that are strongly coupled through the dependence to the field (or state) variables of the material properties, boundary conditions, and source terms. An example of this model has been previously examined and solved [1,2] as a *deterministic or direct problem* according to the definition given by Beck [3]. Similar models have been also proposed by several researchers [4,5], following the same approach, by simplified coupling of heat conduction equation and Maxwell's equations. However, except that for simple mono-dimensional problems the computational requirements are significantly costly. Moreover, the input magnetic data necessary for the model definition are very difficult to obtain experimentally. These drawbacks have urged to search for a more practicable approach. In previous works [6,7] the non linear transient induction heating problem has been solved by the use of the standard heat conduction equation coupled with the proximity-skin electro-magnetic model [8]. As a result only the thermal and the current density fields have to be computed simultaneously, thus avoiding the calculation of the nonessential and costly magnetic field. This model is characterized by the use of two new quantities, that are the coil current and the coupling coil-workpiece parameter. The former allows the optimal design of inductors and a better control of the thermal cycles. The latter allows the estimate of the efficiency of the heating process and the calculation of the instantaneous thermal power supplied to the workpiece during heating. Both these parameters are of fundamental importance for induction heating equipment design and process optimization, however they are not easy to measure practically. Therefore an inverse procedure has been adopted to solve the inherent *parameter estimation problem*. Unfortunately, due to the relatively low amount of experimental data available, no information about the accuracy on the parameter estimation could be given.

The aim of this work is to investigate on the performance of the inverse procedure adopted with respect to the coil current parameter estimation. To validate the parameter estimation procedure actual temperature and parameter measurements have been also performed.

2. THE PHYSICAL PROBLEM

The system investigated is a cylindrical bar, longitudinally fixed, that rotates about its axis to allow an uniform heating. A two-turn coil, coaxial with the bar, is vertically mounted in the heat treatment apparatous. An AC current flowing through the coil (which is cooled by water) heats the surface of the bar by the proximity-skin effect. Both the bar and the coil are fixed during the thermal cycle (single shot heating). Fig.1 shows a scheme of the heating apparatous. The bar is surrounded by air and is cooled by convection and radiation (lines HE, EF). A fixed temperature boundary condition is applied on the upward cylinder base (line HG).

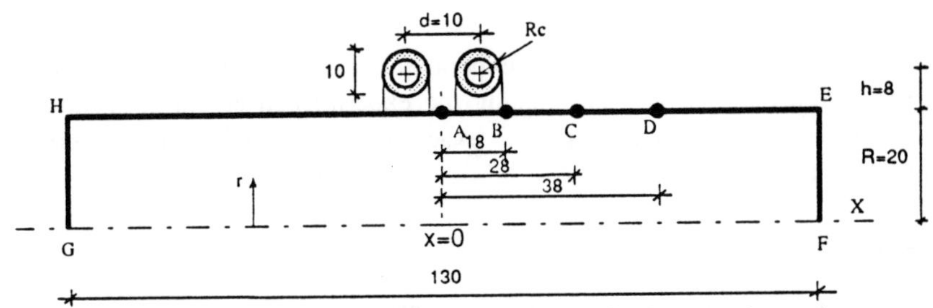

Figure 1 - Scheme of the system analysed (sizes are in mm)

3. THE INVERSE PROBLEM

The methods proposed by Beck for the analysis of heat conduction problems [3] and by Tarantola for the analysis of geological problems [9] are here adopted to solve the parameter estimation problem. Before going into details some definitions are necessary.

A set of observed temperatures **Y** is introduced corresponding to M points on the surface (as A, B, C, and D in fig. 1).

$$\mathbf{Y}^{\mathbf{T}} = \left[\mathbf{Y}_1^{\mathbf{T}} \cdots \mathbf{Y}_j^{\mathbf{T}} \cdots \mathbf{Y}_M^{\mathbf{T}}\right]^{\mathbf{T}} \tag{2a}$$

For the jth element of **Y,** N discrete components representing the N measurements in time are defined, the resulting vector contains N by M elements:

$$\mathbf{Y}_j^{\mathbf{T}} = \mathbf{Y}^{\mathbf{T}}(t_j) = \left[Y_1(t_j) \cdots Y_i(t_j) \cdots Y_N(t_j)\right]^T \tag{2b}$$

Analogously, a vector $\hat{\mathbf{T}}$ is referred to the calculated temperatures corresponding to the same M points and N sampling times:

$$\hat{\mathbf{T}}^{\mathbf{T}} = \left[\hat{\mathbf{T}}_1^{\mathbf{T}} \cdots \hat{\mathbf{T}}_i^{\mathbf{T}} \cdots \hat{\mathbf{T}}_M^{\mathbf{T}}\right]^{\mathbf{T}} \tag{3a}$$

where

$$\hat{\mathbf{T}}_j^{\mathbf{T}} = \hat{\mathbf{T}}^{\mathbf{T}}(t_j) = \left[\hat{T}_1(t_j) \cdots \hat{T}_i(t_j) \cdots \hat{T}_N(t_j)\right]^T \tag{3b}$$

If the material properties and the boundary conditions are the same in the model and in the actual case, then the measured temperatures differ from the calculated ones of a quantity which is a function of measurement errors and errors inherent to the model representation. To account for these disprepancies a *residual* is introduced:

$$s(I) = \hat{T}(I) - Y \tag{4}$$

If the errors associated with the experimental system are known, a suitable error model may be built introducing the experimental errors as a difference between an hypothetical observed exact value $\mathbf{Y}_e$ and that measured by instrument $\mathbf{Y}$:

$$\mathbf{e} = \mathbf{Y}_e - \mathbf{Y} \tag{5}$$

For convenience, **e** may be assumed with a normal statistical distribution, whose average value is zero and the standard deviation σ is known. If the measurement errors are not correlated, the covariance matrix $\mathbf{C}_c$ is diagonal. Each element represents the standard deviation of the related measurement:

$$C_{ii} = \sigma_i^2 \tag{6}$$

Thus, an optimal estimate of the unknown parameter I may be obtained by minimizing with respect to I the following function S (see Appendix B for details):

$$S(I) = \mathbf{s}^{\mathrm{T}}(I)\,\mathbf{C}_c^{-1}\,\mathbf{s}(I) \tag{7}$$

However, since S depends on I through a non-linear relationship, an iterative procedure is required. The parameter is provided by the equation:

$$I^{(k+1)} = I^{(k)} + P^{-1}(I)^{(k)} \cdot H(I)^{(k)} \tag{8}$$

The parameter $I^{(k+1)}$ is re-calculated at each iteration until a suitable convergency criterion is satisfied.

$$\left|I^{(k+1)} - I^{(k)}\right| / \left|I^{(k)}\right| \leq \delta \quad \text{with } \delta \text{ a previously defined tolerance.} \tag{9}$$

At the end of the iteration process, either the standard deviation or the variance of the parameter estimated σ_I may be computed to measure the accuracy of the estimation procedure.

4. THE INVERSE-FEM MODEL

The physical model is complicated by the intrinsic interactions among the fields involved. The non-linearities are due to the dependence on the temperature field of the physical properties and the heat generation. When a parameter estimation problem is solved, a new non linearity effect appears in the model due to the dependence of the state variables on the unknown parameter.

On the other hand phase changes may have a great influence on the results, but in the case of the aluminum the alloy presents the same crystallographic structure up to the fusion temperature.

The standard heat conduction equation with the proper boundary conditions, written in tensorial notation, is:

$$\left[k(T)\cdot T_{,i}\right]_{,i} + q_v(T,I) = \rho\cdot c(T)\cdot \partial_t T \qquad (10)$$

$$-k(T)\cdot T_{,i} = \alpha\cdot(T - T_a) \qquad (11)$$

The magnetic effects are taken into account by the extension to cylindrical geometries [6,7] of the original proximity-skin theory [8]. As a result the spatial current density distribution is determined and its contribution is introduced in the q_v term of eq.(10). Even complex coil-workpiece configurations may be effectively modeled due to the fact that multi-turn effects may be taken into account by application of the superposition principle [11], (see Appendix A).

From equations (10) and (11) the well-known FE equation is obtained:

$$\mathbf{C}(\mathbf{T})\,\dot{\mathbf{T}} + \mathbf{K}(\mathbf{T})\,\mathbf{T} + \mathbf{F}(\mathbf{T},I) = 0 \qquad (12)$$

where **C** and **K** matrices are temperature dependent only, while **F** vector is also dependent on the coil current through the heat generation rate q_V.

Equation (12) can be conveniently discretized in time by a predictor-corrector scheme [10].

The inverse solution algorithm is summarized in the following steps:

1. *Set Data and Boundary/Initial Conditions*
2. *Set Initial parameter $\bar{I}$ and counter k*
3. *Increment counter k*
4. *Set parameter to value I^k in the direct model*
5. *Initialize time loop: $\tau = 0$*
6. *Increment time step $\tau = \tau + \Delta\tau$*
7. *Update electrical- magnetic properties ρ_e, μ*
8. *Compute electric density field $\ddot{J}(r,x)$, and q_V*
9. *Update thermal properties k, ρ, c*
10. *For each time step solve the thermal field $\ddot{T}(r,x)$*
11. *If $\tau < \tau_H$ go to 6*
12. *Set perturbated parameter to $\bar{I}^k + \delta I^k$ in the direct model*
13. *Repeat steps 5 through 11*
14. *Calculate of H,P,X,S,C_e*
15. *Check for convergency criterion eq.(9), if true go to 15*
16. *Compute a new estimate for $\bar{I}$ and go to 3*

17. Print Output: I ; $\ddot{J}(r,x)$; $\ddot{T}(r,x)$; σ_S, σ_I
18. END

5. COMPUTATIONAL RESULTS AND DISCUSSION

An axisymmetric geometry has been investigated in order to simplify the numerical tests, although the model can handle more complex configurations. The IFEM (Inverse-FEm Model) is applied to analyse how different levels of accuracy in the measurements may affect the estimated parameter. Such a problem can arise in practice when the most suitable instrument class has to be compatible with the desired accuracy of the measured data. The optimal experiments design (with particular interest in the choice of the instrument class) can be achieved choosing the appropriate path, a) or b), in fig.2. The main difference between the two paths is in the provision of the input data that have to be supplied to the model (i.e. the coil current I an temperatures **Y**). Usually data come from experiments performed on the real systems or from simulated experiments as shown from path (b) and (a) respectively of fig.2.

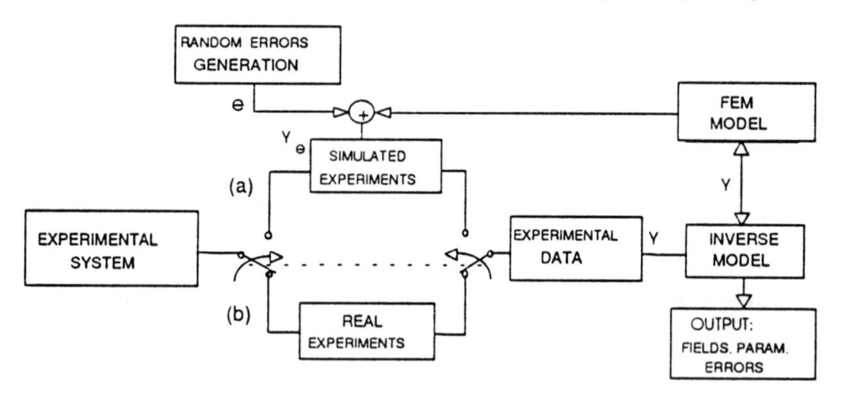

Fig. 2 - Scheme for optimal experiments design

The path (a) is first considered. A hypothetical *true value* for the coil current, I=4920A is assumed in the case of a induction heating process. The duration of the thermal cycle is set to 6 s at the working frequency of 6 kHz. Then the finite element solver is executed (in one iteration) to provide the temperature field into the workpiece. For this test (and the followings) a mesh of 153 nodes and 128 elements has been built. Referring to fig.1, a Dirichlet boundary condition is set on the left side of the cylinder with a temperature of 25 °C. The surfaces exchange by convection and radiation with air whose mean equivalent convection coefficient is 30 W/m^2 K. The initial temperature of the material is set to 25 °C, equal to the environmental air temperature. The thermal conductivity, specific heat, density and electrical resistivity for the given material (an aluminum alloy) are assumed to be variable with the temperature [8]. The average time step used in the calculations is about 0.01 s.

Conventionally the temperatures obtained by the finite element model are designated as "*exact*" Y_e. From the nodal vector Y_e a subset of sample N·by M (=7·4) values has been extracted, corresponding to temperature readings at N selected locations along the surface (as shown in fig.1) and M discrete times. Three sets of random temperature errors (normally distributed) of variance σ_{SIM} (= 4, 5, 15 °C, respectively) are generated by computer [3] and added to Y_e. As a result three vectors **Y** are obtained as input data. These data are successively used in the Ordinary Least Square inverse model (OLS) to study the influence of random errors on estimation of the parameter. In a real optimization or design process many runs would be necessary for each case to have a more reliable *average* parameter estimation. Because of the long calculations needed, the average over three runs are here reported. About three-four iterations are necessary for convergency. The 6th column of table I (tests A1, A2, A3) shows that the parameter can be estimated within an absolute error of about ±30 A. This error value is in agreement with that found in real experiments. Two accuracy indexes (suitably rounded), σ_I and σ_T are provided in the 7th and 8th columns; the former gives the accuracy associated to the estimation procedure, the latter gives the goodness index of the temperature fitting.

Test	N·M	f [kHz]	τ_H [s]	Meth.	σ_{SIM} [°C]	I [A]	σ_I [A]	σ_T [°C]
A0	28	6	6	exact	0	4920	0	0
A1	28	6	6	OLS	5	4891	2.5	5.5
A2	28	6	6	OLS	15	4900	3	10.5
A3	28	6	6	OLS	4	4910	2	4.5
A4	7	6	6	OLS	5	4891	3.5	5
A5	12	6	6	OLS	5	4924	3	3.5
A6	28	6	6	WLS	5	4920	2	2.5
A7	2	6	6	OLS	5	4893	4	4.5
A8	2	6	6	WLS	5	4891	4	3
B1	36	8	12	OLS	-	3966	1.0	16
B2	36	8	12	WLS	-	3961	6.0	15

Table I - Parameter estimation for simulated (A) and real experiments (B) in medium frequency induction heating process.

Further tests (A4-A8) show how the amount of input information influences the parameter estimation. In case A4, only input data at the maximum temperature location (i.e. at point A, in fig.1) are considered; while in the case A5, only data at two time levels are recorded (i.e. the initial, the final and one intermediate). In case A6 a Weighted Least Squares (WLS) inverse model is used, itroducing a suitable covariance matrix C_e, to check how the model behaves when some errors at high temperatures are damped. A significant improvement results in the estimate of the parameter, moreover a faster convergency is noticed (one or two iterations less). In the last cases the effect of only two input data (i.e. the initial and the final temperature, at point A of fig.1) has been investigated with and without the covariance matrix C_c (see cases A7 and A8). Even if σ_T increases, the absolute error of the parameter remains within the range of 30 A. This means that in actual experiments the history at point A is the most critical, therefore it has to be measured accurately.

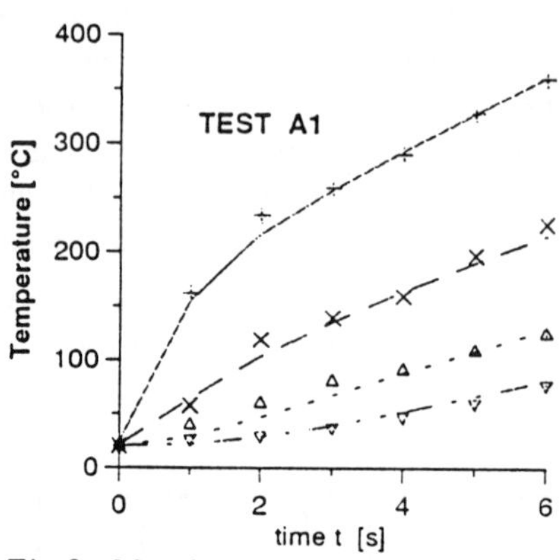

Fig.3. Matching between simulated temperature data (symbols) and model values (lines), in the case of $\sigma_{SIM} = 5°C$.

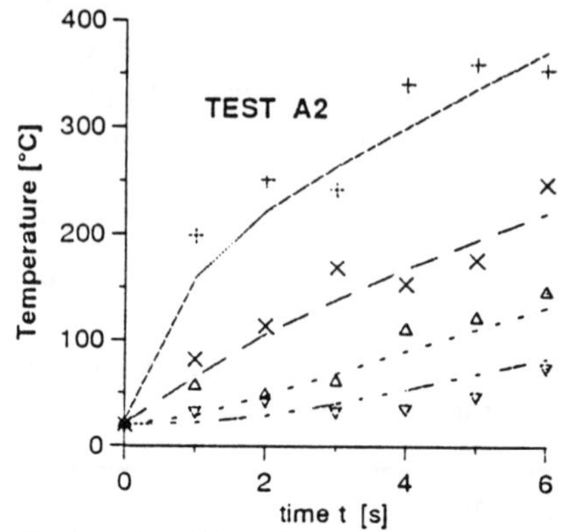

Fig4. Matching between simulated temperature data (symbols) and model values (lines), in the case of $\sigma_{SIM} = 15°C$.

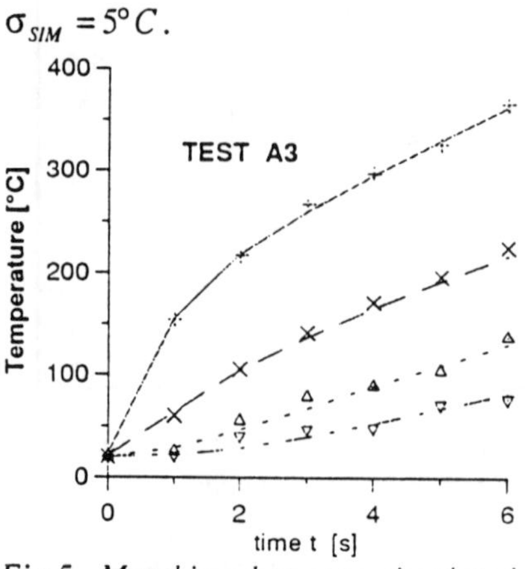

Fig.5. Matching between simulated temperature data (symbols) and model values (lines), in the case of $\sigma_{SIM} = 4°C$.

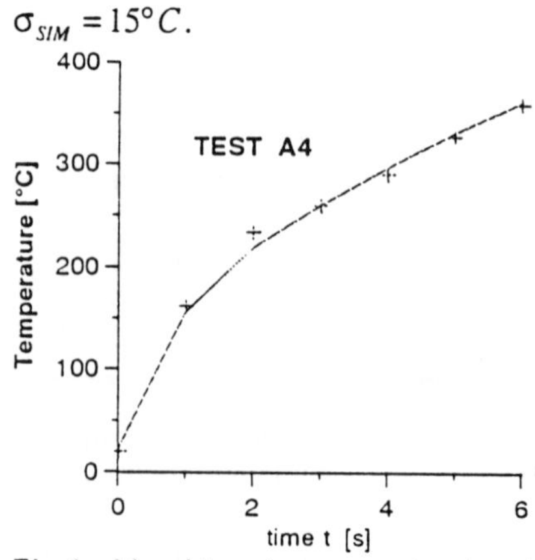

Fig.6. Matching between simulated temperature data (symbols) and model values (lines), with data recorded at point A only.

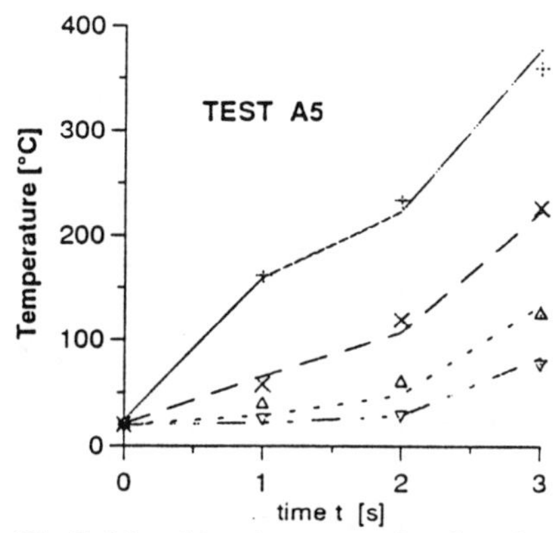

Fig.7. Matching between simulated temperature data (symbols) and model values (lines), with reduced time levels data.

In some cases the only observations at the point A would be enough for a reasonable estimate of the parameter, thus reducing the experimental efforts considerably. Figures 3-7 show how the model values fit the temperature data. To check the validity of the inverse procedure here proposed, actual measurement for the coil current has been carried out. It has resulted about 3930±30 A, under a 8 kHz frequency for a duration of 12 s.

time [s]	x_1 A	x_2 B	x_3 C	x_4 D
0	25	25	25	25
2	174	75	40	27
4	250	123	70	44
6	314	186	105	65
8	378	225	123	78
9	397	238	142	87
10	429	263	161	92
11	455	289	171	110
12	474	301	187	139

Table II. Temperature data in [°C] for an aluminum billet heated by induction for 12 s, at frequency of 8 kHz.

Obviously temperatures are easy to measure by means of radiation pyrometers, if problems due to the cooling media such as vapors are avoided (as in this case). Furthermore the use of pyrometer avoids the need of placing sensors or probes on the workpiece that may cause undesirable disturbances on the electro-magnetic field and viceversa. Hence, temperature histories have been measured at M=4 locations along the surface and N=9 times. Temperatures below 150°C have been measured by a contact thermometer. The data are reported in table II.

In table I there are the results of case B1 obtained from IFEM with an OLS estimator. The case B2 refers to the same problem but with the use of the covariance matrix C_c (WLS), which is built by using error data of the pyrometer employed.

The results show that the parameter estimated is in agreement with that measured with the uncertainty of 30 A. Although a better estimate is obtainable from the model, however it is not recommended since computational times would become prohibitive.

In figures 8 and 9 is shown the fitting between the experimental temperatures and those calculated by the inverse model.

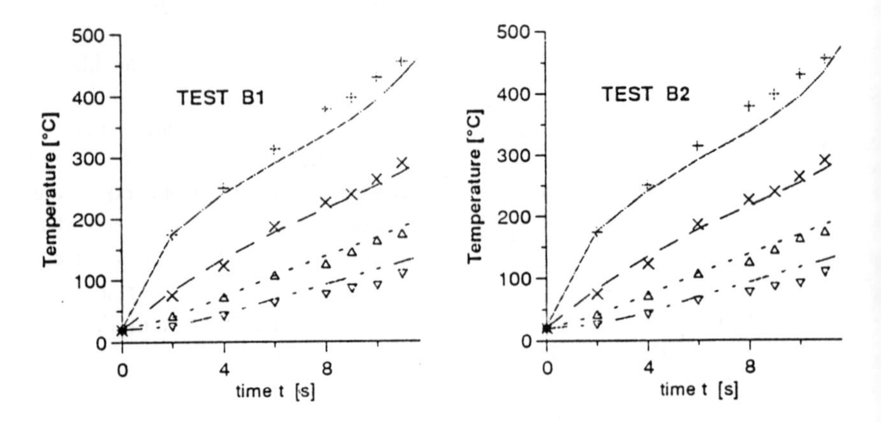

Figure 8-9 - Comparison between experimental and model temperatures in the case of a induction heating of an aluminum billet at frequency of 8kHz for a duration of 12 s: by OLS and WLS estimation respectively.

6. CONCLUSIONS

The results demonstrate that non-linear transient induction heating problem can be effectively modeled by conduction and proximity-skin equations.

On-line and easy-to-measure input data of temperature are used to solve the parameter estimation problem. Simulated experiments show that a good estimation of the coil current can be obtained by relatively low amount of input data.

The experimental validation of the estimation procedure demonstrates that the more accurate is the maximum temperature measurement the more reliable will result the parameter estimation.

If the Monte Carlo approach is used, the parameter estimation will be more accurate if many runs are available for each case; however the long calculation times and the instabilities shown in some case, probably due to ill-posed nature of the problem are critical.

An optimal process technology or an optimal experiment design can take advantage of this approach, especially when more than one parameter has to be estimated simultaneously.

REFERENCES

1. MELANDER M., - Computer Predictions of Progressive induction harden,ing of cylindrical components, Material Science and Technology, V.1, pp.877, (1985).

2. CALÍ M., B. DEBENEDETTI and A. DEMISS, - Non-linear Thermal Problems in Progressive Induction Hardening, Proc. of the AST World Conf. on Advances in Surface Treatment and Surface Finishing, Paris, 3-4 Dec., pp.17, (1986).
3. BECK J.V., ARNOLD K.J., - Parameter estimation in Engineering Science, J.Wiley & Sons, New York, 1977.
4. LAVERS J.D., - Numerical Solution methods for Electroheat Problems, IEEE Trans. on Mag., MAG-19, N.6, pp.2566-2572, (1983).
5. MARCHAND Ch. and A. FOGGIA, - 2-D Finite Element for Program Magnetic Induction Heating, IEEE Trans. on Magn. MAG-19, N. 6, (1983).
6. G. MAIZZA, M. KITAJIMA and B. DEBENEDETTI, - *FEAHT* A PC -based Package for Solving Coupled Problems on Induction Heat Treatments, Proc. of Int. Conf. on HEAT & SURFACE '92, Kyoto 17-20 Nov. (1992).
7. B. DEBENEDETTI and G. MAIZZA, - Metodo Inverso per la Risoluzione di Problemi di Riscaldamento ad Induzione impiegando la Tecnica degli Elementi Finiti, Proc. of the SIMAI '92 Conf., Florence 1-5 June, (1992).
8. J. DAVIES, P. SIMPSON, - Induction Heating Handbook, McGraw Hill, London, 1979.
9. A. TARANTOLA, - Inverse Problem Theory, Elsevier, Amsterdam, 1987.
10. ZIENKIEWICZ O.C., W.L. WOOD, N.W. HINE, - A Unified Set of Single Step Algorithms. Part 1: General Formulation and Applications, Int. J. Num. Meth. Engng. V. 20, pp. 1529-1552, (1984).
11. AI-SHAIKHLI A.K.M. and L. HOBSON, - IEE Proc., V.133, Pt.B, N.5, pp.323, (1986).

NOTATIONS

c Specific heat, **J/(kg K)**

h Distance between bar and coil **m**

H, P Working variables

I Coil current, **A**

$\tilde{I}$ Guess value for coil current, **A**

I_{ML} Coil current estimated by maximum likelihood criterion

$J(r)$ Current density at radius r, **A/m^2**

J_R Current density at radius R, **A/m^2**

k Thermal conductivity, **W/(m K)**

M Number of discrete sample times

N Number of discrete sample locations

$\mathcal{P}_M$ Power density per unit of length, **W/m^3**

q_v Internal heat generation rate, **W/m^3**

S Weighted sum of square residuals

t time, s

T Temperature, K or °C
R_C Coil radius
T_a Ambient temperature, K or °C
α Convection coefficient, $W/(m^2\ K)$
δ Skin depth, m
μ Magnetic permeability,
ρ Density, kg/m^3
ρ_e Electrical resistivity, Ω/m
σ_I Estimated current variance, A
σ_T Variance of T values, °C
τ_H Heating time, s

MATRICES
C Capacitance matrix
$\mathbf{C}_c$ Covariance matrix
d Experimental model data
e Experimental error
F Load vector
K Conductance matrix
m Model parameters
s Residuals
T Temperature vector
$\hat{\mathbf{T}}$ Temperatures calculated at the same nodes and times of **Y**
Y Observed temperatures
$\mathbf{Y}_e$ Exact observed temperatures
X Sensitivity coefficients

APPENDIX A. ELECTRO-MAGNETIC MODEL

The proximity-skin effect theory is described by Davies et al. [3]. This theory is here extended to a cylindrical coil-workpiece geometry. The electro-magnetic field effects produced by a coaxial cilindrical coil depend on the current density distribution. The spatial distribution of the current density field $J(r,x)$ is deduced as follows.

According to the proximity theory, a single-turn coil induces a current density distribution (along x) at maximum radius R, given by:

$$J_R(x) = I/(\pi h)\cdot(1+(x/h))^{-2} \tag{A.1}$$

if the vicinity effects of the coil to the workpiece are also considered a corrected $h' = \sqrt{h^2 - R^2_c}$ has to replace h in (A.1), giving:

$$J_R(x) = \left[I/(\pi\cdot h)\cdot\sqrt{1+(R_c/h)^2}\right]\left[1-(R_c/h)^2+(x/h)^2\right]^{-1} \tag{A.2}$$

where R_C is the coil radius. In eq. (A.2) the dependence of J from the coil current parameter I is clear. The current in the coil has to be estimated over all the thermal cycle. The superposition principle may be applied to eq. (A.1) as demonstrated by Ai-Shaikhili et al. [11] to take into account a two-turn coil:

$$J_R(x) = I/(\pi h)\cdot(1+((x-d/2)/h)^2)+(1+((x+d/2)/h)^2 \tag{A.3}$$

where d is the distance between the coil turns (see fig.1).

The maximum power density (per unit of length), that causes the maximum temperature peak in the cycle, is given by:

$$\mathcal{P}_M = \rho_e/\delta \cdot [I/(\pi h)]^2 = \vartheta_m \cdot R \cdot \rho \cdot c(\vartheta_m)/(2\tau_H) \text{ with } \vartheta_m = T_{MAX} - T_a \quad (A.4)$$

where the left hand side comes from proximity theory and the right from empirical considerations.

Equations (A.4) are used in the model to predict an initial value for I which starts the iterative procedure. The more accurate is the measure of the maximum temperature, the closer will result the initial value to the optimal one. The skin effect is taken into account by the variation of the current density function with radius r:

$$\frac{J(r)}{J(R)} = \frac{\mathrm{ber}'\left(\sqrt{2r/\delta}\right) + j\,\mathrm{bei}'\left(\sqrt{2r/\delta}\right)}{\mathrm{ber}'\left(\sqrt{2R/\delta}\right) + j\,\mathrm{bei}'\left(\sqrt{2R/\delta}\right)} \quad (A.5)$$

where the skin depth is given by

$$\delta = [\rho_e/(\pi \cdot \mu \cdot f)]^{0.5} \quad (A.6)$$

Hence in eq.(A.5) $J(r,x)$ are implicitly function of time and temperature, through the temperature dependence of the physical properties.
Finally the specific rate of heat generation in the volume is calculated by:

$$q_v(r,x) = \rho_e \cdot |J(r,x)|^2 \quad (A.7)$$

APPENDIX B. THE INVERSE METHOD

A general physical problem may be modeled by a minimum set of quantities designated as *model parameters* [3]. Some of these quantities, designated as *model data* **d**, have the property of being measurable; other quantities designated as *model parameters* **m**, are not measurable. A general model may be written as:

$$F(\mathbf{d},\mathbf{m}) = 0 \quad (B.1)$$

Two approaches are possible, i.e. a *direct* and *inverse*. In the *direct models* the equation (B.1) is solved in the form:

$$\mathbf{d} = f(\mathbf{m}) \quad (B.2)$$

in which the measurable quantities **d** are calculable provided that the non-measurable **m** are known. This usually happens when, for example, temperatures are calculated by FEM, starting from the prior knowledge of thermophysical properties and boundary/initial conditions. This type of problems are extensively solved in the past in many engineering fields [1,2,4].

In the *inverse models* equation (B.1) is solved in the form:

$$\mathbf{m} = f(\mathbf{d}) \tag{B.3}$$

where, the non-measurable quantities **d** are calculated from the prior knowledge of the measurable **m**. The latter approach has been successfully applied in many engineering fields [3,9] since it allows the evaluation of system features with the aid of measurements. However it has the disadvantage of adding to the ordinary mathematical complexities of models also the incertainties due to measurement errors. As a result, models of inverse type have to rely on mixed *deterministic* and *probabilistic* techniques.

Specifically, if measurement errors are distributed according to a Gaussian law, the probability density of the unknown parameter I may be expressed by:

$$f(I) = \mathrm{K} \times \exp\left(-\frac{1}{2} S\right) \tag{B.4}$$

K is a suitable constant, while S is a function of the square residuals. If the **d** and **m** are replaced by **Y** and I respectively, and **T** are the calculated temperatures, S is:

$$S(I) = \mathbf{s}^{\mathrm{T}}\, \mathrm{C}_{\mathrm{c}}^{-1}\, \mathbf{s} = \left[\hat{\mathrm{T}}(I) - \mathrm{Y}\right]^{\mathrm{T}} \mathrm{C}_{\mathrm{c}}^{-1} \left[\hat{\mathrm{T}}(I) - \mathrm{Y}\right] \tag{B.5}$$

Tarantola [9] demonstrates that it is possible the decribe adequately the probability density of the unknown parameter by calculating the point of maximum likelihood. It can be obtained by searching for the maximum of (B.4), or by searching for the minimum of S with respect to the parameter I:

$$\mathrm{d}S = \frac{\mathrm{d}S}{\mathrm{d}I}\mathrm{d}I = 2\mathrm{X}^{\mathrm{T}}\, \mathrm{C}_{\mathrm{c}}^{-1}\left[\hat{\mathrm{T}}(I) - \mathrm{Y}\right]\mathrm{d}I = 0; \qquad \mathrm{X} = \frac{\mathrm{d}\hat{\mathrm{T}}(I)}{\mathrm{d}I} \tag{B.6-B.7}$$

where **X** is the sensitivity coefficients vector.

If $\tilde{I}$ is an estimate of I, then the model values $\hat{\mathrm{T}}$ may be obtained by linearization about the parameter I in its neighbourhood:

$$\hat{\mathrm{T}}(I) = \hat{\mathrm{T}}(\tilde{I}) + \mathrm{X}(I - \tilde{I}) \tag{B.8}$$

If equation (B.8) is substituted in (B.4) a non-linear equation will result:

$$\mathrm{X}^{\mathrm{T}}\, \mathrm{C}_{\mathrm{c}}^{-1}\left[\hat{\mathrm{T}}(\tilde{I}) + \mathrm{X}\cdot(I - \tilde{I}) - \mathrm{Y}\right] = 0 \tag{B.9}$$

Solution of (B.9) can be achieved numerically by applying the following iterative scheme:

$$I^{k+1} = I^{k} + (P^{k})^{-1} \cdot H^{k} \quad \text{where} \tag{B.10}$$

$$P^k = \mathbf{X}^{\mathrm{T}}(I^k)\cdot \mathbf{C}_{\mathrm{c}}^{-1}\cdot \mathbf{X}(I^k)$$
$$H^k = \mathbf{X}^{\mathrm{T}}(I^k)\cdot \mathbf{C}_{\mathrm{c}}^{-1}\cdot \left[\hat{\mathbf{T}}(I^k) - \mathbf{Y}\right] \qquad \text{(B.11)}$$

The standard deviation of the parameter I can be estimated as a measure of the procedure accuracy according to:

$$\sigma_I^2 = \mathbf{X}^{\mathrm{T}}(I_{ML})\cdot \mathbf{C}_{\mathrm{c}}^{-1}\cdot \mathbf{X}(I_{ML}) \qquad \text{(B.12)}$$

where I_{ML} is the estimate of the parameter by the maximum likelihood approach.

Finally, the accuracy index of the method to model induction heating experiments is estimated by [3]:

$$\sigma_S^2 = \frac{\mathbf{X}^{\mathrm{T}}(I_{ML})\ \mathbf{X}(I_{ML})}{(N\cdot M)} \qquad \text{(B.13)}$$

RESOLVING POWER OF THE INVERSE PROBLEM ABOUT PALEOCLIMATE[I]

V.I.DMITRIEV[II] , S.G.KOSTYANEV[III]

ABSTRACT

It has long been recognized that terrestrial heat flow density observations from boreholes may be biased due to non-steady surfase temperature conditions. This calls for paleoclimatic corrections which have conventionally been obtained by solving a forward problem based on the equation of heat conduction and models of the surface temperature history and the thermal properties structure. While computationally convenient, the forward method fails (1) to provide a stringent quantitative representation of all available data with confidence limits, (2) to extract all available information in an optimum manner and (3) to provide confidence limits on the parameters of interest. An inverse method in heterehenеous medium which overcomes these shortcomings and an estimation of the resolving power of the inverse problem about the paleoclimate are described in this paper.

1. INTRODUCTION

The problems of the past climate extraction from the geothermal measurements have been attracting ever increasing attention for more than a decade. Tichonov (1935) was the first to pay attention to the possibility of the determination of the alterations of the surface temperature in the past by present observations of temperature alteration in depth. He has proved the uniqueness of the inverse problem as aforesaid upon a specified stratified distribution of thermal and physical parameters. The first evaluation of the paleoclimate has been carried out. The interest toward the paleoclimate problem has increased recently because of the possibility

---

[I] Paleoclimate - climate of previous geological epochs

[II] Academician, Moscow University, Moscow, Russia

[III] Professor, University of Mining and Geology, Sofia, Bulgaria

of a PC-solution. In this connection such papers as Beck [1977, 1982]; Shen & Beck [1983]; Vasseur et al., [1983]; Lachenbruch & Marshal (1986); Nielsen and Beck (1989) and many others can be mentioned.

The present paper submits the genegal setting of the inverse problem in a heterogeneous medium and an estimation of the resolving power of the inverse problem about the paleoclimate. The investigation has taken into consideration the depths, at which the measurements are to be carried out, as well as the problems that may be resolved with these measurements at the specified precision and the models of paleoclimate which are advisable to consider.

2. FORMULATION OF THE INVERSE PROBLEM

The inverse problem of the paleoclimate is set upon the assumption of a specified distribution of thermal and physical parameters, whereas at depth z≥3 km there is a stationary distribution of temperature

$$T_s(z) = T_0 + \frac{q}{\lambda} z, \quad z \geq H, \tag{1}$$

where T_0 is the average surface temperature till the last glacial period, q-terrestrial heat flow, and λ-coefficient of thermal conductivity. The deviation of temperature from the stationary distribution (1) is considered, hereinafter named abnormal distribution and designated by $T_a(x, y, z, t)$. The abnormal temperature is the solution of the boundary value poblem in the strata $0 < z < H$

$$\begin{cases} c\rho \dfrac{\partial T_a}{\partial t} = \operatorname{div}(\lambda \operatorname{grad} T_a), \quad t > 0, \\ T_a(x, y, z, t = 0) = 0, \\ T_a(x, y, z = 0, t) = \varphi(t), \\ \left.\dfrac{\partial T_a}{\partial z}\right|_{z = H} = 0, \end{cases} \tag{2}$$

where c - is the specific heat capacity, and ρ - is the density.
The reverse problem consists of the determination of the temperature distribution on the earth's surface

$$T(z)\Big|_{z=0} = T_0 + \varphi(t)$$

in the past at $0 < t < t_1$ upon a specified abnormal temperature distribution in depth

$$T_a(x_0, y_0, z, t = t_1) = \psi_1(z).$$

This way by $\psi_1(z)$ we have to find out $\varphi(t)$.

3. REDUCTION OF THE INVERSE PROBLEM TO AN INTEGRAL EQUATION OF THE FIRST KIND

The resolution of the non-stationary boundary value problem of heat conductivity (2) according to the principle of Dugamel, may be represented in the following form:

$$T_a(x,y,z,t) = \int_0^t \varphi(t) \frac{\partial U(x,y,z,t-\tau)}{\partial t} d\tau, \tag{3}$$

where the $U(x,y,z,t)$ function is the solution of the boundary value problem

$$\begin{cases} c\rho \dfrac{\partial U}{\partial t} = \operatorname{div}(\lambda \operatorname{grad} U), \quad 0 < t, \ 0 < z < H, \\ U(x,y,z,t=0) = 0, \\ U(x,y,z=0,t) = 1, \\ \left.\dfrac{\partial U}{\partial z}\right|_{z=H} = 0. \end{cases} \tag{4}$$

As long as the thermal and physical parameter distribution in the strata $0 < z < H$ may be considered specified, we may always calculate the distribution U in depth in the drill-hole, as well as in time

$$\frac{\partial U(x_0,y_0,z,t)}{\partial t} = K(z,t) \tag{5}$$

In the case where $K(z,t)$ is specified, and making use of the experimental observations of the abnormal temperature field in the drill-hole at $t = t_1$:

$$T_a(x_0,y_0,z,t=t_1) = \psi_1(z), \tag{6}$$

pursuant to (3), we shall obtain:

$$A\varphi = \int_0^{t_1} \varphi(\tau) K(z,t_1-\tau) d\tau = \psi(z), \ 0 < z < H. \tag{7}$$

Thus we have obtained an integral equation (7) for $\varphi(\tau)$, where $K(z,t)$ and $\psi(z)$ are specified. This integral equation is a Fredholm one, of the first order, the solution of which is unstable. Therefore its solution makes use of the regularization method of (Tikhonov, Arsenin 1979), founded on the use of apriori information, the solution $\varphi_g(t)$. Assuming that the hypothetical temperature distribution on the earth's surface $\varphi(\tau)$ is specified apriori, the solution of the equation (7) may be obtained by the variational problem:

$$\min_{\varphi(t)} \left\{ \|A\varphi - \psi\|^2 - \alpha\|\varphi - \varphi_g\|^2 \right\} \tag{8}$$

where $\| \|^2$ - designates the square of the average square norm, i.e.

$$\|\varphi - \varphi_g\|^2 = \int_0^{t_1} [\varphi(t) - \varphi_g(t)]^2 dt, \tag{9}$$

$$\|A\varphi - \psi\|^2 = \int_0^H \left\{ \int_0^{t_1} \varphi(\tau) K(z, t - \tau) d\tau - \psi(z) \right\}^2 dz, \tag{10}$$

while the constant α designates the regulation parameter and determines the meaning of the apriori information about the solution, Euler's equation for the variational problem (8) is the integral equation of Fredgolm of second order

$$\alpha(\varphi(\tau) - \varphi_g(\tau)) + \int_0^{t_1} \varphi(\tau_0) M(\tau - \tau_0) d\tau_0 = f(\tau), \tag{11}$$

whre

$$M(\tau, \tau_0) = \int_0^H K(z, t_1 - \tau) K(z, t_1 - \tau_0) dz, \tag{12}$$

$$f(\tau) = \int_0^H \psi(z) K(z, t_1 - \tau) dz. \tag{13}$$

Equation (11) is resolved by standard methods, whereas the regulation parameter α is chosen amongst the conditions of execution of the equation (7) with an accuracy equal to the accuracy of $\psi(z)$ measurement.

4. INVESTIGATION OF THE RESOLVING POWER OF THE METHOD

An investigation of the resolving power of the method has to consider the depths at which the measurements are to be taken, the kind of problems that may be solved by these measurements at the specified precision, and the kind of paleoclimate models to be investigated. In order to answer these questions, one has only to consider the problem of penetration into the homogeneous environment of the temperature impulse, specified on the earth surface. Thus we obtain the boundary value problem

$$\begin{cases} \dfrac{\partial U}{\partial t} = a^2 \dfrac{\partial^2 U}{\partial z^2}, & 0 < z, \quad 0 < t, \\ U(z,t)\Big|_{t=0} = T_0 + \dfrac{q}{\lambda} z, & \\ U(z,t)\Big|_{z=0} = \varphi(t) = \begin{cases} T_1, & 0 < t < t_0, \\ T_0, & t > t_0, \end{cases} \end{cases} \tag{14}$$

where $a^2 = \lambda / c\rho$ - coefficient of temperature conductivity.

The abnormal temperature

$$T_a(z,t) = U - T_0 - \frac{q}{\lambda} z$$

is expressed analytically by an integral of probability

$$T_a(z,t) = (T_0 - T_1)\left(\Phi\left(z / 2\sqrt{a^2 t}\right) - \Phi\left(z / 2\sqrt{a^2 (t - t_0)}\right)\right), \tag{15}$$

where

$$\Phi(x) = \frac{2}{\sqrt{\pi}} \int_0^x e^{-y^2} dy. \tag{16}$$

Table 1 contains the relative abnormal temperature calculation

$$\delta U = \frac{T_a(z,t)}{T_0 - T_1} (\Phi(\xi) - \Phi(\xi, \nu)) \tag{17}$$

is dependant on the parameters

$$\xi = \frac{z}{2\sqrt{a^2 t}}; \qquad \nu = \sqrt{\frac{t}{t - t_0}}. \tag{18}$$

Table 1. Relative abnormal temperature calculation

$\xi \backslash \nu$	1.1	1.25	1.5	1.75	2	2.5	3
0.1	-0.011	-0.028	-0.056	-0.083	-0.110	-0.164	-0.216
0.2	-0.022	-0.054	-0.106	-0.157	-0.206	-0.298	-0.381
0.3	-0.031	-0.075	-0.147	-0.224	-0.275	-0.389	-0.468
0.4	-0.038	-0.092	-0.176	-0.259	-0.314	-0.414	-0.482
0.5	-0.043	-0.103	-0.191	-0.264	-0.322	-0.402	-0.446
0.6	-0.045	-0.107	-0.193	-0.258	-0.306	-0.362	-0.385
0.7	-0.046	-0.106	-0.185	-0.239	-0.274	-0.309	-0.319
0.8	-0.045	-0.100	-0.168	-0.210	-0.234	-0.253	-0.257
1.0	-0.038	-0.080	-0.123	-0.144	-0.152	-0.157	-0.157
1.2	-0.028	-0.056	-0.079	-0.083	-0.089	-0.090	-0.090
1.4	-0.018	-0.034	-0.045	-0.048	-0.048	-0.048	-0.048

Fig.1 shows the graphs $\delta U(\xi,\nu)$. It is quite easy to observe that the abnormal temperature field may be disregarded when one of two conditions are satisfied:

$$1.\ t \geq 3t_0 \quad \text{at every} \quad \xi \tag{19}$$

$$2.\ \xi \geq 1,4 \quad \text{at every} \quad \nu \tag{20}$$

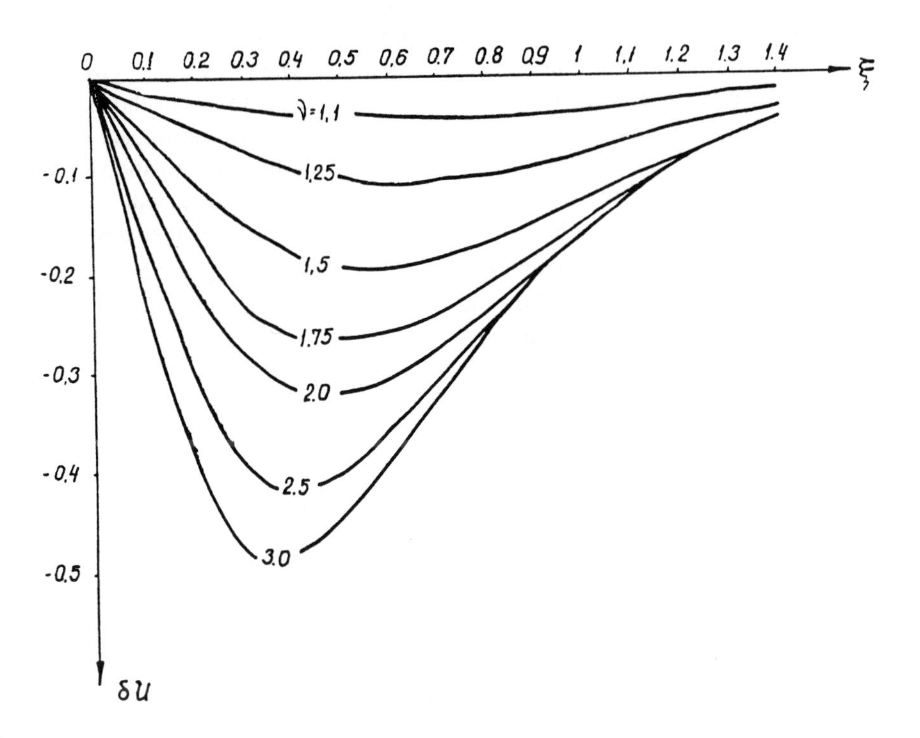

Fig.1. The graphs $\delta U(\xi, \nu)$

The second condition may be determined as a correlation between the depth z and the time t, specified $a^2 = 8.10^{-7}\ m^2/sec$, $t(sec) = 3{,}15.10^7.t_1(years)$. Then (20) will change as follows:

$$z(M) \geq 14\sqrt{t_1(years)}, \tag{21}$$

which means that the abnormal temperature may be disregarded at a depth of more than $14\sqrt{t_1(years)}$, where t_1 - is the time from the initiation of the temperature impulse. The obtained conditions (19) and (20), the abnormal temperature field, allow us to estimate the resolving power of the reverse problem about paleoclimate. Condition (19) shows that if we come out of the termination of the impulse at time $(t - t_0)$, two-times longer than the t_0 impulse, i.e.

$$t - t_0 \geq 2t_0 \tag{22}$$

then the influence of the thermal impulse may be disregarded. The time $t_r = t - t_0 = 2t_0$ may be named relaxation of the thermal impulse. The evaluation of temporary relaxation (22) enables us to evaluate the temporary resolving power of the reverse problem about paleoclimate. This way the last glacial impulse has terminated before 10 thousand years, whereas it has lasted for about 60 thousand years, hence its impact must have been preserved in the depth distribution. The preceding glacial impulse terminated 130 thousand years ago, and lasted about 65 thousand years which means that condition (22) is satisfied in fact its impact has been lost in the depth of present temperature. Thus the investigations on the problem of paleoclimate may consider the temperature distribution on the surface in the recent 100 thousand years only, assuming that beyond that time temperature has a certain mean value of T_0, and there is a mean depth thermal flow q. Therefore the paleoclimate model of temperature distribution has the form as follows:

$$\begin{cases} T(z,t)\Big|_{z=0} = T_0 + T_1(t), \\ T_1(t) = \begin{cases} 0, & t > 0, \\ \varphi(t), & 0 < t < 100 \text{ th. years} \end{cases} \end{cases} \tag{23}$$

The initial moment of time is taken 10^5 years before the present moment. Now we shall consider the degree of elaboration to investigate the function of abnormal temperature of surface $\varphi(t)$ with. The initiation of the glacial period dates back about 70 thousand years from the present moment, therefore the 10 to 20 - thousand year long impulses can not be solved there. This means that we shall have to model the initiation of the glacial period by way of a temperature degree. The end of the glacial period was 10 thousand years ago, hence we may solve with an accuracy of 5 thousand years the effect of gradual transition from the glacial temperature to the present. Summarizing the aforesaid, we may draft the model of the temperature impulse of the glacial period as follows:

$$\varphi(t) = \begin{cases} 0, & 0 < t < t_1, \\ -\dfrac{T_1}{2}\left(1 - \text{th}\dfrac{t - t_2}{t_0}\right), & t_1 < t < 99.9 \text{ th. years} \\ \varphi_1(t), & 99.9 \text{ th. years} < t < 100 \text{ th. years} \end{cases} \tag{24}$$

where T_1 - is the amplitude of the temperature impulse: t_1 - the time of initiation of the latest glacial temperature impulse: t_2 - mean time for the termination of the latest glacial temperature impulse, when the amplitude of the temperature impulse has decreased two times: τ_0 - determines the increasing speed of temperature at the end of the glacial period; $\varphi_1(t)$ - a function determining the change in temperature on the earth surface in the latest 100 years with the variation cycles of the solar activity and changes in climate on Earth through centuries.

Fig.2 shows the typical models of paleoclimate at $t_1=30$ thousand years, $t_2=70$ and 80 th.y. and $\tau_0=5$ and 10 th.y. Thus the reverse problem of paleoclimate is reduced to determination of parameters T_1, t_1, t_2, τ_0 and the function $\varphi_1(t)$. The influence of the $\varphi_1(t)$ function is easy to deal with.

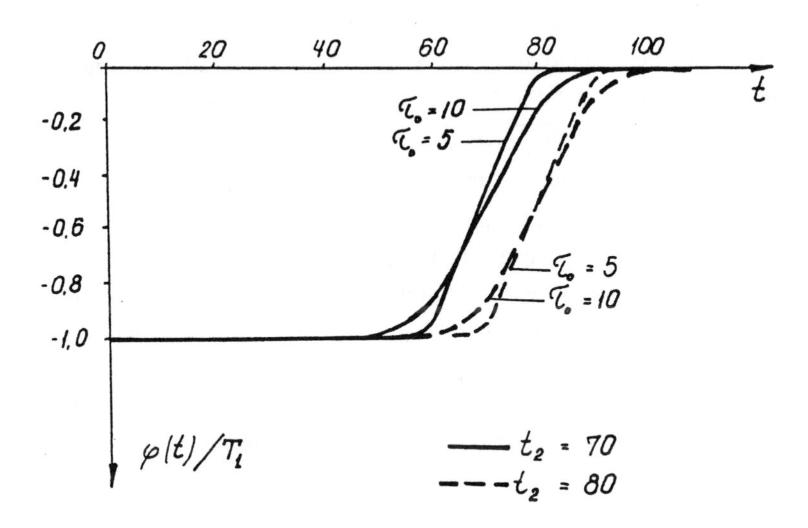

Fig.2.Typical models of paleoclimate

For pursuant to (21) the temperature impulse of 100 years length has totally died away at a depth of 140 meters, and taking measurements at $z \geq 140$ m we may consider $\varphi_1 = 0$. Thus the paleoclimate problem is reduced to the determination of four constants T_1, t_1, t_2, τ_0. Pursuant to (8) we may reduce all these to the problem of minimizing:

$$\min_{T_1, t_1, t_2, \tau_0} \left\{ \int_h^H \left[\int_{t_1}^{t} \varphi(\tau) K(z, t-\tau) d\tau - \psi(z) \right]^2 dz + \right.$$

$$\left. +\alpha \left[\left(t_1 - t_1^g\right)^2 + \left(t_2 - t_2^g\right)^2 + \left(\tau_0 - \tau_0^g\right)^2 \right] \right. \tag{25}$$

where t_1^g, t_2^g, τ_0^g ,are the relevant hypothetical parameters, $t=100$ th.y., $h=140$ m, and

$$\varphi(\tau) = -\frac{T_1}{2}\left(1 - th\frac{t - t_2}{\tau_0}\right). \tag{26}$$

yet there is the problem of the resolving power's dependence on the depth of measurements z, in order to determine the needed depth of investigation H and the step of measurements in depth. For this purpose to investigate we need the evaluation (21) determining the depth, at which the impact of the impulse may be disregarded as well as the evaluation of the depth, at which the impact is maximum. For the glacial impulse

$$\nu = \sqrt{\frac{t}{t - t_0}} \approx 2.5$$

and therefore max δU is obtained at $\xi=0.4$. That means the maximum impact of the latest glacial impulse is approximately equal to

$$z_{max} = 4\sqrt{t(years)}$$

here t - is the time from the initiation of the glacial impulse to present days ($t \approx 70$ th.y.), hence, $z_{max} \approx 1060$ m. The influence of the glacial impulse shall totally die away at a depth of H=3000÷3500 m. Yet the impact of the impulse is well preserved at $\xi=0.1$, which determines the minimum depth of measurements $z_{min} \approx 250$ m. On the grounds of all evaluations we may set out the approximate network of measurements: $z_1 = 250$ m, $z_2 = 750$ m, $z_3 = 1250$ m, $z_4 = 1750$ m, $z_5 = 2250$ m and $z_6 = 2750$ m, which will enable us to analyze an interval of time - 600 to 90000 years.

Fig.3 shows the present temperature distribution in depth at $T_1 = 10°C$, $t_1 = 30$ th.y., $t_2 = 80;90$ th.y. and respectively $\tau_0 = 10;5$ th.y.

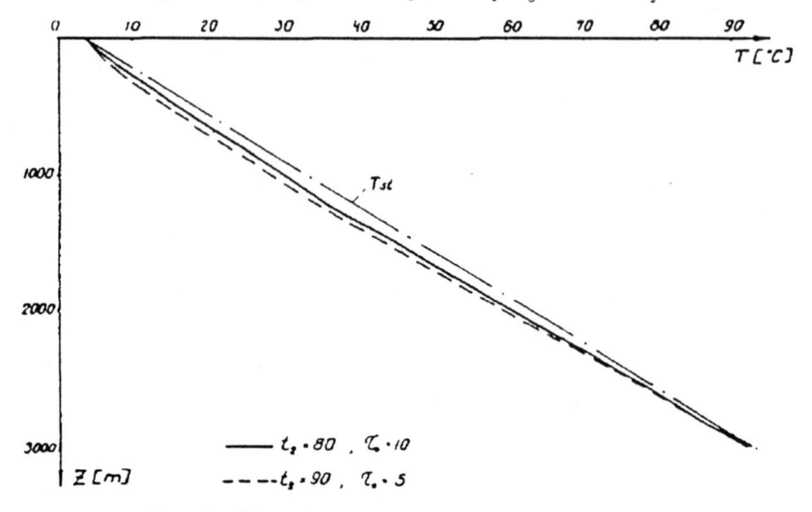

Fig. 3a. The distribution of the total temperature

Fig.3b shows the distribution of the abnormal temperature $T_a(z)$, while fig.3a shows the distribution of the total temperature, whereas the stationary distribution has been taken into consideration

$$T_s = T_0 + \beta z,$$

where $T_0 = 3°C$, $\beta = 0,03°/m$.

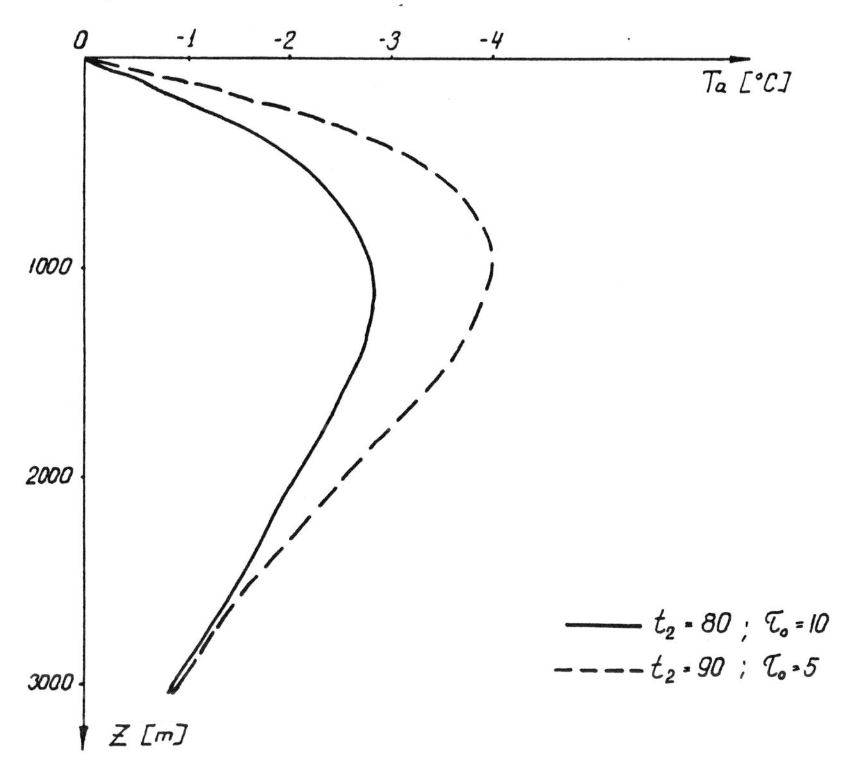

Fig 3b. The distribution of the abnormal temperature

It is evident that the measurement points at 2250 and at 2750 m depth are needed only for the determination of the stationary temperature. The basic information about the glacial temperature impulse is carried by points at 250, 750 and 1250 and 1750 m. These four points are enough for the description of the abnormal field, for by the measurements of the temperature field at these depths, it is always possible with the help of a spline-approximation, to make a continuous distribution of the abnormal temperature. The interpretation of data from geothermal investigations in a drill-hole is carried out in the following way. The changes at depths up to 3 km being specified, we shall draw a linear asymptotic upon the depth points, and the stationary distribution of temperature $T_s(z)$ is to be determined. Reading the $T_s(z)$ from the total field, we shall find out the abnormal distribution of temperature $T_a(z)$. To decrease error in the measurements of $T_a(z)$, a spline-approximation of the curve $T_a(z)$ is drawn. This continuous change in temperature is used in variation problem (25-26) for the determination of the parameters of the glacial impulse T_1, t_1, t_2 and τ_0. Having solved the problem of paleoclimate we may determine the average temperature at small depths by the equation:

$$\overline{T}(z) = T_s(z) + T_a(z).$$

If there is a distribution of temperature, measured in the drill-hole T(z) at small depths - $10\ \text{m} \le z \le 200\ \text{m}$, then by way of

$$\overline{\psi}(z) = T(z) - \overline{T}(z)$$

we may consider the reverse problem of the average temperature change for the last 100 years on the earth surface. For this purpose it is necessary to solve the integral equation (11). This is how the analysis shows that with measurements of temperature in the drill-hole with a precision greater than 5°C it is possible to draw a model of the temperature impulse for the last glacial period, and with more precise measurements at small depths it is possible to restore the average temperature on the earth surface in the course of the last 100 years.

5. CONCLUSIONS

The inverse problem of paleoclimate in a heterogeneous medium is set upon the assumption of a speicfied distribution of thermal and physical parameters, whereas at depth $z \ge 3$ km there is a stationary distribution of temperature. An investigation of the resolving power of the method has to consider the depths at which the measurements are to be taken, the kind of problems that may be solved by these measurements at the specified precision, and the kind of paleoclimate models to be investigated.

Interpretation of the data from geothermal investigations in a drill-hole can be carried out. To decrease error in the measurements of anomalous temperature, a spline-approxination is drawn. This continuous change in temperature is used in corresponding variation problem for the determination of the parameters of the glacial impulse.

6. ACKNOWLEDGMENTS

We gratefully acknowledge the Ministry of education and science - Bulgaria for financial support.

REFERENCES

1. BECK, A.E. - Climatically perturbed temperature gradients and their effect on regional and continental heat-flow means. Tectonophysics, 41:17-19, 1977.
2. BECK, A.E. - Precision logging of temperature gradients and the extraction of past climate. Tectonophysics, 83, 1-11, 1982.
3. LACHENBRUCH, A.H., MARSHALL B.V. - Changing climate: Ceothermal Evidence from Permafrost in the Alaskan Arctic, Science, vol. 234, pp. 689-696, 1986.
4. NIELSEN, S.B. and BECK, A.E. - Heat-flow density values and paleoclimate determined from stochastic inversion of four temperature-depth profiles from the Superior Province of the Canadian Shield, Tectonophysics, 164, 345-359 1989.
5. SHEN, P.Y. and BECK, A.E. - Determination of surface temperature history from bore-hole temperature gradients. J.Geophysics Res., 88, 7485-7493, 1983.
6. TIKHONOV, A.N. - Dokladi AN SSSR, 294-300, V.I.1935.
7. TIKHONOV, A.N. and ARSENIN, V.J. - Metodi reshenija nekorektnih zadach. Moscow, Nauka, 1979.
8. VASSEUR, G. at al., - Holocene paleotemperatures deduced from geothermal measurements. Palaeogeogr., Palaeoclimatol., Palaeoecol., 43, pp. 237-259, 1983.

DETERMINATION OF TRANSFER FUNCTIONS IN MULTIDIMENSIONAL HEAT CONDUCTION BY MEANS OF A FINITE ELEMENT TECHNIQUE.

F.Marcotullio and A.Ponticiello
Dipartimento di Energetica-Università di L'Aquila
Monteluco di Roio-67100 L'Aquila-Italy

SUMMARY

The Z-transfer function method is an efficient way of predicting thermal histories in multidimensional linear systems. A new general procedure to evaluate the heat conduction transfer functions (HCTF) by means of a finite element technique is discussed in this paper. A major feature is that the proposed procedure deals only with thermophysical data processing rather than input/output identification techniques. Thermal response to triangular pulse perturbation related to two-dimensional systems with different boundary conditions is also examined. The results show a good agreement with Ansys code numerical analysis.

1. INTRODUCTION

The prediction of temperature-history in multidimensional heat conduction generally requires a strong computation effort in terms of both time processing and large-scale computer implementation. However, in many applications, an efficient way to perform thermal analysis is to use *input/output* mathematical representations, especially when simulations under various input conditions must be carried out. This approach, which is known as heat conduction transfer functions (HCTF) method, takes advantage of the Z-transformation in order to obtain the transfer functions as the ratio of two polynomials with argument $Z^{-1}=\exp(-s\Delta\tau)$, where s is the Laplace parameter and $\Delta\tau$ the sampling time interval. Once the polynomial coefficients are determined at any selected point of the thermal field, the time-series of the output (temperature response) can be related to any kind of time-series of the input (transient boundary conditions). This method has been successfully employed in the past in one-dimensional applications[1,2] and very recently it also has been applied to multidimensional fields [3,4,5].

Unfortunately, it is not easy to determine Z-transfer functions. In

[4,5] they are identified through the knowledge of the system thermal response to a double ramp perturbation which is determined either by numerical analysis (finite-difference) or by an experimental approach in controlled regime chambers.

In this study we have developed an improved technique for the direct evaluation of the multidimensional HCTF coefficients whose values and number depend only on the system properties and time step $\Delta\tau$. For most applications of practical interest the number of coefficients is generally lower then six in each series.

2. THE PROCEDURE PROPOSED

To set up the proposed procedure we start from the well-known finite element heat conduction NxN equations [6]:

$$\mathbf{C}\dot{\mathbf{a}} + \mathbf{K}\mathbf{a} - \mathbf{f} = 0 \tag{2.1}$$

in which $\mathbf{a}$ is the vector of unknown nodal temperatures of the system and all the matrices are assembled from element submatrices in the standard manner with submatrices $\mathbf{K}^e$, $\mathbf{C}^e$ and $\mathbf{f}^e$ given by [7]:

$$K_{ij}^e = \int_{V^e} \left(k_x \frac{\partial N_i}{\partial x}\frac{\partial N_j}{\partial x} + k_y \frac{\partial N_i}{\partial y}\frac{\partial N_j}{\partial y} + k_z \frac{\partial N_i}{\partial z}\frac{\partial N_j}{\partial z} \right) dV^e + \int_{S^e} h N_i N_j \, dS^e \tag{2.2a}$$

$$C_{ij}^e = \int_{V^e} \rho c N_i N_j \, dV^e \tag{2.2b}$$

$$f_i^e = -\int_{V^e} \dot{q} N_i \, dV^e + \int_{S_1^e} q N_i \, dS_1^e - \int_{S_2^e} h N_i \bar{a} \, dS_2^e \tag{2.2c}$$

For the sake of simplicity we only consider convective boundary conditions so $q=\dot{q}=0$. Then vector $\mathbf{f}$ of eq. 2.1 holds only the m terms of the global domain with surrounding temperatures $\bar{a}$. Now by incorporating in $\mathbf{a}$ the m forcing temperatures $\bar{\mathbf{a}}$, eq.2.1 may be rewritten as:

$$[\mathbf{C}\ \ 0] \frac{d}{d\tau}\begin{Bmatrix}\mathbf{a}\\ \bar{\mathbf{a}}\end{Bmatrix} + [\mathbf{K}\ \ \mathbf{K}_1]\begin{Bmatrix}\mathbf{a}\\ \bar{\mathbf{a}}\end{Bmatrix} = \bar{\mathbf{C}} \frac{d}{d\tau}\begin{Bmatrix}\mathbf{a}\\ \bar{\mathbf{a}}\end{Bmatrix} + \bar{\mathbf{K}}\begin{Bmatrix}\mathbf{a}\\ \bar{\mathbf{a}}\end{Bmatrix} = \mathbf{0} \tag{2.3}$$

where all matrices are $Nx(N+m)$.

Since attention is focused on the study of a selected nodal temperature a, the overall vector of eq.2.3 can be rearranged in order to obtain two separate groups:

- the former, $\mathbf{a}_r$, holding firstly the temperature a of the investigated node followed by the m forcing terms $\bar{\mathbf{a}}$:

$$\mathbf{a}_r^T = \{a \; \bar{\mathbf{a}}\} \tag{2.4}$$

- the latter, $\mathbf{a}_s$, holding the remaining $N\text{-}1$ nodal temperature parameters.

After reordering and partitioning, equation 2.3 can be rewritten as:

$$\begin{vmatrix} \bar{\mathbf{C}}_{1r} & \bar{\mathbf{C}}_{1s} \\ \bar{\mathbf{C}}_{sr} & \bar{\mathbf{C}}_{ss} \end{vmatrix} \begin{Bmatrix} \dot{\mathbf{a}}_r \\ \dot{\mathbf{a}}_s \end{Bmatrix} + \begin{vmatrix} \bar{\mathbf{K}}_{1r} & \bar{\mathbf{K}}_{1s} \\ \bar{\mathbf{K}}_{sr} & \bar{\mathbf{K}}_{ss} \end{vmatrix} \begin{Bmatrix} \mathbf{a}_r \\ \mathbf{a}_s \end{Bmatrix} = \mathbf{0} \tag{2.5}$$

where the subscripts indicate the matrices dimensions. The previous equations are equivalent to:

$$\bar{\mathbf{C}}_{1r}\,\dot{\mathbf{a}}_r + \bar{\mathbf{C}}_{1s}\,\dot{\mathbf{a}}_s + \bar{\mathbf{K}}_{1r}\,\mathbf{a}_r + \bar{\mathbf{K}}_{1s}\,\mathbf{a}_s = \mathbf{0} \tag{2.6a}$$

$$\bar{\mathbf{C}}_{sr}\,\dot{\mathbf{a}}_r + \bar{\mathbf{C}}_{ss}\,\dot{\mathbf{a}}_s + \bar{\mathbf{K}}_{sr}\,\mathbf{a}_r + \bar{\mathbf{K}}_{ss}\,\mathbf{a}_s = \mathbf{0} \tag{2.6b}$$

Solving eq.2.6b, we obtain:

$$\mathbf{a}_s = -\bar{\mathbf{K}}_{ss}^{-1}\left(\bar{\mathbf{C}}_{sr}\,\dot{\mathbf{a}}_r + \bar{\mathbf{C}}_{ss}\,\dot{\mathbf{a}}_s + \bar{\mathbf{K}}_{sr}\,\mathbf{a}_r\right) \tag{2.7a}$$

and, operating the time-derivative:

$$\dot{\mathbf{a}}_s = -\bar{\mathbf{K}}_{ss}^{-1}\left(\bar{\mathbf{C}}_{sr}\,\ddot{\mathbf{a}}_r + \bar{\mathbf{C}}_{ss}\,\ddot{\mathbf{a}}_s + \bar{\mathbf{K}}_{sr}\,\dot{\mathbf{a}}_r\right) \tag{2.7b}$$

If $\ddot{\mathbf{a}}_r$ and $\ddot{\mathbf{a}}_s$ are neglected in eq.2.7b, then:

$$\dot{\mathbf{a}}_s = -\bar{\mathbf{K}}_{ss}^{-1}\,\bar{\mathbf{K}}_{sr}\,\dot{\mathbf{a}}_r \tag{2.7c}$$

and eq.2.7a yields:

$$\mathbf{a}_s = -\bar{\mathbf{K}}_{ss}^{-1}\left[\left(\bar{\mathbf{C}}_{sr} - \bar{\mathbf{C}}_{ss}\,\bar{\mathbf{K}}_{ss}^{-1}\,\bar{\mathbf{K}}_{sr}\right)\dot{\mathbf{a}}_r + \bar{\mathbf{K}}_{sr}\,\mathbf{a}_r\right] \tag{2.7d}$$

The substitution of eq. 2.7c,d in eq.2.6a finally gives:

$$\left(\bar{\mathbf{C}}_{1r} - \bar{\mathbf{C}}_{1s}\,\bar{\mathbf{K}}_{ss}^{-1}\,\bar{\mathbf{K}}_{sr} - \bar{\mathbf{K}}_{1s}\,\bar{\mathbf{K}}_{ss}^{-1}\,\bar{\mathbf{C}}_{sr} + + \bar{\mathbf{K}}_{1s}\,\bar{\mathbf{K}}_{ss}^{-1}\,\bar{\mathbf{C}}_{ss}\,\bar{\mathbf{K}}_{ss}^{-1}\,\bar{\mathbf{K}}_{sr}\right)\dot{\mathbf{a}}_r + \left(\bar{\mathbf{K}}_{1r} - \bar{\mathbf{K}}_{1s}\,\bar{\mathbf{K}}_{ss}^{-1}\,\bar{\mathbf{K}}_{sr}\right)\mathbf{a}_r = 0 \tag{2.8}$$

or in a compact form:

$$\mathbf{M}_1\,\dot{\mathbf{a}}_r + \mathbf{M}_0\,\mathbf{a}_r = 0 \tag{2.9}$$

in which $\mathbf{a}_r$ is expressed by eq.2.4 and all matrices are $1x(m+1)$.
The set of eq.2.9 can be solved by performing a time domain discretization. Setting:

$$\mathbf{a}_r \approx \sum N_i\,\mathbf{a}_{ri} \tag{2.10}$$

and chosing the shape function $N_i(\tau)$ linear we obtain the following matrix recurrence formula [6]:

$$\left(\frac{\mathbf{M}_1}{\Delta\tau} + \mathbf{M}_0\,\theta\right)\mathbf{a}_{r(n+1)} + \left(-\frac{\mathbf{M}_1}{\Delta\tau} + \mathbf{M}_0(1-\theta)\right)\mathbf{a}_{r(n)} = 0 \tag{2.11}$$

or equivalently:

$$\mathbf{U}_{n+1}\,\mathbf{a}_{r(n+1)} + \mathbf{U}_n\,\mathbf{a}_{r(n)} = 0 \tag{2.12}$$

After operating matrix products we obtain:

$$U_{n+1,1}\,a_{n+1} + \sum_{i=1}^{m} U_{n+1,i+1}\,\bar{a}_{n+1,i} + + U_{n,1}\,a_n + \sum_{i=1}^{m} U_{n,i+1}\,\bar{a}_{n,i} = 0 \tag{2.13}$$

Eq.2.13 can be solved by a step-by-step procedure starting from an initial known pattern. The nodal temperature a_{n+1} at time $n+1$ is obtained from the values of the forcing terms $\bar{\mathbf{a}}_{n+1}$ at the same time instant $n+1$ and from the analogous values of the nodal temperature a_n and forcing terms $\bar{\mathbf{a}}_n$ at an earlier time instant n. As a result:

$$a_{n+1} = -\sum_{i=1}^{m} \frac{U_{n+1,i+1}}{U_{n+1,1}}\,\bar{a}_{n+1,i} - \frac{U_{n,1}}{U_{n+1,1}}\,a_n - \sum_{i=1}^{m} \frac{U_{n,i+1}}{U_{n+1,1}}\,\bar{a}_{n,i} \tag{2.14}$$

and, by introducing appropriate coefficients (HCTF's coefficients):

$$d_1\, a_{n+1} = d_0\, a_n + \sum_{i=1}^{m} (b_{0,i}\, \bar{a}_{n,i} + b_{1,i}\, \bar{a}_{n+1,i}) \tag{2.15}$$

being:

$$d_1 = 1 \; ; \; d_0 = -\frac{U_{n,1}}{U_{n+1,1}} \; ; \qquad b_{0,i} = -\frac{U_{n,i+1}}{U_{n+1,1}} \; ; \; b_{1,i} = -\frac{U_{n+1,i+1}}{U_{n+1,1}} \tag{2.16}$$

In conclusion, for any selected nodal temperature we can associate the following set of coefficients whose values depend on the physical properties of the system and the time interval $\Delta\tau$:

$$\left.\begin{matrix} d_0 \;,\; d_1 = 1 \\ b_{0,1} \;,\; b_{1,1} \\ \vdots \\ b_{0,m} \;,\; b_{1,m} \end{matrix}\right\} \; (m+1) \textit{ series two coefficients each}$$

A different result would have been achieved if in eq.2.7b truncation had been performed to the third time-derivatives instead of $\ddot{\mathbf{a}}_r$ and $\ddot{\mathbf{a}}_s$. In this case, the procedure would have led to an equation of the type:

$$\mathbf{M}_2\, \ddot{\mathbf{a}}_r + \mathbf{M}_1\, \dot{\mathbf{a}}_r + \mathbf{M}_0\, \mathbf{a}_r = 0 \tag{2.17}$$

The recurrence formula would have also increased its order involving two recurrence parameters, three time-series ($n+1$, n, $n-1$) and, correspondingly, three coefficients for each of the $m+1$ series.
At this stage, the problem is how to determine proper values for the recurrence parameters in order to ensure convergence and stability. The criterion followed was that of matching the free response ($\bar{\mathbf{a}}=0$) of the system through the numerical solution arising from the recurrence scheme. For each of the natural modes, the time-series a^*_{n+1}, a^*_n, a^*_{n-1}.... are correlated to each other by the characteristic decay term $y_i = e^{-\Delta\tau/\lambda_i}$, being λ_i the eigenvalues of the matrix $\mathbf{K}^{-1}\mathbf{C}$ obtained from eq.2.1:

$$a^*_{n+1} = y_i\, a^*_n = y_i^2\, a^*_{n-1} = ... = y_i^p\, a^*_{n+1-p} \tag{2.18}$$

By substituting eq.2.18 in the corrisponding ($p+1$)-point recurrence scheme we obtain a set of p algebraic equations for the p unknown recurrence parameters. The p decay terms that should be considered depend on their importance in determining the free response. Thus, eq.2.17 will retain as many time-wise derivatives as the eigenvalues which are greater than a certain cut-off value. This will also apply to the recurrence parameters. A special algorithm was developed for two-dimensional domains using linear triangular elements. The code was written in FORTRAN linked to mathematical SFUN/library routines and implemented on a PC-DOS computer.

3. APPLICATIONS AND RESULTS

To verify the accuracy of the HCTF determined through the procedure described, three types of two-dimensional systems with different boundary conditions have been considered:

a. a plate of hollow square profile;
b. a composite plate of solid rectangular profile;
c. a homogeneous plate of solid rectangular profile.

The first two cases undergo convective boundary conditions, in the third case half of the boundary region is surrounded by air while half is held at a known temperature (Fig.3.1).

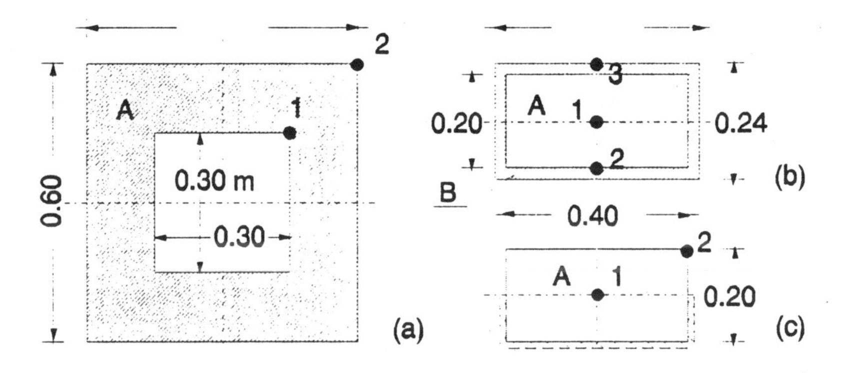

Fig.3.1-Type and sizes of the selected systems.

The chosen input function for the air temperature is a triangular pulse 2-hours duration and 10°C amplitude and both the temperature of the inside fluid of case (a) and the half boundary region of case (c) are kept at a constant uniform value of 0°C. The physical properties of the materials used are listed in Table1, while unit surface conductance is $h=23$ W/m^{2}K and time interval $\Delta\tau=1$ hour. The prescribed temperature at the edges is obtained from the same convection boundary condition letting $h\rightarrow\infty$. Fig3.2 shows the resulting mesh for each system which is set by expanding the domain discretization step-by step and by checking, at each step, the trace of the matrix $\mathbf{K}^{-1}\mathbf{C}$ (sum of all eigenvalues) until its variation becomes negligible.

Tab.I - Thermophysical properties of materials used.

Material	k (W/mK)	ρ (kg/m³)	c (J/kgK)
A (Tile)	0.500	800	1470
B (Insulation)	0.054	15	1220

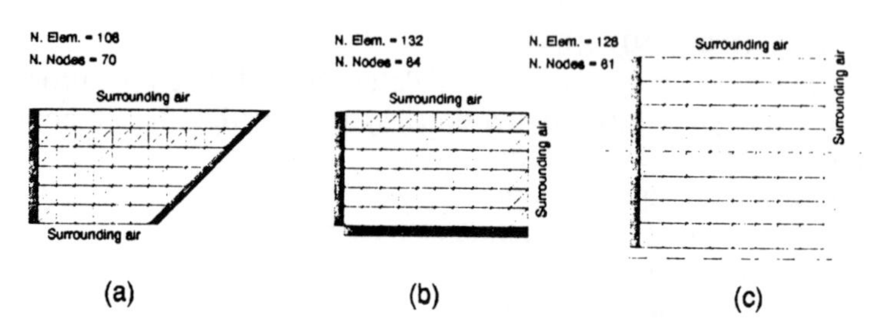

Fig.3.2-Network controlled by significant eigenvalues of matrix $K^{-1}C$

For cases (a) and (c) we obtained three series ($m=2$) of five and four coefficients each for each node, whilst in case (b) there were only two series ($m=1$) with five coefficients (Tab.2,4). The significant eigenvalues are presented in Tab.5 in a decreasing order starting from the dominant.

Tab.2-HCTF coefficients for system (a)

Node	Coeff.	n+1	n	n-1	n-2	n-3
1	b1	-0.001151	0.021934	0.027087	-0.026052	0.003656
1	b2	0.418793	-0.426063	0.029397	0.081514	-0.021348
2	b1	0.777042	-0.989348	0.293031	0.050801	-0.024595
2	b2	0.000052	-0.000340	0.001189	0.000179	-0.000243
	d	-1	1.547318	-0.810660	0.167159	-0.011584

Tab.3-HCTF coefficients for system (b)

Node	Coeff.	n+1	n	n-1	n-2	n-3
1	b1	-0.000745	0.032812	0.000430	-0.009365	0.001739
2	b1	0.110962	-0.092227	-0.035527	0.052573	-0.010911
3	b1	0.905387	-1.664598	0.974167	-0.203438	0.013354
	d	-1	1.853401	-1.097581	0.235987	-0.016677

Tab.4-HCTF coefficients for system (c)

Node	Coeff.	n+1	n	n-1	n-2	n-3
1	b1	-0.01924	0.082850	-0.000378	-0.006111	--------
1	b2	0.007911	0.128940	-0.042525	0.001886	--------
2	b1	0.796789	-0.819942	0.181695	0.008264	--------
2	b2	0.000181	0.006036	-0.002120	-0.000253	--------
	d	-1	1.231859	-0.454817	0.052308	--------

Tab.5-Significant eigenvalues

Case (a)	2.58164	1.38344	0.68179	0.53158	0.44207
Case (b)	10.66330	1.89940	0.58104	0.57064	0.44214
Case (c)	2.48461	0.86400	0.71906	0.44786	-------

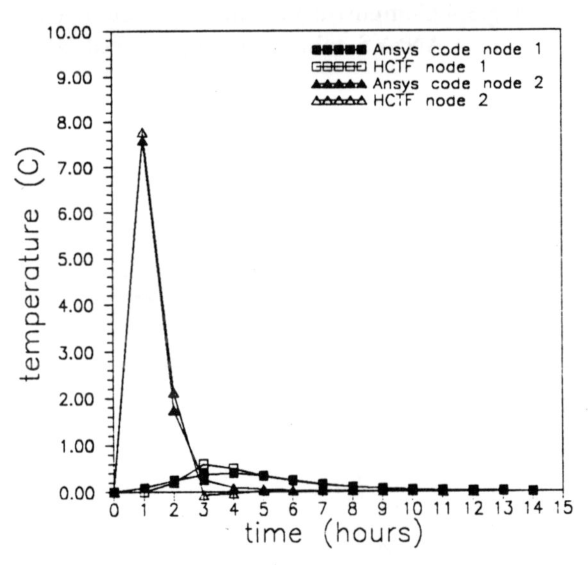

Fig.3.3-Plate of hollow square profile-Comparison of ANSYS solution with proposed HCTF.

The output is represented by the nodal temperature evolutions at the selected points indicated in Fig.3.1 which have been calculated through the convolution equation 2.15. The results were checked against an Ansys Code solution and plotted in Fig.3.3-5. As we can see, the two methods lead to very close agreement with maximum deviations smaller than 0.4°C (at node 2 of the hollow plate).

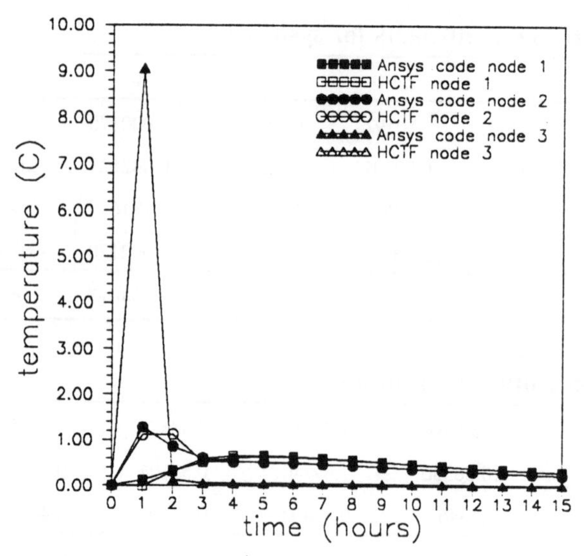

Fig.3.4-Composite plate of solid rectangular profile-Comparison of ANSYS solution with proposed HCTF .

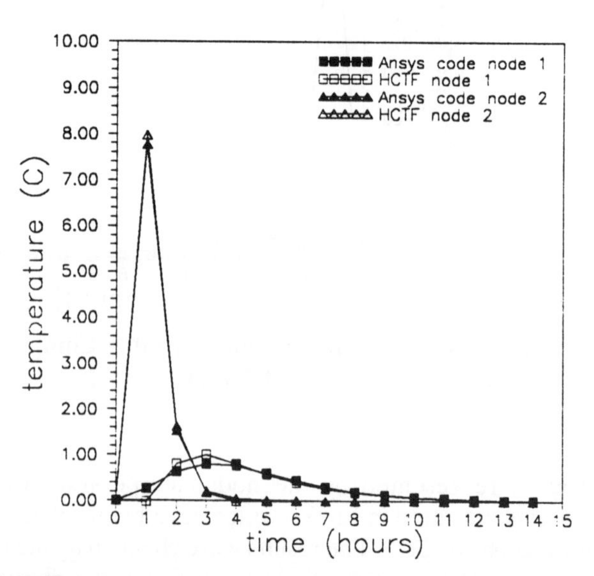

Fig.3.5-Homogeneous plate of solid rectangular profile-Comparison of ANSYS solution with proposed HCTF .

4. CONCLUSIONS

From the above results the following conclusions can be drawn:

• the evaluation of multidimensional HCTF by means the described technique differs from other procedures not only from a methodologic point of view, but also for the improved precision achieved even with a very limited number of coefficients. The main reason is that the procedure only deals with thermophysical data processing rather than input-output identification techniques;

• another interesting advantage is that the procedure needs a small-scale computer implementation (PC) and, once the HCTFs are determined for a preselected node, any further transient analysis can be carried out using a very simple convolution scheme (eq.2.15);

• the method proposed seems to be general and could be applyed in a wide range of physical phenomena.

5. REFERENCES

1. STEPHENSON, D.G. and MITALAS, G.P. - Calculation of Heat Conduction Transfer Function for Multilayer Slabs, ASHRAE Transactions, 1971

2. ASHRAE Handbook of Foundamentals, 1977, Chap. 25

3. SANZ, P.D., Análisis de las funciones de transferencia para la resolu ción de problemas de transmisión de calor en régimen variable en cámeras adiabáticas y de régimen controlado (Analysis of Transfer Function for the Solution of Heat Transfer Problems under Variable Conditions in Adiabatic or Controlled Regime Chambers), 1984, Thesis, Universidad Complutense, Madrid.

4. MASCHERONI, R.H., SANZ, P.D. and DOMINGUEZ, M. - A New Way to Predict Thermal Histories in Multidimensional Heat Conduction: the Z-Transfer Function Method, Int. Comm. Heat & Mass Transfer, 14, pp. 81-90, 1987

5. SALVADORI, U.O., REYNOSO, R.O. and MASCHERONI, R.H. - The Use of Z-Transfer Functions to Evaluate Temperature and Quality Changes in Refrigerated Storage, Proc. of meetings of Comm. B2, C2, D1, D2/3, Int. Institute of Refrigeration, Paris, pp. 695-700, 1990

6. ZIENKIEWICZ, O.C. -The Finite Element Method, Third Edition, McGraw-Hill, London, 1977

7. HUEBNER, K.H. - The Finite Element Method for Engineers, J.

Wiley&Sons, New York, 1975

8. RAO, S.S -The Finite Element Method in Engineering, Second Edition, Pergamon Press, Oxford, 1989

6. LIST OF SYMBOLS

C, **Ce**	Global, element heat capacity matrix.
K, **Ke**	Global, element thermal conductivity matrix.
Ni	Interpolation function associated with the *i*-th nodal degree of freedom.
S^e	Surface of element *e*.
V^e	Volume of element *e*.
a	Vector of global temperature.
$\bar{\mathbf{a}}$	Vector of surronding temperature.
f, $\mathbf{f}^e$	Global, element forcing function vector.
k_x, k_y, k_z	Thermal conductivity along *x, y, z* axes.
h	Unit surface conductance.
q	Rate of heat flow.
$\dot{q}$	Rate of heat generation per unit volume.
c	Specific heat.
m	Number of forcing terms.
ρ	Density
τ	Time.
Δτ	Time step.

The Comparison of the Few Methods of Numerical Solution of Heat Conduction Problem

Radoslaw GRZYMKOWSKI[†]
Malgorzata BIEDRONSKA[‡]

SUMMARY

In the paper there is presented the method of numerical solution of one-dimensional heat conduction equation with boundary conditions of the second and third kind, constructed with the application of interpolating spline function and then it is compared with the scheme of finite differences and Crank-Nicholson's method.

1. INTRODUCTION

The determination of the temperature field in the elements of construction for non-stationary heat conduction has the decisive meaning in the many movement problems and technological processes. In this paper there is limited to one-dimensional problems. The temperature field in the one-dimensional area is described by the Fourier equation which determines $T = T(x,t)$ function:

$$\partial_t T(x,t)=a(x,t)\partial_{xx}T(x,t)+b(x,t)T(x,t)+f(x,t) \tag{1}$$

where a, b, f are the known functions.
The initial conditions:

$$T(x,0) = \psi(x), \tag{2}$$

describe the temperature field in the initial moment.

[†] Dr, D.Sc., eng., Silesian Technical University, Institute of Mathematics, PL-44-100 Gliwice, ul. Zwyciestwa 42.
[‡] D.Sc., Silesian Technical University, Institute of Mathematics, PL-44-100 Gliwice, ul. Zwyciestwa 42.

The boundary condition may be assumed in different way. In the most problems there are considered the conditions of the second kind (if the heat stream is defined on the border) or the third kind (if the exchange of heat with the environment exists). The both conditions may be written by the equation:

$$\partial_x T(X_j,t)=\alpha_j(t)T(X_j,t)+\beta_j(t),\ j=0,1, \tag{3}$$

where X_0, X_1 are respectively the left and right border, $\alpha_j(t)$, $\beta_j(t)$ are the known functions.

The cubic interpolating splines will be used to find the approximation solution of the equation (1) with the boundary-initial conditions (2)-(3).

2. SPLINE FUNCTIONS

Let the mesh Δ_N will be constructed on the interval $<X_0, X_1>$:

$$\Delta_N:\ X_0 = x_0 < x_1 < \dots < x_{N-1} < x_N = X_1$$

The function $S(x) = S(x;\Phi)$ is called the cubic interpolating spline if [1, 4]:

1° $S(x;\Phi) \in P^3(x),\ x \in <x_i, x_{i+1}>,\ i=0,1,\dots,N-1,$

2° $S(x;\Phi) \in C^2<X_0, X_1>,$

3° $S(x_i;\Phi) = \Phi(x_i),\ i=0,1,\dots,N,$

where $P^3(x)$ means a set of polynomials of order not higher that 3. To unique identification there is indispensable to add two boundary conditions. In this paper they are:

$$d_x S(X_j) = \phi_j. \tag{4}$$

Assumed the notations:

$$h_i = x_{i+1} - x_i,\ i=0,1,\dots,N-1, \tag{5}$$

$$M_i = d_{xx} S(x_i;\Phi),\ i=0,1,\dots,N. \tag{6}$$

the spline may be written in form:

$$S(x;\Phi) = M_{i-1}\frac{(x_i - x)^3}{6h_{i-1}} + (\Phi(x_{i-1}) - \frac{M_{i-1}h_{i-1}^2}{6})\frac{x_i - x}{h_{i-1}} +$$

$$+M_i\frac{(x - x_{i-1})^3}{6h_{i-1}} + (\Phi(x_i) + \frac{M_i h_{i-1}^2}{6})\frac{x - x_{i-1}}{h_{i-1}}, \qquad (7)$$

for $x \in \langle x_{i-1}, x_i\rangle$, $i=1,2,\dots,N$.

The coefficients M_i from formula (7) are appointed from the system of equations:

$$2M_0 + M_1 = \frac{6}{h_0}(\frac{\Phi(x_1) - \Phi(x_0)}{h_0} - \phi_0), \qquad (8)$$

$$v_i M_{i-1} + 2M_i + u_i M_{i+1} = w_i, \quad i=1,2,\dots,N-1, \qquad (9)$$

$$M_{N-1} + 2M_N = \frac{6}{h_{N-1}}(\phi_1 - \frac{\Phi(x_N) - \Phi(x_{N-1})}{h_{N-1}}), \qquad (10)$$

as well as:

$$v_i = \frac{h_{i-1}}{h_{i-1} + h_i}, \quad i=1,2,\dots,N-1, \qquad (11)$$

$$w_i = 6\xi_i(\Phi(x_{i+1}) - \Phi(x_i)) + 6\mu_i(\Phi(x_{i-1}) - \Phi(x_i)), \qquad (12)$$

$$u_i = \frac{h_i}{h_{i-1} + h_i}, \quad i=1,2,\dots,N-1, \qquad (13)$$

where:

$$\xi_i = \frac{1}{h_i(h_{i-1} + h_i)}, \qquad (14)$$

$$\mu_i = \frac{1}{h_{i-1}(h_{i-1} + h_i)}. \qquad (15)$$

3. METHOD OF SOLUTION

The solution of the equation (1) with conditions (2)-(3) will be researched in the space-time area:

$$D = \{(x,t):\ X_0 \le x \le X_1,\ 0 \le t \le \Xi < \infty\}.$$

The space-time mesh is constructed on the considered area D:

$$\Omega_{NxK} = \Delta_N x \Delta_K,$$

where Δ_K is time mesh:

$$\Delta_K:\ 0 = t_0 < t_1 < \ldots < t_{K-1} < t_K = \Xi.$$

There is assumed that numerical solution of problem (1)-(3) has the form of the sequence of spline functions $\{S^p(x)\}_0^K$, from which each of them approximates the solution $T(x,t_p)$, $p=0,1,\ldots,K$. Splines functions S^p are generated by the recurrent way by the equation:

$$S^0(x_i) = \psi(x_i),\quad i=0,1,\ldots,N, \tag{16}$$

$$\frac{S^p(x_i) - S^{p-1}(x_i)}{t_p - t_{p-1}} = a(x_i,t_p)\,d_{xx}S^p(x_i) + b(x_i,t_p)S^p(x_i) +$$

$$+ f(x_i,t_p),\quad i=0,1,\ldots,N-1,\quad p=1,2,\ldots,K, \tag{17}$$

$$d_x S^p(X_j) = \alpha_j(t_p)S^p(X_j) + \beta_j(t_p),\ j=0,1,\ p=0,\ldots,K \tag{18}$$

Assumed the notations:

$$\tau_p = t_p - t_{p-1},\quad S_i^p = S^p(x_i),\quad \phi_j^p = d_x S^p(X_j),$$

$$a_i^p = a(x_i,t_p),\quad b_i^p = b(x_i,t_p),\quad f_i^p = f(x_i,t_p),$$

$$\alpha_j^p = \alpha_j(t_p),\quad \beta_j^p = \beta_j(t_p),\quad \psi_i = \psi(x_i),$$

the equations (16)-(18) may be written in form:

$$S_i^o = \psi_i, \tag{19}$$

$$S_i^p - S_i^{p-1} = \tau_p(a_i^p M_i^p + b_i^p S_i^p + f_i^p), \tag{20}$$

$$\phi_j^p = \alpha_j^p S_j^p + \beta_j^p, \quad j=0,1, \tag{21}$$

where $i=1,2,\ldots,N-1$, $p=1,2,\ldots,K$.

Using the formulas (19)-(21) to the system of equations (8)-(10), after transforming, there is obtained for $p=1,\ldots,K$:

$$(2m_0^p + 6h_0^{-2}(1 + h_0\alpha_0^p))S_0^p + (m_1^p - 6h_0^{-2})S_1^p =$$
$$=2d_0^p + d_1^p - 6h_0^{-1}\beta_0^p, \tag{22}$$

$$(v_i m_{i-1}^p - 6\mu_i)S_{i-1}^p + 2(m_i^p + 3\mu_i + 3\xi_i)S_i^p + (u_i m_{i+1}^p - 6\xi_i)S_{i+1}^p =$$
$$=v_i d_{i-1}^p + 2d_i^p + u_i d_{i+1}^p, \quad i=1,2,\ldots,N-1, \tag{23}$$

$$(m_{N-1}^p - 6h_{N-1}^{-2})S_{N-1}^p + (2m_N^p + 6h_{N-1}^{-2}(1 - h_{N-1}\alpha_1^p))S_N^p =$$
$$=d_{N-1}^p + 2d_N^p + 6h_{N-1}^{-1}\beta_1^p \tag{24}$$

where there are the following notations:

$$m_i^p = a_i^p \tau_p^{-1}(1 - \tau_p b_i^p), \quad i=0,\ldots,N, \quad p=1,\ldots,K, \tag{25}$$

$$d_i^p = a_i^p \tau_p^{-1}(S_i^{p-1} + \tau_p f_i^p), \quad i=0,\ldots,N, \quad p=1,\ldots,K. \tag{26}$$

The equation (22)-(24) let us for determination of numerical solution on the all space-time mesh Ω_{NxK}. Moreover, the equations (8)-(10), (7) let determinate the values $S^p(x)$ for the any x from the interval $\langle X_0, X_1 \rangle$. Finally, the obtained solution is the function defined on the set of levels:

$$D_p = \{(x,t): X_0 \le x \le X_1, \ t = t_p\}, \quad p=0,1,\ldots,K,$$

at the same time it is the function of class C^2 on the each level.

To solution of the system of equations (22)-(24) it will be used three-diagonal simplification of the Gauss method, known in the literature as Thomas algorithm [2].

4. THE ANALYSIS

The essential marks of the each difference scheme are compatibility (approximation) and stability [5, 6]. Putting to the system (22)-(24) in the place of S_i^p the values of the exact solution $T(x_i, t_p)$ there is obtained the following result:

The difference scheme (19)-(24) approximates the problem (1)-(3) with order $O(\tau+h)$, where $h = \max(h_i)$. If mesh Δ_N is uniform:

$$h_0 = h_1 = h_2 = \ldots = h_{N-1} = h,$$

then the order of approximation is $O(\tau+h^2)$.

Problem of the stability is more difficult. There is assumed that the coefficients $a(x,t)$, $b(x,t)$ and the functions $\alpha_j(t)$ are constant, and the meshes Δ_N, Δ_K are uniform:

$$h_i = h, \quad i=0,1,\ldots,N-1, \qquad \tau_p = \tau, \quad p=1,2,\ldots,K,$$

If there is assumed that:

$$A = 6h^{-2}\tau a \;, \quad B = 1 - \tau b \;, \quad C_i^p = S_i^p + \tau f_i^p \;,$$

then the system of equations (22)-(24) will have the form:

$$(2B+A(1+h\alpha_0^p))S_0^{p+1} + (B-A)S_1^{p+1} = 2C_0^p + C_1^p - Ah\beta_0^p, \tag{27}$$

$$(B-A)S_{i-1}^{p+1} + 2(2B+A)S_i^{p+1} + (B-A)S_{i+1}^{p+1} = C_{i-1}^p + $$
$$+ \; 4C_i^p + C_{i+1}^p, \quad i=1,2,\ldots,N-1, \tag{28}$$

$$(B-A)S_{N-1}^{p+1} + (2B+A(1-h\alpha_1^p))S_N^{p+1} = C_{N-1}^p + 2C_N^p + Ah\beta_1^p. \tag{29}$$

Using the matrix method of stability analysis [5] it will be shown, that if $f(x,t) \equiv 0$ and $\beta_0 \equiv 0$, $\beta_1 \equiv 0$ then scheme (19) - (21), (22) - (24) is unconditional stable. In the general case it is known, that the necessary conditions are satisfied only.

The obtained order of approximation is the same as in the case of the "ordinary" difference scheme:

$$\frac{T_i^p - T_i^{p-1}}{\tau} = a_i^p \frac{T_{i+1}^p - 2T_i^p + T_{i-1}^p}{h^2} + b_i^p T_i^p + f_i^p, \qquad (30)$$

with the respective boundary conditions.

The scheme (22)-(24) (for $h_i = h$, $i=0,1,\ldots,N-1$) uses the certain global properties of splines, so it is possible expect the qualitatively different results in the certain cases. The more precise analysis of this problem demands still another ways than the methods using in the difference schemes.

5. EXAMPLES OF CALCULATION

$$1^\circ \quad \partial_t T(x,t) = \partial_{xx} T(x,t) + xe^{-t}, \quad 0 \le x \le 1, \quad t \ge 0, \qquad (31)$$

$$T(x,0) = x, \qquad (32)$$

$$\partial_x T(0,t) = \partial_x T(1,t) = 2 - e^{-t}. \qquad (33)$$

There is assumed that $h = 0.125$, $\tau = 0.03125$. The following Table 1 presents results after 12 iterations ($\Xi = 0.375$).

Table 1. Comparison of solutions

x	Spline Method	Real Solution	Difference Method
0	.0017784800500	0	.0017645808500
.125	.1595419195399	.1613617271954	.1595521945331
.25	.3207944812626	.3227234543906	.3208012329530
.375	.4820016702341	.4840851815861	.4820050108135
.5	.6431861371303	.6454469087815	.6431861371336
.625	.8043706040260	.8068086359769	.8043672634539
.75	.9655777929975	.9681703631723	.9655710413153
.875	.1268303547190	.1295320903680	.1268200797350
1	.2881507543100	.2908938175630	.2881368551230

$$2^\circ \quad \partial_t T(x,t) = 5 \cdot 10^{-6} \partial_{xx} T(x,t), \quad 0 \le x \le .1, t \ge 0 \qquad (34)$$

$$T(x,0) = 1450, \qquad (35)$$

$$\partial_x T(0,t) = 10T, \qquad (36)$$

$$\partial_x T(1,t) = -20T. \qquad (37)$$

Table 2. Comparison of solutions

x	Difference Method	Spline Method	Crank-Nicholson Method
0	836.690978150	836.959516877	836.3916154186
.005	876.244636485	876.488464895	875.9291098517
.01	911.017686258	911.236731140	910.6835495834
.015	940.810824849	941.005234055	940.4557133508
.02	965.445430290	965.615629657	965.0672013089
.025	984.764866815	984.911619698	984.3617434466
.03	998.635926880	998.760394568	998.2066414838
.035	1006.950387567	1007.054187857	1006.4943208740
04	1009.626650320	1009.711910467	1009.1439606010
.045	1006.611424168	1006.680823560	1006.1031597550
.05	997.881404224	997.938201563	987.3495920776
.055	983.444889932	983.492929578	982.8925926974
.06	963.343281510	963.386974068	962.7746161744
.065	937.652389092	937.696662322	937.0725015662
.07	906.483487418	906.533705074	905.8984794236
.075	869.984050181	870.045898395	869.4008574103
.08	828.338102461	828.417445559	827.7643259852
.085	781.766137178	781.868847233	781.2098332536
.09	730.524552118	730.656318797	729.9939885915
.095	674.904577443	675.070706671	674.4079676197
.1	615.230679211	615.435890586	614.7759060739

In this example $h = 0.005$, $\tau = 2$. The number of iterations was 51 (Table 2). The calculations were done using method (30) scheme (22)-(24) and using Crank-Nicholson's method (with the order of approximation $O(\tau^2 + h^2)$ [2, 3]).

Comparing results obtained in both examples, it may be seen the great compatibility between classic method of differences and scheme (22)-(24).

3° There is considered the following two-layer initial-boundary problem with the fixed boundary of the division of the layers $X_{0,1}$:

$$\partial_t T_1(x,t)=a_1\partial_{xx}T_1(x,t),\quad X_0\le x\le X_{01}, \tag{38}$$

$$\partial_t T_2(x,t)=a_2\partial_{xx}T_2(x,t),\quad X_{01}\le x\le X_1, \tag{39}$$

$$T_j(x,0)=\Psi_j(x),\quad j=1,2, \tag{40}$$

$$-\lambda_1\partial_x T_1(X_0,t)=\alpha[T_1(X_0,t)-T_\infty], \tag{41}$$

$$\partial_x T_2(X_1,t)=0, \tag{42}$$

$$T_1(X_{01},t)=T_2(X_{01},t), \tag{43}$$

$$-\lambda_1\partial_x T_1(X_{01},t)=-\lambda_2\partial_x T_2(X_{01},t), \tag{44}$$

where a_j, λ_j, $j=1,2$, mean the thermal diffusivity and coefficient of the conduction, α - the coefficient of heat penetration, and T_∞- the temperature of the environment.

The numerical solution of this problem is researched in the space-time area:

$$D^*=\{(x,t):X_0\le x\le X_{01}\cup X_{01}\le x\le X_1,\ 0\le t\le \Xi<\infty\},$$

The mesh Ω^*_{NxK} will be constructed on the area D^*:

$$\Omega^*_{NxK}=\Delta^*_N x\Delta_K,$$

as well as:

$$\Delta^*_N=\Delta^n_0\cup\Delta^N_n,$$

where:

$$\Delta^n_0:\ X_0=x_0<x_1<\ ...\ <x_{n-1}<x_n=X_{01},$$

$$\Delta^N_n:\ X_{01}=x_n<x_{n+1}<\ ...\ <x_{N-1}<x_N=X_1.$$

On the each time layer there is recurrently defined the pair of splines S^p_1, S^p_2 by the formulas:

$$S^0_1(x_i)=\psi_1(x_i),\ i=0,1,...,n, \tag{45}$$

$$S^0_2(x_i)=\psi_2(x_i),\ i=n,n+1,...,N, \tag{46}$$

$$\frac{S^p_1(x_i)-S^{p-1}_1(x_i)}{t_p-t_{p-1}}=a_1 d_{xx}S^p_1(x_i),\ i=1,...,n-1, \tag{47}$$

$$\frac{S^p_2(x_i)-S^{p-1}_2(x_i)}{t_p-t_{p-1}}=a_2 d_{xx}S^p_2(x_i),\ i=n+1,...,N-1 \tag{48}$$

$$-\lambda_1 d_x S_1^p(x_0) = \alpha[S_1^p(x_0) - T_\infty], \tag{49}$$

$$d_x S_2^p(x_N) = 0, \tag{50}$$

$$S_1^p(x_n) = S_2^p(x_n), \tag{51}$$

$$-\lambda_1 d_x S_1^p(x_n) = -\lambda_2 d_x S_2^p(x_n). \tag{52}$$

There is assumed the notations:

$$r_i^p = \frac{t_p - t_{p-1}}{(h_i)^2}, \qquad \sigma_j = \frac{\lambda_j}{a_j}, \qquad S_{j,i}^p = S_j^p(x_i).$$

Using the system of equations (8)-(10), the system of equations (45)-(52) may be written in the form:

$$[2\sigma_1 + 6\lambda_1 r_1^p(1+h_1\alpha)]S_{1,0}^p + [\sigma_1 - 6\lambda_1 r_1^p]S_{1,1}^p =$$
$$= 2\sigma_1 S_{1,0}^{p-1} + \sigma_1 S_{1,1}^{p-1} - 6\alpha r_1^p h_1 T_\infty, \tag{53}$$

$$[\sigma_1 - 6\lambda_1 r_i^p]u_i S_{1,i-1}^p +$$
$$+2[\sigma_1 + 3\lambda_1(r_i^p u_i + r_{i+1}^p v_i)]S_{1,i}^p + [\sigma_1 - 6\lambda_1 r_{i+1}^p]v_i S_{1,i+1}^p =$$
$$= \sigma_1(u_i S_{1,i-1}^{p-1} + 2S_{1,i}^{p-1} + v_i S_{1,i+1}^{p-1}), \quad i=1,\dots,n-1, \tag{54}$$

$$[\sigma_1 - 6\lambda_1 r_n^p]u_n S_{1,n-1}^p +$$
$$+ [(\sigma_1 + 6\lambda_1 r_n^p)u_n + (\sigma_2 + 6\lambda_2 r_{n+1}^p)v_n]S_{1,n}^p +$$
$$+ [\sigma_2 - 6\lambda_2 r_{n+1}^p]v_n S_{2,n+1}^p = \sigma_1 u_n S_{1,n-1}^{p-1} +$$
$$+ 2(\sigma_1 u_n + \sigma_2 v_n)S_{1,n}^{p-1} + \sigma_2 v_n S_{1,n+1}^{p-1}, \tag{55}$$

$$S_{1,n}^p = S_{2,n}^p, \tag{56}$$

$$[\sigma_2 - 6\lambda_2 r_i^p]u_i S_{2,i-1}^p +$$
$$+2[\sigma_2 + 3\lambda_2(r_i^p u_i + r_{i+1}^p v_i)]S_{2,i}^p + [\sigma_2 - 6\lambda_2 r_{i+1}^p]v_i S_{2,i+1}^p =$$
$$= \sigma_2(u_i S_{2,i-1}^{p-1} + 2S_{2,i}^{p-1} + v_i S_{2,i+1}^{p-1}), \quad i=n+1,\dots,N-1, \tag{57}$$

$$[\sigma_2 - 6\lambda_2 r_N^p] S_{2,N-1}^p + [2\sigma_2 + 6\lambda_2 r_N^p] S_{2,N}^p =$$

$$= \sigma_2 S_{2,N-1}^{p-1} + 2\sigma_2 S_{2,N}^{p-1}. \qquad (58)$$

The above system of equations may be written in the matrix form:

$$\mathbb{A} \circ \mathbb{S}^p = \mathbb{S}^{p-1} + \mathbb{R}^p, \quad p=1,2,\dots,K, \qquad (59)$$

where the matrix $\mathbb{A}$ is created by the coefficients appeared at the components of unknown vector:

$$\mathbb{S}^p = [S_{1,0}^p, S_{1,1}^p, \dots, S_{1,n}^p, S_{2,n}^p, \dots, S_{2,N}^p], \quad p=1,\dots,K,$$

in the system of equations (53)-(58), as well as:

$$\mathbb{S}^0 = [\Psi_{1,0}, \Psi_{1,1}, \dots, \Psi_{1,n}, \Psi_{2,n}, \dots, \Psi_{2,N}],$$

where:

$$\Psi_{j,i} = \Psi_j(x_i), \qquad (60)$$

however:

$$\mathbb{R}^p = [-6\alpha r_1^p h_1 T_\infty, 0, 0, \dots, 0].$$

After the evaulation of the values of the approximate solution on the mesh, it is possible to determinate the values of spline S_j^p in the arbitrary point x from the area. It necessary to take advantage of formulas (47),(48),(7), in which the values of splines $S_{j,i}^p$ are instead of $\Phi(x_i)$.

REFERENCES

1. ALBERG J. H., NILSON E. N., WALSH J. L.: The Theory of Splines and Their Applications, Academic Press, New York-London, 1967.
2. ROSENBERG D. V.: Methods for the Numerical Solution of Partial Differential Equations, American Elsewier Publ. Company, N.Y.,1969.
3. SMITH G. D. : Numerical Solution of Partial Differential Equations, Oxford Univ. Press, Oxford, 1978.
4. BIEDRONSKA M., MOCHNACKI B.:The Application of Collocation Method and Spline Functions to Linear and Non-Linear Problems of Stationary Heat Conduction, Bulletin of the Polish Academy of Sciences (Technical Sciences), Vol. 32, No. 5-6, 1984 (297-316).
5. GODUNOV S.: Schemas aux differences, Nauka, Moskva, 1977.

NOMENCLATURE LIST

X - space coefficient,
t - time coefficient,
T(x,t) - the temperature at the point (x,t),
∂_x - the partial derivative on account x,
∂_t - the partial derivative on account t,
d_x - derivative on account x,
S(x) = S(x;Φ) - interpolating spline for the function Φ,
$C^2\langle X_0,X_1\rangle$ - the set of the function which have the continuous of second order,
$P^3(x)$ - the polynomial of third order,
D - the space-time area,
Ω_{NxK} - the space-time mesh.

UNSTEADY HEAT CONDUCTION WITH THE INTERNAL HEAT SOURCE FROM THE NEUTRON FLOW

Justín Murín, Radovan Benkovič, Vladimír Míchal

Department of Mechanics, Faculty of Electrical Engineering of the Slovak Technical University, Ilkovičova 3, 812 19 Bratislava, Slovak Republic

SUMMARY

The heat-hydraulic computation determines the internal heat source in the fuel from the neutron flow in the fuel element of the fuel assembly of nuclear reactors. The internal heat source, the constants of heat conductivity, thermal physical parameters and the initial and boundary conditions are expressed as functions of the neutron flow density.

In addition to the calculation of these parameters, we developed a PC program and solved the radial distribution of temperature fields in the fuel element in various regimes by the finite elements method.

Numerical results are in very good agreement with the analytical solution of the temperature fields in selected points of the fuel element.

1. INTRODUCTION

Computer mechanics is widely applied to solve approximately various tasks in the theory of fields. Currently much attention is paid to bond (or weakly bond) problems, in which for example force and temperature fields or electromagnetic and temperature fields are affecting each other. This may induce changes in the behaviour of the material or of the initial conditions of the problem. In our case the flow of neutrons brings about generation of heat in the fuel assembly which in turn affects the temperature regime. The magnitude and distribution of the temperature field (stationary or non-stationary) in the fuel assembly provide important information for a correct operation of the whole nuclear energy system.

The aim of our work was to carry out the calculation of the non-stationary temperature field by the finite element method (FEM) with considering the internal source of heat depending on the neutron flux density, to develop a program for PC and to verify its function on a computation of the radial temperature field in a fuel assembly. The internal source of heat, the coefficient of heat conductivity and the conditions of heat output by convection are derived from a thermo-hydraulic solution of the reactor.

2. NON-STATIONARY HEAT FLOW WITH AN INTERNAL SOURCE

The problem of non-stationary heat conduction has been treated in detail e.g. in [1,2], therefore we will not discuss the topic. The program TEPLO2D [3] solves planar and axially symmetrical problems of stationary and non-stationary heat conduction by FEM for boundary conditions defining the temperature, heat flux or heat exchange through convection.

The boundary conditions for temperature and for the heat flux can be time dependent. As initial conditions of the problem, the reference ambient temperature and the initial distribution of temperature in the studied body have to be entered. The studied material can be isotropic or anisotropic and is characterised by its heat conductivity λ, density ρ, specific heat c, and by the source of heat q_V.

In the particular problem solved in this paper the program TEPLO2D has been completed by a subroutine providing the calculation of the internal source of heat from the flow of neutrons.

We will only briefly describe the fundamental relations and equations on which the program for PC has been built.

Starting from the Laplace equation of heat conduction and taking into account the initial condition for temperature and boundary conditions for the heat flux and heat transfer by convection, the following functional is obtained, see figure 1.

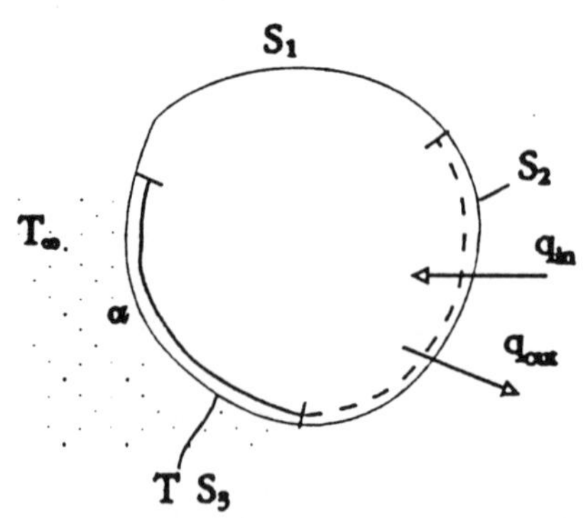

Figure 1. Explanation of the fundamental equation of heat conduction

$$I[T(x,y,z,t)] = \frac{1}{2}\int_V \left[\lambda_x.\left(\frac{\partial T}{\partial x}\right)^2 + \lambda_y.\left(\frac{\partial T}{\partial y}\right)^2 + \lambda_z.\left(\frac{\partial T}{\partial z}\right)^2 - 2\left(\dot{q}-\rho.c\frac{\partial T}{\partial t}\right).T\right]dV$$

$$+\int_{S_2} q.T.dS_2 + \frac{1}{2}\int_{S_3} \alpha(T-T_\infty)^2 dS_3, \qquad (1)$$

where $\lambda_x, \lambda_y, \lambda_z$ are coefficients of heat conductivity [$W.m^{-1}.K^{-1}$], $T=T(x,y,z,t)$ is the temperature field in the studied body at instant t[°C],q_V is the source of heat per unit volume [$W.m^{-3}$], ρ. c is the product of the density and specific heat [$J.m^{-3}.K^{-1}$], α is the coefficient of heat transfer by convection [$W.m^{-2}.K^{-1}$], q_S is the heat flow into the considered region [$W.m^{-2}$], S_2,S_3 denote that part of the body's surface through which there is an inlet or outlet of heat [m^2], and T_∞ is the ambient temperature.

By implementing isoparametric finite elements a system of algebraic equations in the matrix form is obtained

$$[\mathbf{K}].\mathbf{T}+[\mathbf{K}_3].\dot{\mathbf{T}}=\mathbf{P}, \tag{2}$$

where [**K**] is the matrix of heat conductivity, [$\mathbf{K}_3$] is the matrix of specific heat, **T** is the vector of temperatures in the nodes ($\dot{\mathbf{T}}$ is its derivative), **P** is the right hand side comprising the outlet (inlet) of heat by conduction and convection into (from) the ambient and the internal source of heat.

The system of equations is solved by the well known methods used for describing temporally dependent phenomena. When solving practical problems of nuclear energy systems it is furthermore necessary to create a subroutine which in the given step of the solution evaluates the thermophysical properties of the material, the coefficients of heat transfer by convection and the generated heat due to the neutron flow.

The solved planar region may consist of up to 500 elements, 500 nodes, 25 material groups, 500 edges with convection and 25 time sequences, each of them of 50 equal steps. The program consist of a pre-processor (input of data), of the main routine computing the temperature field and of a post-processor (processing of computed results).

The program TEPLO2D, including graphical pre- and post-processor, has been written in C language.

3. THERMO-HYDRAULIC ANALYSIS OF THE FUEL ASSEMBLY

The quantity to begin with is the mean flux of thermal neutrons in the reactor core (RC) $\bar{\varphi}_T$. The following relation is utilised to calculate the internal source of heat

$$q_V \approx E_f.\Sigma_f^5.\bar{\varphi}_T = E_f.N_5.\sigma_f^5.\bar{\varphi}_T, \tag{3}$$

where $E_f = 3{,}09.10^{-11}$ J is the thermal energy which is, on the average, released in one process of nuclear fission, Σ_f^5 is the macroscopic effective cross-section of the fission of U^{235} [m^{-1}], σ_f^5 is the microscopic effective cross-section of the fission of U^{235} [m^2].

Thermal neutrons have a distribution which is very similar to the Maxwell spectrum and the distribution along the radius and height can be written as

$$\varphi_T(r,z) = \varphi_{T0}.J_0.\left(\frac{2,405.r}{R_e}\right).\cos\left(\frac{\pi.z}{H_e}\right), \tag{4}$$

where J_0 is the 1st kind Bessel function of zero-th order, R_e and H_e are the extrapolated radius and height of the RC, respectively, $\varphi_{T0} = k_V.\bar{\varphi}_T$ is the thermal neutron flux in the middle of the RC, and k_V is the volume peaking factor.

The actual shape of function $\varphi_T(r,z)$ is affected by peaking factors. For the whole reactor these can be written as [4]

k_r - radial peaking factor,
maximum value of $k_r = 2{,}31$ actual value $k_r \sim (1{,}8$ to $2{,}1)$,
k_z - axial peaking factor,
maximum value of $k_z = 1{,}57$ actual value $k_z \sim (1{,}4$ to $1{,}5)$,
k_V - volume peaking factor $k_V = k_r \,.\, k_z$.

Since we know the type of reactor, the temperature t_{in} [°C] and pressure p_{in} [MPa] of the cooling liquid at the inlet to the RC are known. For evaluating the enthalpy of water, however, the temperature t_0 and pressure p_0 at the outlet must be known. In the first approach the rise of temperature of the cooling liquid can be estimated as 10 % and the decrease of pressure δ_p at 1 % of their initial values. After calculating of the actual value of $t_0 = t_{H_2O}(H/2)$ from Eqn. (10) and the loss of pressure from Eqn. (11), an iterative calculation of the block can be performed, Eqns. (5 to 11).

The throughput G^{RC} [$kg.s^{-1}$] of the cooling liquid in the RC can be calculated as

$$G^{RC} = \frac{Q_T^{RC}}{h_o - h_{in}}, \tag{5}$$

where h_0, h_{in} are the enthalpies of the liquid at the outlet and inlet of the RC, respectively.

The thermal power of the reactor is calculated from

$$Q_T^{RC} = q_V.V_{RC}, \tag{6}$$

where V_{RC} is the volume of the RC [m^3].

The flux of the cooling liquid through the fuel assembly is

$$G^{fa} = \frac{G^{RC}}{n_{fa}}, \tag{7}$$

where n_{fa} is the number of fuel assemblies in the reactor.

Further, the RC must be divided into several parts and in each of them the temperature and pressure of the liquid must be determined. The calculation is performed for fuel assembly of average and maximum powers. Let us assume that the change of the heat flux per unit of length for a fuel assembly of average power is described by relation

$$\bar{q}_l(z) = \bar{q}_{l0}.\cos\left(\frac{\pi.z}{H_e}\right), \tag{8}$$

where $\bar{q}_{l0} = \dfrac{Q_T^{RC}.k_Z}{n_{fa}.H}$ is the thermal power per unit length for a fuel assembly of average power in the middle of the RC [W.m^{-1}], and H [m] is the height of the RC.

For the fuel assembly of maximum power the value q_{l0}^{max} is substituted for the quantity $\bar{q}_{l0}$, hereby Eqn. (8) becomes

$$q_l^{max}(z) = q_{l0}^{max}.\cos\left(\frac{\pi.z}{H_e}\right), \tag{9}$$

where $q_{l0}^{max} = \dfrac{Q_T^{RC}.k_V}{n_{fa}.H}$ is the thermal power per unit of length for the fuel assembly of maximum power in the middle of the RC.

The change of temperature of the cooling liquid is described by relation

$$t_{H_2O}(z) = t_{in} + \frac{q_{l0}.H_e}{G^{fa}.c_p(z).\pi}.\left(\sin\left(\frac{\pi.z}{H_e}\right) + \sin\left(\frac{\pi.H}{2.H_e}\right)\right), \tag{10}$$

where c_p is the specific heat of water [J.kg^{-1}.K^{-1}].

The hydraulic resistance of the fuel assembly is given by

$$\delta_p = \delta_{p_{FF}} + \delta_{p_L} + \delta_{p_H}, \text{ [Pa]} \tag{11}$$

where $\delta_{p_{FF}}$ is the decrease of pressure due to friction forces in the liquid, δ_{p_L} are local pressure losses, and δ_{p_H} is the hydrostatic loss of pressure.

For determining the temperature of the zircaloy cladding of the fuel element, the coefficient of heat transfer α [W.m^{-2}.k^{-1}] must be known. There are a number of methods and empirical equations for calculating this coefficient [3, 4, 5]. For our purpose the following simple method is sufficiently accurate :

1. The coefficient α_{sb} is determined at surface boiling.
2. The value of α_{vb} is determined after the boiling is developed in the volume .
3. If $\dfrac{\alpha_{vb}}{\alpha_{sb}} \geq 3$, then $\alpha(z) = \alpha_{vb}.0,95$. (12)

$$\text{If } 3 > \frac{\alpha_{vb}}{\alpha_{sb}} \geq 0,5, \text{ then } \alpha(z) = \alpha_{sb}.\left(1 + 0,9025.\left(\frac{\alpha_{vb}}{\alpha_{sb}}\right)\right)^{-1} \tag{13}$$

$$\text{If } 0,5 > \frac{\alpha_{vb}}{\alpha_{sb}} \geq 0, \text{ then } \alpha(z) = \alpha_{sb}. \tag{14}$$

The temperature of the external surface of the cladding of the fuel element is given by

$$t_{Zr}^{ex}(z) = t_{H_2O}(z) + \frac{\kappa.q_l(z)}{\pi.d_{fe}.n_{fe}.\alpha(z)}, \tag{15}$$

where κ is the ratio of the heat released in the fuel and of the total generated heat, $\kappa = 0,92$, d_{fe} is the external diameter of the fuel element [m], and n_{fe} is the number of fuel elements in the fuel assembly.

The temperature of the internal surface of the cladding of the fuel element is

$$t_{Zr}^{in}(z) = t_{Zr}^{ex}(z) + \frac{\kappa . q_l(z) . \delta_{Zr}}{\pi . d_{Zr} . n_{fe} . \lambda_{Zr}(z)}, \tag{16}$$

where δ_{Zr} is the thickness of the clearance of the fuel element, λ_{Zr} [W.m^{-1}.s^{-1}] is the thermal conductivity of the cladding, and d_{Zr} is the internal diameter of the cladding of the fuel element.

The temperature of the external surface of the fuel is

$$t_{UO_2}^{ex}(z) = t_{Zr}^{in}(z) + \frac{\kappa . q_l(z) . \delta_{He}}{\pi . d_{odf} . n_{fe} . \lambda_{He}(z)}, \tag{17}$$

where d_{odf} is the external diameter of the fuel, δ_{He} is the thickness of the clearance between the fuel and the cladding, and λ_{He} is thermal conductivity of the clearance.

The temperature of the internal surface of the fuel is

$$t_{UO_2}^{in}(z) = t_{UO_2}^{ex}(z) + \frac{\kappa . q_l(z)}{4 . \pi . n_{fe} . \lambda_{UO_2}} . \left(1 - \frac{2 . d_{ch}^2}{d_{odf}^2 - d_{ch}^2} . \ln \frac{d_{odf}}{d_{ch}} \right), \tag{18}$$

where d_{ch} is the diameter of the central hole and λ_{UO_2} is the thermal conductivity of the fuel.

In VVER-type reactors, the regime of boiling of liquid is established in the thermally most loaded regions, the liquid not being heated to the temperature of saturation. In this regime there is no danger of damaging the Zirconium cladding of the fuel element, nevertheless at least a simple calculation of the critical heat flux of the first kind, q_l^{CR} , is carried out and the obtained result is compared with the actual heat flux [6,7].

It is advantageous to introduce the emergency coefficient as

$$K_{CR}(z) = \frac{q_l^{CR}(z)}{q_l(z)}. \tag{19}$$

Because of the low accuracy of empirical equations used to calculate the flux q_l^{CR}, the value of the coefficient of safety below 1,3 is unacceptable and the calculation of the most loaded fuel assembly must be carried out a new for other initial parameters.

The results of the testing thermo-hydraulic calculation for medially and maximally loaded fuel assembly are shown in figure 2. The presented result were obtained for a VVER-1000 reactor. The initial average flux of thermal neutrons was $\bar{\varphi}_T = 7.10^{16}$neutron.m^{-2}.s^{-1} at average fuel enrichment of 3,5 %. It is to be noted that the calculated results agree well with respective parameters of a real reactor.

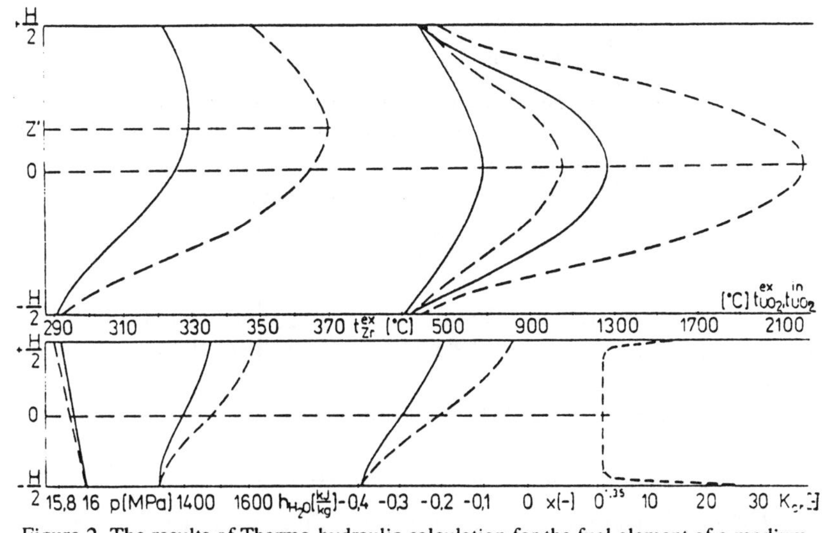

Figure 2. The results of Thermo-hydraulic calculation for the fuel element of a medium (-) and maximum (--) fuel assembly powers, where p - pressure of the cooling liquid, h - enthalpy of the cooling liquid, x - dryness of the cooling liquid, K_{CR} - emergency coefficient to origin critical heat flux of the first kind.

4. CALCULATION OF THE THERMAL FIELD BY WAY OF TEPLO2D PROGRAMME

By using the developed programme we will make a calculation of the radial division of the thermal field in the fuel element of the fuel assembly (Figure 3.). For practical purposes the calculation of the thermal field is justified in the point of maximum temperature of the external surface of the fuel element z' and in the point of maximum temperature of the fuel z = 0. The fuel element is of medium or maximum powers. FEM model is in figure 4.

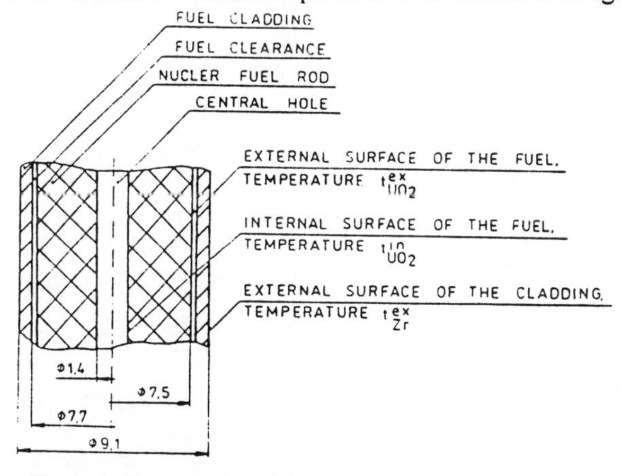

Figure 3. Cross-section of the fuel element of VVER-1000 reactor.

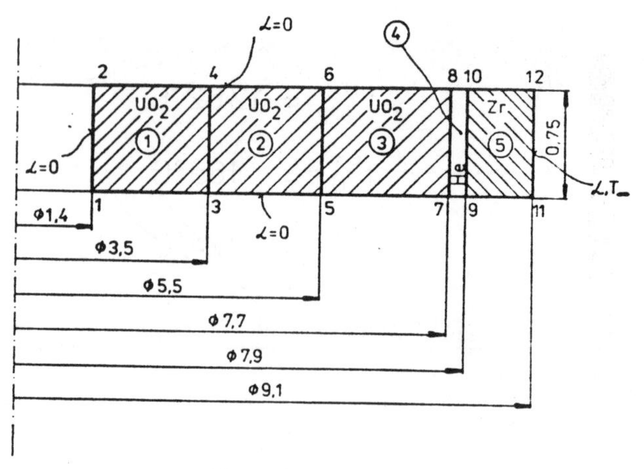

Figure 4. FEM model.

The initial quantities for the calculation of the radial thermal field are coefficients of the thermal conductivity of the fuel, of the contact clearance and the cladding [8], then it is the specific power of the internal sources of heat per fuel capacity unit for the fuel element of medium and maximum powers fuel assembly, coefficient of heat transfer $\alpha(z)$ (17) - (19) and the temperature of the cooling liquid (14). The surface of the model limited by points 11 - 1 - 2 - 12 has been isolated ($\alpha = 0$) and on the external surface of the fuel element (points 12 -11) there is a heat transfer by convection with a known value α and the surrounding temperature T_∞.

Figure 5 shows graphical output of the programme displaying radial division of the thermal field of the fuel element of the medium output fuel assembly.

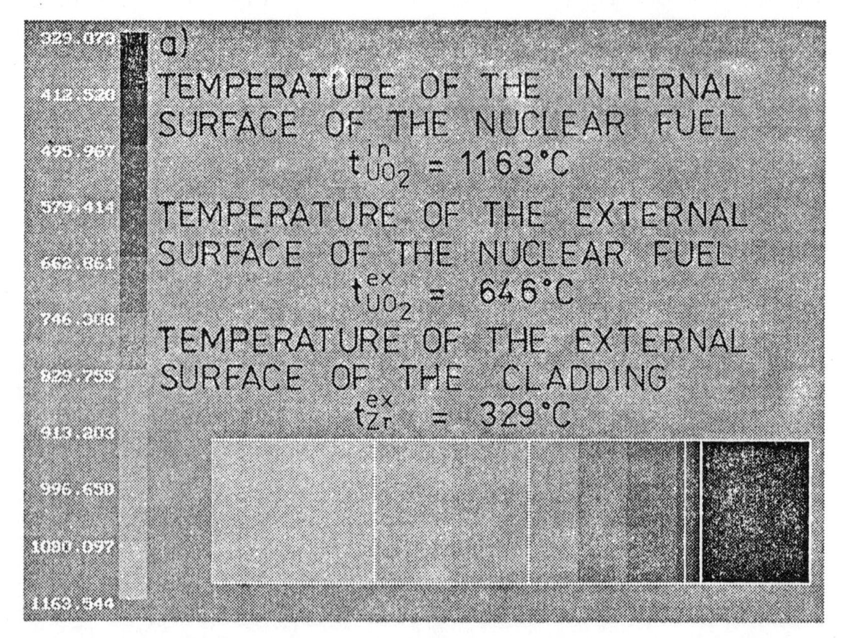

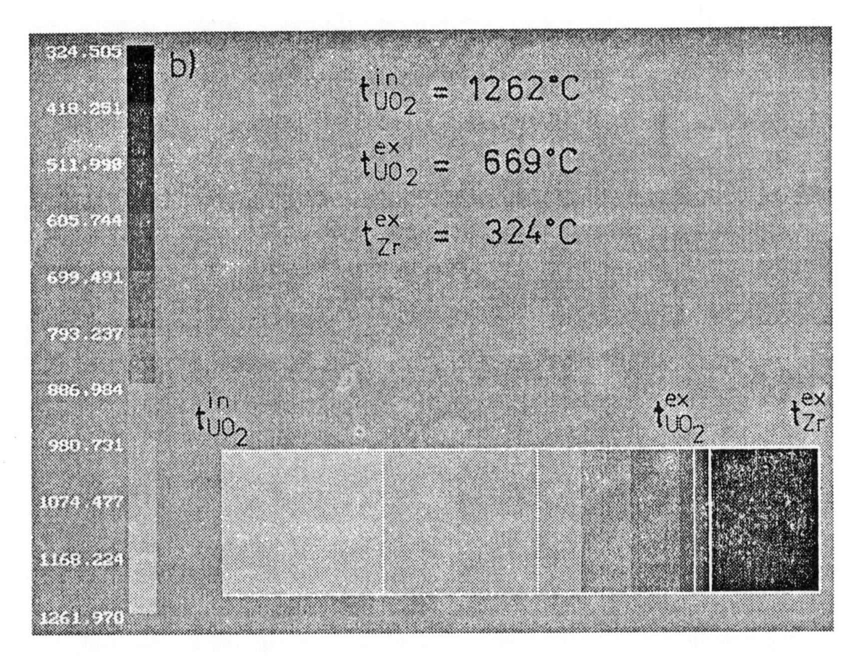

Figure 5. Radial division of thermal field of the fuel element of medium output fuel assembly with the given medium density of thermal neutrons flow $\varphi_T = 7.10^{16}$ neutron.m^{-2}.s^{-1}. a) z' = 0,5m b) z = 0m.

Figure 6 shows a graphical output from the programme displaying radial division of the thermal field of the fuel element of the maximum power fuel assembly.

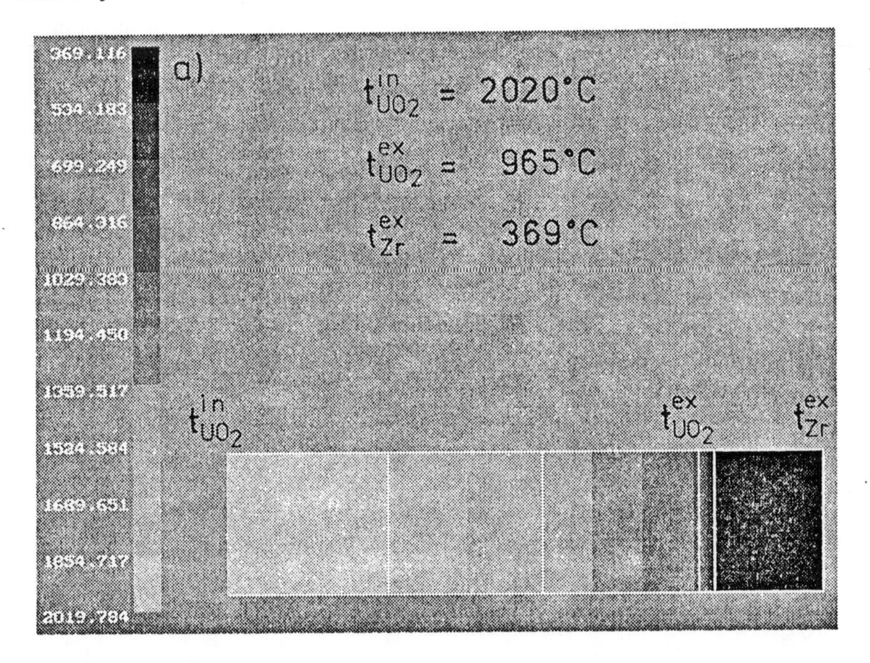

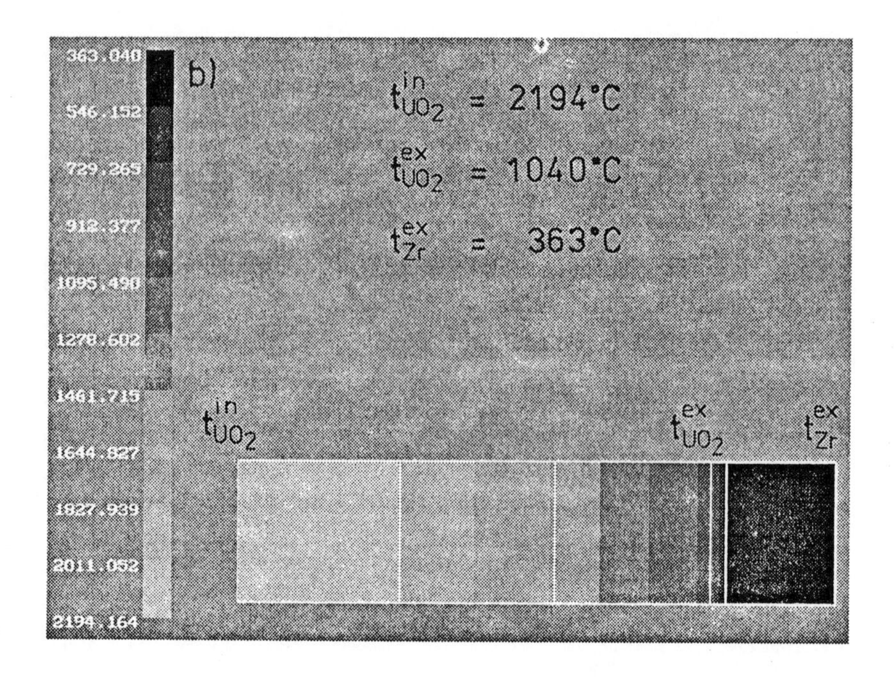

Figure 6. The significance of this figure is identical with figure 5, but for the fuel element of maximum power fuel assembly

When we compare the results of FEM with analytical solution (figure 2) we will find the correspondence of temperatures on the surface of the fuel element to be very good.

Figure 7 shows the results of calculation of maximum temperatures of fuel and of the external surface of the cladding of the fuel element of medium and maximum powers fuel assembly with densities of thermal neutrons flow φ_T = 1; 3; 5; 7.10^{16} neutron.m^{-2}.s^{-1}.

The calculation that was made supposed the inlet temperature of cooling liquid t_{in} to be constant. Certain non-linearity of temperature dependence on neutron flow is due to non-linear dependence of coefficients of thermal conductivity of fuel, contact clearance, the cladding of the fuel element and thermo-physical quantities of water, on temperature.

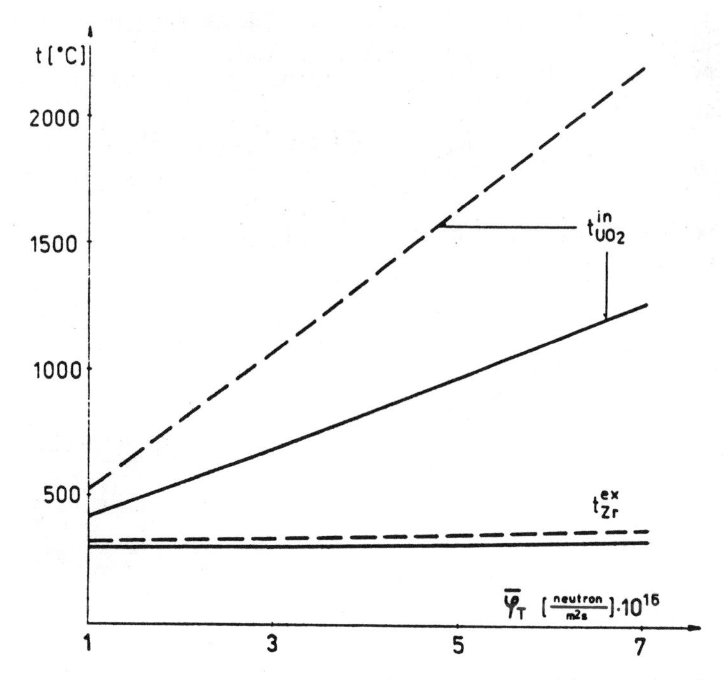

Figure 7. Maximum fuel temperatures and temperatures of external surface of the cladding of the fuel element of medium and maximum powers fuel assembly with various densities of thermal neutrons flow.

5. CONCLUSION

The submitted article deals with the problem of radial division of thermal field in a fuel element, while the generated heat in fuel and other corresponding input parameters of the task are bound to the density of neutron flow. We have reached good accordance of results of the testing task with its analytical solution. The results we have got suggest that making use of this programme for similar tasks in the area of nuclear-energy equipment and other tasks with inward sources of heat, feasible.

REFERENCES

1. RAO, S.S. - The Finite element method in engineering, Pergamon Press, New York, 1982.
2. HINTON, E., OWEN, D.R.J. - An introduction to Finite Element Computations, Pineridge Press Limited, Swansea, U.K., 1979.
3. BENKOVIČ, R. - Diploma work, EF STU Bratislava, 1992.
4. BARTOLOMEJ, G.G., BAT, G.A., BAJBAKOV, V.D., ALCHUTOV, M.S. - Theory of nuclear reactors, Energoatomizdat, Moscow, 1989.
5. MÍCHAL, V. - Diploma work, MEI Moscow, 1992.

6. DEMENTJEV, B.A. - The nuclear power reactors, Energoatomizdat, Moscow, 1990.
7. GALIN, N.M., KIRILOV, P.L. - Heat and mass transmission, Energoatomizdat, Moscow, 1987.
8. PETUCHOV, B.S., GENIN, L.G., KOVALOV, S.A. - Heat and mass transmission in nuclear power stations, Energoatomizdat, Moscow, 1986.
9. HEŘMANSKÝ, B. - Thermomechanics of nuclear reactors, Academia, Prague, 1986.

APPLICATION OF R-FUNCTIONS METHOD TO NUMERICAL SOLUTION OF NON-STEADY THERMAL PROBLEMS

Andrzej WAWRZYNEK [1)]

SUMMARY

In the paper a generalization of R - functions method (RFM), applied to approximate solution of non-steady heat transfer problem is presented. In particular, the so called R-functions are used in order to describe the considered spatial - time domains. Up to now, the RFM was adapted only for numerical computations of steady problems, whereas in the case of non-steady ones, it was connected with Laplace's transformation (the transformation concerns a time-variable). As an example, the problem of cooling in the domain of infinite plate with different boundary conditions is solved. The numerical solutions are compared to analytical ones.

1. INTRODUCTION

The basis of R - function theory and its applications are widely presented in the Rvachev's monographs [2,3] and also, among others, in the papers [1,4,5].

The essence of this method consists in:

- construction of an analytical equation describing a geometry of the whole considered domain in the form $\omega = \omega(x_1, x_2, x_3) = 0$;
- construction of analytical equations concerning the parts of boundary $\omega_i = \omega_i(x_1, x_2, x_3) = 0$, for which the different boundary conditions are assumed;
- construction of general structure of solution (GSS), it means a searching for this class of function which satisfies, for the considered boundary, all boundary conditions in exact way. The GSS in this form contains unknown parameters - these are found in the numerical part of algorithm.

[1)] dr, Silesian Technical University,
ul.Krzywoustego 7, PL-44-121 Gliwice

The author of this paper discussed the problem of RFM applications in the range of thermal conductivity in previous researches [1,4,5]. In particular, the functional for linear non-steady problems applying R - functions and differential time discretization is presented in [1]. The paper [4] deals with a solidification problem; at the same time a certain method of its linearization (see [6]) supplements the basic RFM algorithm. The article presented here constitutes an extension of the paper [5].

As it was emphasized, the prevailing results in the range of RFM application for numerical simulation of heat transfer processes were obtained in such way that the time-variable was treated separately, whereas in this paper the same functions are adapted for spatial - time domain.

In particular, the considered geometrical area Ω is extended on domain $\Omega^* = \Omega \times <0, \theta>$, and, using R - functions, the analytical description of this area is derived. Next, the GSS is constructed and numerical algorithm of unknown parameters determination is described. The solution obtained is smooth and continuous in the whole spatial - time domain. At the same time all boundary and initial conditions are satisfied in exact way. The shape of the area is taken into account in exact way, too. The boundary conditions and kinds of them can change with time.

2. GOVERNING EQUATIONS

Let a considered object occupy an area $\Omega \subset R^3$ (or R^2 or R^1) and let us assume that its boundary $\partial\Omega$ can be divided into n parts. Thermophysical parameters λ, c, ρ (thermal conductivity, specific heat, mass density) of the object are constant. Non-steady temperature field $T(\boldsymbol{x},t)$ in $\Omega(\boldsymbol{x})$ describes the Fourier's equation in the form

$$\frac{\partial T}{\partial t} = \frac{\lambda}{c\rho} \sum_{i=1}^{3} \frac{\partial^2 T}{\partial x_i^2} \quad \text{for } \boldsymbol{x} = (x_1, x_2, x_3) \in \Omega \ , \ t \in (0, \infty) \ ; \tag{1}$$

Energy equation is supplemented by the following boundary conditions given for successive parts of $\partial\Omega$

$$T|_{\partial\Omega_i} = T_{oi} = f_i, \quad i = 1, ..., k \ ; \ (\boldsymbol{x}, t) \in \partial\Omega_i \times (0, \infty) \ ; \tag{2}$$

$$\left(\frac{\partial T}{\partial \nu} + h_j T\right)\Bigg|_{\partial\Omega_j} = f_j, \quad j = k+1, ..., n \ ; \quad (\boldsymbol{x}, t) \in \partial\Omega_j \times (0, \infty) \ ; \tag{3}$$

($f_j = \alpha_j T_\infty / \lambda$ and $h_j = \alpha_j / \lambda$, α_j – heat transfer coefficient for $\partial\Omega_j$), and initial condition

$$T(\boldsymbol{x}, t)|_{t=0} = T_0(\boldsymbol{x}) = f_0 \ , \quad \boldsymbol{x} \in \Omega \ . \tag{4}$$

3. ANALYTICAL DESCRIPTION OF SPATIAL-TIME DOMAIN

The Cartesian product $\Omega^* = \Omega \times <0, \theta)$ can be treated as finite cylinder Ω^* which base is Ω and height is equal to θ (see Fig.1). The boundary of Ω^* consists of the following surfaces: $\Omega_0'' = \Omega$, for $t=0$, $\Omega_{-1}'' = \Omega'$ for $t=\theta$ and $\Omega'' = \partial\Omega \times (0, \theta) = \Omega_1'' \cup \ldots \cup \Omega_n''$.

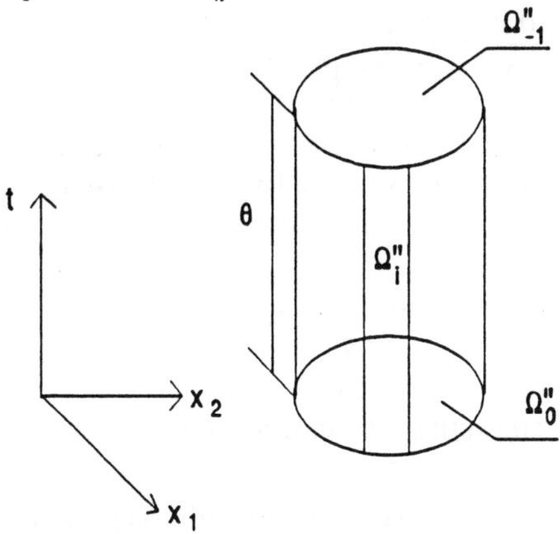

Figure 1. The spatial–time domain Ω^*

Introducing R–functions $\wedge_0$, $\vee_0$, $(^-)$ defined as below

$$
\begin{aligned}
P \wedge_0 Q &= P + Q - \sqrt{P^2 + Q^2} \\
P \vee_0 Q &= P + Q + \sqrt{P^2 + Q^2} \\
\overline{P} &= -P
\end{aligned}
\tag{5}
$$

where P and Q are the arguments, we find the analytical description of domain Ω^* in the form

$$
\omega^* = \omega^*(x, t) = [\omega(x)] \wedge_0 [(\theta - t)t/\theta] \; . \tag{6}
$$

Function $\omega(x, t)$ has a property

$$
\omega(x, t) \begin{cases} <0 \; , & (x, t) \notin \Omega^* \; , \\ =0 \; , & (x, t) \in \partial\Omega^* , \\ >0 \; , & (x, t) \in \Omega^* \; . \end{cases} \tag{7}
$$

Now the boundary and initial conditions are as follows

$$T\Big|_{\Omega_0''} = f_i\,, \quad i = 0, \ldots, k; \qquad \text{for } (x, t) \in \Omega_i\,; \tag{8}$$

$$\left(\frac{\partial T}{\partial \nu} + h_j T\right)\Bigg|_{\Omega_j''} = f_j, \quad j = k+1, \ldots, n\,; \qquad \text{for } (x, t) \in \Omega_j\,; \tag{9}$$

and

$$T(x, t)|_{\Omega_0''} = f_0\,, \quad \text{for } x \in \Omega\,. \tag{10}$$

4. GENERAL STRUCTURE OF SOLUTION

The class of functions satisfying conditions (8-10) given for $\partial\Omega^*$ is introduced

$$T(x, t) = F_0 - \omega\, D_1(F_0) + \omega F_1 + \Phi F_2 - \omega\, D_1(\Phi F_2) - \omega\, \Phi\, H_1 \tag{11}$$

where

$$F_0 = \frac{\sum_{i=0}^{k} \frac{f_i}{\omega_i}}{\sum_{i=0}^{n} \frac{1}{\omega_i}}, \quad F_1 = \frac{\sum_{j=k+1}^{n} \frac{f_j}{\omega_j}}{\sum_{j=k+1}^{n} \frac{1}{\omega_j}},$$

$$F_2 = \frac{\sum_{j=k+1}^{n} \frac{1}{\omega_j}}{\sum_{j=0}^{n} \frac{1}{\omega_j}}, \quad H_1 = \frac{\sum_{j=k+1}^{n} \frac{h_j}{\omega_j}}{\sum_{j=k+1}^{n} \frac{1}{\omega_j}}$$

In formula (11), D_1 is a differential operator

$$D_1(.) = \sum_{i=1}^{3} \frac{\partial \omega}{\partial x_i} \frac{\partial (.)}{\partial x_i} + \frac{\partial \omega}{\partial t} \frac{\partial (.)}{\partial t} \tag{12}$$

and

$$D_1(\omega)\Big|_{\partial\Omega^*} = 1 ,$$

$$D_1(.)\Big|_{\Omega_j''} = \sum_{i=1}^{3} \frac{\partial\omega}{\partial x_i}\frac{\partial(.)}{\partial x_i} = \frac{\partial(.)}{\partial \nu} \tag{13}$$

$$\text{for } j=1,\, 2,\, \ldots,\, n\, ;$$

where $\partial(.)/\partial\nu$ is the normal derivative. At the same time $\Phi = \Phi(x, t)$ is a function of the form

$$\Phi(x, t) = \sum a_{ij}\, \varphi_{ij}(x, t) \tag{14}$$

where a_{ij} are unknown parameters and φ_{ij} are given functions (see [1,4,5]). Since $\omega = 0$ for $(x, t) \in \partial\Omega^*$ we can check that

$$T(x, t)\Big|_{\partial\Omega^*} = F_1 + \Phi F_2 \tag{15}$$

and

$$\begin{aligned} T(x, t)\Big|_{\partial\Omega^*} &= f_i\, , \quad i = 0,\, 1,\, \ldots,\, k\, ; \\ T(x, t)\Big|_{\Omega_j''} &= \Phi\, , \quad j = k+1,\, \ldots,\, n\, : \\ D_1(T)\Big|_{\partial\Omega^*} &= F_1 - \omega\, \Phi\, H_1\, ; \\ D_1(T)\Big|_{\Omega_j''} &= f_j - \Phi h_j\, , \quad j = k+1,\, \ldots,\, n. \end{aligned} \tag{16}$$

so all boundary conditions (8-10) are satisfied.

5. SOLUTION OF THE PROBLEM

In order to solve the problem described above (1-3), it is necessary to determine the unknown parameters a_{ij}. It can be done on the basis of collocational condition (function $T = T(x, t)$ satisfies equation (1)), or the stationary condition of the least square criterion can be taken into account; it means

$$J[a_{ij}] = \int_{\Omega^*} \left(\frac{\partial T}{\partial t} - \frac{\lambda}{c\rho} \sum_{i=1}^{3} \frac{\partial^2 T}{\partial x_i^2} \right)^2 d\Omega^* = \min_{a_{ij} \in R} \tag{17}$$

In this way, one can obtain a system of linear algebraic equations determining the searched parameters a_{ij}

6. NUMERICAL EXAMPLES

The algorithm presented above was applied for the problem of temperature field identification in infinite plate area.

It was assumed that heat exchange between object and environment is described by the third kind of boundary condition, at the same time
- for the case (a): $h_i=h=2.0$ [m^{-1}] and $f_i=f=0.666$ [m^{-1}] ($\alpha=100$ [W/m^2deg]);
- for the case (b): $h_i=h=10.0$ [m^{-1}] and $f_i=f=3.333$ [m^{-1}] ($\alpha=500$ [W/m^2deg]);
- for the case (c): $h_i=h=20.0$ [m^{-1}] and $f_i=f=6.667$ [m^{-1}] ($\alpha=1000$ [W/m^2deg])

and the initial temperature $T_0=f_0=500$ [deg], the ambient temperature T_∞ = 30 [deg].

The function Φ (see formula (14)) was accepted in the form of power series

$$\Phi = \Phi(x, t) = \sum_{\substack{i,j=0 \\ i+j<N}}^{N} a_{ij} \left(\frac{x}{g}\right)^i \left(\frac{t}{\theta}\right)^j \tag{18}$$

where g is the thickness of the plate. The unknown coefficients a_{ij} were determined on the basis of functional minimalization condition (17) (in numerical realization the Gauss method of double integration was used). Satisfactory results were obtained even in the case of rare discretization of area Ω^* and for N = 5 (see the formula (18)) and then the main matrix of algebraic equations system is of the range 21 x 21.

Figures 2-4 illustrate the course of temperature field in half thickness of plate for distinguished moments of time whereas Figures 5-7 the cooling curves at selected points in the plate domain.

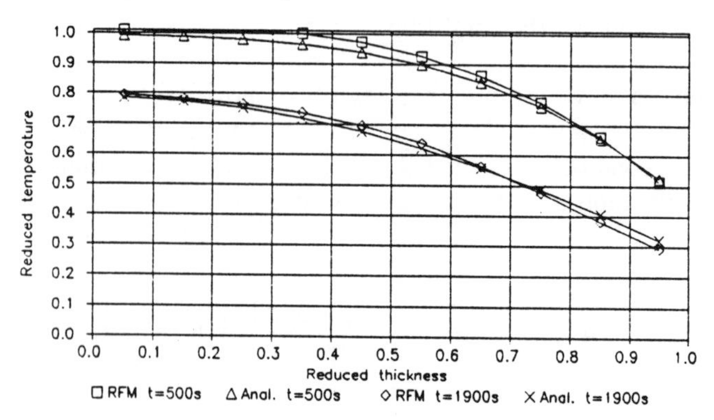

Figure 3. Temperature field for case (b): t=500 [s] and t=1900 [s]

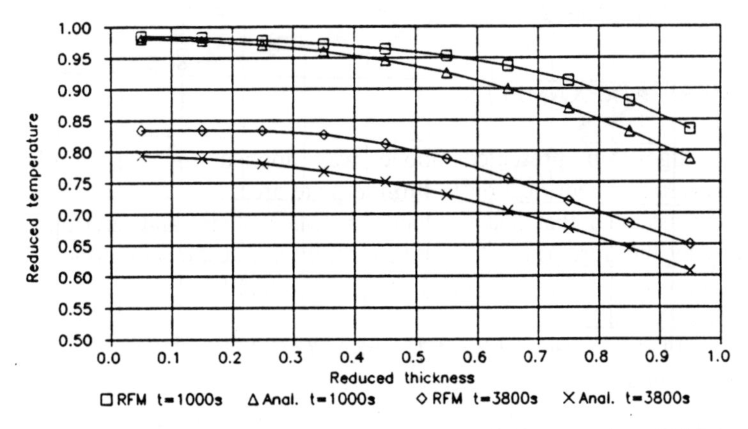

Figure 2. Temperature field for case (a): $t=1000$ [s] and $t=3800$ [s]

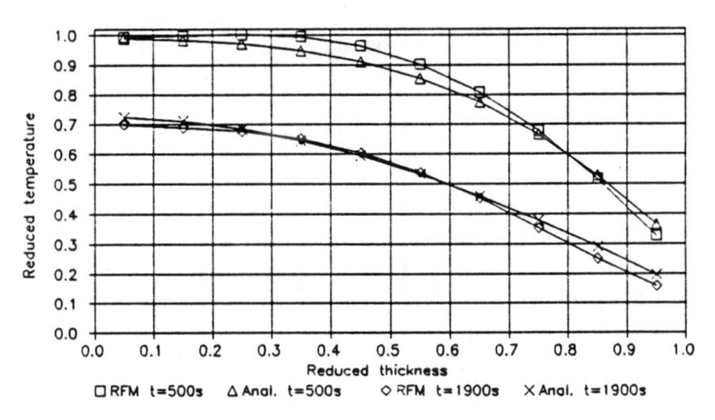

Figure 4. Temperature field for case (c): $t=500$ [s] and $t=1900$ [s]

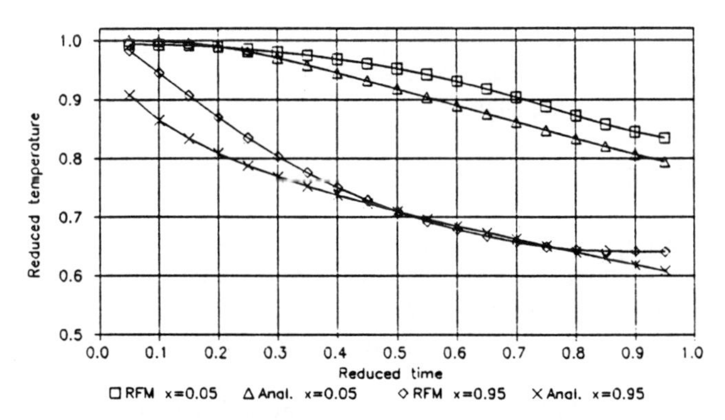

Figure 5. Temperature field for case (a): $x=0.05$ and $x=0.95$

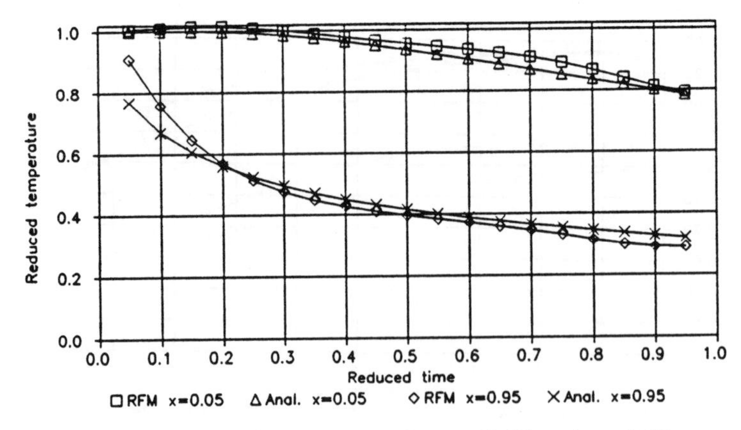

Figure 6. Temperature field for case (b): x=0.05 and x=0.95

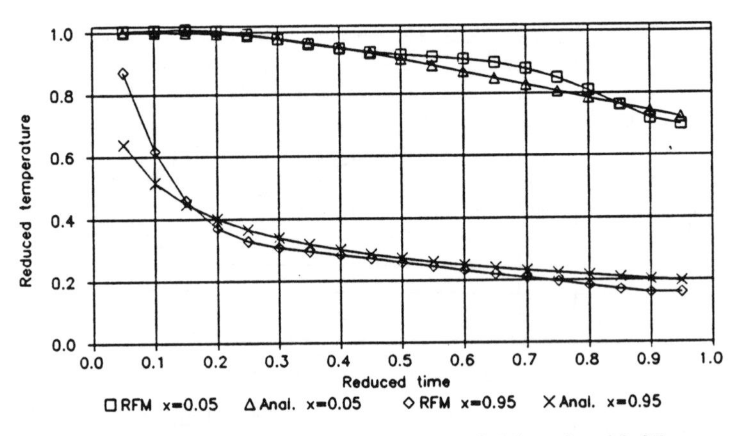

Figure 7. Temperature field for case (c): x=0.05 and x=0.95

A comparison of numerical and analytical solutions (see Figs 2-7) confirms the effectiveness of the proposed algorithm (the error of temperature field identification is not more than 7%).

7. FINAL REMARKS

The most important advantages of the presented algorithm are follows
- an exact approximation of the shape of considered spatial-time area,
- an exact identification of boundary conditions,
- small number of unknown parameters describing temporary temperature field,
- the temperature field is a smooth and continues function.

REFERENCES

1. WAWRZYNEK, A. - Application of R-Function Method for Computation of the Temperature Field in the Volume of the Continuous Casting Mould. Bull.Pol.Acad.Sc.,Tech.Sc., Vol.39, No. 4, pp.633-645, 1991.
2. RVACHEV, V.L. & SLESARENKO, A.P. - Algebra of Logic and Integral Transformations in Boundary Value Problems (in Russian), Ed. Naukova Dumka, Kiev, 1976.
3. RVACHEV, V.L. & RVACHEV, V.A. - Nonclassical Methods of Approximation Theory in the Boundary Value Problems (in Russian), Ed. Naukova Dumka, Kiev, 1979.
4. MAJCHRZAK, E. & WAWRZYNEK, A. - Utilization of R-Functions Method in Numerical Modelling of Solidification Process, Advanced Computational Methods in Heat Transfer II, Vol.1: Conduction, Radiation and Phase Change, Ed. Wrobel, L.C., Brebbia C.A. and Nowak, A.J., Comp. Mech. Publ. & Elsevier, 1992.
5. WAWRZYNEK, A. & GRZYMKOWSKI, R. - Nonclassical Approach to the Solution of the Initial Problem in Arbitrary Domain, Numerical Methods in Engineering '92, Ed. Hirsch Ch., Zienkiewicz, O.C. and Onate E., Elsevier, 1992.
6. MAJCHRZAK, E. - Utilization of Boundary Element Method in the Thermal Theory of Foundry, Proceedings of the 1st Int. Conf., Vol.2 Heat Transfer, Southampton, Ed. Wrobel, L.C. and Brebbia, C.A., Comp. Mech. Publ. de Gruyer, 1991.

NOMENCLATURE LIST

$x = (x_1, x_2, x_3)$	- spatial variable
t, θ	- time variable and time duration of the process
T, T_0, T_∞	- temperature field, initial and ambient temperature
$\Omega, \partial\Omega$	- geometrical area and its boundary
$\Omega^*, \partial\Omega^*$	- spatial-time domain and its boundary
ω, ω_i	- analytical form of area Ω, Ω_i
λ, c, ρ	- thermophysical parameters
h_i, f_i	- boundary and initial functions or constants
a_{ij}, φ_{ij}	- unknown coefficients and given function describing temperature field
$\wedge_0, \vee_0, (^-)$	- Rvachev's operations

ACKNOWLEDGMENT : This research is a part of the Project No 3 3623 91 02 sponsored by KBN.

SECTION 3

SOLIDIFICATION AND CASTING PROCESSES

An efficient finite element method for mould filling simulation in metal casting

R.W.Lewis A.S.Usmani
J.T.Cross
Institute for Numerical methods in Engineering
University College of Swansea

Abstract

The process of mould filling in casting has been modelled. At this stage the molten metal flow is assumed to be laminar. The flow calculations are performed by a semi-implicit solver for the incompressible Navier-Stokes equation using a velocity-pressure segregated approach. The moving front of the molten metal is determined by an explicit solution of a pure advection equation. The temperature field is determined by an explicit solution of the energy equation. The advection terms in all the equations are discretized using a Taylor-Galerkin procedure. The interface between the metal and mould is modelled using special interface elements. The model is tested by applying it to a practical example.

1 Introduction

The primary benefit of a mould filling simulation in casting is that it can provide a more realistic, non-uniform initial temperature distribution for the solidification stage and produce an initial velocity field for the modelling of inertial and natural convection. Examination of the flow patterns during filling can reveal a wide range of potential difficulties with the height, dimensions and positioning of risers or the gating system.

The method used for the modelling of mould filling is based upon the *volume of fluid* method (VOF)[1] (also called the pseudo-concentration method [2] or saturation function method [3, 4]) which is, in turn, based upon the well known *marker and cell* (MAC) technique. The VOF method retains the conceptual simplicity of the MAC technique but demands considerably less computer effort. The marker particles are replaced by a nodal variable which is advected using the velocities from the solution of the Navier-Stokes equations. The main advantage of the method is that the moving front does not have to be explicitly determined, neither do the front boundary conditions have to be satisfied explicitly.

Many finite element and control volume finite difference models exist for numerical simulation of the mould filling process. However, considerable scope remains for further research. The main aim of this work has been to further develop the finite element modelling of this process towards faster solution times and better simulation quality. The Taylor-Galerkin discretization of the governing equations offers major gains on both these fronts. The former is due to the explicit nature of the resultant equation systems and the latter to the excellent stability and smoothness of the advected variable field that is achieved with negligible false diffusion, which is a major drawback of many existing models.

The flow of molten metal into a casting is invariably turbulent, however this has not been incorporated in the model at this stage as this is still under development.

2 Governing differential equations and boundary conditions

The conservation equations and their boundary conditions used in the present model are as follows.

Conservation of mass

$$\nabla \cdot \mathbf{v} = 0 \tag{1}$$

in which $\mathbf{v}$ is the velocity vector.

Conservation of momentum

An expanded version of the stress divergence form is used.

$$\rho \left(\frac{\partial \mathbf{v}}{\partial t} + \nabla \cdot \mathbf{v} \right) = -\nabla P + \nabla \cdot \mu \left(\nabla \mathbf{v} + (\nabla \mathbf{v})^T \right) + \rho \mathbf{g} \tag{2}$$

in which P is the pressure, μ is the dynamic viscosity, ρ is the density and $\mathbf{g}$ is the gravitational acceleration.

The essential or Dirichlet boundary conditions for the Navier-Stokes equations are specified in terms of the velocities at the boundaries. Pressure may not be specified at the boundary as it is an implicit variable in an incompressible flow [5] which propagates at infinite speed to deliver a solenoidal velocity field. The natural or Neumann boundary conditions may be applied as normal and/or tangential traction forces,

$$f_n = -P + 2\mu \frac{\partial v_n}{\partial n} \tag{3}$$

$$f_\tau = \mu \left(\frac{\partial v_n}{\partial \tau} + \frac{\partial v_\tau}{\partial n} \right) \tag{4}$$

where n and τ are the unit normal and tangent vectors respectively.

Conservation of the metal front position

A pseudo-concentration function is used to track the free fluid surface. If this function is represented by $F(x, y, t)$ for 2-D flow, the first order pure advection equation, which conserves the function $F(x, y, t)$, is written as,

$$\frac{\partial F}{\partial t} + \mathbf{v} \cdot \nabla F = 0 \tag{5}$$

A particular value of the pseudo-concentration function, F_c, is associated with the free fluid surface [6], and it can be tracked in time by simply plotting the contour of F_c at each timestep. The value of $F(x, y, t) = 0.0 = F_c$ is used to mark the free surface, while $F(x, y, t) > 0.0$ indicates the fluid region and $F(x, y, t) < 0.0$ the empty region. As this is a hyperbolic or pure advection equation, the boundary values of F are required only at the nodes where the fluid enters the cavity.

Conservation of energy

Finally, the heat transfer is controlled by the advective-diffusive energy equation, which is,

$$\rho c \left(\frac{\partial T}{\partial t} + \mathbf{v} \cdot \nabla T \right) = \nabla \cdot k \nabla T \tag{6}$$

where c and k are the specific heat and thermal conductivity respectively, and T is the temperature.

Dirichlet boundary conditions for this equation consist of specified temperature at the boundaries. The general Neumann boundary condition for the energy equation may be written as,

$$k \frac{\partial T}{\partial n} + q + h(T - T_a) = 0 \tag{7}$$

where q, h and T_a are specified boundary heat flux, convective heat transfer coefficient and the ambient temperature respectively.

3 Description of the algorithm

The Navier-Stokes equations are discretized using the fractional-step method of Donea *et. al.*[7]. The advection part of the Navier-Stokes equations is discretized via the Taylor-Galerkin procedure of Laval and Quartapelle [8]. The Taylor-Galerkin procedure of Donea *et. al.*[9, 10] has been used to discretize the pseudo-concentration equation and the energy equation. An unstructured mesh of four noded quadrilateral elements has been used for the example problem. Linear shape functions have been employed for all the nodal variables except pressure which has been assumed to be constant over an element.

A detailed description of the full computational algorithm follows.

3.1 The flow field

The Navier-Stokes equations are solved in three separate stages: advection, viscous diffusion and incompressibility.

Advection

$$\frac{\partial \mathbf{v}}{\partial t} + \mathbf{v} \cdot \nabla \mathbf{v} = 0 \tag{8}$$

This equation is first discretized in time using the first two terms of the Taylor series expansion for the temporal derivative. The temporally discretized form obtained is given in reference [8] as,

$$\frac{\mathbf{v}^* - \mathbf{v}_n}{\Delta t} = -(\mathbf{v}_n \cdot \nabla)\mathbf{v}_n + \frac{\Delta t}{2}\left(\left(\left(\mathbf{v}_n \cdot \nabla\right)\mathbf{v}_n\right) \cdot \nabla\right)\mathbf{v}_n + \frac{\Delta t}{2}\left(\mathbf{v}_n \cdot \nabla\right)\left(\mathbf{v}_n \cdot \nabla\right)\mathbf{v}_n \tag{9}$$

where $\mathbf{v}^*$ is an intermediate velocity field, and $\mathbf{v}_n$ is the velocity at timestep n. This equation is now spatially discretized using the Galerkin form of the finite element method (GFEM). The fully discretized system results in a matrix system of the form,

$$\mathbf{M}\left(\frac{\mathbf{v}^* - \mathbf{v}_n}{\Delta t}\right) = \mathbf{S}_1(\mathbf{v}_n) \tag{10}$$

This can be solved explicitly if $\mathbf{M}$ is converted to a 'row-sum type' lumped mass matrix. However, a lumped mass approach will always degrade advection solutions [11]. Donea *et. al.*[7] proposed an iterative explicit procedure which uses both the lumped and consistent mass matrices to obtain a significantly improved solution. This procedure has been used here to obtain the intermediate step velocity $\mathbf{v}^*$ from equation (10). Specified velocities may be imposed on $\mathbf{v}^*$ to satisfy the boundary conditions. The matrix $\mathbf{S}_1$ in equation (10) is the advection matrix which includes terms arising from the Taylor-Galerkin discretization.

Viscous diffusion

$$\rho \frac{\partial \mathbf{v}}{\partial t} = \nabla \cdot \mu \left(\nabla \mathbf{v} + (\nabla \mathbf{v})^T \right) + \rho \mathbf{g} \tag{11}$$

This equation is discretized by using a standard GFEM procedure to obtain,

$$\mathbf{M} \left(\frac{\mathbf{v}^{**} - \mathbf{v}^*}{\Delta t} \right) = \mathbf{K}_1(\mathbf{v}^*) + \mathbf{F} \tag{12}$$

which is solved explicitly using a lumped form of $\mathbf{M}$. An iterative solution is not necessary as the lumped explicit solutions for diffusion remain smooth if the timestep size is within specified stability limits. Specified boundary values of velocity may again be imposed on $\mathbf{v}^{**}$. In equation (12), $\mathbf{K}_1$ is the standard viscous diffusion matrix and $\mathbf{F}$ is the force vector containing contributions from gravity and/or other sources of applied loading.

Incompressibility

The velocity field obtained above ($\mathbf{v}^{**}$) does not, as yet, satisfy conservation of mass; the remaining equations to be dealt with are

$$\rho \frac{\partial \mathbf{v}}{\partial t} = -\nabla P \tag{13}$$

$$\nabla \cdot \mathbf{v} = 0 \tag{14}$$

If equations (13) and (14) are discretized via GFEM, we obtain,

$$\mathbf{M} \left(\frac{\mathbf{v}_{n+1} - \mathbf{v}^{**}}{\Delta t} \right) = \mathbf{C} P_{n+1} \tag{15}$$

$$\mathbf{C}^T(\mathbf{v}_{n+1}) = 0 \tag{16}$$

where $\mathbf{C}$ is a gradient and $\mathbf{C}^T$ a divergence matrix. Taking the divergence of both sides of equation (16) (after multiplying by $\mathbf{M}^{-1}$) and cancelling the $\mathbf{v}_{n+1}$ term using equation (16), we have,

$$\mathbf{C}^T \mathbf{M}^{-1} \mathbf{C} P_{n+1} = -\frac{\mathbf{C}^T \mathbf{v}^{**}}{\Delta t} \tag{17}$$

This equation is used to calculate the pressure, P_{n+1}. The final velocities may then be obtained from equation (15).

Further details and analysis of segregated velocity pressure type formulations for finite element modelling of the incompressible Navier-Stokes equations may be obtained in Gresho [12].

3.2 Pseudo-concentration function for front tracking

The pseudo-concentration equation (5) is a hyperbolic or pure advection equation. If the pseudo-concentration function is advected using conventional GFEM severe oscillations are generated. This problem can be remedied by a Taylor-Galerkin discretization of equation (5) as in reference [6]. If such a discretization is performed the following equation is obtained,

$$\frac{F_{n+1} - F_n}{\Delta t} = \left(-(\mathbf{v}_{n+\frac{1}{2}} \cdot \nabla) + \frac{\Delta t}{2} (\mathbf{v}_n \cdot \nabla)^2 \right) F_n \tag{18}$$

where $\mathbf{v}_{n+\frac{1}{2}}$ represents $\frac{\mathbf{v}_{n+1}+\mathbf{v}_n}{2}$. The fully discretized equations result in a matrix system of the form,

$$\mathbf{M} \left(\frac{F_{n+1} - F_n}{\Delta t} \right) = \mathbf{S}_2(F_n) \tag{19}$$

This equation is solved using the iterative explicit procedure mentioned earlier. The advection matrix, $\mathbf{S}_2$, includes the diffusion-like terms arising from the Taylor-Galerkin discretization.

3.3 Heat transfer during filling

A heat transfer analysis requires that both advection and diffusion are dealt with. To this end equation (6) is broken down into convection and thermal diffusion stages.

Convection

$$\frac{\partial T}{\partial t} + \mathbf{v} \cdot \nabla T = 0 \tag{20}$$

This equation is exactly the same as equation (5) for the pseudo-concentration function. A Taylor-Galerkin discretization of equation (20) results in an expression for an intermediate temperature field,

$$\frac{T^* - T_n}{\Delta t} = \left(-(\mathbf{v}_{n+\frac{1}{2}} \cdot \nabla) + \frac{\Delta t}{2} (\mathbf{v}_n \cdot \nabla)^2 \right) T_n \tag{21}$$

which is similar to equation (19). Spatial discretization of the above, using standard GFEM, results in a matrix system of the form,

$$\mathbf{M}\left(\frac{T^{*}-T_{n}}{\Delta t}\right)=\mathbf{S}_{3}(T_{n}) \tag{22}$$

Again this equation is solved using the iterative explicit procedure. The advection matrix, $\mathbf{S}_3$, includes the Taylor-Galerkin terms.

Thermal diffusion

$$\rho c\frac{\partial T}{\partial t}=\nabla\cdot k\nabla T \tag{23}$$

This part is discretized using standard GFEM and results in,

$$\mathbf{M}\left(\frac{T_{n+1}-T^{*}}{\Delta t}\right)=\mathbf{K}_{2}(T^{*}) \tag{24}$$

from which the temperatures at the end of the stage are calculated by the lumped explicit procedure, and $\mathbf{K}_2$ is the standard heat diffusion matrix.

4 Metal-mould interface element

During the filling process (and during solidification after filling), thermal barriers exist at the metal/mould interface in the form of die coatings and air gaps. If the heat transfer at this interface is to be accurately modelled, the heat transfer coefficient between the two faces must be known for all possible conditions. This information is not readily available. However, the present algorithm incorporates an interface element, the stiffness of which depends upon the heat transfer coefficient. The element stiffnesses from the interface elements are incorporated in matrix $\mathbf{K}_2$ in equation (24). The details of this element are available in reference [13].

5 A test example of filling with curved flow boundaries

Figure 1 shows the problem definition for this example, while Figures 2, 3 and 4 show the mesh, the initial position of the fluid front and a typical velocity field respectively. Figure 5 shows the front profiles and temperature contours at various times. The boundary conditions for the flow allow free slipping at the walls. The properties used for the Navier-Stokes solver

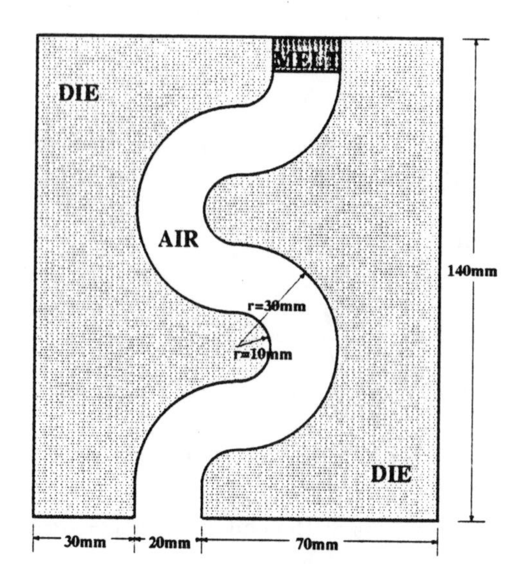

Figure 1: Problem definition for the curved cavity example

Figure 2: Mesh used for the curved cavity example

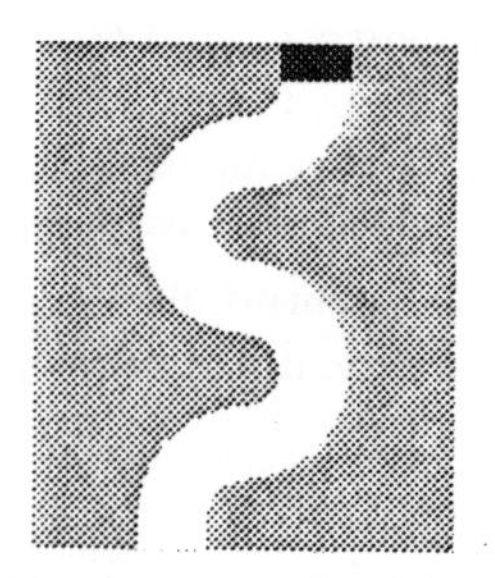

Figure 3: Starting position for the curved cavity example

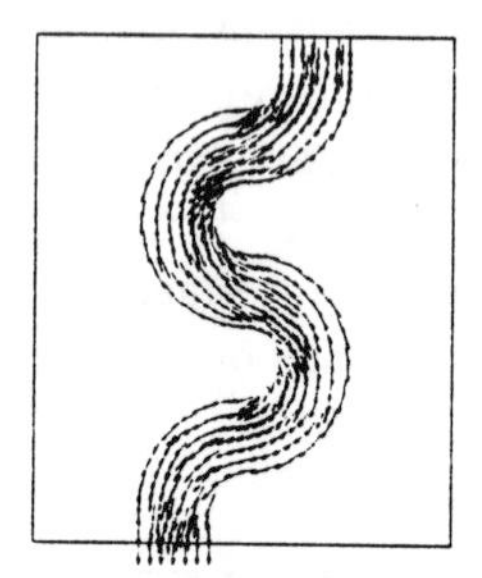

Figure 4: Typical velocity vectors for the curved cavity example

correspond to an Re of approximately 100.0. For the thermal analysis the real properties of metal (aluminium) and die (steel) have been used.

$$\rho_{met} = 2700.0\ kg/m^3 \qquad k_{met} = 200.0\ W/mK \qquad c_{met} = 900.0\ J/kgK$$

$$\rho_{die} = 7800.0\ kg/m^3 \qquad k_{die} = 48.0\ W/mK \qquad c_{die} = 450.0\ J/kgK$$

The coefficient of heat transfer for the metal-mould interface was assumed to be 30000 W/Km^2. The metal is assumed to flow by gravity from a fixed initial position as shown in Figure 3.

The mesh contained 1483 elements and 1543 nodes and was run for 850 timesteps.

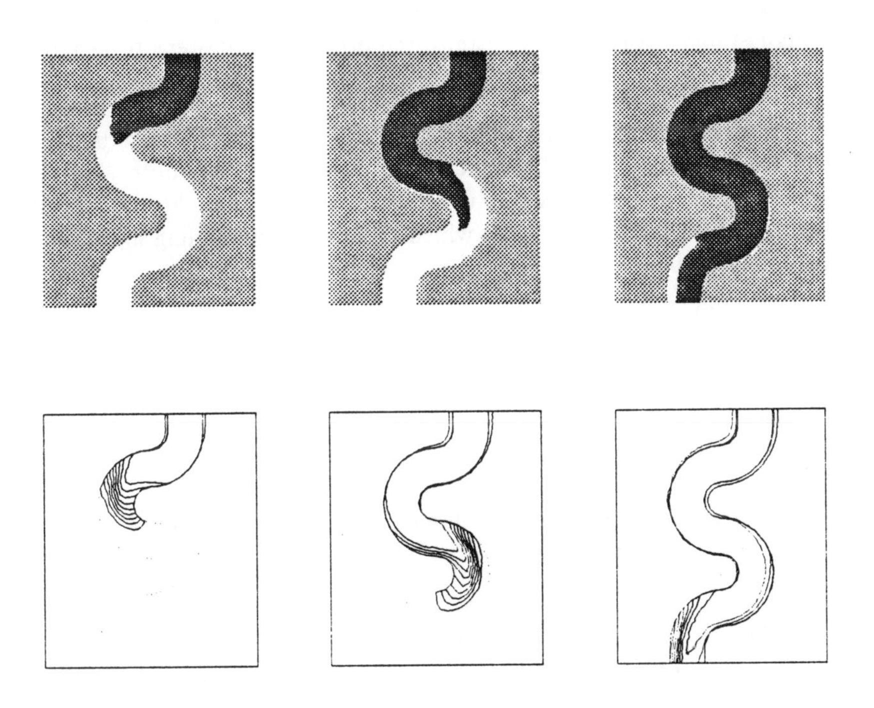

Figure 5: Front positions and temperature contours at 0.4, 0.6 and 0.85 seconds for the curved cavity example

6 Conclusions

One of the problems with mould filling analyses is the length of time required to achieve a solution. The fact that the results shown in Section 5 were obtained in only 10 minutes on a SUN SPARCstation 2 means that the model can now be used in conjunction with heat transfer and post-analysis defect prediction modules, to provide a fully integrated casting analysis package. The temperature profiles show little false diffusion and the interface elements performed well, as can be seen from the discontinuity of the temperature field at the metal mould interface.

The explicit nature of this model means that timestep size limitations must be adhered to. This would normally make a model more expensive than an implicit one. However, for advection dominated flows, the timestep size is merely a reflection of the physics of the problem, and implicit algorithms, although stable with larger timesteps, require small timesteps to achieve acceptable accuracy.

The proposed model provides an efficient and reliable solution to a difficult numerical problem. With the inherent advantages of the finite element method in terms of geometry resolution, it is expected that this model will provide an excellent foundation on which a realistic mould filling simulation facility may be built.

References

[1] C.W.Hirt and B.D.Nichols. Volume of fluid (vof) method for the dynamics of free boundaries. *Journal of Computational Physics*, 39:201–225, 1981.

[2] E.Thompson. Use of pseudo-concentrations to follow creeping viscous flows during transient analysis. *International Journal for Numerical Methods in Fluids*, 6:749–761, 1986.

[3] T.J.Smith and D.B.Welbourn. The integration of geometric modelling with finite element analysis for the computer-aided design of castings. *Applied Scientific Research*, 44:139–160, 1987.

[4] R.W.Lewis, K.Morgan, E.D.L.Pugh, and T.J.Smith. A note on discontinuous numerical solutions of the kinematic wave equation. *Inter-*

national Journal for Numerical Methods in Engineering, 20:555–563, 1984.

[5] P.M.Gresho, R.L.Lee, and R.L.Sani. On the time-dependent solution of the incompressible Navier-Stokes equations in two and three dimensions. In *Recent Advances in Numerical Methods in Fluids*, volume 1. Pineridge Press Limited, Swansea, 1980.

[6] A.S.Usmani, J.T.Cross, and R.W.Lewis. A finite element model for the simulation of mould filling in metal casting and the associated heat transfer. *International Journal for Numerical Methods in Engineering*, 35:787–806, 1992.

[7] J.Donea, S.Giuliani, H.Laval, and L.Quartapelle. Finite element solution of the unsteady Navier-Stokes equations by a fractional step method. *Computer Methods in Applied Mechanics and Engineering*, 30:53–73, 1982.

[8] H.Laval and L.Quartapelle. A fractional-step Taylor-Galerkin method for unsteady incompressible flows. *International Journal for Numerical Methods in Fluids*, 11:501–513, 1990.

[9] J.Donea. A Taylor-Galerkin method for convective transport problems. *International Journal for Numerical Methods in Engineering*, 20:101–119, 1984.

[10] J.Donea, S.Giuliani, H.Laval, and L.Quartapelle. Time-accurate solution of advection-diffusion problems by finite elements. *Computer Methods in Applied Mechanics and Engineering*, 45:123–145, 1984.

[11] P.M.Gresho, R.L.Lee, and R.L.Sani. Advection-dominated flows with emphasis on the consequences of mass lumping. In *Finite Elements in Fluids*, volume 3. John Wiley and Sons, 1978.

[12] P.M.Gresho. On the theory of semi-implicit projection methods for viscous incompressible flow and its implementation via a finite element method that also introduces a consistent mass matrix. part 1: Theory and part 2: Implementation. *International Journal for Numerical Methods in Fluids*, 11:587–659, 1990.

[13] A.S.Usmani. *Finite Element Modelling of Convective-Diffusive Heat Transfer and Phase Transformation with Reference to Casting Simulation - Ph.D. Thesis*. University of Wales, Swansea, 1991.

NUMERICAL MODELLING OF FILLING AND SOLIDIFICATION IN METAL CASTING PROCESSES : A UNIFIED APPROACH

C. R. Swaminathan* and V. R. Voller**

*Department of Mechanical Engineering
**Department of Civil and Mineral Engineering
University of Minnesota, Minneapolis, MN 55455, U.S.A.

SUMMARY

On appealing to an analogy with the numerical modelling of phase change, an algorithm for modelling the filling of metal into a mould is developed. This algorithm closely resembles the basic and well known Volume of Fluid (VOF) method. The key difference is that the modified algorithm can allow for an implicit time solution, free of the Courant criteria. The modified filling algorithm is incorporated into a unified numerical model for filling and solidification heat transfer. This unified model is applied to the two-dimensional filling and solidification of a wheel casting.

1. INTRODUCTION

Recent studies on the performance of metal shape castings have emphasized the need for prediction and control of defects. Computational modelling has taken a significant role in these studies [1]. Three key phenomena that need to be modelled are :

1. The filling of the hot molten metal into the mould shape, in particular the tracking of the free surface between the metal and the air.

2. The subsequent or concurrent heat transfer and solidification of the metal.

3. The development of the stress field as the metal cools.

Computational models for each of these phenomena have been reported in the literature, indeed preliminary models that have investigated the coupled nature of the phenomena have also been reported [2-5]. In this paper we will focus on the coupling between the filling and the heat transfer.

Models and numerical algorithms for solving filling and solidification in shape castings are common place and for a wide range of problems commercial codes are available. In terms of run times, the computation involved in predicting the filling of the mould cavity is typically an order of magnitude larger than the computation time required to solve a solidification problem alone. The reasons for this are simple. There are about four decades worth of literature on solving solidification problems and as a result there is wide choice of efficient methodologies (see [6] for a summary of these methods). On the other hand numerical algorithms for filling are much more limited, with the vast majority of algorithms based on the so called 'Volume of Fluid' (VOF) method [7]. The VOF method is explicit in time and therefore subject to the Courant criteria. This severely limits the time step size and thereby controls the run time.

The filling algorithm presented in this paper will be based on a modification [8] of the VOF approach. In essence we will develop a 'numerical analogy' between filling and solidification. This approach allows us to employ an implicit solution procedure free of the Courant criteria. The resulting discrete equations can be solved by techniques identical to those used in solving solidification problems. A major advantage of this approach is that it can easily accommodate a unified solution of filling and solidification. This unification will be the central focus in the current work.

2. NUMERICAL MODELLING OF FILLING

The objective of a filling model is to track the liquid/air free surface and thereby predict the filling pattern in a mould. The filling pattern has a direct influence on the subsequent heat transfer and also plays a key role in determining the location of defects such as porosity. An ideal filling algorithm should be compatible with general purpose codes, must preserve a sharp interface between the two fluids, must be free of any time step restrictions and must be independent of the numerical discretization method.

A basic governing equation describing the filling process can be obtained on considering a spatially fixed representative elementary volume (REV) which at some point in time is undergoing filling, see Fig. 1. On assuming negligible mass transfer between liquid and gas phases, a balance of the liquid mass alone over the REV can be written as

$$\frac{d}{dt}\int_V \rho_1 \, G \, dV = -\int_S \rho_1 \, G \, \vec{u}.\vec{n} \, dS \qquad (1)$$

where ρ_1 is the density of the liquid, $\vec{u}$ is the velocity, $\vec{n}$ is the unit outward normal to the surface S and G is the volume fraction of the metal. On defining

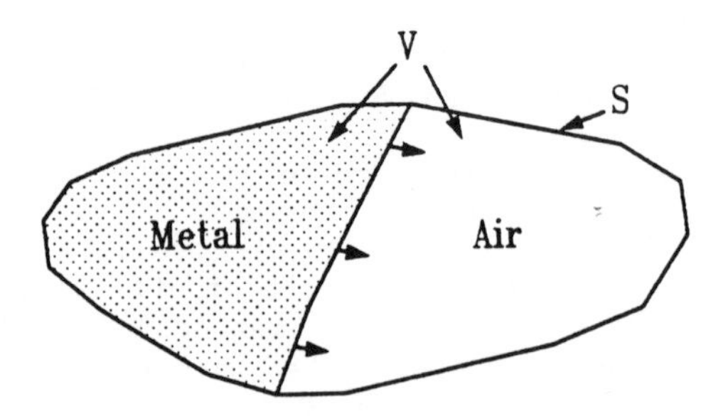

Fig. 1 - A spatially fixed representative elementary volume.

$$F = \int_V G \, dV \tag{2}$$

to be the volume averaged liquid fraction over the REV and on recognizing that the volume V is fixed, Eq. (1) can be rewritten as

$$\frac{\partial}{\partial t}(\rho_1 F) = -\int_S \rho_1 G \, \vec{u}.\vec{n} \, dS \tag{3}$$

This equation can be immediately used in a numerical solution on associating the REV with a control volume (or cell) defined on the numerical mesh.

2.1 The Volume of Fluid (VOF) method

Note that on setting G to be equal to F, Eq. (3) can be rewritten in the form

$$\frac{\partial}{\partial t}(\rho_1 F) + \nabla.(\rho_1 \vec{u} F) = 0 \tag{4}$$

This is the equation modelled in the well known Volume of Fluid (VOF) method of Hirt and Nichols [7]. In solving Eq. (4) numerically, a challenging task is to obtain a smear-free interface. Hirt and Nichols propose a donor-acceptor type approximation for the fluxes at the interface to achieve this objective. The only drawback is that an explicit time stepping scheme has to be used so that not more than one control volume (or cell) fills over a particular time step.

In this work, we draw upon an analogy with the numerical modelling of phase change, to obtain an alternative relationship between F and G, a relationship that allows for a smear free yet implicit solution of the resulting equations. This analogy with phase change will be presented next.

2.2 Analogy with phase change : a new F .vs. G relationship [8]

A common method for solving phase change problems is to use the so called enthalpy method. In this approach the progress of the phase change is tracked on the introduction of a local liquid fraction associated with each control volume in the numerical discretization. An often solved test problem involves the heat conduction controlled melting of a pure material initially at the phase change temperature. In the numerical solution of this problem, based on an enthalpy method, when a cell is undergoing the phase change, the temperature of that cell remains fixed at the phase change temperature. Essentially a phase change cell can only receive heat from the fully liquid cells and heat cannot be transmitted to neighbouring cells. In our treatment of filling we will develop a model that will have an analogous response, i.e.,

> when a computational cell is filling it can only receive fluid ; it cannot empty (transmit fluid) until it is completely filled

This can be achieved on using an upwind principal [9] for evaluating G at the control volume surfaces and on relating F to G in the following manner

$$\begin{aligned} G &= 0 \quad \text{while} \quad 0 \le F < 1 \\ F &= 1 \quad \text{while} \quad 0 < G \le 1 \end{aligned} \qquad (5)$$

This analogy with melting is illustrated in Fig. 2 ; in effect filling can be modelled as a melting problem with zero specific heat in the liquid.

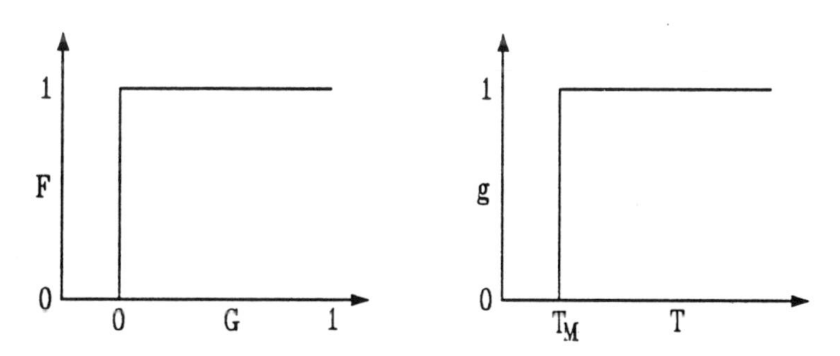

Fig. 2 - Analogy between filling and melting phase change.

3. GOVERNING EQUATIONS FOR COMBINED FILLING AND SOLIDIFICATION

In deriving the governing equations for combined filling and solidification, we shall consider a phase change system with a columnar dendritic morphology with the solid being stationary. This greatly simplifies the derivation of the governing equations. In such systems, we can define a system velocity $\bar{u}$, such that

$$\vec{u} = \vec{u}_a = g\ \vec{u}_1 \qquad \text{and} \qquad \vec{u}_s = 0 \tag{6}$$

where g is the volume fraction of the liquid metal in the combined solid plus liquid region and the subscripts $[\,]_s$, $[\,]_1$ and $[\,]_a$ represent the solid, liquid and air phase, respectively.

Based on these discussions, the mass conservation in the combined solid and liquid phase takes the form

$$\frac{\partial}{\partial t}(\bar{\rho}\ F) = -\int_S \rho_1\ G\ \vec{u}.\vec{n}\ dS \tag{7}$$

where

$$\bar{\rho} = g\ \rho_1 + (1-g)\ \rho_s \tag{8}$$

For the case of equal densities in the solid and liquid phases, Eq. (7) simplifies to

$$\frac{\partial F}{\partial t} = -\int_S G\ \vec{u}.\vec{n}\ dS \tag{9}$$

In addition to the mass conservation equation, one can derive a volume continuity equation

$$\nabla.\vec{u} + \frac{1}{\rho_1}\left\{\frac{\partial}{\partial t}(F\,[g\ \rho_1 + (1-g)\ \rho_s]) + G\ \vec{u}.\nabla\rho_1\right\} + \frac{1}{\rho_a}\left\{\frac{\partial}{\partial t}([1-F]\ \rho_a) + [1-G]\ \vec{u}.\nabla\rho_a\right\} = 0 \tag{10}$$

by addition of the single phase equations. If we assume a constant density in each phase Eq. (10) reduces to

$$\nabla.\vec{u} + \frac{\partial}{\partial t}\left(F\ [1-g]\left[\frac{\rho_s}{\rho_1} - 1\right]\right) = 0 \tag{11}$$

For the case of equal densities in the solid and liquid phases, Eq. (11) simplifies to

$$\nabla.\vec{u} = 0 \tag{12}$$

The momentum balance in the combined liquid and air phase can be written as

$$\frac{\partial}{\partial t}(\bar{\rho}_F\ \vec{u}) + \nabla.(\bar{\rho}_G\ \vec{u}\ \vec{u}) - \nabla.(\bar{\mu}_G\ \nabla\vec{u}) + \vec{S} = 0 \tag{13}$$

where

$$\overline{\rho}_F = F\rho_1 + (1 - F)\rho_a$$
$$\overline{\rho}_G = G\rho_1 + (1 - G)\rho_a \tag{14}$$
$$\overline{\mu}_G = G\mu_{eff} + (1 - G)\mu_a$$

The source term S takes the form,

$$\vec{S} = (F\rho_1 + [1 - F]\rho_a)\vec{gr} \tag{15}$$

where $\vec{gr}$ is the acceleration due to gravity. Buoyancy effects in the liquid are modelled via the Boussinesq approximation, i.e.,

$$\rho_1 = \rho_{ref}(1 - \beta[T - T_{ref}]) \tag{16}$$

Eq. (13) is applied over the entire solid, liquid and air region and the velocity in the solid is switched to zero using a viscosity model [10],

$$\mu_{eff} = \mu_1\left(\frac{\mu_s}{\mu_1}\right)^{(1-g)} \tag{17}$$

where the solid viscosity μ_s is given a large value (e.g., 10^6). Alternatively, one could also use the Darcy source approach [11] to model the flow in the mushy region.

The conservation of energy can be written down in terms of the mixture enthalpy as

$$\frac{\partial\overline{H}}{\partial t} + \nabla.(\overline{u\,H}) - \nabla.(\overline{k}_G\nabla T) = 0 \tag{18}$$

In Eq. (18), the mixture quantities are defined as

$$\overline{H} = \overline{(\rho\,c)}_F T + F\rho_1 L g \tag{19}$$

$$\overline{u\,H} = \overline{(\rho\,c\,u)}_G T + G\rho_1 L\vec{u}_1 g \tag{20}$$

$$\overline{k}_G = G(g k_1 + (1 - g)k_s) + (1 - G)k_a \tag{21}$$

where

$$\overline{(\rho\,c)}_F = F(g\rho_1 c_1 + (1 - g)\rho_s c_s) + (1 - F)\rho_a c_a \tag{22}$$

$$\overline{(\rho\,c\,u)}_G = G(g\rho_1 c_1\vec{u}_1 + (1 - g)\rho_s c_s\vec{u}_s) + (1 - G)\rho_a c_a\vec{u}_a \tag{23}$$

Eq. (18) has been derived by adding the single phase energy balance equations. For the phase change system under consideration, on using Eq. (6), Eq. (18) simplifies to

$$\frac{\partial}{\partial t}\left[(\overline{\rho\, c})_F\, T\right] + \nabla.\left[(\overline{\rho\, c})_G\, \vec{u}\, T\right] - \nabla.(\overline{k}_G\, \nabla T) + S = 0 \qquad (24)$$

where

$$(\overline{\rho\, c})_G = G\, \rho_1\, c_1 + (1 - G)\, \rho_a\, c_a \qquad (25)$$

The source term S in Eq. (24) takes the form,

$$S = \frac{\partial}{\partial t}(F\, \rho_1\, L\, g) + \nabla.(G\, \rho_1\, L\, \vec{u}) \qquad (26)$$

On using the mass continuity equation, Eq. (26) simplifies to

$$S = \frac{\partial}{\partial t}(\rho_s\, L\, F\, [g - 1]) \qquad (27)$$

To summarize, Eqs. (9), (12), (13) and (24) represent the governing equations that will be used in the current work to model the combined filling and solidification phenomena.

4. NUMERICAL SOLUTION

The coupled equations derived in the previous section are discretized using the control volume based finite difference method of Patankar [9]. A fully implicit time integration scheme is used for all the equations. The convective terms are discretized using the power law approximation. The velocity pressure coupling is achieved using a staggered grid arrangement [9]. The resulting discrete equations are solved using the SIMPLER algorithm [9]. Within a given time step, the following iterative solution procedure is adopted :

1. Solve the momentum and volume continuity equations (Eqs. (12) and (13)) using the SIMPLER algorithm to obtain the velocity and pressure fields.
2. Solve the mass continuity equation (Eq. (9)) to obtain the F and G fields.
3. Solve the energy equation (Eq. (24)) to obtain the temperature and liquid fraction fields.
4. Repeat steps 1-3 until convergence is achieved.

In the present work convergence is measured in terms of three quantities, namely, the volume imbalance, the mass imbalance and the energy imbalance. The nonlinearity between g and T (or equivalently between F and G) is handled using the predictor-corrector approach outlined in [6].

5. RESULTS AND DISCUSSIONS

The filling and solidification of a Al-4.5% Cu alloy wheel casting in both sand and steel moulds is studied. The

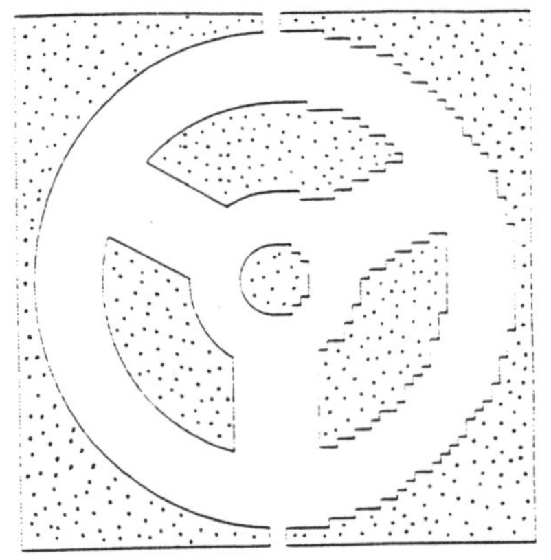

$L = 0.1905$ m
$R_1 = 0.1778$ m
$R_2 = 0.1270$ m
$R_3 = 0.0635$ m
$R_4 = 0.0254$ m

$Th_{cast} = 0.015875$ m
$Th_{sand} = 0.0254$ m (front and
$Th_{steel} = 0.00635$ m back)

Fig. 3 - Three spoke wheel casting : the left half shows the actual casting while the right half shows the discrete approximation.

dimensional details of the casting are shown in Fig. 3. It is bottom filled against gravity with molten metal at a superheat of 50 C through the 6.35 mm ingate. The filling stage lasts for about 5.4 s. The inlet velocity is assumed to vary linearly with time. Air is allowed to escape out of the mould through the 6.35 mm riser at the top of the mould. The mould is assumed to be at a preheat of 150 C. The bottom face of the mould is assumed to be insulated. All the other faces including the front and back faces lose heat to the ambient (25 C) via convection and radiation. An effective heat transfer coefficient of 15 $W/m^2/K$ is used. A perfect contact between the solidified shell and the mould walls is assumed. The thermophysical properties of the alloy, air, sand, and the steel mould are summarized in Table 1. The g vs T relationship for the aluminium alloy is given as

$$g = 1 \quad \text{if} \quad T \geq T_1$$

$$g = \left(\frac{T_M - T}{T_M - T_1}\right)^{-\frac{1}{1-\kappa}} \quad \text{if} \quad T_E < T < T_1 \tag{28}$$

$$0 < g \leq g_E \quad \text{if} \quad T = T_E$$

$$g = 0 \quad \text{if} \quad T < T_E$$

where T_1 is the liquidus temperature, T_M is the melting point of pure aluminium, T_E is the eutectic temperature and ρ is the partition coefficient of the alloy.

Exploiting symmetry, the solution is carried out in one half the domain using a 30 X 60 grid of uniform control volumes. Variable time stepping is used such that during the filling stage, 5 cells are filled over each time step. The momentum equations are under-relaxed by a factor of 0.25.

Table 1 - Thermophysical properties for the problem.

Al-4.5% Cu alloy			Air		
ρ =	2500	kg/m^3	ρ =	0.35	kg/m^3
μ =	0.025	kg/m/s	μ =	4 X 10^{-5}	kg/m/s
β =	4 X 10^{-5}	/K	C_p =	1000	J/kg/K
C_p =	1000	J/kg/K	k =	0.025	W/m/K
k_l =	90	W/m/K		Sand	
k_s =	200	W/m/K	ρ =	1500	kg/m^3
L =	3.9 X 10^5	J/kg	C_p =	1046	J/kg/K
T_l =	647	C	k =	1.116	W/m/K
T_M =	660	C		Steel	
T_E =	548	C	ρ =	7800	kg/m^3
κ =	0.14		C_p =	450	J/kg/K
			k =	48	W/m/K

Fig. 4 shows the results in the case of a sand mould. Figs. 4(a) - 4(k) depict the filling pattern in equal intervals of 10 % full, while Figs. 4(l) - 4(s) depict the solidification pattern in terms of the fraction liquid again in intervals of 10 % solid phase. The solidification contours are 0.25 units apart. During the relatively short filling stage (5.4 seconds) little heat transfer occurs due to the insulating effect of the sand. The metal loses most of its superheat during this stage and is close to being isothermal at the end of filling. The subsequent cooling to complete solidification requires an additional 575 seconds. Cooling during the post filling stage is primarily conduction controlled and very close to the one that would have been obtained in an analysis that neglected the filling stage and convection effects. The slight asymmetry is due to the adiabatic conditions that prevails at the lower ingate face.

Fig. 5 shows the results in the case of a metal mould. Figs. 5(a) - 5(k) depict the filling pattern in intervals of 10 % full. The time for filling is the same as in the sand mould case (5.4 seconds). The filling pattern, particularly in the hub region (see Figs. 5(f) - 5(h)) is different as compared to the sand casting. Due to the rapid cooling as the metal reaches the hub, a reasonable amount of solidification has occurred (see Figs. 5(l) - 5(n)). This restricts the filling of the hub from the central spoke and the rim becomes a preferential filling path. As a result filling of the hub is completed via the side spokes and not from below. At the end of the filling stage a major portion of the metal is in the mushy state (approximately 45% of the casting is in the solid phase, see Fig. 5 (o)). Cooling is much more rapid in this case and the solidification is complete in 85 seconds. In the post filling stage the hottest metal located near the ingate is unable to rise and move away up the spokes due to the resistance to flow in the mushy region. This can be noted on observing that the 'hot spot' remains in the gate area (see Figs. 5 (p) - 5(t)).

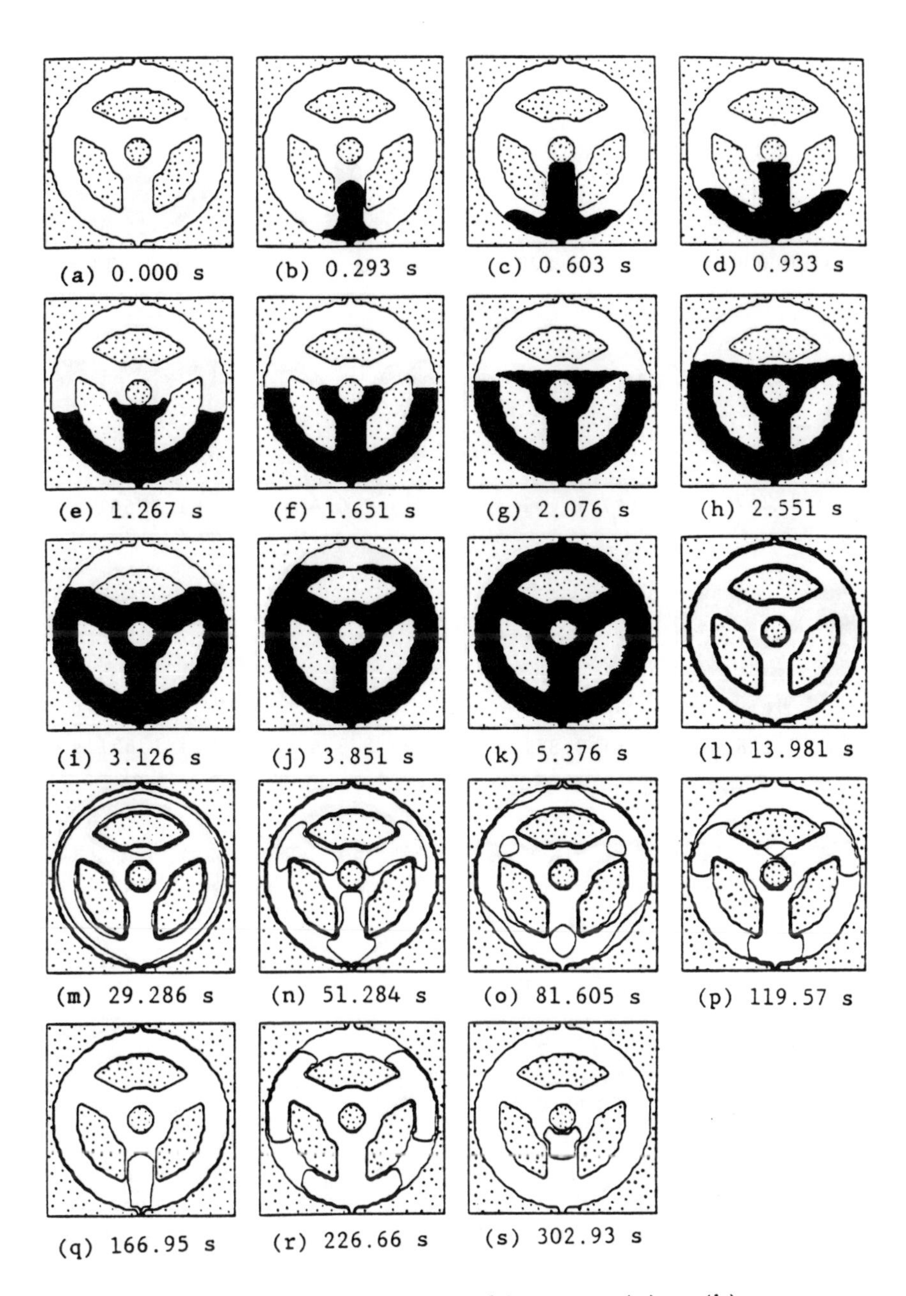

Fig. 4 - Results for the sand mould case : (a) - (k) progress of filling (F = 0.5), and (l) - (s) progress of solidification (g = 0.25, 0.5, 0.75).

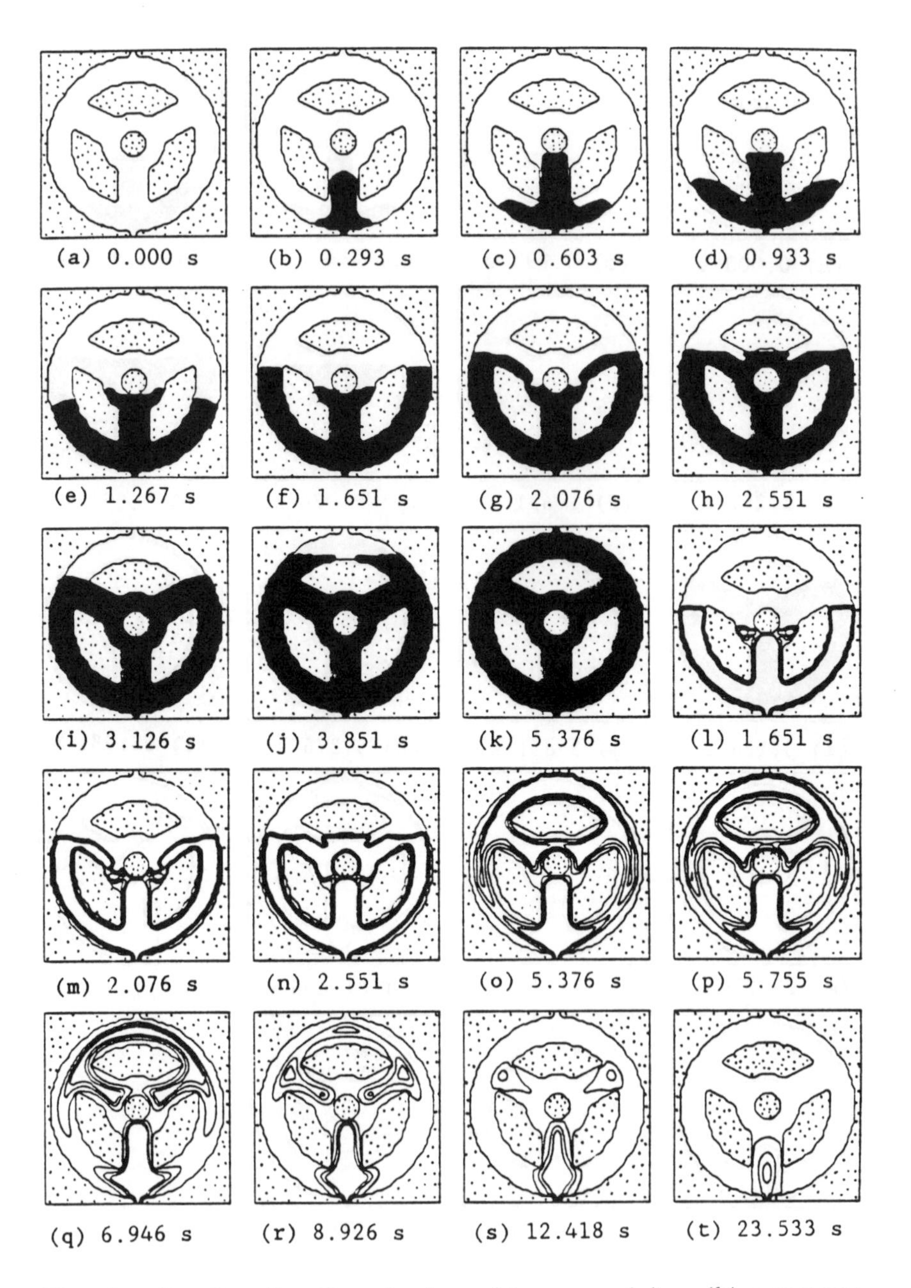

Fig. 5 - Results for the steel mould case : (a) - (k) progress of filling (F = 0.5), and (l) - (t) progress of solidification (g = 0.25, 0.5, 0.75).

6. CONCLUSIONS

The objective of this paper has been to develop and present a unified model of filling and solidification heat transfer of metal in a mould. A key feature of the work has been the introduction of a filling algorithm which, in a numerical setting, has identical characteristics to the solidification algorithm. This allows for the construction of a unified approach which can be based on traditional heat and mass transfer techniques. In particular, an implicit time approximation can be used and the resulting equations are amenable to standard control volume or finite element discretization techniques.

Emphasis in this work is based on the development of the appropriate equations and the numerical solution procedures. Preliminary results, however, show promise and clearly indicate the need to carry out a coupled filling and solidification heat transfer analysis.

8. ACKNOWLEDGEMENTS

This work was supported by the National Center for Excellence in Metalworking, operated by Concurrent Technologies Corporation, Johnstown (Pennsylvania, U.S.A.) under contract to the U.S.Navy as part of the U.S.Navy Manufacturing Technology Program. Computational time on the CRAY-YMP was provided by the Minnesota Supercomputer Institute (MSI), Minneapolis (Minnesota, U.S.A.).

9. REFERENCES

1. Modeling of Casting, Welding and Advanced Solidification Processes - VI, Conference Proceedings, eds. T. S. Piwonka, V. Voller, and L. Katgerman, TMS-AIME, Warrendale, PA, 1993.
2. R. A. Stoehr and C. Wang, Coupled Heat Transfer and Fluid Flow in the Filling of Castings, AFS Trans., vol. 96, pp. 733-740, 1988.
3. G. Dhatt, D. M. Gao, and A. Ben Cheikh, A Finite Element Simulation of Metal Flow in Moulds, Int. J. Numer. Methods Eng., vol. 30, pp. 821-831, 1990.
4. A. S. Usmani, J. T. Cross, and R. W. Lewis, A Finite Element Model for the Simulations of Mould Filling in Metal Casting and Associated Heat Transfer, Int. J. Numer. Methods Eng., vol. 35, pp. 787-806, 1992.
5. B. G. Thomas, Stress Modeling of Casting Processes : An Overview, Modeling of Casting, Welding and Advanced Solidification Processes - VI, pp. 519-534, Conference Proceedings, eds. T. S. Piwonka, V. Voller, and L. Katgerman, TMS-AIME, Warrendale, PA, 1993.
6. C. R. Swaminathan and V. R. Voller, On the Enthalpy Method, Int. J. Numer. Methods Heat Fluid Flow, in print,

1993.

7. C. W. Hirt, and B. D. Nichols, Volume of Fluid (VOF) Method for the Dynamics of Free Boundaries, J. Comput. Phys., vol. 39, pp. 201-225, 1981.
8. C. R. Swaminathan and V. R. Voller, A Time Implicit Filling Algorithm, to appear in Appl. Math. Model., 1993.
9. S. V. Patankar, Numerical Heat Transfer and Fluid Flow, Hemisphere, Washington D.C., 1980.
10. A. S. Usmani, R. W. Lewis, and K. N. Seetharamu, Natural Convection Controlled Change of Phase, Int. J. Numer. Methods Fluids, vol. 15, pp. 1023-1035, 1991.
11. A. D. Brent, V. R. Voller, and K. J. Reid, Enthalpy Porosity Technique for Modelling Convection-Diffusion Phase Change : Application to the Melting of a Pure Metal, Numer. Heat Transfer, vol. 13B, pp. 297-318, 1988.

THERMO-ELASTO-VISCOPLASTIC MODEL FOR METAL CASTING PROCESSES

P. M. M. Vila Real [1]; C. A. Magalhães Oliveira [2]
1 Lecturer at the Dept. of Mech. Engineering, University of Porto
2. Ass. Prof. at the Dept. of Mech. Engineering, University of Porto

ABSTRACT

The Finite Element Method is used to model the thermomechanical behaviour of ductile cast iron using metallic moulds. Heat conduction is assumed for the heat transfer analysis and an elasto-viscoplastic model is employed to predict the development of thermal stresses and strains. Special finite elements with coincident nodes are used to model the heat transfer and the mechanical contact at the metal-mould interface. The local heat transfer coefficient between the casting and the mould may be dependent on the air gap formation. The latent heat evolution effect is taken into account by the use of the enthalpy method. An iterative procedure is required to take into account the material and the contact non-linearity. A real casting has been modelled and numerical results are compared with the experimental measurements.

1. INTRODUCTION

Stresses in casting occur during the manufacturing process usually because of volume changes on solidification, non-uniform temperature distributions within the casting body during cooling, and/or mechanical constraint between the mould and the casting. Elastic or plastic constitutive relations are sufficient for engineering purposes for most metallic materials at low temperatures. However, at elevated temperatures time dependent plastic behaviour cannot be ignored, wherefore the material should be treated as elasto-viscoplastic. In this paper a viscoplastic constitutive relation of Perzina type [1,12,13,15,21,22] is used to describe the mechanical behaviour of ductile cast iron.

2. FORMULATION OF THERMO-ELASTO-VISCOPLASTIC BEHAVIOUR

For most casting processes, with the exception of processes like roll casting [8], where large plastic deformation is a significant part of the process, the heat generated during the deformation can be neglected, making the thermal and the mechanical problems uncoupled. In this work the thermal and mechanical problem are uncoupled but they cannot be resolved separately, because the local heat transfer coefficient at the metal-mould interface may be dependent on the air gap formation. The technique involves concurrently solving an uncoupled set of equations (the transient heat conduction equation and the incremental equilibrium equation, assuming an elasto- viscoplastic behaviour) within a time interval, but at the beginning of each new time step the heat transfer coefficient between the casting and the mould is updated along the interface. The same finite element formulation (finite element mesh, shape function, etc.) and the same equation solution technique, are used in both the thermal and stress analysis.

2.1 Thermal analysis

The governing equation for non-linear, transient heat conduction in the domain Ω, takes the form

$$\nabla\cdot(k\nabla T)=\rho c\frac{\partial T}{\partial t} \tag{1}$$

where k is the thermal conductivity, ρ the density, c the specific heat T the temperature and t the time. The temperature field which satisfies eq. (1) in Ω must satisfy the following boundary conditions: prescribed temperatures $\overline{T}$ on a part Γ_T of the boundary, specified heat flux $\overline{q}$ and heat flux by convection and/or radiation or a combination of these conditions on a part Γ_q of the boundary and also flux through the metal-mould interface, which can be given as, [19]

$$q_i = h_{int}(T_P - T_M) \qquad \text{on } \Gamma_i \tag{2}$$

where h_{int} is the heat transfer coefficient at that interface, T_P and T_M are the casting and the mould temperature at the interface, respectively.

Using finite elements Ω^e to discretize the domain Ω, a weak formulation and the Galerkin method for choosing the weighting functions, we obtain [18] the following system of differential equations:

$$\underset{\sim}{K}\underset{\sim}{T}+\underset{\sim}{C}\dot{\underset{\sim}{T}}=\underset{\sim}{F} \tag{3}$$

where

$$K_{ij}=\sum_{e=1}^{E}\int_{\Omega^e}\nabla N_i k\nabla N_j d\Omega^e+\sum_{e=1}^{Q}\int_{\Gamma_q^e}h_{cr}N_iN_jd\Gamma_q^e+\sum_{e=1}^{I}K_{ij}^{int} \tag{4}$$

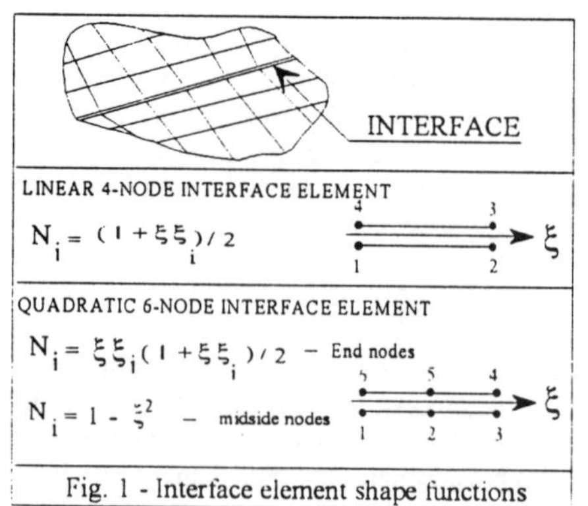

Fig. 1 - Interface element shape functions

$$C_{ij}=\sum_{e=1}^{E}\int_{\Omega^e}\rho c N_iN_jd\Omega^e \tag{5}$$

$$F_i=\sum_{e=1}^{Q}\int_{\Gamma_q^e}(h_{cr}T_\infty-\overline{q})N_id\Gamma_q^e \tag{6}$$

where E is the total number of element, Q is the number of elements with boundary type Γ_q, I is the number of elements of interface type Γ_i, N_i and N_j are shape functions, h_{cr}, is the combined convection and radiation heat-transfer coefficient, K_{ij}^{int} is the interface element contribution to the matrix $\underset{\sim}{K}$ and is defined as [19]

$$K_{ij}^{int} = \pm \int_{\Gamma_i^e} h_{int} N_i N_j d\Gamma_i^e \tag{7}$$

where N_i and N_j are shape functions defined in figure 1. The sign + must be used if i and j belong to the same side of the element and the sign - in the opposite case. It should be emphasised that the interface elements have no contribution for $\underset{\sim}{C}$ and $\underset{\sim}{F}$. The complete resolution of the problem needs the integration of the set of ordinary differential equations (3), which can be found in reference [18].

2.1.1 Numerical modelling of phase change

The latent heat evolution effect is taken into account by the use of the enthalpy method. In this work, at integration points, the specific heat c is computed as

$$c = \sqrt{\frac{\nabla H \cdot \nabla H}{\nabla T \cdot \nabla T}} \tag{8}$$

where H is the specific enthalpy. The matrix $\underset{\sim}{C}$ (5) can then be computed. This technique prevents the underestimation of the capacity matrix if the solidification interval is found inside one element with no integration points in the mushy zone, where the specific heat is dramatically increased [6].

2.2 Mechanical analysis

The mechanical behaviour of the casting process has been modelled using a thermo-elasto-viscoplastic constitutive law, together with a small strain theory. It is assumed that the total strain rate is the sum of the elastic strain rate $\dot{\underset{\sim}{\varepsilon}}_e$, viscoplastic strain rate $\dot{\underset{\sim}{\varepsilon}}_{vp}$, thermal strain rate $\dot{\underset{\sim}{\varepsilon}}_{th}$, plus the liquid-solid transformation strain rate $\dot{\underset{\sim}{\varepsilon}}_{tr}$ (which is specific of the solidification of most of metal alloys [3]):

$$\dot{\underset{\sim}{\varepsilon}} = \dot{\underset{\sim}{\varepsilon}}_e + \dot{\underset{\sim}{\varepsilon}}_{vp} + \dot{\underset{\sim}{\varepsilon}}_{th} + \dot{\underset{\sim}{\varepsilon}}_{tr} \tag{9}$$

The elastic strain rate depends on the total stress rate according to

$$\dot{\underset{\sim}{\varepsilon}}_e = \underset{\sim}{D}^{-1} \dot{\underset{\sim}{\sigma}} \tag{10}$$

in which $\underset{\sim}{D}$ represents the elastic matrix. Assuming the Von Mises yield criterion the onset of viscoplastic flow is governed by

$$F = \sqrt{3J'_2} - Y(T) = 0 \tag{11}$$

J'_2 being the second deviatoric stress invariant, and Y(T) the uniaxial yield stress defined as a function of the temperature. For associated viscoplasticity the viscoplastic strain rate is assumed to be given by [1, 10-17,20-22]

$$\dot{\underset{\sim}{\varepsilon}}_{vp} = \gamma \langle \Phi(F) \rangle \frac{\partial F}{\partial \underset{\sim}{\sigma}} \tag{12}$$

where γ is a fluidity parameter and $\langle\Phi\rangle$ is the flow function taken as non-zero for positive values of the yield function F, only. In this work as in the references [11,21], we used Φ=F.

The thermal strain $\underset{\sim th}{\varepsilon}$ is given by

$$\underset{\sim th}{\varepsilon} = \left[\int_{T_{ref}}^{T} \alpha(T)dT \right] \underset{\sim}{I} \tag{13}$$

in which $\alpha(T)$ is the thermal expansion coefficient dependent on temperature.

The transformation strain can be calculated [3] as follows:

$$\underset{\sim tr}{\varepsilon} = f_s(T)\frac{1}{3}\frac{\Delta V}{V}\underset{\sim}{I} = f_s(T)\beta\underset{\sim}{I} \tag{14}$$

where $f_s(T)$ is the volume solid fraction, $\Delta V/V$ the relative volume change due to the total liquid-solid transformation [3] and β is the linear expansion coefficient in accordance with the transformation dilatation. In this work we treat the transformation strain in the same way as the thermal strain. The combined effect of thermal and transformation strain given as

$$\underset{\sim tt}{\varepsilon} = \underset{\sim tr}{\varepsilon} + \underset{\sim th}{\varepsilon} = \left[f_s(T)\beta + \int_{T_{ref}}^{T} \alpha(T)dT \right] \underset{\sim}{I} \tag{15}$$

can be better understood by figure 2. For the definition of the solid fraction f_S we used a pseudo-lever rule [4] given by

$$f_s(T) = \begin{cases} 0 & ; \quad T > T_l \\ \dfrac{T_l - T}{T_l - T_s} & ; \quad T_s \le T \le T_l \\ 1 & ; \quad T < T_s \end{cases} \tag{16}$$

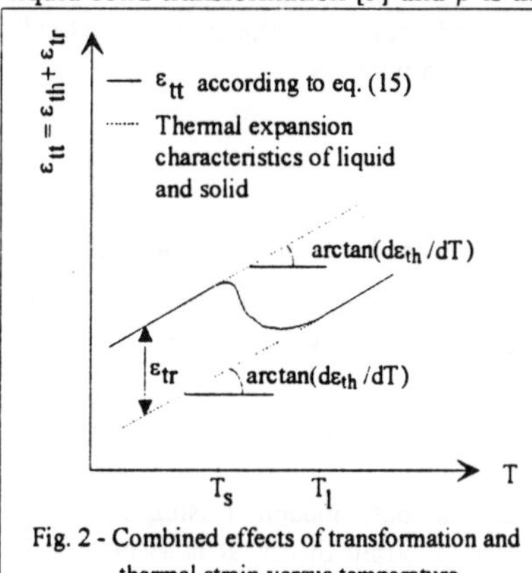

Fig. 2 - Combined effects of transformation and thermal strain versus temperature.

where T_s and T_l are the solidus and liquidus temperatures respectively.

2.2.1 The viscoplastic strain increment

In the time interval $\Delta t_n = t_{n+1} - t_n$ the viscoplastic strain increment $\Delta\underset{\sim vp}{\varepsilon}^n$ may be expressed as

$$\Delta\underset{\sim vp}{\varepsilon}^n = \Delta t_n \left[(1-\theta)\underset{\sim vp}{\dot{\varepsilon}}^n + \theta\underset{\sim vp}{\dot{\varepsilon}}^{n+1} \right] \qquad 0 \le \theta \le 1 \tag{17}$$

The viscoplastic strain rate $\underset{\sim vp}{\dot{\varepsilon}}^{n+1}$ at time station t_{n+1} can be approximated by the first terms of a Taylor's expansion as [13]

$$\underset{\sim vp}{\dot{\varepsilon}}^{n+1} = \underset{\sim vp}{\dot{\varepsilon}}^n + \left(\frac{\partial \underset{\sim vp}{\dot{\varepsilon}}}{\partial \underset{\sim}{\sigma}} \right)^n \Delta\underset{\sim}{\sigma}^n = \underset{\sim vp}{\dot{\varepsilon}}^n + \underset{\sim}{H}^n \Delta\underset{\sim}{\sigma}^n \tag{18}$$

where $\Delta\underset{\sim}{\sigma}^n$, is the stress change occurring in the time interval Δt_n. Substituting from (18) in (17) the viscoplastic strain increment can be written as:

$$\Delta\underset{\sim vp}{\varepsilon}^n = \Delta t_n \underset{\sim vp}{\dot{\varepsilon}}^n + \theta\Delta t_n \underset{\sim}{H}^n \Delta\underset{\sim}{\sigma}^n = \Delta t_n \underset{\sim vp}{\dot{\varepsilon}}^n + \underset{\sim}{C}^n \Delta\underset{\sim}{\sigma}^n \tag{19}$$

2.2.2 The stress increment

The incremental stress change occurring in time interval Δt_n is given by

$$\Delta \underset{\sim}{\sigma}^n = \underset{\sim}{D}^n \Delta \underset{\sim e}{\varepsilon}^n = \underset{\sim}{D}^n \left(\Delta \underset{\sim}{\varepsilon}^n - \Delta \underset{\sim}{\varepsilon}^n_{vp} \right) \tag{20}$$

in which $\underset{\sim}{D}^n$ is the elasticity matrix. Expressing the total strain in terms of the displacement increment, $\Delta \underset{\sim}{u}^n$ and substituting from (19) in (20), then

$$\Delta \underset{\sim}{\sigma}^n = \left(\left[\underset{\sim}{D}^n \right]^{-1} + \underset{\sim}{C}^n \right)^{-1} \left(\underset{\sim}{B} \Delta \underset{\sim}{u}^n - \Delta t_n \dot{\underset{\sim vp}{\varepsilon}}^n \right) = \hat{\underset{\sim}{D}}^n \left(\underset{\sim}{B} \Delta \underset{\sim}{u}^n - \Delta t_n \dot{\underset{\sim vp}{\varepsilon}}^n \right) \tag{21}$$

in which B is the usual strain-displacement matrix.

2.2.3 Incremental equation of equilibrium

The incremental form of the equilibrium equation is

$$\int_\Omega \underset{\sim}{B}^T \Delta \underset{\sim}{\sigma}^n d\Omega = \Delta \underset{\sim}{f}^n = \Delta \underset{\sim mc}{f}^n + \Delta \underset{\sim tt}{f}^n \tag{22}$$

in which $\Delta \underset{\sim mc}{f}^n$ represents the change in equivalent nodal loads resulting from mechanical loads occurring in the time interval Δt_n, and $\Delta \underset{\sim tt}{f}^n$ is the combination of the thermal and transformation load increment. Substituting from (21) in (22), results in the following expression for the displacement increment

$$\Delta \underset{\sim}{u}^n = \left[\underset{\sim}{K}^n \right]^{-1} \Delta \underset{\sim}{V}^n \tag{23}$$

in which

$$\underset{\sim}{K}^n = \int_\Omega \underset{\sim}{B}^T \hat{\underset{\sim}{D}}^n \underset{\sim}{B} d\Omega \tag{24}$$

and the incremental pseudo loads, $\Delta \underset{\sim}{V}^n$ are given by

$$\Delta \underset{\sim}{V}^n = \int_\Omega \underset{\sim}{B}^T \hat{\underset{\sim}{D}}^n \dot{\underset{\sim vp}{\varepsilon}}^n \Delta t_n d\Omega + \Delta \underset{\sim}{f}^n \tag{25}$$

The numerical stability of the algorithm is strongly related to the time step length [22].

2.2.4. Equilibrium correction

Due to the linearisation implied in (17), equilibrium equation will generally not be exactly satisfied and residual forces

$$\underset{\sim}{\psi}^{n+1} = \underset{\sim}{f}^{n+1} - \int_{\Omega} \underset{\sim}{B}^{T} \left(\underset{\sim}{\sigma}^{n+1} - \underset{\sim tt}{\sigma}^{n+1} \right) d\Omega \tag{26}$$

will exist at the end of the time step. This requires the inclusion of an equilibrium correction process. The simples and most economic approach is to include the residual or out-of-balance forces (26) in the pseudo loads to be employed for the next time step

$$\Delta \underset{\sim}{V}^{n+1} = \int_{\Omega} \underset{\sim}{B}^{T} \underset{\sim}{\hat{D}}^{n+1} \underset{\sim vp}{\dot{\varepsilon}}^{n+1} \Delta t_{n+1} d\Omega + \Delta \underset{\sim}{f}^{n+1} + \underset{\sim}{\psi}^{n+1} \tag{27}$$

Such a technique avoids an iterative process. However, in this work, due to the contact non-linearity at the metal-mould interface the following iterative procedure within time increments, based on the references [14,17,20] has been used: Suppose that at time $t = t_n$ we have an equilibrium situation and $\underset{\sim}{u}^n$, $\underset{\sim}{\sigma}^n$, $\underset{\sim}{\varepsilon}^n$, $\underset{\sim}{\varepsilon}^n_{vp}$, $\underset{\sim}{\dot{\varepsilon}}^n_{vp}$ e $\underset{\sim}{f}^n$ are known.

1) Compute the displacement increments associated with time step Δt_n for the iteration i

$$\Delta \underset{\sim i}{u}^{n} = \left[\underset{\sim i}{K}^{n} \right]^{-1} \Delta \underset{\sim i}{V}^{n} \tag{28}$$

- for the first iteration, i=1

$$\Delta \underset{\sim 1}{V}^{n} = \int_{\Omega} \underset{\sim}{B}^{T} \underset{\sim}{\hat{D}}^{n} \underset{\sim vp}{\dot{\varepsilon}}^{n} \Delta t_n d\Omega + \Delta \underset{\sim}{f}^{n} + \underset{\sim}{\psi}^{n} \tag{29}$$

- for the subsequent iteration i>1

$$\Delta \underset{\sim i}{V}^{n} = \underset{\sim i}{\psi}^{n+1} \tag{30}$$

2) Evaluate the total displacements and stresses as

$$\underset{\sim i}{u}^{n+1} = \underset{\sim i-1}{u}^{n+1} + \Delta \underset{\sim i}{u}^{n} \tag{31}$$

$$\underset{\sim i}{\sigma}^{n+1} = \underset{\sim}{\hat{D}}^{n} \left(\underset{\sim}{B} \underset{\sim i}{u}^{n+1} - \underset{\sim vp}{\varepsilon}^{n} - \Delta t_n \underset{\sim vp}{\dot{\varepsilon}}^{n} + \Delta t_n \theta \underset{\sim}{H}^{n} \underset{\sim}{\sigma}^{n} \right) + \underset{\sim tt}{\sigma}^{n+1} \tag{32}$$

3) Calculate the following vectors

$$\Delta \underset{\sim i}{\sigma}^{n} = \underset{\sim i}{\sigma}^{n+1} - \underset{\sim}{\sigma}^{n} \tag{33}$$

$$\underset{\sim vp,i}{\varepsilon}^{n+1} = \underset{\sim vp}{\varepsilon}^{n} + \Delta t_n \underset{\sim vp}{\dot{\varepsilon}}^{n} + \Delta t_n \theta \underset{\sim}{H}^{n} \Delta \underset{\sim i}{\sigma}^{n} \tag{34}$$

$$\underset{\sim vp,i}{\dot{\varepsilon}}^{n+1} = \gamma \langle \Phi(F) \rangle \frac{\partial F}{\partial \underset{\sim i}{\sigma}^{n+1}} \tag{35}$$

4) Compute the residual forces

$$\underset{\sim i}{\psi}^{n+1} = \underset{\sim}{f}^{n+1} - \int_{\Omega} \underset{\sim}{B}^{T} \left(\underset{\sim i}{\sigma}^{n+1} - \underset{\sim tt}{\sigma}^{n+1} \right) d\Omega \tag{36}$$

and check for convergence according to $\left\| \underset{\sim i}{\psi}^{n+1} \right\| / \left\| \underset{\sim}{f}^{n+1} \right\| < TOL$

- if the convergence has been reached go to step 5.
- if not, set $\Delta \underset{\sim i}{V}^{n} = \underset{\sim i}{\psi}^{n+i}$ and go to step 1.

5) If convergence has been reached, consider $\underset{\sim}{\sigma}^{n+1} = \underset{\sim i}{\sigma}^{n+1}$, $\underset{\sim vp}{\varepsilon}^{n+1} = \underset{\sim vp,i}{\varepsilon}^{n+1}$, $\underset{\sim vp}{\dot{\varepsilon}}^{n+1} = \underset{\sim vp,i}{\dot{\varepsilon}}^{n+1}$,

$\underset{\sim}{u}^{n+1} = \underset{\sim i}{u}^{n+1}$, $\underset{\sim}{\psi}^{n+1} = \underset{\sim i}{\psi}^{n+1}$, and define the pseudo load vector for the next time step as

$$\Delta \underset{\sim}{V}^{n+1} = \int_{\Omega} \underset{\sim}{B}^{T} \underset{\sim}{\hat{D}}^{n+1} \underset{\sim vp}{\dot{\varepsilon}}^{n+1} \Delta t_{n+1} d\Omega + \Delta \underset{\sim}{f}^{n+1} + \underset{\sim}{\psi}^{n+1} \tag{37}$$

To reduce the residuals several iterative schemes can be used [14,17]. Here we adopted the Newton-Raphson method.

2.2.5. Numerical treatment of contact problems

The interface problem is decomposed into a pure contact problem in the normal direction and the frictional resistance in the tangential direction of the interface. The behaviour of the interface elements is characterised by the relation between the relative displacements of the surfaces in contact and the normal and shear stresses at the interface. An isoparametric formulation is used and the elements are zero thickness with coincident nodes. In our work the interface element is based on joint element developed in ref. [5] and later used in ref. [7,15,16,20]

2.2.5.1. Slip criterion

The onset of irreversible relative slip between casting and mould is governed by the following slip criterion

$$F = |\tau| - \chi = 0 \tag{38}$$

where τ is the shear stress and the state variable χ is defined by

$$\chi = \mu |\sigma_n| \tag{39}$$

in which μ is the Coulomb's coefficient of friction and σ_n is the stress in the normal direction of the interface.

2.2.5.2. Displacement definition

The relative displacement at any point along the element represented in figure 3, is given by

$$\underset{\sim}{\varepsilon}^{*} = \begin{Bmatrix} \gamma^{*} \\ \varepsilon_n^{*} \end{Bmatrix} = \begin{Bmatrix} t_{x'} - b_{x'} \\ t_{y'} - b_{y'} \end{Bmatrix} \tag{40}$$

where γ^{*} and ε_n^{*} are the relative displacements in the direction of the local axis x' and y' respectively, and the letter t refers to the top side of the element and the letter b to the bottom side. So $t_{x'}$ represents the displacement in the local x'-axis direction at the top side of the element, etc..

Using the usual finite element representation, the displacements at any point of the 6-node element represented in figure 3 are given by

$$t_{x'} = \sum_{i=4}^{6} N_i u'_i\,; \qquad t_{y'} = \sum_{i=4}^{6} N_i v'_i\,; \qquad b_{x'} = \sum_{i=1}^{3} N_i u'_i\,; \qquad b_{y'} = \sum_{i=1}^{3} N_i v'_i \qquad (41.a;b;c;d)$$

where u'_i and v'_i are displacements of node i in the local x' and y' axes respectively, and N_i are shape functions listed in figure 1 for both linear 4-node and quadratic 6-node interface elements.

Substituting from (41) in (40) the relative displacements can be written as

$$\underset{\sim}{\varepsilon}^{*} = \underset{\sim}{N}' \underset{\sim}{d}' \qquad (42)$$

in which

$$\underset{\sim}{N}' = \begin{bmatrix} -\underset{\sim}{N}_1 & -\underset{\sim}{N}_2 & -\underset{\sim}{N}_3 & \underset{\sim}{N}_4 & \underset{\sim}{N}_5 & \underset{\sim}{N}_6 \end{bmatrix} \qquad (43)$$

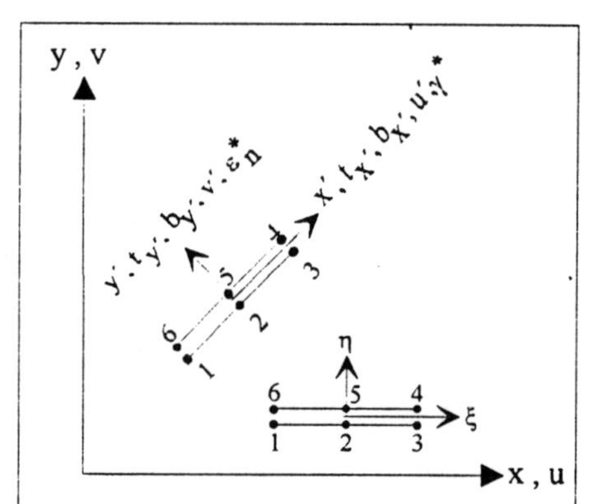

Fig. 3 - Schematic representation of global, local and parent axes for the interface element.

and

$$\underset{\sim}{d}' = \begin{Bmatrix} \underset{\sim}{d}'_1 & \underset{\sim}{d}'_2 & \underset{\sim}{d}'_3 & \underset{\sim}{d}'_4 & \underset{\sim}{d}'_5 & \underset{\sim}{d}'_6 \end{Bmatrix}^T \qquad (44)$$

where $\underset{\sim}{N}_i = N_i \underset{\sim}{I}$, with $\underset{\sim}{I}$ being a 2x2 identity matrix and $\underset{\sim}{d}'_i = \{u'_i \quad v'_i\}^T$. Relating the local displacements u', v' with the global ones by

$$\begin{Bmatrix} u' \\ v' \end{Bmatrix} = \underset{\sim}{T} \begin{Bmatrix} u \\ v \end{Bmatrix} \qquad (45)$$

or

$$\underset{\sim}{d}' = \underset{\sim}{T}\underset{\sim}{d} \qquad (46)$$

where $\underset{\sim}{T}$ is the well known transformation matrix [5], the eq. (42) becomes

$$\underset{\sim}{\varepsilon}^{*} = \underset{\sim}{N}' \underset{\sim}{d}' = \underset{\sim}{N}' \underset{\sim}{T}\underset{\sim}{d} = \underset{\sim}{B}\underset{\sim}{d} \qquad (47)$$

where $d_i = \{u_i \quad v_i\}^T$ are the global displacements of node i and $\underset{\sim}{B}_i = \underset{\sim}{N}'_i \underset{\sim}{T}$.

2.2.5.3. Stress-relative displacement relationship

The normal stress is given by

$$\sigma_n = k_n \varepsilon_n^{*} \qquad (48)$$

where k_n is the normal stiffness, which is chosen as an arbitrary large number for numerical convenience. Note that only compressive normal stresses are allowed, i. e. $\sigma_n \le 0$. The incremental relation form of the shear stress-tangential relative displacement relationship is given by

$$\Delta\tau = k_t \Delta\gamma^{*} \qquad (49)$$

where k_t is the tangential stiffness. The frictional shear force, however, is limited by the slip criterion (38). In two dimensional cases the preceding relationship is defined as shown in figure 4 [15], where the state variable χ is considered to be a known state variable (39).

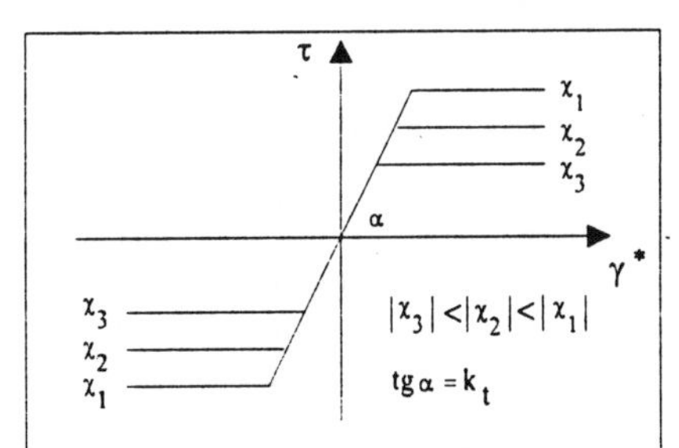

Fig. 4 - Frictional shear-relative displacent relationship

2.2.5.4. Stiffness matrix

The stiffness matrix of an interface element is defined in the standard finite element manner as

$$\underset{\sim}{K} = \int_{\Gamma^e} \underset{\sim}{B}^T \underset{\sim}{D} \underset{\sim}{B} d\Gamma \qquad (50)$$

where $\underset{\sim}{D}$ is the matrix which relates the stresses and the relative displacements in the local co-ordinates system, so that

$$\Delta \underset{\sim}{\sigma} = \begin{Bmatrix} \Delta\tau \\ \Delta\sigma_n \end{Bmatrix} = \begin{bmatrix} \bar{k}_t & 0 \\ 0 & k_n \end{bmatrix} \begin{Bmatrix} \Delta\gamma^* \\ \Delta\varepsilon_n^* \end{Bmatrix} = \underset{\sim}{D} \Delta \underset{\sim}{\varepsilon}^* \qquad (51)$$

The non-linear tangent behaviour is modelled by the appropriate variation of $\bar{k}_t$ as described in the next section.

2.2.5.5. Computational procedure

According to our previous discussion, a computational procedure based on the Newton Raphson method can be set up [15,20]. This can be summarised as follows:

- for each time step $\Delta t_n = t_{n+1} - t_n$:
  Define the active interface elements and corresponding integration (or Gauss) points.
  - for each iteration i within the time step and for each active Gauss point:

    1. Evaluate the stiffness matrix $\underset{\sim}{K}_i$ using the appropriate tangential stiffness $\bar{k}_t$

       (for the first time step and first iteration set $\bar{k}_t = k_t$. For subsequent iterations $\bar{k}_t$ is defined in computational step 6.).
    2. Solve the global system of equations defined in (28).
    3. Update the nodal incremental displacements, $\Delta \underset{\sim}{u}^i = \Delta \underset{\sim}{u}^{i-1} + \Delta \underset{\sim}{u}^i$
    4. Evaluate $\Delta \varepsilon_i^{*n+1}$ and the total relative displacements $\underset{\sim}{\varepsilon}_i^{*n+1} = \underset{\sim}{\varepsilon}^{*n} + \Delta \underset{\sim}{\varepsilon}_i^{*n+1}$
    5. Calculate the stress in the normal direction according to (48) and evaluate χ_{n+1} defined in (39). Note that only non-positive values of normal stress are allowed, otherwise set $\sigma_n = \tau = 0$
    6. Evaluate the shear stress τ and the stiffness $\bar{k}_t$ for the next iteration

       $$\tau_i^{n+1} = \tau^n + \bar{k}_t \Delta\gamma_i^{*n+1} \qquad (52)$$

       If $|\tau| > \chi$ at t_{n+1} correct the current values of τ and $\bar{k}_t$ according to

$$\tau = \chi \frac{\Delta\gamma^*}{|\Delta\gamma^*|}; \qquad \bar{k}_t = \frac{\chi}{|\Delta\gamma^*|} \tag{53}$$

7. Evaluate the out of balance (or residual) forces.

Computational steps 1. to 7. are repeated until the convergence be reached.

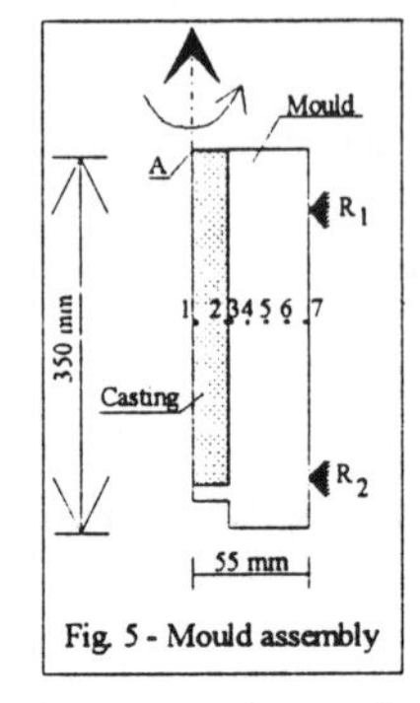

Fig. 5 - Mould assembly

3. NUMERICAL SIMULATION AND RESULTS

A cylindrical casting with 30 mm of diameter poured in a cylindrical mould as shown in figure 5, has been studied. A set of seven thermocouples were used to record the temperature versus time data from the casting as well as from the mould. In the thermo-elasto-viscoplastic analysis, we used a Bingham model suggested in ref. [11] which uses (13) with Φ=F and $\gamma = Ce^{-A/T}$, where T is in Kelvin, A=4800K and C=0.0025. Figure 6 shows, for the thermocouples located at the mould, a good agreement between the experimental and the numerical results. For the other two thermocouples located on the casting, the differences may have been due to an incomplete modelling of the phase-change process or the location of the thermocouples considered in the numerical analysis was not exactly coincident with their real position inside the casting.

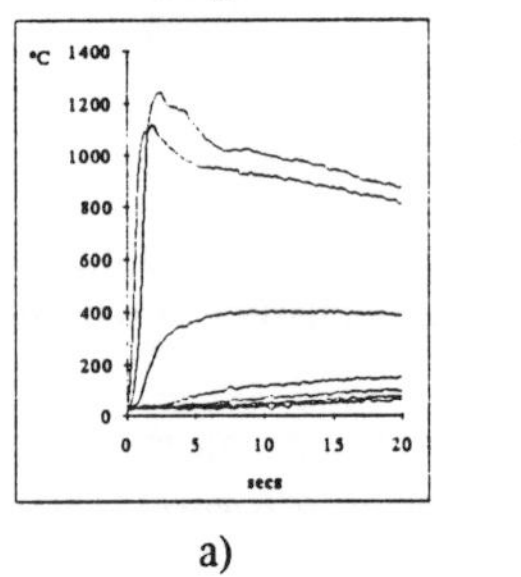

a)

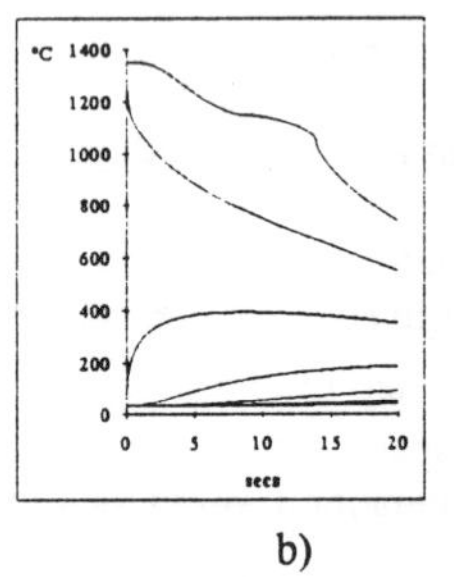

b)

Fig. 6 - Time history of temperatures at thermocouples location.
a) experimental results; b) numerical results.

Figure 7 shows a radial temperature profile at the central zone of the mould assembly, after 15.0 seconds. It can be also seen in that figure the temperature drop at the metal-mould interface. In figure 8 it is shown a radial profile of simulated hoop stress σ_θ after 15.0 seconds. In figure 9 a) and 9 b) we can see the effect of the expansion of the ductile iron due to the precipitation of graphite, which was taken into account through the thermal dilatation coefficient, defined as a function of the iron specific volume and temperature [8]

$$\alpha(T) = \frac{1}{3V} \cdot \frac{V - V_{ref}}{T - T_{ref}} \tag{54}$$

where V_{ref} and T_{ref} are the specific volume of reference and the temperature of reference respectively. In fig. 9 b) we can verify the effect of the transformation strain ε_{tr} given by eq. (14). Here it has been considered a linear expansion coefficient resulting from liquid-solid transformation β=.38% [9]. In figures 10 and 11 we can see the effect of the actualisation of the heat transfer coefficient between the casting and the mould, $h_{int,}$ function of the possible air gap formation. In fig. 10 it was used a coefficient of heat transfer at the interface constant and equal to 5000 w/m^2°C, and the result of fig. 11 has been obtained using h_{int} = 5000 w/m^2°C, if there is contact between de casting and the mould and h_{int} = 2500 w/m^2°C in the opposite case. Figure 12 shows the discontinuity of the temperature field at the metal-mould interface, and a deformed mesh is shown in figure 14, where it can be clearly seen the gap between the casting and the mould.

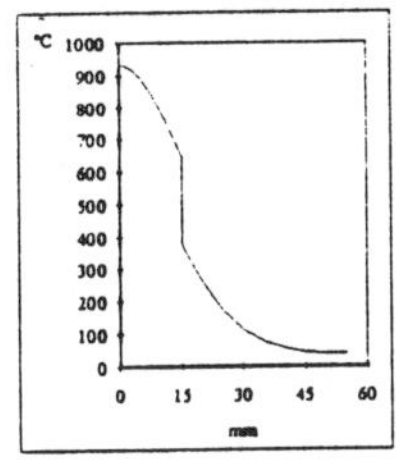

Fig. 7 - Radial temperature profile.

Fig. 8 - Radial hoop stress σ_θ profile.

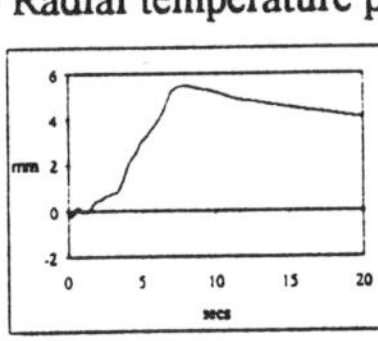

a)

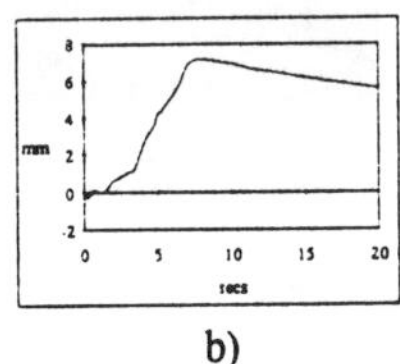

b)

Fig. 9 - Time history of displacement of point A. a) Without considering the transformation dilatation; b) Considering the transformation dilatation.

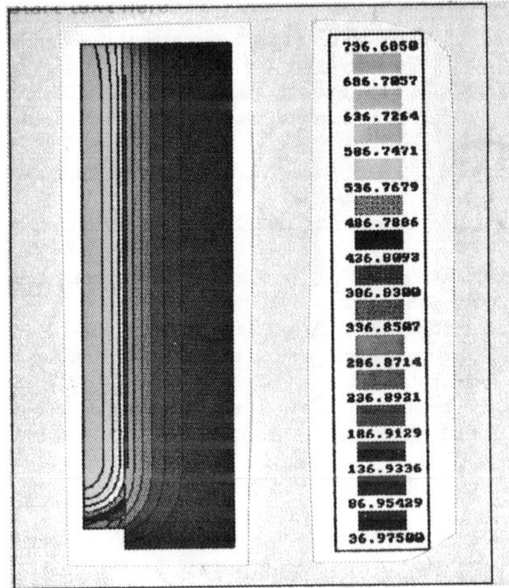

Fig. 10 - Temperature field after 20.0 secs, simulated with constant h_{int}.

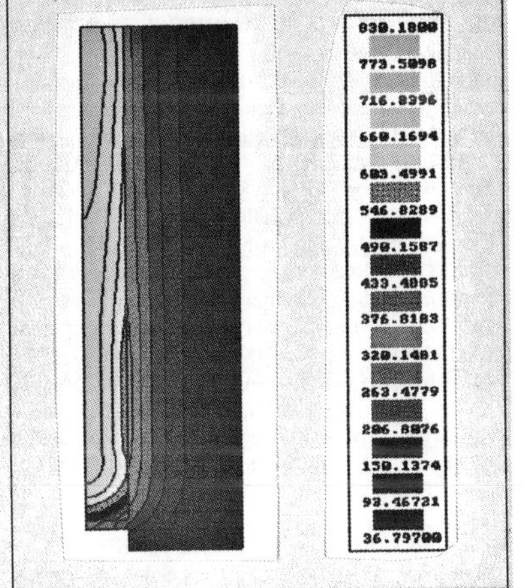

Fig. 11 Temperature field after 20.0 secs, simulated with variable h_{int}.

4. CONCLUSIONS

The Theoretical basis for the thermomechanical analysis of casting process has been presented. A coupled thermo-elasto-viscoplastic model for estimating the temperature, stress and deformation fields has been developed. The effect of considering the volume expansion due to the precipitation of graphite as well as due to the liquid-solid transformation has been shown. Thermal results were compared with experimental measurements.

5. ACKNOWLEDGEMENTS

The authors are grateful to Prof. Barbedo de Magalhães for the experimental results, to the INIC and JNICT for the financial support.

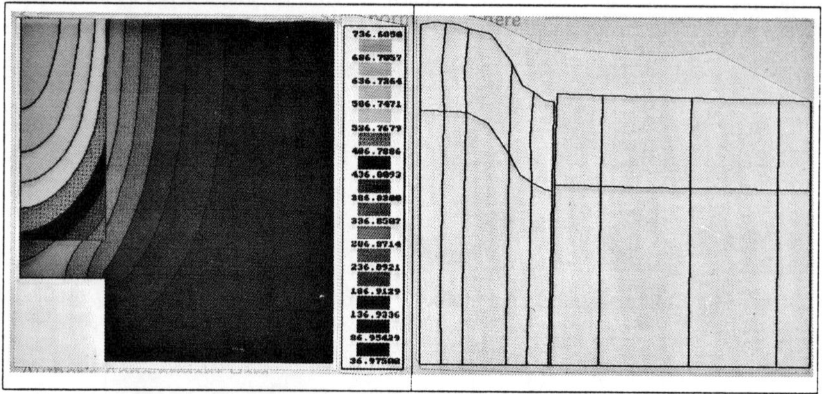

Fig. 12 - Zoom of temperature distribution. Fig. 13 - Zoom of the deformed mesh.

6. REFERENCES

1. BARROS, M. H., "Thermo-Viscoplastic Analysis of Concrete Structures by the Finite Element Method", PhD Thesis (in Portuguese), Coimbra 1987.
2. BEER, G. "An Isoparametric Joint/Interface Element for Finite Element Analysis", Int. J. Num. Meth. Engng., Vol. 21, 585-600, 1985.
3. BELLET, M.; BAY, F.; BRIOIST, J. J.; CHENOT, J. L., "Modelling of the thermomechanical coupling in the cooling stage of the casting process", NUMIFORM92, pp. 771-776, France, Set. 1992.
4. BRADLEY, F. J.; SAMONDS, M. "On the Application of Source Term, Enthalpy, and Micro modelling Algorithms to the Numerical Simulation of Near-eutectic Ductile Iron Solidification and Shrinkage Prediction", submitted to the Journal of Applied Mathematical Modelling, July (1991).
5. CÉSAR DE SÁ, J. M. A.; OWEN, D.R.J., "A study of Grain Boundary Sliding Mechanism by the Finite Element Method", Report C/R/449/83, University College of Swansea, U. K., 1983.
6. COMINI, G.; DEL GUIDICE, S.; ARO, O., "A Conservative Algorithm for Multidimensional Conduction Phase Change", Int. J. Num. Meth. Engng., Vol. 30, 697-709, 1990.
7. COUTO MARQUES, J., "Development of a Finite Element Package for the Solution of Non-linear Dynamic Problems With Application to the Seismic Analysis of Dams", MSc. Thesis, Porto, FEUP 1985.
8. DANTZIG, J. A., "Thermal Stress development in metal casting process", Met. Sci. Tech. vol. 7 (3), 133-178,1989.
9. KAMAMOTO, S.; NISHIMORI, T.; KINOSHITA, S. "Analysis of Residual Stress and Distortion Resulting from Quenching in Large Low-Alloy steel shafts", Mat. Sci. Tech., Vol. 1, 798-804798-804, 1985.
10. LEWIS, R. W.; MORGAN, K.; ROBERTS, P. M.,"Determination of Thermal Stresses in Solidification Problems", Numerical Analysis of Forming Processes, edited by J. F. T. Pittman, O. C. Zienkiewicz,, R. D. Wood, J. M. Alexander, John Wiley & Sons Ltd., pp. 405-431, 1984.
11. MARCELIN, J.L.; ABOUAF M.; CHENOT, J.L., "Analysis of Residual Stresses in Hot-Rolled Complex Beams", Comp. Meth. Appl. Mech. Engin., Vol 56, pp. 1-16, 1986.
12. MARTINS, R. A. F.; BARROS, M. HELENA; DINIS, LUCIA M. J. S., "Non-Associated Viscoplasticity-A Symmetric Algorithm for Implicit Solution". Proc. of Num. Meth. for Nonlinear Problems, Ed: C. Taylor, D. R. Owen, E. Hinton, F. B. Damjanic. Pineridge Press Ltd, U. K. pp. 130-139,1986.
13. OWEN, D. R. J., HINTON, E., "Finite Elements in Plasticity: Theory and Practice", Pineridge Press, U. K., 1980.
14. REIS GOMES, C. M. B.,"Matrix Update Methods in Non-linear Analysis of Two Dimensional and Thin Shell Problems", PhD Thesis, University College of Swansea, U. K. 1982.
15. RODIC, I. T., "Numerical Analysis of Thermomechanical Processes During Deformation of Metals at High Temperatures", PhD Thesis, University College of Swansea, U. K., 1989.
16. SOUSA, LUÍSA C.; CÉSAR DE SÁ, J. "Finite Element Modelling of Thermo-Mechanical Forming Processes Using a Mixed Method", NUMIFORM 92, pp.229-235, France, Set. de 1992.
17. TRIGO BARBOSA, J., "Finite Element Non-linear Analysis of Reinforced Plates and Shells", PhD Thesis (in Portuguese) University of Porto, 1993.
18. VILA REAL, P. M. M., OLIVEIRA, C. A. M., "Finite Element Modelling of Thermo-Elastic Behaviour of Solids With High Thermal Gradients", MSc. Thesis (in Portuguese), Porto, FEUP,1988.
19. VILA REAL, P. M. M.; OLIVEIRA, C. A. M., "Formulation of a Finite Element with Coincident Nodes to Model the Thermal Interface Between Two Distinct Bodies", Revista de ENGENHARIA, FEUP, pp. 53-57,Jan. 1991.
20. VILA REAL, P. M. M.; OLIVEIRA, C. A. M., "Finite Element Study of the Thermo-Elasto-Viscoplastic Behaviour of Metal Casting Processes", (in Portuguese) proceedings of 2ª Jornadas de Mecânica Computacional, Coimbra, 28-30 de Setembro de 1992, M 15.
21. ZHIGANG WANG; INOUE, T., "Viscoplastic Constitutive Relation Incorporating Phase Transformation-Application to Welding", Mat. Sci. Tech., Vol. 1, 899-903, 1985.
22. ZIENKIEWICZ, O. C., CORMEAU, I. C., "Viscoplasticity, Plasticity and Creep in Elastic Solids - A Unified Numerical Solution Approach", Int. J. Num. Meth. Engng., vol. 8, pp. 821-845, 1974.

FLUID FLOW AND HEAT TRANSFER MODELLING OF MOULD FILLING IN CASTING PROCESSES

M.R. Tadayon, J.A. Spittle, S.G.R. Brown

IRC in Materials for High Performance Applications,
Material Engineering Department, University College of Swansea,
Swansea SA2 8PP, U.K.

SUMMARY

This paper describes a numerical model to simulate fluid flow with free surface and heat transfer during the filling of two-dimensional mould cavities in casting processes. The fluid flow model is an extended SOLA-VOF type for incompressible viscous flow using a uniform finite difference grid. The heat transfer model is based on the energy type formulation in which the net thermal energy in a cell is used to compute the cell temperature. The model caters for various flow and thermal boundary and interface conditions encountered in mould filling problems. The simulated results of some practical mould filling examples are demonstrated.

1 INTRODUCTION

In casting processes many complex physical phenomena such as transient fluid flow, heat transfer and solidification may occur simultaneously or in rapid succession. The relationship between these phenomena and how they affect the macro/micro-structures, quality and cost of the casting are thus the major concerns of the foundry engineers.

Many of the previous efforts at modelling casting processes have concentrated on fluid flow and/or solidification aspects after mould filling and neglecting the effects resulting from the filling stage. However, the nature of the fluid flow during the mould filling could have a significant influence on as cast structure, the extent and forms of both porosity (gas and shrinkage) and segregation distributions and consequently the mechanical properties of the final product. In fact, mould filling probably represents the most critical stage of the casting process. Particularly, insufficient attention to the design of the running system and the filling pattern of the mould can lead to high levels of casting rejects due to entrained inclusions and gas porosity.

Computer modelling is still a long way from being able to address most of the effects associated with fluid flow and it is only within the last five years that computer models for mould filling have emerged which also include heat transfer and simplified treatments of solidification. Most of these models are based on the Solution Algorithm-Volume of Fluid (SOLA-VOF) finite difference technique [1]. This is one of a number of techniques capable of analysing highly transient free surface flow problems. The VOF formulation [2] is also being incorporated into finite element models of mould filling [3] to track the free surface of the liquid. At the present time, mould filling analyses are primarily being performed to study the filling patterns, flow conditions (laminar and/or turbulent), and also to predict the temperature field distribution in the casting system during the filling stage. The latter obviously permits a more accurate prediction of the subsequent progress of solidification.

In this paper, a model of fluid flow and heat transfer for analysis of mould filling in casting processes is presented. The fluid flow model is an extended SOLA-VOF type for incompressible viscous flow using a uniform finite difference grid. The heat transfer model employs the principle of the conservation of thermal energy where the net thermal energy of a cell is used to compute its temperature [4]. The advantage of this model is that it uses the information obtained from the SOLA-VOF analysis to calculate the advective thermal energy in and out of a cell. The models are capable of treating a multitude of flow and thermal boundary and interface conditions encountered in mould filling problems. Finally, some realistic cases of mould filling are simulated and the results are illustrated.

2 MATHEMATICAL MODELS

A system of conservation equations of mass, momentum and energy is derived via local volume averaging [5]. The equations are then solved numerically by the finite difference method to calculate velocity, pressure and temperature fields. The numerical solution is formulated using a uniform staggered grid, *i.e.* velocities at cell sides and pressure and temperature at cell centres. The evolution of free surface is monitored by solving a transport equation for the volume of fluid in each cell using the calculated velocity field.

2.1 Fluid Flow Calculation

The fluid dynamics principles of conservation of mass and momentum for incompressible viscous flow can be expressed respectively as the Continuity and Navier-Stokes equations as follows:

Continuity equation:

$$D = \frac{\partial u}{\partial x} + \frac{\partial v}{\partial y} = 0 \tag{1}$$

Navier-Stokes equations:

$$\frac{\partial u}{\partial t} + u\frac{\partial u}{\partial x} + v\frac{\partial u}{\partial y} = -\frac{\partial P}{\partial x} + \nu\left(\frac{\partial^2 u}{\partial x^2} + \frac{\partial^2 u}{\partial y^2}\right) \tag{2}$$

$$\frac{\partial v}{\partial t} + u\frac{\partial v}{\partial x} + v\frac{\partial v}{\partial y} = -\frac{\partial P}{\partial y} + \nu\left(\frac{\partial^2 v}{\partial x^2} + \frac{\partial^2 v}{\partial y^2}\right) + g \tag{3}$$

where u and v are the velocity components in the x and y directions respectively, while D is the divergence, P the ratio of pressure to density, ν the kinematic viscosity, g the gravitational constant and t the time.

Initially, a tentative (initial or present time step) velocity and pressure field is introduced into the Navier-Stokes equations to calculate a new (predicted) velocity field which does not satisfy the zero divergence requirement for each cell. Consequently, a pressure adjustment term, δP is derived as,

$$\delta P = -D/(\partial D/\partial P) \tag{4}$$

and the new velocity field is then corrected by adjusting the pressure in each cell as follows:

$$\begin{aligned} P_{i,j} &\rightarrow P_{i,j} + \delta P \\ u_{i\pm1/2,j} &\rightarrow u_{i\pm1/2,j} \pm \delta t\delta p/\delta x \\ v_{i,j\pm1/2} &\rightarrow v_{i,j\pm1/2} \pm \delta t\delta p/\delta y \end{aligned} \tag{5}$$

where δt is the time step and δx and δy are the cell sizes in x and y directions respectively. The adjustment process is iterated until zero divergence is satisfied for all fluid cells. The convergence rate of the iteration process can be accelerated by implementing a Successive Over-Relaxation (SOR) method [6].

The numerical procedure described above could suffer from instability if system parameters such as time step and cell size are not chosen properly. The stability analysis of such numerical formulations are usually very complex, however, a heuristic stability analysis has been proposed by Hirt [7]. Furthermore, the convective terms of the Navier-Stokes equations require some form of *upwinding* to stabilise the computation. Here, the stability of the convective terms have been retained by using a weighted combination of the upstream (donor-cell) and centred-difference approximation [1].

2.2 Free Surface Calculation

An important problem in modelling mould filling is to monitor the evolution of the fluid free surface. For this purpose, a VOF function F is defined where $F = 1$ denotes a full cell, $F = 0$ an empty cell and $0 < F < 1$ a surface cell. The function F is transported with the velocity field and it behaves as a conserved property. The transient behaviour of F is governed by the equation,

$$\frac{\partial F}{\partial t} + u\frac{\partial F}{\partial x} + v\frac{\partial F}{\partial y} = 0 \tag{6}$$

When the free surface equation (6) is integrated over a computational cell, the changes of F in a cell reduce to fluxes of F across the cell faces. Therefore, special care must be taken in computing these fluxes to preserve the sharp definition of free surfaces. The method employed here, uses a type of *donor-acceptor* flux approximation [8]. The essential idea is to use information about F downstream as well as upstream of a flux boundary to establish a crude interface shape, and then utilise this shape in computing the flux. Finally, the orientation of the free surface is determined from the values of F in a group of adjacent cells.

2.3 Heat Transfer Calculation

The conservation equation of thermal energy governing the transfer of heat in the metal and the mould can be presented in terms of enthalpy, H as follows:

$$\frac{\partial H}{\partial t} + u\frac{\partial H}{\partial x} + v\frac{\partial H}{\partial y} = k\left(\frac{\partial^2 T}{\partial x^2} + \frac{\partial^2 T}{\partial y^2}\right) \tag{7}$$

where T is the temperature and k the conductivity. The advantage of the enthalpy formulation is that the evolution of latent heat, *i.e.* solidification, can easily be accounted for.

The thermal energy in and out of each face of a cell is transported through conduction/convection. The conductive heat flux is calculated according to Fourier's first law and the convective heat transfer between the metal and the mould and their surroundings is evaluated with the Newtonian cooling law. The advective heat transfer through bulk fluid motion can only be calculated when the amount of fluid in and out of a cell is determined. The advected thermal energy can be expressed in mathematical form as,

$$Q = H \cdot \Delta V \cdot \delta F \tag{8}$$

where Q is the thermal energy advected through a cell side, δF the amount of F transported. Also, H and ΔV are the enthalpy and the volume of the donor cell respectively. The advantage of this formulation is that δF is already evaluated in the free surface calculation.

As the thermal energy in and out of every cell is calculated, the net thermal energy of a cell is then used to compute the cell temperature.

3 NUMERICAL SIMULATION AND RESULTS

The capabilities of the model, to analyse the flow pattern and temperature variation during the mould filling, are demonstrated by simulating two mould filling examples.

The first example is to simulate flow in a complete casting system, consisting of a pouring basin, down sprue, sprue base, horizontal runner and vertical square mould cavity as shown in Figure 1. The liquid metal enters the casting system with a constant velocity from an inlet on the top left hand side of the pouring basin. The simulated result, illustrated in Figure 1(a) shows that initially, the pouring basin becomes partially full and liquid metal is free falling through the sprue. As the sprue base is filled and the liquid metal enters the horizontal runner, see Figure 1(b), the down sprue is pressurised forcing the pouring basin to become full. At a later time, shown in Figure 1(d), the liquid metal enters the vertical cavity and the in-rushing metal forms a wave against the opposite wall. The liquid metal wave gradually develops and rolls over itself to meet the inflow stream into the cavity. Figure 1(f) shows the rotational flow field in the filled square cavity and, also, the last regions to fill in the casting system.

The second example represents a geometry similar to many encountered in gravity sand cast iron components. The casting system shown in Figure 2, consists of a large feeder connected to the casting via a side gate. In practice, many casting defects are discovered at or near gate locations and casting quality is often found to be sensitive to the gate location and its geometry. This is generally due to the significant localised heating of the mould near the gate during the filling.

To investigate the effect of the gate size on the filling and mould heating in the gate region, two hypothetical cases of gravity sand casting have been modelled. The two cases are identical except that one has a gate twice the height of the other. The liquid metal is poured from the top of the feeder with a constant velocity. The simulated results for the gravity sand casting with small and large gates are illustrated in Figures 2 and 3 respectively.

In the case of the casting with the small gate, the result in Figure 2(a) indicates that flow through the gate has become pressurised, leading to larger flow velocities. This also encourages the filling of the feeder before the casting cavity is filled. At the same time, the mould around the gate region is considerably heated by the liquid metal in the gate and the feeder. Figure 2(b) shows that the feeder is almost full before the casting cavity becomes half full. This encourages the heating of the mould area above the gate even more.

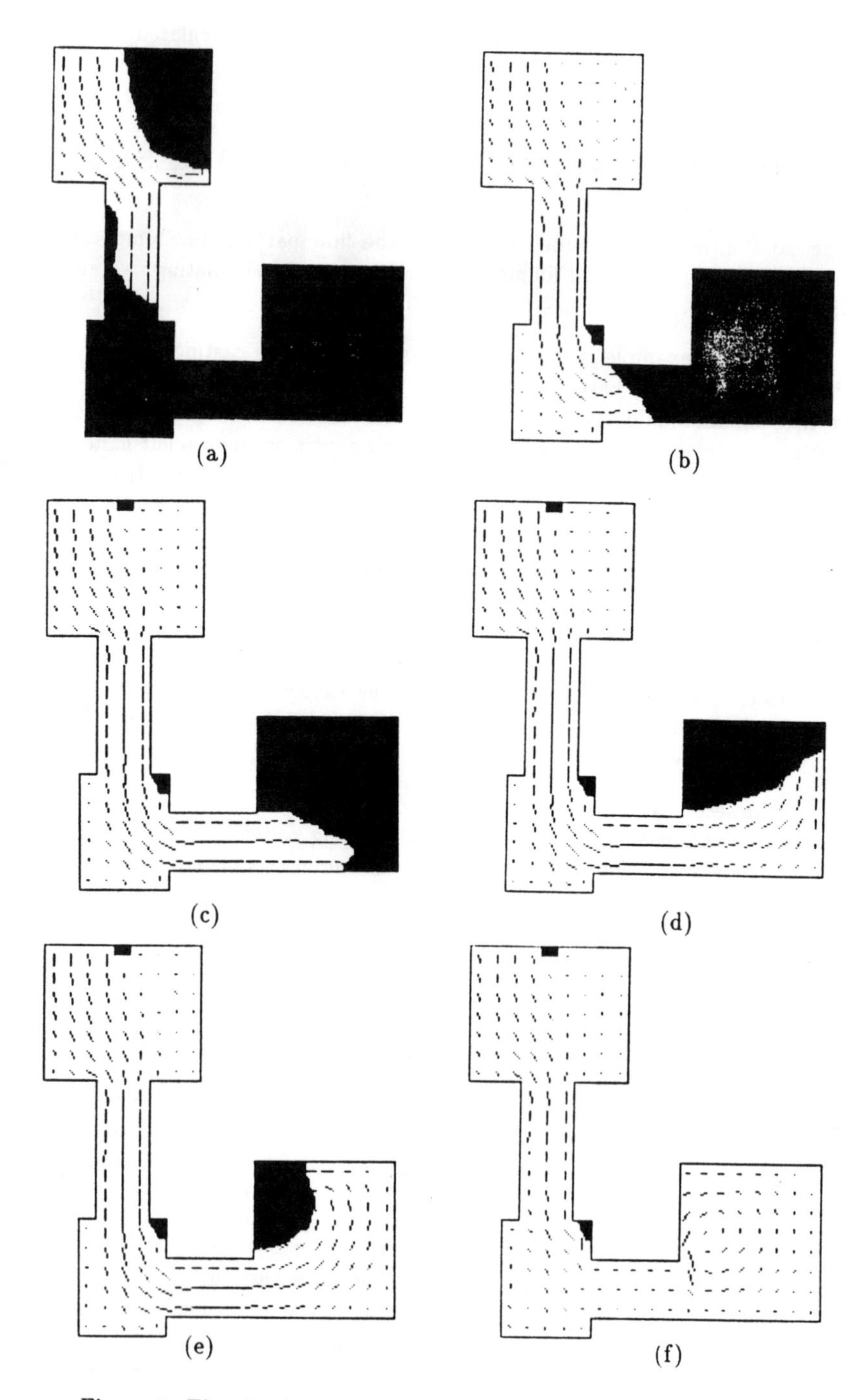

Figure 1. The simulated filling pattern for a complete casting system.

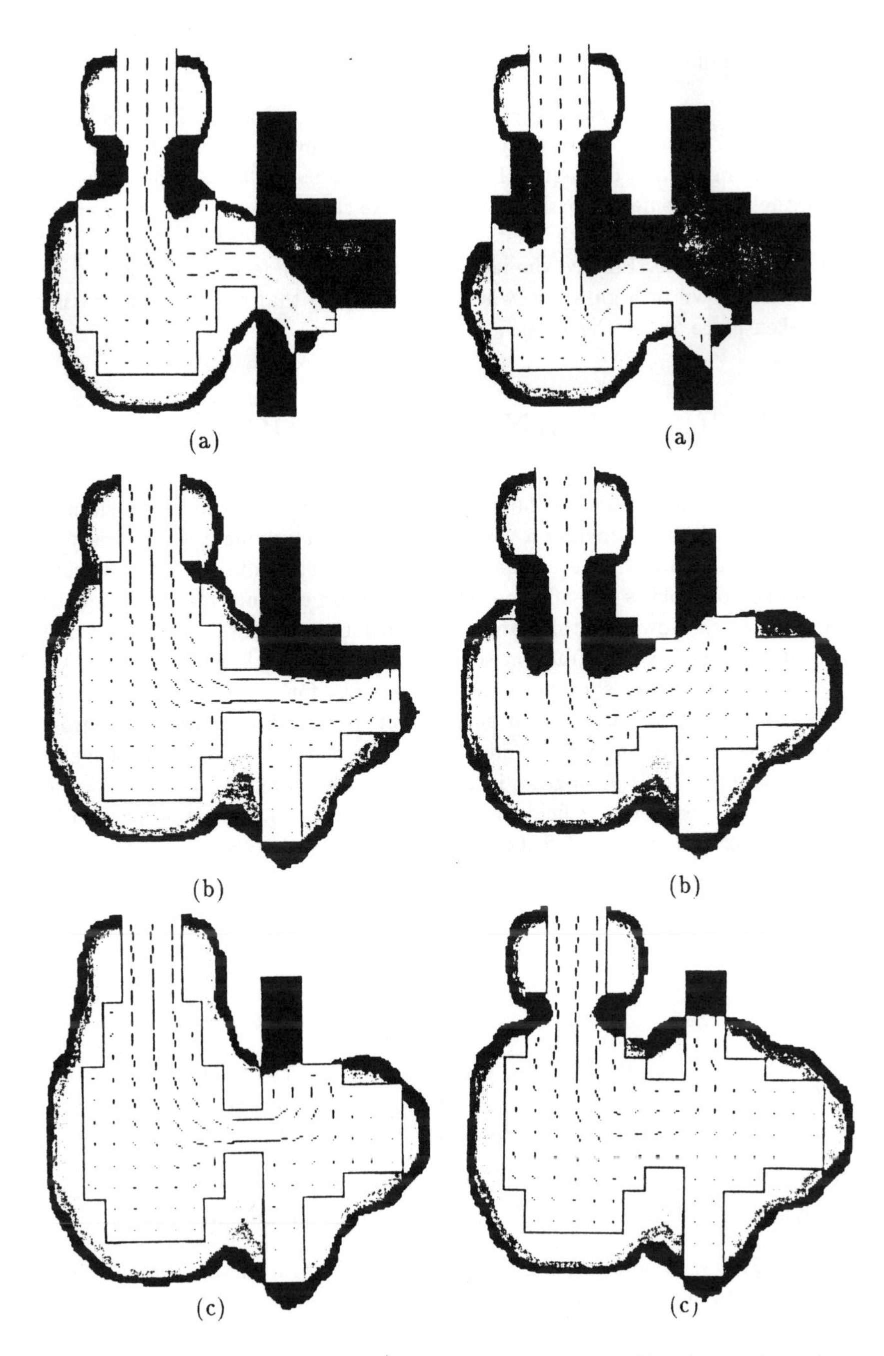

Figure 2. Gravity sand casting with a small gate.

Figure 3. Gravity sand casting with a large gate.

Figure 2(c) shows the last area of the mould to fill with the feeder being completely full.

In the case of the casting with the large gate, the simulated results imply a different scenario. Figure 3(a) shows that the liquid metal is free flowing through the gate and consequently only heats the mould area below the gate. Also, the feeder remains partially full while the casting cavity fills up to about 80%, see Figure 3(b), and the metal just comes into contact with the mould at the top of the gate. Finally, the last areas to fill in the feeder and the casting cavity are shown in Figure 3(c) which also indicates that some mould heating has taken place above the gate.

4 DISCUSSION AND CONCLUSIONS

The numerical model of fluid flow and heat transfer for mould filling simulation, presented here, provides a very useful tool for studying a wide variety of casting phenomena which are usually examined experimentally. In the present model the flow field is not affected by solidification of liquid metal during the filling process. However, this can be included by coupling temperature and viscosity or velocity and fraction solid of the melt. With the aid of a CAD system which includes such mathematical models, foundry designers are able to minimise turbulence, avoid air entrapment, provide for separation of dross and inclusions from the metal and produce the desired distribution of metal flow through the multiple ingates in various casting processes. They could also determine the temperature fields in the metal and mould, the extent of solidification and certain casting defects. This ultimately leads to a more economical production of higher quality castings with improved macro/micro-structures and mechanical properties.

References

[1] NICHOLS, B.D., HIRT, C.W. and HOTCHKISS, R.S. - SOLA-VOF:A Solution Algorithm for Transient Fluid Flow with Multiple Free Boundaries, *Los Alamos Sci. Lab., Rept. LA-8355*, August 1980.

[2] HIRT, C.W. and NICHOLS, B.D. - Volume of Fluid (VOF) Method for Dynamics of Free Boundaries, *J. Comp. Phys.*, **39**, pp. 201-225, 1981.

[3] SAMONDS, M. and WAITE, D. - 3D Finite Element Simulation of Filling Transients in Metal Castings, *Mathematical Modelling for Materials Processing*, Eds. M. Cross, J.F.T. Pittman and R.D. Wood, Clarendon Press, 1993.

[4] LIN, H.J. and HWANG, W.S. - Combined Fluid Flow and Heat Transfer Analysis for the Filling of Casting, *Modeling and Control of Casting and Welding Processes IV*, Eds. A.F. Giamei and G.J. Abbaschian, The Minerals, Metals and Materials Society, 1988.

[5] BIRD, R.B., STEWART, W.E. and LIGHTFOOT, E.N. - *Transport Phenomena*, John Wiley & Sons, 1960.

[6] PATANKAR, S.V. - *Numerical Heat Transfer and Fluid Flow*, Hemisphere Publishing Corp., 1980.

[7] HIRT, C.W. - Heuristic Stability Theory for Finite-Difference Equations, *J. Comp. Phys.*, **2**, pp. 339-355, 1968.

[8] RAMSHAW, J.D. and TRAPP, J.A. - A Numerical Technique for Low-Speed Homogeneous Two-Phase Flow with Sharp interfaces, *J. Comp. Phys.*, **21**, pp. 438-453, 1976.

PREDICTIONS OF IMPLICIT AND EXPLICIT FINITE DIFFERENCE MODELS FOR THE SOLIDIFICATION OF LONG FREEZING RANGE ALLOYS

S.G.R. Brown, M.R. Tadayon and J.A. Spittle*

SUMMARY

At the present time there is considerable international interest in the establishment of 'criteria functions' that can be empirically related to microstructural features of as-cast alloys. These functions are thermal/temporal parameters, the values of which can be predicted using numerical models. From experimental observations over many years, local solidification time, t_s, has become an accepted function for assessing and anticipating dendrite arm spacings in long freezing range aluminium alloys. In this paper, the authors examine and compare the predictions of 1D implicit and explicit finite difference models for the solidification of an Al-4wt.% Cu alloy. It has been observed that the local solidification time values predicted by the two models at specific locations can be significantly different, particularly at low superheats.

1. INTRODUCTION

Research groups throughout the world are engaged in trying to develop 'second generation' numerical solidification simulation software capable of predicting microstructural evolution and, in turn, in-service mechanical properties and safe lifetimes of components. An indirect approach to this problem involves the establishment of criteria functions which can be quantitatively related to particular experimentally measured microstructural features such as dendrite arm spacings, grain size, amount of dispersed porosity etc. These criteria functions, which are thermal/temporal parameters, can be determined experimentally, e.g. from thermal analysis studies, and computationally from numerical heat transfer/solidification models. The alternative direct approach is to attempt to model the kinetics of the transformations and those factors influencing solid/liquid interface morphology and scale of structural features.

* The authors are in the IRC in Materials for High Performance Applications
Department of Materials Engineering, University College, Swansea SA2 8PP

In the short term, the indirect approach to microstructural prediction is the most attractive although it is not yet accepted that the scale (size, amount) of a particular microstructural feature can be singularly related to a thermal/temporal parameter defining a local solidification environment. Grain size and pore size, for example, will also be dependent on the density of potential substrates for heterogeneous nucleation. Porosity levels might also be influenced by the gas content of the melt or gas pick-up from the mould.

Although laboratory experiments can demonstrate clear correlations between microstructure and properties, it is normally extremely difficult to anticipate the structural details (volume fraction of phases, size scale of constituents, morphologies etc) that will arise at specific positions in commercially produced complex-shaped components such as castings. Numerical computer simulation therefore offers the possibility of predicting such aspects prior to processing and of design of components for durability. As the drive towards weight reduction in transport increases, particularly in the automotive industry, research groups around the world are assessing the 'criterion function' approach to predicting microstructures and, the low density, high strength, long freezing range aluminium alloys are being subjected to extensive investigation. It is extremely important to determine the possibilities and limitations of this approach and to establish international procedures for incorporating such functions into models.

1.1 Important Microstructural Features

Although the overall microstructure of a cast alloy is responsible for the alloy's mechanical properties, when modelling microstructure individual structural features have usually been emphasised. The most important of these are dendrite arm spacing of the primary solidifying phase and porosity.

Dendrite arm spacing reflects the cooling rate which also influences all other transformations that the alloy may undergo. Dendrite arm spacing is therefore often used as an indicator of the total microstructure. Considerable experimental data exists concerning the influence of cooling rate, R, or local solidification time, t_s, on dendrite arm spacings in long freezing range aluminium alloys [1]. t_s is the time taken for the temperature at a given location to fall from the liquidus to the solidus. Within the range of solidification conditions normally occurring in commercial castings, R and t_s are therefore widely accepted as 'criteria functions' for dendrite arm spacings in aluminium alloys. Dendrite arm spacings in turn have been used to predict tensile properties of aluminium alloys [2]. However, it is impossible to use the knowledge of these interrelationships in any predictive manner unless the data is incorporated into numerical models.

Porosity in castings arises from solidification shrinkage, gas or a combination of both. Porosity is obviously an undesirable defect and its presence can therefore be used to assess the quality of a casting. In the case of long freezing range aluminium alloys, porosity levels have been related to tensile properties [3] and fatigue lives [4].

In the absence of gas, the type and location of solidification shrinkage depends significantly on the mode of freezing of the alloy. Various forms include centreline macroshrinkage and dispersed microporosity. Gas precipitation is not a problem in all metal alloys but where found, e.g. in aluminium alloys, is usually combined with shrinkage. In determining criteria functions for porosity, care must be taken in experimentation regarding gas content and inclusion levels [5]. Inclusions are important since they influence the nucleation of pores. In the past few years, a variety of functions have been established either experimentally or using numerical models for describing porosity distribution in long freezing range aluminium alloys [6-10].

1.2 Criteria Functions

As mentioned above, these functions are thermal/temporal parameters. Such parameters can be R, t_s, G (temperature gradient), V_s (solidus velocity) or combinations of these variables e.g. $G/\sqrt{R}$. Certain structural features such as microporosity are likely to be determined by events that take place close to the solidus temperature. Others like dendrite arm spacings are governed by the conditions that exist over the complete freezing range of the primary phase. It is therefore critical when determining the values of criteria functions at specific locations in castings to decide (a) at what precise stage during solidification the parameters will be determined and (b) exactly how the parameters will be calculated. Comparison of data of different research workers is impossible unless the procedures relating to (a) and (b) have been clearly specified.

1.3 Establishment of Criteria Functions

(i) Experimental Determination

Experimentation is one method of establishing criteria functions for microstructural features and indeed is the only way in which the true quantitative relationship between a feature and a function can be obtained. This involves extensive thermal analysis of castings and relating the solidification conditions obtained from the thermal analysis data at specific locations to the observed microstructures at those locations. As mentioned earlier, this approach has led to the establishment of local solidification time as a function which can be correlated with dendrite arm spacing. Calculating the values of functions does present problems because the cooling rate at a specific location changes continuously with time. Cooling rate also influences the nucleation and growth temperatures of the constituents. Care must therefore be exercised as indicated in (a) and (b) in 1.2 when determining the values from the thermal analysis curves.

(ii) Computer Modelling

In this case, experimentally determined dimensions of microstructural features at specific locations in a solidified material are related to the values of functions (e.g. local solidification time) at those locations determined from a numerical heat transfer/solidification model. For any particular function, e.g. G or t_s, the values calculated will obviously depend on the materials physical properties data and boundary conditions used in the model. The modelled values at specific locations would only correspond with the experimentally derived values in (i) for a well validated model. Using numerical models, it is certainly easier to examine a variety of functions and to establish which one/s can be qualitatively correlated with the form of variation of the microstructural feature of interest.

2. PRESENT INVESTIGATION

In the present study, the influence of formulation on the computed values of local solidification times is examined for 1D heat transfer/solidification finite difference models. The predictions of implicit and explicit schemes are examined for the solidification of a 50mm long Al-4wt.% Cu alloy charge for two different melt superheats of 0.5K and 53K.

The model assumes that heat is removed from a chilled surface that is maintained at a constant temperature of 298K and that the heat transfer coefficient at the metal/chill surface interface is 1000 W/m^{2}K. Latent heat evolution during freezing was handled assuming relationships between fraction solid, f_s, and local temperature T. During primary phase solidification, the Scheil equation [11] below was used. This assumes complete solute mixing in the liquid and no solute diffusion in the solid.

$$f_s = 1 - \left(\frac{T_m - T}{T_m - T_l} \right)^{\frac{1}{k_0 - 1}} \tag{1}$$

where T_m is the melting point of aluminium (933K), T_l the liquidus of the alloy (920K) and k_0 the equilibrium partition coefficient (0.18). It was assumed that the solidification of the non-equilibrium eutectic portion of the alloy occurred over a 2K temperature range (821K-819K) during which time the fraction solid varied linearly with temperature. The overall f_s vs. T relationship for the freezing of the alloy is therefore as shown in Fig. 1.

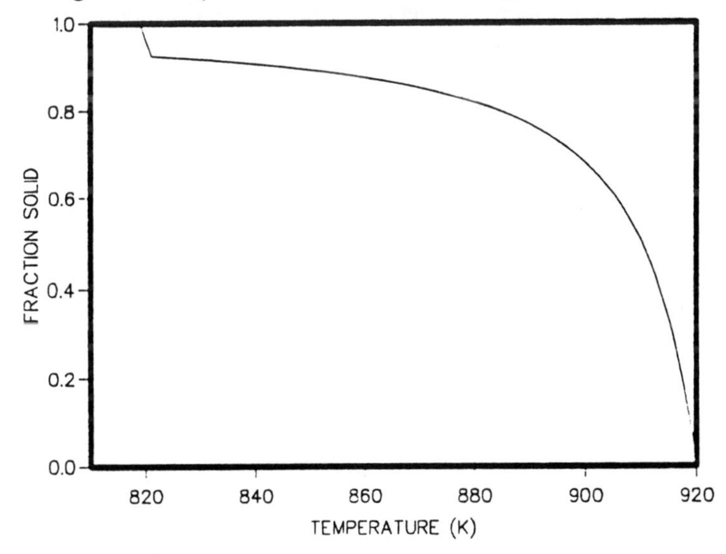

Figure 1 Fraction solid - Temperature relationship used in the present models for Al-4wt.%Cu.

2.1 Implicit and Explicit Finite Difference Models

The field equation governing the solidification of an alloy in a general casting domain Ω whose thermophysical properties are highly nonlinear can be expressed as,

$$\frac{\partial H}{\partial T}\dot{T} - \nabla.[k\nabla T] = 0 \qquad \text{in } \Omega \tag{2}$$

subject to the relevant initial and boundary conditions of

$$T(\mathrm{x},0) = T_0(\mathrm{x}) \qquad \text{in } \Omega \tag{3}$$

$$T(\mathrm{x},t) = \hat{T}(\mathrm{x},t) \qquad \text{on } \Gamma_1 \tag{4}$$

$$k\nabla T.\hat{n} + q + q_c = 0 \qquad \text{on } \Gamma_2 \tag{5}$$

where H is enthalpy, k the conductivity, t the time, $\hat{T}$ the specified temperature, $\hat{n}$ the unit normal to the boundary, x the coordinate vector, q the specified heat flux and q_c the convective heat flux defined as

$$q_c = h_c(T - T_c) \tag{6}$$

in which h_c is the convective heat transfer coefficient and T_c the temperature at which there is no convection.

The term $\frac{\partial H}{\partial T}$ in equation (2) may be approximated by the effective heat capacity method [12] in which the effective heat capacity C is defined as,

$$C(T) = \rho\left[c(T) - L\frac{df_s}{dT}\right] \tag{7}$$

where ρ is the density, c the specific heat and L the latent heat.

The discretisation of equation (2) in 1D space and time, using finite difference [13] in space and time and the θ-method [14] as the time integration scheme results in,

$$\begin{aligned}\left[C_i^{n+\theta} + r\theta\left(k_{i+\frac{1}{2}}^{n+\theta} + k_{i-\frac{1}{2}}^{n+\theta}\right)\right]T_i^{n+1} \;=\;& C_i^{n+\theta}T_i^n \\ &+r[(1-\theta)\,k_{i+\frac{1}{2}}^{n+\theta}\,(T_{i+1}^n - T_i^n) \\ &-\;(1-\theta)\,k_{i-\frac{1}{2}}^{n+\theta}\,(T_i^n - T_{i-1}^n) \\ &+\theta\left(k_{i+\frac{1}{2}}^{n+\theta}\,T_{i+1}^{n+1} + k_{i-\frac{1}{2}}^{n+\theta}\,T_{i-1}^{n+1}\right)]\end{aligned} \tag{8}$$

where $r = \Delta t/\Delta x^2$, Δt is the time step and Δx the grid size.

By changing the value of θ in equation (8), different members of θ-method can be produced. For example,

$\theta = 0$ - Forward Difference or Euler-Forward (Explicit)

$\theta = 1/2$ - Crank Nicolson (Implicit)

$\theta = 2/3$ - Galerkin (Implicit)

$\theta = 1$ - Backward Difference or Euler-Backward (Implicit)

The parameter θ also controls the stability and accuracy of the equation (8). The values of $1 \geq \theta \geq \frac{1}{2}$ lead to an unconditionally stable scheme, but for $\theta < \frac{1}{2}$, the scheme is conditionally stable only if

$$\Delta t < \frac{\Delta x^2}{2(k/\rho c)}$$

It should also be noted that in the case of the explicit scheme, the effective heat capacity needs to be expressed with a backward difference approximation.

The implicit schemes of equation (8) produce a nonlinear system of equations which may be solved using an iterative method of solution. Here, a simple Successive Over-Relaxation (SOR) [13] is employed.

3. RESULTS

Figs. 2 and 3 illustrate the predicted thermal analysis curves at various locations from the chill using the implicit and explicit models for a superheat of 0.5K. The two sets of curves appear to be virtually identical even though, as discussed in the next paragraph, predictions of parameters relating to the solid plus liquid region using the two models can be significantly different.

Figs. 4 and 5 show the rates of movement of the liquidus and solidus (819K) isotherms respectively for the two superheats using the implicit and explicit models. For each superheat, the distance - time plots for the solidus are almost identical for the two models. However, the equivalent plots for the liquidus are significantly different for the two models and this difference increases with decreasing superheat. At both superheats, the predicted length of time that it takes the liquidus to reach the end of the charge is shorter in the case of the explicit model. The distance from the chill at which the liquidus advance curves for the two models become different increases with increasing superheat.

Fig. 6 illustrates the calculated local solidification times for both superheats for each of the models. The curves highlight two features. Firstly, in accordance with the predicted movements of the liquidus and solidus isotherms, at each superheat the predictions of t_s for the two models differ. The difference increases with distance from the chill and is exacerbated by a decrease in superheat.

Secondly, the curves exhibit a distinctive shape; the t_s values initially increase with distance to a position beyond which they suddenly decrease (and in the case of the implicit model continuously decrease with distance).

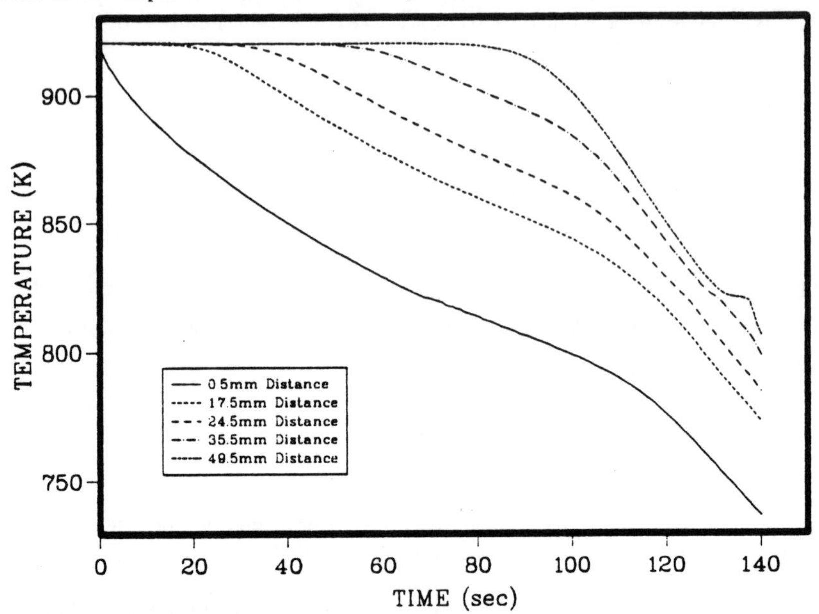

Figure 2 Predicted cooling curves at various distances from the chill using the implicit model

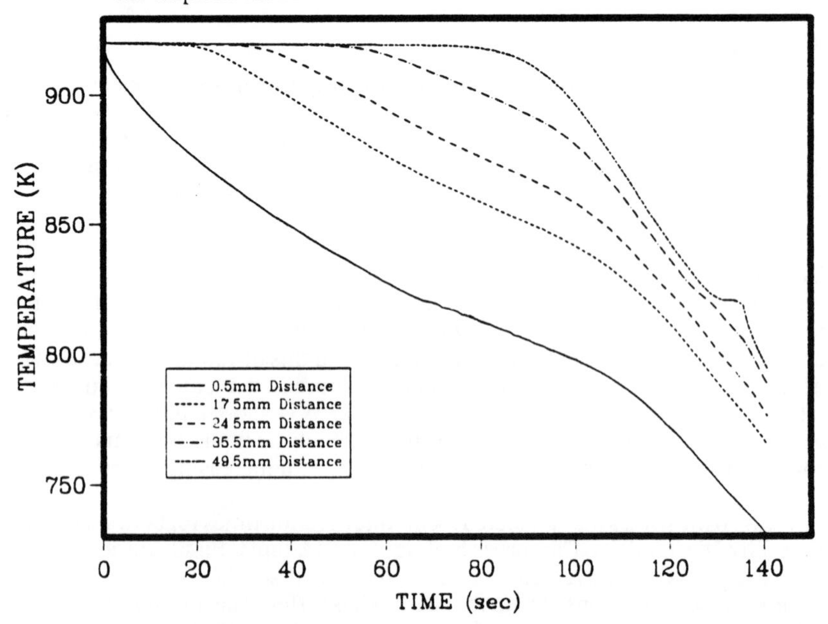

Figure 3 Predicted cooling curves at various distances from the chill using the explicit model

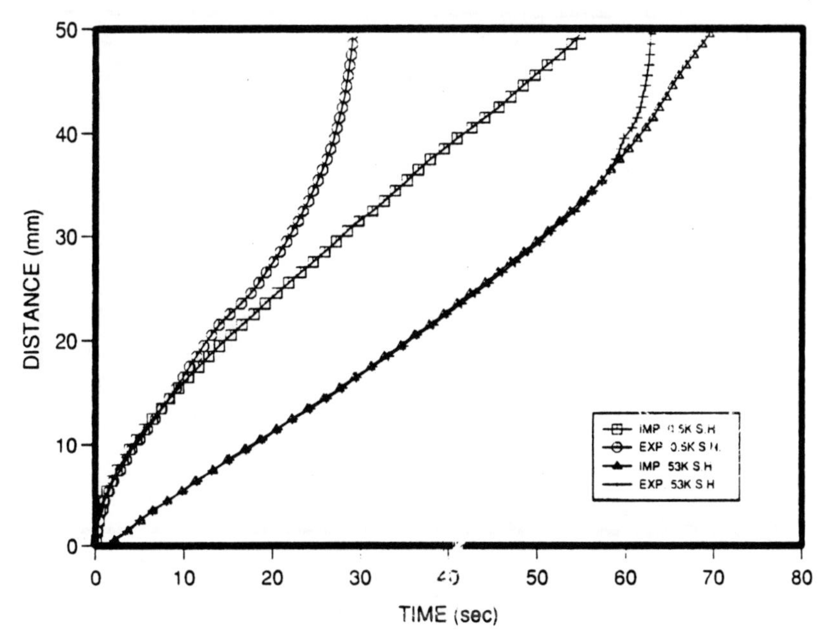

Figure 4 Predicted liquidus position as a function of time for superheats of 0.5K and 53K using the implicit and explicit models.

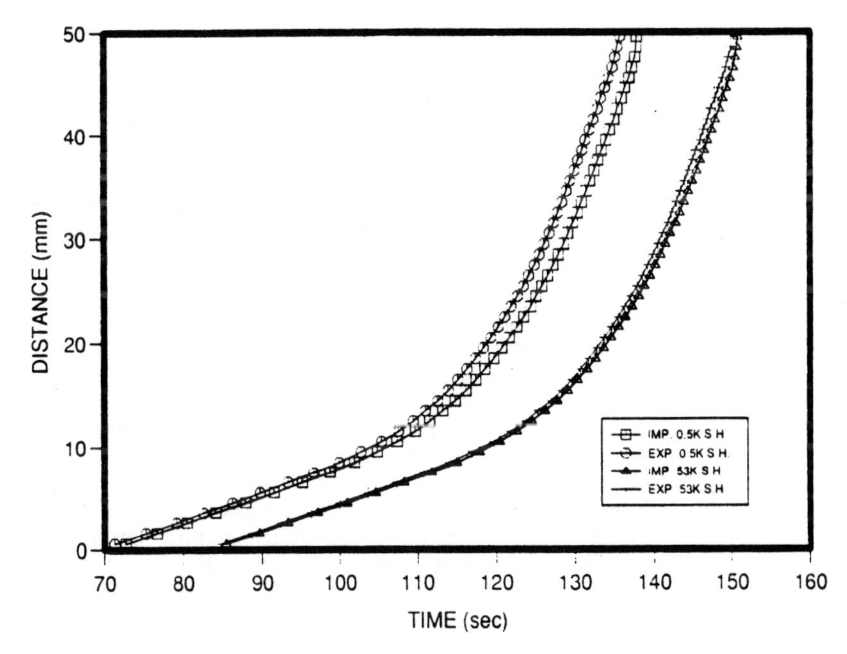

Figure 5 Predicted solidus position as a function of time for superheats of 0.5K and 53K using the implicit and explicit models.

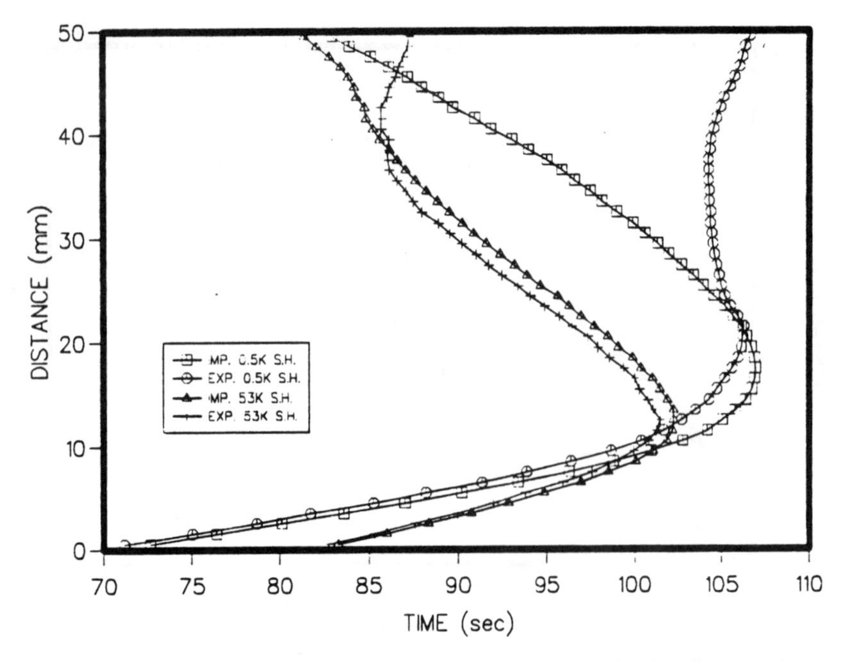

Figure 6 Predicted local solidification times along the alloy bar for superheats of 0.5K and 53K using the implicit and explicit models.

4. DISCUSSION

The results illustrate that implicit and explicit finite difference models can give different predictions of the form of the variation of local solidification time with increasing distance from the chill surface for the 1D freezing of a long freezing range aluminium alloy. The difference in the predictions becomes greater as the melt superheat decreases. This discrepancy is possibly due to the inability of the explicit formulation to model the true rate of advance of the liquidus when the temperature gradient ahead of the liquidus becomes very shallow. Under this shallow gradient condition, the explicit model predicts an unrealistically high rate of advance of the liquidus which results in larger predicted t_s values. In the high superheat case, the t_s values predicted by the two models are very similar until near the end of the charge when, again, the temperature gradient ahead of the liquidus falls to a very low value.

The curves of local solidification time versus distance along the charge indicate that t_s initially continuously increases until a position is reached beyond which t_s starts to fall again, Fig. 6. This position corresponds with that at which the solidus isotherm velocity displays a marked increase, see Fig. 5. It is thought that these phenomena are related to the shape of the fraction solid - temperature curve predicted by the Scheil equation. As can be seen in Fig. 1., on cooling there is initially a very high rate of solid formation with decreasing temperature but at lower temperatures the curve is very shallow.

5. CONCLUSIONS

Modelling groups around the world are currently evaluating the application of numerical modelling to the prediction of microstructures in as-solidified materials using 'criteria functions'. The latter are thermal/temporal parameters the values of which can be determined from the models. The present data relates to the values of the parameter 'local solidification time', t_s, determined at specific locations in a unidirectionally solidified Al-4wt.% Cu alloy charge using finite difference models. The data demonstrate that

(a) the values of t_s obtained from implicit and explicit models can differ, particularly at low melt superheats.

(b) the method of handling latent heat evolution will also influence the predicted values and form of variation of t_s with distance along the charge.

6. REFERENCES

[1] BOWER, T.F., BRODY, H.D. and FLEMINGS, M.C. - Measurements of Solute Distribution in Dendritic Solidification. Trans. Met. Soc. AIME, Vol. 236, pp. 624-634, 1966.

[2] OSWALT, K.J. and MISRA, M.S. - Dendrite Arm Spacing (DAS): A Nondestructive Test to Evaluate Tensile Properties of Premium Quality Aluminium Alloy (Al-Si-Mg) Castings. AFS Int. Cast Metals J., Vol. 1, pp. 23-40, 1981.

[3] McLELLAN, D.L. and TUTTLE, M.M. - Aluminium Castings - A Technical approach, AFS Trans., Vol. 91, pp. 243-252, 1983.

[4] WICKBERG, A., GUSTAFSSON, G. and LARSSON, L-E. - Microstructural Effects on the Fatigue Properties of a Cast Al7SiMg Alloy, SAE Technical Paper Series, No. 840121, pp. 1-8, 1984.

[5] SURI, V.K., HUANG, H., BERRY, J.T. and HILL, J.L.- Applicability of Thermal Parameter Based Porosity Criteria to Long Freezing Range Aluminium Alloys. AFS Trans., Vol. 100, pp. 92-106, 1992.

[6] LEE, Y.W., CHANG, E. and CHIEU, C.F. - Modeling of Feeding Behaviour of Solidifying Al-7Si-0.3Mg Alloy Plate Casting. Metall. Trans. B, Vol. 21B, pp. 715-722, 1990.

[7] PAN, E.N., LIN, C.S. and LOPER, C.R. - Effects of Solidification Parameters on the Feeding Efficiency of A356 Aluminium Alloy. AFS Trans., Vol. 98, pp. 735-746, 1990.

[8] BRODY, H.D. VISWANATHAN, S. and STOEHR, R.A. - Predicting Shrinkage Microporosity in Aluminium Copper Alloys, Modeling of Casting, Welding and Advanced Solidification Processes V, Eds. Rappaz, M., Osgu, M.R. and Mahin, K.W., The Minerals, Metals and Materials Society, 1991.

[9] VENKATARAMANI, R., MADHUSUDANA, K., PALANIAPPAN, K.R. and PRABHAKAR, O. - Soundness Profiles in Aluminium Alloy Castings, Ibid.

[10] BROWN, S.G.R. and SPITTLE, J.A. - Finite Element Simulation of Solidification of Aluminium Casting Alloy LM25, Mater. Sci. Technol., Vol. 6, pp. 543-547, 1990.

[11] SCHEIL, E. - Zeik. f. Met., Vol. 34, p. 70, 1942.

[12] CLYNE, T.W. - The use of heat flow modelling to explore solidification phenomena, Metall. Trans. B, Vol. 13B, pp. 471-477, 1982.

[13] PATANKAR, S.V. - Numerical Heat Transfer and Fluid Flow, Hemisphere Publ. Corp. 1980.

[14] HUGHES, T.J.R. - Unconditionally Stable Algorithms for Nonlinear Heat Conduction, Comp. Meth. Appl. Mech. Eng., Vol. 10, pp. 135-139, 1977.

SOLIDIFICATION PROBLEM. NUMERICAL MODEL ON THE BASIS OF GENERALIZED FDM

Bohdan MOCHNACKI[1] and Marek BALCER[2]

SUMMARY

Generalized FDM [1] can be treated as a certain algorithm joining the typical features of FDM (theoretical approach) and FEM or BEM (the way of domain discretization). The method is peculiarly effective in the case of linear problems even so the non-linear problems can be taken into account too (e.g. [2]). In this paper the solidification and cooling processes in casting with complex shape domain are analyzed. The basic algorithm of GFDM for casting and mold sub-domains supplemented by a procedure 'linearizing' considered problem is discussed. In numerical realization a certain improvements of Gauss elimination method presented by M.Balcer in [3] have been used.

1. GOVERNING EQUATIONS

The spatial heterogenous domain Ω of the casting and mold in which the solidification and cooling processes are taking place constitutes a sum of sub-domains $\Omega_m \in \Omega$, $m=1, 2, \ldots, M$ at the same time indexes $m>3$ identify the mold sub-areas. Considering the solidification process it is possible to distinguish the following sub-domains in the casting volume (comp. Fig. 1)

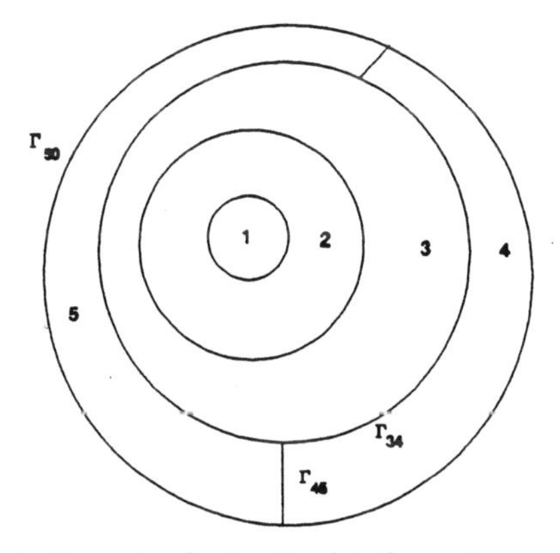

Fig. 1. Domain Ω, 1–liquid, 2-mushy zone, 3–solid, 4, 5–mold

[1] Prof., Silesian Technical University, Krzywoustego 7, 44–121 Gliwice
[2] Dr., Silesian Technical University, Krzywoustego 7, PL–44–121 Gliwice

$$X \in \Omega_1(t) \rightarrow T(X, t) > T_L, \qquad S(X, t) = 0$$
$$X \in \Omega_2(t) \leftrightarrow T(X, t) \in \langle T_S, T_L \rangle, \qquad S(X, t) = \varphi(T) \qquad (1)$$
$$X \in \Omega_3(t) \rightarrow T(X, t) < T_S, \qquad S(X, t) = 1$$

where T, X, t denote a temperature, spatial co-ordinates and time, $\langle T_S, T_L \rangle$ – interval of temperature in which a solidification process proceeds and latent heat L_V is evolved, $S(X, t)$ – volumetric fraction of solid state at the neighbourhood of point X, $\varphi(t)$ – a function determining the emanation of L_V.

A non-steady temperature field in considered domain is described by a system of partial differential equations of the type

$$C_m(T)\, \partial_t T(X, t) = \text{div}\left[\lambda_m \,\text{grad}\, T(X, t)\right] + q_{Vm}, \quad m = 1, \ldots, M \qquad (2)$$

where C_m is a specific heat related to an unit of volume, q_{Vm} – capacity of internal heat sources in Ω_m. One can notice that $q_{Vm}=0$ for $m \neq 2$ whereas for $m=2$ $\quad q_{Vm} = L_V\, \partial_t S = L_V\, \varphi'(T)\, \partial_t T$ [4]. Finally the following system of equations is considered

$$C_m(T)\partial_t T(X, t) = \text{div}\left[\lambda_m \,\text{grad}\, T(X, t)\right], \quad m = 1, \ldots, M \qquad (3)$$

where

$$C_m(T) = \begin{cases} C_1, & T > T_L, \\ C_2 - L_V\, \varphi(T), & T \in \langle T_S, T_L \rangle, \\ C_3, & T < T_S, \\ C_m, & m > 3. \end{cases} \qquad (4)$$

It should be pointed that by choosing properly the function $\varphi(T)$ it is possible to obtain correct (from a technological point) model of macroscopic solidification for wide class of metals and alloys [4]. Assuming according to Viejnik (e.g. [5]) that $\varphi(T)$ is a linear function of the form $\varphi(T) = (T_L - T)/(T_L - T_S)$ one obtains

$$C_m(T) = \begin{cases} C_1, & T > T_L, \\ C_2 - \dfrac{L_V}{T_L - T_S}, & T \in \langle T_S, T_L \rangle, \\ C_3, & T < T_S, \\ C_m, & m > 3. \end{cases} \qquad (5)$$

A substitute thermal capacity C_m, $m \leq 3$ changes in a 'jumping' manner at points T_S and T_L and this parameter (or enthalpy of metal [6]) should be in special way 'smoothed' but taking into account the method presented in this paper it is possible to obtain the proper results even in the case of step functions – on condition that certain correcting procedures will be introduced.

Boundary conditions supplementing the mathematical model are given in the form

$$X \in \Gamma_{kl} : \quad \begin{cases} -\lambda_k n \cdot \operatorname{grad} T_k(X, t) = -\lambda_l n \cdot \operatorname{grad} T_l(X, t) \\ T_k(X, t) = T_l(X, t) , \quad k \neq l \end{cases} \tag{6}$$

where Γ_{kl} are the contact surfaces between casting and mold sub-domains and mold sub-areas. On the outer surface of the system it is assumed that

$$X \in \Gamma_{k0} : \quad -\lambda_k n \cdot \operatorname{grad} T_k(X, t) = \alpha (T - T_\infty) \tag{7}$$

where α is a heat transfer coefficient, T_∞ – ambient temperature. For $t=0$

$$T_m(X, 0) = T_{m0}(X) , \quad \Omega_1(0) = \Omega_2(0) = \varnothing \tag{8}$$

2. CORRECTION OF TEMPERATURE FIELD IN THE CASTING AREA

A procedure of temperature field correction can be joined with optional method of Fourier's equation numerical solution in which computations are realized 'step by step' for successive levels of time. The basic idea of this algorithm is a 'rebuilding' of temporary solution for considered level of time it means that a pseudo-initial condition for next transition from t^{f+1} to t^{f+2} is in a certain way transformed. The adequate formulae result from energy balance for neighbourhood of considered point or element - but they can be also educed in analytical way [7]. Presented below approach to the problem of numerical simulation of solidification process constitutes a certain generalization of Temperature Recovery Method - in particular the adaptation of TRM to the problem of solidification proceeding in an interval of temperature. The method was presented by Szargut J.and Mochnacki B. in [8], next by Hong C.P., Umeda T., and Kimura Y. [10] and generalized by Majchrzak E. [8]

So, in the numerical realization the casting-mold system is reduced to the system molten metal - mold. In this way the casting area becomes a homogenous one. On the basis of knowledge of temperature field at t^f moment of time the new temperature distribution for t^{f+1} can be found according to the algorithm for linear Fourier's equations. Next the correction of obtained results is introduced. In discussed case one ought to consider the following transitions - Figure 2:

$$\begin{array}{ll}(A) & T_i^f > T_L\,,\quad T_i^{f+1} > T_L\,,\\ (B) & T_i^f > T_L\,,\quad T_i^{f+1} \in \langle T_S,\, T_L\rangle\,,\\ (C) & T_i^f > T_L\,,\quad T_i^{f+1} < T_S\,,\\ (D) & T_i^f \in \langle T_S,\, T_L\rangle\,,\quad T_i^{f+1} \in \langle T_S,\, T_L\rangle\,,\\ (E) & T_i^f \in \langle T_S,\, T_L\rangle\,,\quad T_i^{f+1} < T_S\,,\\ (F) & T_i^f < T_S\,,\quad T_i^{f+1} < T_S\end{array} \tag{9}$$

where T_i is a temperature at point $X_i \in \Omega_m$, $m=1, 2, 3$.

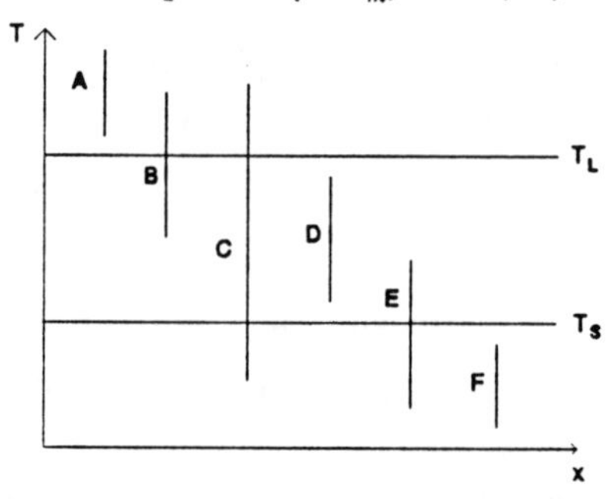

Fig. 2. Correction of temperature field

For the point which fulfills condition (A) value of temperature is not corrected. In cases (B)−(F) the following formulas allow to find the proper values of searched function

$$\begin{array}{ll}(B): & \hat{T}_i^{f+1} = T_L - \dfrac{C_1\left(T_L - T_i^{f+1}\right)}{C_2}\,,\\[2ex] (C): & \hat{T}_i^{f+1} = T_S + \dfrac{L_V}{C_3} - \dfrac{C_1\left(T_L - T_i^{f+1}\right)}{C_3}\,,\\[2ex] (D): & \hat{T}_i^{f+1} = T_i^f - \dfrac{C_1\left(T_i^{f+1} - T_i^f\right)}{C_2}\,,\\[2ex] (E): & \hat{T}_i^{f+1} = T_S + \dfrac{L_V}{C_3} - \dfrac{C_2\left(T_L - T_i^f\right)}{C_3} - \dfrac{C_1\left(T_i^f - T_i^{f+1}\right)}{C_3}\,,\\[2ex] (F): & \hat{T}_i^{f+1} = T_i^f - \dfrac{C_1\left(T_i^f - T_i^{f+1}\right)}{C_3}\end{array} \tag{10}$$

where $\hat{T}_i^{f+1}$ is the corrected value of temperature.

3. THE GENERALIZATION OF FDM

The spatial differential mesh Ω_h for Ω domain is defined a set of points $X_1, X_2, \ldots, X_n$ in which the inner nodes and boundary ones are distinguished. The cartesian product of Ω_h and time mesh:

$0=t^0<t^1< \ldots <t^f<t^{f+1}< \ldots <t^F<\infty$ with step $\Delta t^f=t^{f+1}-t^f$ creates a spatially−temporal mesh $\Omega_{h,\Delta}$.

In domain Ω_h we choose a subset of nodes and in the local (connected with central node) co-ordinate system they will be identified by index 0 (central node), the next ones forming a star by 1, 2, ..., k (Fig. 3 – 2D problem, $X=\{x, y\}$ is here considered).

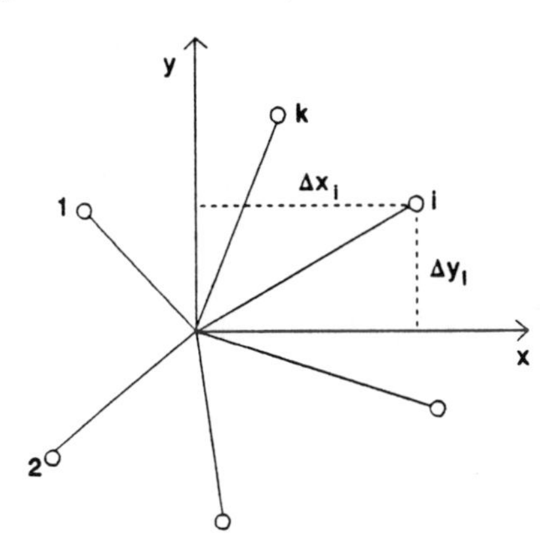

Fig. 3. k – points internal star

If the function $T\ (X,\ t)$ is developed in Taylor's power series in the neighbourhood of point X_0 then taking into account the terms of second grade we obtain

$$\begin{aligned} T_i^{f+1} &= T_0^{f+1} + \partial_x T_0^{f+1}\Delta x_i + \partial_y T_0^{f+1}\Delta y_i + \\ &+ 0.5\cdot\left(\partial_{xx} T_0^{f+1}\Delta x_i^2 + \partial_{yy} T_0^{f+1}\Delta y_i^2 + 2\partial_{xy} T_0^{f+1}\Delta x_i \Delta y_i\right) \end{aligned} \tag{11}$$

where $T_0^{f+1}=T\ (x_0,\ y_0,\ t^{f+1})$, $T_i^{f+1}=T\ (x_i,\ y_i,\ t^{f+1})$ etc.

The functional F in the form

$$F = \sum_{i=1}^{k}\left[\left(T_0^{f+1} - T_i^{f+1} + \partial_x T_0^{f+1}\Delta x_i + \ldots + \partial_{xy} T_0^{f+1}\Delta x_i \Delta y_i\right)w_i\right]^2 \tag{12}$$

is considered; at the same time w_i constitute the weights and they are selected in this way, in order to take into account the rejected terms of power series.

For instance one can assume that $w_i = 1/\sqrt{\left(\Delta x_i^2 + \Delta y_i^2\right)^3}$.

The necessary condition of functional F minimum gives the system of algebraic equations

$$\boldsymbol{A} \cdot \boldsymbol{D} = \boldsymbol{E} \cdot \boldsymbol{T} \tag{12}$$

where

$$\boldsymbol{A} = \begin{bmatrix} d_{2,0} & d_{1,1} & 0.5\,d_{3,0} & 0.5\,d_{1,2} & d_{2,1} \\ d_{1,1} & d_{0,2} & 0.5\,d_{2,1} & 0.5\,d_{0,3} & d_{1,2} \\ 0.5\,d_{3,0} & 0.5\,d_{2,1} & 0.25\,d_{4,0} & 0.25\,d_{2,2} & 0.5\,d_{3,1} \\ 0.5\,d_{1,2} & 0.5\,d_{0,3} & 0.5\,d_{2,2} & 0.25\,d_{0,4} & 0.5\,d_{1,3} \\ d_{2,1} & d_{1,2} & 0.5\,d_{3,1} & 0.5\,d_{1,3} & d_{2,2} \end{bmatrix} \tag{13}$$

and $d_{s,l} = \sum_i \Delta x_i^s \, \Delta y_i^l \, w_i^2$, $s = 0, \ldots 4$; $l = 0, \ldots, 4$.

The matrix $\boldsymbol{E}$ has a form

$$\boldsymbol{E} = \begin{bmatrix} \Delta x_1 w_1^2 & \Delta x_2 w_2^2 & \ldots & \Delta x_k w_k^2 \\ \Delta y_1 w_1^2 & \Delta y_2 w_2^2 & \ldots & \Delta y_k w_k^2 \\ 0.5\,\Delta x_1^2 w_1^2 & 0.5\,\Delta x_2^2 w_2^2 & \ldots & 0.5\,\Delta x_k^2 w_k^2 \\ 0.5\,\Delta y_1^2 w_1^2 & 0.5\,\Delta y_2^2 w_2^2 & \ldots & 0.5\,\Delta y_k^2 w_k^2 \\ \Delta x_1 \Delta y_1 w_1^2 & \Delta x_2 \Delta y_2 w_2^2 & \ldots & \Delta x_k \Delta y_k w_k^2 \end{bmatrix} \tag{14}$$

whereas

$$\begin{aligned} \boldsymbol{T} &= \left\{ T_1 - T_0 , T_2 - T_0 , \ldots, T_k - T_0 \right\}^{\mathrm{T}} \\ \boldsymbol{D} &= \left\{ \partial_x T_0^{f+1} , \partial_y T_0^{f+1} , \ldots, \partial_{xy} T_0^{f+1} \right\}^{\mathrm{T}} \end{aligned} \tag{15}$$

and one can obtain the approximation of derivatives for central node of the star:

$$\boldsymbol{D} = \boldsymbol{A}^{-1} \cdot \boldsymbol{E} \cdot \boldsymbol{T} \tag{16}$$

In the case of boundary nodes the procedure of differential operators construction is the same, but it is possible to take into account the less number of terms of the Taylor's series because

$$n \cdot \text{grad}\, T_0^{f+1} = \partial_x T_0^{f+1} \cos\alpha_0 + \partial_y T_0^{f+1} \sin\alpha_0 \tag{16}$$

The typical for FDM approach leads to linear system of equations (here implicit differential scheme is considered) and the set of temperature values for $f+1$ level of time can be found.

The systems of linear algebraic equations, used to solving the problem, have a general band form and this part of the paper concerns some remarks about construction of effective algorithm based on the Gauss elimination which used the certain properties of general band matrixes and computer data type architecture.

One of the most important aspects in effective algorithm design is the use of appropriate data structure. It has an essential influence on computing time and on making better the operating memory managing, for instance it is possible to increase the range of solved problem for fixed numerical method without using the external memory. So, many problems lead to the working out the system of equations with the band form main matrix. That system could be described in memory in array data type form. It is not efficient method because on the one hand the majority of matrix elements are equal to zero and, on the other hand computational process has a long access time to double indexed array.

Let us construct the following data structure. Each equation can be represented by a record of the form

```
eq = record
        il : integer;
        w : array [1..maxil} of ax;
        b : real;
     end;
where
ax = record
        nr : integer;
        v : real;
     end;
```

Used variables have the following interpretation: il - number of non-zero components in equation, w - record of pair (nr, v) at the same time nr - index of position in equation, v - value of element corresponding to above index, b - value of free component.

Whole system of equations can be described by a vector of records, but the most effective and natural data form for using an external memory is the file of records.Introduced above the variable **Maxil** corresponds to the supremum of non-zero elements in rows of considered matrix. For example, it is easy to notice that for typical FDM algorithm for 2D problems **Maxil**=5, whereas for GFDM this variable is not greater then maximum number of nodes creating the stars in considered spatial mesh.

In the case of exact methods using the elimination algorithm each step of computations changes the form of equations contained the eliminated variable and the number of non-zero components in these equations as a rule increases. That, for the systems of equations exists a problem how to limit range of **Maxil** - except for a natural limit (the number of all variables). It can be shown that in case of systems with band form structure the upper limit of **Maxil** corresponds to double width of band. This property was a basis of algorithm called Band Elimination Algorithm which was used for solving the problem presented in next chapter. Moreover, the BEA for $2k+1$ band width is characterized by $0(nk^2)$ computational complexity and it is a feature which should be emphasized.

4. EXAMPLE OF NUMERICAL COMPUTATIONS

The casting with complex shape (lateral section of pouring spout) made from cast carbon steel has been considered. The cast is produced in typical molding sand. The following thermophysical parameters have been assumed:

C_1 = 6150 [kJ/m^3 K]	C_2 = 56000	C_3 = 4875	C_4 = 1200
λ_1 = 0.03 [kW/mK]	λ_2 = 0.03	λ_3 = 0.03	λ_4 = 0.001
T_S= 1470 [deg C]	T_L =1505	$T_1(0)$=1550	$T_4(0)$=50

The shape of considered section is shown in Fig. 4. The mesh assumed in casting area was irregular at the same time the nodes in mold sub-domain created a rectangular mesh with variable step (the outer contour of mold was rectangular one).

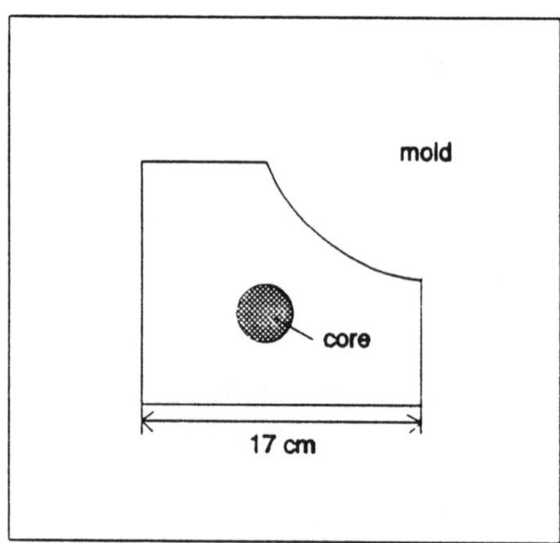

Fig. 4. Casting and mold sub−domains

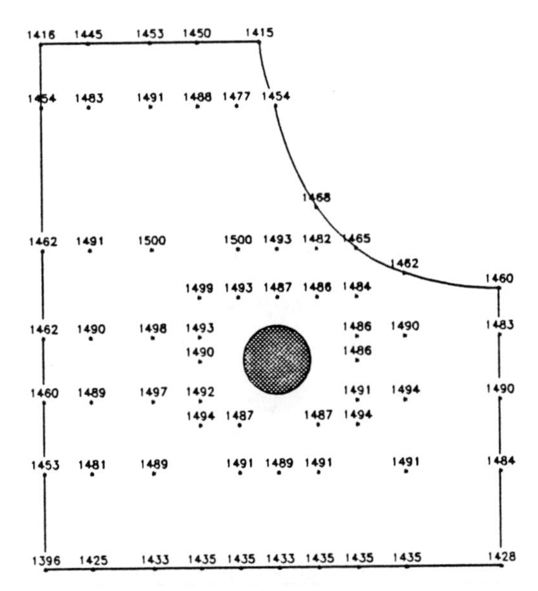

Fig. 5. Temperature field after 300 s

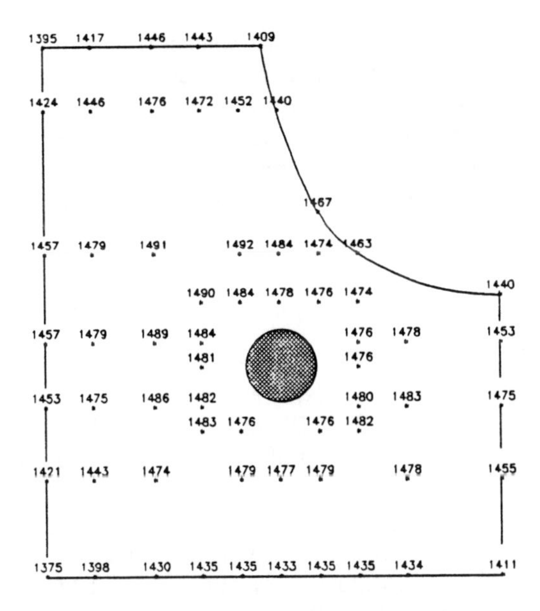

Fig. 6. Temperature field after 600s

Figures 5–7 illustrate the examples of obtained results (the boundary temperatures along the core surface and in mold domain are not marked).

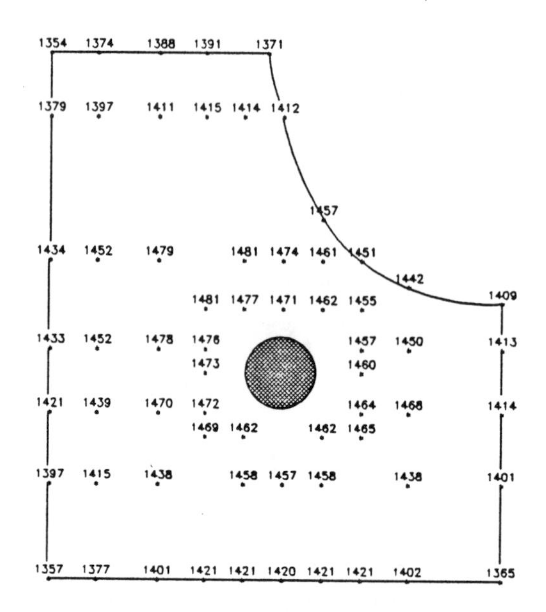

Fig. 7. Temperature field after 900s

For boundary nodes it was assumed that contact temperature at the moment $t=0^+$ results from the Schwarz's analytical solution and the temperature field for $t=t^1$ was found separately for casting and mold sub-domains (along the contact surface the 1st kind condition was given). The second and next steps of computations were initiated by determining of boundary temperature values and in the same way temperature field in casting and mould areas have been calculated.

REFERENCES

1. MOCHNACKI, B. & SUCHY, J. – Modelling and Simulation of Casting Solidification (in Polish), Ed. PWN, Warsaw, 1993.
2. MOCHNACKI, B. & BIEDRONSKA, M. – A Generalized FDM Application for Numerical Resolving of the Stefan Problem, Bull. Pol. Acad. Sc., Techn. Sc., No 7–8, pp. 459–466, 1989.
3. BALCER, M. – The Numerical Aspects of Heat Transfer Problems Modelling (in Polish), Doctor's Thesis, Gliwice, 1992.
4. MOCHNACKI, B. – Substitute Thermal Capacity of Metal Solidifying in an Interval of Temperature, Bull. Pol. Acad. Sc., Techn. Sc., No 3–4, pp. 127–143, 1984.
5. LONGA, W. – Casting Solidification (in Polish), Ed. Silesia, Katowice, 1986.
6. BUDAK, B.M. et al. – FDM with Smoothing of Parameters in Stefan

Problem Solution, Journal of Applied Mathematics and Physics, No 5, pp. 828–840, 1965.
7. MAJCHRZAK, E. – Utilization of BEM for Numerical Analysis in the Casting–Mold System, Computational Modeling of Free and Moving Boundary Problems, Vol. 2, Ed. Computational Mechanics Publications, de Gruyter, 1991.
8. SZARGUT, J. & MOCHNACKI, B. – Differential Model of Steel Ingot Solidification, Arch. of Metal., No 3, pp. 270–289, 1971.
9. HONG, C.P., UMEDA, T. & KIMURA, Y. – Numerical Models for Casting Solidification. Part I., Metall. Trans. B, Vol. 15B, pp. 91–99, 1984.

NOMENCLATURE LIST

T, $X=\{x, y\}$, t	- temperature, spatial co-ordinate and time
T_i^f	- temperature at node X_i for t^f level of time
C, λ	- specific heat and thermal conductivity
T_L, T_S, L_V	- interval of alloy solidification and latent heat
S, q_V	- fraction of solid state, capacity of heat sources
∂_t, ∂_x, ∂_{xx}	- time and spatial derivatives
Δx_i, Δy_i, w_i	- parameters of mesh.

ACKNOWLEDGMENT
This research is a part of the Project No 3 3623 91 02 sponsored by KBN.

NUMERICAL MODEL OF SOLIDIFICATION AND CRYSTALLIZATION IN THE DOMAIN OF CASTING

Ewa MAJCHRZAK[1] & Ryszard SKOCZYLAS[2]

SUMMARY

The solidification process proceeding in the volume of pure metal or alloy can be treated as a macroscopic one, but more exact approach requires to regard the microscopic aspects of considered phenomena it means the course of nucleation and nuclei growth. In particular the component of energy equation called source function should be determined on the basis of microscopic crystallization model.

In the paper a certain algorithm applying the Boundary Element Method for parabolic equation with additional term describing a capacity of heat sources is presented. The source function results from the considerations concerning the crystallization process and can be determined by a certain iterative procedure. In the final part of the paper the results of numerical computations are shown.

1. INTRODUCTION

Boundary Element Method is one of the most effective ways of boundary–initial problems solution but the problem of casting solidification and crystallization modelling on the basis of the BEM is a rather difficult one. Yet, essential advantages of the BEM suggest that the application of this method for numerical simulation of Stefan's problem should be a subject of numerous scientific investigations concerning the effective algorithms construction. At present the majority of fundamental for theory and practice results have been obtained by combination of typical BEM algorithm for linear parabolic equations with the procedures 'linearizing' a solidification problem (for example Alternating Phase Truncation Method – refs 1 and 2, Temperature Recovery Method – refs 3 and 4, and their generalizations –

---

[1] Dr Sc, Silesian Technical University, Gliwice, Krzywoustego 7

[2] Dr Sc, Academy of Mining and Metallurgy, Cracow, Reymonta 23

refs 5 and 6). It should be pointed that this approach to the solidification problem allows to analyze the course of process only in macroscopic scale. In this paper an other approach will be presented. In particular energy equation with the term called 'source function' will be introduced at the same time an 'action' of this term is determined by the course of crystallization process.

2. GOVERNING EQUATIONS

We consider the solidification and cooling processes in domain of pure metal (e.g. aluminium) which in equilibrium conditions solidifies in constant temperature T^* (solidification point). Non−steady temperature field in considered domain $D=D_1(t) \sqcup D_2(t)$ describes the following equation

$$x \in D: \quad c(T)\frac{\partial T(x,\, t)}{\partial t} = \mathrm{div}[\lambda(T)\,\mathrm{grad}\,T(x,\, t)] + q_V(x,\, t) \qquad (1)$$

where

$$c = \begin{cases} c_1, & T>T^* \\ c_2, & T<T^* \end{cases}, \quad \lambda = \begin{cases} \lambda_1, & T>T^* \\ \lambda_2, & T<T^* \end{cases} \qquad (2)$$

and c is a specific heat related to an unit of volume for molten metal $D_1(t)$ and solidified part of casting $D_2(t)$, λ − thermal conductivity of $D_1(t)$ and $D_2(t)$, respectively, q_V − source function, T, x, t − temperature, spatial co-ordinates and time.

The last component in equation (1) is equal to

$$q_V(x,\, t) = L\frac{\partial S(x,\, t)}{\partial t} = -L\frac{\partial \hat{S}(x,\, t)}{\partial t} \qquad (3)$$

where S is a solid state fraction in the neighbourhood of considered point $x \in D$ whereas $\hat{S}\,(x,\, t)=1-S\,(x,\, t)$.

The distribution of $S\,(x,\, t)$ is determined by mathematical model of crystallization process. In this paper the following description of this process taken into account (see refs 7 and 8)

− The number of nuclei is proportional to the second power of undercooling

$$N(x,\, t) = \Psi\, \Delta T^2(x,\, t) \qquad (4)$$

where Ψ is a nucleation coefficient, ΔT − undercooling below equilibrium temperature. It was assumed that the nucleation stops at maximum undercooling it means if $\Delta T\,(x,\, t+\Delta t) < \Delta T\,(x,\, t)$ then

$$N(x,\, t+\Delta t) = N(x,\, t)\,. \qquad (5)$$

– The solid phase growths (equiaxial grains) is determined by following formula

$$\frac{\mathrm{d}R(x,\, t)}{\mathrm{d}t} = \mu_1 \Delta T^2(x,\, t) + \mu_2 \Delta T^3(x,\, t) \tag{6}$$

where R is a grain radius and μ_1, μ_2 – growth coefficients.

– Relation between linear growth rate and the liquid fraction is given by the Johnson–Mehl–Avrami–Kolmogorov type equation

$$\hat{S}(x,\, t) = \exp\left\{ -\frac{4}{3}\pi N(x,\, t) \left[\int_0^t \mathrm{d}R(x,\, t) \right]^3 \right\} . \tag{7}$$

To sum up, the source function describing solidification process is of the form

$$q_V(x,\, t) = -L\, \frac{\partial}{\partial t} \exp\left\{ -\frac{4}{3}\pi\, \Psi \Delta T^2 \left[\int_0^t \left(\mu_1 \Delta T^2 + \mu_2 \Delta T^3 \right) \mathrm{d}t \right]^3 \right\} . \tag{8}$$

Taking into account the thermophysical properties of considered material (Al) it can be assumed that $c_1 = c_2 = c = \mathrm{const}$, $\lambda_1 = \lambda_2 = \lambda = \mathrm{const}$ and finally one obtains the following energy equation

$$\frac{\partial T(x,\, t)}{\partial t} = a\, \mathrm{div}[\mathrm{grad}\, T(x,\, t)] + \frac{q_V(x,\, t)}{c} . \tag{9}$$

Above equation is supplemented by boundary conditions of the form

$$x \in \Gamma : \quad \Phi[T(x,\, t),\, \boldsymbol{n} \cdot \mathrm{grad}\, T(x,\, t)] = 0 \tag{10}$$

and initial condition

$$t = 0 : \quad T(x,\, 0) = T_0(x) . \tag{11}$$

It should be pointed that presented mathematical model is very convenient as a base of the boundary element method application because one has not problems with fundamental solution determination and effective numerical algorithm construction, trough, a certain iterative procedure in order to find temporary value of source function must be introduced.

3. BOUNDARY ELEMENT METHOD FOR PARABOLIC EQUATIONS DESCRIBING TEMPERATURE FIELD OF SOURCES

Applying a weighted residual formulation to eqn (9) one obtains

$$\int_{t^0}^{t^F} \int_D \left[a\, \mathrm{div}(\mathrm{grad}\, T) - \frac{\partial T}{\partial t} - \frac{q_V}{c} \right] T^*\, \mathrm{d}D(x)\, \mathrm{d}t = 0 . \tag{12}$$

A tapering function T^* of the form

$$T^*(\xi, x, t^F, t) = \frac{1}{4\pi a(t^F - t)^{d/2}} \exp\left[\frac{-r^2}{4a(t^F - t)}\right] H(t^F - t) \quad (13)$$

is the fundamental solution of the following equation

$$a\,\mathrm{div}\left[\mathrm{grad}\,T^*(\xi, x, t^F, t)\right] + \frac{\partial T^*(\xi, x, t^F, t)}{\partial t} = -\Delta(\xi, x)\,\Delta(t^F, t) \quad (14)$$

where r is the distance from considered point x to the point ξ where concentrated heat source is applied, $[t^0, t^F]$ is the considered interval of time, $\Delta(\xi, x)$, $\Delta(t^F, t)$ are the delta functions and $H(t^F - t)$ Heaviside's function, d is the dimension of problem, at the same time a heat flux can be found

$$q^*(\xi, x, t^F, t) = -\lambda \frac{\partial T^*(\xi, x, t^F, t)}{\partial n}\,. \quad (15)$$

Using twice the Gauss Ostrogradski theorem for $\int\limits_{t^0}^{t^F}\int\limits_{D} \mathrm{div}[\mathrm{grad}\,T]\,T^*\,\mathrm{d}D\,\mathrm{d}t$

and integrating the component $\int\limits_{t^0}^{t^F}\int\limits_{D} \frac{\partial T}{\partial t} T^*\,\mathrm{d}D\,\mathrm{d}t$ by parts we obtain

$$\int\limits_{t^0}^{t^F}\int\limits_{D}\left[a\,\mathrm{div}(\mathrm{grad}\,T^*) + \frac{\partial T^*}{\partial t}\right] T\,\mathrm{d}D\,\mathrm{d}t = \int\limits_{D} T T^*\,\mathrm{d}D\Big|_{t^0}^{t^F} + \\ + \frac{1}{c}\int\limits_{t^0}^{t^F}\int\limits_{\Gamma}(q\,T^* - q^*T)\,\mathrm{d}\Gamma\,\mathrm{d}t + \frac{1}{c}\int\limits_{t^0}^{t^F}\int\limits_{D} q_V T\,\mathrm{d}D\,\mathrm{d}t\,. \quad (16)$$

Taking into account the properties of the fundamental solution (comp. ref 9) the integral equation for $\xi \in D$ is of the form

$$T(\xi, t^F) = \frac{1}{c}\int\limits_{t^0}^{t^F}\int\limits_{\Gamma}\left[q^*(\xi, x, t^F, t)T(x, t) - T^*(\xi, x, t^F, t)q(x, t)\right]\mathrm{d}\Gamma\,\mathrm{d}t + \\ - \int\limits_{t^0}^{t^F}\int\limits_{D}\frac{q_V(x, t)}{c} T^*(\xi, x, t^F, t)\,\mathrm{d}D\,\mathrm{d}t + \int\limits_{D} T^*(\xi, x, t^F, t^0)\,T(x, t^0)\,\mathrm{d}D$$

(17)

at the same time if $\xi \rightarrow \Gamma$ then the boundary integral equation has got the same form as presented above but in the place of component $T(\xi, t^F)$ appears the component $C(\xi)T(\xi, t^F)$ where $C(\xi)=0.5$ for a smooth boundary point ξ.

In this way discussed in the chapter 2 mathematical model of solidification and cooling process can be replaced by the integral equation (17) in which the given initial and boundary conditions should be taken into account and a source function $q(x, t)$ is determined on the basis of microscopic model.

In order to obtain a solution of eqn (17) a time mesh is defined

$$0 = t^0 < t^1 < \ldots < t^{f-1} < t^f < \ldots t^F < \infty \tag{18}$$

and using the 1st scheme of BEM the temperature field for the moment t^f is obtained only on the basis of known temperature values at the moment t^{f-1}.

The boundary Γ in divided into N boundary elements Γ_j, $j=1, 2, \ldots N$ and the interior D is divided into L internal cells D_l, $l=1, 2, \ldots L$. Then a numerical approximation of eqn (17) is of the form

$$\begin{aligned} C(\xi^i)T(\xi^i, t^f) = & \frac{1}{c}\sum_{j=1}^{N} \int_{\Gamma_j} \left[\int_{t^{f-1}}^{t^f} q^*(\xi^i, x^j, t^f, t^{f-1})\,\mathrm{d}t \right] T(x^j, t^f)\,\mathrm{d}\Gamma_j + \\ & - \frac{1}{c}\sum_{j=1}^{N} \int_{\Gamma_j} \left[\int_{t^{f-1}}^{t^f} T^*(\xi^i, x^j, t^f, t^{f-1})\,\mathrm{d}t \right] q(x^j, t^f)\,\mathrm{d}\Gamma_j + \\ & - \frac{1}{c}\sum_{l=1}^{L} \int_{D_l} \left[\int_{t^{f-1}}^{t^f} T^*(\xi^i, x^l, t^f, t^{f-1})\,\mathrm{d}t \right] q_V(x^l, t^f)\,\mathrm{d}D_l + \\ & + \sum_{l=1}^{L} \int_{D_l} T^*(\xi^i, x^l, t^f, t^{f-1})\,T(x, t^{f-1})\,\mathrm{d}D_l \,. \end{aligned} \tag{19}$$

Time integration is possible in an analytic way whereas the integrals with respect to Γ_j and D_l must be calculated numerically. The final system of algebraic equations can be written in the form

$$\begin{aligned} \sum_{j=1}^{N} H_{ij}\,q(x^j, t^f) = & \sum_{j=1}^{N} G_{ij}\,T(x^j, t^f) - \sum_{l=1}^{L} Z_{il}\,q_V(x^l, t^f) + \\ & + \sum_{l=1}^{L} P_{il}\,T(x^l, t^{f-1}) \,, \quad i=1, 2, \ldots, N \,. \end{aligned} \tag{20}$$

From the system (20) unknown values of 'missing' temperature or heat fluxes in the boundary nodes are found using Gauss elimination method with pivoting and next the temperature field for internal nodes can be calculated from the following formula

$$T(\xi^i, t^f) = \sum_{j=1}^{N} G_{ij} T(x^j, t^f) - \sum_{j=1}^{N} H_{ij} q(x^j, t^f) - \sum_{l=1}^{L} Z_{il} q_V(x^l, t^f) + \\ + \sum_{l=1}^{L} P_{il} T(x^l, t^{f-1}), \qquad i = N+1, N+2, \ldots, N+L. \tag{21}$$

The determined temperature distribution constitutes a pseudo−initial condition for calculations concerning the next time step.

As it was mentioned the source function is specified by an iteration process. Let us assume that during an interval of time $t^f - t^{f-1}$ the temperature field in central point of internal element D_l decreases below the solidification point T^*. In the area of considered element the undercooling $\Delta T_l^f = T^* - T_l^f$ appears and a source function is generated. The number of nuclei and the rate of solid phase growth found on the basis of (4) and (6) are introduced to eqn (7), in this way the first approximation of $q_V(x, t)$ is found – comp. Fig. 1.

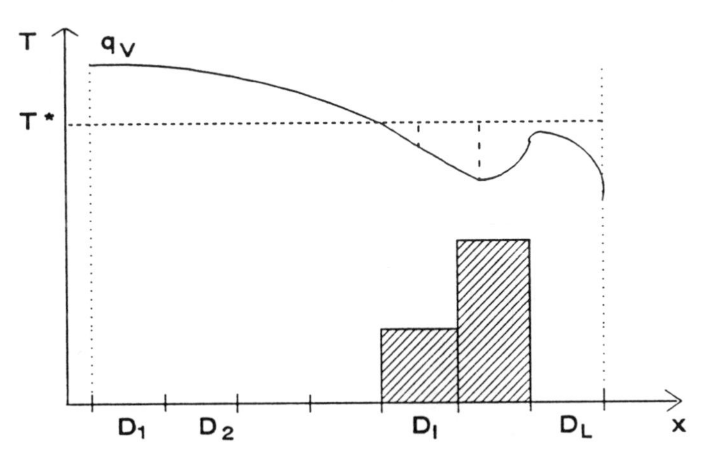

Fig. 1. Identification of source function

The transition from t^{f-1} to t^f is calculated again at the same time the distribution of q_V obtained in previous step is taken into account. In this way one can find the new values of undercooling ΔT_l^f and to correct the course of q_V in casting domain. The iterative process is stoped when the assumed output criterion is satisfied. The testing calculations show that presented above iterative algorithm can be divergent. The only way to avoid this situation is a reduction of time interval or the acceptation of source functions assigned for previous interval of time (the function keeps up with temperature field).

4. EXAMPLE OF NUMERICAL SIMULATION

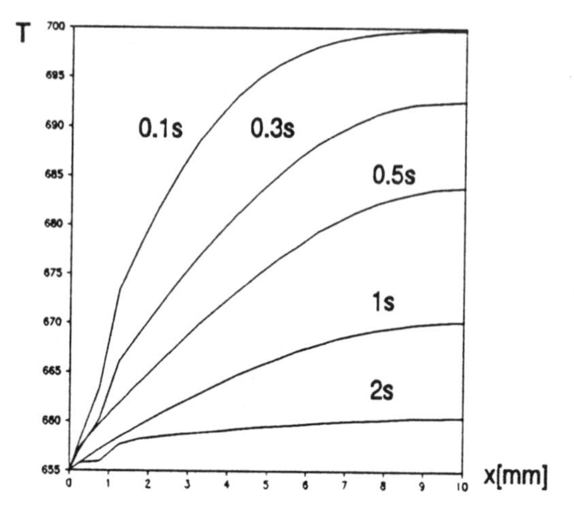

Fig. 2. Temperature field in plate domain

The plate with thickness 2cm made from aluminium solidifies in typical moulding sand. The inital temperature of molten metal is equal to T_0=700°C, solidification point: T^*=660°C, the temperature of the outer surface of casting was found on the basis of well known Schwarz's solution and the first kind boundary condition in the form $T(g)$=655°C. It should be pointed that this approximation of real boundary condition is quite satisfactory – Ref 10. Thermophysical parameters of metal are the following λ=150[W/mK], c=2880[kJ/m^{3}K], L=936000[kJ/m^3], $\mu_1=3\cdot10^{-6}$[m/sK^2], $\mu_2=0$, $\Psi=10^{10}$[1/m^{3}K^2]. 1D internal elements with thickness 0.1cm are used whereas Δt=0.004s. The part of obtained results is shown in Fig. 2, 3, 4.

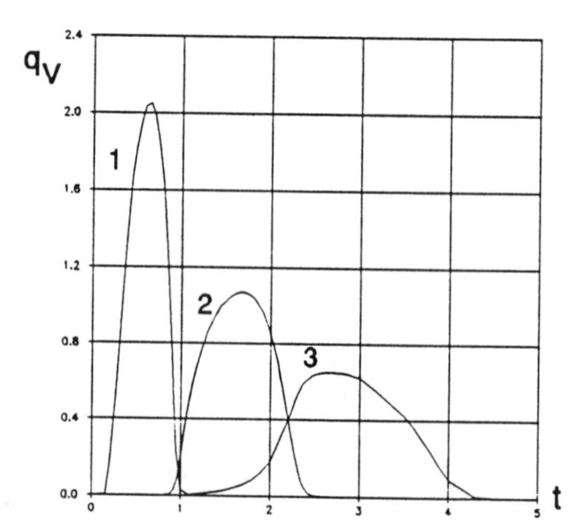

Fig. 3. Source function [kW/cm^3] at selected points
1–x=0.025cm, 2–x=0.075cm, 3–x=0.125cm

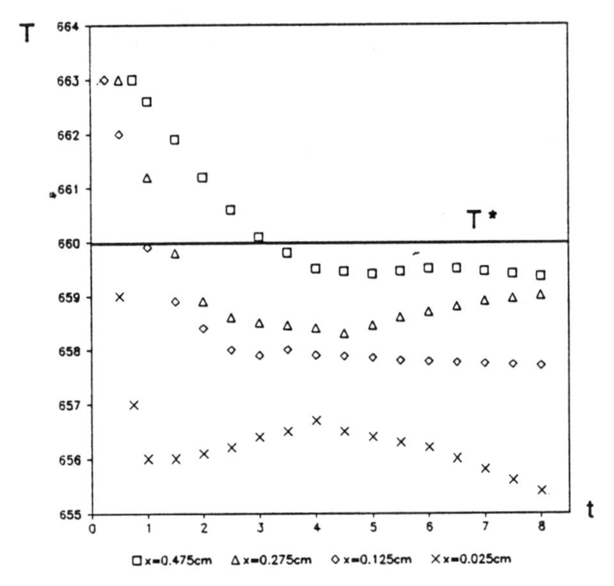

Fig. 4. Undercooling curves

REFERENCES

1. ROGERS, J., BERGER, A. & CIMENT, M. – The Alternating Phase Truncation Method for a Stefan Problem. SIAM J. Num. Anal., No 16, pp. 565–587, 1979.

2. MAJCHRZAK, E. – Numerical Simulation of Continuous Casting Solidification by Boundary Element Method. Engineering Analysis with Boundary Elements, Vol. 18, 1993.

3. HONG, C.P., UMEDA, T. & KIMURA, Y. – Numerical Models for Casting Solidification. Part I. Metall. Trans. B, Vol. 15B, pp.91–99, 1984.

4. MOCHNACKI, B. & SUCHY, J. Modelling and Simulation of Casting Solidification (in Polish), Ed. PWN, Warsaw, 1993.

5. MAJCHRZAK, E. – Utilization of BEM for Numerical Analysis of Thermal Processes in the Casting–Mould System, Computational Modelling of Free and Moving Boundary Problems, Vol. 2, Ed. Computational Mechanics Publications, de Gruyter, 1991.

6. MAJCHRZAK, E. & WAWRZYNEK, A. – Utilization of R–Functions Method in Numerical Modelling of Solidification Process, Advanced Computational Methods in Heat Transfer, Vol. 1, Ed. Computational Mechanics Publications, Elsevier, pp. 307–319, 1992.

7. THEVOZ, Ph., DESBIOLLES, J.L. & RAPPAZ, M. – Modeling of Equiaxed Microstructure Formation in Casting, Metall. Trans. A, Vo. 20A, pp. 311–321, 1989.

8. SKOCZYLAS, R. – Micro–Macroscopic Modelling of Solidification of Complex Shaped Castings Using PC, Solidification of Metals and Alloys, No 17, pp. 145–153, 1992.
9. BREBBIA, C.A., TELLES, J.C.F. & WROBEL, L.C. – Boundary Element Techniques, Ed. Springer–Verlag, 1984.
10. LONGA, W. – Casting Solidification (in Polish), Ed. Silesia, Katowice, 1986.

NOMENCLATURE LIST

T, x, t – temperature, spatial co-ordinate, time
D Γ – considered domain and its boundary
T^*, L, S – solidification point, latent heat, solid state volumetric fraction
c, λ, a – specific heat, thermal conductivity, diffusion coefficient
Φ, ΔT, μ_1, μ_2 – nucleation coefficients, undercooling, growth coefficients
q_V – source function
$<t^0, t^F>$ – considered interval of time
$T^*(\xi, x, t^F, t)$, q^* – fundamental solution and heat flux
t^{f-1}, t^f – two successive levels of time
Γ_j, $j=1, 2, \ldots N$, D_l, $l=1, 2, \ldots L$ – boundary elements and internal cells.

ACKNOWLEDGMENT
This research is a part of the Projects No 3 3623 91 02 and No 3 0139 91 01 sponsored by KBN

A Numerical Model for the Steel Thin-Slab Casting Process

A.G.Gerber and A.C.M.Sousa

Dept. of Mechanical Engineering, University of New Brunswick
Fredericton, N.B., E3B 5A3, Canada

ABSTRACT

A parametric study is presented for the Hazelett thin-slab steel casting process. The paper investigates the influence of the belt heat transfer, steel superheat, casting speed, and nozzle diameter on the solidification rate within the caster. Numerical results compare well with available literature and show that the belt heat transfer and casting speed are the most influential process parameters on the solidification rate. A fixed-grid enthalpy formulation is used allowing, with a single set of equations, the description of the liquid region, the two-phase mushy zone, and the solid region.

1. INTRODUCTION

Over the last decade steel producers from the United States, Europe and Japan have invested considerable research into the development of direct-casting processes. The interest has stemmed out of a desire to reduce production and investment costs by simplifying the overall steel-making process. The simplification results from the casting of a steel product close to or at the necessary thickness for cold rolling. The elimination of most if not all process steps between casting and cold rolling results in a much less complicated process, with energy savings and lower investment costs. Direct-casting research has focused primarily on three traditional process configurations: the twin-belt, twin-roll, and single-roll arrangements. In addition considerable research is being directed toward newer innovations such as spray forming and electromagnetic levitation casting processes. The twin-belt process produces thin slabs approximately 25 to 75 mm in thickness while the double and single roll processes are used for producing thin strip sections typically 1 - 6 mm and 20 - 500 μm respectively [1,2].

Thin slab product is typically not thin enough for direct introduction into cold rolling facilities but requires some additional hot rolling.

Of the twin-belt processes being studied the Hazelett process has received the most attention. The high casting speeds of the process, its successful history in the non-ferrous industry, low investment costs, and the ability to cast sections of different sizes has made it very popular. This paper presents a parametric study for the Hazelett caster, in which the process is examined in terms of belt heat transfer coefficient, steel superheat, belt speed and inlet diameter. Farouk et al. [3] have presented numerical results for the Hazelett caster, but a systematic analysis of the relevant process parameters was not performed.

2. PROCESS DESCRIPTION

The Hazelett process, depicted in Fig. 1a), incorporates two parallel moving belts each intensely cooled by high velocity water jets. Metal block side dams move along with the bottom belt to contain the molten charge transversely. The side dams can be adjusted to cast sections of different widths, and the belts adjusted to vary the thickness. The process can be operated at high casting speeds up to 8 m/min allowing for the possibility of direct rolling, and can be set up to cast at varying angles including vertical.

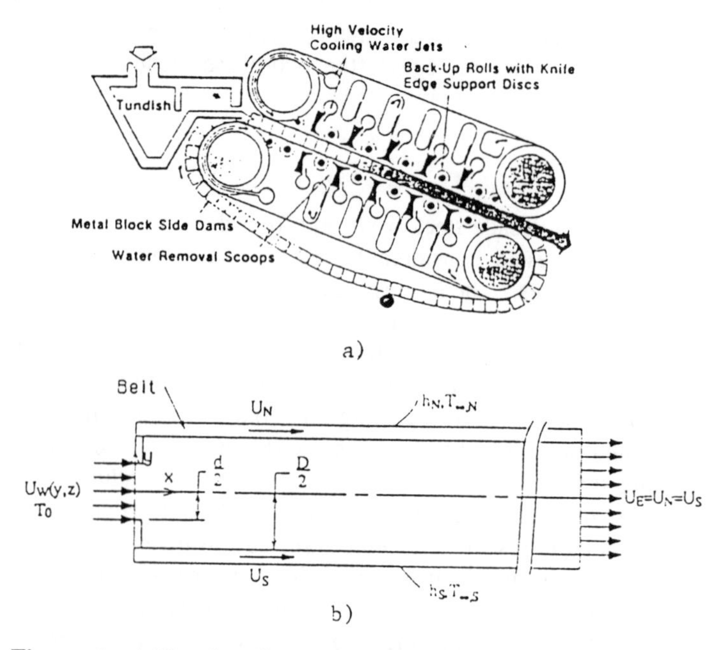

Figure 1. a) Hazelett Caster [4] ; b) schematic ($U_N=U_S=U_b$).

The modelling of the process can be simplified by assuming that heat transfer is the same on the top and bottom of the cast, allowing for symmetry about the centreline. In addition, since the length to thickness ratio for the caster is very high, fluid recirculation is limited to a few hundred millimetres from the inlet after which the fluid flow becomes highly streamlined. As a result, the outflow velocity is equal to the belt velocity, U_b, and the exiting temperature profile does not need to be specified due to the parabolic nature of the flow. Two-dimensional flow was assumed for this study as the width, 18 to 140 cm depending on the design of the caster, is typically much larger than the thickness. In addition no slip is allowed between the belt and the solidified shell. A simplified schematic of the process is shown in Fig.1b).

The velocity profile at the inlet is assumed to be uniform for this study, an assumption that has little influence on the overall solidification characteristics of the process which is the aim of this study. The magnitude of the inlet velocity is set to maintain mass continuity with the caster exit conditions. In addition the flow is assumed laminar with no buoyancy effects in all regions, an assumption justifiable considering that the entrance Reynolds numbers for this study are close to the laminar/turbulent transition, and because of the 60:1 length to width ratio of the solution domain. The long narrow mold, with its moving walls, quickly streamlines the flow isolating the turbulence to a narrow region at the entrance.

Critical to the process are the overall heat transfer coefficients, h, used to characterize heat flow along the belts. The heat transfer coefficients incorporate the thermal resistances at the cooling water and belt interface, the belt, the coating layer on the belt, and the air gap between the coating and solidified shell. The thermal resistance of the air gap is the single most important factor in determining the overall heat transfer coefficient. The cooling water for this study is assumed to be at 300 K.

3. PROBLEM FORMULATION

3.1 Governing Equations

In studying solidification problems the physical model must be comprehensive enough to fully describe the liquid region, the solid-liquid mushy zone, and the solidified material. To meet this requirement the paper uses a fixed-grid approach proposed in [6] to solve the governing conservation equations for mass, momentum, and energy. The governing equations used in the present study, in tensorial notation, are as follows:

$$\frac{\partial}{\partial x_j}(\rho u_j)=0, \qquad j=1,2. \tag{1}$$

$$\frac{\partial}{\partial x_j}(\rho u_j u_i)=\frac{\partial}{\partial x_j}(\mu\frac{\partial u_i}{\partial x_j})+\frac{\partial}{\partial x_j}(\mu\frac{\partial u_j}{\partial x_i})-\frac{\partial P}{\partial x_i}, \qquad i=1,2;\ j=1,2. \tag{2}$$

$$\frac{\partial}{\partial x_j}(\rho u_j i)=\frac{\partial}{\partial x_j}(\frac{k}{c_p}\frac{\partial i}{\partial x_j})-\frac{\partial}{\partial x_j}(\rho u_j \Delta H); \qquad j=1,2. \tag{3}$$

Equation 3 is based on a total enthalpy formulation where the total enthalpy, H, is defined as the sum of the sensible heat, $i=c_pT$ and the latent heat ΔH, or simply $H=c_pT + \Delta H$. Such a formulation allows for the proper accounting of latent heat transport and diffusion in the two-phase region.

There are many factors that can influence the rate at which latent heat is released, such as temperature, cooling rate, nucleation rate, and for alloys, solute distribution. In general however, the problem is simplified by setting the latent heat release as a function of temperature (or enthalpy) alone. For a pure metal the solidification event is an isothermal one while for alloys the event occurs over a distinct temperature range. In this study a linear phase change was implemented so that with $\Delta H=f(i)$ yields:

$$\begin{array}{ll} \Delta H=0 & for\ \ i\le c_pT_s \\ 0\le\Delta H\le L & for\ \ c_pT_s\le i\le c_pT_l \\ \Delta H=L & for\ \ i\ge c_pT_l \end{array} \tag{4}$$

where L is the total latent heat of fusion, and T_s and T_l are the solidus and liquidus temperatures respectively. The functional relationship between ΔH and i for a linear phase change can be stated as:

$$\Delta H=\frac{L}{2}+L\frac{i-\frac{1}{2}c_p(T_l+T_s)}{c_p(T_l-T_s)} \tag{5}$$

Important to the solution of the governing equations in this problem is the coupling of the momentum and energy equations through viscosity. In solid-liquid phase change problems using a fixed-grid the presence of a solid matrix and its influence on the fluid flow must be accounted for either through the viscosity, or, through an appropriate source term in the momentum equations. In this paper the presence of the solid matrix is simulated by gradually increasing the viscosity over the mushy region until bulk fluid flow stops, at which point the viscosity is set to a large value.

Bulk fluid flow in this paper was assumed to end at 0.67 solid fraction as suggested in [3], while an exponential relation is used to approximate the viscosity variation over the mushy region [7], namely:

$$\mu = \mu_\ell e^{Bs} \tag{6}$$

In this equation μ_l is the viscosity in the fully liquid region, B is an arbitrary constant set to 5, and s, the solid fraction, is defined with the equation $s=1-\Delta H/L$. In order to computationally handle the large variations in viscosity a harmonic mean of the diffusion coefficients is used at the grid interfaces [5].

3.2 Numerical Formulation

The formulation of the discretized equations was based on a control-volume approach [5]. In this approach, when evaluating the interface energy fluxes, the latent heat terms must be isolated and added to the discretized source term. Since the latent heat is advected from one control-volume to the next an upwind differencing scheme is used to approximate the interface value. The solution of the system of discretized equations for momentum and energy is achieved using the SIMPLER [5] method, with the updating of the velocity field carried out by employing the SIMPLEC procedure [8].

A common problem in obtaining convergence in fixed-grid solidification problems is the oscillation of the latent heat from 0 to L, especially when the phase-change region is small. To avoid this the current study uses an iterative approach suggested in [9] for updating the latent heat field. The approach computes the latent heat for the next iteration, ΔH^{k+1}, using the formula $\Delta H^{k+1}=\Delta H^k+i^k-f^{-1}(\Delta H^k)$, subject to the limits of $0\leq\Delta H^{k+1}\leq L$ at each node. The function $f^{-1}(\Delta H^k)$ is found by rearranging Eq.(5) to isolate the enthalpy term i. The latent heat field is swept once using this approach following a reasonable convergence of the enthalpy field.

4. PROGRAM VERIFICATION

To determine the consistency of the computer code a grid dependency test was performed for grids 16x80, 16x100, 26x80 and 26x100 in the x and y - directions respectively. The results show for the solidification length that differences less than 3% are found between the coarsest and finest mesh sizes. Where differences are noticeable it is a result of the reduction in grid density in the y-direction which lessens the ability to resolve the large temperature gradients (and thus the solidification front) in that direction. Due to the length of the solution domain (1.5 m long for a width of 25 mm) the grid spacing must be expanded exponentially in the x-direction according to the relation $\Delta x = ae^{b(i-1)}$, where for the 26x100 grid density 'a' is set to 0.001 mm and 'b' to 0.043. The integer 'i' is the array index in the x-direction. The mesh is refined at the entrance where the highest solidification rate occurs and where recirculation takes place due to the contracted nozzle inlet. As the flow becomes more streamlined the grid spacing is expanded exponentially up to the caster exit, while in the y-direction a uniform grid density is retained for the entire length.

The present predictions for specific cases were compared against published experimental and numerical results on the solidification and heat transfer phenomena in a Hazelett caster. As seen in Fig.2 these predictions are in very good agreement with the solidification rate obtained experimentally in [10]. From this test it is noticeable that the heat transfer coefficient is much higher during the first 160 mm (2 seconds of mold residence time) to take into account the high temperature gradients in the molten metal close to the moving belt. As the solid shell develops the increased resistance of the solid shell and air gap (formed as the solid contracts) reduces the heat transfer coefficient. In Fig.3 the program results are in excellent agreement with the numerical results presented in [3] for a value of $h = 1395$ W/m^2.

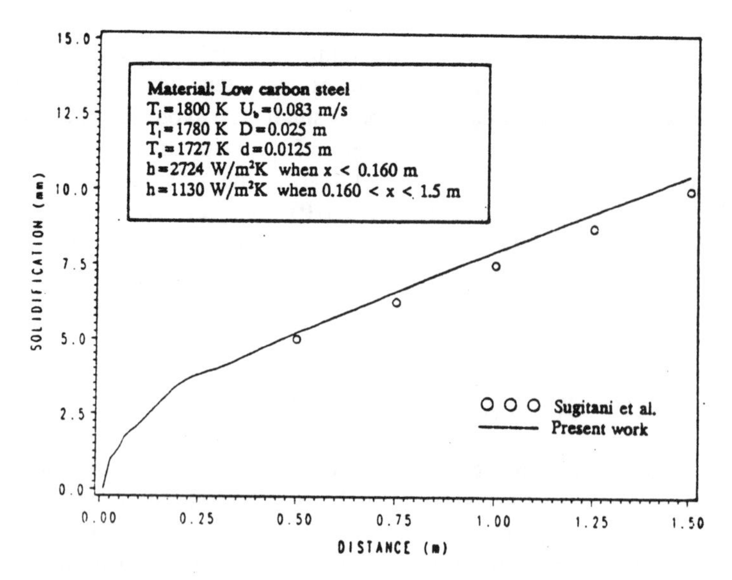

Figure 2. Comparison with experimental results in [10]

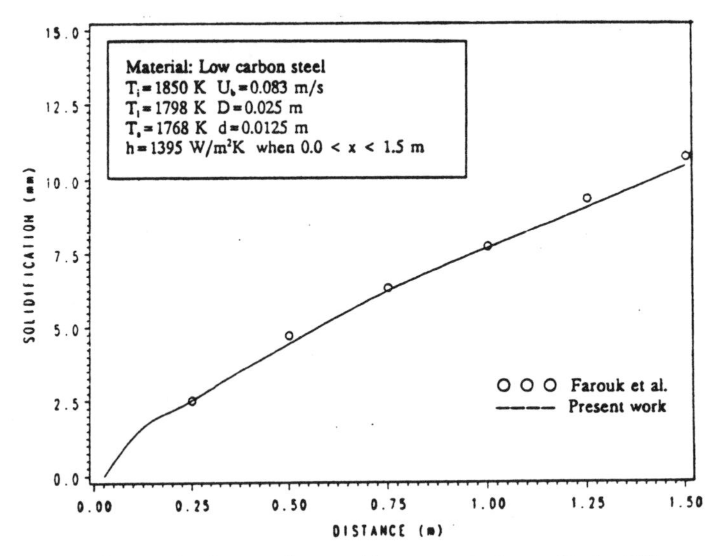

Figure 3. Comparison with numerical results in [3]

5. PROGRAM RESULTS

The program was run to test the influence of various process parameters on the operation of the Hazelett caster. The parameters tested were the overall heat transfer coefficient, the belt speed, molten metal superheat, and the influence of the nozzle diameter. The caster width remained constant at 25 mm for all tests since the width does not significantly influence the solidification rate although it is important in obtaining higher production capacities. The material studied was low carbon steel with a solidius temperature of $T_s = 1780$ K and a liquidus temperature of $T_l = 1727$ K.

Figures 4 through 6 present results that are typical of all the simulations conducted, and are shown to provide a qualitative description of the process not readily observed in the parametric results to be described later. As can be seen in Fig.4 the temperature remains quite high in the casting core, while significant cooling occurs in the solidified shell. The low thermal diffusivity of steel inhibits heat removal so that full solidification is not reached, for an $h = 1395$ W/m^{2}K, until the end of the solution domain. The streamlines for the first 100 mm past the inlet (Fig.5) show the streamlining of the flow due to the viscous influence of the moving walls, which, combined with the increasing viscosity in the mushy zone, isolate the recirculation to a narrow region near the inlet. Fig.6 tracks the growth of the two-phase mushy zone over the entire caster length with bulk fluid flow persisting up to a solid fraction of 0.67 as noted previously.

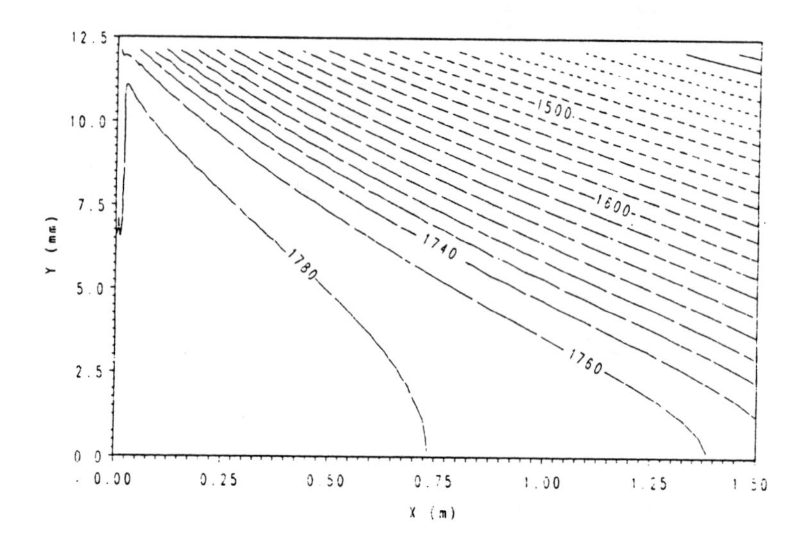

Figure 4. Temperature field over the caster length.
(h=1500 W/m^2K, U_b=0.08 m/s, T_i=1800 K, D=0.0125 m, d=0.00625 m)

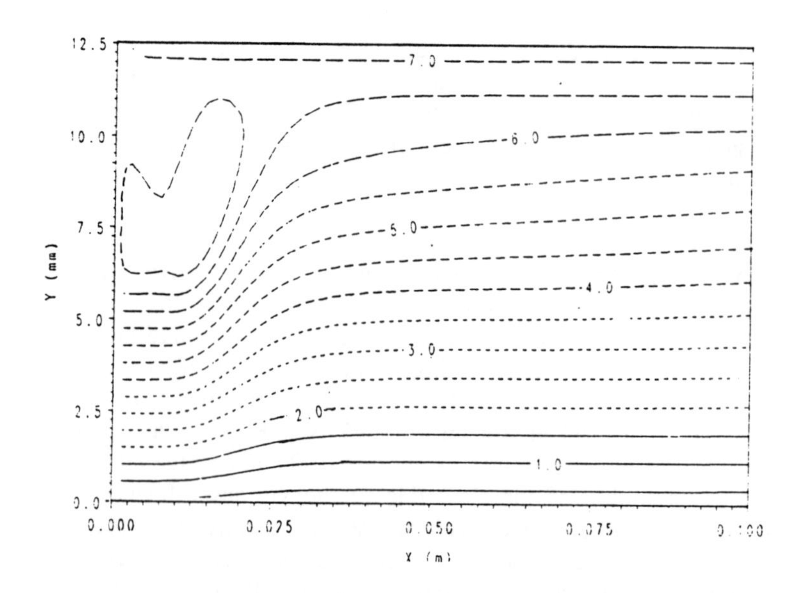

Figure 5. Flow streamlines at caster inlet

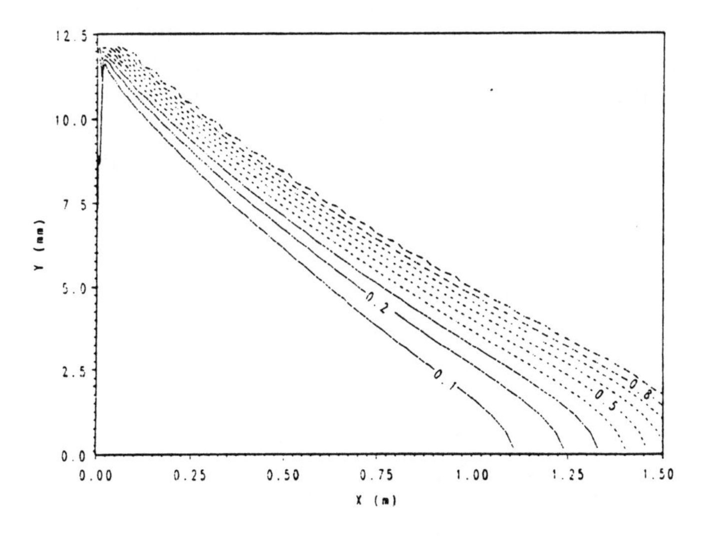

Figure 6. Growth of the mushy zone over the caster length.

5.1 Parametric Study

Table 1. Experimental heat transfer coefficients for the Hazelett caster

Author	h(W/m²K)
Itoyama [11]	1395
Spitzer [12]	1200*
Farouk [3]	1131
Sugitani [10]	975*
Whitmore [13]	581
*After 2 seconds in the mold	

(a) Belt Heat Transfer: In testing the influence of the belt heat transfer coefficient the tests were conducted over a range of values broad enough to include those values typically found experimentally. In Table 1 various experimentally determined heat transfer values are tabulated for Hazelett-type processes. The numerical predictions shown in Fig.7 highlight the importance of accurately determining the belt heat transfer coefficient. A significant difference occurs in the rate of solidification between 500 W/m^2K and 1500 W/m^2K which, considering the range of the experimental values reported, could lead to significant overestimation or underestimation of the shell thickness. The caster load is fully solidified at a shell thickness of 12.5 mm as symmetry about the centreline is assumed.

(b) Caster Belt Speed: The belt speed also has a significant effect on the overall solidification rate in the caster. Slower belt velocities increase the residence time of the load allowing for additional cooling, but also slower production rates. Fig.8 shows the solidification rate at various speeds with slower solidification rates at higher speeds and solidification occurring very rapidly at the lower belt speeds. The need for the support provided by a moving mold is seen at the higher belt speeds where a significant portion of the cast has not yet solidified.

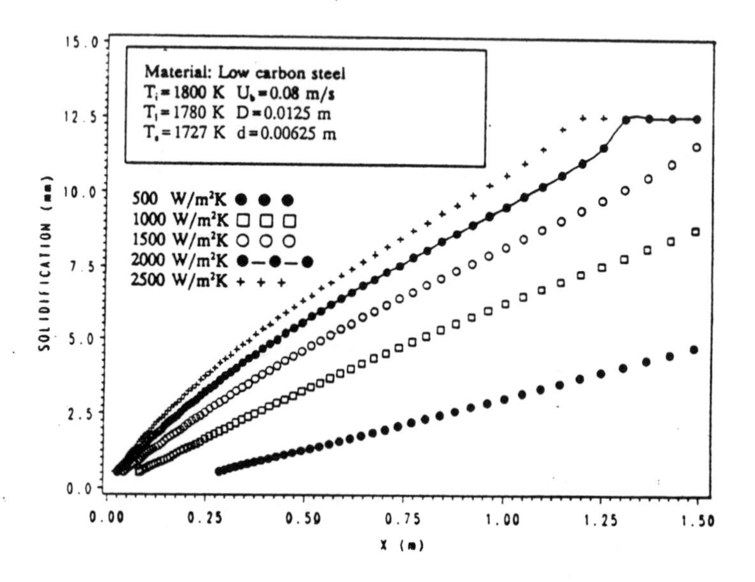

Figure 7. Effect of belt heat transfer, h, on solidification rate.

(c) Steel Superheat: The influence of the superheat of the molten metal at the entrance into the caster was also analyzed and results shown in Fig.9. As expected higher superheats resulted in reduction of solidification rates. The additional sensible heat that must be removed delays the commencement of solidification, while the amount of latent heat to be removed remains the same. The influence of this parameter is however, less important than the belt speed and heat transfer coefficient.

(d) Nozzle Diameter: The influence of the nozzle diameter upon solidification was tested for values of 0.0125 m, 0.00625 m and 0.003125 m, respectively. The present findings indicate that for the range considered this influence is negligible (maximum departure of solidification for diameters 0.0125 m and 0.003125 m, respectively, at outflow, is less than 1%). Since the length of the cooling surface is much larger than the entrance width the entrance conditions have little effect on the overall solidification rates. The entrance conditions, however, may have a significant effect on the microstructure of the material.

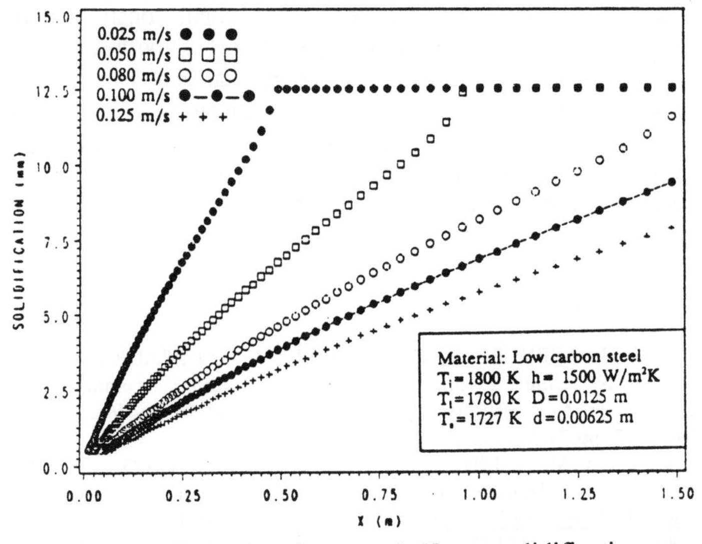

Figure 8. Effect of casting speed, U_b, on solidification rate.

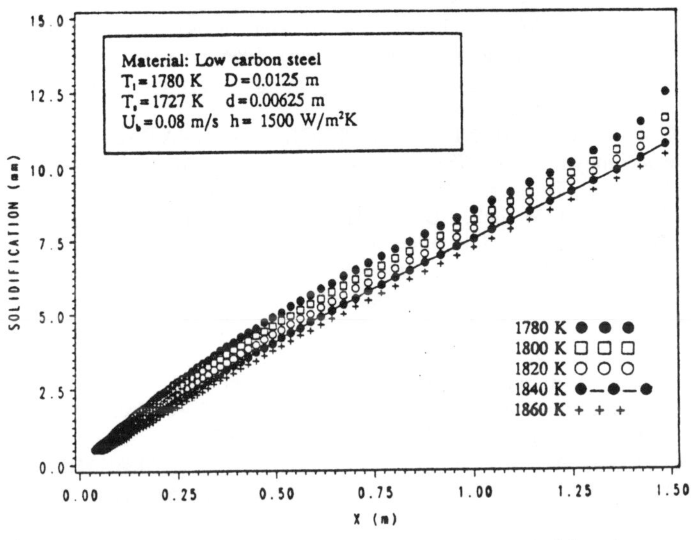

Figure 9. Effect of steel superheat, T_i, on solidification rate.

6. CONCLUSIONS

From this study it can be concluded that the solidification rate for the cast is influenced significantly by the belt overall heat transfer coefficient, and the casting speed. This finding has some practical implications regarding control of the process. Under actual conditions, the overall heat transfer coefficient can be varied only for a very limited range.

leaving the casting speed, which is to a great extent constrained by the production rates, as the only controlling parameter. Superheat and nozzle diameter, as the study points out for the range analyzed, have little or no effect upon the solidification rate. Entrance effects such as turbulence, recirculation, and nozzle diameter are limited to a narrow region near the entrance and therefore have little influence on the overall solidification rate in the caster. Their influence upon product microstructure may be of importance.

7. REFERENCES

[1] MARR, H., SPENCELEY, G., LUDLOW, V., and PEACE J. - Horizontal Continuous Casting of Thin Slabs at the British Steel Corporation - Teesside Laboratories, Near Net Shape Casting, Iron and Steel Society Inc., Warrendale, PA, 1987, pp.21-30.

[2] SOUSA, A.C.M., ŠELIH, J., GERBER, A.G., and LENARD, J.G. - Heat and Fluid Flow Simulation of the Melt-Drag Single-Roll Strip Casting Process, J. of Materials Processing Technology, Vol.34, pp.473-480, 1992.

[3] FAROUK, B., APELIAN, D., and KIM, Y.G. - A Numerical and Experimental Study of the Solidification Rate in a Twin-Belt Caster, Metallurgical Transactions, Vol.23B, pp.477-492, 1992.

[4] HAZELETT, R.W. - The Present Status of Continuous Casting Between Moving Flexible Belts, Iron and Steel Engineer, June, 1966, pp.105-110.

[5] PATANKAR, S.V. - Numerical Heat Transfer and Fluid Flow, Hemisphere Publishing Corporation, New York, 1980.

[6] PRAKASH, C., SAMONDS, M. and SINGHAL, A.K. - A Fixed-Grid Numerical Methodology for Phase Change Problems Involving a Moving Heat Source, Int. J. Heat Mass Transfer, Vol.30, No.12, pp.2690-2694, 1987.

[7] SALCUDEAN, M. and ABDULLAH, Z. - On the Numerical Modelling of Heat Transfer During Solidification Processes, Int. J. Numerical Methods in Engineering, Vol.25, pp.445-473, 1988.

[8] VAN DOORMAAL, J.P., and RAITHBY, G.D. - Enhancements of the Simple Method for Predicting Incompressible Fluid Flows, Numerical Heat Transfer, Vol.7, pp.147-163, 1984.

[9] VOLLER, V.R. and PRAKASH, C. - A Fixed Grid Numerical Modelling Methodology for Convection-Diffusion Mush Region Phase-Change Problems, Int. J. Heat Mass Transfer, Vol.30, pp.1709-1719, 1987.

[10] SUGITANI, Y., NAKAMURA, M., SHIRAI, Y., OKAZAKI, T., and YOSHIHARA, M. - Solidification and Heat Transfer Phenomena in the Twin Belt Caster, Trans. ISIJ, Vol.26, pp.153-154, 1986.

[11] ITAYAMA, S., NAKATO, H., NOZAKI, T. and HABU, Y. - Development and Solidification Characteristics of Horizontal Thin-Slab Caster, Tetsu-to-Hagane, Vol.71, pp.S272, 1985.

[12] SPITZER, K.H. - Investigation of Heat Transfer Between Metal and a Water-Cooled Belt Using a Least Square Method, Int. J. Heat Mass Transfer, Vol.34, No.8, pp.1969-1974, 1991.

[13] WHITMORE, B.C. and HLINKA, J.W. - Continuous Casting of Low-Carbon Steel Slabs by the Hazelett Strip-Casting Process, J. Metals, August, 1968, pp.68-73.

Potential Applications of Intelligent Preprocessing in the Numerical Simulation of Castings

Rajesh S. Ransing, Yao Zheng and Roland W. Lewis

Institute for Numerical Methods in Engineering
University College of Swansea, University of Wales, Swansea, UK

Summary

The areas of integration between the numerical simulation of castings and artificial intelligence (AI) techniques are discussed at the pre-processing stage. The major areas of possible integration which are identified are firstly, the interfacial boundary conditions and secondly, in the determination of density points in the mesh generation. It has been observed experimentally that for pure aluminium castings, depending upon the various process, material and geometrical parameters, the interfacial heat transfer coefficient can vary in the range of 2000 - 16000 W/m^2K. A scheme based on heuristics is discussed for the quantitative estimation of the interfacial heat transfer coefficient values with the simultaneous consideration of various influencing parameters. Also, the scope of AI techniques in the mesh generation for thermal problems has been reviewed.

1 Introduction

User friendly preprocessing for FEM analysis has always been of major concern to numerical analysts. In the thermal analysis of the casting process, difficulties in the mesh generation for highly intricate casting geometries are experienced as well as allowing the casting designer the flexibility to interact with the software in the foundrymans language. This paper addresses the possible preprocessing areas where artificial intelligence can contribute constructively. Over the past few years, efforts have been reported in the literature regarding the applications of artificial intelligence techniques to finite element methods in general, which has aroused considerable interest in the AI community. [1]. Various promising areas have been explored at the preprocessing stage in this direction in order to alleviate the efforts required by a casting designer in particular. The objective of using AI techniques at the preprocessing stage is to enable a foundryman to use FEM software in a more realistic and user friendly way and also to minimising the learning period.

The success of any finite difference or finite element analysis, also depends greatly on the precise estimation of the pertaining physical conditions. A typical foundrymen with years of experience is hesitant to acquire a technical understanding of the details involved in heat transfer problems. However, this knowledge is essential for accurate modelling purposes. The solution to this problem has not as yet, been clearly addressed in the literature. The accuracy of the simulation of the solidification process of metal castings inside metal moulds mainly depends on the rate of heat removal from the metal to the mould. The air gap, which usually develops at the interface between the solidifying metal and the surrounding mould or chill, influences magnitude of the heat transfer coefficient at the interface which is a function of casting geometry, mould and metal material, chills used and also the time elapsed. [3] More importantly the determination of this interfacial heat transfer coefficient is purely based on experimental data. Current data is available only for a few typical cases, which need to be extrapolated for more general geometries. Our efforts are directed towards enabling the foundrymen to use this current technical information in his realistic simulation of castings based on heuristic techniques. A scheme for integrating this knowledge with the finite element program is also illustrated.

In the next section a brief understanding of AI and expert systems as applicable to this domain is given. Then, its application to predicting boundary conditions and appropriate mesh generation is discussed.

2 Artificial Intelligence and Finite Element Methods

2.1 Conceptual Background

In simple words, artificial intelligence is an area of computer science which enables computers to mimic human thinking and reasoning processes using a collection of different programming techniques and programming languages. Many other definitions are reported in the AI literature but similar ideas are conveyed in each one of them. Expert systems - a branch of AI - uses these concepts to enable computers to function in decision support roles as advisors, personifying human expert decision-making capabilities. Earlier expert systems have mostly relied on rule based architecture, representing knowledge in the IF - THEN format. The decision is given by inferencing these IF - THEN statements in a logical manner. Recently, other representation schemes such as graphs, frames, object oriented paradigm have also been reported[4]. A neural network - another branch of AI - is an assembly of a large number of highly interconnected simple processing units. The connections between two neurons have a strength referred to as a weight. The knowledge is stored in the form of these weights. Neural networks are best applied for mapping problems which are tolerant of a high error rate and have lots of example date available, but to which hard and fast rules can not be easily applied.

Identification of the appropriate areas for the application of AI techniques is very important for its success. Normally, small domains of knowledge where enough expertise is essential for problem solving, are considered for AI applications.

2.2 Review of Previous Work

With the commercial success of AI related programs in the late eighties, many applications, including finite element methods, were explored for the use of AI tools such as expert systems. A sudden surge of papers relating finite element methods with expert systems appeared in this period. Initial applications of such expert systems were in the field of structural engineering. [5–7] The objective during this period was to demonstrate the applicability of AI systems for finite element analysis. Most of these systems advised the user on node selection, selection of elements, mesh generation *etc.*, for a particular type of loading. Breitkopf *et al.* [8] developed a knowledge based system for algorithm selection in the case of nonlinear finite element structural analysis. Yeh *et al.*[9] reported the use of this technology for debugging FEM input data.

Apart from applications in structural analysis, Carey and Patton[1] have explored the various possible areas integrating expert systems with finite element methods. The nature of an expert system in finite element analysis tends more towards a consultation mode rather than the problem solving. Such a consultancy mode can be useful in taking decisions such as choice of elements, mesh refinement, size of incremental time steps, decisions on whether to remesh the domain and also whether to repeat a calculation at a given level with a different increment *etc.* during the FE analysis. Analysis and interpretation of the results stemming from the outcome of the calculation can also be handled by an expert system. No work has yet been reported in this area of analysis and interpretation. The use of expert system for determining appropriate time steps in dynamic finite element programs has been illustrated by Ramirez[20]

Interestingly, during this period thermal problems remained un-investigated until very recently. Expert system aid in generating the initial meshes for thermal problems has been discussed by Kang *et al.*[2]

In this paper, attention is focussed more on thermal analysis problems in the numerical simulations of castings, particularly at the interface between seperate domains.

3 Preprocessing in Casting Simulation

The major tasks in the preprocessing preparation of casting problems are of inputting the geometry, generating a mesh and inputting the boundary conditions and related material data base. During this exercise, human expertise is required only in the quantitative estimation of these values. Boundary conditions are considered along the boundary elements as well as at the interface elements. The scope for estimating the boundary conditions based on heuristics is discussed in the next section.

3.1 Boundary Conditions

For thermal as well as casting analyses, the problem of specifying one or more boundary conditions involves deciding between fixed temperature values or insulated boundaries or constant flux temperature conditions along with their relative magnitudes. For a casting process, expertise is required in arriving at these values, but once defined initially, then for further cases the estimation is straightforward which may not necessarily require experts guidance.

In the simulation of aluminium die castings, previously interfacial elements, utilised a value in the range of 2500 to 5000 W/m^2K before the formation of an air gap and a constant value of 400 W/m^2K after the formation the gap [10]. However, with the current results available from experimental work on aluminium die castings, the heat transfer coefficients can vary from 2500 to 16000 W/m^2K depending upon various parameters. Moreover, these parameters differ from casting to casting, hence for each type of casting expert guidance is necessary for the precise estimations of these values. This analysis is discussed in more detail in the next section, from an expert systems applications point of view.

3.1.1 Interfacial Heat Transfer

In permanent mould or die castings, the resistance to heat flow at the metal - mould interface has a significant influence on the solidification rate of the casting. Until recently, in finite element simulation of castings, an implicit assumption of a perfect contact between metal and mould was considered for the problem of heat transfer during filling. As these heat transfer coefficients are input value to the finite element program, and are usually extremely difficult to assess correctly. The accuracy of the input data is also equally important in the finite element analysis in order to precisely estimate the output parameters. In reality most foundrymen are unaware of the appropriate values of such heat transfer coefficients and tend to use values available in handbooks, which can be too general for a given application. Where as, in this paper with the primary study of the past literature, it has been shown that even for a

given material and its surface finish, numerous other factors such as casting geometry, design and process parameters can also greatly influence its value. This problem can be subdivided into two sub problems as:
1. Arriving at an appropriate value of the interfacial heat transfer coefficient which is a function of temperature for each element.
2. Integrating a suitable and efficient interfacial model with finite element analysis.

The second problem of developing an interfacial model with finite element analysis has been discussed in the past. Samonds[11] suggested two interface models based on the coincident node technique and the other on a thin element technique. A model based on the coincident node technique was presented by Lewis *et al.*[12]. Recently, Usmani[13] proposed an interface element of zero thickness based on a Newtonian heat transfer model which is also simple to implement. This type of interface element was found to be more convenient for integrating with user defined interfacial heat transfer coefficients.

In the next section, the influence of various parameters on the heat transfer coefficient is depicted graphically.

3.1.2 Factors Governing Interfacial Heat Transfer Coefficients

The judgements and the inferences presented in this paper are based on the number of experimental results published in related technical journals. There has been a great interest on the metallurgical side in determining experimentally the metal - mould interfacial heat transfer coefficient. Often, the experiments are done to study the influence of only one or two parameters. Wheras, for practical applications, influence of all the parameters need simultaneous consideration. In this section the experimental results published over the last decade are compared together and also analysed. The use of heuristic methods have been explored for the quantitative estimation of heat transfer coefficient values under various conditions.

In the metallurgical literature, the variation of the interfacial heat transfer coefficients has been studied and emphasised with respect to time. As the solidification time is very much dependent on casting geometry and experimental conditions, for the generalised analysis its variation with respect to surface temperature has been investigated. This variation with respect to temperature can be easily implemented at the element level as the value of temperature at the previous instance is readily available.

Ho and Phelke[14] have determined the interfacial heat transfer coefficient values where a copper chill was placed on the top of a cylindrical pure aluminium casting and also at the bottom. The variation of interfacial heat transfer coefficient and surface temperature with respect to time was investigated. The graphs showing this variation were digitised with the help of a digitiser

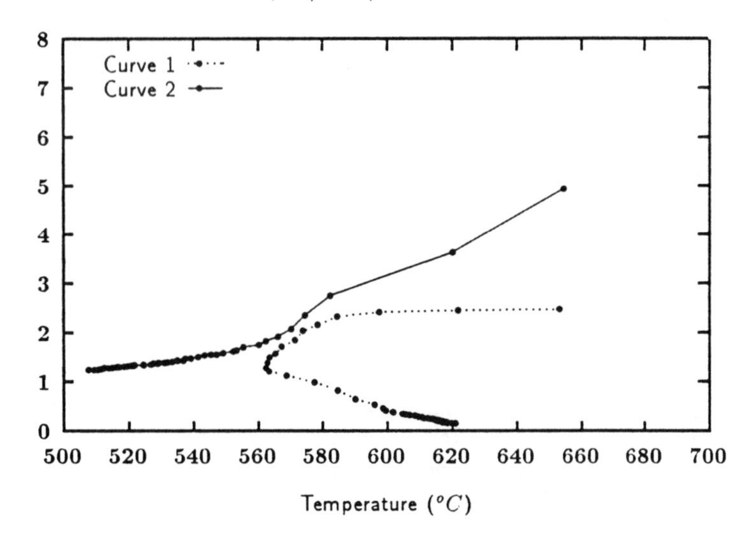

Figure 1: Heat transfer coefficient *vs* surface temperature for pure aluminium cylindrical casting with copper chill at the top (Curve 1) and at the bottom (Curve 2)

and the variation of interfacial heat transfer coefficient with respect to surface temperature was investigated. All the graphs which are based on the past experimental results in this paper are drawn by digitising the original graphs and reploting them as interfacial heat transfer coefficient Vs surface temperature. Fig. 1 shows the respective variation when the chill is on the top surface and also when the chill is at the bottom surface. Tadayon and Lewis[15] have also proposed an interface model considering the effect of a metallostatic head alone. The interfacial heat transfer coefficient was assumed to very linearly with the metallostatic head.

The formation of an air gap and the heat transfer mechanism through a gap was studied by Nishida and Engler[16] for cylindrical and flat castings. Experimental results for thermal resistance *vs* time and the surface temperature *vs* time were given for a pure aluminium. The variation of the interfacial heat transfer coefficient with respect to temperature was again derived from these charts by digitising them and is analysed herein (Fig. 2). It can be noted that for flat shapes the interfacial heat transfer coefficient values are greater than those for cylindrical shapes. This trend, can also be extended logically, to consider the cores. The thermal expansion of the core interferes with the thermal contraction of the casting, leading to zero gap width development. This behaviour will always tend to result in higher interfacial heat transfer coefficient values as compared to other shapes. The nature of gap width development for various shapes has been shown by Mahallawy and Assar[19]. If

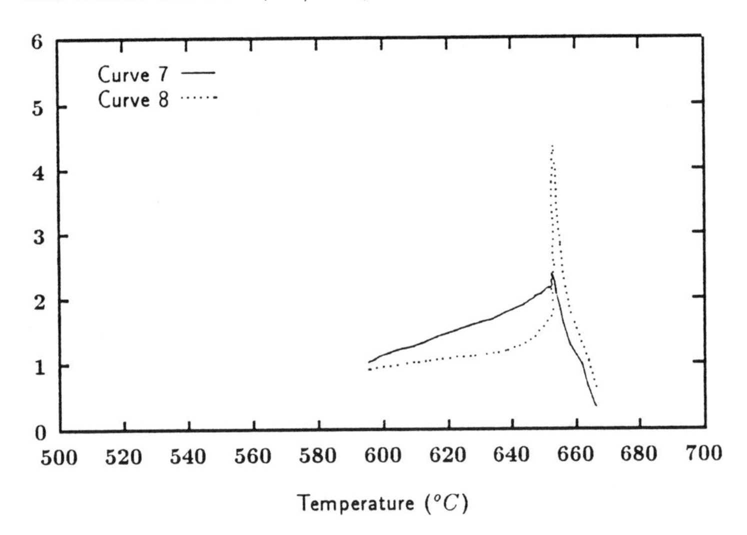

Figure 2: Heat transfer coefficient *vs* surface temperature for pure aluminium casting without a chill for cylindrical shape (Curve 7) and flat shape (Curve 8)

the convexity of the shape is defined as the reciprocal of the radius of curvature, it may be possible to capture these variables with the help of heuristics. It is interesting to note that as the convexity reduces from a positive value to zero and then to a negative value, there is an increase in the magnitude of the interfacial heat transfer coefficient profile with respect to temperature.

Further experiments were reported by Mahallawy *et al.*[17] indicating the influence of superheat on the interfacial heat transfer coefficient variation. Again, plots of the interfacial heat transfer coefficient with respect to surface temperature are regenerated here with the help of a digitiser (Fig. 3) to highlight the influence of superheat. It can be observed that as the superheat of the molten metal increases, there is a substantial increase in the magnitude of the interfacial heat transfer coefficient for all values of surface temperatures. The type of chill, or the length of the casting also influences the interfacial heat transfer coefficient variation. Fig. 4 is reproduced from Taha *et al.*[18] for sand and copper chills for cylindrical shaped pure aluminium castings. It is obvious that the cooling conditions have a significant influence on the heat transfer coefficient variation. However, with the aid of experimental results, the order of influence can now be determined. Further, to add to the complexity of the quantitative analysis of heat transfer coefficient values, Assar[3] highlights the difference in the variation in the gap width formation for horizontal and vertical surfaces of a cylindrical casting Fig. 5. Taha[18]also shows the experimental variation of the interfacial heat transfer coefficient for

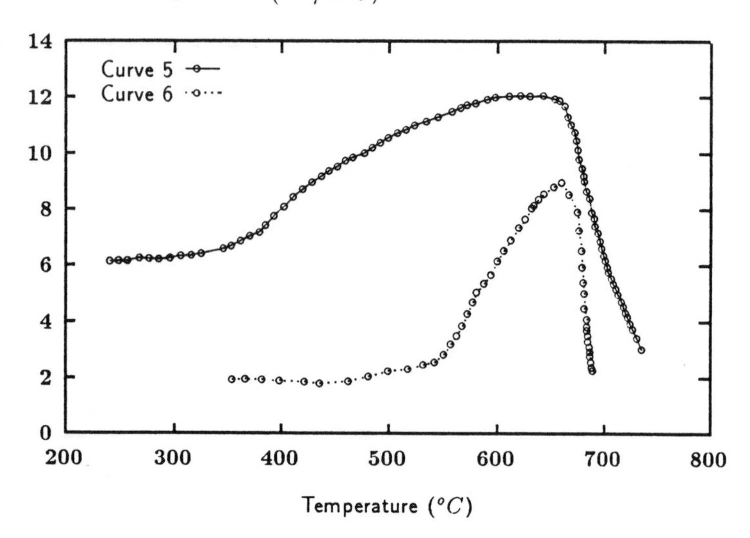

Figure 3: Heat transfer coefficient *vs* surface temperature for pure aluminium cylindrical casting with copper chill and 115 ^oC superheat (Curve 5), and 40 ^oC superheat (Curve 6)

different lengths for a Al-4.5

3.1.3 Interpretation of Results

The variation of the heat transfer coefficient with respect to temperature for various parameters for a pure aluminium casting are superimposed to generate Fig. 6. It is important to note the consistency of the shape in various plots. As far as the qualitative analysis is concerned a similar trend is followed in all the diagrams. Following points can be observed from Figure. 6.

a. The interfacial heat transfer coefficient is always a maximum at the liquidous temperature.
b. The influence of superheat on the magnitude of interfacial heat transfer coefficient values is more predominant than that of the weight of a casting. From curves 4 and 5, it can be seen that 65% increase in the weight and 30% decrease in the superheat has resulted in only 16% overall increase in the interfacial heat transfer coefficient.
c. Curves 3 and 4 indicate that a 160% increase in the conductivity of the mould surface at the interface can alter the heat transfer coefficient values by as much as 263%.
d. When the shape of the castings changes from cylindrical to flat, the maximum value of the interfacial heat transfer coefficients can increase as much as

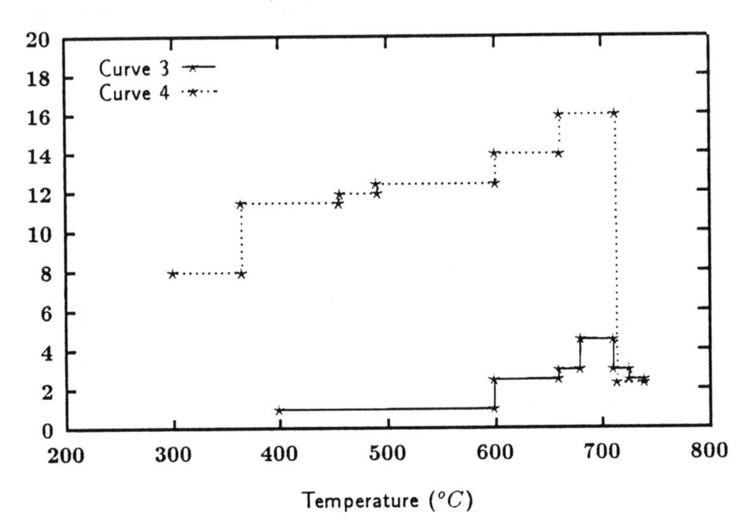

Figure 4: Heat transfer coefficient *vs* surface temperature for pure aluminium cylindrical casting with sand chill (Curve 3) and copper chill (Curve 4) with 80 ^oC superheat

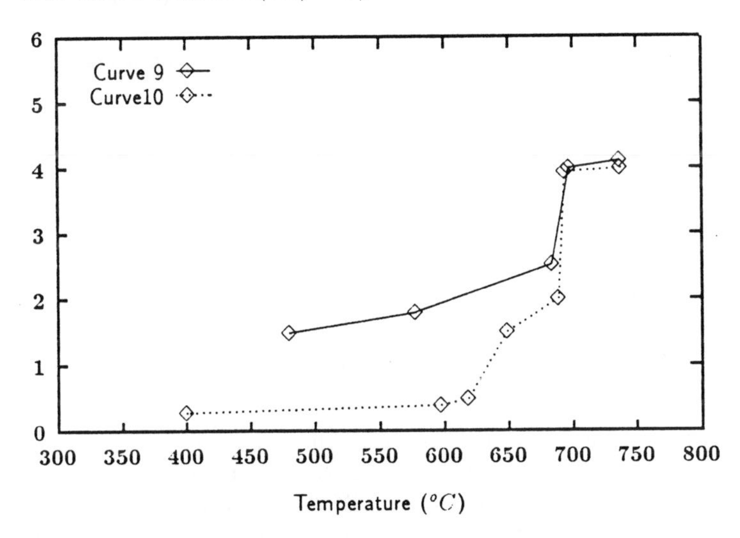

Figure 5: Heat transfer coefficient *vs* surface temperature for Zinc with horizontal surface (Curve 9) and vertical surface (Curve 10)

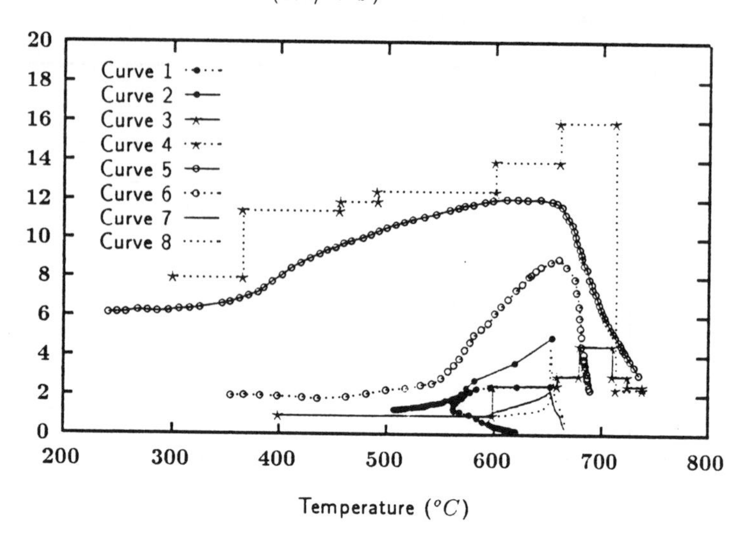

Figure 6: Superimposed heat transfer coefficient *vs* surface temperature variations for pure aluminium

100%.
e. With the introduction of the copper chill, an increase of around 120% can be expected in the value of the interfacial heat transfer coefficient (Curves 2 and 7).
f. Curves 5 and 6 indicate that during solidification an average increase of 100% can be achieved in the values of the interfacial heat transfer coefficient if the superheat is increased by around 175%.
g. With vertical or horizontal orientation of the surfaces, an average increase of 100% can be achieved in the values (Fig. 5).

Also, it can be seen that the basic shape of the curve remains similar for the various combinations of process, material and geometric parameters. Hence, if the curve is defined with a few control points such as shown in Fig. 7, its shape can be manipulated intelligently with the aid of expert systems technology to yield the appropriate values. The nature of the expert system is also described in Fig. 7.

A typical representative example is indicated in Fig. 8 to emphasis the need for considering differential heat transfer coefficients. The interfacial boundary conditions at surfaces AB, BC, DE, EF, FG, GA and the curved surface H should be different in magnitude. Therefore, appropriate heat transfer coefficients need to be chosen to represent a realistic simulation.

Moreover, such an analysis improves the accuracy of the results when different exothermic or endothermic cooling aids, shapes, addition or deletion of

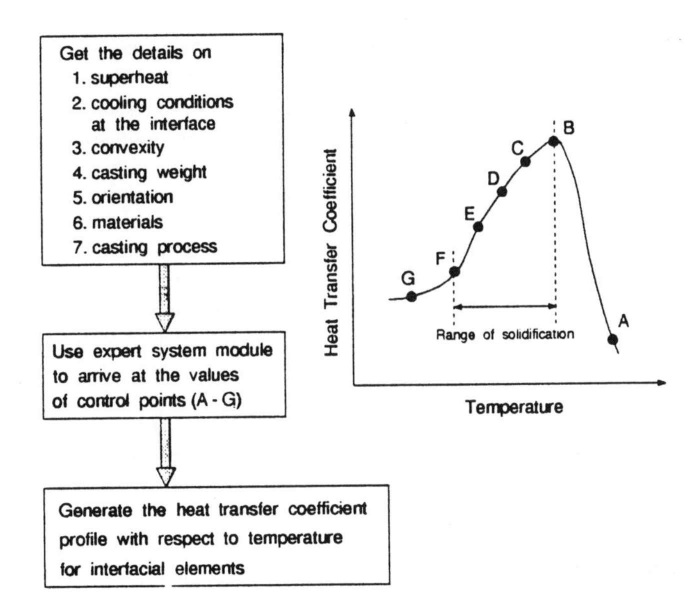

Figure 7: Expert system module for incorporating boundary conditions at interfacial elements

some features such as cores *etc.* are used, enabling a more realistic concurrent design of castings to be achieved.

3.2 Mesh Generation

3.2.1 General View

In the finite element simulation of casting problems, the end result may be greatly affected by the discretization chosen for domain. Considerable expertise is normally required in choosing is control parameters. These meshes

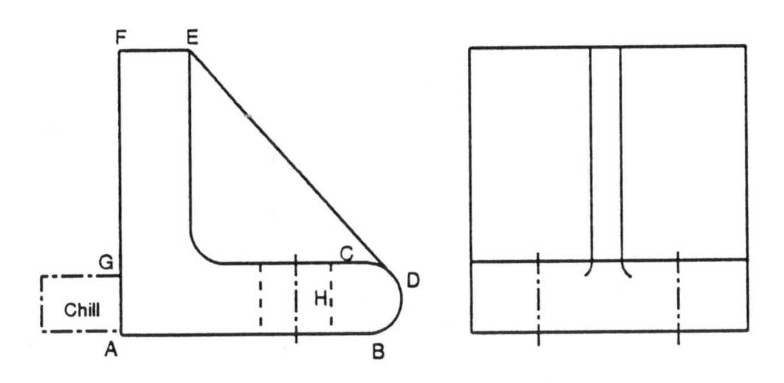

Figure 8: Representative casting geometry

of differing density will result in different accuracies achieved in simulation. An excessively fine mesh leads to unnecessary computational costs whereas a highly mesh produces intolerable approximation errors. As far as adaptive mesh generation with error estimation is concerned, the main difficulty is in establishing initial mesh requirements.

For a mesh generator with a variable density facility, the dominant factors of the mesh topology are density distribution and boundary description. There are different ways to define domain boundaries, some approaches can be set out to describe the geometry boundaries artificially. However, currently, it would appear that the best possible utilization of AI would be in the controlling of mesh density distribution.

Dolšak *et al.*[21] have presented a framework for a mesh generation expert system and demonstrated that rules for deciding appropriate control values for meshes can be inductively constructed from expert provided examples.

3.2.2 Knowledge-Based Approach

Kang *et al.*[2] presented a new approach to mesh generation, the main point was to incorporate the information about the object geometry as well as the boundary and loading conditions thus generating an *a-priori* mesh which will be more refined around the critical regions of the domain concerned. The expert system generated was able to intelligently identify critical regions and choose a proper mesh size by performing an approximate heat flux calculation.

According to the experience of the authors, a mesh can be properly constructed based on essential information for critical regions, a mesh density intensity distribution and a basic mesh size. A schematic illustration is given in Fig. 9. The critical region (or points, or surfaces) can be determined by the geometrical features (Critical Region A), and boundary conditions (Critical Region B). The critical regions will be identified together with the mesh density intensity distribution. The basic mesh size is governed by both geometrical features and the numerical accuracy desired.

3.2.3 Neural Network Techniques

Ahn *et al.*[22] demonstrated a self-organizing mesh based on neural network (self-organizing feature mapping). With user-supplied mesh density functions and a boundary mesh the mesh generator was able to provide a graded mesh with similar asymptotic characteristics to a weighted Dirichlet tessellation and dual Delaunay triangulation. The mesh generation included local mesh restrictions such as a fixed boundary and/or internal meshes.

Dyck *et al.*[23] presented a system which predetermined the mesh density by using a neural network. The system can be trained by incorporating exam-

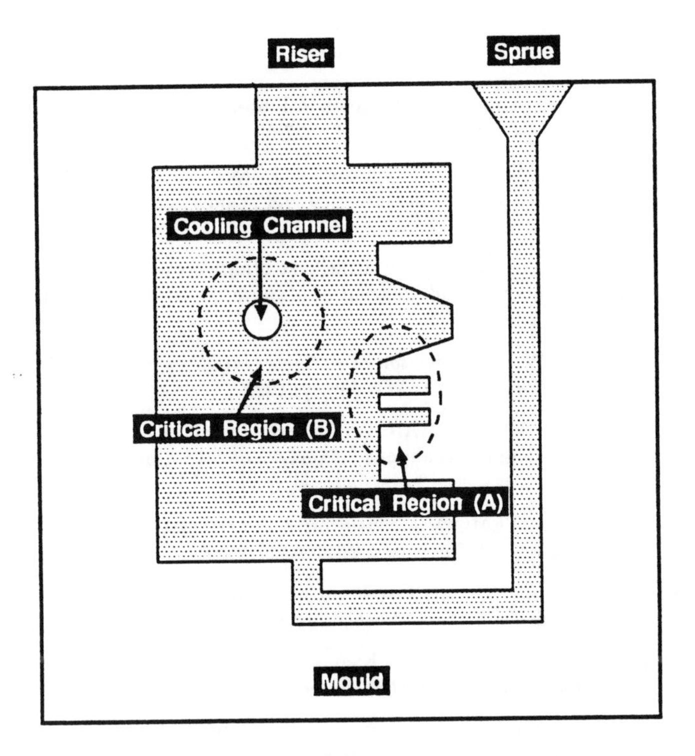

Figure 9: The schematic diagram of casting example

ples of ideal meshes. As stated in [23], the initial mesh should be very close to the final, optimal mesh.

4 Conclusions

In this paper, the possibilities of applying AI techniques are emphasized which would lead to more computationally efficient preprocessing, and the realistic simulation of castings. With the development of AI as a branch of computer science, the applications for such a technique in engineering computation will expand rapidly. Although only a few potential applications have been mentioned, the area has received considerable interest, and will hopefully play a valuable role in the numerical simulation of castings.

Acknowledgements

This work has been carried out under partial financial support by the COST504 program, which is greatfully acknowledged.

References

[1] CAREY, C. F. – Toward Expert Systems in Finite Element Analysis, *Communications in Applied Numerical Methods*, **3**, 527-533,1987.

[2] KANG, E. and HAGHIGHI, K. – A Knowledge-Based A-Priori Approach to Mesh Generation in Thermal Problems, *Inter. J. for Numer. Methods in Engng.*, **35**, 915-937, 1992.

[3] ASSAR, A. W. M. – On the interfacial heat transfer coefficient for cylindrical ingot casting in a metal mould, *Journal of Material Science Letters*, **11**, 601-606, 1992.

[4] PATTERSON, D. W. – *Introduction to Artificial Intelligence and Expert Systems*, Prentice Hall, Englewood Cliffs, New Jersey, 1990.

[5] CAGAN, J. *et al.* – PLASHTRAN: An Expert Consultant on Two-Dimensional Finite Element Modelling Techniques, *Engineering with Computers*, **2**, 199-208, 1987.

[6] CHEN, J. L. *et al.* – FEMOD: A consultative Expert System for Finite Element Modelling, *Computer and Structures*, **29**, 99-109, 1988.

[7] FENVES, S. J. *et al.* – On the Development of an Integrated Knowledge-Based Finite Element Analysis System, *Mathl. Comput. Modelling*, **14**, 124-129, 1990.

[8] BREITKOF, P. *et al.* – Knowledge Engineering Enhancement of Finite Element Analysis, *Communications in Applied Numerical Methods*, **3**, 359-366, 1987.

[9] YEH, Y. *et al.* – Building an Expert System for Debugging FEM Input Data With Artificial Neural Networks, *Expert Systems with Applications*, **5**, 59-70, 1992.

[10] SAHM, P. R. and HANSEN, P. N. – Numerical Simulation and Modelling of Casting and Solidification Processes for Foundry and Cast House, *International Committee of Foundry Technical Associations*, 1984.

[11] SAMONDS, M. T. – Finite Element Simulation of Solidification in Sand Mould and Gravity Die Castings, Ph D Thesis, University of Wales, Swansea, 1985.

[12] LEWIS, R. W. *et al.* – Solidification in Castings by the Finite Element Method, *Material Science and Technology*, **6**, 482-489, 1990.

[13] USMANI, A. S. – Finite Element Modelling of Convective-Diffusive Heat Transfer and Phase Transformation with Reference to Casting Simulation, Ph D Thesis, University of Wales, Swansea, 1991.

[14] HO, K. and PEHLKE, R. D. – Metal-Mould Interfacial Heat Transfer, *Metallurgical Transactions B*, **16 B**, 585-594, 1985.

[15] TADAYON, M. R. and LEWIS, R. W. – A Model of Metal-Mold Interfacial Heat Transfer for Finite Element Simulation of Gravity Die Castings, *Cast Metals*, **1**, 24-28, 1988.

[16] NISHIDA, Y. *et al.* – The Air Gap Formation Process at the Casting-Mould Interface and the Heat Transfer Mechanism through the Gap, *Metallurgical Transactions B*, **17 B**, 833-844, 1986.

[17] MAHALLAWY, N. A. *et al.*, – Effect of Melt Superheat on Heat Transfer Coefficient for Aluminium Solidifying against Copper Chill, *Journal of Materials Science*, **26**, 1729-1733, 1991.

[18] TAHA M. A. *et al.*, – Effect of Melt Superheat and Chill Material on Interfacial Heat-Transfer Coefficient in End-Chill Al and Al-Cu alloy castings, *Journal of Materials Science*, **27**, 3467-3473, 1992.

[19] MAHALLAWY, N. A. and ASSAR, A.M., – Metal-Mould Heat Transfer Coefficient using end-chill experiments, *Journal of Materials Science Letters*, **7**, 205-208, 1988.

[20] RAMIREZ, M. R., – An Expert System for Setting Time Steps in Dynamic Finite Element Programs, *Engineering with Computers*, **5**, 205-219, 1989.

[21] DOLŠAK, B. and JEZERNIK, A., – Mesh Generation Expert System for Engineering Analysis with FEM, *Computers in Industry*, **17**, 309-315, 1991.

[22] AHN, C., *et al.*, – A Self-Organizing Neural Network Approach for Automatic Mesh Generation, *IEEE Trans. Magnetics*, **27**, 4201-4204, 1991.

[23] DYCK, D. N., *et al.*, – Determining an Approximate Finite Element Mesh Density Using Neural Network Techniques, *IEEE Trans. Magnetics*, **28**, 1767-1770, 1992.

ASPECTS OF DIRECT, HEURISTIC AND INVERSE MODELLING FOR CONTINUOUS CASTING PROCESSES

Boehmer, J.R. and Fett, F.N.

Institute for Energy Technology, Department of Mechanical Engineering, University of Siegen, Paul-Bonatz-Str. 9-11, D-W-5900 Siegen, GERMANY

ABSTRACT

Three different approaches of modelling in the wider sense, i.e. direct, heuristic and inverse modelling, are discussed with regard to their potential to contribute to actual problem solving in the domain of continuous casting processes. It is shown that synergy effects can be gathered standardizing their common ground and principles. Starting from a modular modelling concept, a methodology of knowledge and model-library based problem solution support is proposed. This concept allows a problem-specific model utilization with knowledge support during all phases of the modelling process. Model simulation, knowledge-based (and in special case, rule-based) systems as well as inverse modelling can be used effectively in combination, i.e. both in a competitive and complementary way.

1. INTRODUCTION

The continuous casting process in its various realizations for different ferrous and nonferrous metals, and for different shapes like billets, tubes, blooms, slabs or compact strips, is determined by a multitude of degrees of freedom, the combination of which conditions the quality of the products. The major principle of all variations of the basic continuous casting process [1] is to supply molten metal into a mould, and to cool down the metal in a controlled way. The primary task of mould cooling is to shape the metal and to remove enough energy from the melt to form an at least strength-bearing strand shell. To compensate the shrinkage due to the strand's cooling, the mould is tapered. Below the mould further cooling of the strand is controlled by water, sprayed directly onto the strand surface. The goal is to adjust a cooling water distribution and intensity that causes a certain desired cooling behaviour, avoiding dangerous gradients.

The different variations of the continuous casting process take into account the peculiarities of different materials and shapes as well as 'design philosophies'. Today, there are great demands on operating safety and reliability of the continuous casting process just as on the quality of the products. Object of intensive efforts are measures for cost-saving and productivity improvements, optimizing plant engineering and process control. This prerequisites reliable knowledge about the process determining factors and their interaction.

Process modelling in its widest sense means to deal on a theoretical level with abstract and manageable representations of reality, regarding only those (few) aspects of the real systems and processes that are considered to be relevant and necessary for a successful solution of a given problem situation. The initiating goals may be clustered to five different application areas:

- defect diagnosis and avoidance,
- efficiency and quality improvement,
- furtherdevelopment of plant construction and/or processing,
- casting new materials, and
- off-line staff training.

The goal of this article is to discuss three different approaches of modelling in the wider sense, i.e. model simulation, expert systems, and inverse modelling, and their potential to contribute to actual problem resolving, and to show that they can be used both in a competitive and complementary way.

2. MODULAR MODELLING CONCEPT

Basic interest of mathematical modelling of technical systems is always the need to understand the process, to describe existing interactions and to explain observed phenomena, in order to be able to make decisions in the end. In the course of this one transforms the problem situation into a comprehensive and manageable representation, abstracting from all unessential situation features. Mathematical models then combine cause-effect interrelations which are formalized by systems of governing equations. Heuristic models deal also with cause-effect interrelations, however formalized by systems of heuristic rules. The usual procedure of modelling (see Table 1) is to subdivide the overall problem into partial problems, regarding clear and manageable subsystems and process units. Taking into account the physical background of involved effects, the analysis of influential factors and the distinction between independent and dependent variables leads to the construction of cause-effect interrelations. The mathematical as well as the heuristic, i.e. rule-based modelling of them induce each a coupled system of separate and self-contained partial models with defined interfaces. I.e., the most suitable way is to formulate partial models for partial phenomena, and to combine them to an overall process model. For the usually computerized modelling process, a standardized form of these computer models is of advantage. Exceeding a critical quantity, these models should be organized in a model library [2]. Then, the modelling procedure can be reduced to the selection, connection, and parametrization of pre-formulated partial models.

2.1 Problem Delimitation and Modelling

General objective for all continuous casting processes is to control and to improve the production rates ensuring a safe and robust operation, and keeping a defined minimum product quality. This prerequisites a concrete idea of the meaning of the terms 'safety' and 'quality'. It should be noted that these terms are not independent from each other. Let 'quality', above all, stand for the absence of crack formation or shrink holes. So, a defect-free product may be existent if no cracks or cavities were ascertainable with the naked eye or at a closer look. Therefore, the term 'quality' refers to certain defined quality features of the

1. problem analysis 1.1. defining the relevant system and subsystems 1.2. looking at process units 1.3. distinguishing influential factors, independent and dept. variables	problem level
2. physical background 2.1. separating involved effects 2.2. describing physical principles 2.3. finding relevant properties 2.4. formulating cause-effect interrelations	physical level
3. applying operational expert knowledge *and/or* 4. applying previous computation results *and/or* 5. applying simulation models	tool selection level
5.1. partial model formulation 5.2. assembling relevant calculation components 5.3. interconnecting the partial models to an overall model 5.3.1. reproducing the cause-effect interrelation *and/or* 5.3.2. building a transformed hierarchical framework (concluding from effects to causes by inverse modelling) 5.4. model calibration	programming level
6. qualitative simulation *and/or* 7. quantitative simulation, parameter optimization *and/or* 8. reverse calculation	utilization level
9. systemization of results, interpretation and assessment 10. generation of process advices	harvesting level

Table 1 - Levels of problem solution support using direct, heuristic and/or inverse modelling

products. On the other hand, the term 'safety' can be put down to the sense of reliability and reproducibility of the process. This refers not only to adjustable entities but particularly to output properties, i.e. some characteristics of product quality. Following this argumentation, the demand on 'safety' becomes clear to be a preliminary stage to keep, and further, to improve quality. This prerequisites, that all influencing factors and their effectiveness must be known to be able to avoid, to clear or to compensate disturbances.

At the beginning, looking at the real process one finds an incomplete and shadowy view at a tremendous jumble of influencing factors. Analyzing the scenario in view of the actual problem, it condenses e.g. to the scenario of Figure 1. A consequent application of steps (1.1) to (2.4) in Table 1 may lead to the cause-effect interrelation depicted in Figure 2. In that case especially temperatures and thermal stresses are of interest, because the reason for crack formation is an overstraining of the material´s ability to react pliable on existing thermal loads, mainly due to locally different material contraction. Accordingly, thermal, material property, and thermo-mechanical models have been developed. Within the blocks 'thermal model' or 'contraction model' further couplings between the partial systems strand / mould / coolant are performed as well as for an evaluation of link variables like solidification geometry, local solidification enthalpy fraction, convection-modulated heat conductivity coefficients of the melting bath, surface heat transfer coefficients, pressure and friction forces, displacements, etc.

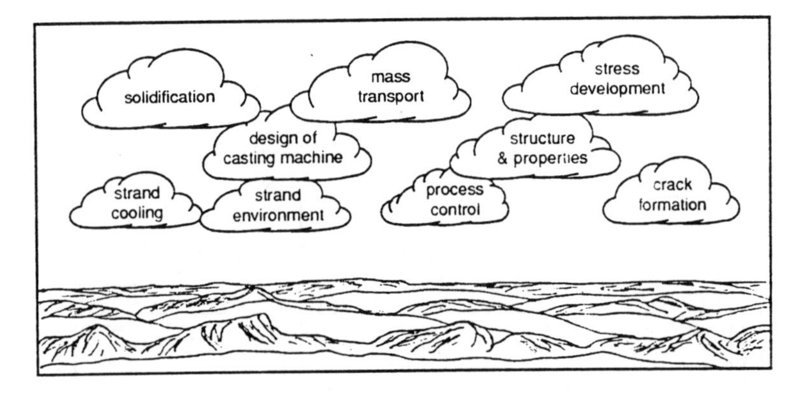

Figure 1 - Jumble of influencing factors on the quality of a continuously cast product

With this, the couplings between mass flow (strand movement), temperature distribution, metallic structure, stresses and strains, and the influences of the outside world have been reduced, such that separate, self-contained partial models can be treated *sequentially* (regarding each constant conditions), and balanced by iteration, to obtain a self-consistent solution in the end. Emphasis is on the two way coupling between the heat transfer and the thermal distortion of the strand. The effective heat transfer between the strand and the mould is largely determined by the size of the gap between them, and the size of the gap is governed by the cooling conditions.

The interrelation shown in Figure 2 is the first stage of a qualitative model. It can serve as a basis both for quantitative modelling (see chapters 2.2 and 2.3) and heuristic, i.e. rule-based modelling within the framework of an expert system (see chapter 3). Transforming it to an effect-causes hierarchy, it becomes the basis for inverse modelling (chapter 4).

2.2 Formulation of Partial Process Models

Already in the beginnings of industrial utilization of continuous casting processes the predictability, especially of invisible states, became a central question. Fundamental work has been done investigating the principles of solidification [3] and describing the influence of solidification parameters on structural properties [4]. Early papers (e.g. [5]) already dealt with the formation of residual stresses in continuously cast billets. The application of general theories in mathematical form (i.e. balance equations [6], state equations, phenomenological transport equations) allows to describe the cooling performance during the casting process in an efficient and illustrative way [7]. So, standard works like [8] form partial model libraries *in formulae*.

Owing to the development of computer hard- and software, the requirements on meaningfulness and usefulness of modelling and simulation grew. The application of finite difference and finite element methods as standard methods of numerical analysis [9] lead to sophisticated mathematical models. This evolution was also of important influence on the modelling of continuous casting processes. Concerning our own work in this field, in Table 2 partial process models for quantitative modelling, developed in Siegen during the last

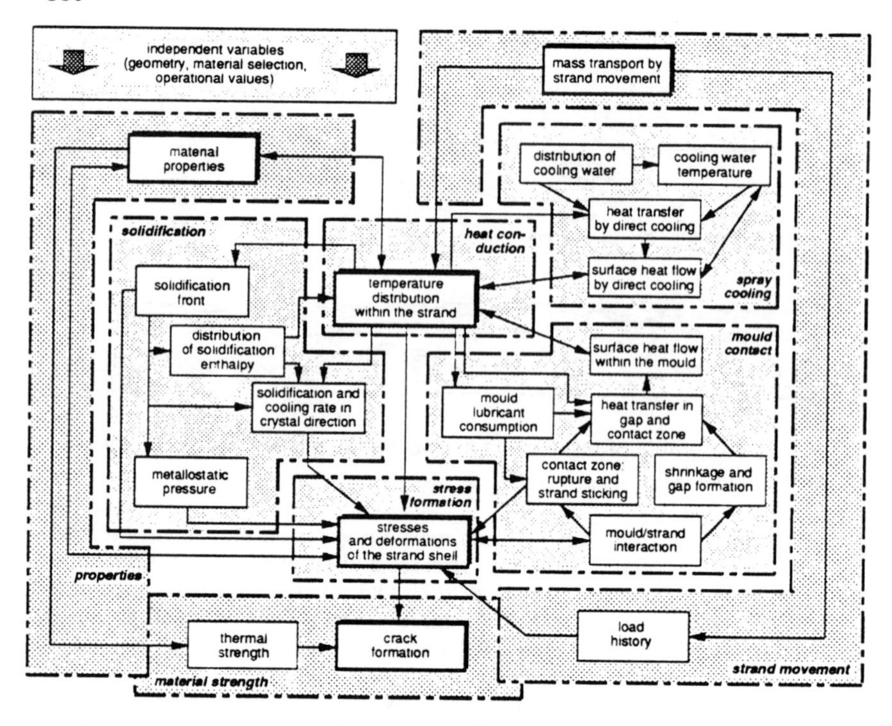

Figure 2 - Partial model interconnection to represent the physics of continuous steel bloom casting

decade, are summarized. Depending on the shape of the products, cylindrical or Cartesian coordinate formulations are used. In all cases where unification is reasonable, generalized coordinate formulations are used. Depending on the specific conditions, one-, two- or fully three-dimensional investigations can be realized both for steady-state operational points and transient processes.

The metal strand to be cast is in the centre of attention. Therefore, all plant configuration is concentrated in the boundary conditions of the strand, which themselves are explained by separate models. For example, if the geometrical arrangement of cooling water slots in the mould walls affect on mould cooling performance, a detailed model of mould heat conduction has to be taken into account. Or a spray distribution function has to be calculated considering the kinetics of water droplets sprayed via nozzles. Also, there may be an interaction between strand behaviour and boundary conditions via strand surface temperatures, or the contact problem in the mould.

The totality of all partial models form a model library to support modelling of a whole family of continuous casting processes. Prerequisites to take advantage from it are the conditions for modular programming:

- existence of standardized module characterization profiles describing application area, functionning and limitations,
- guaranteed model validity, and control of upper and lower limits of model variables,
- standardized interfaces with regard to calling syntax, parameter transfer and connection to external variables and files.

sector	sub-sector	model	code	options	identifier
shapes		product shapes	SC	billets / tubes / blooms / slabs / thin slabs	R / T / B / S / C
		coordinate systems		cylindrical, rectangular, generalized	C / R / G
thermal sector	strand modelling	heat conduction models for given initial and boundary conditions	HC	1D / 2D / 3D	'000' ... '111'
				full / half / quarter cross-section	F / H / Q
				steady-state / transient	S / T
				constant / variable properties	C / V
		mass transport by strand movement	MT	continuous enthalpy transport	C
				stepwise shifting procedure	S
		solidification models	SM	heat source method	SM - 001
				equivalent specific heat method	SM - 002
				enthalpy method	SM - 003
				separation & phase boundary approach	SM - 004
		effects of electromagnetic stirring	ES	determination of effected zone	ES - D
				effective heat conductivity method	ES - 001
	transfer models	heat transfer models	BC	direct mould contact: optional with casting flux / contact pressure / depending on reciprocation parameters	BC - 001 ... BC - 005
				gap heat transfer (for different media)	BC - 010 / 011
				surface radiation models	BC - 020 / 021
				free and forced convection	BC - 025 / 026
				direct heat transfer to cooling water	BC - 030 / 038
				heat transfer to support / transport rolls	BC - 050
	boundary modelling	mould models	MD	round, square and rectangular shapes	R / B / S
				1D / 2D / 3D	'000' ... '111'
				arrangement: inside / outside moulds	I / O
				multi-layered	L / M
				multi-tapered	T
				diff. cooling water slot configurations	H / S
		mould oscillation models	MO	geometry / velocities / heat transfer	MO - 001 / 002
		water spray distribution models	SD	1D / 2D / 3D, diff. nozzles / angles	SD - 001 / 002
		cooling water models	CW	heating	CW - 001
				boiling	CW - 010
				vaporization	CW - 100
material sector		crystal growth models	CG	orientation	CG - 001
				grain size	CG - 005
		TTT, TTA, and tempering characteristics	PC	microstructure	PC - 001
				properties	PC - 010
				microstresses	PC - 030
		segregation of alloying elements	SE		SE - 001
		cavity models	CP	center porosity / gas holes	CP - 001
		thermophys. material properties	PR		PR - 001 / 030
		material strength models	MS	depending on: temperature, structure, deformation rate	MS - 001, MS - 002
stress / contraction sector		stress / strain analysis	SA	1D / 2D / 3D	'000' ... '111'
				full / half / quarter / eigth cross-section	F / H / Q / E
				thermo-elasto-plastic / creep	E / P / C
				considering the material´s load history	S / H
				considering mushy zone (optional)	S / M
				considering phase dilatation changes	N / D
				mould or roll contact / friction	F / C / R
		assessment of resulting stresses	CC	crack criteria	CC - 001 / 005
		mould deformation models	DM	1D / 2D	'000' ... '111'
				rigid / yielding	R / Y
				thermo-elasto-plastic	E / P
				slip joint of mould parts considered	N / S
		contact models	CM		CM - 001 / 002

Table 2 - Partial process models available in the model library

2.3 Problem-Specific Model Synthesis

The interrelations network once described, each block of it (for an example see again Figure 2) is to be represented by a partial model from the library. Corresponding to this relationship, i.e., depending on the modelling purpose, the partial models of interest are selected and put together to an entire simulation model. It is put in concrete terms with respect to coordinate system, dimension, symmetry, time dependence and solution procedures for the resulting equation system. The degree of model extent and precision determines the 'grain size' of the models and the number of internal links.

Table 3 shows some examples for problem-specific modelling for different continuous casting processes. According to the modelling goal and the peculiarities of the special casting process to be considered (especially material and geo-

modelling goal	type of casting process	mathematical model	model type
cooling performance, material structure and final properties, stress/ crack formation, mould contact, (for given conditions)	(round) billet casting, axisymmetric boundary conditions	2D, steady-state, coupled thermal and mechanical models with sub-models, e.g. for solidification and crystal growth, strand movement, heat transfer through mould walls, mould contact and gap formation, spray cooling, material strength	D
time until complete solidification, and averaged temperature course	axisym. billets (steel)	1D, transient, thermal, diff. boundary models	D
	axisym. copper billets	2D, transient, thermal, diff. boundary models	D
	steel blooms	2D, transient, thermal, moving strand slice	D
optimization of mould taper	slab casting	2D/3D steady-state thermal and (feedback coupled) contraction model; mould stability and contact model	D
development of strand shell stresses	(round) billets	2D axisym. analysis of incremental thermal strain due to incremental temperature change	D
	steel blooms	2D plane strain or 3D incremental analysis due to incremental temperature change	D
stresses in the spray cooling zones	bloom or slab casting, if boundary conditions change (abruptly)	3D transient analysis, uncoupled thermal model with given boundary conditions; followed by separate stress analysis	D
temperature distribution	copper tubes, homog. boundary conditions	2D axisymmetric thermal model with inner and outer mould	D
	copper tubes, heterog. boundary conditions	3D thermal model with inner and outer mould, cooling water distribution model	D
thermal contact estimation	bloom casting (steel)	mould heat conduction model, contact/gap heat transfer model	I
adaption of spray cooling intensity	steel bloom or billet casting	1D/2D transient inverse thermal model with a constraining stress/strain model	I
optimization of mould taper	steel bloom or billet casting	1D/2D transient inverse thermal model with constraining stress/strain and mould model	I
influence of mould layers and/or design of cooling water slots	slab casting	e.g. steady-state thermal model both for strand and mould, coupled contraction model	D
temperature distribution	thin slab casting	2D, thermal model both for strand and mould, coupled contraction model	D

Table 3 - Examples for problem-specific modelling for continuous casting processes (model type: D=direct, I=inverse modelling)

model specification	model performance
SC - RC	shape is 'round billets', cylindrical coordinate system
HC - 5HSV	2D (r,z) coordinates: '5' decimal is '101' dual, i.e. from the chosen (r,φ,z) system are r and z used; half cross-section, steady-state analysis, temperature-dependent properties
MT - C	continuous enthalpy method
SM - 003	solidification enthalpy is included in overall enthalpy flux balance
ES - 001	electromagnetic stirring effects only via local heat conductivities
BC - 005 BC - 010 BC - 020 BC - 025 BC - 032	boundary conditions are: - mould contact with casting flux layer - gap heat transfer in the mould - surface radiation - free and forced convection to air - direct heat transfer to cooling water, model 'REINERS/JESCHAR'
MD - R5OMTH	the thermal mould model for round shapes with (r,z) coordinates considers a multi-layered, multi-tapered outside mould with fully homogeneous water cooling
CW - 111	the cooling water model includes heating, boiling and vaporization
CG - 001	crystal orientation normal to the solidification front. solidification rate taken into account
PR - 001 PR - 002 PR - 003 PR - 004 PR - 006 PR - 010	thermophysical material properties for: - thermal conductivity - thermal heat capacity - density - modulus of elasticity - thermal expansion coefficient - nonlinear temperature-stress-strain relationship
MS - 001	temperature-dependent material strength
SA - 5HPHSNC	stress-strain analysis in (r,z) cylindrical coordinate system for a half longitudinal cross-section, thermo-plastic material, analysis traces the material's load history, no mushy zone, no phase changes in the solid, mould contact constraints considered
CC - 003	crack criterion no.3 : thermal-strength standardized maximum stresses
DM - 1YPS	mould deformation model in (r)-direction, thermo-elastoplastic, slip joint of mould parts
CM - 002	iterative balancing of strand and mould deformation analysis, coupled via external forces (i.e. not displacements)

Table 4 - Partial models (ref. Table 2) used for the billet model around the kernel '*SC-RC*' (first example in Table 3)

metry peculiarities), an appropriate mathematical model is formed, using partial models out of the library (Table 2) and a standardized program kernel as basic framework. As an example, the model for axisymmetric billet casting (first example in Table 3) consists of the shape/coordinate kernel '*SC-RC*' and the partial models summarized in Table 4. With this specific model a whole class of problems out of the billet casting domain can be supported, but the model is optimized with regard to program size and computing efficiency. Particularly the latter topic is of importance when dealing with complex models such as 3D contact models (e.g. example 5: "optimization of mould taper" in Table 3) which require extensive programming and computation efforts.

3. SUPPLEMENTING MODELLING BY EMPIRICAL KNOWLEDGE

Expert systems [10], also called knowledge-based systems, capture non-algorithmic knowledge of experts in their software and make it available to any user capable of running the system. The basic idea behind expert systems is to provide a computer with the knowledge and inference procedures needed to solve problems that would normally require significant human expertise for their solution [11]. The application domain for expert systems arises especially in heuristic or symbolic contexts, e.g. for selection problems, assessment or decision-making. Looking at empirical knowledge available, one can distinguish between two different areas:

- knowledge about proper model formulation, model synthesis, and decision-supplying utilization of models,
- long-term operational experience, and from in-plant experiments.

3.1 Knowledge Support for Simulation Processing

Dealing with mathematical models requires a methodology for the formulation, organization and coordination of all possible plan scenarios suitable to execute a requested job [12]. On the *organization or preprocessing level* the real problem situation has to be analyzed and formalized to a demand profile, compiling geometrical, process and material parameters (see Figure 3). Additionally, process constraints have to be taken into account, such as primary and secondary product requirements, and also equipment restrictions like, e.g., capacity or safety relevant limitations. On the other hand there are model constraints, concerning modelbase capacity, characteristics and limitations of available models, the relationship between the models each other as well as the relationship between problem and model frames. Furthermore, experience about previous simulations as well as empirical knowledge may be provided.

Coordination level: As a result of preprocessing, different options for problem-solution support may be offered like in a display window: application of operational expert knowledge (see also chapter 3.2), browsing and evaluating adequate results of previous computations, proposing and assembling relevant calculation components. In the latter case, available informations can be used either to interconnect the relevant partial models to an overall model to perform model simulations, parametric investigations or parameter optimizations, or for inverse modelling (see chapter 4).

Activities on the *execution level* shall be discussed for the case of direct modelling: Making available simulation models to persons who perhaps are

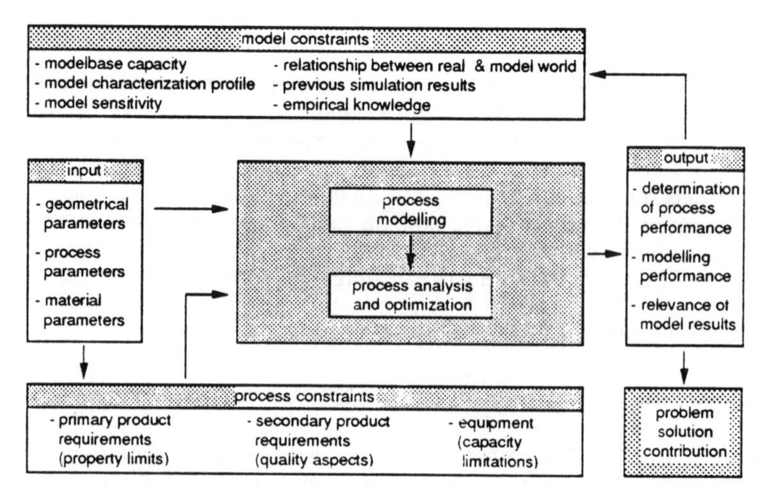

Figure 3 - Block structure of process design and simulation

domain experts but inexperienced simulation users, there is need to support the use, and to guard against the misuse of such models. Knowledge-based simulation management covers an input check for a given model configuration, regarding upper and lower limits of the design variables as well as physical and technological constraints, and improper parameter combinations. Furthermore, it supplies model use, indicating and visualizing important results, commenting on result confidence, pointing to parameter sensitivities or applying optimization strategies.

On the *postprocessing level* there is the opportunity for knowledge support processing the simulation results. This concerns the compression, systemization and interpretation of the results both for single simulation runs and comparative simulations as well as the assessment of the practical relevance of model results. It covers also the recording of modelling performance and result quality, and the derivation of consequences for further modelling.

3.2 Defect Diagnosis and Avoidance by Heuristic Modelling

The cause-effect interrelation shown in Figure 2 can also serve as a basis for heuristic modelling. In this context, Table 5 shows the structure of knowledge-based defect diagnosis and avoidance, representing two ways of consultations:

- *Running a defect diagnosis:* A certain casting defect has been detected (for defect catalogues see, e.g. [13,14]). Now the question is about the steps necessary to take remedial action. The system inquires all values of the known influencing factors, asking "what has been done anyway?". Then it tries to identify applicable rules by variance comparison of the conditional parts of the stored rules, and to present operation instructions attached to the rules found.
- *Performing process advices:* Here the task is to check the operability of an intended parameter set, and to indicate possible risks. Performing again a variance comparison of the conditional sides of the stored rules the system tries to find and to display those rules which indicate appropriate consequences to be expected. To avoid probable defects the system suggests corresponding operating instructions for improvement.

Figure 4 shows the beginnings of an overall expert system for the field of continuous casting of high-grade steels [14], which has been applied to the bow-type continuous casting. As a first result, there is a knowledge base with about 2000 production rules for 9 different material groups considering 33 influencing factors and taking into account 14 different defect types. Fuzzy-logic

	running a defect diagnosis	*performing process advices*
input	the casting defects detected	an intended parameter set
goal	advice for remedial action	operability check, indication of possible risks
action	check of influencing factors, identification and utilization of applicable rules, data or simulation models	check of influencing factors and rules/data available, model computation
output	instructions for operation	assessment comment(s), proposal for a parameter change

Table 5 - Structure of knowledge-based defect diagnosis and avoidance

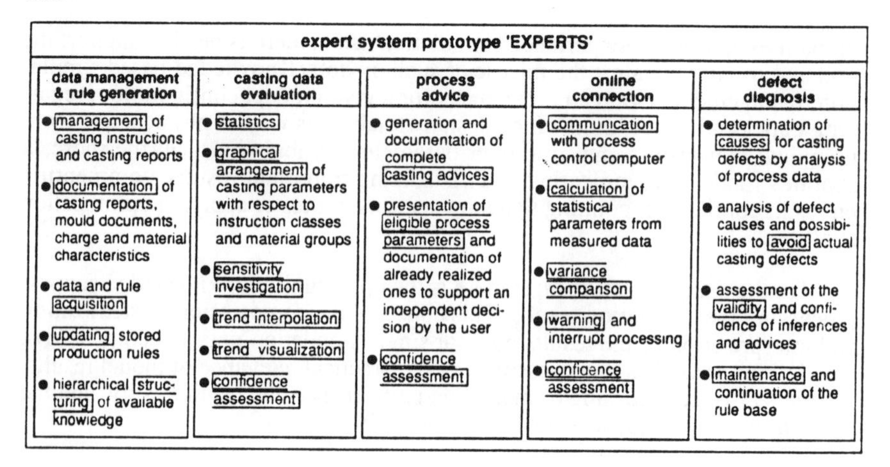

Figure 4 - Main branches of the 'EXPERTS' system for the continuous casting of different steel grades

parameters, rule priorities and confidence classes facilitate to qualify the operating instructions inferred and to avoid inconsistencies.

Although knowledge *acquisition* for such a knowledge-based system is mainly based on casting reports, specialists´ interviews and questionnaires, the most promising way of knowledge *representation* is again a well-ordered cause-effect structure, at least in the form outlined in Figure 5. The tree structure there contains separate partial elements. I.e., all rules gained from knowledge acquisition and stored in the rule-base are not *a priori* related to each other. Rules consist of a conditional and a consequential part, each possibly consisting of different sub-elements. They may be formulated in variable form, e.g.:

'if <*condition*> is <*A*> then <*consequence*> is <*B*>',

'if <*condition*> is <*C*> then <*consequence*> is <*D*>', etc.

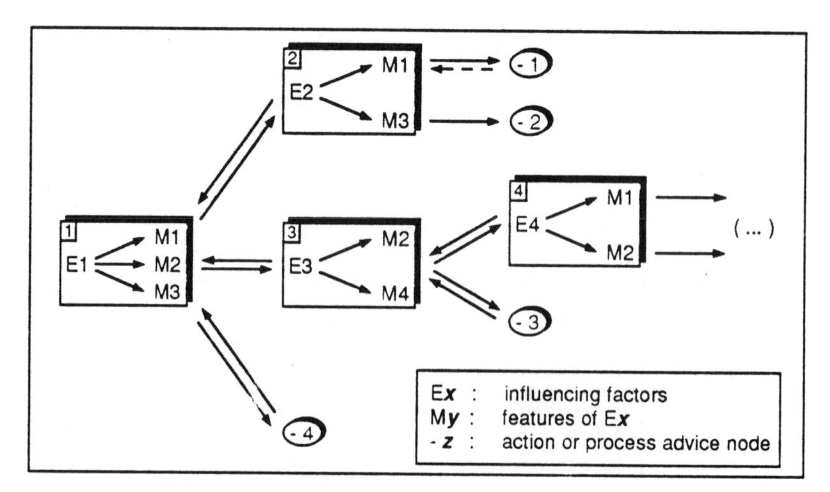

Figure 5 - Schematic interconnection of knowledge elements; the framework can be used in both directions

In terms of Figure 5 this means:
'if <*influencing factor E1*> is <*M1*> this is caused by <*E2*>',
'if <*influencing factor E1*> is <*M3*> follow advice <*-4*>', etc.
In this structure, also more complex influences can be represented. Depending on the type and value of the applicable feature (e.g. *M1*) of a certain factor (e.g. *E1*), other knowledge elements are searched to explain the feature, or an advice is given. The same structure can be used processing rules like, e.g.:
'if <*E1*> is <*M1*> and <*E3*> is <*M2*> and <*E4*> is <*M1*> then <...>'.

4. INVERSE MODELLING

Notion and application fields of inverse modelling of continuous casting processes have been described earlier (e.g. [15]). At this place rather the correlation between direct, heuristic and inverse modelling shall be indicated, regarding their common basis of cause-effect interrelations. As well, the reduction possibilities and synergy effects using standardized model libraries are adressed. 'Inverse modelling' means to find the causing conditions for given effects. This effects may be certain values of state variables, entire field distributions or transient courses, for example. In this context, the use of similarity characteristics [16] is of great advantage. Applied successfully, inverse modelling gives a chance to obtain problem solutions without the obligation to perform extensive model simulations for different parameter settings. The basic idea of inverse modelling is to transpose dependent and independent process variables on a theoretical level.

The methodology of inverse modelling is quite similar to that of direct modelling discussed in chapter 2. The strategy is, to define several abstraction levels with such partial models that can be inverted easily, and to calculate back from given and as feasible or even as optimal found field distributions to the causative conditions on a first abstraction level. Then, taking these causes themselves as dependent effects, the causative conditions on the next (lower) abstraction level are to be estimated. The procedure is repeated until geometric values or adjustable operational parameters are obtained.

The chance to build reverse solvable models depends on the possibility to reduce the complexity of the governing model equations. So, an essential facility for inverse modelling is first a complexity reduction outlining cause-effect interrelation diagrams and preparing partial models or keeping them in a model library. What follows, is a procedure which can be called 'hierarchical modelling'. Figure 6 shows a simplified example. Starting from the relatively untidy relationship model in the left part, a goal-oriented transformation and rearrangement leads to a *hierarchical* framework with refining abstraction levels. The decisive thing here is, that within the hierarchy dependent and independent variables are exchanged by analytical or numerical inversion of an appropriate direct model. If model inversion is not feasible, the corresponding direct model is object of a minimization algorithm, i.e. the input of a non-reversible direct model is estimated by minimizing the error between actual and desired model output. The simplest and most efficient method to do that, is a root finding method, provided that for the estimation error a unique root exists. Then, usually 4 to 5 iteration steps are sufficient for an adequate solution. Out of this, two application examples shall be mentioned briefly:

- Coupling a purely thermal model with a constraining stress/strain-model, and computing the strand shrinkage within the mould, due to cooling, opti-

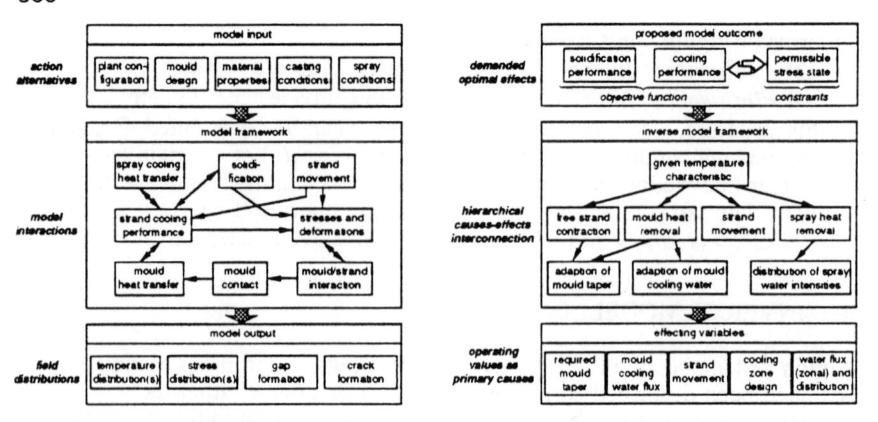

Figure 6 - Simplified cause-effect framework of a continuous casting process model, and its transformed hierarchical representation

mum mould taper(s) can be fitted to the estimated heat transfer conditions, otherwise requiring a multi-field analysis.

In a similar manner the secondary cooling conditions can be obtained: From estimated heat transfer coefficients and strand surface temperatures the unknown amount and distribution of cooling water, as an operational variable, can be obtained with the aid of inverse formulated heat transfer models.

To exclude ambiguous solutions, additional engineering knowledge is needed. It concerns the properties of design variables and their limits, a weighing of physical effects, the feasibility of technological realization, and the relation between model sensitivity and plant sensitivity to model parameters. However, this shows parallels to direct and heuristic modelling.

CONCLUSIONS

Disregarding formalization aspects, process modelling focusses on a reducing and therefore manageable representation of reality. Depending on the goals initiating modelling, and also on the structure of available knowledge, three different types of modelling, i.e. direct, heuristic and inverse modelling can be distinguished but found related descendants of one origin. Applying the methodology of modelling in a consequent manner, a common basis is found in a graphical formalization of cause-effect interrelations. Starting from this, each block may be directly filled by a computer model so that the overall model representation is quite modular. Different models may be derived and used without violation of the overall system structure. Exceeding a critical quantity, the partial models and tools should be organized in a model library.

The model library concept is not revolutionary but evolutionary. It evolves directly from previous process modelling, fueled by actual needs for standardization, generalization, documentation and effective computing. Inverting appropriate direct partial models and re-arranging them in a hierarchical framework of cause-effect interrelations, the same model library is basis for inverse modelling. The model library concept can also be applied to the formalization of heuristic knowledge, treating partial elements of empirical knowledge like partial mathematical models. Concequently, the knowledge and model library concept forms a platform to shorten the difficulties and time requirements of the modelling process as well as to improve the quality of model utilization. Given a certain problem, the user may choose between different options, i.e. applying operational expert knowledge, applying previous computation results, performing model simulations, performing reverse calculations, or may pursue different ways to compare and to verify the results.

Further investigations will focus on the minimization of supervision and interaction with a human operator being both modelling and domain specialist, as well as to enlarge modelbase capabilities with additional partial models. However, this should not give the impression that knowledge-based modelling could ever *totally* replace human imaginativeness and skill. It only takes *routine* tasks from the user to get more time for more intelligent tasks.

REFERENCES

1. JUNGHANS, H. - Deutsches Reichspatent DRP 510361, 1927.
2. BOEHMER, J.R. - Integration wissensbasierter Systeme bei Modellbildung und Simulation, dargestellt für das Beispiel Stranggußsimulation, Mathematische Modellbildung für Energieumwandlungsprozesse, Ed. Boehmer, J.R., VDI Verlag, 1992.
3. CHALMERS, B. - Principles of Solidification, John Wiley & Sons, New York, 1964.
4. BOLLING, G.F. - Manipulation of Structure and Properties, Solidification, Ed. Hughel, T.J. and Bolling, G.F., American Society for Metals, Ohio, 1971.
5. ROTH, A., WELSCH, M. and ROERIG, H. - Über die Eigenspannungen in Strangguß-Blöcken aus einer eutektischen Al-Si-Legierung, Aluminium, Vol. 24, pp.206-209, 1942.
6. FINE, H.A. and GEIGER, G.H. - Handbook on Material and Energy Balance Calculations in Metallurgical Processes, The Metallurgical Society of AIME, 1979.
7. DARDEL, Y. - La Transmission de la Chaleur au Cours de la Solidification, du Réchauffage et de la Trempe de lAcier, Editions de la Revue de Métallurgie, 1964.
8. RUDDLE, R.W. - The Solidification of Castings, The Institute of Metals, London, 1957.
9. CIARLET, P.G. and LIONS, J.L. (eds.) - Handbook of Numerical Analysis, Elsevier Science Publ., 1990.
10. WATERMAN, D.A. - A Guide to Expert Systems, Addison-Wesley, 1986.
11. FEIGENBAUM, E.A. - Expert Systems in the 1980s, Machine Intelligence, State of Art Report, ser. 9, no. 3, Ed. Bond, A., Pergamon Infotech Ltd., Maidenhead, 1981.
12. VALAVANIS, K.P. and SARIDIS, G.N. - A Review of Intelligent Control Based Methodologies for Modelling and Analysis of Hierarchically Intelligent Systems, Intelligent Control 1990, Los Alamitos IEEE Comput. Soc. Press, 1990.
13. NEWTON, R.L. - Definitions and Causes of Continuous Casting Defects, ISI Publ. 106, The Iron and Steel Institute, London, 1967.
14. BOEHMER, J.R., FETT, F.N., POEPPEL, M. and HENTRICH, R. - EXPERTS: An Expert System for the Continuous Casting Process (Development and Applications), Advances in Continuous Casting: Research and Technology, Ed. Taha, M.A. and El-Mahallawy, N.A., Woodhead, Cambridge, 1992.
15. BOEHMER, J.R. and FETT, F.N. - Feasibility and Limits of Inverse Modelling with Regard to the Continuous Casting Process, Numerical Methods in Thermal Problems, Vol. VII, Part 1, Ed. Lewis, R.W., Chin, J.H. and Homsy, G.M., Pineridge, 1991.
16. JESCHAR, R. - Anwendung der Ähnlichkeitstheorie in der Metallurgie, Kinetik metallurgischer Vorgänge bei der Stahlherstellung, Ed. Dahl, W., Lange, K.W. and Papamantellos, D., Verlag Stahleisen, 1972.

A NEW APPROACH TO SOLIDIFICATION AND MATERIAL STRAIN HISTORY DURING THE CONTINUOUS CASTING OF METALS

Boehmer, J.R., Fett, F.N. and Jordan, M.

Institute for Energy Technology, Department of Mechanical Engineering, University of Siegen, Paul-Bonatz-Str. 9-11, D-W-5900 Siegen, GERMANY

ABSTRACT

Coupled thermal and thermo-mechanical models are well-established to explain some aspects of crack formation during the continuous casting of metals. Overcoming the disadvantages of static formulations especially for the stress state, a more realistic approach to model the material's load history is proposed. For each strand shell element its experiences passing all zones of a continuous casting machine are traced, i.e. cooling down slowly or abruptly, perhaps reheating, and being compressed or drawn by neighbouring elements. The approach concerns the formulation of moving coordinate systems and moving boundary conditions, splitting the entire cooling process to a sequence of load steps. Emphasis of the nonlinear thermo-elastoplastic model, basing on temperature-dependent stress-strain relationships is on phase change and material properties. To explain the relationship between shrinkage, mould stability and gap formation, and their effect on the cooling conditions in the mould, modelling of strand contact performance in the mould is an integral part of the model.

1. INTRODUCTION

During the last decade efforts have been intensified to attend the continuous casting of billets, tubes, blooms, slabs and compact strips of steel and nonferrous metals by mathematical modelling and computation. The main interest is focused on solidification and cooling conditions, microstructure, strand contact and strand contraction as well as on thermal stresses of the material. Objective is the preparation of effective, realistic and reliable software models, to support the deepening of insights in general, to reduce the amount of expensive in-plant experiments, to explain the appearance of certain casting defects as well as to furtherdevelop process technology.

The idea to investigate and to predict the stress state during the continuous casting of metals is not new. Already in the beginnings of industrial utilization of the continuous casting process Roth et. al. [1] were interested in residual stresses in Al-Si billets. Twenty years later Weiner and Boley published their fundamental work on elasto-plastic thermal stresses in a solidifying body [2]. Figure 1 outlines the principle structure of a continuous casting machine for

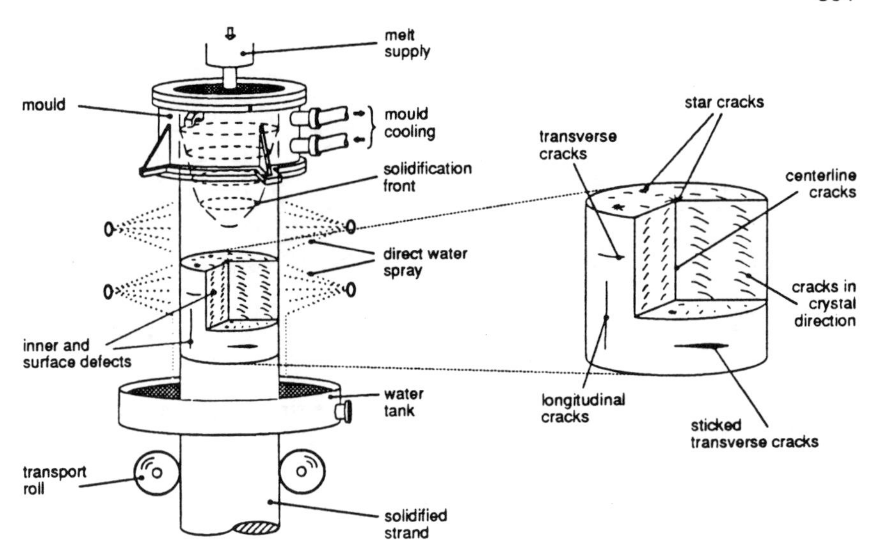

Figure 1 - Principle structure of a continuous casting machine for copper billets, and some crack types occuring

nonferrous metal billets with its different cooling zones, and an overview of some of most important crack types occuring.

Steinkamp [3] explains the crack formation during cooling by too strong temperature gradients along the billet radius. Consequently, both too high and too low cooling rates in the secondary cooling zone may cause damage. This shall be explained by a simple two-layer model (see Figure 2). Starting from a stress-free state (temperature T_0), in the left part of Figure 2 the surface layer is cooled more intensive than the inner one (cooling I). This causes radial and hoop stresses due to different shrinkage. The result are compressive tangential stresses in the centre, tensile tangential stresses on the surface, and compressive stresses in radial direction. On the other hand, if the inner parts are stronger cooled than the outer (or the surface is reheated), in tangential direction there is tension in the centre and compression on the surface, and in radial direction the two layers tend to pull from each other.

The goal of mathematical modelling of strand shell stresses, caused by the cooling and moulding conditions, is to understand the effects of the casting conditions on crack formation, to avoid critical states and to find optimal casting parameters in the end. Until now, quite a number of more or less complex mathematical models for continuous casting processes has been developed, to describe and to explain especially the occurance of high-temperature cracks. Starting from the fundamental equations of classical continuum mechanics usually the residual stress state of the solidified casting material is investigated, sometimes fully coupled with the thermal conditions, but always neglecting both cooling history and strand movement, and its direct transitory influence on the stresses. However, the mathematical, and especially the numerical, representation of the real integration path, taking into account nonlinear material behaviour, phase transformation and surface contact (by transient loadings and moving boundary conditions) make the model quite complex.

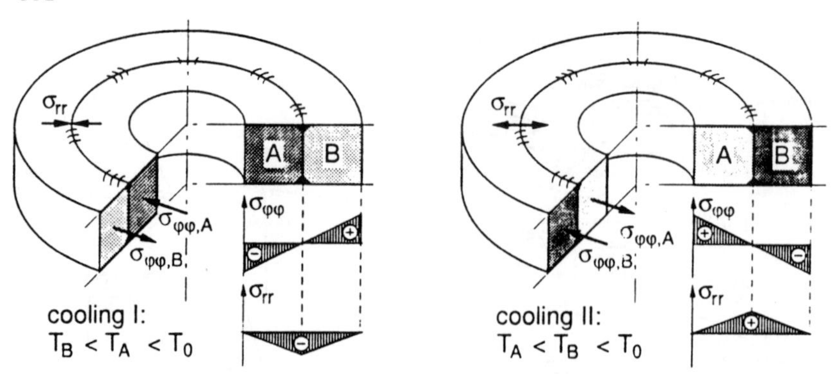

Figure 2 - Development of tensile and compressive stresses in two ideally connected concentric rings for different cooling strategies

2. COUPLED THERMAL AND MECHANICAL ANALYSIS

The mechanical model to be described here is part of a modular computer model to simulate solidification, heat removal and thermal strain during the continuous casting of round shapes [4], which itself is an integral part of a more general model library concept for continuous casting processes [5]. Figure 3 shows the principle structure of the embedding of the stress/strain model within the overall model. The reason for crack formation can be seen in an overstraining of the material´s ability to react pliable on existing thermal loads, which themselves are caused by locally different material contraction. Accordingly, thermal, material property, and thermo-mechanical models have been developed. Within the blocks 'thermal model' or 'contraction model' further couplings between the partial systems strand/mould/coolant are performed as well as for an evaluation

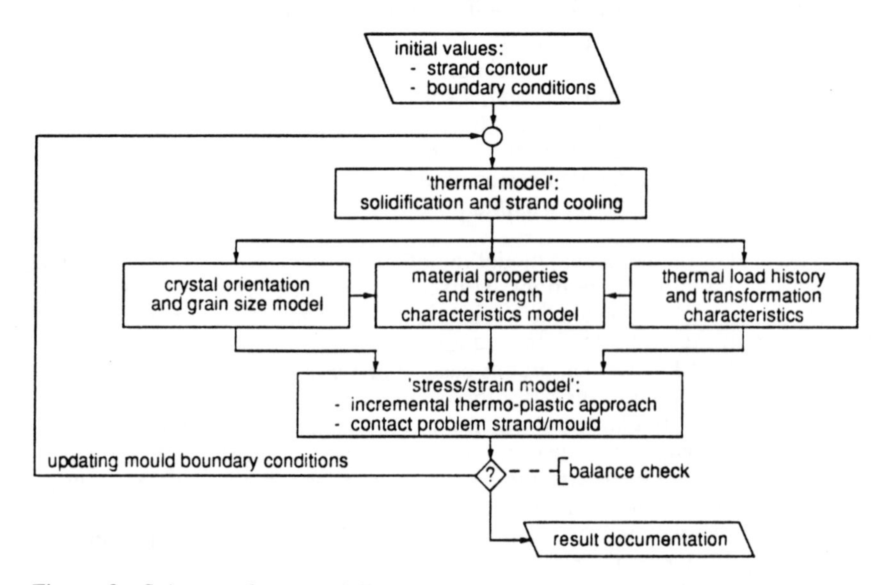

Figure 3 - Scheme of sequentially coupled calculation of temperatures, material properties, deformations and stresses to simulate the physics of the cc process

of link variables like solidification geometry, local solidification enthalpy fraction, convection-modulated heat conductivity coefficients of the melting bath, surface heat transfer coefficients, pressure and friction forces, displacements, etc. With this, the couplings between mass flow (strand movement), temperature distribution, metallic structure, stresses and strains, and the influences of the outside world have been reduced, such that separate, self-contained partial models can be treated sequentially (regarding each constant conditions), and balanced by iteration. Emphasis is on the two way coupling between the heat transfer and the thermal distortion of the strand. The effective heat transfer between the strand and the mould is largely determined by the size of the gap between them, and the size of the gap is governed by the cooling conditions. Because the present analysis focuses on the steady-state case, the balancing iteration ends if mould boundary conditions, i.e. contact and gap geometry, and the accompanying heat transfer coefficients become steady-state.

As a result of thermal analysis, Figure 4 shows in its upper part a calculated temperature distribution in a half longitudinal cross-section of a copper billet. The cooling strategy is to spray water onto the strand surface after leaving the mould. So, only four different boundary models (mould cooling, radiation to air, spray cooling and, not indicated in Figure 4, a collecting tank) are to be applied. The most sophisticated model is the mould cooling model, being feedback coupled with contraction analysis. In the lower part of Figure 4 the axial and, correlated with the constant casting speed, at the same time temporal temperature courses at four different locations, tracking the signals of fictitious inner thermocouples, are outlined. One can see the strong temperature gradients necessary to remove the solidification enthalpy and also effects on the strand surface, such as reheating of the strand shell for cooling breakdown and the nonlinear temperature dependency of spray cooling heat transfer.

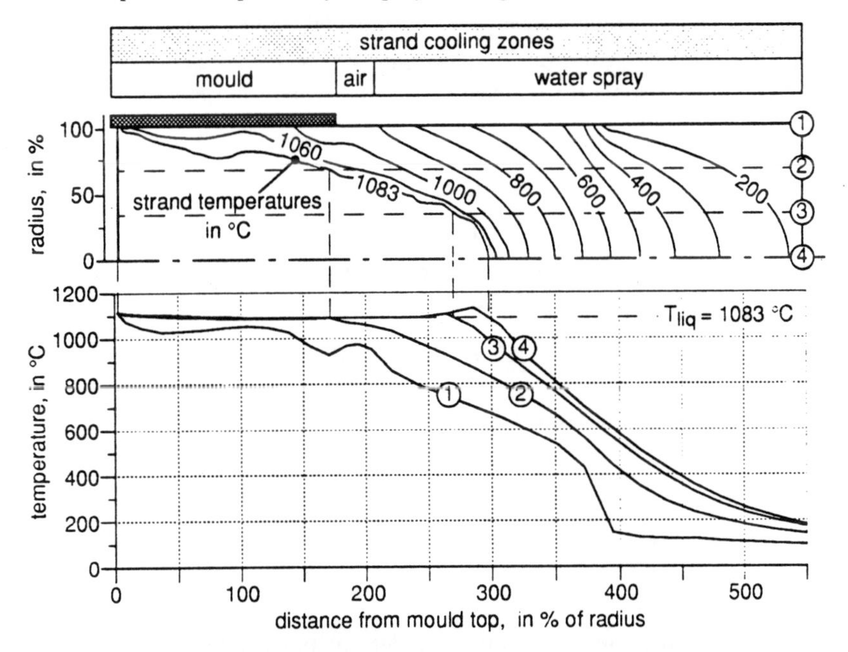

Figure 4 - Calculated temperature distribution in half a longitudinal cross-section of a copper billet, and temperature courses at four different locations

3. OVERCOMING STATIC FORMULATIONS

In this section at first the 'usual' approach of continuum mechanics to the calculation of thermal induced stresses shall be outlined. Although moving, the solidified strand shell is considered to be stationary, and the thermal load to each observed strand volume element is calculated with regard to the temperature change relative to a global reference temperature. Because this reference temperature holds for the stress-free state, in the case of a solidification process the solidus temperature is taken as reference. With this, residual stresses can be calculated, and qualitative indications of defect potentials or possibilities for process improvement can be extracted, especially regarding and comparing stress change rates but which are a posteriori determined. Nevertheless, some model results like, e.g. the prediction of unplausible compressive radial stresses in the strand centre just after final solidification, remained unsolved and weakened the overall performance of coupled modelling of the continuous casting process. Switching from a stationary to a moving coordinate system in the mathematical representation and including the real integration path in the calculation of thermal stresses these open questions could be cleared. The underlying method is the central topic of this article.

3.1 Calculation of Residual Stresses

The load collective of the strand contains metallostatic pressure of the melt, the contact pressure between strand and mould, the temperature gradients within the solidified strand shell as well as strand transport forces. The description of contraction behaviour and stress formation is performed by a finite element model for the axisymmetric case. The standard equations [6] for the relation between the unknown displacements $\underline{\delta} = \{u, v\}^T$, the nodal forces $\underline{F} = \{F_r, F_z\}^T$, the stresses $\underline{\sigma} = \{\sigma_{zz}, \sigma_{rr}, \sigma_{\varphi\varphi}, \tau_{rz}\}^T$, the strains $\underline{\varepsilon} = \{\varepsilon_{zz}, \varepsilon_{rr}, \varepsilon_{\varphi\varphi}, \gamma_{rz}\}^T$ and the mechanical properties, for the moment represented by the elasticity matrix $\underline{\underline{D}}$ are

$$\underline{\underline{K}} \cdot \underline{\delta} = \underline{F} \quad , \tag{1}$$

$$\underline{\varepsilon} = \underline{\underline{B}} \cdot \underline{\delta} \quad , \tag{2}$$

$$\underline{\sigma} = \underline{\underline{D}} \cdot (\underline{\varepsilon} - \underline{\varepsilon}_0) + \underline{\sigma}_0 \quad , \tag{3}$$

where $\underline{\underline{B}}$ is the strain-displacement matrix. The global stiffness matrix $\underline{\underline{K}}$ is composed of element stiffness matrices

$$\underline{\underline{K}}^e = \int_V \underline{\underline{B}}^T \cdot \underline{\underline{D}} \cdot \underline{\underline{B}} \, dV \quad . \tag{4}$$

For isotropic material and axisymmetric stress state the elasticity matrix is

$$\underline{\underline{D}} = \frac{E}{(1+\nu)\cdot(1-2\nu)} \cdot \begin{bmatrix} 1-\nu & \nu & \nu & 0 \\ \nu & 1-\nu & \nu & 0 \\ \nu & \nu & 1-\nu & 0 \\ 0 & 0 & 0 & \frac{1-2\nu}{2} \end{bmatrix} \quad . \tag{5}$$

The vector of initial strains $\underline{\varepsilon}_0$ contains all deformations not depending on the stress state, i.e. transformation shrinkage or dilatation $\underline{\varepsilon}_0^{tr}$ and especially thermal strains $\underline{\varepsilon}_0^{th}$:

$$\underline{\varepsilon}_0 = \underline{\varepsilon}_0^{tr} + \underline{\varepsilon}_0^{th} = \underline{\varepsilon}_0^{tr} + \begin{Bmatrix} \alpha \\ \alpha \\ \alpha \\ 0 \end{Bmatrix} \cdot (T - T_0) \quad , \tag{6}$$

where T_0 is the zero-stress reference temperature mentioned above. Considering the temperature dependence of material properties, an averaged element temperature $(T+T_0)/2$ is taken into account. For the discretization of the strand shell continuum the use of constant strain triangular ring elements is of advantage for an excellent represenation of strand shell geometry, and it is sufficient with regard to calculation precision. On the basis of a detailed description of solidification shrinkage, the model further allows to predict the occurance of shrink holes.

To model thermo-elastoplastic material behaviour using *original* material characteristics (σ-ε-T relations) the method of initial stresses [6] has been applied. Using the term $\underline{\sigma}_0$ of relation (3), the solution $\underline{\sigma}^{el}$ of linear thermo-elastic analysis can be corrected if the yield stress has been exceeded locally:

$$\underline{\sigma}_0 = \underline{\sigma}^{el} - \underline{\sigma}^{corr} \quad . \tag{7}$$

The method is to introduce fictitious corrective forces

$$\underline{F}^{\sigma_0} = \int_V \underline{\underline{B}}^T \cdot \underline{\sigma}_0 \, dV \quad , \tag{8}$$

which are added to the nodal forces $\underline{F}$ in (1). Basis for the calculation of $\underline{\sigma}^{corr}$ is the von Mises material model [7,8,9].

The thermal distortion of the mould is modelled by a separate module which is explained in detail in [9]. Figure 5 shows the three residual thermo-elastoplastic principal stresses (σ_1, σ_2, σ_3) for the temperature course shown in Figure 4, using the 'static' approach (1) to (8) and taking into account the contact problem between strand and mould, so that (1) becomes:

$$\underline{\underline{K}} \cdot \underline{\delta} = \underline{F} + \underline{F}^{\varepsilon_0} + \underline{F}^{\sigma_0} + \underline{F}^{cont} \quad . \tag{9}$$

Looking at the stresses which, for better interpretation, are standardized by the local temperature-dependent material strength $\sigma_{th,ul}$, one can note extreme contact responses in the strand shell at the mould exit. Taking into account the temperature profile (Figure 4), it can be seen that the reheating of the strand shell, which begins in the middle of the mould, has an expansive effect. But this thermal expansion trend is restricted by a strong mould taper, so that the strand shell is squeezed through the mould's cone. To avoid this effect, the mould taper must be stronger in the upper part and slighter at the mould's end. Comparing the temperature gradients and the circumferential stresses on the surface in the spray cooling zone it can be seen that, in the present case, intensive surface cooling can compensate high surface stresses. After final solidification one can find quite high compressive stresses in all directions in the centre. Looking again at the temperatures it can be seen that the centre is first stronger cooled than the surface, but after one radius length after final solidifica-

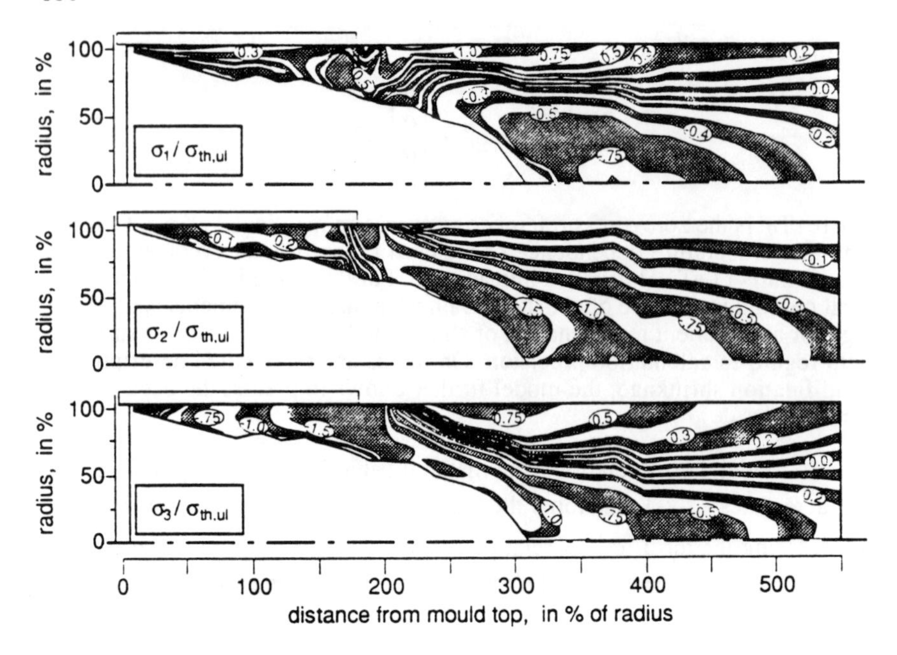

Figure 5 - Thermal-strength standardized thermo-elastoplastic principal stresses using the static stress analysis approach with a global reference temperature

tion the surface is stronger cooled than the centre. Going back to the considerations of section 1, one would expect first tensile hoop stresses in the centre and compressive hoop stresses on the surface as well as tensile radial stresses, and then an increasing trend to reverse these conditions. The load step approach to be discussed in the following is supposed to clear present contradictions to plausibility.

3.2 Moving Coordinate System and Load Step Formulation

Regarding the fact that each just solidified strand volume element, frozen and wedged in the matrix of neighbouring elements, is drawn in line to the casting direction, its thermal load history can be obtained by tracking its temperature as it is moved down through the casting machine. In Figure 4 such cooling histories are outlined for four di ferent locations in the strand. The total thermal load associated with the temperature change relative to the zero-stress reference temperature T_0, in our case the solidification temperature, is:

$$\underline{F}^{\varepsilon_0} = -\int_V \underline{\underline{B}}^T \cdot \underline{\underline{D}} \cdot \underline{\varepsilon}_0 \, dV \quad , \tag{10}$$

where $\underline{\varepsilon}_0$ is defined by (6). Applying the total thermal load between the beginning of solidification (temperature T_0) and the actual thermal state (temperature $T(r,z)$) in only one load step (as described in section 3.1) assumes a *linear* cooling history, i.e. constant temporal gradients for each considered volume element. To approximate the real temporal integration path for the calculation of thermal stresses, $\underline{F}^{\varepsilon_0}$ is split up to a number L of incremental loads which are applied sequentially:

$$^{(ls)}\Delta\underline{F}^{\varepsilon_0} = -\int_V {}^{(ls)}\underline{\underline{B}}^T \cdot {}^{(ls)}\underline{\underline{D}} \cdot {}^{(ls)}\Delta\underline{\varepsilon}_0 \, dV$$

$$= -\int_V {}^{(ls)}\underline{\underline{B}}^T \cdot {}^{(ls)}\underline{\underline{D}} \cdot [{}^{(ls)}\Delta\underline{\varepsilon}_0^{tr} + \underline{\alpha} \cdot ({}^{(ls)}T - {}^{(ls-1)}T)] \, dV \quad , \tag{11}$$

where ${}^{(0)}T = T_0 = T_{sol}$, ${}^{(L)}T = T(r,z)$ and

$$\sum_{ls=1}^{L} ({}^{(ls)}T - {}^{(ls-1)}T) = T - T_0 \quad . \tag{12}$$

Taking into account that each strand volume element is moved downstream, the incremental loads ${}^{(ls)}\Delta\underline{F}^{\varepsilon_0}$ are calculated for a short time interval ${}^{(ls)}\Delta t$, in which the considered elements are not only cooled down by $({}^{(ls)}T - {}^{(ls-1)}T)$ but also moved by

$$^{(ls)}\Delta z = {}^{(ls)}v_C \cdot {}^{(ls)}\Delta t \quad , \tag{13}$$

according to the actual casting speed ${}^{(ls)}v_C$. This results in a moving coordinate procedure as shown in Figure 6. For clarification, the axial increments ${}^{(ls)}\Delta z$ there are chosen quite rough; in fact they have to be considerably smaller to represent cooling nonlinearities (e.g. the strand shell reheating in the mould) sufficiently.

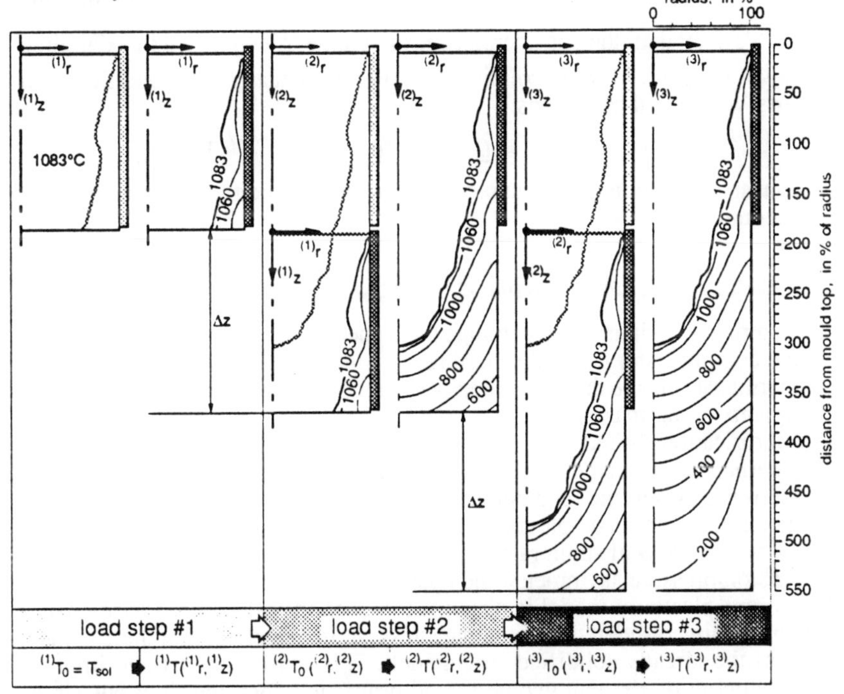

Figure 6 - Moving coordinate approach for three different load steps; shown are each initial (T_0) and final $({}^{(ls)}T)$ strand temperatures

The incremental formulation of the standard FE equations leads to a modified set of equations which have to be evaluated for all L load steps:

$$^{(ls)}\underline{\underline{K}} \cdot {}^{(ls)}\Delta\underline{\delta} = {}^{(ls)}\Delta\underline{F} \ , \tag{1'}$$

$$^{(ls)}\Delta\underline{\varepsilon} = {}^{(ls)}\underline{\underline{B}} \cdot {}^{(ls)}\Delta\underline{\delta} \ , \tag{2'}$$

(the $\underline{\underline{B}}$ matrix has to be updated for each load step due to the change in geometry each triangular element is subject to),

$$^{(ls)}\Delta\underline{\sigma} = {}^{(ls)}\underline{\underline{D}} \cdot ({}^{(ls)}\Delta\underline{\varepsilon} - {}^{(ls)}\Delta\underline{\varepsilon}_0) + {}^{(ls)}\Delta\underline{\sigma}_0 \ . \tag{3'}$$

The total displacements, strains and stresses for each moment (ls) result from superposition:

$$^{(ls)}\underline{\delta} = \sum_{\ell=1}^{ls} {}^{(\ell)}\Delta\underline{\delta}({}^{(\ell)}r, {}^{(\ell)}z) \ , \tag{14}$$

$$^{(ls)}\underline{\varepsilon} = \sum_{\ell=1}^{ls} {}^{(\ell)}\Delta\underline{\varepsilon}({}^{(\ell)}r, {}^{(\ell)}z) \ , \tag{15}$$

$$^{(ls)}\underline{\sigma} = \sum_{\ell=1}^{ls} {}^{(\ell)}\Delta\underline{\sigma}({}^{(\ell)}r, {}^{(\ell)}z) \ . \tag{16}$$

If ℓ becomes L, the final displacements, strains and stresses are obtained. Intermediate solutions after a load step $(ls\text{-}1)$ are, according to the strand movement, mapped from the coordinate system $({}^{(ls-1)}r, {}^{(ls-1)}z)$ to the new one $({}^{(ls)}r, {}^{(ls)}z)$. By that, the considered control volume is lengthened in axial direction, giving room for further solidification. For the actually new solidified volume parts, zero initial conditions

$$^{(ls-1)}\underline{\delta}^0 = \underline{0} \ ; \quad {}^{(ls-1)}\underline{\varepsilon}^0 = \underline{0} \ ; \quad {}^{(ls-1)}\underline{\sigma}^0 = \underline{0} \ ; \quad {}^{(ls-1)}T = T_{sol} \tag{17}$$

are assumed. Additionally, shrinkage effects during solidification may be taken into account.

3.3 Phase Change and Material Properties

The material behaviour of the cooling strand shell is characterized by

- the cooling history, initially starting from the melt,
- thermal shrinkage due to cooling with perhaps temperature-dependent coefficients of thermal expansion,
- solidification shrinkage,
- a perhaps disjointed dilatation course due to solid phase transformations,
- thermo-elastoplastic (and perhaps creep) strains,
- material strength and plasticity, dependent on local temperature, amount of working and local strain rate,
- tension/pressure asymmetry and hysteresis.

The thermo-plasticity model is described in sections 3.1 and 5. Phase change phenomena are considered twice in the stress/strain model. At first there is movement of the already solidified strand shell. This causes at least two implications: mapping of element geometry and properties from one fictitious strand slice to the next following, and initializing the properties of the just solidifying volume parts; in addition, solidification shrinkage may be taken into account.

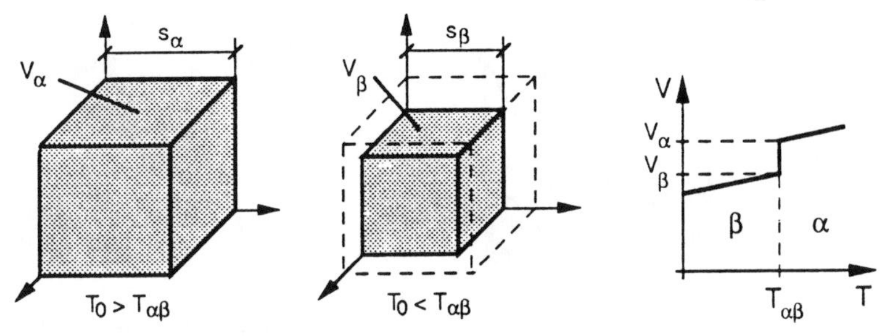

Figure 7 - Shrinkage of a cube due to phase change from phase α to phase β

On the other hand, abrupt density changes can be modelled with $\underline{\varepsilon}_0^{tr}$ in (6):

$$\underline{\varepsilon}_0^{tr} = \begin{Bmatrix} 1 \\ 1 \\ 1 \\ 0 \end{Bmatrix} \cdot \varepsilon_{\alpha\beta} = \begin{Bmatrix} 1 \\ 1 \\ 1 \\ 0 \end{Bmatrix} \cdot \left(\sqrt[3]{1 + \frac{V_\beta - V_\alpha}{V_\alpha}} - 1 \right) \quad . \tag{18}$$

For the nomenclature see Figure 7.

Figure 8 shows the consequences of the moving coordinate approach considering solidification to and phase transformation in the solid. It refers to the change in conditions during a single load step. The phase states considered are:

- the initial phase state after load step *(ls-1)*,
- the same state after performing incremental strand movement,
- the transformations during load step *(ls)*,
- the final phase state after load step *(ls)*.

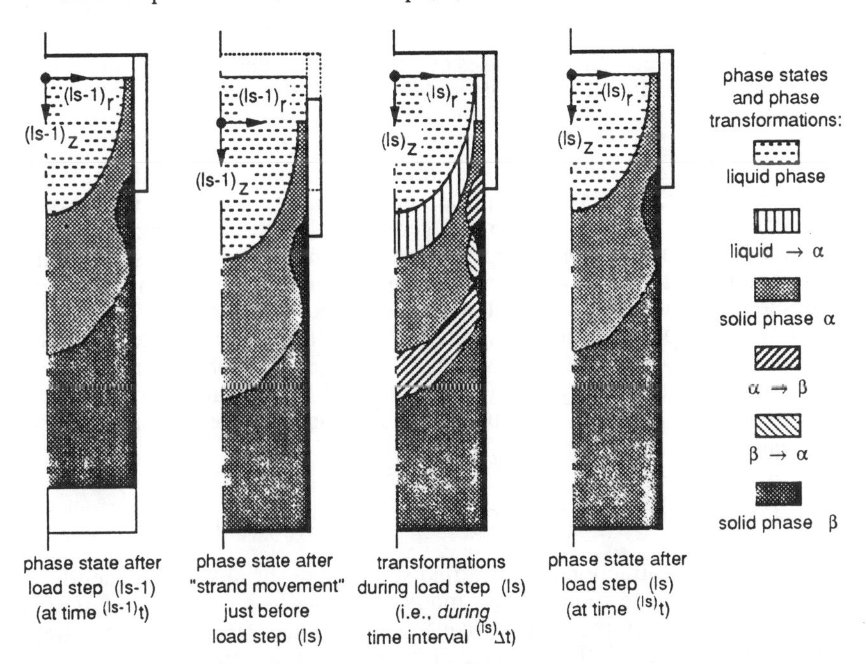

Figure 8 - Consideration of phase transformations during one time step

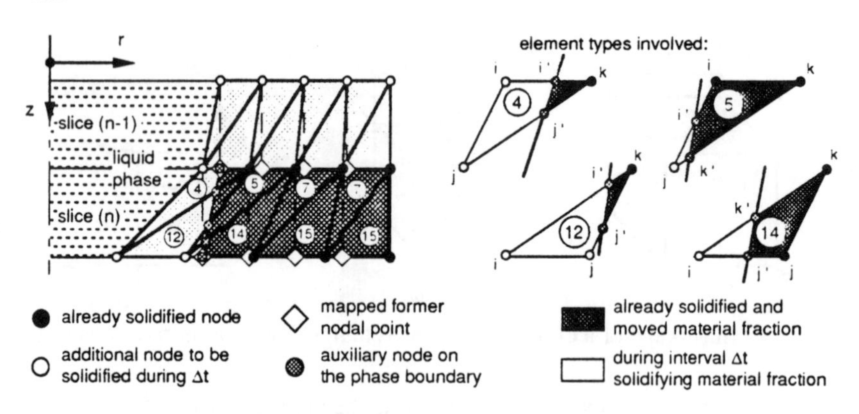

Figure 9 - Element geometry and property mapping due to strand movement between two load steps

For the stress/strain analysis the axisymmetric strand shell is discretized to ring elements with triangular cross-section. The implications involved in geometry and property mapping during the moving coordinate system and load-step approach shall be discussed regarding Figure 9. Let us consider slice *(n-1)*, moved downstream to its new position as slice number *(n)*. Let us further assume that slice *(n-1)* had been just solidified from the melt before cooling during the first load step. Regarding the course of the solidification front, it becomes clear that, after moving slice *(n-1)*, its former position is again filled with solidifying and cooling material. As well, in position *(n)* the slice *(n-1)* is growing to a larger thickness, as shown in Figure 9. Therefore, the dark shaded area represents older material than the light ones. Regarding the triangular discretization, the former course of the solidification front causes a differentiation between sixteen different cases of new element intersection by the former phase boundary. In our example only six different cases from them are outlined.

The mapping procedure prerequisites a method to realize an exact superposition of stresses, strains and displacements according to (14) to (16), taking into account the variation of element geometry. For the mapping procedure, only nodal point values are treated. I.e., the element-related stresses and strains are transformed to nodal values to perform the superposition. Mapping these values to the next following nodal point line, the former discretization in radial direction does not meet the actual one. Actual values are calculated by linear interpolation between the former values. To transfer transformation strains, the volume fraction of already transformed material is calculated and taken into account. During the load step calculation all properties are evaluated for the temperature level $(^{(ls)}T + ^{(ls-1)}T)/2$.

4. MODELLING STRAND CONTACT IN THE MOULD

Due to thermal contraction the solidified strand shell shrinks. Consequently, to guarantee an adequate cooling in the mould, the inner face of the mould must be tapered. A too strong mould taper causes high contact forces between strand and mould, leading at last to sticking or breaking off the strand shell. If, on the other hand, mould taper is not sufficient to compensate strand shrinkage, mould cooling breaks down which may cause breakouts. Finally, mould contact need not to be stationary: First, the decrease of strand surface cooling as a result of gap formation may cause a re-heating of the surface layers from the hot strand centre, what decreases the gap width owing to thermal expansion of the strand. The gap once closed forces again strand shrinkage and gap formation. On the other hand, mould oscillation causes in time local changes in friction forces, depending on the relative movement between strand and mould.

In the mechanical analysis the strand surface is first supposed to be free. The deformed structure is then compared to the inner geometry of the mould. If *negative* gap widths are detected, a method has to be applied to limit strand extension to the mould surface ('mould restriction'). In principle, there are two methods to realize this: prescribed displacements or corrective forces. For the first method the procedure is to determine and to apply all locations and values of prescribed displacements. The second method is to determine and then to apply those contact pressure forces which satisfy the mould restrictions. This has the advantage, that, leaving the mould, the strand surface elements keep only their irreversible plastic deformations when the corrective forces disappear.

According to (1') the deformation increments are calculated assuming the contact conditions of the last load step. In case of $ls=1$ the strand surface is supposed to be initially free. Being a component in the displacement vector (14) the total *radial* displacements are then:

$$^{(ls)}u\,(r,z) \;=\; \sum_{\ell=1}^{ls} {}^{(\ell)}\Delta u\,(r,z) \;. \tag{19}$$

With this, the new strand surface coordinates result in

$$^{(ls)}r_S\,(z) \;=\; {}^{(0)}r_S\,(z) + {}^{(ls)}u\,({}^{(0)}r_S, z) \;. \tag{20}$$

This is to be compared with the inner radius of the mould (which itself is a result of a mould deformation model):

$$^{(ls)}r_S\,(z) \;\leq\; {}^{(ls)}r_{mld}\,(z) \tag{21}$$

for all surface nodes with $0 \leq z \leq z_{mld}$. If condition (21) is violated at a position z_C, the last radial deformation increments are set to

$$^{(ls)}\Delta u_S\,(z_C) \;=\; {}^{(ls)}r_{mld}\,(z_C) \;-\; {}^{(ls\text{-}1)}r_S\,(z_C) \;, \tag{22}$$

and, after checking all axial nodes (1') has to be modified to

$$^{(ls)}\underline{\underline{\hat{K}}} \cdot {}^{(ls)}\Delta\underline{\hat{\delta}} \;=\; {}^{(ls)}\Delta\underline{\hat{F}}^{pl} \;, \tag{23}$$

where the prescribed displacements (22) are included in the following form (the example shows only the consideration of *one* prescribed displacement, and omits the load step and plasticity indices):

$$\begin{bmatrix} K_{11} & \cdots & 0 & \cdots & K_{1n} \\ \vdots & \ddots & 0 & \cdot^{\cdot^{\cdot}} & \vdots \\ 0 & 0 & 1 & 0 & 0 \\ \vdots & \cdot^{\cdot^{\cdot}} & 0 & \ddots & \vdots \\ K_{n1} & \cdots & 0 & \cdots & K_{nn} \end{bmatrix} \cdot \begin{Bmatrix} \Delta\delta_1 \\ \vdots \\ \Delta\hat{\delta}_j \\ \vdots \\ \Delta\delta_n \end{Bmatrix} = \begin{Bmatrix} \Delta F_1 - K_{1j} \cdot \Delta\hat{\delta}_j \\ \vdots \\ \Delta\hat{\delta}_j \\ \vdots \\ \Delta F_n - K_{nj} \cdot \Delta\hat{\delta}_j \end{Bmatrix} \tag{24}$$

and solved again for the displacement vector ${}^{(ls)}\Delta\hat{\underline{\delta}}$, taking into account the correction of plastic stresses, so that (23) becomes

$$ {}^{(ls)}\hat{\underline{\underline{K}}} \cdot {}^{(ls)}\Delta\hat{\underline{\delta}}^{pl} = {}^{(ls)}\Delta\hat{\underline{F}}^{pl} + {}^{(ls)}\Delta\hat{\underline{F}}^{\sigma_0} , \tag{25} $$

where ${}^{(ls)}\Delta\hat{\underline{F}}^{\sigma_0}$ are the plastic corrections due to additional displacements in the mould. The final result is then ${}^{(ls)}\Delta\hat{\underline{\delta}}^{pl}$.

Whereas the method of prescribed displacements stops here, the more sophisticated contact force method is to calculate the force equivalents of the fixed displacements due to mould restriction. They follow from

$$ {}^{(ls)}\Delta\underline{F}^{pl}_{mld} = {}^{(ls)}\underline{\underline{K}} \cdot {}^{(ls)}\Delta\hat{\underline{\delta}}^{pl} - {}^{(ls)}\Delta\underline{F}^{pl} , \tag{26} $$

where ${}^{(ls)}\underline{\underline{K}}$ is the original stiffness matrix from (1'). From ${}^{(ls)}\Delta\underline{F}^{pl}_{mld}$ only *compressive* forces are taken into account, but keeping the force equilibrium. This has the following background: Physically, it is not necessary to force *all* mould penetrating nodes separately onto the mould surface. Because they are connected to each other, restraining one node will move the adjoining nodes too, possibly forming a gap there. This effect would result in a tensile mould restriction force ${}^{(ls)}\Delta\underline{F}^{pl}_{mld}$ there. Therefore, resulting tensile forces are omitted, and the remaining compressive forces are decreased by the same amount. After correcting ${}^{(ls)}\Delta\underline{F}^{pl}_{mld}$ in this way to ${}^{(ls)}\Delta\tilde{\underline{F}}^{pl}_{mld}$, the resulting system equation becomes

$$ {}^{(ls)}\underline{\underline{K}} \cdot {}^{(ls)}\Delta\tilde{\underline{\delta}}^{pl} = {}^{(ls)}\Delta\underline{F}^{pl} + {}^{(ls)}\Delta\tilde{\underline{F}}^{pl}_{mld} , \tag{27} $$

Its solution ${}^{(ls)}\Delta\tilde{\underline{\delta}}^{pl}$ approximately satisfies the contact problem.

The procedure has to be integrated within the strand movement procedure. While all state variables of the strand are tied to a moving coordinate system, the mould boundary conditions (and force increments) are tied to the fixed global coordinate system. So, for a strand surface element moving downstream through the mould there are variable boundary conditions as a result of own deformation and change in mould width. Leaving the mould, the boundary conditions lapse for this element, but the (plastic) deformations of the strand remain.

5. INCLUSION OF THERMO-PLASTICITY

Some methodological aspects of the inclusion of thermo-plasticity within the linear standard FEM calculation have been discussed in section 3.1 for the static case. The advantage of the applied method is that *original* stress-strain diagrams can be input to the model without modification. Tracking the load history of the cast material, besides the solidification of new material at the inner surface of the strand shell, already pre-loaded and pre-strained material is subject to further load increments. In contrast to the static approach outlined in section 3.1, for each element during the number of load steps *different* stress-strain relationships (for different temperature levels) are to be evaluated, considering each different initial conditions (initial strains). Therefore, a method has to be applied

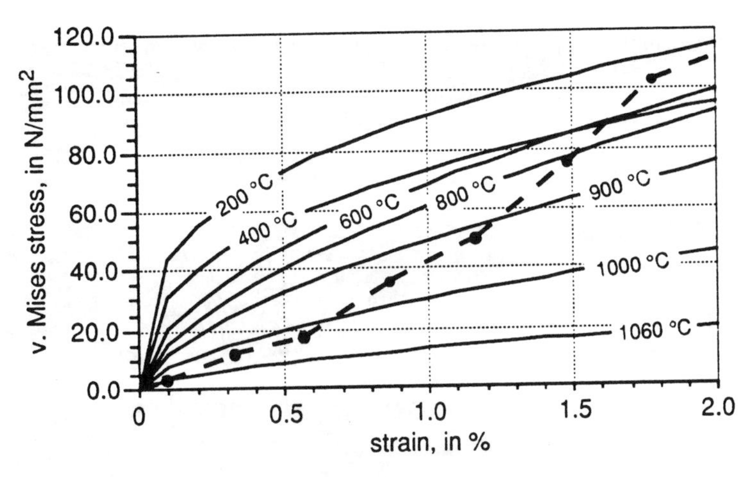

Figure 10 - Stress-strain diagrams for different temperature levels, and a characteristic loading history for a near-surface element

to switch from one σ-ε-curve to another but taking into account the remaining (i.e. the plastic) pre-strains. Figure 10 traces the variation of effective stresses and strains for a near-surface element which underlies the most eventful thermal history. The same property transformation can be performed for strain-rate dependent data, if available.

6. NUMERICAL SIMULATION AND RESULTS

According to the modular modelling concept the block 'stress/strain model' in Figure 3 has been programmed and tested separately before including into the entire simulation model. As a calculation example, Figure 11 shows calculation results of a 16-load-step approach. For comparison reasons the residual thermo-elastoplastic principal stresses (σ_1, σ_2, σ_3) are outlined in the same form as in Figure 5. Because a constant casting velocity was assumed, the moving procedure covers at least one element row (strand slice). Selecting 32 axial element rows, one can perform 1-, 2-, 4-, 8-, 16- and 32-load-step calculations. For the present case it can be shown that there is no significant change in the results of 16- and 32-load-step analysis. First, in the centre after final solidification one can find high tensile stresses in all directions. The solidified strand shell in the mould underlies the 'mould restriction', what causes high radial and circumferential compressive stresses as well as tensile stresses in the axial direction. Regarding the principal stress components σ_1 and σ_2, one can note that σ_1, mainly represents stresses in axial, and σ_2 represents stresses mainly in the radial direction. But there is one exception: In the spray cooling area the principal directions are rotated by 90° so that the zero radial stress condition on the free surface is reflected by σ_1, and not by σ_2 here. Regarding the circumferential stresses, the expectations gained from the simple layer model (Figure 2) can be confirmed. The spray water cooling following the mould causes a stronger cooling of the surface layers than in the inner parts. This results in tensile stresses in the surface layers. The situation is reversed after final solidification of the strand which is accompanied by extensive temperature gradients

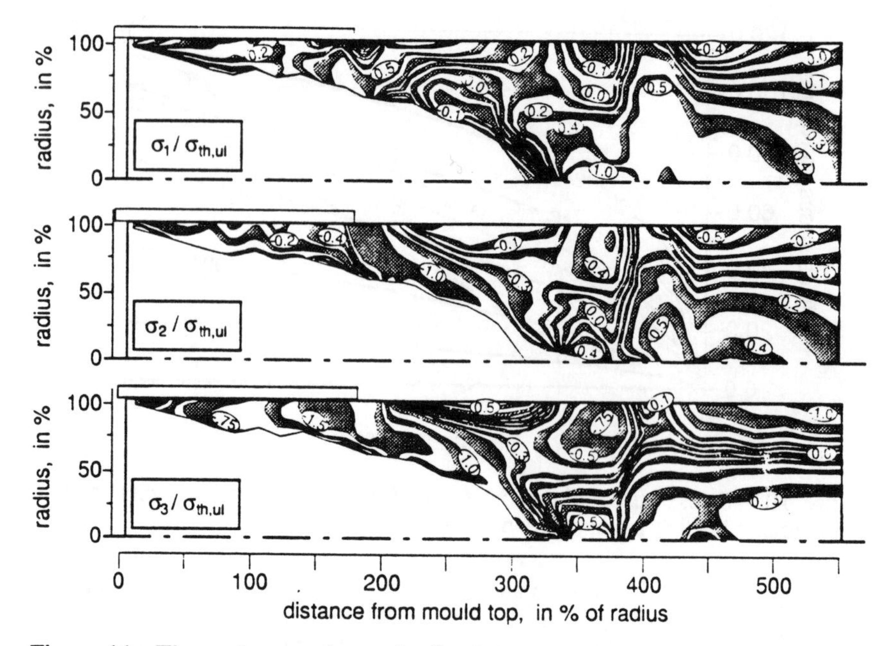

Figure 11 - Thermal-strength standardized thermo-elastoplastic principal stresses using a 16-load-step approach

and shrinkage. Therefore, circumferential tensile stresses can be found here, in contrast to compressive ones in the surface layers. But there is another tensile stress region downstream on the surface. Going back to Figure 4, one can see high temperature gradients in this region, because the burnout temperature for spray cooling heat transfer has been exceeded, so that again the surface is locally stronger cooled than the centre. Finally, and according to the temperature course, inner tensile and outer compressive stresses are obtained.

REFERENCES

1. ROTH, A., WELSCH, M. and ROERIG, H. - Über Eigenspannungen in Strangguß-Blöcken aus einer eutektischen Al-Si-Legierung, Aluminium, Vol.24, 206-209, 1942.
2. WEINER, J.B. and BOLEY, B.A. - Elasto-Plastic Thermal Stresses in a Solidifying Body, J.mech.Phys.Solids, Vol.11, pp.145-154, 1963.
3. STEINKAMP, W. - Abschätzung der Sekundärkühlleistung zum rißfreien Stranggießen niedriglegierter Kupferwerkstoffe, Z.Metallkunde, Vol.68, pp.470-477, 1977.
4. BOEHMER, J.R. - Ein modulares PC-Modell zur Berechnung von Erstarrung, Wärmeabfuhr und Materialbelastung beim Stranggießen von Metall-Rundformaten, Z.Metallkunde, Vol.84, pp.91-99, 1993.
5. BOEHMER, J.R. and FETT, F.N. - A Modular Framework to Utilize Mathematical Modelling for the Continuous Casting Process, Advances in Continuous Casting: Research and Technology, Ed. Taha, M.A. and El-Mahallawy, N.A., Woodhead, Cambridge, 1992.
6. ZIENKIEWICZ, O.C. - The Finite Element Method, McGraw-Hill, London, 1977.
7. RECKLING, K.A. - Plastizitätstheorie und ihre Anwendung auf Festigkeitsprobleme, Springer Verlag, Berlin, 1967.
8. MENDELSON, A. and MANSON, S.S. - NACA Techn. Note 4088, 1957.
9. BOEHMER, J.R. - Mathematisches Modell für das System Strang-Strangkühlung beim Strangguß von Metallen, Diss. Univ. Siegen, 1987.

SECTION 4

NATURAL AND/OR FORCED CONVECTION

STABILITY AND NUMERICAL DIFFUSION OF FINITE ELEMENT SCHEMES FOR CONVECTION-TYPE PROBLEMS

G. Comini†, M. Manzan‡, and C. Nonino§

SUMMARY

Various finite element schemes of the Bubnov-Galerkin and Taylor-Galerkin types are analysed, both in the transient and in the steady state, to obtain the expressions of numerical diffusion. Then, with reference to the transient advection-diffusion equation, stability limits are determined by means of a general Von Neumann procedure. Finally, the operational equivalence between Taylor-Galerkin methods, utilised for pseudo-transient calculations, and Petrov-Galerkin methods, derived for the steady state forms of the advection-diffusion equation is illustrated. Practical guidelines to identify the most suitable algorithm for any given convection-type problem are provided.

1. INTRODUCTION

The finite element method, based on Galerkin formulations, has become a well-established procedure for the solution of convection-type problems, where both advection and diffusion mechanisms must be accounted for in the mathematical model. On the other hand, for advection-dominated problems, standard Bubnov-Galerkin procedures can lead to space-oscillations, or "wiggles", in the numerical solution unless the mesh is refined to reduce the local Peclet number [1,2]. The use of "upwinding", or Petrov-Galerkin types of discretization, as an alternative treatment for node-to-node oscillations, has been investigated by many authors [1-3]. However, Petrov-Galerkin methods have been derived for the steady-state forms of the transport equations and their extension to the transient solutions is not always possible [1,2]. Instead, for the transient advection-diffusion equations, alternative discretization procedures, connected with special time integration schemes, have gained wide acceptance in recent years [1,2,4,5].

The most popular, among the new methods, are the balancing tensor diffusivity, the characteristic-based Galerkin and the Taylor-Galerkin methods [1-5]. Despite their widely different mathematical and physical foundations, all these methods lead to discretized equations that are either very similar, when the source terms are variable, or identical, when the source terms are

† Professor, ‡ Researcher, § Associate Professor
Istituto di Fisica Tecnica e di Tecnologie Industriali, Università di Udine,
Viale Ungheria 41, 33100 Udine, ITALY

constant. In fact, the discretized equations always contain upwinding terms that are dependent on the time increment used and are characterised by effects that do not disappear when the steady state has been reached. Thus, even if we are only interested in steady state solutions, we can still use one of these algorithms to find pseudo-transient solutions without "wiggles". On the other hand, with sufficiently refined meshes, standard Bubnov-Galerkin discretizations and implicit time integration schemes can still be considered a viable choice for the solution of convection problems. In fact, the standard Crank-Nicolson scheme is not plagued by numerical diffusion, while the standard fully implicit scheme allows the use of very large time steps and, even if it is characterised by a numerical diffusion in the transient, it does not present any numerical diffusion when the steady state has been reached [5].

At this point, the global picture can be rather confusing: we appreciate unconditional stability but we want zero numerical diffusion and, occasionally, we like to suppress wiggles without having to refine the mesh. We can have, at a price, any one of these nice features but we cannot have all of them built into a single algorithm. Obviously, at the end, we will have to compromise, but not before having evaluated very carefully all the possible approaches. In this paper several popular finite element procedures of the Bubnov-Galerkin and Taylor-Galerkin types are investigated, with the aim of finding numerical diffusion errors and stability limits. For the first time, in the context of the finite element method, numerical diffusion errors are evaluated both for transient and steady state situations. Actually, it has been established long ago that numerical diffusion errors can be different for transient and pseudo-transient calculations but, until now, to the authors' knowledge this distinction has been made only in the context of finite difference methods [6]. In the text, we also find the stability limits of the investigated algorithms by means of a very general procedure which is normally employed in the context of the finite difference method [7]. Finally, well-known theoretical analyses [1,2] are invoked to demonstrate the operational equivalence between Taylor-Galerkin methods, utilised for pseudo-transient calculations, and Petrov-Galerkin methods, derived for the steady state forms of the advection-diffusion equations.

In the next sections, the results obtained are discussed in detail, with the aim of establishing practical guidelines to identify the most suitable algorithm for any problem of the convection type.

2. NUMERICAL DIFFUSION ERRORS

The effect of diffusion is to smooth out gradients and any differences in the independent variable. Therefore, we can say that an algorithm is affected by artificial, or numerical, diffusion when the computed amplitudes of the elementary wave components are always smaller than the corresponding exact, or analytical, values and, consequently, perturbations are damped more than the physics requires. On the other hand, numerical schemes can also show a non-physical, "negative" diffusion and, in such a case, perturbations are amplified, sometimes to an extent that leads to numerical instability. The performances of finite element schemes with respect to diffusion errors are usually evaluated with reference to the pure advection equation to ensure that all diffusion effects have a non-physical origin. For a constant property, incompressible fluid, the pure advection equation can be written in the form

$$\frac{\partial T}{\partial t} = T_t = -u\,\frac{\partial T}{\partial x} = -u\,T_x \qquad (1)$$

where T is the temperature, t is the time and u is the velocity. If u is time independent, from Eq. (1) we obtain immediately

$$T_{tt} = \frac{\partial T_t}{\partial t} = -\frac{\partial}{\partial t}\left(u\,\frac{\partial T}{\partial x}\right) = -u\,\frac{\partial T_t}{\partial x} = u^2\,\frac{\partial^2 T}{\partial x^2} = u^2\,T_{xx} \tag{2}$$

The time discretization of Eq. (1) can be achieved by computing the value of the unknown T, at the time level $(n+1)\Delta t$, by means of "exact" Taylor series expansions of initial point $n\Delta t$

$$T^{n+1} \cong T^n + \Delta t\,T_t^{n+\alpha} \tag{3}$$

and

$$T^{n+1} \cong T^n + \Delta t\,T_t^n + \frac{\Delta t^2}{2}\,T_{tt}^{n+\gamma} \tag{4}$$

where α and γ are weighting parameters, suitably chosen in the interval between 0 and 1. Obviously, for α and γ arbitrarily chosen, the above equations are not exact any more and Eq. (3) yields an accuracy of the first order for the time expansion, while Eq. (4) leads to an accuracy of the second order. From Eqs. (3) and (1), we obtain the algorithms of the Bubnov-Galerkin type (BG) for the time discretization of the pure advection equation

$$\frac{T^{n+1} - T^n}{\Delta t} = T_t^{n+\alpha} = -u\,T_x^{n+\alpha} \cong -u\left[\alpha\,T_x^{n+1} + (1-\alpha)\,T_x^n\right] \tag{5}$$

In Eq. (5), the advection term can be evaluated anywhere in the interval between n and $(n+1)$ and, accordingly, for $\alpha = 0$ we obtain the BG explicit algorithm (BG–E), for $\alpha = 1/2$ we obtain the BG Crank-Nicolson algorithm (BG–C) and for $\alpha = 1$ we have the BG fully implicit algorithm (BG–I). Similarly, Eqs. (4), (1) and (2), yield the algorithms of the Taylor-Galerkin type (TG) for the partial time discretization of the pure advection equation

$$\frac{T^{n+1} - T^n}{\Delta t} = T_t^n + \frac{\Delta t}{2}\,T_{tt}^{n+\gamma} \cong -u\,T_x^n + \frac{u^2\,\Delta t}{2}\left[\gamma\,T_{xx}^{n+1} + (1-\gamma)\,T_{xx}^n\right] \tag{6}$$

In Eq. (6) the advection term is always evaluated at the level n, while the diffusion-like term can be evaluated anywhere in the interval between n and $(n+1)$. Accordingly, for $\gamma = 0$ we obtain the TG fully explicit algorithm (TG–E), for $\gamma = 1/2$ we obtain the TG Crank-Nicolson algorithm (TG–C) and for $\gamma = 1$ we have the TG implicit algorithm (TG–I).

Equations (5) and (6) can be discretized, with respect to the space coordinates, using a standard Bubnov-Galerkin process. For the typical node i, using elements of equal size Δx and linear shape functions, we obtain the assembled finite element equations [8]

$$\begin{aligned}
&\frac{1}{6\,\Delta t}\left[(T_{i-1}^{n+1} + 4\,T_i^{n+1} + T_{i+1}^{n+1}) - (T_{i-1}^n + 4\,T_i^n + T_{i+1}^n)\right] \\
&\quad = -\frac{u}{2\,\Delta x}\left[\alpha\,(T_{i+1}^{n+1} - T_{i-1}^{n+1}) + (1-\alpha)\,(T_{i+1}^n - T_{i-1}^n)\right] \\
&\quad + \beta\,\frac{u^2\,\Delta t}{2\,\Delta x^2}\left[\gamma\,(T_{i-1}^{n+1} - 2\,T_i^{n+1} + T_{i+1}^{n+1}) + (1-\gamma)\,(T_{i-1}^n - 2\,T_i^n + T_{i+1}^n)\right]
\end{aligned} \tag{7}$$

and

In determining the accuracy of the various algorithms, it is worth noting that Eqs. (7) and (8) can be considered numerical representations of the following modified transport equation [6,9]

$$T_t + u\,T_x = a_{et}\,T_{xx} + \epsilon_t^* = \epsilon_t \tag{9}$$

where a_{et} is the transient equivalent numerical diffusion, ϵ_t^* is the residual transient truncation error term and ϵ_t is the global transient truncation error term. Obviously, in the steady state we have

$$T_{i\pm1}^{n+1} = T_{i\pm1}^{n} = T_{i\pm1} \tag{10}$$

and thus Eqs. (7) and (8) reduce to the same form

$$\frac{u}{2\,\Delta x}\,(T_{i+1} - T_{i-1}) = \beta\,\frac{u^2\,\Delta t}{2\,\Delta x^2}\,(T_{i-1} - 2\,T_i + T_{i+1}) \tag{11}$$

which can be considered the numerical representation of the transport equation

$$u\,T_x = a_{es}\,T_{xx} + \epsilon_s^* = \epsilon_s \tag{12}$$

where a_{es} , ϵ_s^* and ϵ_s are the steady-state equivalent numerical diffusion, the residual truncation error and the global truncation error, respectively.

For all the schemes referred to in Table 1, from the general Taylor series expansion around grid point i and one of the two time levels $n\Delta t$ or $(n+1)\Delta t$

$$\begin{aligned} T(x+h,t+k) &= T(x,t) + h\,T_x(x,t) + k\,T_t(x,t) \\ &+ \frac{h^2}{2}\,T_{xx}(x,t) + h\,k\,T_{xt}(x,t) + \frac{k^2}{2}\,T_{tt}(x,t) \\ &+ \frac{h^3}{6}\,T_{xxx}(x,t) + h^2\,k\,T_{xxt}(x,t) + h\,k^2\,T_{xtt}(x,t) + \frac{k^3}{6}\,T_{ttt}(x,t) + \ldots \end{aligned} \tag{13}$$

we obtain the equivalent numerical diffusion coefficients and the truncation errors. In Eq. (13) h and k are the space and time increments, respectively. Similar analyses have been carried out by many authors, in the context of the finite difference method [6], and the results obtained are widely accepted as being applicable to multidimensional problems, with and without physical diffusion terms. Thus, it is quite reasonable to assume a general validity also for the present results since, in one-dimensional problems, with equal elements, linear shape functions and lumped capacity matrices, finite element algorithms lead to the same discretized equations which are obtained by using finite differences.

From Table 1 we can see that the BG–C schemes alone are characterised by truncation errors that do not include a numerical diffusion coefficient, either in the transient or in the steady-state analyses. Instead, the other BG algorithms exhibit numerical diffusion in the transient analysis but not in the steady-state analysis, while TG algorithms do not exhibit numerical diffusion in the transient analysis, but present numerical diffusion in the steady-state analysis. In particular, BG–E algorithms show a negative numerical diffusion in the transient analysis and, consequently, these algorithms are unstable if the partial differential equation does not include a physical diffusion term. Besides, it is worth noting that consistent capacity matrices always yield better space accuracies than lumped capacity matrices. Finally, it must be pointed out that the absence of numerical diffusion is not the only criterion to judge the performance of an algorothm. In fact, also dispersion errors may play an important role in determining the overall accuracy of a numerical scheme [2,6-8]. However, the present analysis is limited to diffusion errors.

3. STABILITY ANALYSIS

In analysing the numerical stability, we consider the transient advection-diffusion equation, since we are interested in the solution of convection-type problems. For a constant property, incompressible fluid the transient energy equation can be written as

$$\frac{\partial T}{\partial t} = T_t = -u\,\frac{\partial T}{\partial x} + a\,\frac{\partial^2 T}{\partial x^2} = -u\,T_x + a\,T_{xx} \tag{14}$$

where a is the thermal diffusion coefficient. The time discretization of Eq. (14) can be achieved by following the same steps outlined in the previous section. In fact the only addition, with respect to Eq. (1), is the physical diffusion term that, on the other hand, can be discretized in the same way as the numerical diffusion term. Thus, for the typical node i , we obtain the assembled finite element equations

$$\begin{aligned}
&\frac{1}{6\,\Delta t}\left[(T_{i-1}^{n+1} + 4\,T_i^{n+1} + T_{i+1}^{n+1}) - (T_{i-1}^{n} + 4\,T_i^{n} + T_{i+1}^{n})\right] \\
&\quad = -\frac{u}{2\,\Delta x}\left[\alpha\,(T_{i+1}^{n+1} - T_{i-1}^{n+1}) + (1-\alpha)\,(T_{i+1}^{n} - T_{i-1}^{n})\right] \\
&\quad + \beta\,\frac{u^2\,\Delta t}{2\,\Delta x^2}\left[\gamma\,(T_{i-1}^{n+1} - 2\,T_i^{n+1} + T_{i+1}^{n+1}) + (1-\gamma)\,(T_{i-1}^{n} - 2\,T_i^{n} + T_{i+1}^{n})\right] \\
&\quad + \frac{a}{\Delta x^2}\left[\vartheta\,(T_{i-1}^{n+1} - 2\,T_i^{n+1} + T_{i+1}^{n+1}) + (1-\vartheta)\,(T_{i-1}^{n} - 2\,T_i^{n} + T_{i+1}^{n})\right]
\end{aligned} \tag{15}$$

and

$$\begin{aligned}
\frac{1}{\Delta t}\,(T_i^{n+1} - T_i^{n}) &= -\frac{u}{2\,\Delta x}\left[\alpha\,(T_{i+1}^{n+1} - T_{i-1}^{n+1}) + (1-\alpha)\,(T_{i+1}^{n} - T_{i-1}^{n})\right] \\
&+ \beta\,\frac{u^2\,\Delta t}{2\,\Delta x^2}\left[\gamma\,(T_{i-1}^{n+1} - 2\,T_i^{n+1} + T_{i+1}^{n+1}) + (1-\gamma)\,(T_{i-1}^{n} - 2\,T_i^{n} + T_{i+1}^{n})\right] \\
&+ \frac{a}{\Delta x^2}\left[\vartheta\,(T_{i-1}^{n+1} - 2\,T_i^{n+1} + T_{i+1}^{n+1}) + (1-\vartheta)\,(T_{i-1}^{n} - 2\,T_i^{n} + T_{i+1}^{n})\right]
\end{aligned} \tag{16}$$

for consistent and lumped capacity matrices, respectively. Clearly, the values of parameters α, β, γ and ϑ determine the particular BG or TG algorithms, as summarised in Table 2 where the first three letters and the subscript in the abbreviation identifying the numerical scheme have the same meaning illustrated in the previous section. The last letter, instead, refers to the time discretization algorithm utilized for the diffusion term: E stands for explicit, C for Crank-Nicolson, and I for fully implicit. Again, we can point out that in one-dimensional problems, with elements of equal size, linear shape functions, and lumped capacity matrices, finite element algorithms lead to the same discretized equations that are obtained by using finite differences.

Equations (15) and (16) can be conveniently rewritten as

$$b_1\,T_{i-1}^{n+1} + (1-b_1-b_2)\,T_i^{n+1} + b_2\,T_{i+1}^{n+1} = d_1\,T_{i-1}^{n} + (1-d_1-d_2)\,T_i^{n} + d_2\,T_{i+1}^{n} \tag{17}$$

Table 1 - Values of the parameters α, β, and γ to be used in Eqs. (7), (8), and (11) to obtain various finite element schemes for the solution of the advection equation (1). Symbols a_{et} and a_{es} denote the transient and the steady-state equivalent diffusion coeffecients, while symbols ϵ_t and ϵ_s indicate the transient and the steady-state trucation errors.

Scheme	α	β	γ	a_{et}	ϵ_t	a_{es}	ϵ_s
BG_C–E	0	0	–	$-\frac{u^2\,\Delta t}{2}$	$O(\Delta t, \Delta x^4)$	0	$O(\Delta x^2)$
BG_L–E	0	0	–	$-\frac{u^2\,\Delta t}{2}$	$O(\Delta t, \Delta x^2)$	0	$O(\Delta x^2)$
BG_C–C	$\frac{1}{2}$	0	–	0	$O(\Delta t^2, \Delta x^4)$	0	$O(\Delta x^2)$
BG_L–C	$\frac{1}{2}$	0	–	0	$O(\Delta t^2, \Delta x^2)$	0	$O(\Delta x^2)$
BG_C–I	1	0	–	$\frac{u^2\,\Delta t}{2}$	$O(\Delta t, \Delta x^4)$	0	$O(\Delta x^2)$
BG_L–I	1	0	–	$\frac{u^2\,\Delta t}{2}$	$O(\Delta t, \Delta x^2)$	0	$O(\Delta x^2)$
TG_C–E	0	1	0	0	$O(\Delta t^2, \Delta x^4)$	$\frac{u^2\,\Delta t}{2}$	$O(\Delta x^2)$
TG_L–E	0	1	0	0	$O(\Delta t^2, \Delta x^2)$	$\frac{u^2\,\Delta t}{2}$	$O(\Delta x^2)$
TG_C–C	0	1	$\frac{1}{2}$	0	$O(\Delta t^2, \Delta x^4)$	$\frac{u^2\,\Delta t}{2}$	$O(\Delta x^2)$
TG_L–C	0	1	$\frac{1}{2}$	0	$O(\Delta t^2, \Delta x^2)$	$\frac{u^2\,\Delta t}{2}$	$O(\Delta x^2)$
TG_C–I	0	1	1	0	$O(\Delta t^2, \Delta x^4)$	$\frac{u^2\,\Delta t}{2}$	$O(\Delta x^2)$
TG_L–I	0	1	1	0	$O(\Delta t^2, \Delta x^2)$	$\frac{u^2\,\Delta t}{2}$	$O(\Delta x^2)$

$$\frac{1}{\Delta t}\left(T_i^{n+1} - T_i^n\right) = -\frac{u}{2\,\Delta x}\left[\alpha\left(T_{i+1}^{n+1} - T_{i-1}^{n+1}\right) + (1-\alpha)\left(T_{i+1}^n - T_{i-1}^n\right)\right] \tag{8}$$
$$+\,\beta\,\frac{u^2\,\Delta t}{2\,\Delta x^2}\left[\gamma\left(T_{i-1}^{n+1} - 2\,T_i^{n+1} + T_{i+1}^{n+1}\right) + (1-\gamma)\left(T_{i-1}^n - 2\,T_i^n + T_{i+1}^n\right)\right]$$

for consistent and lumped capacity matrices, respectively. Clearly, the values of parameters α, β, and γ determine the particular BG or TG algorithms, as summarised in Table 1 where the subscripts in the abbreviation identifying the numerical scheme refer to the use of consistent (C) or lumped (L) capacity matrices.

Table 2 - Values of the parameters α, β, and γ to be used in Eqs. (15) and (16) to obtain various finite element schemes for the solution of the energy equation (14). The various schemes are characterized by different stability limits.

Scheme	α	β	γ	ϑ	Stability limits
BG_C–EE	0	0	–	0	$Co \leq \frac{1}{6} Pe$; $Co \leq \frac{2}{Pe}$
BG_L–EE	0	0	–	0	$Co \leq \frac{1}{2} Pe$; $Co \leq \frac{2}{Pe}$
BG_C–CC	$\frac{1}{2}$	0	–	$\frac{1}{2}$	unconditionally stable
BG_L–CC	$\frac{1}{2}$	0	–	$\frac{1}{2}$	unconditionally stable
BG_C–II	1	0	–	1	unconditionally stable
BG_L–II	1	0	–	1	unconditionally stable
TG_C–EE	0	1	0	0	$Co \leq -\frac{1}{Pe} + \sqrt{\frac{1}{3} + \left(\frac{1}{Pe}\right)^2}$
TG_L–EE	0	1	0	0	$Co \leq -\frac{1}{Pe} + \sqrt{1 + \left(\frac{1}{Pe}\right)^2}$
TG_C–EC	0	1	0	$\frac{1}{2}$	$Co \leq \frac{\sqrt{3}}{3}$
TG_L–EC	0	1	0	$\frac{1}{2}$	$Co \leq 1$
TG_C–EI	0	1	0	1	$Co \leq \frac{1}{Pe} + \sqrt{\frac{1}{3} + \left(\frac{1}{Pe}\right)^2}$
TG_L–EI	0	1	0	1	$Co \leq \frac{1}{Pe} + \sqrt{1 + \left(\frac{1}{Pe}\right)^2}$
TG_C–CC	0	1	$\frac{1}{2}$	$\frac{1}{2}$	unconditionally stable
TG_L–CC	0	1	$\frac{1}{2}$	$\frac{1}{2}$	unconditionally stable
TG_C–II	0	1	1	1	unconditionally stable
TG_L–II	0	1	1	1	unconditionally stable

where

$$
\begin{aligned}
b_1 &= \frac{1}{6} - \frac{1}{2}\,\alpha\,Co - \beta\,\gamma\,Fo_e - \vartheta\,Fo \\
b_2 &= \frac{1}{6} + \frac{1}{2}\,\alpha\,Co - \beta\,\gamma\,Fo_e - \vartheta\,Fo \\
d_1 &= \frac{1}{6} + \frac{1}{2}\,(1-\alpha)\,Co + \beta\,(1-\gamma)\,Fo_e + (1-\vartheta)\,Fo \\
d_2 &= \frac{1}{6} - \frac{1}{2}\,(1-\alpha)\,Co + \beta\,(1-\gamma)\,Fo_e + (1-\vartheta)\,Fo
\end{aligned}
\tag{18}
$$

are the coefficients of the general scheme (15) with consistent capacity matrices, while

$$
\begin{aligned}
b_1 &= -\frac{1}{2}\,\alpha\,Co - \beta\,\gamma\,Fo_e - \vartheta\,Fo \\
b_2 &= \frac{1}{2}\,\alpha\,Co - \beta\,\gamma\,Fo_e - \vartheta\,Fo \\
d_1 &= \frac{1}{2}\,(1-\alpha)\,Co + \beta\,(1-\gamma)\,Fo_e + (1-\vartheta)\,Fo \\
d_2 &= -\frac{1}{2}\,(1-\alpha)\,Co + \beta\,(1-\gamma)\,Fo_e + (1-\vartheta)\,Fo
\end{aligned}
\tag{19}
$$

are the coefficients of the general scheme (16) with lumped capacity matrices

$$Co = u\,\frac{\Delta t}{\Delta x} \tag{20}$$

is the cell Courant number related to the advection term,

$$Fo = a\,\frac{\Delta t}{\Delta x^2} \tag{21}$$

is the cell Fourier number related to the physical diffusion term, and

$$Fo_e = \left(\frac{u^2\,\Delta t}{2}\right)\,\frac{\Delta t}{\Delta x^2} = \frac{Co^2}{2} \tag{22}$$

is the equivalent cell Fourier number related to the diffusion-like term.

Following the Von Neumann method for stability analysis, we assume that the numerical solution can be expressed by means of a Fourier series [7], whose typical term is

$$T^n_{i\pm 1} = F^n\,e^{I\,\sigma\,(x_i \pm \Delta x)} \tag{23}$$

where $I = \sqrt{-1}$ is the imaginary unit and σ is the wave number. Substituting definition (23) into Eq. (17) and using the identities

$$e^{\pm I\,\sigma\,\Delta x} = \cos(\sigma\,\Delta x) \pm I\,\sin(\sigma\,\Delta x) \tag{24}$$

we obtain the amplification factor

$$G = \frac{F^{n+1}}{F^n} = \frac{1 - (d_1 + d_2)\,[1 - \cos(\sigma\,\Delta x)] - I\,(d_1 - d_2)\,\sin(\sigma\,\Delta x)}{1 - (b_1 + b_2)\,[1 - \cos(\sigma\,\Delta x)] - I\,(b_1 - b_2)\,\sin(\sigma\,\Delta x)} \tag{25}$$

The modulus of the amplification factor G must be lower than or equal to one if the solutions are to remain bounded. Thus multiplying by the conjugate of G we obtain the condition which must be satisfied to have stability

$$|G|^2 = |G\,\overline{G}| = \frac{D_1 \sin^4\left(\frac{\sigma\,\Delta x}{2}\right) + D_2 \sin^2\left(\frac{\sigma\,\Delta x}{2}\right) + 1}{B_1 \sin^4\left(\frac{\sigma\,\Delta x}{2}\right) + B_2 \sin^2\left(\frac{\sigma\,\Delta x}{2}\right) + 1} \leq 1 \tag{26}$$

where

$$\begin{aligned} D_1 &= 16\, d_1\, d_2 \\ D_2 &= 4\left[(d_1 - d_2)^2 - (d_1 + d_2)\right] \\ B_1 &= 16\, b_1\, b_2 \\ B_2 &= 4\left[(b_1 - b_2)^2 - (b_1 + b_2)\right] \end{aligned} \tag{27}$$

It has been shown in Reference 7, that the necessary and sufficient conditions to satisfy Eq. (26) are

$$D_1 - B_1 + D_2 - B_2 \leq 0 \tag{28}$$

and

$$D_2 - B_2 \leq 0 \tag{29}$$

Condition (28) leads to the diffusion limit for stability, which is of the form

$$Fo = \frac{Co}{Pe} \leq K \tag{30}$$

where K is a suitable numerical constant and Pe is the cell Peclet number

$$Pe = \frac{Fo}{Co} = \left(\frac{u\,\Delta t}{\Delta x}\right)\left(\frac{\Delta x^2}{a\,\Delta t}\right) = \frac{u\,\Delta x}{a} \tag{31}$$

It must be pointed out that, in Eq. (30), the velocity u does not play any role. Condition (29) leads to the advection limit for stability, which is of the form

$$Co \leq f\left(\frac{1}{Pe}\right) \tag{32}$$

and depends on the velocity. Limits (30) and (32) must hold simultaneously and, for the schemes referred to in Table 2, they can be expressed by means of simple algebraic relations. If both limits are always satisfied, the algorithm is said to be unconditionally stable.

Among the Bubnov-Galerkin schemes, the implicit procedures are unconditionally stable, while the BG–EE schemes are characterised both by a diffusion and an advection limit. Taylor-Galerkin algorithms instead are either unconditionally stable or are characterised only by an advection limit.

4. STEADY-STATE PROBLEMS

It has already been pointed out that steady-state solutions can be found by means of a pseudo-transient simulation. Thus, it is worth analysing the behaviour of the finite element schemes for the advection-diffusion equation (14) when the steady state has been reached. In such a case, Eq. (14) becomes

$$u\,\frac{\partial T}{\partial x} - a\,\frac{\partial^2 T}{\partial x^2} = u\,T_x - a\,T_{xx} \tag{33}$$

while, because of Eq. (14), both Eqs. (15) and (16) reduce to

$$\frac{u}{2\,\Delta x}\,(T_{i+1} - T_{i-1}) - \frac{1}{\Delta x^2}\left(a + \beta\,\frac{u^2\,\Delta t}{2}\right)(T_{i-1} - 2\,T_i + T_{i+1}) = 0 \tag{34}$$

or, in the dimensionless form, to

$$\begin{aligned} -\left(\frac{1+\beta\,Co}{2} + \frac{1}{Pe}\right)\,T_{i-1} + \left(\beta\,Co + \frac{2}{Pe}\right)\,T_i \\ -\left(\frac{1-\beta\,Co}{2} - \frac{1}{Pe}\right)\,T_{i+1} = 0 \end{aligned} \tag{35}$$

It can be shown that, with a given mesh, wiggles can be avoided and temperature gradients can be correctly represented if the conduction resistance is smaller than the advection resistance and, consequently, the cell Peclet number does not become so large to change the sign of the coefficients of T_{i+1} in Eq. (35). This requirement yields

$$\frac{1-\beta\,Co}{2} - \frac{1}{Pe} \geq 0 \tag{36}$$

Therefore, for Bubnov-Galerkin schemes ($\beta = 0$), we obtain

$$Pe \leq 2 \tag{37}$$

while, for Taylor-Galerkin schemes ($\beta = 1$), we obtain

$$Pe \leq \frac{2}{1 - Co} \tag{38}$$

i.e. the usual conditions to avoid, in the steady-state, the typical "2 Δx" wave patterns.

Obviously, in the steady state, we can also use Petrov-Galerkin formulations [2,3,10,11] with various forms of weighting functions that lead to different "upwind" finite element schemes. In one dimension, with elements of equal size, upwinding can be accounted for by means of a weighting parameter δ in the assembled equation for the typical node i

$$\begin{aligned} \frac{u}{2\,\Delta x}\,[\delta\,(T_i - T_{i-1}) + (1-\delta)\,(T_{i+1} - T_{i-1})] \\ + \frac{a}{\Delta x^2}\,(T_{i-1} - 2\,T_i + T_{i+1}) = 0 \end{aligned} \tag{39}$$

We have full upwinding for $\delta = 1$, and a central difference representation of the advection term for $\delta = 0$. Finally, if we write, Eq. (39) in dimensionless form we obtain

$$-\left(\frac{1+\delta}{2} + \frac{1}{Pe}\right)\,T_{i-1} + \left(\delta + \frac{2}{Pe}\right)\,T_i - \left(\frac{1-\delta}{2} - \frac{1}{Pe}\right)\,T_{i+1} \tag{40}$$

For $\beta = 1$, equation (40) and Eq. (35) are identical, provided that the cell Courant number is equal to the weighting parameter δ. This proves the identity between steady-state, Petrov-Galerkin solutions and pseudo-transient Taylor-Galerkin solutions. Since in one-dimensional problems, with elements of equal size, Petrov-Galerkin schemes yield solutions that are exact at nodes when an optimal weighting parameter

$$Co = \delta_{opt} = \coth(Pe) - \frac{1}{Pe} \tag{41}$$

is used [2,3,10,11], also Taylor-Galerkin schemes can be optimized by assuming $Co = \delta_{opt}$. Actually, if we use the full explicit Taylor-Galerkin scheme (TG–EE), we do not even have to utilise Eq. (41) because it has been shown that nearly-optimal, pseudo transient solutions can be obtained by choosing a cell Courant number as close as possible to the limiting, or critical, value for stability [2]. These considerations cannot be directly extended to multi-dimensional problems, with irregular meshes. However, in pseudo-transient solutions of multidimensional problems, the time increment used plays an important role, and nearly optimal pseudo-transient solutions can be obtained by choosing, also in this case, a time step as close as possible to the critical value for stability [2].

5. CONCLUSIONS

The results outlined in the previous sections can lead to criteria for the choice of the most suitable algorithm for any given task, even if the analysis does not take into account the effects of dispersion errors. Clearly, the authors' preferences do not include Petrov-Galerkin schemes because they cannot be easily interpreted from either a physical or mathematical point of view. Thus, the guidelines we can offer are limited to Bubnov-Galerkin and Taylor-Galerkin schemes.

Pseudo-transient simulations. In the solution of steady-state problems by means of pseudo-transient simulations, stability and the absence of stationary, numerical diffusion are characteristics of great interest. Therefore, fully implicit Bubnov-Galerkin (BG–I I) schemes can be used confidently since they allow the use of very large time steps to reach convergence and, in the steady-state are not affected by numerical diffusion. The only reason not to use BG–I I schemes for pseudo-transient simulations can be the appearance of wiggles, connected with cell Peclet numbers that are too large, i.e. with mesh sizes that are too large with respect to the local velocity values. In these cases the mesh should be refined but, if the refinement is too expensive, we can consider the alternative use of one of the Taylor-Galerkin schemes, characterised by a stationary numerical diffusion that plays the role of an upwinding parameter.

Transient simulations. In truly transient simulations, we can use Bubnov-Galerkin, Crank-Nicolson schemes (BG–CC) which are unconditionally stable and are not affected by numerical diffusion. In the transient, consistent capacity matrices perform better than lumped capacity matrices, especially if the mesh is irregular and, consequently, the algorithm of choice might be the BG_C-CC. As an alternative, Taylor-Galerkin schemes might also be considered since they are not affected by numerical diffusion in transient calculations.

Taylor-Galerkin schemes. Many schemes of this type have been proposed and, therefore, their respective ranges of applicability are still the object of research. Here we can only say that fully explicit Taylor-Galerkin algorithms (TG–EE) can be very convenient with lumped capacity matrices since do not require a matrix inversion to march ahead in time. However, to have better stability characteristics, the TG–EI and the TG–I I schemes can also be considered. The TG–EI schemes perform better than the TG–EE schemes with respect to stability but are not unconditionally stable as the TG–I I schemes.

On the other hand, with TG–EI schemes we do not have to perform the system matrix factorization every time step because only the constant physical diffusion term is dealt with implicitly and, in the computer programs, we can use a re-solution facility. Instead with TG–I I schemes, also the variable diffusion-like term is dealt with implicitly and, consequently, a new matrix factorization must be performed every time step. However, as a final remark, it must be pointed out that Taylor-Galerkin algorithms can be affected by serious dispersion errors which can deteriorate the overall accuracy.

Acknowledgements: The support to this research by M.U.R.S.T. and C.N.R. is gratefully acknowledged.

REFERENCES

1. ZIENKIEWICZ, O.C., LOEHNER, R., MORGAN, K., and NAKAZAWA, S. - Finite Elements in Fluid Mechanics - A Decade of Progress, *Finite Elements in Fluids*, Ed. Gallagher, R.H. *et al.*, Vol. 5, 1-26, Wiley, London, 1984.
2. ZIENKIEWICZ, O.C. and TAYLOR, R.L. - *The Finite Element Method*, Vol. 2, 4th Ed., Mc Graw-Hill, London, 1991.
3. HUGHES, T.J.R. and BROOKS, A. - A Theoretical Framework for Petrov-Galerkin Methods with Discontinuous Weighting Functions: Application to Streamline-Upwind Procedure, *Finite Element in Fluids*, Ed. Gallagher, R.H. *et al.*, Vol. 4, 47-65, Wiley, London, 1982.
4. GRESHO, P.M., CHAN, S.T., LEE, R.L. and UPSON, C.D. - A Modified Finite Element Method for Solving the Time-Dependent, Incompressible Navier-Stokes Equations. Part 1: Theory, *Int. j. numer. methods fluids*, Vol. 4, 557-598, 1984.
5. COMINI, G., DEL GIUDICE, S., and NONINO, C. - Taylor-Galerkin Algorithms for Convection-Type Problems, *Advanced Computational Methods in Heat Transfer II*, Ed. Wrobel, L.C. *et al.*, Vol.1, 535-554, Computational Mechanics Publications, Southampton, 1992.
6. ROACHE, P.J. - On Artificial Viscosity, *Journal of Computational Physics*, Vol. 10, 169-184, 1972.
7. HIRSCH, C. - *Numerical Computation of Internal and External Flows*, Vol. 1, Wiley, Chichester, UK, 1988.
8. PEPPER, D.W. and BAKER, A.J. - Finite Differences Versus Finite Elements, *Handbook of Numerical Heat Transfer*, Ed. Minkowycz W.J. *et al.*, Chapter 13, 519-577, Wiley, New York, 1988.
9. CHEN, Y. and FALCONER, R.A. - Advection-Diffusion Modelling Using the Modified Quick Scheme, *Int. j. numer. methods fluids*, Vol. 15, 1171-1196, 1992.
10. CHRISTIE, J., GRIFFITHS, D.F., MITCHELL, A.R., and ZIENKIEWICZ, O.C. - Finite Element Methods for Second Order Differential Equations with Significant First Derivatives, *Int. j. numer. methods eng.*, Vol. 10, 1389-1396, 1976.
11. HEINRICH, J.C., HUYAKORN, P.S., and ZIENKIEWICZ, O.C. - An 'Upwind' Finite Element Scheme for Two-Dimensional Convective Transport Equation, *Int. j. numer. methods eng.*, Vol. 11, 131-143, 1977.

NUMERICAL STUDIES OF THREE-DIMENSIONAL EFFECTS IN NATURAL CONVECTION: THE DOUBLE-GLAZING PROBLEM

W. J. Decker(I),(II) and J. J. Dorning(I),(II)

SUMMARY

The effects of the third dimension, the span-wise dimension, on flow in the double-glazing problem of natural convection have been studied numerically. The investigations were carried out using a recently developed block iterative nodal integral method for coarse-mesh calculations to determine the steady-state and time-dependent three-dimensional flows at various Rayleigh numbers. Analogous two-dimensional calculations also were done, and the results of the two sets of calculations were compared to identify the basic effects of the third dimension on the asymptotic steady flows and on the transient flows. The salient three-dimensional effects are discussed in some detail. Euler-Newton parameter continuation methods were used to calculate the steady-state solutions at various values of the Rayleigh number and to identify spurious singular points that occur when excessively coarse meshes are used. The final calculations were then done for various Rayleigh numbers using meshes such that the problem of nearby spurious singular points does not arise.

1 INTRODUCTION

Natural convection, the flow driven by buoyancy that results from the thermal expansion of a fluid subjected to heating, is crucial to system dynamics in a broad spectrum of scientific and technological areas — from the development of meteorological flows, to the cooling of high-tech electronic components, to the passive removal of decay heat from advanced so-called inherently safe nuclear reactors. Relevant to many of these systems in a fundamental way is the archetypical problem of a rectangular parallelepiped of fluid with one side wall heated and the opposite wall cooled, and with the other two walls and bottom and top insulated. This so called double-glazing or hot wall problem has been studied fairly extensively in two-dimensional settings. (See the survey article by DE VAHL DAVIS and JONES[1], for example.) Motivated by the numerous important contemporary applications in areas such as those mentioned above, in which three-dimensional effects often

(I)Engineering Physics Program, University of Virginia, Charlottesville, VA 22903-2442.

(II)This research was supported by the U. S. Army Research Office under Grant No. DAAL03-90-G-0175 and by NASA under Grant No. NAGW-3021.

are crucial, studies of this now-classic computational fluid dynamics problem have been carried out in a three-dimensional setting and are reported here. The relevant equations and the geometric and physical details of the problem are given in the next section. In the third section the basic ideas of the coarse-mesh computational method used to do the calculations, the nodal integral method [2–4], are summarized, and the moving-template overlapping block method developed recently [4] for the memory efficient solution of the discrete-variable nodal integral method equations is described very briefly. Some details on the Euler-Newton parameter continuation and the convergence criteria used in the calculations also are given.

It is well known that spurious singular points, usually related to spurious bifurcation points, can be encountered when an excessively low order approximation is used in the numerical solution of a complex nonlinear problem, e.g. one described by nonlinear partial differential equations. Such spurious bifurcation points have been studied in some detail in fluid mechanics, in the context of finite difference solutions to the classic two-dimensional driven cavity problem [5] where it was found, using arc-length continuation methods [6], that spurious turning points (saddle-node bifurcation points) occurred at Reynolds numbers in the laminar flow region when 30×30, 40×40 and 50×50 meshes were used (but not when 100×100 and 120×120 meshes were used). Analogous studies, also done using arc-length continuation methods, of the same two-dimensional driven cavity problem, but in the context of the nodal integral method, showed that this method also led to spurious turning points, but these occurred only for the extremely coarse 5×5, 6×6 and 7×7 meshes [7]. With this background on the driven cavity problem in mind, simple numerical studies to estimate the values of the Rayleigh number as a function of mesh at which spurious singular points occur for nodal integral method solutions to the double-glazing problem were carried out before embarking on the fairly large three-dimensional steady-state and time-dependent calculations done at $\mathrm{Ra} = 10^4$ to study the effect of the third dimension on the flow. The simple calculations done to estimate these bifurcation points are summarized in Sec. 4.1. The results of the calculated three-dimensional and two-dimensional flows at $\mathrm{Ra} = 10^4$ are compared in Sec. 4.2 where the salient effects of the third dimension on the flow are discussed, the most important of which is the weakening of the main vortex due to the drag that results from the front and rear span end walls and the freedom of motion provided by the third dimension which allows fluid rising along the hot wall and falling along the cold wall to move in the span direction thereby decreasing the force with which it drives the main vortex. The development of this flow in time is discussed briefly in Sec. 4.3. Finally, the conclusions are summarized in Sec. 5, the last section.

2 THE DOUBLE-GLAZING PROBLEM

The equations used to describe the thermally driven flow are the time-dependent compressible Navier-Stokes equations represented using the Boussinesq approximation to the equation of state, coupled to the energy conservation equation, which in dimensionless form are

$$\frac{\partial u}{\partial x} + \frac{\beta_x}{\beta_y}\frac{\partial v}{\partial y} + \beta_x \frac{\partial w}{\partial z} = 0, \tag{1}$$

$$\frac{\partial u}{\partial t}+\frac{\partial u^2}{\partial x}+\frac{\beta_x}{\beta_y}\frac{\partial uv}{\partial y}+\beta_x\frac{\partial uw}{\partial z} = -\frac{\partial T_x}{\partial x}+\left(\frac{\beta_x}{\beta_y^2}\frac{\partial^2 u}{\partial y^2}+\beta_x\frac{\partial^2 u}{\partial z^2}\right), \quad (2)$$

$$\frac{\partial v}{\partial t}+\frac{\partial vu}{\partial x}+\frac{\beta_x}{\beta_y}\frac{\partial v^2}{\partial y}+\beta_x\frac{\partial vw}{\partial z} = -\frac{\partial T_y}{\partial y}+\left(\frac{1}{\beta_x}\frac{\partial^2 v}{\partial x^2}+\beta_x\frac{\partial^2 v}{\partial z^2}\right), \quad (3)$$

$$\frac{\partial w}{\partial t}+\frac{\partial wu}{\partial x}+\frac{\beta_x}{\beta_y}\frac{\partial wv}{\partial y}+\beta_x\frac{\partial w^2}{\partial z} = -\frac{\partial T_z}{\partial z}+\left(\frac{1}{\beta_x}\frac{\partial^2 w}{\partial x^2}+\frac{\beta_x}{\beta_y^2}\frac{\partial^2 w}{\partial y^2}\right) - \beta_x\frac{\mathrm{Ra}}{\mathrm{Pr}}\theta, \quad (4)$$

$$\frac{\partial \theta}{\partial t}+\frac{\partial \theta u}{\partial x}+\frac{\beta_x}{\beta_y}\frac{\partial \theta v}{\partial y}+\beta_x\frac{\partial \theta w}{\partial z} = \left(\frac{1}{\mathrm{Pr}}\right)\left(\frac{1}{\beta_x}\frac{\partial^2 \theta}{\partial x^2}+\frac{\beta_x}{\beta_y^2}\frac{\partial^2 \theta}{\partial y^2}+\beta_x\frac{\partial^2 \theta}{\partial z^2}\right); \quad (5)$$

where u, v, and w are the x, y, and z components of the velocity; θ is the temperature; $T_x = \nu\frac{\partial u}{\partial x} - p$, $T_y = \nu\frac{\partial v}{\partial y} - p$ and $T_z = \nu\frac{\partial w}{\partial z} - p$ are the normal stresses; β_x is the width-to-height aspect ratio; β_y is the depth-to-height span aspect ratio; Pr is the Prandtl number; and Ra is the Rayleigh number. Throughout the studies reported here $\beta_x = \beta_y = 1$ (a cube) and $\mathrm{Pr} = 0.71$ (air), while Ra is set at various values. This set of dimensionless variables, found by a vertical viscous scaling, differs from those used by DE VAHL DAVIS and JONES [1] by a factor of Pr in all the dependent variables.

The boundary conditions for the double-glazing problem are that the right side wall ($x = 1.0$) is held at (dimensionless) temperature $\theta = 1$, the left side wall ($x = 0.0$) is held at (dimensionless) temperature $\theta = 0$, while the front, back, top and bottom are insulated. The walls of the cube have no-slip boundary conditions for the velocity, so that $u = v = w = 0$ on all six walls. Since the pressure is determined only to within an arbitrary constant for a closed system such as this, it (or more precisely the normal stress) is specified at one point on the front wall to eliminate that arbitrary constant and "set the gauge" of the pressure. The above equations and boundary conditions specify well-posed problems to which the approximate solutions described here were found by discretization using a nodal integral method.

3 THE NUMERICAL METHODS

The numerical method used to study these three-dimensional, time-dependent and stationary flows was a nodal integral method [2–4]. A very short descriptive summary of the method for three-dimensional, time-dependent convective flows follows; a more detailed discussion can be found in Refs. [2–4].

3.1 The Nodal Integral Method

To develop a nodal integral method, the original set of partial differential equations is "transverse-averaged" locally within a space-time computational volume element (or "node" in the nomenclature) sequentially over all but one space-time coordinate, yielding a separate set of ordinary differential equations in each space-time variable. The total number of these ODEs associated with each space-time node is equal to the number of space-time coordinates multiplied by the number of original partial differential equations. This set of equations is made up of first- and second-order

ODEs that result from the first- and second-order partial derivatives in the original equations. The forcing functions $S(\eta)$, $\eta = x$, y, z, or t (which include unknown transversed-averaged partial derivatives, etc.) in these ODEs temporarily are taken as known and approximated by a Legendre expansion within the node (computational element). The first order equations have the generic form $d\psi(\eta)/d\eta = S(\eta)$. Solving these first order equations results in expressions for $\overline{S}^{\eta}$ and for the node-average of the unknown over all coordinates. Solving the second order equations, which are of the generic form $d^2\psi(\eta)/d\eta^2 = S(\eta)$, yields expressions for the first derivative of the unknown and for its average over a node. Using these quantities and enforcing: (1) Continuity of unknowns and their derivatives across node interfaces, (2) Equality of the separately calculated quadruple space-time averages of the velocity components over a node, (3) Equality of the same averages (over all 3+1 coordinates) of the pressure, and (4) Conservation of all four PDEs over a node, a set of nonlinear algebraic equations is developed, with the unknowns being the node-surface-averaged velocity components, normal stresses and temperatures. Additional equations for the insulated boundaries and the no-slip boundary conditions must be added. Finally, owing to the fact that the cavity is closed, the sum of the quadruply integrated mass conservation equations is not linearly independent of the normal velocity boundary conditions; hence, one of these mass conservation equations must be eliminated and replaced by an equation "setting the gauge" of the pressure, or more precisely of the normal stress. Steady-state solutions are obtained simply by requiring that the same solution be given at two successive time levels, which is algebraicly equivalent to using a nodal integral method to discretize the steady-state equations. The whole procedure yields a set of nonlinear discrete-variable equations, and creates a well-posed algebraic problem for the node-surface-averaged quantities, which are the final discrete-variable unknowns and which are averaged over all but one coordinate and evaluated at the surface value of that coordinate (e.g., $\overline{\overline{\overline{v}}}^{xt}(+b)$). These equations are solved via Newton iterations carried out using the moving template overlapping block iterative method [8] recently implemented [4] to invert the Jacobian of the nonlinear algebraic equations. This overlapping block iterative method makes possible the Newton's method solution of the final nonlinear equations without requiring the storage or direct inversion of large matrices. Hence, it is extremely memory efficient, and makes it possible to carry out many large-scale time-dependent calculations on systems with modest memory sizes, including workstations and even *i*486 PCs.

Nodal integral methods, thus developed, have the characteristic of being accurate even on coarse meshes [2–4]. For example, in the two-dimensional double-glazing problem, the flow for Ra $= 10^4$, calculated using nodal integral methods on a uniform 10×10 mesh, was found to be as accurate as a finite element method solution calculated using a non-uniform 12×12 mesh and as accurate as a finite difference method solution calculated using a uniform 21×21 mesh [4]. In computational cost comparisons that have been made in the past, the method has been shown to reduce significantly, in comparison with traditional methods, the computational time and cost for a given desired accuracy [2]. Nodal integral methods also have an additional advantage. They lead to a Jacobian matrix in the Newton's method solution that is dominated by an overlapping block diagonal structure. This can be exploited, as described in Ref. 4, in such a way that the matrices that actually must be inverted never exceed those associated with a 2×2 ($\times$ 2) grid in the solution to a two- (three-) dimensional problem using an $N \times N$ ($\times N$) grid. Thus, the new moving [2×2 ($\times$ 2)] template overlapping block iterative nodal integral method is computationally efficient, and is well suited for solving computationally taxing three-dimensional flows.

3.2 Parameter Continuation

The steady-state solutions were obtained over a range of Ra using Euler-Newton parameter continuation [9]. The equations were solved using Newton iterations, starting from a zero initial guess, for the first parameter value (usually Ra $= 10^3$) to obtain, at that value, the complete solution vector $U(\mathrm{Ra})$ to the nonlinear system $F(U(\mathrm{Ra}); \mathrm{Ra})$. Then a new initial guess was generated for the solution vector at a new Rayleigh number via Euler extrapolation

$$U^*(\mathrm{Ra} + \Delta\mathrm{Ra}) = U(\mathrm{Ra}) - J^{-1}(U(\mathrm{Ra}); \mathrm{Ra})\frac{dF}{d\mathrm{Ra}}(U(\mathrm{Ra}); \mathrm{Ra})\Delta\mathrm{Ra}, \quad (6)$$

after which the Newton iterations were converged at the new parameter value. Automatic control was used on the Euler steps in Ra in order to optimize the number of Newton iterations required at each parameter value [6]. The optimum number was set at three.

Convergence of both the Newton iterations and the moving template overlapping block iterations done to invert the Jacobian at each Newton iteration was based on the ℓ_1-norm of the relative change in the total solution vector from one iteration to the next. That is, the iterations were terminated when

$$\sum_{m=1}^{M} \left| \frac{U_m^i - U_m^{i-1}}{U_m^{i-1}} \right| < \epsilon, \quad (7)$$

where ϵ was set to 10^{-12} for the overlapping block iterations used to invert the Jacobian, and to 10^{-6} for the Newton iterations.

4 NUMERICAL RESULTS

The numerical results reported here are in three different categories. The first are those of calculations done using Euler-Newton continuation to locate spurious singular points in the three-dimensional steady-state solution branch, and to establish that they are indeed spurious. The second are those of the steady-state calculations, done to identify three-dimensional effects, carried out at Ra $= 10^4$ using a grid such that nearby singular points do not occur; and the third are those of the time-dependent calculations carried out at the same Ra and using the same grid to study the time evolution of the flow.

4.1 Identification of Singular Points

In order to determine how fine a mesh would be required to do reliable three-dimensional calculations for Rayleigh numbers of the order of 10^4, i.e. that would not be plagued by nearby spurious points corresponding to singular inverses of

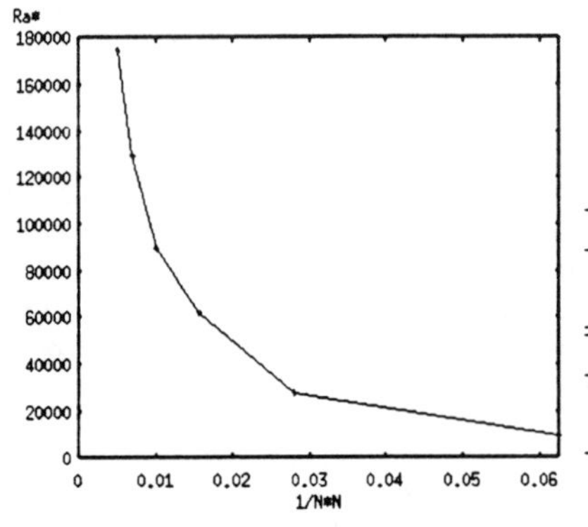

Figure 1: Ra vs. $1/N^2$

Grid (3-D)	Singular Point
$4 \times 4 \times 4$	$\simeq 1.1 \times 10^4$
$6 \times 6 \times 6$	$\simeq 3.0 \times 10^4$
Grid (2-D)	**Singular Point**
4×4	$\simeq 1.0 \times 10^4$
6×6	$\simeq 2.8 \times 10^4$

Table 1

the Jacobian, Euler-Newton parameter continuation calculations of the steady-state solutions were carried out for various grids. These sequences of calculations were started from Ra $= 10^3$ for successively less coarse grids for both the three-dimensional flows and the two-dimensional flows and continued until the behavior of the iterations done to invert the Jacobian indicated that a singular point was being approached. The value of the Rayleigh number Ra* at which such a point occurred for a very coarse grid was estimated, the grid was refined and process was repeated. If a bifurcation were associated with such a singular point and it were a physical bifurcation, this type of sequence of critical Rayleigh numbers as a function of grid number Ra$^*(N)$ typically would converge to a limit Ra$^*(\infty)$ which would be an estimate of the physical bifurcation value. Conversely, if there exists a spurious singular point, and possibly an associated spurious bifurcation point, that is merely an artifact of the grid, the sequence Ra$^*(N)$ typically diverges, suggesting that the critical parameter value Ra$^*(N)$ moves off to infinity as the mesh is refined. In this case, if a calculation is to be done at a specific Rayleigh number, it is important to use a grid such that there are no spurious singular values Ra$^*(N)$ (already diverged to infinity), or in a more practical vein, such that Ra$^*(N)$ is much greater than the Rayleigh number of interest.

There are other ways to estimate bifurcation points (both spurious and physical) that are more explicit, more precise, and far more elegant. Smooth continuation through turning points is straightforward via arc-length continuation, an extension of Euler-Newton continuation in which the system equations are supplemented by the fixed-point branch arc length equation, lifting the branch out of the bifurcation diagram so that the new fixed point (which then includes the parameter value) becomes a regular function of the arc-length parameter [5–7]. A more general and more direct method is to supplement the equations with additional equations corresponding to the conditions for various types of bifurcations — requiring that the Jacobian have a zero eigenvalue, the corresponding null-space eigenvector be normalized, etc. [10]. These methods directly yield both the solution at the bifurcation point and the bifurcation value of the parameter; however, they typically are computationally intensive and very costly to carry out for large-scale problems, since they at least double the number of unknowns and the size of the nonlinear system to be solved for a fixed grid. Hence, in order to avoid these costs, and since the important objectives here were first to determine whether singular points occur and second to establish that they indeed are spurious, these extended system approaches were not used. Rather, it was sufficient to interpret the failure of the iterative inversion of the Jacobian to converge, at the first Newton iteration for a new parameter value following an Euler-Newton continuation step, as an indication that a singular point was being approached. This gave an estimate of the singular point. Of course, if these estimates converged as the grid was refined, a more careful analysis of the corresponding physical bifurcation point would have been essential. However, this was not the case. The singular points were thus estimated by Euler-Newton continuation for a sequence of grids first for the two-dimensional double-glazing problem, and then for two grids for the three-dimensional version of this problem. Figure 1 shows the resulting estimated values of $\mathrm{Ra}^*(N)$ vs. $1/N^2$ from the 4×4, 6×6, 8×8, 10×10, 12×12 and 14×14 grid calculations of the two-dimensional flows. The corresponding points are connected by straight lines. The plot indicates that the singular point is spurious and is moving off to infinity very rapidly as the grid is refined from its initial very coarse 4 x 4 structure. Analogous estimates were made by Euler-Newton continuation from $\mathrm{Ra} = 10^3$ for the three-dimensional problem, but only for the two coarsest grids $4\times4\times4$ and $6\times6\times6$. These estimates are compared in Table 1 with the corresponding estimates for the two-dimension problem, i.e. those for the 4×4 and 6×6 grids. The close agreement, combined with the fact that the values resulting from the three-dimensional calculations are larger, suggests that a spurious bifurcation which is fundamentally a two-dimensional phenomenon may be occurring and that the estimates of $\mathrm{Ra}^*(N)$ based on the two-dimensional calculations also might provide reasonable, perhaps even conservative, estimates for those appropriate to the three-dimensional problem. Thus it was concluded that a $10\times10\times10$ grid was sufficient for the calculation of the flow at $\mathrm{Ra} = 10^4$ in the three-dimensional double-glazing problem. Nevertheless, the $10\times10\times10$ steady-state calculations for $\mathrm{Ra} = 10^4$ discussed below were carried out via Euler-Newton continuation starting from $\mathrm{Ra} = 10^3$; no singular points were encountered during this sequence of calculations, which is consistent with a value for the Rayleigh number at the spurious singular point that is of the order of 9×10^4 as suggested by Fig. 1 and Table 1.

4.2 The Effects of the Third Dimension

One of the purposes in studying this problem was to determine how the fluid motion in the span direction changes the flow in the symmetric midplane ($y = 0.5$) as compared to the two-dimensional flow. This indicates how well the results of two-dimensional calculations approximate the more realistic three-dimensional flow. The flow outside the symmetric midplane also was examined to determine how it deviates from the flow in the midplane. This included observations of how the main vortex deforms as it approaches the span end walls, and an examination of the flow in the span direction that is related to the strength of the main vortex. All the flows discussed in this subsection are steady-state flows computed on a 10×10 ($\times$ 10) grid at $\mathrm{Ra} = 10^4$ following Euler-Newton continuation from $\mathrm{Ra} = 10^3$. Figure 2(a) shows the computed three-dimensional flow in the midplane of symmetry. Comparison with the corresponding two-dimensional flow, shown in Fig. 2(b), indicates that the main vortex in the three-dimensional flow is not as strong as that in the two-dimensional flow. This is clearer from Fig. 3 which provides a quantitative comparison — the difference between the three-dimensional midplane flow and the

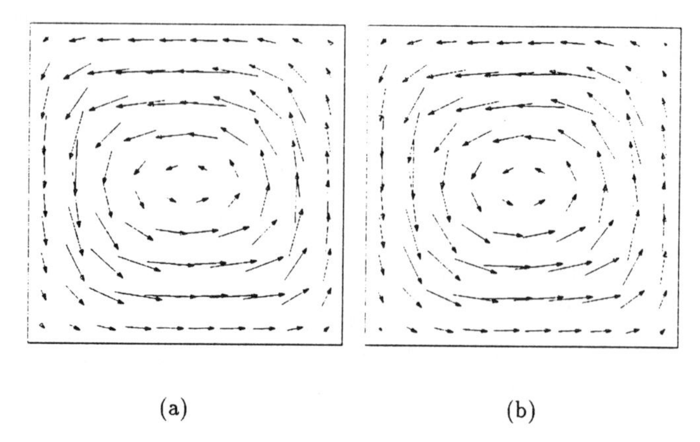

(a) (b)

Figure 2: (a) Flow in the Midplane of Symmetry $y = 0.5$ for Ra = 10^4 Calculated Using a $10 \times 10 \times 10$ Mesh. (b) Two-Dimensional Flow for Ra = 10^4 Calculated Using a 10×10 Mesh.

two-dimensional flow, scaled by a factor of 10 over that in Fig. 2(a). That the main vortex in the three-dimensional flow is weaker is a result of two effects. The first is the drag effect of the span end walls which act as a momentum sink because of the no-slip boundary conditions. The second is the freedom for motion in the span direction which permits the fluid rising along the hot wall and falling along the cold wall to move away from the symmetric midplane and therefore drive the main vortex less hard than it would if it were confined to the midplane. Figure 3 also shows that the largest difference between the flows occurs where the hot fluid first approaches the cold wall in the upper left corner, and where the cold fluid first approaches the hot wall in the lower right corner. This greater motion of the three-dimensional flow in the midplane of symmetry also results from the freedom the fluid has to move in the third dimension, and will be discussed in more detail at the end of this subsection.

Figure 4(a) shows how the fluid circulates in the $x = 0.5$ plane, a plane bounded by the four insulated walls. The u-component of the velocity is suppressed; however, the flow is out of the figure and away from the hot wall in the upper half, and is into the figure and toward the hot wall in the lower half. The direction of gravity is downward causing the vertical asymmetry. The magnitude of the vectors is scaled by a factor of 10 over that in Fig. 2(a). It is clear that the flow in the span direction produces longitudinal vortices that intersect each quadrant of the plane. These vortices, whose axes are locally in the x-direction, circulate the fluid from the span end walls at $y = 0.0$ and $y = 1.0$ toward the midplane of symmetry. This circulation of course is weaker than the main vortex by which it is driven. The four locally longitudinal vortices shown in Fig. 4(a) most likely are cross sections of two side-by-side tori standing on end. However, this is not suggested by Fig. 4(b) which shows the flow in the horizontal midplane at $z = 0.5$ with the hot wall at the top of the figure and the cold wall at the bottom. The absence of clearly defined vortices in the four quadrants of this figure suggests that those shown in Fig. 4(a) do not connect to form tori, which necessarily would intersect this plane. However, they may connect and intersect this plane with the corresponding vortices obscured in

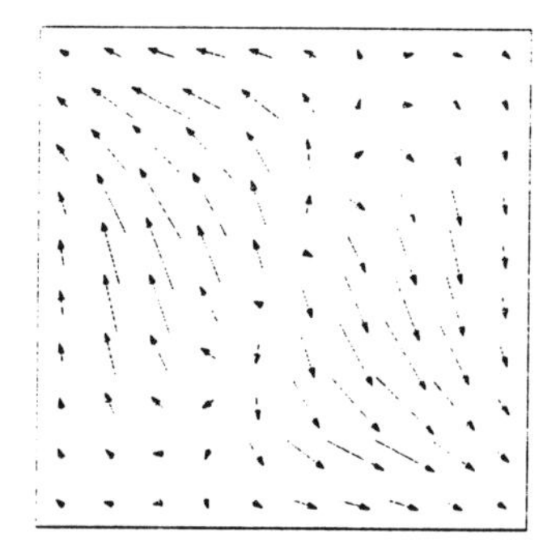

Figure 3: Difference Between the Three-Dimensional Flow in the Midplane (Figure 2(a)) and the Two-Dimensional Flow (Figure 2(b)): $\vec{v}_{3D} - \vec{v}_{2D}$.

Fig. 4(b) because they are very small and weak, and are located extremely close to the span end walls near the corners of Fig. 4(b). (This is suggested, albeit not conclusively, by the flows in the $z = 0.4$ and $z = 0.6$ planes, which are not shown here.) Otherwise, the two pairs of vortices could merge in a topologically very complicated way into the upward flow of the main vortex along the hot wall and into its downward flow along the cold wall. The magnitude of the vectors in this figure is scaled the same as that of the vectors in Fig. 4(a).

The flow in the back half of the cavity is indicated in Fig. 5(a) by the components of the velocity vectors in the planes of constant y ($y = 0.6, 0.7, 0.8, 0.9$). Clearly the main vortex weakens as the distance in the span direction from the symmetric midplane ($y = 0.5$) is increased. The center of that vortex also appears to move a little downward and slightly to the right (toward the hot wall) in the sequence of constant-y planes as the far span wall ($y = 1.0$) is approached. In this figure the black vectors indicate flow toward the symmetric midplane at $y = 0.5$, and the gray vectors indicate flow away from it. Hence, the span direction flow in the interior of the main vortex is away from the far wall toward the symmetric midplane. The main vortex acts like a funnel cloud pulling fluid away from the far (and near) span end walls depositing it (from both sides) spread out about the symmetric midplane. The gray vectors throughout much of the outer region of the vortex, principally near the corners and the bottom and top, indicate that the return flow from the symmetric midplane to the span walls is through the co-rotating outer part of this same vortex. This return flow in the outer part of the vortex near the bottom and top of the cavity also was very evident in Fig. 4(a). That it does not also occur in a significant amount along the hot and cold walls, at least at the horizontal midplane ($x = 0.5$) is clear from Fig. 4(b) which is consistent with the flows indicated along the hot and cold walls in Fig. 5(a). The stronger return flows (from the symmetric midplane to the span end walls) in the upper left and lower right corners, most evident in the $y = 0.6$, 0.7 and 0.8 planes, provides the path away from the symmetric midplane that leads to the largest difference between the flow in that plane and the corresponding calculated two-dimensional flow. This difference appears as the largest velocity differences shown in the upper right and lower left regions of Fig. 3 which were mentioned above. In fact there is a clear and

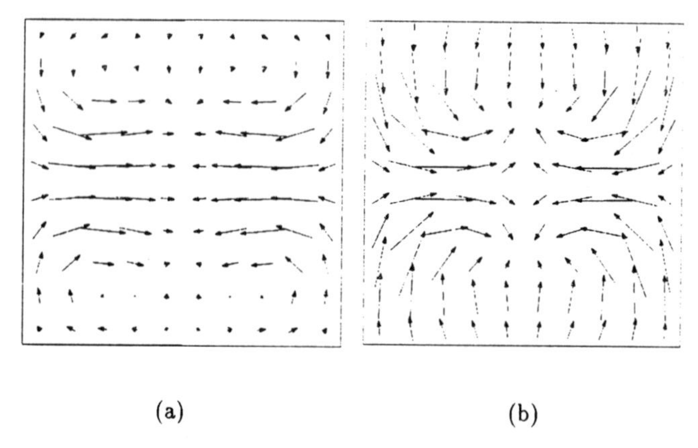

Figure 4: (a) Flow in the Plane $x = 0.5$, for Ra = 10^4 Calculated Using a $10 \times 10 \times 10$ Mesh. (b) Flow in the Plane $z = 0.5$, for Ra = 10^4 Calculated Using a $10 \times 10 \times 10$ Mesh.

logical correspondence between the velocity difference vectors indicated in Fig. 3 and the gray return flow vectors in Fig. 5(a).

4.3 The Time Evolution of the Three-Dimensional Flow

Time-dependent calculations of the evolution of the three-dimensional flow in the double-glazing problem also were carried out using the same moving template overlapping block iterative solution of the nodal integral method equations. These were calculated on a $10 \times 10 \times 10$ mesh using a time step $\Delta t = 0.0025$ (in dimensionless time units) starting at zero initial velocity and zero temperature field (equal to the cold wall temperature) in the fluid, and a hot wall dimensionless temperature of unity. As in the steady-state calculations no-slip velocity boundary conditions were used on all six walls.

Not unlike the time-evolution of the two-dimensional flow, the main vortex initially develops very close to the hot wall and initially is very elongated vertically (since the initial fluid temperature was equal to that of the cold wall). As time evolves it moves outward toward the center of the cavity with this motion not surprisingly progressing more rapidly at the symmetric midplane than near the span end walls where their drag effects inhibit it. This is clear from the flow fields computed early in the transient which are shown on constant-y surfaces in Fig. 5(b). At long times this flow evolves to the steady flow shown in Fig. 5(a) where the center of the main vortex in the $y = 0.6$ plane has moved to the left and slightly downward from its position at $t = 0.025$ shown in Fig. 5(b) and its center in the $y = 0.9$ plane has moved to the left and slightly upward from its position at $t = 0.025$. The longitudinal vortices — not shown, but analogous to those in Fig. 4(a) for the steady flow — also gradually develop and strengthen as time evolves and the main vortex develops fully and establishes the final form of both the inflow to the symmetric midplane, which it forces from its slower moving ends at the span end walls, and the return outflow, which it ejects outward and back along the four corners and the

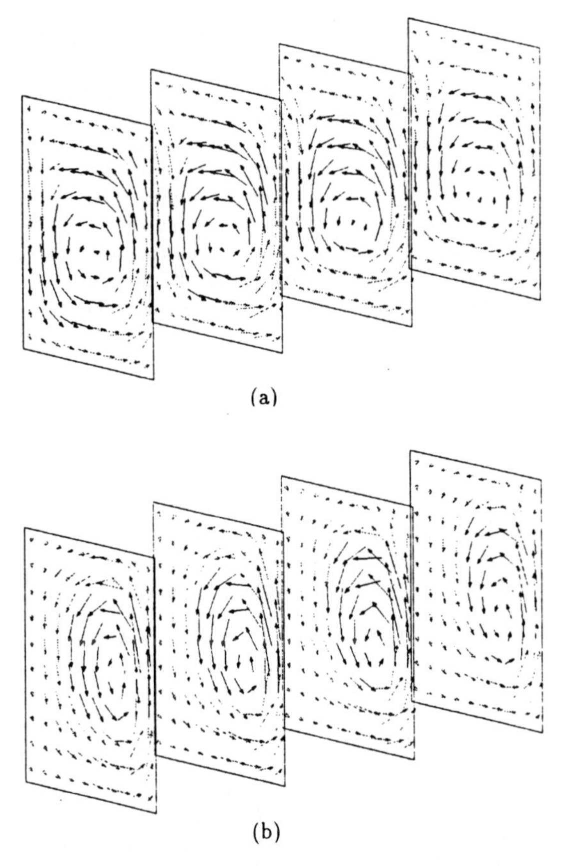

(a)

(b)

Figure 5: Flows in the Planes $x = 0.6, 0.7, 0.8$ and 0.9 for Ra $= 10^4$ Calculated Using a $10 \times 10 \times 10$ mesh: (a) The Steady-State Flow; (b) The Time-Dependent Flow Early in the Transient at $t = 0.025$, Calculated Using a Time Step $\Delta t = 0.0025$.

bottom and top toward the span end walls.

5 CONCLUSION

The recently developed moving-template overlapping block iterative nodal integral method was applied to the study of three-dimensional effects in a fundamental natural thermal convection problem. Steady-state and time-dependent two- and three-dimensional calculations were done to determine the transient and steady structure of the flow. Not surprisingly, the main vortex in the three-dimensional flow was not quite as strong as the one in the corresponding two-dimensional flow. This is due both to the drag from the span end walls in the three-dimensional case, and to flow away from the symmetric midplane along the hot and cold walls which occurs in the three-dimensional case and reduces the buoyancy forces that drive the main vortex. The flow, that the main vortex drives in its interior, from its slower rotating ends at the span end walls to its faster rotating portion at the symmetric midplane was clearly seen, as was the return flow from the symmetric midplane to the span end walls in the outer part of the main vortex along the four corners and the bottom and top of the cavity. The longitudinal vortices, associated with the fundamentally three-dimensional effects corresponding to the span-direction flows driven by the main vortex, were clearly identified. However, whether they close to form two side-by-side tori or are part of a complex topology in which their axes inject into the main vortex flow is not completely clear; most likely this can be resolved by larger scale calculations. Finally, the time-dependent calculations showed the transient form of the three-dimensional effects, and the development of the resulting three-dimensional structures from no-flow initial conditions.

References

[1] G. de Vahl Davis and I. P. Jones. Natural Convection in a Square Cavity: A Comparison Exercise. *Internat. J. Numer. Methods Fluids*, 3:227–248, 1984.

[2] Y. Y. Azmy and J. J. Dorning. A Nodal Integral Approach to the Numerical Solution of Partial Differential Equations. In *Advances in Reactor Computations*, pages 893–909. American Nuclear Society, LaGrange Park, IL, 1983.

[3] G. L. Wilson, R. A. Rydin, and Y. Y. Azmy. Time-Dependent Nodal Integral Methods in Fluid Dynamics. In *Proceedings of the International Topical Meeting on Advances in Reactor Physics, Mathematics, and Computation*, pages 1681–1696. Section Francaise de L'American Nuclear Society. Paris, France, 1987.

[4] W. J. Decker and J. J. Dorning. A Block Iterative Nodal Integral Method for Fluid Dynamics Problems. In *Proc. of the Joint Int. Conf. on Math. Methods and Supercomputing in Nuclear Appl.*. American Nuclear Society/European Nuclear Society, Karlsruhe, Germany, 1993.

[5] R. Schreiber and H. B. Keller. Spurious Solutions in Driven Cavity Problems. *J. Comp. Physics*, 49:165–172, 1983.

[6] H. B. Keller. *Lectures on Numerical Methods in Bifurcation Problems.* Springer-Verlag, New York, 1987.

[7] Y. Y. Azmy and J. J. Dorning. Arc-Length Continuation through Limit Points of Nodal Integral Method Solutions to the Navier-Stokes Equations. In *Numerical Methods for Nonlinear Problems* [C. Taylor, E. Hinton, D. R. J. Owen and E. Oñate, editors], volume 2, pages 672–687. Pineridge Press. Swansea, U. K., 1984.

[8] G. Radicati and Y. Robert. Vector and Parallel CG-like Algorithms for Sparse Non-Symmetric Systems. Report 681–M, IMAG/TIM3, Grenoble, France, 1987.

[9] J. M. Ortega and W. C. Rheinboldt. *Iterative Solution of Nonlinear Equations in Several Variables.* Academic Press, New York, 1970.

[10] G. Moore and A. Spence. The Calculation of Turning Points of Nonlinear Equations. *SIAM Journal of Numerical Analysis*, 17(4):567–576, 1980.

HIGH RAYLEIGH NUMBER NATURAL CONVECTION IN A RECTANGULAR ENCLOSURE CONTAINING WATER NEAR ITS MAXIMUM DENSITY

Tatsuo Nishimura
Dept. of Mech. Eng. Yamaguchi Univ. Ube 755 Japan
Akio Wake
Shimizu Corporation, Tokyo 135, Japan
and
Eiji Fukumori
Toyoda Automatic Loom Works, Obu 474, Japan

ABSTRACT

Natural convection of pure water in a rectangular enclosure with vertical endwalls differentially heated near the temperature of maximum density was investigated by a time-dependent penalty finite element model. Attention is focussed on the main features of the flow at high Rayleigh numbers not considered previously ($Ra=10^5-10^8$). Water in a rectangular enclosure for aspect ratio of 1.25 is initially kept at 4 °C. The enclosure is then suddenly heated and cooled on the opposing vertical walls, i.e., 8°C and 0°C. Rayleigh number is varied by the size of the enclosure.

In the Rayleigh number range considered here, steady state is reached in time $2t_f$, where t_f is the time to steady state suggested by Patterson & Imberger. The steady-state flow field and temperature structure are symmetric about the midpoint of the enclosure and a stable sinking jet is formed in the interior of the enclosure due to the density inversion. Laminar boundary layer approximation is applicable near the vertical walls.

1. INTRODUCTION

Many natural phenomena involve buoyancy-induced flows of cold water close to its freezing point. The mechanism of such flows is considerablly complicated by the fact that its density reaches a maximum value at about 4°C, giving rise to a variety of intriguing phenomena. There have been numerous experimental and numerical studies concerned with the effect of the density inversion of water.

Watson [15] first presented a numerical study of the flow in a square enclosure where one vertical wall was kept at 0°C, while the opposite wall was heated to temperatures above 0 °C. Seki et al. [12] reported a combined experimental and numerical investigation of natural convection in rectangular enclosures.

The cold vertical wall was kept at 0 °C, while the temperature of the hot wall was varied from 1 to 12 °C. Inaba and Fukuda [6] investigated natural convection in an inclined rectangular enclosure where the temperature of the one wall was maintained at 0 °C, while the opposite hot wall temperature was varied from 2 to 20 °C. Lin and Nansteel [9] studied numerically the effect of Rayleigh number over the range 10^3 to 10^6 on Nusselt number in a square enclosure. In these studies, the range of Rayleigh numbers considered is relatively low ($Ra < 10^6$).

Only three papers related to natural convection at high Rayleigh numbers ($Ra > 10^6$) were identified during the literature search. Lankford and Bejan [8] conducted experiments of a high aspect ratio (A=5.05) enclosure where a constant heat flux boundary condition was imposed on the hot vertical wall, while the opposite wall was cooled at temperatures below 4 °C. A scale analysis was used to correlate average Nusselt number measured experimentally in the Rayleigh number range 10^8 to 10^{11}. Ivey and Hamblin [7] performed experiments of natural convection in low aspect ratio (A=0.059-0.265) enclosures filled with water at high Rayleigh numbers ($Ra = 10^5 - 10^8$). The vertical hot wall temperature was 8 °C and the cold wall temperature was 0 °C. The flow field consisted of a double cellular circulation, but the flow was not always stable. That is, a sinking jet formed at the central portion of the enclosure undergoes a series of large-scale meanders for $Ra > 10^7$. An apparently similar feature was reported by Lankford and Bejan [8]. Braga and Viskanta [2] investigated experimentally and numerically transient natural convection in a rectangular enclosure (A=0.5) in the Rayleigh number range 10^7 to 10^8. The fluid was initially stagnant and the temperature of one vertical wall was suddenly lowered to 0 °C. Attention was focussed on the development of the flow in the presence of the density inversion. Since steady state flow and temperature fields are not described, the onset of the unstable sinking jet is not evident in their experiments.

Thus, the experimental studies suggest that the flow in the presence of the density inversion becomes unstable at high Rayleigh numbers. However, the mechanism of unstable flow has not been explained, and also the effect of unstable flow on the heat transfer of the system have not been clarified. These aspects have motivated the present investigation. As an initial step, we performed numerical calculations of natural convection with the density inversion of water in a rectangular enclosure in the Rayleigh number range 10^5 to 10^8, and examined the induced temperature and velocity fields as well as the heat transfer of the system, during both transient and steady state.

2. MATHEMATICAL MODEL AND NUMERICAL SOLUTION

We consider transient and steady natural convective motion of water contained in a rectangular, two-dimensional enclosure (A=1.25) shown in Fig. 1. Initially ($t<0$) the water is stagnant at a uniform temperature 4 °C. At time $t \geq 0$ a uniform temperature

8 °C is imposed in the left vertical wall (x=0), while the right vertical wall (x=L) is kept at 0°C. The top and bottom connecting walls are insulated.

The following simplifying assumptions are made in the analysis :(1) the flow is two-dimensional, laminar and incompressible, and (2) the thermophysical properties are independent of temperature, except for the density in the buoyancy force. These simplifications were made in order to reduce the computational effort during transient and steady state.

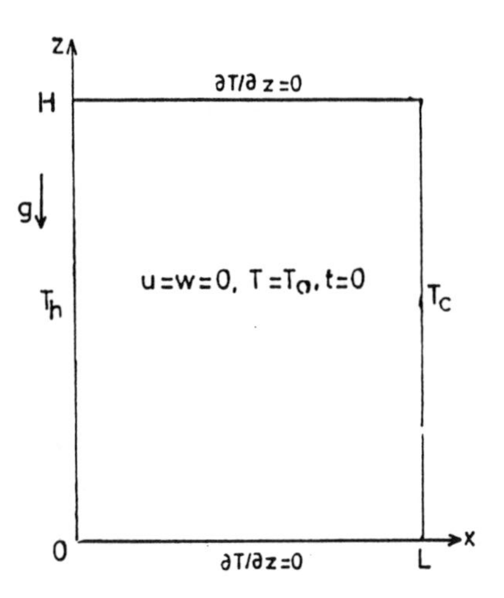

Fig. 1 Physical model and boundary conditions.

The appropriate two-dimesional conservation of mass, momentum and energy equations are, respectively

$$\frac{\partial u}{\partial x} + \frac{\partial w}{\partial z} = 0 \tag{1}$$

$$\rho_o\left(\frac{\partial u}{\partial t} + u\frac{\partial u}{\partial x} + w\frac{\partial u}{\partial z}\right) = \mu\frac{\partial}{\partial x}\left(\frac{\partial u}{\partial x} + \frac{\partial u}{\partial x}\right) + \mu\frac{\partial}{\partial z}\left(\frac{\partial u}{\partial z} + \frac{\partial w}{\partial x}\right) - \frac{\partial p}{\partial x} \tag{2}$$

$$\rho_o\left(\frac{\partial w}{\partial t} + u\frac{\partial w}{\partial x} + w\frac{\partial w}{\partial z}\right) = \mu\frac{\partial}{\partial x}\left(\frac{\partial w}{\partial x} + \frac{\partial u}{\partial z}\right) + \mu\frac{\partial}{\partial z}\left(\frac{\partial w}{\partial z} + \frac{\partial w}{\partial z}\right) - \frac{\partial p}{\partial z} - \rho g \tag{3}$$

$$\frac{\partial T}{\partial t} + u\frac{\partial T}{\partial x} + w\frac{\partial T}{\partial z} = \alpha_o\left(\frac{\partial^2 T}{\partial x^2} + \frac{\partial^2 T}{\partial z^2}\right) \tag{4}$$

The initial and boundary conditions are

$$u=w=0 \text{ and } T=T_o=4\,^\circ C \text{ for } t < 0 \tag{5}$$

and

$$T=T_h=8\,^\circ C \text{ at } x=0 \text{ and } 0 \leqslant z \leqslant H \tag{6}$$

$$T=T_c=0\,^\circ C \text{ at } x=L \text{ and } 0 \leqslant z \leqslant H \tag{7}$$

$$\partial T/\partial z=0 \text{ at } z=0 \text{ and } 0 \leqslant x \leqslant L \tag{8}$$

$$\partial T/\partial z=0 \text{ at } z=H \text{ and } 0 \leqslant x \leqslant L \tag{9}$$

In equation (3), ρg is the buoyancy force, with ρ being the local density corresponding to the local temperature. Water has a non-linear density-temperature relationship that attains its maximum at about T=4°C. Several investigators have proposed different correlations for the density of water as a function of temperature. In this study, a simplified form of the correlation over the range 0 to 8°C suggested by Simons [13] is used

$$\rho = \rho_o(1 - \beta^\bullet (T - T_o)^2) \tag{10}$$

where β°=6.8x10^{-6} ($^{\circ}$C)$^{-2}$ and T_0 is the reference temperature 4°C. The definition of Rayleigh number is

$$Ra = g\Delta\rho H^3/(\rho_o \nu_o \alpha_o) = g\beta^{\circ}(T_o - T_c)^2 H^3/(\nu_o \alpha_o) \quad (11)$$

The model equations are solved numerically using a penalty finite element method. The penalty function approximation is especially attractive in the solution of confined recirculating flow. Fukumori and Wake [3] indicated that the pressure is approximately represented by the following equation

$$p = p_s - \lambda\left(\frac{\partial u}{\partial x} + \frac{\partial w}{\partial z}\right) \quad (12)$$

where P_s is the hydrostastic pressure and λ is the penalty parameter, assumed to be large.

As a result of the penalty approximation, the pressure and the mass conservation equation are eliminated from the system of equations (1)~(3). The characteristics of the algorithm used in this study are summarized as follows: (1) Four-noded bilinear isoparametric elements are used for spatial discretization. (2) the penalty parameter is fixed at $\lambda = 10^7$. (3) the reduced-integration method is applied to the terms which include the penalty parameter. (4) The Crank-Nicolson method is used for discretization of the time derivative terms. (5) Convective and buoyancy force terms are treated explicitly in time and while other terms remain implicit. (6) The well-known computational difficulty for convection dominated transport problems is controlled by applying the method of optimum-added dissipation [5] to the energy equation (4).

Since determination of an optimum mesh size is of practical importance, three nonuniform meshes (40x32, 48x52 and 60x62 nodal points) were used. The basic mesh consisted of 40x32 nodal points and was increased to 60x62 nodal points for higher Rayleigh numbers. The nonuniform mesh provided a higher concentration of nodes near the central portion and the walls of the enclosure where the velocity and thermal gradients need to be accurately resolved.

The criterion for convergence to a steady state solution was that the maximum change in flow and temperature fields satisfy the following inequality:

$$[\phi^{n+1} - \phi^{n}] < 10^{-6} \quad (13)$$

where ϕ in turn represents the velocities and temperature.

3. RESULTS AND DISCUSSION

The following four types of enclosures were considered in this study. The aspect ratio was fixed (A=1.25) and the height and width of the enclosure were variable (HxL= 5cm x 4cm, 20cm x 16cm, 30cm x 24cm and 40cm x 32cm).

We first compare the numerical calculation with an experiment performed additionaly for a small enclosure (5cm x 4 cm) to confirm the validity of the numerical analysis. Figure 2 shows the steady state streamlines and isotherms at T_c=0 °C and T_h=8 °C. The flow field consists of a double cellular circulation, and a stable sinking jet is formed in the interior of the enclosure. The temperature field is symmetric about the midpoint of the enclosure. The isotherm of 4°C is vertically oriented and is located at the interface of the double cellular circulation. Figure 3 shows a visualization photograph of the flow and temperature fields using liquid crystals under the same thermal conditions. The encapsulated thermochromic liquid crystals have been reported to be useful for flow and temperature visualizations in the density inversion process of water [10]. The agreement with the numerical results is satisfactory, although the circulation at the hot wall is slightly larger than that at the cold wall. This difference may be due to the effect of variable thermophysical properties of the kinematic viscosity and thermal diffusivity of water, which is not considered in the present computation.

Figure 4 shows comaprison between numerical and experimental average Nusselt numbers for

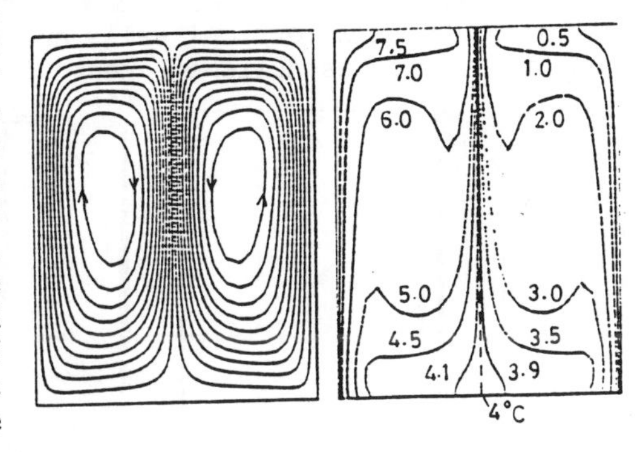

Fig. 2 Streamlines and isotherms (5 cm x 4 cm , T_h=8°C, T_c=0°C).

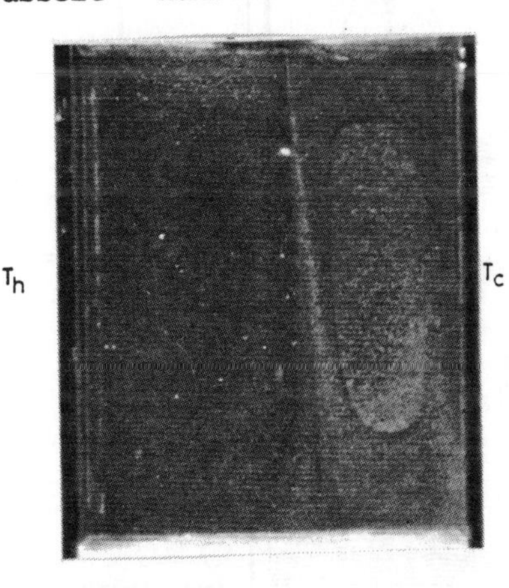

Fig. 3 Visualization photograph by liquid crystals.

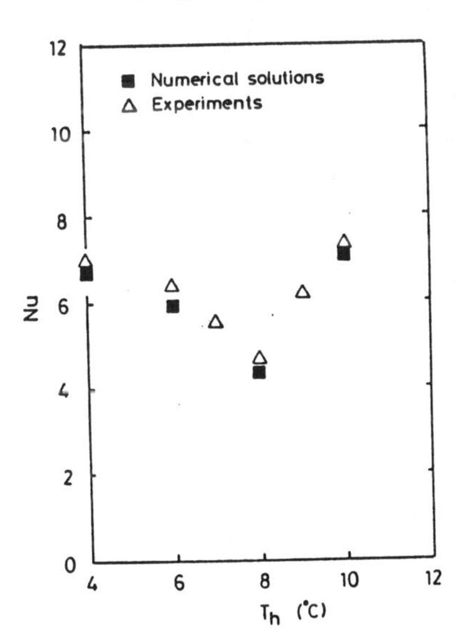

Fig. 4 Average Nusselt numbers for different T_h.

different hot wall temperatures (T_h = 4-10°C). The numerical results are in good agreement with the experimental data, and the minimum average Nusselt number occurs at T_h=8 °C. Thus the above results supports the validity of the numerical analysis.

Next we describe the results for large enclosures at high Rayleigh numbers. In the range of Rayleigh numbers considered here, the flows are stable and steady state is attained, contrary to expectations from the experimental results of Ivey and Hamblin [7].

As an example, Fig. 5 shows the sequence of the transient streamlines and isotherms for different time levels at $Ra=3.2x10^8$ of a 40cm x 32cm enclosure. At the very beginning of the transient process, heat transfer is dominated by conduction (Fig. 5 (a)). As convection becomes dominant, buoyant fluid rises along the two vertical walls and boundary layer type of motion is first established near the walls. The thermal

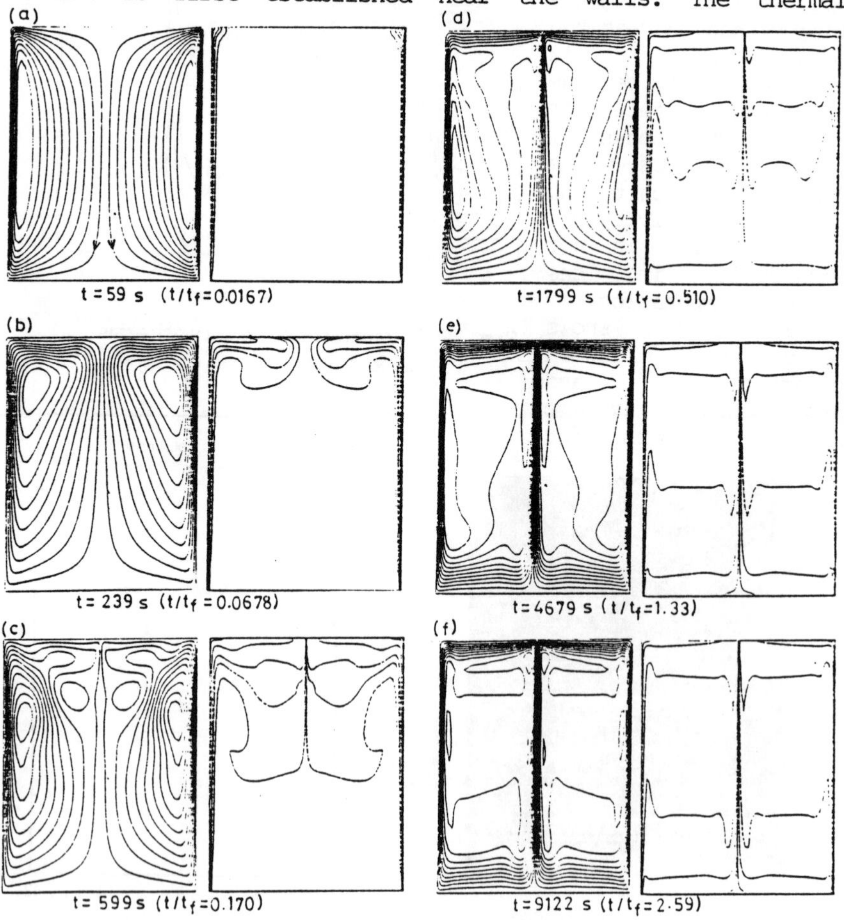

Fig. 5 Transient streamlines and isotherms (40 cm x 32 cm, T_h=8°C and T_c=0°C).

boundary layer is seen clearly from the isotherms (Fig. 5 (b)). The presence of the top horizontal adiabatic wall forces the thermal boundary layers into the center of the enclosure, forming thermal intrusion layers. The two opposing thermal intrusion layers emanating from each vertical wall meet in the center of the enclosure, turn and sink toward the bottom (Fig. 5 (c)). At the bottom, the sinking flow divides symmetrically into two counter flows along the bottom toward the base of the respective vertical walls to complete the circuit. Thus a double-cellular circulation arises.

As the intensity of sinking flow becomes greater, a stable sinking jet with a temperature very close to 4°C is formed (Figs. 5 (d) (f)). Following the variation of the flow, thermal stratification is gradually developed in the core region of the enclosure. Finally the flow and temperature fields reach steady state (Fig. 5 (f)).

The development of the thermal stratification in the core region (x=0.75L) for different sizes of enclosures (different Rayleigh numbers) is shown in Fig. 6, where times levels are nondimensionlized by the time to steady state suggested by Patterson and Imberger [11] of $t_f = HL/(2\alpha Ra^{1/4})$. The figure indicates that the asymptotic approach to steady state has the same trend for any Rayleigh number and that steady state is reached in time $2t_f$. The reason for $2t_f$ is explained as follows. In this system, the thermal stratification begins from the top of the enclosure as shown in Fig. 5. While in the system considered by Patterson and Imberger, the stratification begins from the top and bottom of the enclosure, respectively, and, therefore, the time to steady state is t_f, which is experimentally confirmed by Yewell et al. [16].

Since steady state reaches even at high Rayleigh numbers considered here, we examine the main features of boundary layers near the vertical walls in the steady state velocity and temperature fields. Figure 7 shows the vertical

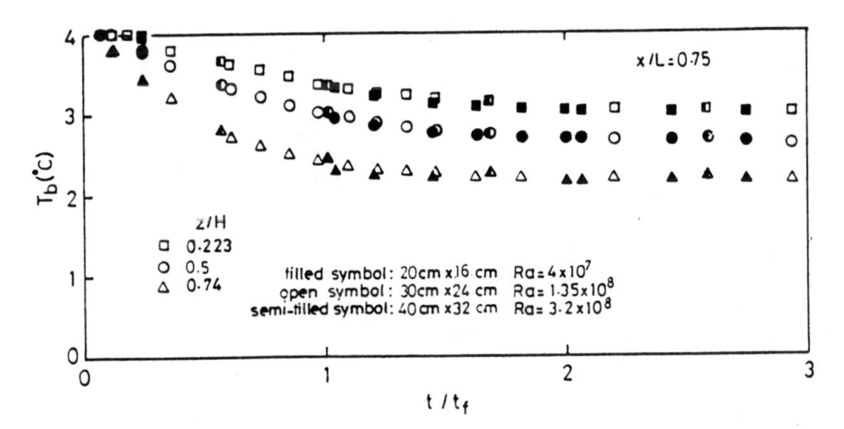

Fig. 6 Time variation of temperatures in the core region for different enclosures or Rayleigh numbers.

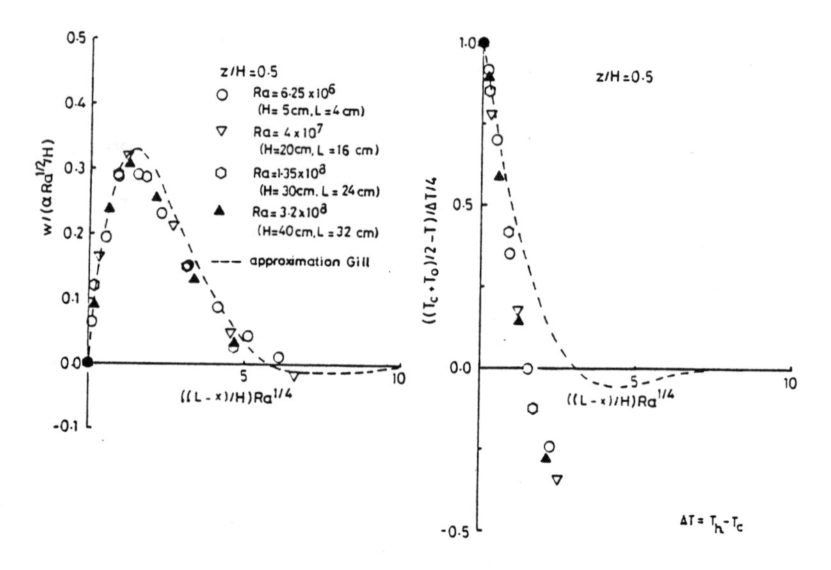

Fig. 7 Vertical velocity and temperature profiles near the cold wall for different enclosures or Rayleigh numbers.

velocity and the temperature profiles at the mid-height z=H/2 near the cold vertical wall. Gill's approximation of the boundary layer solution [4] is also shown in this figure for reference. Gill's approximation is applicable to the velocity and temperature profiles, when a half of the present enclosure contains fluids for which density decreases linearly with increasing temperature and the temperature difference between the vertical walls is kept at 4 C. In the figure, the velocity is nondimensionlized by the buoyant velocity scale $\alpha Ra^{1/2}/H$, and the profile follows with Gill's approximation for any Rayleigh number, while the temperature profiles deviate from the approximation.

Figure 8 shows the thermal stratification in the core region at x=0.75L. The numerical result obtained by Lin and Nansteel [9] is also shown. The present result for a 5cm x 4cm enclosure approximately agrees with their data for a 3.6cm x 3.6cm enclosure. Although the enclosure size is larger, the change of thermal

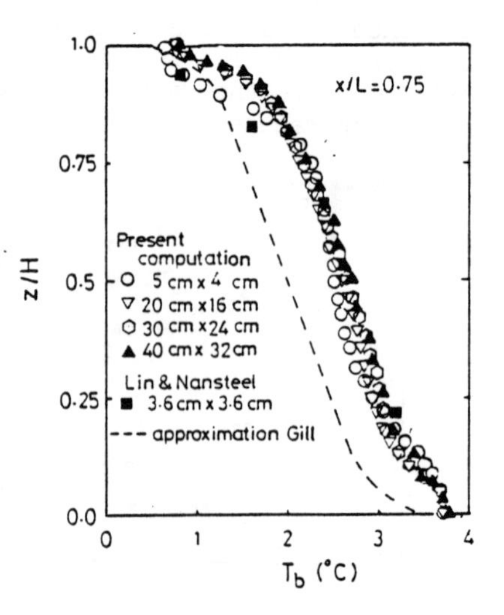

Fig. 8 Thermal stratification in the core region.

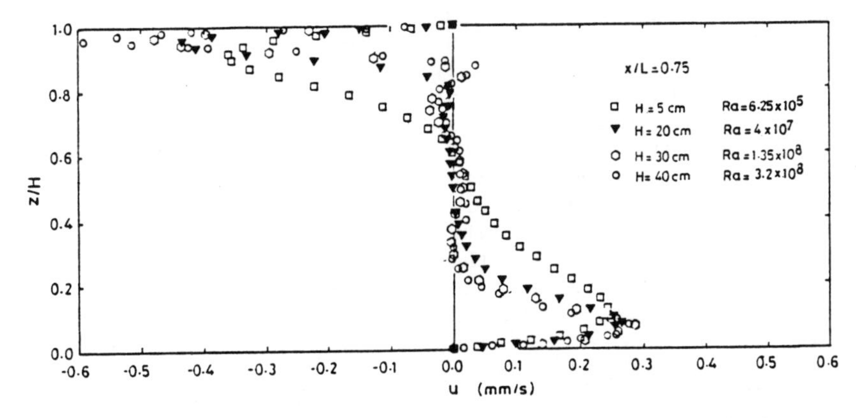

Fig. 9 Horizontal velocity profiles in the core region for different enclosures.

stratification is minute. The core temperature depends only on the vertical coordinate z, but is quite different from Gill's approximation. Thus it is evident that the difference in temperature profile in the thermal boundary layer between the present result and Gill's approximation shown in Fig. 7 is due to the core temperature.

The horizontal velocities in the core region are also estimated. Figure 9 shows the results at x=0.75L for different Rayleigh numbers. The profile is asymmetric with the point of flow reversal for any Rayleigh number, and the degree of asymmetry is more apparent with increasing Rayleigh number. It should be noted that the maximum velocity near the top horizontal wall (z=H) is two times that near the bottom wall (z=0) at $Ra=3.2\times10^8$, which is not identical to the symmetric profile in z obtained by Simpkins and Chen [14].

We now investigate the features of the sinking flow formed in the central portion of the enclosure due to the density

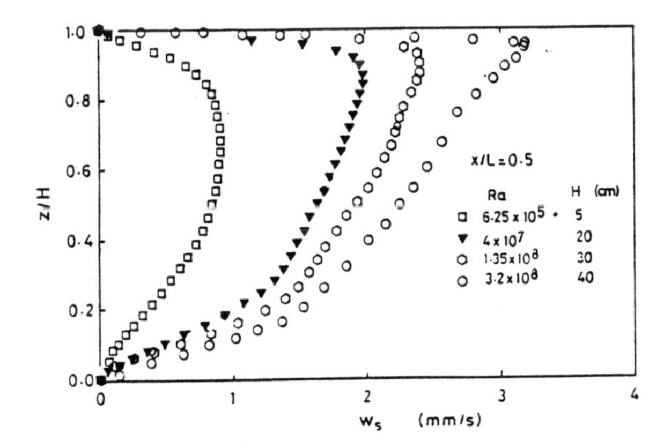

Fig. 10 Sinking velocity profiles for different enclosures.

inversion. Figure 10 shows sinking velocity profiles in z. Although the profile is similar for different Rayleigh numbers, the position of the maximum velocity shifts toward the top of the enclosure with increasing Rayleigh number. The maximum velocity nondimensionlized by the buoyant velocity scale, is found to be almost constant ($w_s/(\alpha Ra^{1/2} H)=0.4$) at Rayleigh numbers considered here, indicating the usefulness of the velocity scale approach.

Figure 11 shows the local Nusselt number profiles at the cold vertical wall. The local Nusselt number is found to be proportional to $Ra^{1/4}$, except near the top and bottom walls, also indicating the validity of laminar boundary layer approximation. The average Nusselt numbers for different Rayleigh numbers are shown in Fig. 12. Numerical results by other investigators [9, 12] are included in the figure. The present results are found between the correlation proposed from Gill's approximation by Bejan [1] and the correlation proposed by Lankford and Bejan [8]. Gill's approximation underpredicts the heat transfer rate because of

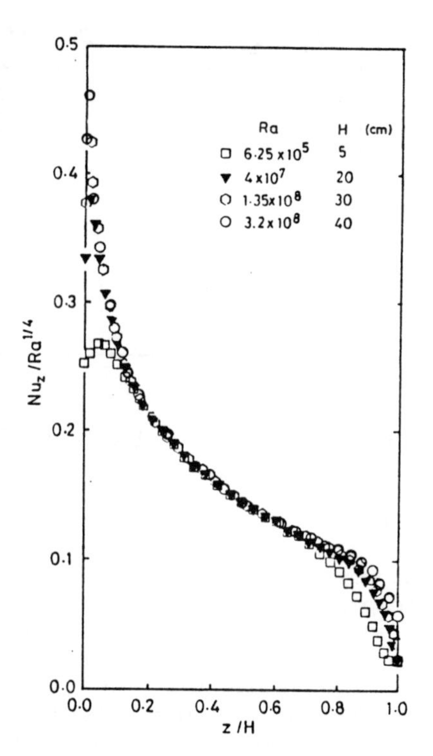

Fig. 11 Local Nusselt numbers.

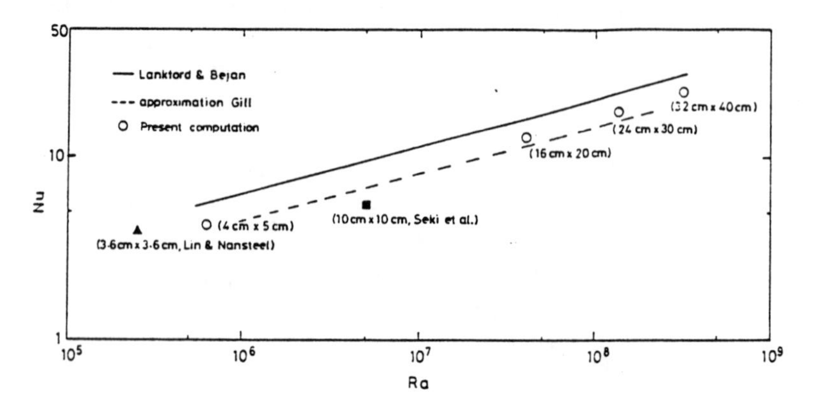

Fig. 12 Average Nusselt number vs. Rayleigh number.

the difference in the core temperature shown in Fig. 8. The correlation by Lankford and Bejan was proposed under the assumption that the temperature in the core region is constant at 4 °C, and that the thermal resistance at the interface of the double cellular circulation is negligible. Since the thermal resistance cannot be ignored even at high Rayleigh numbers in the present system, Nusselt numbers are lower than predicted by the correlation.

4. CONCLUSIONS

Numerical computations were performed to study natural convection in rectangular enclosures containing water near its maximum density. Attention was focussed on the features of the flow at high Rayleigh numbers. The following conclusive remarks are made.

(1) Steady state is gradually reached in time $2t_f$, where t_f is defined as $HL/(2\alpha Ra^{1/4})$.

(2) The velocity profile near the vertical walls agrees well with Gill's approximation of laminar boundary layer solution, but the temperature profile disagrees due to the discrepancy in the core temperature.

(3) The usefulness of time and velocity scales suggested by Patterson and Imberger [11] is verified in the study.

In the present study, the flows at high Rayleigh numbers are stable and symmetric about the midpoint of the enclosure, contrary to expectations from the previous experiments. Asymmetries of temperature and velocity may be needed for the flow to become unstable and oscillatory. Asymmetries can be numerically obtained, by considering the variable thermophysical properties or by using an asymmetric density-temperature relationship proposed by other investigators. This is a future work for the writers.

REFERENCES

1. BEJAN, A. - A Note on Gill's Solution of Free Convection in Vertical Enclosure. J. Fluid Mech., Vol.90, pp.561-568, 1979.

2. BRAGA, S.L. and VISKANTA, R. - Transient Natural Convection of water near Its Density Extremum in a Rectangular Cavity. Int. J. Heat Mass Transfer, Vol.35, No.4, pp.861-875, 1992.

3. FUKUMORI, E. and WAKE, A. - The Linkage between Penalty Function Method and Second Viscosiy Applied to Navier-Stokes Equations. Coputational Mechanics, Cheung, Lee and Leung (eds) Balkema, Rotterdam Holland, pp.1385-1390, 1991.

4. GILL, A.E. - The Boundary-Layer Regime for Convection in a Rectangular Cavity. J. Fluid Mech., Vol.26, pp.515-536, 1966.

5. Hughes, T.J.R. - A Simple Scheme for Developing Upwind Finite Elements. Int. J. Num. Meth. Eng., Vol.12, pp.1359-1365, 1978.

6. INABA, H. and FUKUDA, T. - Natural Convection in an Inclined Square Cavity in Regions of Density Inversion of Water. J. Fluid Mech., Vol.142, pp.363-381, 1984.

7. IVEY, G.N. and HAMBILIN, P.F. - Convection near the Temperature of Maximum Density for High Rayleigh Number, Low Aspect Ratio, Rectangular Cavities. J. Heat Transfer, Vol.111, pp.100-105, 1989.

8. LANKFORD, K.E. and BEJAN, A. - Natural Convection in a Vertical Enclosure Filled with Water near 4 C. J. Heat Transfer, Vol.108, pp.755-763, 1986.

9. LIN, D.S. and NANSTEEL, M.W. - Natural Convection Heat Transfer in a Square Enclosure Containing Water near Its Density Maximum. Int. J. Heat Mass Transfer, Vol.30, No.11, pp.2319-2329, 1987.

10. NISHIMURA, T., FIJIWARA, M., HORIE, N. and MIYASHITA, H. - Temperature Visualizations by Use of Liquid Crystals of Unsteady Natural Convection during Supercooling and Freezing of Water in an Enclosure with Lateral Cooling. Int. J. Heat Mass Transfer, Vol.34, No.10, pp.2663-2668, 1991.

11. PATTERSON, J. and IMBERGER, J. - Unsteady Natural Convection in a Rectangular Cavity. J. Fluid Mech., Vol.100, pp.65-86, 1980.

12. SEKI, N., FUKUSAKO, S. and INABA, H. - Free Convection Heat Transfer with Density Inversion in a Confined Rectangular Vessel. Wärme und Stoffubertragung, Vol.11, pp.145-156, 1978.

13. SIMONS, T.J. - Circulation Models of Lakes and Inland Seas. Can. Bul. Fish. Aquatic. Sci., Vol.203, pp.1-146, 1980.

14. SIMPKINS, P.G. and CHEN, K.S. - Convection in Horizontal Cavities. J. Fluid Mech., Vol.166, pp.21-39, 1986.

15. WATSON, A. - The Effect of the Inversion Temperature on the Convection of Water in an Enclosed Rectangular Cavity. Q. J. Mech. Appl. Math., Vol.15, pp.423-446, 1972.

16. YEWELL, R., POULIKAKOS, D. and BEJAN, A. - Transient Natural Convection Experiments in Shallow Enclosures. J. Heat Transfer, Vol.104, pp.533-538, 1982.

APPLICATION OF NUMERICAL METHODS TO CONVECTION IN A DIFFERENTIALLY HEATED CAVITY WITH HIGH ASPECT RATIO

Nigel G. Wright†, P. Andrew Sleigh‡ and Philip H. Gaskell‡
†Department of Mathematics
City University, London, EC1V 0HB. UK

‡Department of Mechanical Engineering
University of Leeds, Leeds, LS2 9JT. UK

SUMMARY

The paper presents solutions to the flow in shallow laterally heated cavities. The algorithm uses a fully coupled point-by-point solver in conjunction with nonlinear multigrid. A high order bounded discretization is implemented by means of defect correction. Solutions are presented at aspect ratios upto 100 and comparison is made with asymptotic theory. Solutions for one particular aspect ratio, 20, are presented at Rayleigh numbers upto 10^6. The solutions are further enhanced by using locally refined grids.

1 INTRODUCTION

The problem of a differentially heated cavity has received much attention for the case of aspect ratio one[1, 2, 3]. However, most occurrences of this problem involve large or small aspect ratios, such as the temperature control of circuit board components under natural convection, heating and ventilation control in building design and construction, solar-energy collectors and atmospheric & fluvial dispersion in the environment. These cases have received much attention from theoretical fluid dynamicists using asymptotic analysis[4]. The flow is divided into different flow regimes in each of which certain parameters are dominant and an asymptotic analysis is carried out in each. This results in different predictions for the various regions and these are integrated to give an overall prediction of the flow.

So far there has only been a small amount of work in this area using numerical techniques[5]. This has been due to the high gradients involved in the transitions between the different regions of the flow and the difficulty of fitting grids to very long or tall cavities. In this paper we present numerical solutions to these problems with aspect ratios of the order of 10^2. This is done using algorithms based on those developed by us for the case of aspect ratio one. The main features of this algorithm are:

- high order bounded discretizations[6] implemented via defect correction[' for efficiency and stability.
- fully coupled solver for the primitive variables that are used.
- multigrid algorithm allowing for solutions with large numbers of finite difference nodes.
- local grid refinements for the end regions of the flow.

These are particularly beneficial in the flows under consideration here. The discretization allows for accurate resolution of steep gradients without loss of boundedness. The combination of multigrid and local grid refinement allows for high mesh densities to be placed in the areas where they are most required i.e. the boundary layers and end-zones.

The solutions presented cover a range of parameters and demonstrate the successful application of the above techniques to these problems.

2 GOVERNING EQUATIONS

The governing equations for the flow are, in two-dimensional cartesian coordinates:

$$\rho\frac{\partial u}{\partial t}+\rho u\frac{\partial u}{\partial x}+\rho v\frac{\partial u}{\partial y}=-\frac{\partial p}{\partial x}+\mu\nabla^2 u \qquad (1)$$

$$\rho\frac{\partial v}{\partial t}+\rho u\frac{\partial v}{\partial x}+\rho v\frac{\partial v}{\partial y}=-\frac{\partial p}{\partial y}+\mu\nabla^2 v+\rho g \qquad (2)$$

$$\rho\frac{\partial T}{\partial t}+\rho u\frac{\partial T}{\partial x}+\rho v\frac{\partial T}{\partial y}=\rho\alpha\nabla^2 T+q \qquad (3)$$

$$\frac{\partial(\rho u)}{\partial x}+\frac{\partial(\rho v)}{\partial y}=0. \qquad (4)$$

The density variation in the buoyancy term satisfies the equation

$$\rho = \rho_0(1 - \beta(T - T_0)) \tag{5}$$

where ρ_0 is a reference density and T_0 is a reference temperature. In this work we have adopted the Boussinesq approximation that β may be neglected everywhere except in the buoyancy term. A modified pressure p' is introduced to simplify the buoyancy term

$$p' = \rho_0 g y + p \tag{6}$$

this gives

$$\frac{\partial p'}{\partial y} = \rho g + \frac{\partial p}{\partial y}. \tag{7}$$

The constants are defined as follows:

g	acceleration due to gravity
μ	dynamic viscosity
$\nu = \mu/\rho_0$	kinematic viscosity
$\alpha = \kappa/(\rho_0 C_p)$	thermal diffusivity
κ	thermal conductivity
C_p	specific heat capacity
β	coefficient of thermal expansion
L	length of the cavity
H	height of the cavity
A	aspect ratio of the cavity = L/H.

Non-dimensionalization is performed in various ways in asymptotic analysis in order to assist in the comparison of different terms. These non-dimensionalizations are not always the most suitable for numerical methods, in that they can lead to large differences between the terms in the partial differential equations. This in turn can lead to convergence difficulties. We adopt a non-dimensionalization that prevents these problems and this is outlined below.

These equations are non-dimensionalized using:

$$u = u_0 U$$
$$v = u_0 V$$
$$x = LX$$
$$y = LY$$
$$p = \rho u_0^2 P$$
$$T = T_1 + (T_2 - T_1)\theta$$
$$t = \left(\frac{L}{u_0}\right) \tau$$

Prandtl number, $Pr = \frac{\mu C_p}{\kappa} = \frac{\nu}{\alpha}$
Grashof number, $Gr = \frac{\beta g L^3}{\nu^2}(T_2 - T_1)$
Rayleigh number, $Ra = GrPr = \frac{\beta g L^3}{\alpha\nu}(T_2 - T_1)$.

The reference velocity is taken as $u_0 = \sqrt{\beta g L(T_2 - T_1)}$. Neglecting time dependent terms, the equations become:

$$U\frac{\partial U}{\partial X} + V\frac{\partial U}{\partial Y} = -\frac{\partial P}{\partial X} + \sqrt{\frac{Pr}{Ra}}\nabla^2 U \tag{8}$$

$$U\frac{\partial V}{\partial X} + V\frac{\partial V}{\partial Y} = -\frac{\partial P}{\partial Y} + \sqrt{\frac{Pr}{Ra}}\nabla^2 V + \theta \tag{9}$$

$$U\frac{\partial \theta}{\partial X} + V\frac{\partial \theta}{\partial Y} = \frac{1}{\sqrt{PrRa}}\nabla^2 \theta \tag{10}$$

$$\frac{\partial U}{\partial X} + \frac{\partial V}{\partial Y} = 0. \tag{11}$$

3 SOLUTION STRATEGY

The solution strategy contains three stages: discretization of the partial differential equations to give a set of algebraic equations, solution of these equations by an iterative technique and acceleration of this technique by use of a multigrid algorithm. Each of these is described below.

3.1 Discretization

The above partial differential equations are discretized on a mesh of uniform square control volumes. The staggered grid approach is used to prevent pressure oscillations. This approach is supported by earlier work[3]. Each of the partial differential equations is integrated over the surface of the control volume to give an algebraic equation in terms of the face fluxes. It is these face fluxes that must be approximated by interpolation. There are various ways of doing this, as discussed below.

It has been shown in previous work[8] that the use of the traditional hybrid differencing (a combination of zero and first order interpolation) is not satisfactory. This is due to the numerical diffusion introduced by the upwind differencing term that is used in regions of steep gradients. The first extension was to use quadratic interpolation, giving QUICK differencing[9]. Unfortunately, this was only a partial answer as higher accuracy was obtained at the expense of unphysical over/under shoots in the results.

In this work we use the SMART algorithm of Gaskell and Lau[6]. This is unconditionally bounded and has second order accuracy in regions where such a definition is relevant and has been implemented in a number of complex flows[10].

3.2 Coupled Solver

Unlike the widely used SIMPLE technique[11] the equation solver used here treats all the three variables $u, v\&p$ simultaneously. This gives greater stability and faster convergence. Each control volume is visited in turn and at each the four face velocities and the pressure at the central node are updated. This gives a 5x5 matrix that is easily inverted by LU decomposition. Obviously, each velocity is updated twice and this gives the scheme the stability lacked by a single update method. After one iteration over the grid for momentum and continuity, the temperature field is update via the temperature transport equation. Simultaneous updating of temperature has been investigated, but offered no significant advantageous.

Because of the non-linearity the coefficients of the matrix must be calculated using values at the previous iteration. This necessitates under-relaxation of the updates to the variables. The iteration is continued until the residual measure defined by

$$\| r \|_2 = \left(\frac{\sum_{i,j} \left(r^{u^2}_{i,j} + r^{v^2}_{i,j} + r^{t^2}_{i,j} + r^{c^2}_{i,j} \right)}{4 \times n \times n} \right)^{1/2}$$

is less than 10^{-5}.

3.3 Multigrid

A current serious constraint on the application of CFD techniques to slender cavities is that of limited resolution. One solution is to design iterative schemes with higher error reductions. However, these still have the disadvantage of a deteriorating convergence rates as the iteration proceeds. The one idea that has opened the door to high resolution solutions in many flow situations is the concept of multigrids. The method seeks to obtain the initial fast convergence of an ordinary scheme throughout the iteration procedure, thus giving very fast solution techniques.

When observing the convergence of a non-multigrid iterative technique it can be seen that initially convergence is rapid, but that this soon stops and error reduction becomes very slow. This is a manifestation of the efficiency of an iterative solver in eliminating errors of wavelength similar to the mesh size only. In order to maintain this fast convergence and exploit this property of iterative solvers, multigrid methods solve the problem on a hierarchy of coarser grids. This has the added advantage that iterations on coarser grids takes less time. Equations on coarser grids are amended to ensure that they represent the actual solution on the finer grids by the adding of the following source term:

$$F^{k-1} = I_k^{k-1} r^k + L^{k-1}(I_k^{k-1} q^k), \tag{12}$$

where I_k^{k-1} is the restriction from grid level k to grid level k-1. Here, because the Navier-Stokes equations are non-linear we use the Full Approximation Storage (FAS) version of multigrids.

For full details of the theory readers are referred to an earlier text[8]. It has been shown theoretically by Brandt and Dinar[12] (who first proposed these techniques) that such an algorithm will be very efficient.

When smoothing in an multigrid algorithm, all that is required is to smooth the error on the current grid, not eliminate it. So the accuracy of the discretization is not relevant. The only criteria are stability and efficiency. So it would appear natural to use a low order discretization for smoothing and implement a high order discretization through a correction to the source term in the equation. This is the basis of the defect correction method as outlined by Hackbusch[7]. To find a new iteration u^{n+1} from the equation

$$L\underline{u}^{n+1} = \underline{r}^n \tag{13}$$

requires the use of two operators

- L^l - lower order discretization for smoothing
- L^h - high order discretization for residual calculation.

The following system needs to be solved

$$L^l\underline{u}^{n+1} = \underline{r}^n + L^l\underline{u}^n - L^h\underline{u}^n \tag{14}$$

This has been implemented in this case and is very successful in increasing stability whilst maintaining accuracy[13].

4 RESULTS

The algorithm outlined above was implemented on a SUN Sparcstation with a mesh size of 1/32 (that is 32 control volumes vertically) at various values of the Rayleigh number and at Prandtl number of 0.733. Validation was successfully carried for the case of aspect ratio one[2, 3].

A further step, used in asymptotic analysis, is to work with $Ra_l = \frac{Ra}{A}$ as a non-dimensional number where A is the aspect ratio. This is usually achieved by dividing temperature θ by A. However, rather than implementing this in the non-dimensionalization we use a value of $Ra = Ra_l \times A$.

Asymptotic theory predicts that as $A \to \infty$ the flow is characterized by a long central area of parallel flow and two end-zones where flow reversal takes place. To investigate this we solved for a number of aspect ratios ($20 \to 100$) whilst keeping Ra_l constant. Asymptotic theory suggests that the value of the streamfunction at the mid-point of the cavity should tend to a constant as $A \to \infty$. Looking at Table 1 and Figure 1 it can be seen that this is the case from numerical results. Further confirmation should be obtained from solutions at higher aspect ratios, but this was not possible with the present code. This is due to degraded convergence caused by the large number of control volumes in the x-direction on the coarsest grid level. Techniques for overcoming this are under investigation - see Section 5.

Ra	A	Ra_l	ψ_m	ω_m
2.e4	20	1.e3	1.966e-2	3.187e-1
4.e4	40	1.e3	1.423e-2	2.326e-1
6.e4	60	1.e3	1.163e-2	1.923e-1
8.e4	80	1.e3	1.021e-2	1.630e-1
1.e5	100	1.e3	9.670e-3	1.574e-1

Table 1: Value of streamfunction and vorticity at the mid-point for various values of A

Figures 2,3 & 4 show streamfunction, temperature and vorticity for the case of $A = 20$ at Rayleigh numbers of $10^3, 10^4, 10^5 \& 10^6$. These show the development of a central core flow with a stratified temperature field. At higher Ra an extra recirculation is observed in the end-zone (Figure 4). This behaviour is confirmed by comparison with asymptotic results[5].

As stated above a central core region of parallel flow develops as aspect ratio increases. In this region the gradients are much smaller than those in the end-zones where the flow is turning. An extremely fine mesh is

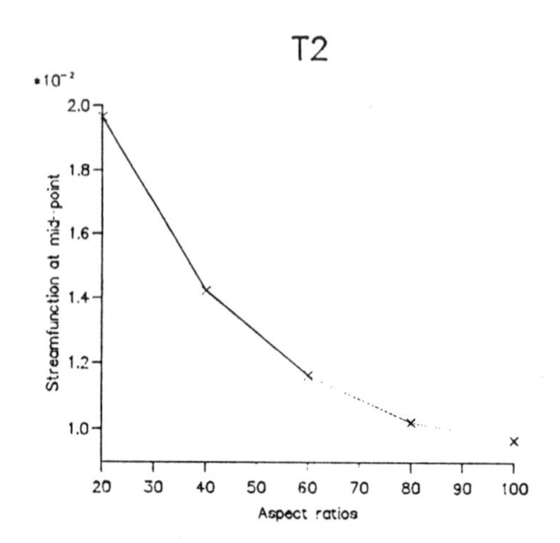

Figure 1: Streamfunction at the mid-point for various aspect ratios. (Only part of the cavity is shown.)

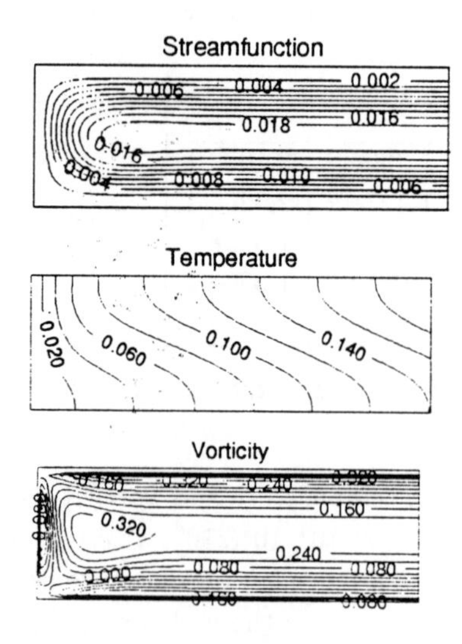

Figure 2: Streamfunction, vorticity and temperature $Ra = 2 \times 10^4$ and $A = 20$. (Only part of the cavity is shown.)

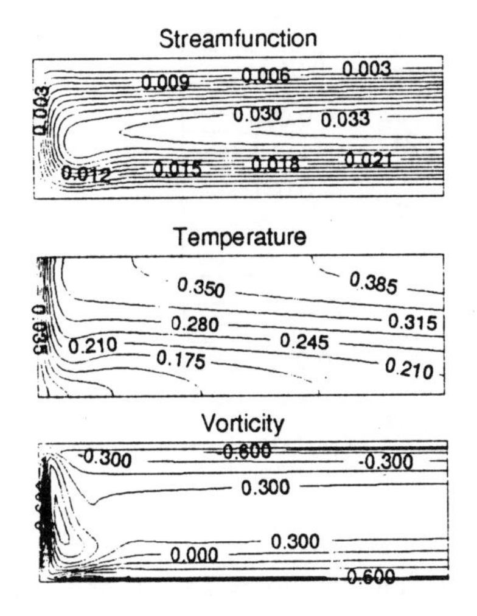

Figure 3: Streamfunction, vorticity and temperature $Ra = 2 \times 10^5$ and $A = 20$. (Only part of the cavity is shown.)

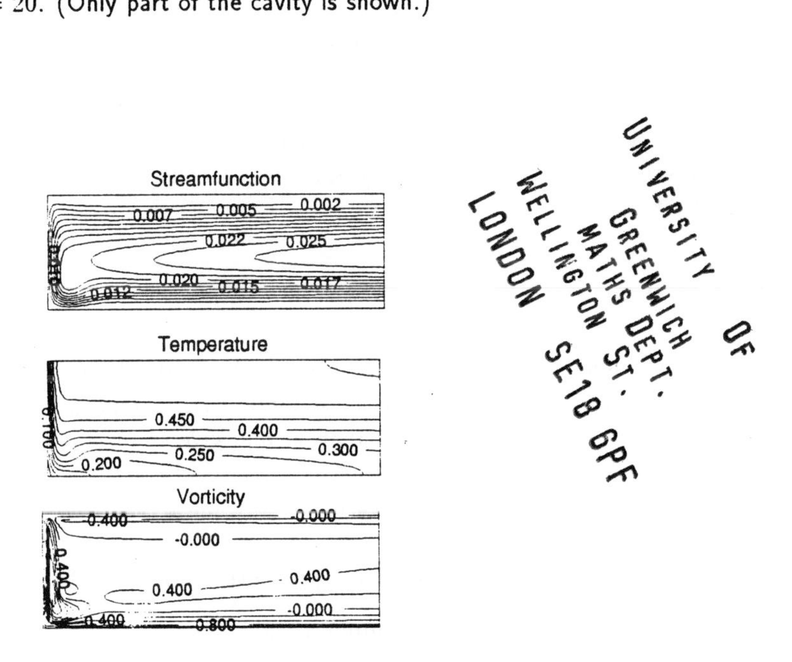

Figure 4: Streamfunction, vorticity and temperature $Ra = 2 \times 10^6$ and $A = 20$.

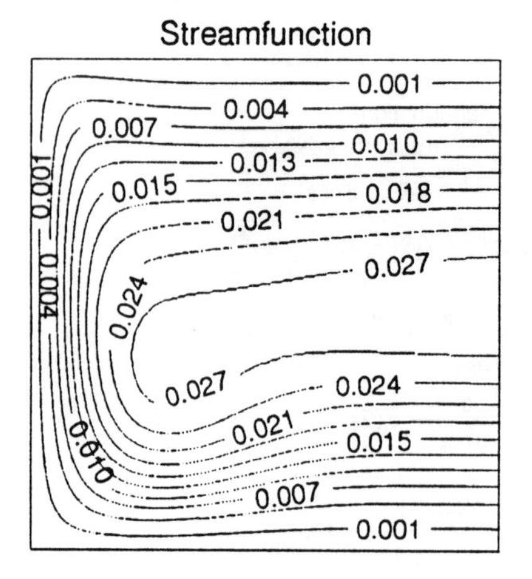

Figure 5: Streamfunction in the end-zone for $Ra = 2 \times 10^5$ and $A = 20$ on a locally refined grid of 128×128 control volumes.

not as important here as it is in the end-zones. In view of this, Figure 5 shows a locally refined mesh solution. This was obtained by taking the full grid solution and refining to a mesh size 1/64 in the region ($0 < x < 2$ & $0 < y < 1$) and then further to a mesh size 1/128 in the region ($0 < x < 1$ & $0 < y < 1$). The new solution gives results similar to the full grid, but obviously the accuracy of the locally-refined solution is higher.

5 CONCLUSION

The use of higher order discretization with fully coupled algorithm has been successfully demonstrated for a horizontal cavity. This gives results that appear to confirm asymptotic analysis and give an insight into such flows.

Further work will involve:

- more detailed comparison with asymptotic work.
- different values of Prandtl number.
- tall cavities.
- overcoming convergence difficulties of multigrid in very slender cavities by use of semi-coarsening.

References

1 G. de Vahl Davis. Natural convection of air in a square cavity: A bench mark numerical solution. *International Journal for Numerical Methods in Fluids*, 3:249–264, 1983.

2 P.H. Gaskell and N.G. Wright. A multigrid algorithm for the investigation of thermal, recirculating fluid flow problems. In *Numerical Methods in Thermal Problems, Proceedings of The Fifth International Conference.* Pineridge Press, July 1987.

3 P.A. Sleigh, P.H. Gaskell, and N.G. Wright. Non-linear coupled multigrid solutions to thermal problems, employing different nodal grid arrangements and convective transport approximations. In *Numerical Methods in Laminar and Turbulent Problems, Proceedings of The Seventh International Conference.* Pineridge Press. 1991.

4 P.G. Daniels, P.A. Blythe, and P.G. Simpkins. High rayleigh number thermal convection in a shallow laterally heated cavity. Technical Report 11523-871019-61, Bell Laboratories, 1987.

5 P. Wang. *Thermal convection in slender laterally-heated cavities.* PhD thesis, City University, 1993.

6 P.H. Gaskell and A.K.C. Lau. Curvature compensated convective transport : SMART, a new boundedness preserving transport algorithm. *International Journal for Numerical Methods in Fluids*. 8:617–641, 1988.

7 W. Hackbusch. *Robust Multigrid Methods*, volume 23 of *Notes on Numerical Fluid Dynamics.* Vieweg-Verlag, 1988.

8 N.G. Wright. *Multigrid Solution of Elliptic Fluid Flow Problems.* PhD thesis, University of Leeds, 1988.

9 B.P. Leonard. A stable and accurate convective modelling procedure based on quadratic upstream interpolation. *Computational Methods in Applied Mechanics and Engineering*, 19:59–98, 1979.

10 P.H. Gaskell and A.K.C. Lau. The method of curvature compensation and its use in the prediction of highly recirculating flows. In *Proceedings of the AIAA/ASME/SIAM/APS 1st National Fluid Dynamics Congress, Part 1*, pages 272–279, 1988.

11 S.V. Patankar and D.B. Spalding. A calculation procedure for heat, mass and momentum transfer in three-dimensional parabolic flows. *International Journal of Heat Mass Transfer*, 15:1787–1806, 1972.

12 A. Brandt and N. Dinar. Multigrid solution to elliptic flow problems. In S. Parter, editor, *Numerical Methods in Partial Differential Equations*, pages 53–147. Academic Press, 1977.

13 N.G. Wright and P.H. Gaskell. An efficient multigrid approach to solving highly recirculating flows. *submitted to Computers and Fluids*, 1993.

UNSTEADY NATURAL CONVECTIVE FLOW IN AN ENCLOSURE WITH A PARTIALLY HEATED WALL WITH A VARYING TEMPERATURE

P. H. Oosthuizen and J. T. Paul
HEAT TRANSFER LABORATORY, Dept. Mechanical Engineering
Queen's University, Kingston, Ontario, Canada K7L 3N6

ABSTRACT

Natural convective flow in a square enclosure with one wall partially heated and the opposite wall cooled to a uniform lower temperature and with the remaining walls adiabatic has been numerically studied. The temperature of the heated wall section, in general, varys sinusoidally with time. The flow has been assumed to be laminar and two-dimensional. Fluid properties have been assumed constant except for the density change with temperature that gives rise to the buoyancy forces, this being treated by means of the Boussinesq approximation. The governing equations, expressed in terms of stream function and vorticity, have been written in dimensionless form. The resultant equations, subject to the assumed boundary conditions, have been solved using the finite-element method. The solution has as parameters the Rayleigh number based on the enclosure width and the mean temperature difference, the Prandtl number, the dimensionless size and position of the heated wall section and the frequency and magnitude of the dimensionless wall temperature variation. Results have been obtained for a Prandtl number of 0.7 for for Rayleigh numbers between between 1,000 and 1,000,000 for various geometrical parameters for various forms of dimensionless wall temperature variation with dimensionless time. These results have been used to study the effects of the governing parameters on the mean and local heat transfer rate variations.

1. INTRODUCTION

The present paper describes a numerical study of two-dimensional, laminar natural convective flow in a square enclosure with one wall partially heated to a uniform temperature, T'_H and the opposite wall cooled to a uniform lower temperature T'_C , which is less than T'_H . The remaining walls are adiabatic. The temperature of the heated wall section is uniform but it is, in general, varying sinusoidally with time.

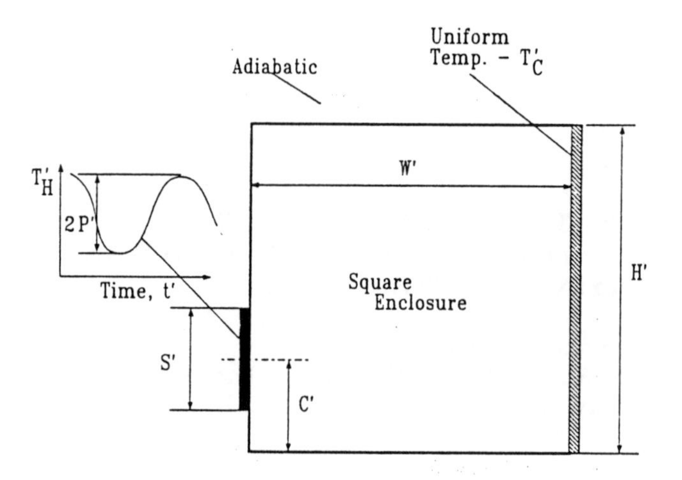

Figure 1. Situation considered in present study.

The flow situation is, thus, as shown in Fig. 1. The situation considered is an approximate model of some situations that occur in the cooling of electronic and electrical equipment.

The flow and heat transfer in enclosures when the temperature of one vertical wall is suddenly increased has been quite widely studied. Most of this work has dealt with the case where the wall is heated to a uniform temperature over its entire height e.g. see [1] to [11]. A few studies of this type of flow have dealt with the situation where the wall is only partially heated e.g. see [12] to [14]. The case where the hot wall temperature is varying with time has received relatively little attention. A study of the effects of a sinusoidlly varying hot wall temperature for the fully heated wall case is described in [15] which also gives a review of past work concerned with the time-varying hot wall temperature case. These past studies of the flow that arises when the hot wall temperature is varying with time have all essentially been concerned with the situation in which the entire hot wall is at a uniform temperature. The present study differs from these past studies in that an enclosure with a partly heated wall has been considered, this situation potentially having significant practical application.

2. GOVERNING EQUATIONS AND SOLUTION PROCEDURE

It has been assumed that the flow is laminar and two- dimensional and that the fluid properties are constant except for the density change with temperature which gives rise to the buoyancy forces. This has been treated by using the Boussinesq approach.

The solution has been obtained in terms of the stream function and vorticity defined, as usual, by:

$$u' = \frac{\partial \psi'}{\partial y'} \quad , \qquad v' = -\frac{\partial \psi'}{\partial x'}$$

$$\omega' = \frac{\partial v'}{\partial x'} - \frac{\partial u'}{\partial y'} \tag{1}$$

The prime ($'$) denotes a dimensional quantity.

The following dimensionless variables have then been defined:

$$\psi = \psi' / \alpha \quad , \qquad \omega = \omega' \, W'^2 / \alpha$$

$$T = (T' - T'_C) / (\overline{T'_H} - T'_C) \quad , \quad t = t' \alpha / W'^2 \tag{2}$$

where $\overline{T'_H}$ is the time averaged temperature of the hot wall section.

In terms of these dimensionless variables, the governing equations are:

$$\frac{\partial^2 \psi}{\partial x^2} + \frac{\partial^2 \psi}{\partial y^2} = -\omega \tag{3}$$

$$\frac{\partial \omega}{\partial t} + \left(\frac{\partial \psi}{\partial y} \frac{\partial \omega}{\partial x} - \frac{\partial \psi}{\partial x} \frac{\partial \omega}{\partial y} \right) - Pr \left(\frac{\partial^2 \omega}{\partial x^2} + \frac{\partial^2 \omega}{\partial y^2} \right)$$

$$= Ra \frac{\partial T}{\partial x} \tag{4}$$

$$\frac{\partial T}{\partial t} + \left(\frac{\partial \psi}{\partial y} \frac{\partial T}{\partial x} - \frac{\partial \psi}{\partial x} \frac{\partial T}{\partial y} \right)$$

$$- \left(\frac{\partial^2 T}{\partial x^2} + \frac{\partial^2 T}{\partial y^2} \right) = 0 \tag{5}$$

The Rayleigh number Ra is defined by:

$$Ra = \beta \, g \, W'^3 (\overline{T'_H} - T'_C) / \nu^2 \tag{6}$$

The boundary conditions on the solution are:

On all walls:

$$\psi = 0 , \quad \frac{\partial \psi}{\partial n} = 0$$

At $x = 0$: on heated section

$$T = T_H$$

At $x = 1$:

$$T = 0$$

On remaining wall segments:

$$\frac{\partial T}{\partial n} = 0$$

where n is the coordinate measured normal to the wall surface considered. T_H is the time varying dimensionless temperature of the heated wall section. It is here assumed to be given by:

$$T_H = 1 + P \sin(2\pi t/p) \tag{7}$$

where P is the dimensionless half amplitude of the hot section temperature fluctuation and p is the dimensionless period of this fluctuation.

It has been assumed that the fluid is at rest and at the cold wall temperature, i.e. it has been assumed that the initial conditions are:

$$t = 0: \psi = 0, \ T = 0$$

The above dimensionless equations, subject to the initial and boundary conditions, have been solved using a finite element procedure.

3. RESULTS

The solution, in general, has the following parameters:

- The Rayleigh number, Ra
- The Prandtl number, Pr
- The aspect ratio of the enclosure, A
- The dimensionless size of the heated wall section, S
- The dimensionless position of the heated wall section, C
- The dimensionless amplitude of the temperature fluctuation, P
- The dimensionless period of the temperature fluctuation, p

Results will only be presented here for the case of a Prandtl number of 0.7 and for a square enclosure i.e. for an aspect ratio of 1. Although results have been obtained for a number of different values of the dimensionless period of the hot wall temperature fluctuation, results will only be presented here for a dimensionless period of 0.02, these being typical of those obtained at other dimensionless period values. The remaining governing parameters are then the Rayleigh number, the size and position of the heated wall section and the amplitude of the hot wall temperature fluctuation.

The main result considered here is the instantaneous mean heat transfer rates at the hot and cold wall. These heat transfer rates have

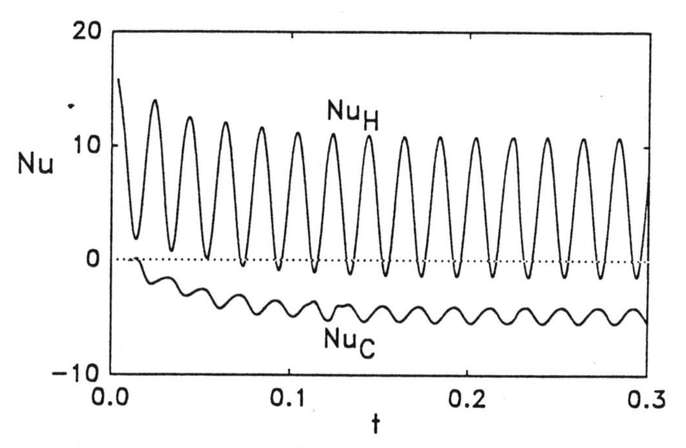

Figure 2. Variations of mean Nusselt numbers on hot and cold walls with dimensionless time for a fully heated wall for P = 0.5 and p = 0.02 and for a square enclosure with Ra = 10^5.

been expressed in terms of mean Nusselt numbers based on the full enclosure width, W' and the mean overall temperature difference, $(\overline{T'_H} - T'_C)$.

Figure 2 shows typical variations of the mean Nusselt numbers at the hot and cold walls for the case of a fully heated wall case. It will be seen that, since the fluid in the enclosure is initially at rest, there is an initial transient phase but by a dimensionless time of 0.2 the variations have become periodic. The initial transient will not be considered here, attention being given to the results in the periodic region. It should be noted that in this periodic region the time-averaged mean values of the Nusselt numbers on the hot and cold walls are equal in magnitude. It will be seen from the results given in Fig. 2 that the amplitude of the hot wall mean Nusselt number variation is much greater than that for the cold wall, the hot wall Nusselt number being negative for part of the cycle. This, together with the fact that the Nusselt number variations on the two wall are somewhat out of phase with each other and with the temperature fluctuation indicates that a pseudo-steady state flow does not exist in the enclosure in the situation considered. The effect of the heated section size on the heat transfer rate variation is illustrated by the results given in Fig. 3. This shows the variations of the hot and cold wall mean Nusselt numbers for two heated section sizes in the periodic region. It will be seen that the amplitude of the mean Nusselt number variations decreases as the size of the heated element is decreased. The effects of the heated section position on the heat transfer rate variation is illustrated by the results given in Fig. 4. This shows the variations of the hot and cold wall mean Nusselt numbers for three heated section positions in the periodic region. It will be seen that the average Nusselt numbers are greatest when the heated section is near the centre of the wall and that the amplitude of the mean Nusselt number variations again decreases as the heat transfer rate from the heated section decreases.

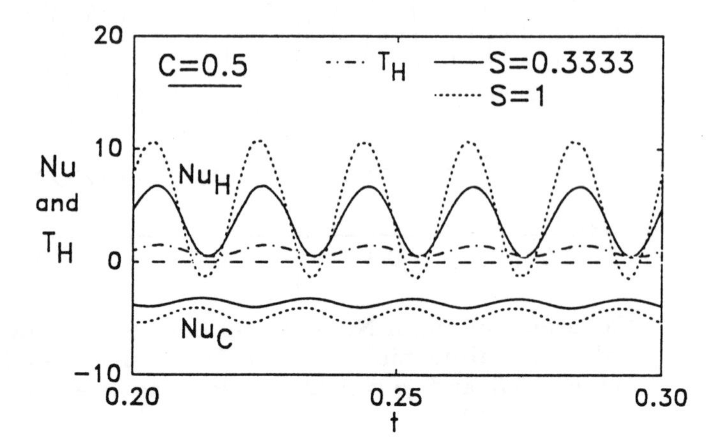

Figure 3. Variations of mean Nusselt numbers on hot and cold walls with dimensionless time for $P = 0.5$ and $p = 0.02$ and for a square enclosure with $Ra = 10^5$. Results are given for heated wall sections that are centred on the wall and have $S = 0.3333$ and 1. The variation of hot wall temperature is also shown.

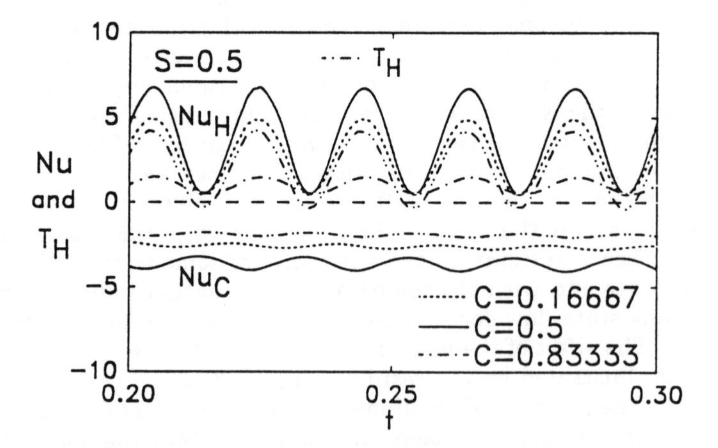

Figure 4. Variations of mean Nusselt numbers on hot and cold walls with dimensionless time for $P = 0.5$ and $p = 0.02$ and for a square enclosure with $Ra = 10^5$. Results are given for heated wall section with $S = 0.5$ that is placed at various positions up the wall. The variation of hot wall temperature is also shown.

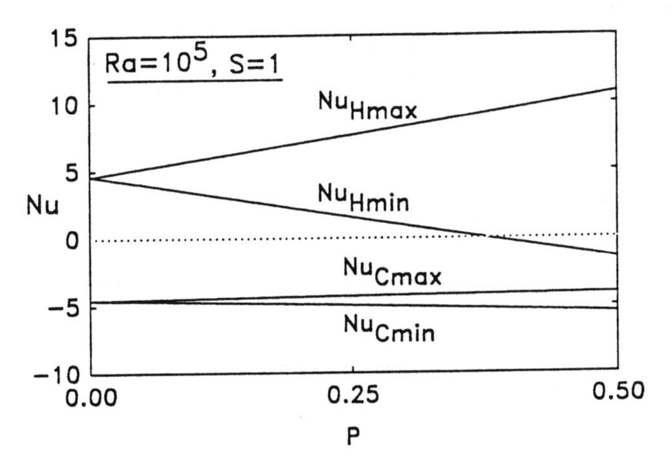

Figure 5. **Effect of dimensionless half-amplitude on the maximum and minimum values of the mean Nusselt numbers on the hot and cold walls for $p = 0.02$ and for a square enclosure with $Ra = 10^5$. The results are for a fully heated wall.**

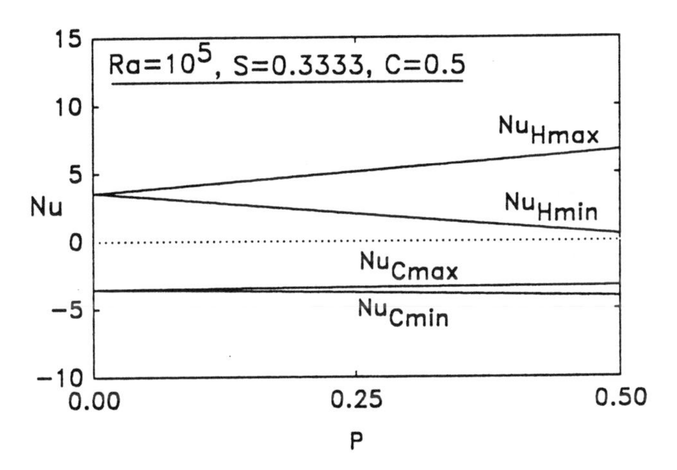

Figure 6. **Effect of dimensionless half-amplitude on the maximum and minimum values of the mean Nusselt numbers on the hot and cold walls for $p = 0.02$ and for a square enclosure with $Ra = 10^5$. The results are for a heated wall section with $S = 0.5$ that is centred on the wall.**

Instead of considering the actual variations of the mean Nusselt numbers in the periodic region, attention will now be restricted to the maximum and minimum values of these Nusselt numbers on the two walls. Figures 5 and 6 show the effect of the half-amplitude of the hot

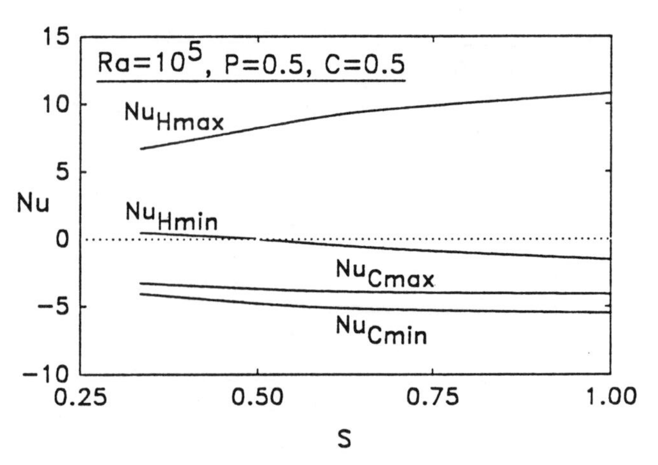

Figure 7. Effect of dimensionless element size on the maximum and minimum values of the mean Nusselt numbers on the hot and cold walls for P = 0.5 and p = 0.02 and for a square enclosure with Ra = 10^5. The results are for a heated wall section that is centred on the wall.

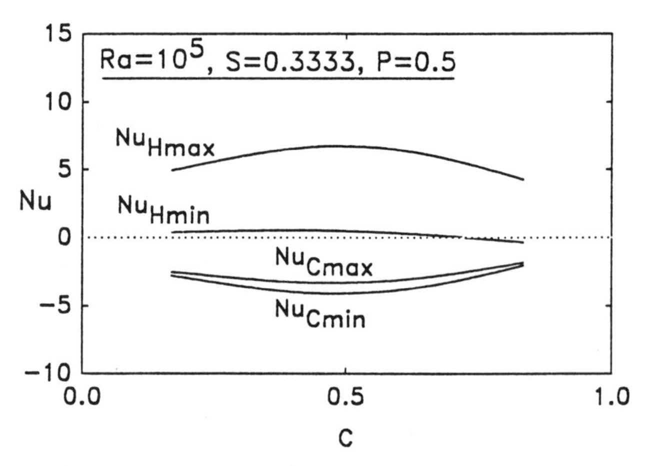

Figure 8. Effect of dimensionless element position on the maximum and minimum values of the mean Nusselt numbers on the hot and cold walls for P = 0.5 and p = 0.02 and for a square enclosure with Ra = 10^5. The results are for a heated wall section with S = 0.3333 .

wall section temperature fluctuation on these Nusselt number values for various hot wall section sizes. It will be seen, as mentioned, before that as the heated section size decreases, the average mean Nusselt number decreases and the difference between the maximum and minimum values of the Nusselt number on a wall decreases. This is further illustrated

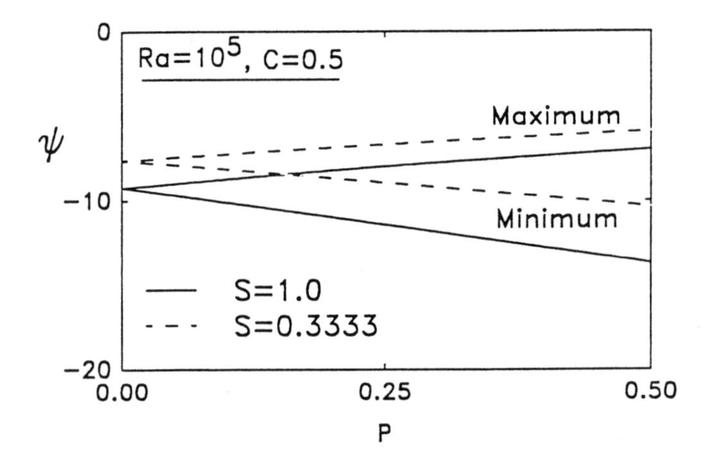

Figure 9. **Effect of dimensionless element size on the maximum and minimum values of the dimensionless stream function for $P = 0.5$ and $p = 0.02$ and for a square enclosure with $Ra = 10^5$. The results are for a heated wall section that is centred on the wall.**

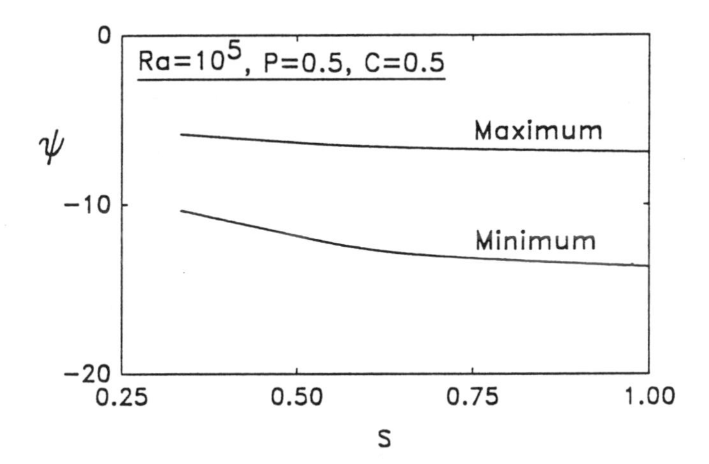

Figure 10. **Effect of dimensionless half-amplitude on the maximum and minimum values of the dimensionless stream function for $p = 0.02$ and for a square enclosure with $Ra = 10^5$. The results are for heated wall sections with $S = 1$ and $S = 0.3333$ that are centred on the wall.**

by the results given in Fig. 7. The effect of the heated section position on the maximum and minimum values of the mean Nusselt numbers is illustrated by the results given in Fig. 8.

Lastly, attention will be given to the minimum value of the dimensionless stream function in the enclosure, this being an indication of the intensity of the fluid motion in the enclosure. Figures 9 and 10 illustrate the effects of the amplitude of the dimensionless temperature fluctuation and the heated section size the maximum and minimum values of this dimensionless stream function.

4. CONCLUSIONS

The difference between the maximum and minimum values of the mean Nusselt numbers on the hot and cold walls in the periodic region decreases as the size of the heated section decreases and as this section is moved away from the centre of the wall.

5. NOMENCLATURE

A = aspect ratio, H' / W'
C = C' / W'
C' = vertical position of centre of heated wall section
g = gravitational acceleration
H' = height of enclosure
k = thermal conductivity
Nu = Mean Nusselt number based on W'
Nu_C = Mean Nusselt number for cold wall
Nu_H = Mean Nusselt number for hot wall
Nu_{Cmax} = Maximum value of mean Nusselt number for cold wall
Nu_{Cmin} = Minimum value of mean Nusselt number for cold wall
Nu_{Hmax} = Maximum value of mean Nusselt number for hot wall
Nu_{Hmin} = Minimum value of mean Nusselt number for hot wall
n = n' / W'
n' = coordinate measured normal to surface
P = dimensionless half amplitude
P' = half amplitude of hot wall temperature fluctuation
Pr = Prandtl number
p = dimensionless period of hot wall temperature fluctuation
Ra = Rayleigh number based on W'
S = S' / W'
S' = size of heated element
T = dimensionless temperature
T' = temperature
T'_H = temperature of hot walls
$\underline{T_H}$ = dimensionless temperature of hot walls
$\overline{T'_H}$ = mean temperature of hot wall
T'_C = temperature of cold surface
t = dimensionless time
t' = time
u = dimensionless velocity component in x' direction
u' = velocity component in x' direction

v = dimensionless velocity component in y' direction
v' = velocity component in y' direction
W' = width of enclosure
x = dimensionless x' coordinate
x' = horizontal coordinate position
y = dimensionless y' coordinate
y' = vertical coordinate position
α = thermal diffusivity
β = bulk coefficient
ν = kinematic viscosity
ψ = dimensionless stream function
ψ' = stream function
ω = dimensionless vorticity
ω' = vorticity

6. ACKNOWLEDGEMENTS

This work was supported by the Natural Sciences and Engineering Research Council of Canada.

7. REFERENCES

1. CHAN, A. M. C. and BANERJEE, S. - Three-Dimensional Numerical Analysis of Transient Natural Convection in Rectangular Enclosures, Journal of Heat Transfer , Vol. 101, pp. 114-119. 1979.
2. PATTERSON, J. and IMBERGER, J. - Unsteady Natural Convection in a Rectangular Cavity, Journal of Fluid Mechanics , Vol. 100, Part 1, pp. 65-86. 1980.
3. KUBLBECK K., MERKER G. P. and STRAUB, J. - Advanced Numerical Computation of Two-Dimensional Time-Dependent Free Convection in Cavities, International Journal of Heat and Mass Transfer , Vol. 23, pp. 203-217. 1980.
4. IVEY, G. N. - Experiments on Transient Natural Convection in a Cavity, Journal of Fluid Mechanics , Vol. 144, pp. 489-401. 1984.
5. NICOLETT, V. F., YANG, K. T. and LLOYD, J. R. - Transient Cooling By Natural Convection in a Two-Dimensional Square Enclosure, International Journal of Heat and Mass Transfer , Vol. 28, pp. 1721-1732. 1985.
6. HALL, J. D., BEJAN, A. and CHADDOCK, J. B. - Transient Natural Convection in a Rectangular Enclosure with One Heated Side Wall, International Journal of Heat and Fluid Flow , Vol. 9, pp. 396-404. 1988.
7. SCHLADOW S. G., PATTERSON J. C. and STREET, R. L. - Transient Flow in a Side-Heated Cavity At High Rayleigh Number: a Numerical Study, Journal of Fluid Mechanics , Vol. 200, pp. 121-148. 1989.

8. HYUN, J. M. and LEE, J. W. - Numerical Solutions for Transient Natural Convection in a Square Cavity with Different Sidewall Temperatures, International Journal of Heat and Fluid Flow , Vol. 10, pp. 146-151. 1989.
9. PATTERSON, J. C. and ARMFIELD, S. W. - Transient Features of Natural Convection in a Cavity Journal of Fluid Mechanics , Vol. 219, pp. 469-497. 1990.
10. JEEVARAJ, C. G. and PATTERSON, J. C. - Experimental Study of Transient Natural Convection of Glycerol-Water Mixtures in a Side Heated Cavity International Journal of Heat and Mass Transfer , Vol. 35, pp. 1573-1587. 1992.
11. OOSTHUIZEN, P. H. and PAUL. J. T. - Unsteady Free Convective Flow in an Enclosure Containing Water Near Its Density Maximum, Fundamentals of Natural Convection , V. S. Arpaci and Y. Bayazitoglu, eds., New York ASME HTD - Vol. 140, pp. 83-91, 1990.
12. KUHN, D. C. S. and OOSTHUIZEN, P. H. - Unsteady Natural Convection in a Partially Heated Rectangular Cavity, Journal of Heat Transfer , Vol. 109, No. 3, pp. 798-801. 1987.
13. KUHN, D. C. S. and OOSTHUIZEN, P. H. - Transient Three-Dimensional Flow in an Enclosure with a Hot Spot on a Vertical Wall, International Journal of Numerical Methods in Fluids , Vol. 8, pp. 369-385. 1988.
14. KUHN, D. C. S. and OOSTHUIZEN, P. H. - Transient Three-Dimensional Natural Convective Flow in a Rectangular Enclosure with Two Heated Elements on a Vertical Wall, Proceedings 5th International Conference on Numerical Methods in Thermal Problems , Vol. V, No. 1, pp. 524-535. 1987.
15. KAZMIERCZAK, M. and CHINODA, Z. - Buoyancy Driven Flow in an Enclosure with Time Periodic Boundary Conditions, International Journal of Heat and Mass Transfer , Vol. 35, pp. 1507-1518. 1992.

NATURAL CONVECTION IN A RECTANGULAR ENCLOSURE WITH A PARTIALLY HEATED WALL AND PARTLY FILLED WITH A POROUS MEDIUM

P. H. Oosthuizen and J. T. Paul
HEAT TRANSFER LABORATORY, Dept. Mechanical Engineering
Queen's University, Kingston, Ontario, Canada K7L 3N6

ABSTRACT

Two-dimensional natural convective flow in a rectangular enclosure with part of one wall heated to a uniform temperature and the opposite wall uniformly cooled to a lower temperature and with the remaining wall portions adiabatic has been considered. The enclosure is partly filled with a fluid and partly filled with a porous medium which is saturated with the same fluid, there being no partition between the fluid and the porous medium layers. It has been assumed that the flow is steady, laminar and two-dimensional and that fluid properties are constant except for the density change with temperature which gives rise to the buoyancy forces. The usual Darcy assumptions have been adopted in the porous layer, except that the viscous shear stress term, i.e. the Brinkman term, has been retained although the inertia term has been neglected. The governing equations have been written in terms of the stream function and vorticity and expressed in dimensionless form. These dimensionless equations, subject to the boundary conditions, have been solved using a finite element procedure. Results have mainly been obtained for a Prandtl number of 0.7 for a wide range of the governing parameters. The main results considered is the mean heat transfer rate across the enclosure. The effects of changes in the governing parameters on this mean heat transfer rate has been considered.

1. INTRODUCTION

The situation considered in the present study is shown in Figure 1. It involves two-dimensional natural convective flow in a rectangular enclosure with part of one wall *BC* (see Figure 2) heated to a uniform temperature, T'_H, and the opposite wall *EF* uniformly cooled to a temperature, T'_C, which is less than T'_H, and with the remaining walls *CDHE* and *BAGF* adiabatic. The enclosure is partly filled with a fluid and partly filled with a porous medium which is saturated with the same fluid, there being no partition between the fluid and the porous medium layers.

The situation under consideration is an approximate model of some situations that occur in building practice. Previous studies of enclosures that are partly filled with a porous medium have indicated that the heat transfer rate across the cavity is very significantly reduced as compared to that for a fluid filled cavity. The main aim of the present study was to determine how this reduction in the heat transfer rate is affected by the position and size of the heated wall section when the wall is only partly heated. The results have application in situations where the heat transfer across an enclosure has to be reduced but cost considerations require that as little insulation material as possible be used.

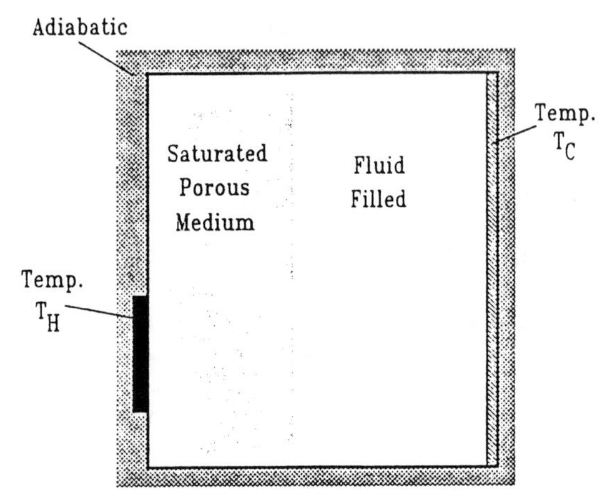

Figure 1. Situation considered in present study.

The flow and heat transfer in enclosures that are partly filled with a fluid and partly filled with a porous medium has been considered by a number of workers both for the case where there is no barrier between the layers and for the case where there is an impermeable barrier between the layers. Typical of these studies are those of Poulikakos and Bejan [1], Lauriat and Mesguich [2], Beckermann et al [3], Oosthuizen and Paul [4], Tong and Subramanin [5], Tong et al [6], Oosthuizen and Paul [7] and Song and Viskanta [8]. These studies were all essentially concerned with a situation in which the entire hot wall was at a uniform temperature. The case of heat transfer across an enclosure partly filled with a porous medium and partly filled with a fluid when the one wall is partly heated and with an impermeable barrier between the porous medium and the fluid has been considered by Oosthuizen and Paul [9]. The present study differs from these past studies in that an enclosure with a partly heated wall and with no barrier between the layers has been considered, this situation potentially having significant practical application.

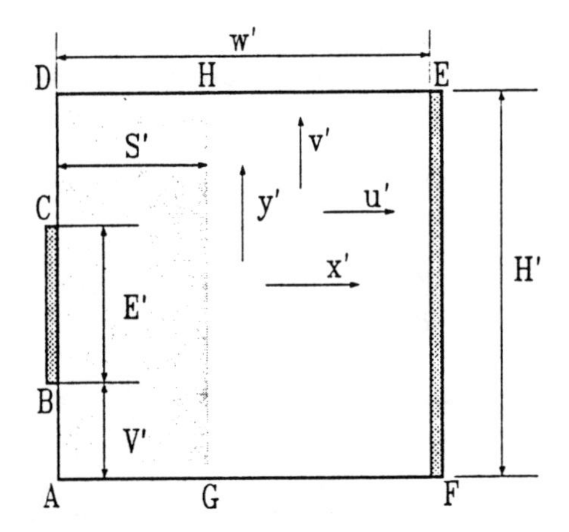

Figure 2. Coordinate system used.

2. GOVERNING EQUATIONS AND SOLUTION PROCEDURE

It has been assumed that the flow is steady, laminar and two-dimensional and that fluid properties are constant except for the density change with temperature which gives rise to the buoyancy forces, this being treated using the Boussinesq approach. It has also been assumed that, in the porous medium, the inertia term in the momentum equation is negligible. The usual Darcy assumptions have then been adopted in the porous layer, except that the viscous shear stress term, i.e. the Brinkman term, has been retained although the inertia term has been neglected.

The solution has been obtained in terms of the stream function and vorticity defined, as usual, by:

$$u' = \frac{\partial \psi'}{\partial y'} \quad , \qquad v' = -\frac{\partial \psi'}{\partial x'}$$

$$\omega' = \frac{\partial v'}{\partial x'} - \frac{\partial u'}{\partial y'} \tag{1}$$

The prime ($'$) denotes a dimensional quantity. In the porous layer, the velocity is, of course, the superficial or Darcian mean velocity.

The following dimensionless variables have then been defined:

$$\psi = \psi' / \alpha_f \quad , \qquad \omega = \omega' \, W'^2 / \alpha_f$$

$$T = (T - T'_C) / (T'_H - T'_C) \tag{2}$$

where $\alpha_f = k_f / \rho_f c_f$ and where the subscript f denotes fluid properties. The cold wall temperature, T'_c, has been taken as the reference temperature. The coordinate system used is shown in Figure 2.

In terms of these dimensionless variables, the governing equations for the porous medium are:

$$\frac{\partial^2 \psi}{\partial x^2} + \frac{\partial^2 \psi}{\partial y^2} = -\omega \qquad (3)$$

$$\left(\frac{\nu}{\nu_f}\right)\left(\frac{\partial^2 \omega}{\partial x^2} + \frac{\partial^2 \omega}{\partial y^2}\right) - \frac{\omega}{Da} = -Ra\frac{\partial T}{\partial x} \qquad (4)$$

$$\frac{\partial^2 T}{\partial x^2} + \frac{\partial^2 T}{\partial y^2} - \left(\frac{k_p}{k_f}\right)\left(\frac{\partial \psi}{\partial y}\frac{\partial T}{\partial x} - \frac{\partial \psi}{\partial x}\frac{\partial T}{\partial y}\right) = 0 \qquad (5)$$

Similarly, the dimensionless governing equations for the fluid layer are:

$$\frac{\partial^2 \psi}{\partial x^2} + \frac{\partial^2 \psi}{\partial y^2} = -\omega \qquad (6)$$

$$\left(\frac{\partial^2 \omega}{\partial x^2} + \frac{\partial^2 \omega}{\partial y^2}\right) - \frac{1}{Pr}\left(\frac{\partial \psi}{\partial y}\frac{\partial \omega}{\partial x} - \frac{\partial \psi}{\partial x}\frac{\partial \omega}{\partial y}\right)$$

$$= -Ra\frac{\partial T}{\partial x} \qquad (7)$$

$$\frac{\partial^2 T}{\partial x^2} + \frac{\partial^2 T}{\partial y^2} - \left(\frac{\partial \psi}{\partial y}\frac{\partial T}{\partial x} - \frac{\partial \psi}{\partial x}\frac{\partial T}{\partial y}\right) = 0 \qquad (8)$$

The boundary conditions on the solution are:

On all walls:

$$\psi = 0, \quad \frac{\partial \psi}{\partial n} = 0$$

At $x = 0$: on BC in Fig. 1

$$T = 1$$

At $x = 1$:

$$T = 0$$

On remaining wall segments:

$$\frac{\partial T}{\partial n} = 0$$

where n is the coordinate measured normal to the wall surface considered. Across the interface between the liquid and porous layers the velocity components, the pressure, the shear stress, the temperature and heat flux have been assumed to be continuous. These conditions lead to the requirement that the stream function be continuous across the interface and to the following conditions on the dimensionless stream function and vorticity across the interface:

$$\left.\frac{\partial \psi}{\partial x}\right|_f = \left.\frac{\partial \psi}{\partial x}\right|_p$$

$$\left.\frac{\partial \omega}{\partial x}\right|_f = \left.\frac{\partial \omega}{\partial x}\right|_p + \left(\frac{1}{Da}\right)\left.\frac{\partial \psi}{\partial x}\right|_p$$

$$\left.\frac{\partial T}{\partial n}\right|_f = \left.\frac{\partial T}{\partial n}\right|_p \left(\frac{k_p}{k_f}\right)$$

where the subscripts f and p refer to conditions on the fluid and porous medium sides sides of the interface respectively.

The above dimensionless equations, subject to the boundary conditions, have been solved using a finite difference procedure. The solutions for the porous medium and fluid layers were obtained simultaneously using the matching conditions across the interface between the layers, nodal points being selected to lie along this interface.

3. RESULTS

The solution has the following parameters (see Figure 2):

- The Rayleigh number, Ra
- The Darcy number, Da
- The Prandtl number, Pr
- The dimensionless thickness of the porous medium layer, S
- The dimensionless size of the heated wall section, E
- The dimensionless position of the heated wall section, V
- The aspect ratio of the enclosure, A
- The conductivity ratio, k_p / k_f
- The viscosity ratio, ν_p / ν_f

In the present study, the conductivity ratio has been taken as 1. This does not correspond exactly to any real physical situation but is closely satisfied in many cases. The viscosity ratio has also been taken as 1 which appears, on the basis of available experimental results, to be a reasonably good assumption. With the conductivity and viscosity ratios taken as 1, the governing parameters reduce to the Rayleigh number, the Darcy number, the Prandtl number and the geometrical arrangement. Most of the results given below have been been obtained for a Prandtl number of 0.7.

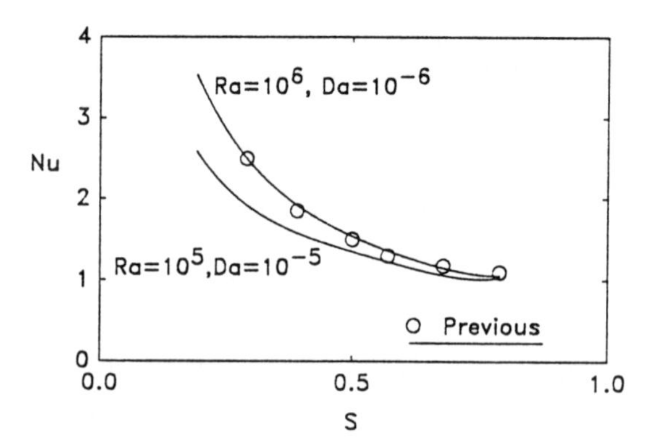

Figure 3. Effect of dimensionless porous layer thickness on mean Nusselt number for a fully heated wall for Pr = 1 and A = 1 , for Ra = 10^6 and Da = 10^{-6} and for Ra = 10^5 and Da = 10^{-5} . Previous results are from Beckermann et al [3].

The main result considered here is the mean heat transfer rate across the enclosure. This heat transfer rate has been expressed in terms of a mean Nusselt number based on the full enclosure width, W' and the overall temperature difference, $(T'_H - T'_C)$.

Although the main focus of the present study was on the case where the wall is partially heated, some results were first obtained for the fully heated wall case. These results have been used in assessing the effect of heated section size on the heat transfer rate. They have also been used to validate the numerical procedure used, previous results only being available for this fully heated wall case. Since these previous results are mainly for a Prandtl number of 1, the present results for the fully heated wall case have been obtained for this value of Prandtl number. Typical variations of Nu with porous layer thickness for this case for two combinations of Ra and Da values are shown in Figure 3. Some previous results from Beckermann et al [3] are also shown in this figure and good agreement between the present results and these previous results will be seen to exist. It will be also be seen from the results given in Figure 3 that the mean heat transfer rate decreases rapidly with increasing S until the porous layer occupies roughly 70% of the enclosure width. For larger values of S the interface position has relatively little effect on the heat transfer rate which tends to the pure conduction value, Nu being equal to one in this limiting case because the conductivity ratio has been assumed to be equal to 1. This indicates that even when the enclosure is only partly filled with a porous medium the convective motion can essentially be suppressed, the heat transfer then effectively being by pure conduction. Typical variations of Nu with Da for the fully heated wall case with Pr = 1 case are shown in Figure 4. These results are for an enclosure that is half filled with a porous medium. These results, which are also in good agreement with

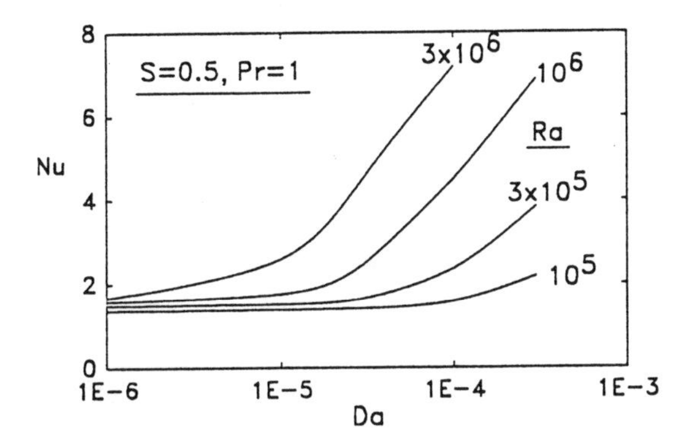

Figure 4. Effect of Darcy number on mean Nusselt number for a fully heated wall for $Pr = 1$, $A = 1$ and $S = 0.5$ for various values of Ra.

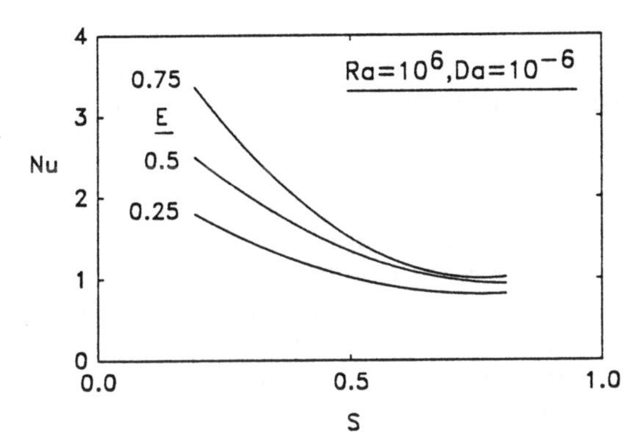

Figure 5. Effect of dimensionless porous layer thickness and element size on mean Nusselt number for a centrally mounted element for $Pr = 0.7$, $A = 1$, $Ra = 10^6$ and $Da = 10^{-6}$.

previous results, show that at small values of the Darcy number, the mean heat transfer rate is almost independent of Da , there being essentially no fluid motion in the porous layer under these circumstances with the result that the thermal resistance of this layer then dominates the heat transfer rate. As the Darcy number increases, a point is reached, however, at which Nu starts to increase quite rapidly with increasing Da as a result of the "penetration" of the fluid motion into the porous layer.

Attention will next be given to the partially heated wall case. As mentioned before, these results have been obtained for a Prandtl number of 0.7. Typical variations of Nu with porous layer thickness for fixed values of Ra and Da for various values of the heated section size for a heated section that is centrally placed on the wall are shown in Figure

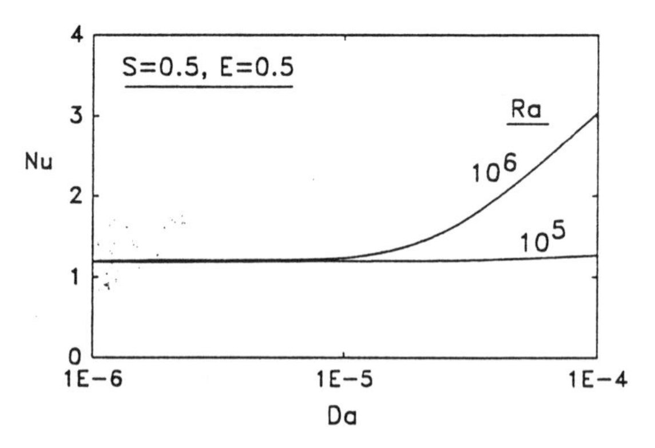

Figure 6. Effect of Darcy number on mean Nusselt number for a centrally mounted element for Pr = 0.7 , A = 1 , S = 0.5 and E = 0.5 for two values of Rayleigh number.

5. As with the fully heated wall case, it will be seen that the mean heat transfer rate decreases rapidly with increasing S until the porous layer occupies roughly 70% of the enclosure width. For larger values of S the interface position has relatively little effect on the heat transfer rate which tends to the pure conduction value, the value of Nu in this limiting case varying with the heated section size. These and other similar results for other conditions indicate that, while the actual heat transfer rates are dependent on heated section size, the thickness of the porous layer needed to suppress the effects of the convective motion is almost the same with a partially heated wall as it is with a fully heated wall under the same conditions. Typical variations of Nu with Da for two values of Ra for a centrally mounted heated section with E = 0.5 are shown in Figure 6. These results are for an enclosure that is half filled with a porous medium. Comparing these results with those given in Figure 4 for the fully heated wall case shows that the value of Darcy number at which Nu starts to increase as a result of the penetration of the fluid motion into the porous layer is almost the same in the two cases. Thus, again, although the size of the heated wall section effects the heat transfer rate, it does not have a strong effect on the conditions at which the rise in the heat transfer rate occurs. Typical variations of Nu with heated section position for two values of Darcy number and fixed values of the other parameters are shown in Figure 7. At the lower Darcy number considered there is essentially no fluid motion in the porous layer and the heated section position has little effect on the mean heat transfer rate. At the higher Darcy number considered, however, the mean heat transfer rate decreases significantly as the heated section is moved up the wall.

Lastly, the effect of enclosure aspect ratio on the mean heat transfer rate will be considered, typical results being given in Figures 8

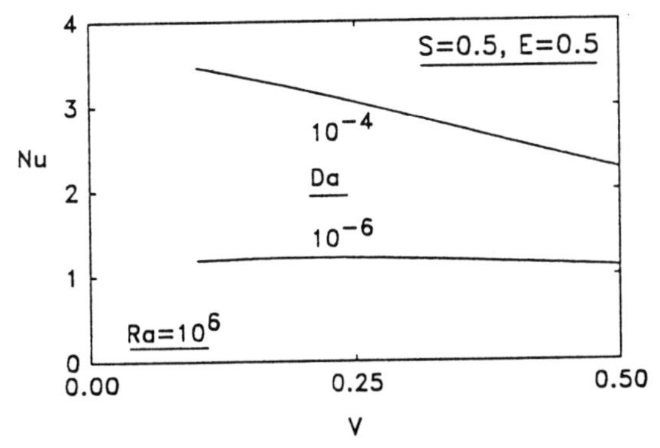

Figure 7. **Effect of dimensionless element position on mean Nusselt number for $Pr = 0.7$ for $Ra = 10^6$, $A = 1$, $S = 0.5$ and $E = 0.5$ for two values of Darcy number.**

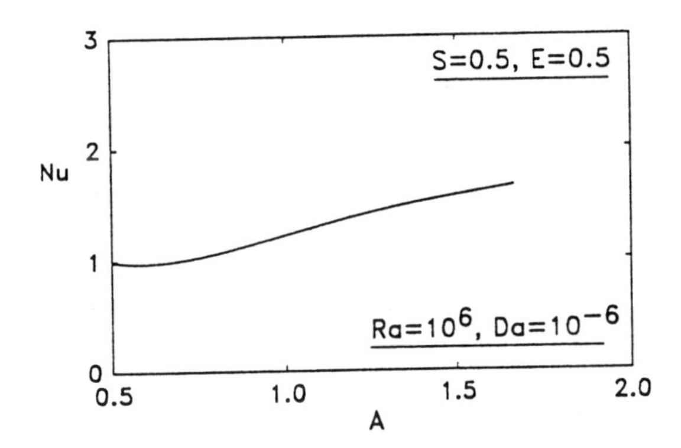

Figure 8. **Effect of enclosure aspect ratio on mean Nusselt number for $Pr = 0.7$ for $Ra = 10^6$, $Da = 10^{-6}$, $S = 0.5$ and for a centrally mounted element with $E = 0.5$.**

and 9. These results are all for a centrally mounted heated section of fixed size. Figure 8 illustrates the effect of aspect ratio on the heat transfer rate for fixed values of Ra and Da. These numbers and Nu, it will be recalled, are here based on the full width of the enclosure. It will be seen from Figure 8 that at low aspect ratios the mean Nusselt number is essentially equal to that with pure conduction but as the the aspect ratio increases, which corresponds to an increase in the height of the enclosure and a corresponding increase in the length of cold wall, the mean Nusselt number increases. Figure 9 shows the effect of the thickness of the porous layer on the mean Nusselt number for two aspect ratios and fixed values of Ra and Da. The value of the porous layer thickness beyond which the convection is essentially suppressed will be seen to be approximately the same for both aspect ratios.

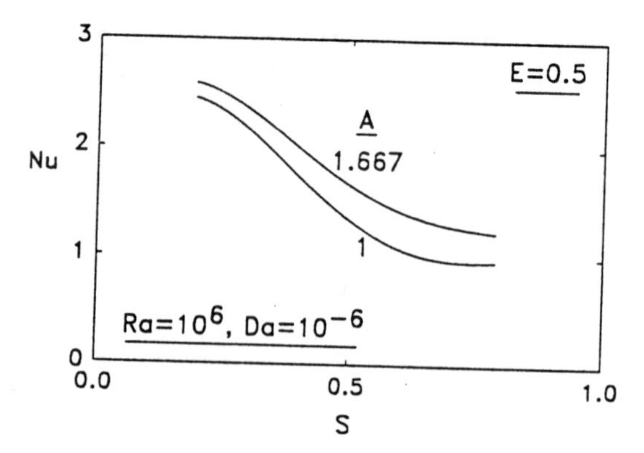

Figure 9. Effect of dimensionless porous layer thickness on mean Nusselt number for a centrally mounted element for $Pr = 0.7$, $Ra = 10^6$, $Da = 10^{-6}$ for two values of aspect ratio.

4. CONCLUSIONS

The position and size of the heated wall section and the aspect ratio of the enclosure have little effect on the thickness of the porous layer required to reduce the heat transfer rate to essentially the value it would have with pure conduction.

5. NOMENCLATURE

A = H' / W' c = specific heat
Da = Darcy number
E = E' / W'
E' = size of heated element
g = gravitational acceleration
H' = height of enclosure
K = permeability of porous medium
k = thermal conductivity
Nu = mean Nusselt number based on W'
n = n' / W'
n' = coordinate measured normal to surface
Pr = Prandtl number
p' = pressure
Ra = Rayleigh number based on W'
S = S' / W'
S' = thickness of porous layer
T = dimensionless temperature
T' = temperature
T'_H = temperature of hot wall
T'_C = temperature of cold wall
u' = velocity component in x' direction

$V \quad = V' / W'$
$V' \quad$ = vertical position of lower edge of heated element
$v' \quad$ = velocity component in y' direction
$W' \quad$ = width of enclosure
$x \quad$ = dimensionless x' coordinate
$x' \quad$ = horizontal coordinate position
$y \quad$ = dimensionless y' coordinate
$y' \quad$ = vertical coordinate position
$\beta \quad$ = coefficient of thermal expansion
$\Delta \quad$ = displacement of barrier
$\nu \quad$ = kinematic viscosity
$\rho \quad$ = density
$\psi \quad$ = dimensionless stream function
$\psi' \quad$ = stream function
$\omega \quad$ = dimensionless vorticity
$\omega' \quad$ = vorticity

Subscripts

$f \quad$ = fluid properties
$p \quad$ = porous media properties

6. ACKNOWLEDGEMENTS

This work was supported by the Natural Sciences and Engineering Research Council of Canada.

7. REFERENCES

1. POULIKAKOS, D. and BEJAN, A. - Natural Convection in Vertically and Horizontally Layered Porous Media Heated From the Side. International Journal of Heat and Mass Transfer , Vol. 26, pp. 1805-1813. 1983.
2. LAURIAT, F. and MESGUICH, F. - Natural Convection and Radiation in an Enclosure Partially Filled With a Porous Insulation, ASME Paper 84-WA/HT-101, 1984.
3. BECKERMANN, C., RAMADHYANI, S. and VISKANTA, R. - Natural Convection Flow and Heat Transfer Between a Fluid Layer and a Porous Layer Inside a Rectangular Enclosure, Natural Convection in Porous Media , V. Prasad and N. A. Hussain, eds., New York, ASME HTD - Vol. 56, pp. 1-12, 1986.
4. OOSTHUIZEN, P. H. and PAUL. J. T. - Free Convective Flow in a Cavity Filled with a Vertically Layered Porous Medium. Natural Convection in Porous Media , V. Prasad and N. A. Hussain. eds., New York ASME HTD - Vol. 56, pp. 75-84, 1986.

5. TONG, T. W. and SUBRAMANIAM, E. - Natural Convection in Rectangular Enclosures Partially Filled With a Porous Medium. International Journal of Heat and Fluid Flow Vol. 7, pp. 3-10. 1986.
6. TONG, T. W., FARUQUE, M. A., ORANGI, S. and SATHE, S. B. - Experimental Results for Natural Convection in Vertical Enclosures Partially Filled with a Porous Medium, Natural Convection in Porous Media V. Prasad and N. A. Hussain, eds., New York, ASME HTD - Vol. 56, pp. 85-93, 1986.
7. OOSTHUIZEN, P. H. and PAUL, J. T. - Natural Convection in an Inclined Cavity Half-Filled with a Porous Medium, ASME Paper 87-HT-15, 1987.
8. SONG, M. and VISKANTA, R. - Natural Convection Flow and Heat Transfer Between a Fluid Layer and an Anisotropic Porous Layer Within a Rectangular Enclosure, Heat Transfer in Enclosures , ASME HTD-Vol. 177, pp. 1-12, 1991.
9. OOSTHUIZEN, P. H. and PAUL, J. T. - Free Convection in an Inclined Cavity with a Partially Heated Wall and Partly Filled With a Porous Medium, Proceedings, 3rd U.K. National Heat Transfer Conference, Vol. 2, pp. 1107-1113, 1992.

NATURAL CONVECTION IN A CAVITY BY THE BEM

Zlatko Rek, Leopold Škerget

University of Maribor, Smetanova 17, 62000 Maribor, Slovenia

1 SUMMARY

The aim of this paper is to use Boundary Element Method (BEM) for coupled problem of fluid flow and heat transfer in the case of natural convection for high Rayleigh's (Ra) numbers. Vorticity-velocity formulation is used. Some improvements to classical BEM are being made to achieve convergence and stability for high Ra numbers, such as using implicit scheme, collecting all fluxes and splitting velocity into constant and variable part. All of the mentioned modification of the original scheme were tested on the well known case of natural convection in a cavity for Ra numbers in the range from 10^3 to 10^7. Results show a very good agreement with benchmark solutions proposed by De Vahl Davis, [2],[3].

2 GOVERNING EQUATIONS

In steady-state conditions, the partial differential equations set governing the transport phenomena in incompressible viscous fluid flow represents the basic conservation balances of mass, momentum and enthalpy written below

$$\nabla \cdot \vec{v} = 0 \tag{1}$$

$$\rho(\vec{v} \cdot \nabla)\vec{v} = -\nabla p + \mu \nabla^2 \vec{v} + \rho \vec{g} \tag{2}$$

$$(\vec{v} \cdot \nabla)T = a\nabla^2 T \tag{3}$$

for primitive variables: velocity $\vec{v}$, pressure p and temperature T. Material properties: density ρ, dynamic viscosity μ, and thermal diffusivity $a = \lambda/\rho c_p$, where λ is conductivity and c_p specific heat, are assumed to be constant. $\vec{g}$ is the gravity acceleration.

The buoyancy body force can be approximated with Boussinesq relation, where the temperature influence on density is used only with body forces, while it is neglected in all other terms. The following relation between the fluid density and temperature may be used $\rho = \rho_0[1 - \beta(T - T_0)]$, where ρ_0 is some reference density at temperature T_0, and β is the volume coefficient of thermal expansion. Using this relation the momentum equation (2) can now be written as

$$\rho_0(\vec{v} \cdot \nabla)\vec{v} = -\nabla P + \mu\nabla^2\vec{v} + \rho_0\beta(T - T_0)\vec{g} \tag{4}$$

P being the modified pressure $P = p - \rho_0\vec{g} \cdot \vec{r}$ with $\vec{r} = (x, y, z)$ being the position vector.

Vorticity-velocity ($\vec{w} - \vec{v}$) formulation is used to eliminate pressure term and divide the fluid flow into its kinematic and kinetic part, see [9],[11],[4]. The boundary-domain integral representation of the partial differential equation set (1),(4),(3) for plane flow, where $\vec{w} = (0, 0, w)$, stands

$$c(\xi)\vec{v}(\xi) + \int_\Gamma (\nabla u^* \cdot \vec{n})\vec{v}\, d\Gamma = \int_\Gamma (\nabla u^* \times \vec{n}) \times \vec{v}\, d\Gamma + \int_\Omega w\vec{k} \times \nabla u^*\, d\Omega \tag{5}$$

$$\begin{aligned} \nu c(\xi)\vec{v}(\xi) + \nu \int_\Gamma wq^*\, d\Gamma &= \nu \int_\Gamma \frac{\partial w}{\partial n} u^*\, d\Gamma - \int_\Gamma [wv_n + \beta g_t T] u^*\, d\Gamma + \\ &\quad \int_\Omega [w\vec{v} + \beta(g_y, -g_x)T]\nabla u^*\, d\Omega \end{aligned} \tag{6}$$

$$ac(\xi)T(\xi) + a\int_\Gamma Tq^*\, d\Gamma = a\int_\Gamma \frac{\partial T}{\partial n} u^*\, d\Gamma - \int_\Gamma Tv_n u^*\, d\Gamma + \int_\Omega T\vec{v} \cdot \nabla u^*\, d\Omega \tag{7}$$

where u^* is fundamental solution, q^* is it's derivative, $\vec{n}$ is unit normal to the boundary, and $c(\xi)$ is geometry coefficient.

2.1 Numerical solution

In order to obtain approximate numerical solutions of the velocity, vorticity and temperature fields, one has to discretize the boundary into N_E boundary elements and domain into N_C internal cells and interpolate the functions and derivatives with appropriate polynomials, [1],[9],[5].

When equation (6) is applied to all nodes, the following system for discrete values can be written when the unknown is vorticity

$$\begin{aligned} \nu[H]\{w\} &= -[G]\{q\} - [G][v_n + \beta g_t T]\{w\} + \\ &\quad \Big([D_x][v_x + \beta g_y T] + [D_y][v_y - \beta g_x T]\Big)\{w\} \end{aligned} \tag{8}$$

$$q = -\nu\frac{\partial w}{\partial n} \tag{9}$$

or if the unknown is vorticity flux

$$\nu[H]\{w\} = -[G]\{\tilde{q}\} + \Big([D_x][v_x + \beta g_y T] + [D_y][v_y - \beta g_x T]\Big)\{w\} \tag{10}$$

$$\tilde{q} = q + v_n w + \beta g_t T \tag{11}$$

After rearranging columns and rows due to the application of boundary conditions, the above system can be written as

$$\Big[[M][N][O]\Big] \begin{Bmatrix} \{w\}_{\Gamma_1} \\ \{\tilde{q}\}_{\Gamma_2} \\ \{w\}_{\Omega} \end{Bmatrix} = \{F\} \tag{12}$$

and submatrices represents

$$[M] = \nu[H] + [G][v_n + \beta g_t T] - \Big([D_x][v_x + \beta g_y T] + [D_y][v_y - \beta g_x T]\Big)_{\Gamma} \tag{13}$$

$$[N] = -[G] \tag{14}$$

$$[O] = \Big([D_x][v_x + \beta g_y T] + [D_y][v_y - \beta g_x T]\Big)_{\Omega} \tag{15}$$

Described scheme for fluid kinetic part is implicit because unknowns are both boundary and domain values, in contrary to classical BEM where unknowns are only boundary values, while domain values are computed explicitly.

Similar system can be written for heat energy transport, bearing in mind that a stands for ν and buoyancy term is canceled, [8],[6]. Fluid kinematic part stays unchanged what means that it is implicit for boundary vorticity or vorticity flux values, and explicit for domain velocities.

When very fine discretization is used or Ra number is high, the system matrix is very ill conditioned ($cond(A) \approx 10^7$), what means that all integrals have to be evaluated carefully. With numerical integration, it is very hard to obtain accurate results, so linear elements and triangular cell are used where analytical integration can be performed.

Assigning unknown flux as total flux (diffusion + convection + generation), the numerical scheme becomes more stable for high Ra numbers, [4]. In the case of natural convection, the vorticity generation is caused by buoyancy term on the boundary. Including this term into system matrix, nonlinearity of the righthand-side vector significantly reduces.

3 GEOMETRY, BOUNDARY CONDITIONS AND DISCRETIZATION

The pure buoyancy driven natural convection is considered. This problem has been proposed by De Vahl Davis et al., [2],[3] as a standard example for comparing different numerical techniques. There is a square

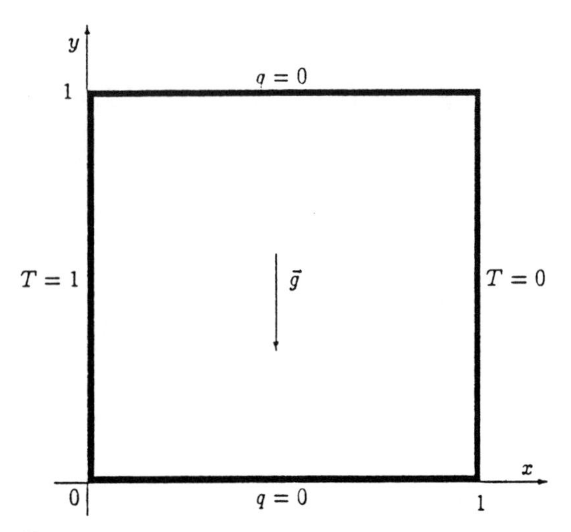

Figure 1: Geometry and boundary conditions.

cavity filled with viscous incompressible fluid. The motion is caused by buoyancy force due to the heated left wall subject to a unit temperature, while the right wall is kept at zero temperature. The top and bottom walls are adiabatic. Geometry of the problem and boundary conditions are presented on Fig. 1. Fig. 2 shows a discretized model. Nonuniform mesh with 41×41 nodes was used to better describe boundary layer at high Ra numbers. Aspect ratio between largest and smallest element was 10. There was 160 boundary elements and 3200 cells, with totally 1681 nodes. The interpolation was linear, due to the analytical evaluation of integrals.

4 RESULTS

Tabels Tab. 1 – Tab. 5 summarize comparison of BEM results and benchmark values for $Ra = 10^3$, 10^4, 10^5, and 10^6. The bencmark solution for $Ra = 10^7$ is not given by De Vahl Davis et al. in [3], so the results of Stevens [10] are used for comparison. The following quantities are given in tabels: the stream function at the mid-point of the cavity $|\Psi|_{mid}$, the maximum absolute value of the stream function $|\Psi|_{max}$ and its location $@x, y$, the maximum horizontal velocity on the vertical mid-plane of the cavity (together with its location) U_{max}, the maximum vertical velocity on the horizontal mid-plane of the cavity (together with its location) V_{max}, the average Nusselt number Nu on the vertical boundary of the cavity at $x = 0$, the maximum value of the local Nusselt number Nu_{max} on the boundary at $x = 0$ (together with its location), and the mini-

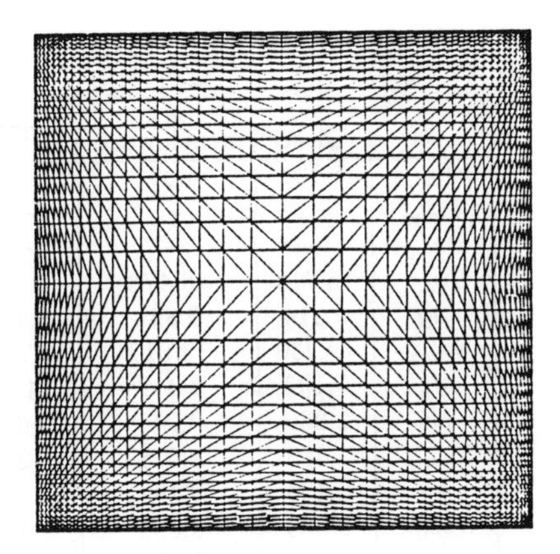

Figure 2: Discretized model: 160 elements, 3200 cells, 1681 nodes.

	BEM	benchmark	$\Delta\%$
$\lvert\Psi_{mid}\rvert$	1.153	1.174	1.8
$\lvert\Psi_{max}\rvert$	-	-	-
$@x, y$	-	-	-
U_{max}	3.595	3.649	1.5
$@y$	0.814	0.813	-0.1
V_{max}	3.624	3.697	2.0
$@x$	0.186	0.178	-4.5
Nu	1.113	1.118	0.4
Nu_{max}	1.496	1.505	0.6
$@y$	0.079	0.092	14.1
Nu_{min}	0.648	0.692	6.4
$@y$	1.000	1.000	0.0

Table 1: Results for $Ra = 10^3$.

mum value of the local Nusselt number Nu_{min} on the boundary at $x = 0$ (together with its location).

It can be seen from tabels that BEM results agree very good with bencmark values. There are some larger differences when coordinates are compared, but it can be explained that BEM results represent nodal values, while bencmark results are obtained with higher order interpolation. Figures Fig. 3 Fig. 6 shows velocity field $\vec{v}$, streamlines Ψ, isotherms T and vorticity lines w for all Ra numbers.

	BEM	benchmark	Δ%
$\lvert\Psi_{mid}\rvert$	5.121	5.071	-1.0
$\lvert\Psi_{max}\rvert$	–	–	–
$@x, y$	–	–	–
U_{max}	16.210	16.178	-0.2
$@y$	0.814	0.823	1.1
V_{max}	19.230	19.617	2.0
$@x$	0.136	0.119	-14.3
Nu	2.230	2.243	0.6
Nu_{max}	3.499	3.528	0.8
$@y$	0.136	0.143	4.9
Nu_{min}	0.550	0.586	6.1
$@y$	1.000	1.000	0.0

Table 2: Results for $Ra = 10^4$.

	BEM	benchmark	Δ%
$\lvert\Psi_{mid}\rvert$	9.191	9.111	-0.9
$\lvert\Psi_{max}\rvert$	9.860	9.612	-2.6
$@x, y$	0.289,0.563	0.285,0.601	-1.4,6.3
U_{max}	33.200	34.730	4.4
$@y$	0.864	0.855	-1.1
V_{max}	67.260	68.590	1.9
$@x$	0.065	0.066	1.5
Nu	4.505	4.519	0.3
Nu_{max}	7.643	7.717	1.0
$@y$	0.096	0.081	-18.5
Nu_{min}	0.702	0.729	3.7
$@y$	1.000	1.000	0.0

Table 3: Results for $Ra = 10^5$.

	BEM	benchmark	Δ%
$\lvert\Psi_{mid}\rvert$	17.510	16.320	-7.3
$\lvert\Psi_{max}\rvert$	17.840	16.750	-6.5
$@x, y$	0.159,0.500	0.151,0.547	-5.3,8.5
U_{max}	64.880	64.630	-0.4
$@y$	0.840	0.850	1.2
V_{max}	219.400	219.360	0.0
$@x$	0.040	0.038	-5.3
Nu	8.836	8.800	-0.4
Nu_{max}	17.037	17.925	5.0
$@y$	0.052	0.038	-36.8
Nu_{min}	1.010	0.989	-2.1
$@y$	1.000	1.000	0.0

Table 4: Results for $Ra = 10^6$.

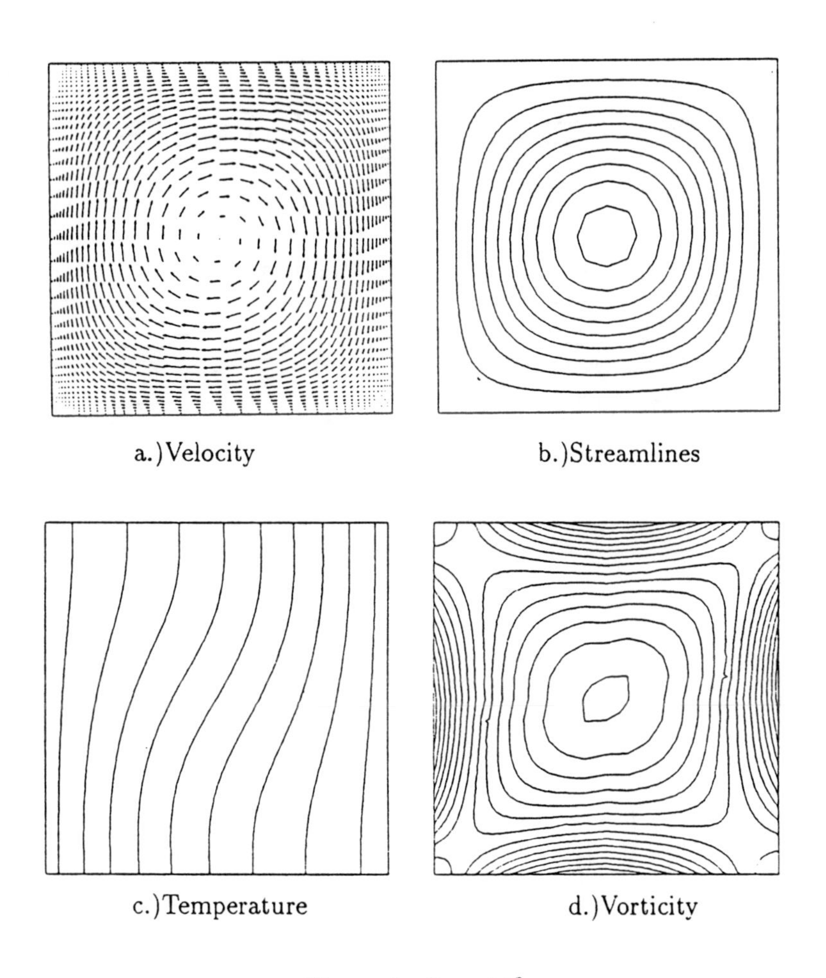

Figure 3: $Ra = 10^3$

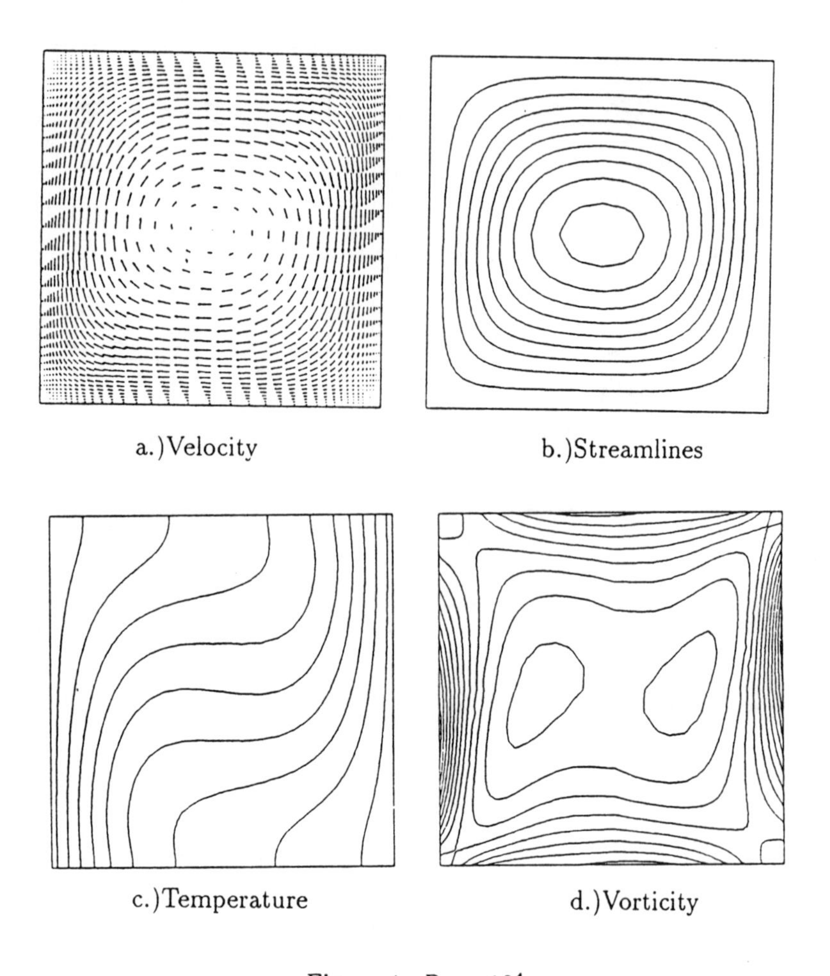

a.)Velocity

b.)Streamlines

c.)Temperature

d.)Vorticity

Figure 4: $Ra = 10^4$

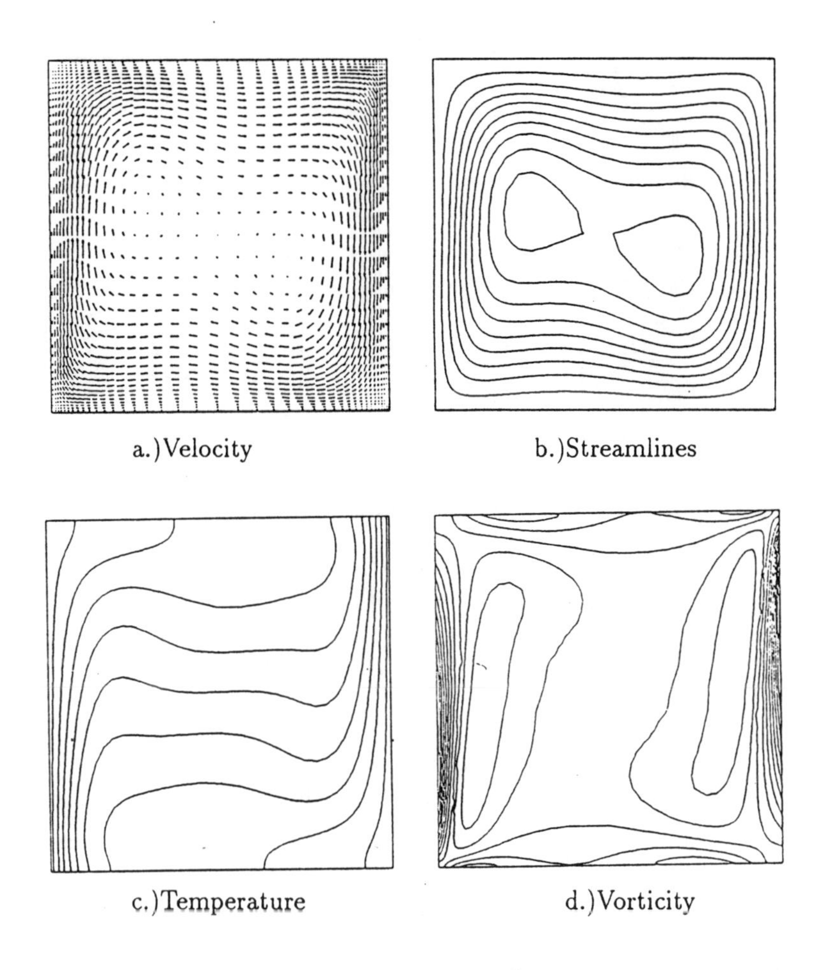

Figure 5: $Ra = 10^5$

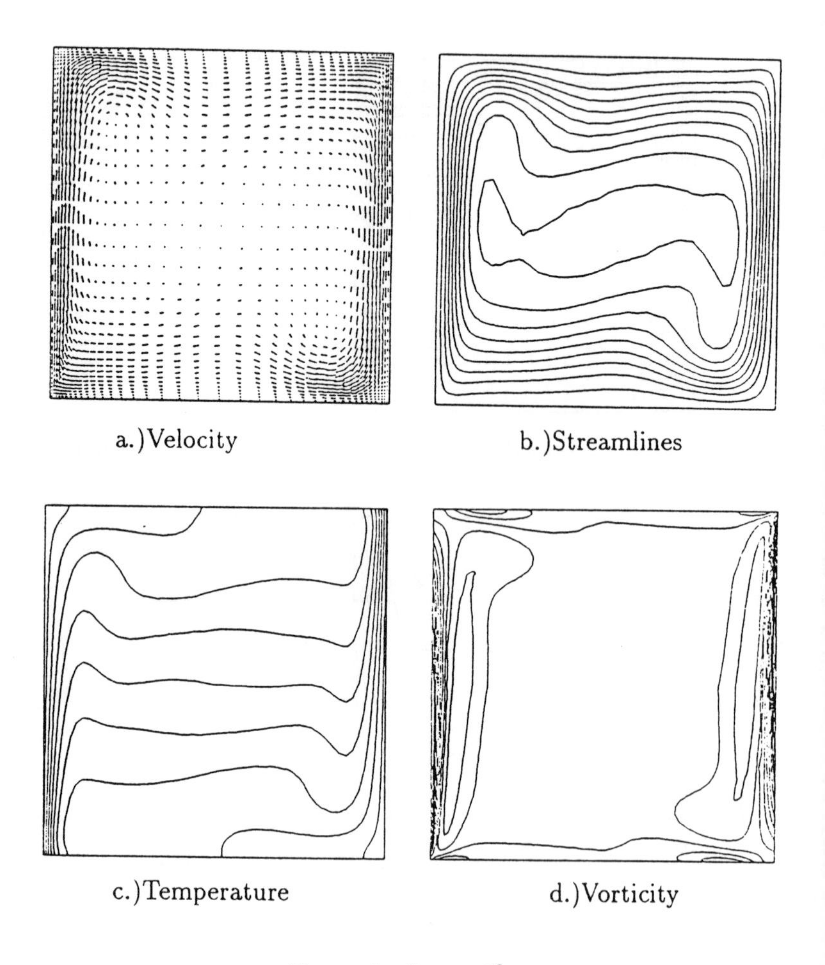

Figure 6: $Ra = 10^6$

	BEM	benchmark	Δ%
$\lvert\Psi_{mid}\rvert$	28.900	29.300	1.4
$\lvert\Psi_{max}\rvert$	31.040	30.300	-2.4
$@x, y$	0.096,0.500	0.080,0.570	20.0,12.2
U_{max}	154.900	–	–
$@y$	0.864	–	–
V_{max}	620.200	710.000	12.6
$@x$	0.021	0.022	4.5
Nu	15.859	16.500	3.8
Nu_{max}	34.051	–	–
$@y$	0.030	–	–
Nu_{min}	1.496	–	–
$@y$	1.000	–	–

Table 5: Results for $Ra = 10^7$.

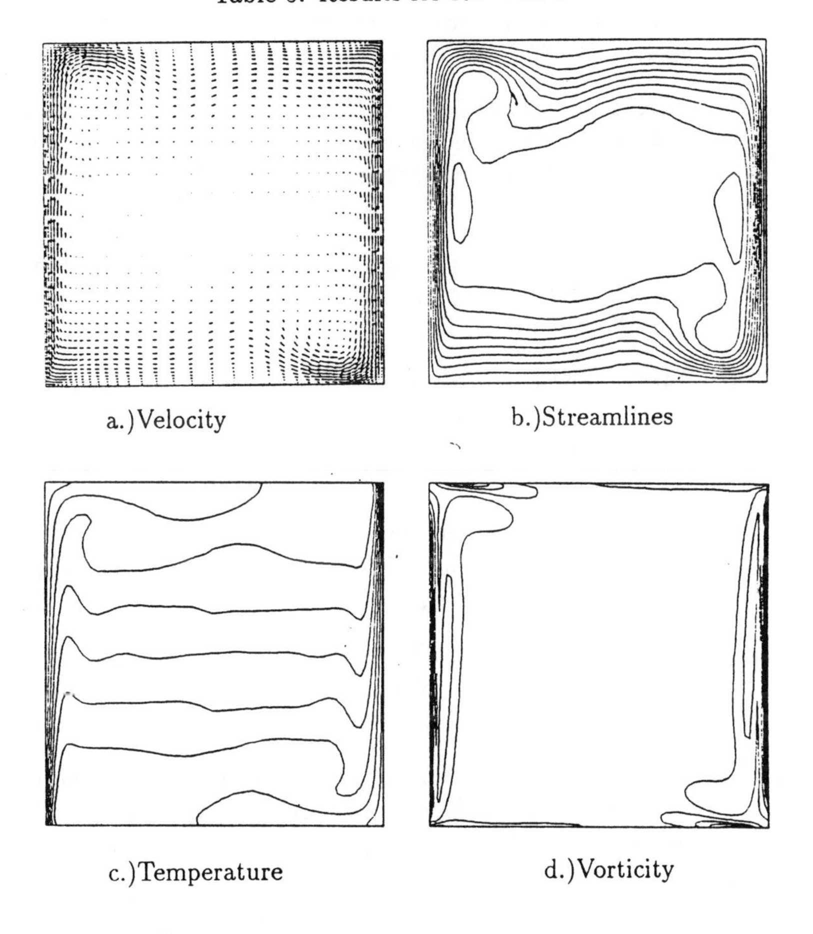

Figure 7: $Ra = 10^7$

References

[1] BREBBIA C.A.: **The Boundary Element Method for Engineers**, Pentech Press, London, Halstead Press, New York, 1978.

[2] DAVIS G.D.V., JONES I.P.: Natural Convection in a Square Cavity: A Comparison Exercise, **Int. Jou. for Num. Meth. in Fluids**, Vol.3, pp.227-248, 1983.

[3] DAVIS G.D.V.: Natural Convection of Air in a Square Cavity: A Bench Mark Numerical Solution, **Int. Jou. for Num. Meth. in Fluids**, Vol.3, pp.249-264, 1983.

[4] ONISHI K., KUROKI T., TANAKA M.: Boundary Element Method for Laminar Viscous Flow and Convective Diffusion Problems, **Topics in Boundary Element Research**, Vol.2, Ch.8, pp.209-229, Springer Verlag, Berlin, 1985.

[5] REK Z., SKERGET P., ALUJEVIC A.: Boundary-Domain Integral Method for Mixed Convection, **International Conference on Boundary Element Methods**, Vol.2, pp.219-230, 1990.

[6] REK Z., SKERGET P., ALUJEVIC A.: Vorticity-Velocity Formulation for Turbulent Flow by BEM, **Boundary Elements in Fluid Dynamics**, pp.123-130, Computational Mechanics Publications, Southampton, 1992.

[7] REK Z., SKERGET P.: Hyper Singular Boundary Element Method in Fluid Dynamics, **Zeitschrift für Angewandte Mathematik und Mechanik**, Vol.?, pp.?-?, 1993.

[8] SKERGET P., ALUJEVIC A., BREBBIA C.A., Kuhn G.: Natural and Forced Convection Simulation Using the Velocity-Vorticity Approach, **Topics in Boundary Element Research**, Vol.5, Ch.4, pp. 49-86, Springer-Verlag, 1989.

[9] SKERGET P., KUHN G., ALUJEVIC A., BREBBIA C.A.: Time dependent transport problems by BEM, **Advances in Water resources**, Vol.12, No.1, pp.9-20, 1989.

[10] STEVENS W.N.R.: Finite Element, Stream Function-Vorticity Solution of Steady Laminar Natural Convection, **Int. Jou. for Num. Meth. in Fluids**, Vol.2, pp.349-366, 1982.

[11] WU J.C.: Problems of General Viscous Flow, **Developments in BEM**, Vol.2, Ch.2, Elsevier Appl.Sci.Publ., London, 1982.

A Numerical Study of non-Darcian Effects on Natural Convection Flows Over a Vertical Cylinder in a Saturated Porous Medium

Cha'o-Kuang Chen, Professor,
Department of Mechanical Engineering,
National Cheng Kung University, Tainan, Taiwan, R. O. C.

Chien-Hsin Chen, Associate Professor,
Department of Mechanical Design Engineering,
National Yunlin Polytechnic Institute, Yunlin, Taiwan, R.O.C.

SUMMARY

This work discusses the influence of non-Darcian flow phenomena on the natural convection flows over a vertical cylinder embedded in a saturated porous medium. The non-Darcian flow effects often involved in fluid flow through porous media include high-flow-rate inertial losses, solid-boundary viscous friction, nonuniform porosity distribution and transverse thermal dispersion. These effects significantly alter the flow and heat transfer characteristics from those predicted by Darcy flow model. The heat transfer results based on different flow models are obtained to demonstrate the importance of the various non-Darcian flow effects.

INTRODUCTION

Heat transfer and fluid flow through porous media has been studied in a broad range of different fields and applications in recent years. A better understanding of porous media heat transfer is required in many thermal engineering disciplines. Some examples of applications are geothermal operations, chemical catalytic reactors, packed-sphere beds, heat exchange systems, and thermal insulation engineering. The problem of natural convection over a vertical cylinder in a saturated porous medium was first analyzed by Minkowycz and Cheng [1] based on the widely used Darcy's flow model. A more accurate numerical solution for this problem has been obtained by Kumari et al. [2]. However, the Darcy's law, whcih was used in most of the existing analytical studies for convective heat transport in porous media, neglects the effects of solid-boundary resistance and the inertial forces on fluid flow and heat transfer through porous media. In some practical systems, the permeability of the porous medium is not uniform. The variation in permeability is due to the nonuniform porosity condition which can be found, for example, in packed beds of spheres. Boundary, inertia and variable-porosity effects have been investigated extensively [3-5] in boundary layer flow and heat transfer along a flat plate in a porous medium. However, these non-Darcian flow effects on vertical-cylinder natural convection in porous media have not been considered.

In the present study we shall consider the boundary, inertia, non-uniform porosity and thermal dispersion effects on natural convection flow over a vertical heated cylinder embedded in a saturated porous medium. Boundary effects can be accounted for through the Brinkman's extension, where a viscous term is added to the the momentum equation. The inertia effects can be modeled through including a velocity squared term in the momentum equation, which has become known as Forchheimer's extension. The variations of porosity in the vicinity of the solid boundary can be approximated by exponential functions. Owing to the porosity variations, it is expected that the thermal conductivity will vary across the porous medium. When the inertia effects are prevalent, the transverse thermal dispersion effect becomes important [6]. This dispersion effect and variable stagnant thermal conductivity are taken into consideration in the energy equation. The governing equations have been solved numerically using an implicite finite-difference scheme. Typical velocity and termperature profiles as well as the local heat transfer rate along the streamwise direction are illustrated.

ANALYSIS

Consider a steady natural convection flow over a vartical cylinder of radius r_0 maintained at a uniform temperature T_w and immersed in a saturated porous medium. We assume that the convective fluid and the porous medium are in local thermal equilibrium. Based on the usual Boussinesq approximation, the boundary-layer equations in a cylindrical coordinate system are [4]

$$\frac{\partial}{\partial x}(ru) + \frac{\partial}{\partial r}(rv) = 0 \tag{1}$$

$$\frac{\mu}{K}u + \rho C u^2 = \rho g \beta (T - T_\infty) + \frac{\mu}{\phi}\frac{1}{r}\frac{\partial}{\partial r}(r\frac{\partial u}{\partial r}) \tag{2}$$

$$u\frac{\partial T}{\partial x} + v\frac{\partial T}{\partial r} = \frac{1}{r}\frac{\partial}{\partial r}(\alpha_e r\frac{\partial T}{\partial r}) \tag{3}$$

where u and v are the components of velocity in the x and r-directions; T and g are the temperature and gravitational constant respectively; ρ, μ, and β are the density, viscosity, and the thermal expansion coefficient of the fluid; K, C, and ϕ are the permeability, inertia coefficient, and porosity of the porous medium; and $\alpha_e = k_e/(\rho c)$ is the effective thermal diffusivity of the porous medium with k_e denoting the effective thermal conductivity of the saturated porous medium and ρc the product of the density and specific heat of the fluid. The appropriate boundary conditions for the problem are

$$u = v = 0, T = T_w \quad \text{at } r = r_o \tag{4}$$

$$u = 0, T = T_\infty \quad \text{as } r \longrightarrow \infty \tag{5}$$

The porosity of the porous medium is assumed to be dependent on the distance from the boundary. The commonly used exponential function can approximate very well the near-wall porosity variation such as in packed beds of spheres [7]

$$\phi = \phi_\infty + (\phi_w - \phi_\infty)\exp[-N(r - r_0)/d] \tag{6}$$

where $\phi_\infty = 0.4$ is the free-stream porosity; $\phi_w = 0.9$ is the porosity at wall; and the constant $N = 6$ is used to represent the porosity decay [7]. It is noted that the oscillations of the porosity, which are considered to be secondary, are neglected in the present analysis. Both the inertia coefficient C and the permeability K of the porous medium depend on the sphere diameter and the porosity and are determined from the experimental results of Ergun [8]

$$K = \frac{d^2\phi^3}{150(1-\phi)^2} \tag{7}$$

$$C = \frac{1.75(1-\phi)}{d\ \phi^3} \tag{8}$$

where d is the particle diameter.

It is known that the effective thermal conductivity k_e of a saturated porous medium is composed of a sum of the stagnant thermal conductivity k_d (due to molecular diffusion) and the thermal dispersion conductivity k_t (due to mechanical dissipation), i.e.,

$$k_e = k_d + k_t \tag{9}$$

The stagnant thermal conductivity of a packed-sphere bed can be given by the following semi-analytical expression [9]

$$\frac{k_d}{k_f} = (1-\sqrt{1-\phi}) + \frac{2\sqrt{1-\phi}}{1-\lambda B}\left[\frac{(1-\lambda)B}{(1-\lambda B)^2}\ln\left(\frac{1}{\lambda B}\right) - \frac{B+1}{2} - \frac{B-1}{1-\lambda B}\right] \tag{10}$$

where $B = 1.25[1-\phi)/\phi]^{10/9}$ and $\lambda = k_f/k_s$ is the ratio of the thermal conductivity of the fluid phase to that of the solid phase. Equation (10) shows that the stagnant thermal conductivity is a function of position for a nonuniform porosity medium. As proposed by Hsu and Cheng [1], it is assumed that the thermal dispersion conductivity is of the form

$$\frac{k_t}{k_f} = D_t\frac{ud}{\alpha_f}\frac{1-\phi}{\phi^2} \tag{11}$$

where D_t is an empirical contant.

The continuity equation is automatically satisfied by introducing the stream function ψ as

$$ru = \frac{\partial\psi}{\partial r} \quad \text{and} \quad rv = -\frac{\partial\psi}{\partial x} \tag{12}$$

The governing equations (1)-(3) are nondimensionalized by introducing the following dimensionless variables:

$$\eta = \frac{r^2-r_0^2}{2r_0}\frac{Ra^{0.5}}{x}, \xi = \frac{2x}{r_0}Ra^{-0.5}$$
$$f(\xi,\eta) = \psi(x,y)/(r_0\alpha_f Ra^{0.5}), \theta = \frac{T-T_\infty}{T_w-T_\infty} \tag{13}$$

where $Ra = \rho g\beta K_\infty x(T_w - T_\infty)/\mu\alpha_f$ is the modified local Rayleigh number. In terms of the new variables, the momentum and energy equations are

$$\frac{DaRa}{\phi}\frac{\partial}{\partial\eta}\left[(1+\xi\eta)\frac{\partial^2 f}{\partial\eta^2}\right]-Gr^*(\frac{C}{C_\infty})(\frac{\partial f}{\partial\eta})^2-(\frac{K_\infty}{K})\frac{\partial f}{\partial\eta}+\theta=0 \quad (14)$$

$$\sigma\frac{\partial}{\partial\eta}\left[(1+\xi\eta)\frac{\partial\theta}{\partial\eta}\right]+\left[\frac{1}{2}f+(1+\xi\eta)\frac{\partial\sigma}{\partial\eta}\right]\frac{\partial\theta}{\partial\eta}=\frac{1}{2}\xi(\frac{\partial f}{\partial\eta}\frac{\partial\theta}{\partial\xi}-\frac{\partial\theta}{\partial\eta}\frac{\partial f}{\partial\xi}) \quad (15)$$

where $Da = K_\infty/x^2$ is the local Darcy number, $Gr^* = g\beta K_\infty^2 C_\infty(T_w - T_\infty)/\nu^2$ is the modified Grashof number, and $\sigma = \alpha_e/\alpha_f$. The dimensionless boundary conditions are

$$f' = 0, f+\xi\frac{\partial f}{\partial\xi}=0, \theta=1 \quad \text{at } \eta=0 \quad (16)$$

$$f' = 0, \theta = 0 \quad \text{as } \eta\longrightarrow\infty \quad (17)$$

where the primes indicate differentiation with respect to η. The thermal dispersion conductivity in terms of the new variables is

$$\frac{k_t}{k_f}=D_t\frac{1-\phi}{\phi^2}Ra_d f' \quad (18)$$

where $Ra_d = g\beta K_\infty d(T_w - T_\infty)/\nu\alpha_f$.

A main objective of this work is to determine the heat transfer rate from the vertical cylinder. Consider first the local heat flux along the vertical cylinder which can be computed from

$$q=-k_e(\frac{\partial T}{\partial r})_{r=r_0} \quad (19)$$

The local heat transfer rate at the wall can be represented in terms of the local Nusselt number Nu, which is defined as $Nu = hx/k_e$, where h is the local heat transfer coefficient. Combining equation (19) with the definition of h, i.e. $q = h(T_w - T_\infty)$, gives the heat transfer parameter

$$Nu/Ra^{0.5} = -\theta'(\xi,\theta) \quad (20)$$

The numerical solutions for equations (14) and (15) with the boundary conditions (16) and (17) have been obtained based on an implicit finite-difference scheme, namely, the Keller Box method [11,12]. This numerical scheme has several very desirable features that make it appropriate for the solution of partial differential equations. The main features of this method involve second-order accuracy with arbitrary ξ and η spacings, allowing very rapid ξ variations, and allowing easy programming of the solution of large numbers of coupled equations. The details of the solution procedure by this method are descrided in the literature (see, for example, [12]). Hence, they are not repeated here in the interest of brevity.

RESULTS AND DISCUSSION

In this section, the main results of the non-Darcian effects on natural convecton about a vertical cylinder in porous media are reported and discussed. The numerical simulations were carried out for 3 mm and 5 mm diameter glass beads ($k_s = 1.05 W m^{-1} K^{-1}$). Since several different non-Darcian effects are taken into consideration, the following legends are used: BIV, which indicates Boundary, Inertia, and Variable porosity effects; nBnIU, which denotes no Boundary, no Inertia, and Uniform porosity effects, etc. It may be remarked that nBnIU, which is the Darcy's flow model, is the case reported previously in Refs. [1,2]. Also the case nBIU with $\xi = 0$ reduces to the flat plate case studied by Plumb and Huenefeld [3]. In order to verify the accuracy of the numerical calculations, we repeated these special cases and compared our results with those in Refs. [2,3]. They were shown to be in good agreement. The comparison is not presented here for brevity.

The focus in this study is first on the boundary, inertia and variable porosity effects with thermal dispersion neglected. This can be done by setting $Ra_d = 0$ in equation (18). Figures 1 and 2 depict the typical velocity profiles resulting from the inclusion of various non-Darcian effects. The boundary and inertia effects are found to decrease the velocity in the thermal boundary layer. Furthermore, the boundary effect is more significant near the leading edge. This is because that the parameter $DaRa$ in equation (14), which characterizes the importance of the boundary effect, is inversely proportional to the downstream distance x. When the effects of nonuniform porosity are taken into consideration, the channeling profiles [5] are created with the occurrence of a maximum velocity close to the solid boundary. This flow-channeling phenomenon is caused by the nonhomogeneous porosity variation with high porosity regions near the wall.

Shown in figures 3 and 4 are the dimensionless temprerature profiles corresponding to the velocity profiles in figures 1 and 2. The inertia effects are found to thicken the thermal boundary layers and to reduce the temperature gradients at the wall. It can also be seen that with porosity variation included the temperature gradient is increased owing to the strong flow-channeling effect.

Figures 5-7 aim to identify the contribution of each of the various effects on the heat transfer rate for the heated cylinder. Inclusion of the boundary friction term (BnIU) yields a decrease in the heat transfer rate when compared with the case where the Darcy's flow model is assumed (nBnIU). The heat transfer rate is further reduced when the non-Darcian inertial term is also included (BIU). The inertia effect appears to be important for porous systems of large modified Grashof numbers. The third non-Darcian effect involves variable-porosity media where the porosity, permeability and inertial coefficient all vary with the distance from the wall. The nonuniform porosity variation causes flow channeling, resulting in a great enhancement in heat transfer.

Figures 8 and 9 illustrate the influence of thermal dispersion on the heat transfer rate. This dispersion effect is expected to be significant as the flow inertia is prevalent. The empirical constant D_t in equation (11) should be determined from experiment. A value of $D_t = 0.02$ is used to examine qualitatively the thermal dispersion effect [10]. It can be summarized that the heat transfer rate is drastically increased by taking into consideration the thermal dispersion effect. This great heat transfer enhancement caused by the dispersive transport can be attributed to the better mixing of the convective fluid within the pores. The thermal dispersion effect is more pronounced at a larger value of Ra_d.

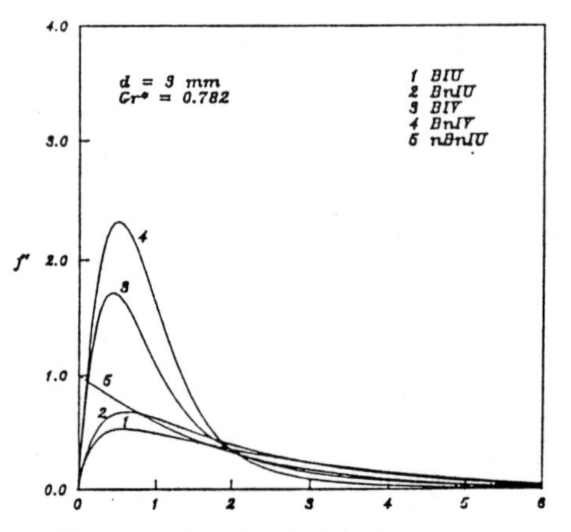

Fig. 1 The velocity distributions at $x = 16mm$.

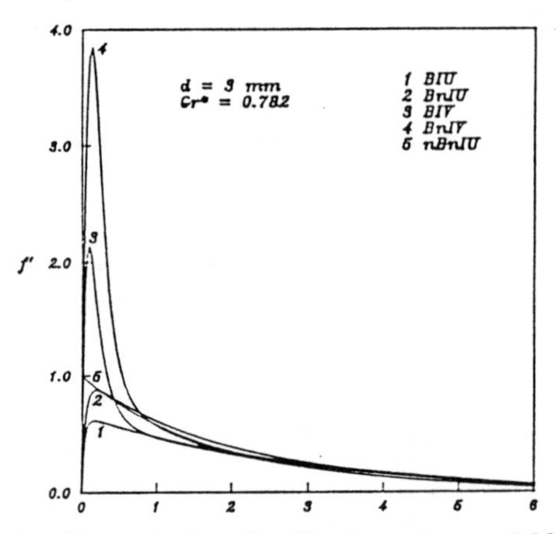

Fig. 2 The velocity distributions at $x = 360mm$.

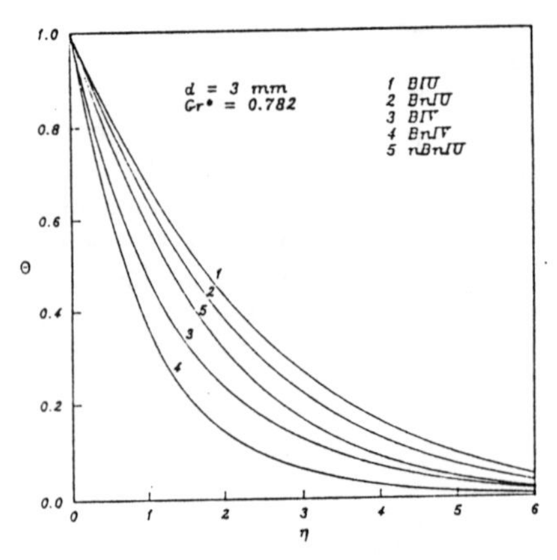

Fig. 3 The temperature distributions at $x = 16mm$.

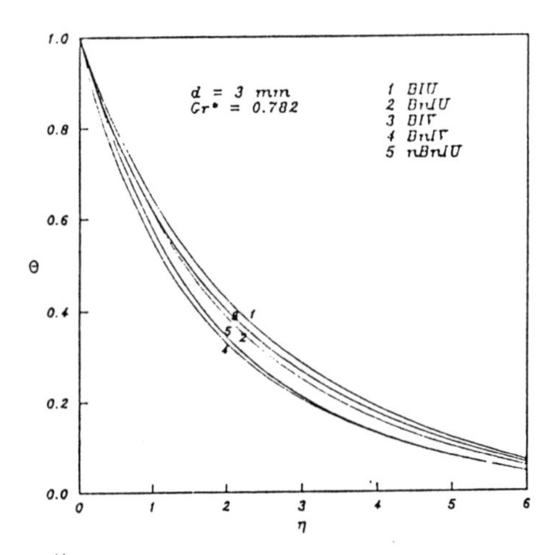

Fig. 4 The temperature distributions at $x = 360mm$.

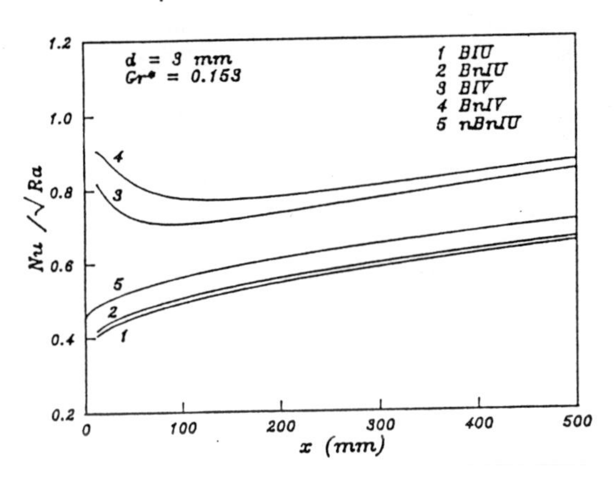

Fig. 5 Local heat transfer for different flow models at $d = 3mm$, $Gr^* = 0.153$.

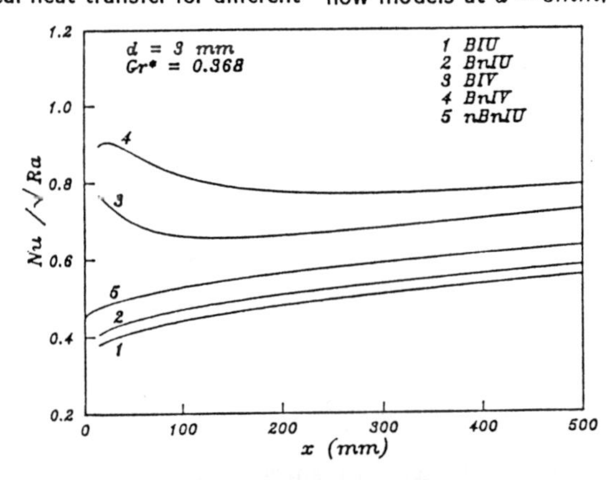

Fig. 6 Local heat transfer for different flow models at $d = 3mm$, $Gr^* = 0.368$.

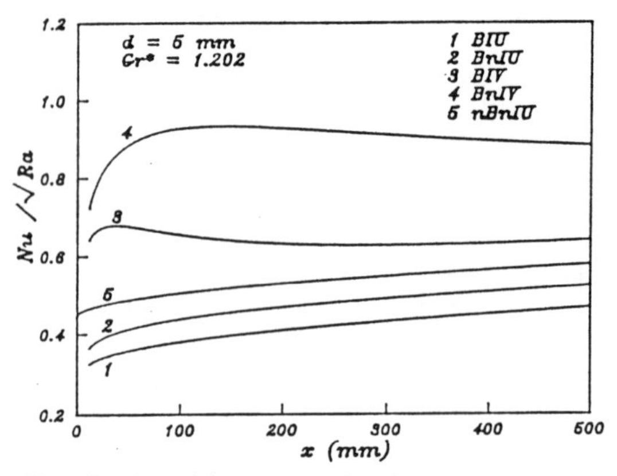

Fig. 7 Local heat transfer for different flow models at $d = 5mm$, $Gr^* = 1.202$.

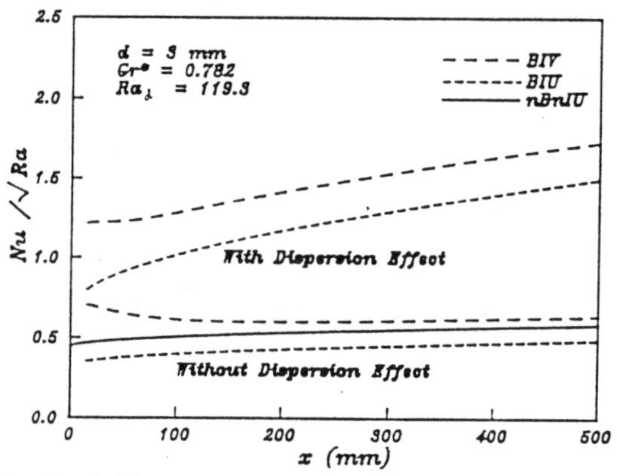

Fig. 8 Influence of thermal dispersion on the local heat transfer parameters.

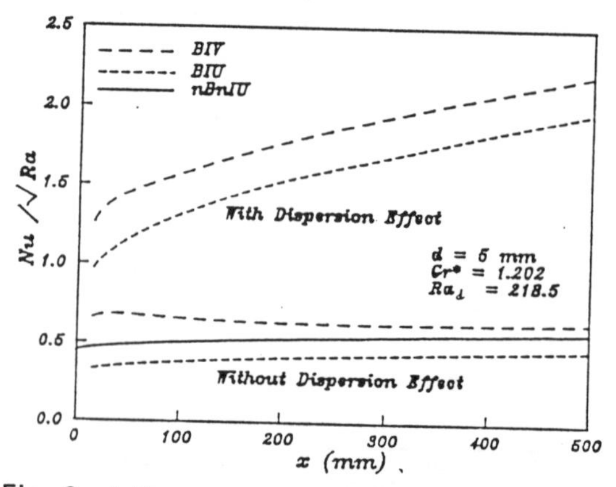

Fig. 9 Influence of thermal dispersion on the local heat transfer parameters.

CONCLUSION

The foregoing analysis demonstrates the importance of the non-Darcian flow effects in modeling natural convection flows along a vertical cylinder embedded in a porous medium. The flow inertia and boundary friction terms are included in the momentum equation with nonuniform porosity variation taken into account. The effects of variable stagnant thermal conductivity and transverse thermal dispersion are taken into consideration in the energy equation. By the inclusion of these non-Darcian effects, it is seen that the velocity and temperature profiles differ significantly from those predicted using the Darcy's law. Both the flow inertia and the no-slip boundary effects are found to reduce the velocity and heat transfer in the boundary layer. The effect of flow-channeling is found to augment dramatically the thermal communication between the porous matrix and the solid boundary. Thermal dispersion in the buoyancy-induced flow is shown to increase the heat transfer considerably.

NOMENCLATURE

B constant defined in equation (10)

C inertia coefficient

c specific heat of the fluid

Da Darcy number, K_∞/x^2

D_t empirical constant defined in equation (11)

d particle diameter

f dimensionless stream function

Gr^* modified Grashof number, $g\beta K_\infty^2 C_\infty (T_w - T_\infty/\nu^2$

g gravitational constant

h heat transfer coefficient

K permeability

k_d stagnant thermal conductivity

k_e effective thermal conductivity

k_f thermal conductivity of fluid

k_s thermal conductivity of particles

k_t thermal dispersion conductivity

N empirical constant in equation (6)

Nu Nusselt number

q local heat flux

Ra modified local Rayleigh number

Ra_d modified Rayleigh number based on particle diameter

r radial coordinate

r_0 radius of cylinder

T temperature

u x-component velocity

v r-component velocity

x streamwise coordinate

Greek symbols

α_e effective thermal diffusivity

α_f thermal diffusivity of fluid

β thermal expansion coefficient

η pseudo-similarity variable

θ dimensionless temperature

λ thermal conductivity ratio of the solid phaase to fluid phase

η viscosity of the fluid

ν kinematic viscosity of fluid

ξ dimensionless streamwise coordinate

ρ density of the fluid

σ α_e/α_f

ϕ porosity

ψ stream function

Subscripts

∞ quantities away from the wall

w quantities at wall

REFERENCES

1. W. J. MINKOWYCZ and P. CHENG, Free convection about a vertical cylinder embedded in a porous medium, Int. J. Heat Mass Transfer, Vol. 19, pp.805-813, 1976.

2. M. KUMARI, I. POP and G. NATH, Finite-difference and improved perturbation solutions for free convection on a vertical cylinder embedded in a saturated porous medium, Int, J. Heat Mass Transfer, Vol. 28, pp.2171-2174, 1985.

3. O. A. PLUMB and J. C. HUENEFELD, Non-Darcy natural convection from heated surfaces in saturated porous media, Int. J. Heat Mass Transfer, Vol. 24, pp.765-768, 1981.

4. K. VAFAI and C. L. TIEN, Boundary and inertia effects on flow and heat transfer in porous media, Int. J. Heat Mass Transfer, Vol. 24, pp.195-203, 1981.

5. K. VAFAI, Convective flow and heat transfer in variable-porosity media, J. Fluid Mech., Vol. 147, pp.233-259, 1984.

6. P. CHENG, Thermal dispersion effects in non-Darcian convective flows in a saturated porous medium, Lett. Heat Mass Transfer, Vol. 8, pp.267-270, 1981.

7. R. F. BENENATI and C. B. BROSILOW, Void fraction distribution in beds of spheres, AIChE., Vol. 8, pp.359-361, 1962.

8. S. ERGUN, Fluid flow through packed columns, Chem. Engng. Progress, Vol. 48, pp.89-94, 1952.

9. P. ZEHNER and E. U. SCHLUENDER, Waermeleitfahigkeit von schuettungen bei massigen temperaturen, Chemie-Ingr-Tech., Vol. 42, pp.933-941, 1970.

10. C. T. HSU and P. CHENG, Closure schemes of the macroscopic energy equation for convective heat transfer in porous media, Int. Comm. Heat Mass Transfer, Vol. 15, pp.689-703, 1988.

11. H. B. KELLER, A new difference scheme for parabolic problems, Numerical solutions of partial differential equations, Vol. 2, (J. Bramble, ed.), Academic, New York, 1970.

12. T. CEBECI and P. BRADSHAW, Physical and computational aspects of convective heat transfer, Springer-Verlag, New York, 1984.

NATURAL CONVECTION HEAT TRANSFER IN VERTICAL CHANNELS WITH ADIABATIC ENTRANCE SECTION

Davor Zvizdic

Faculty of Mechanical Engineering and Naval Architec.
University of Zagreb, Department of Heat engineering,
Salajeva 1, 41000 Zagreb, Croatia

SUMMARY

This paper describes methodology and numerical procedure for calculation of convection heat transfer and fluid flow in vertical, natural convection cooled, parallel-plate channels having, in upper section, heat exchange surfaces and adiabatic lower section. The influence of lower section length on heat transfer in upper section is numerically investigated. Governing equations are parabolic, expressing conservation of mass, momentum and energy for two-dimensional laminar flow with Boussinesq's approximation applied for taking the buoyancy force into account. Numerical procedure for calculation of coupled temperature, velocity and pressure fields involving control volume based marching technique is proposed, beginning at the inlet cross section and proceeding upward to the channel exit. The influence of lower section length on heat transfer in upper section is numerically investigated for a group of air-cooled channels. Heat flux and pressure distributions as well as overall heat transfer parameters are provided in graphical form.

1. INTRODUCTION

Numerical modelling of natural-convection heat transfer and fluid flow in vertical open-ended channels is a necessary step in design stages of apparata that incorporate them as well as building blocks of overall numerical models. Designers of souch equipment faced with opposing space/performance requirements need reliable prediction of their thermal performance. Studies of buoyancy-driven fluid flow in vertical parallel-plate channels have been reported extensively [1,2,3,4], but mostly pertainig to symmetric or asymmetric isoflux or isothermal walls. Solutions for complex temperature profiles along channel walls have been reported by present author in [5]. Special cases, seemingly reported here for the first time, are channels with entrance sections that do not participate in heat transfer (adiabatic). This configuration is encountered in electrical apparata such as dry-type and oil-filled transformers with entrance "collars", electronic equipment as well as in process and heat engineering systems.

2. PROBLEM STATEMENT

Consider a parallel plate vertical channel (Fig.1) of height H and width b consisting lenghthwise of two sections. The top section (length Lq) has two heat exchange surfaces: Lq(y=0) and Lq(y=b) with wall temperatures T(x,y=0) and T(x,y=b) higher then temperature of surrounding fluid (T_a).

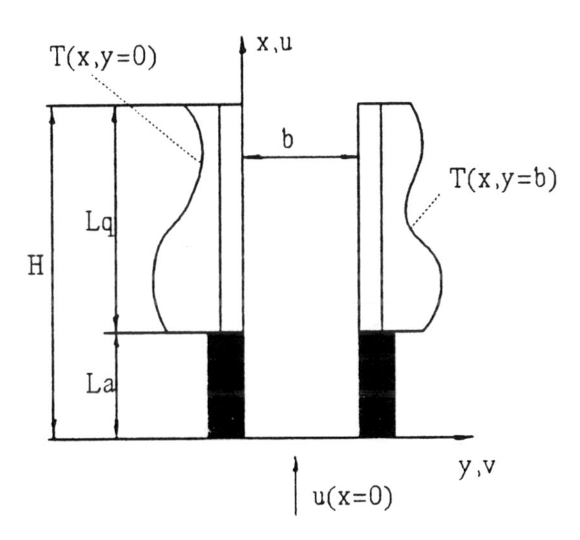

Fig. 1

The lower, entrance section (legth La) is adiabatic. Due to heat exchange in the upper section, vertical buoyancy-driven upflow in the channel is established with velocity bandary layers beginning to develop at x=0 and temperature boundary layers at x=La.

2.1 Governing equations

As a basis for two-dimensional calculation of heat transfer, simultaneous non-linear partial differential equation system in parabolic form is established, describing conservation of mass, momentum and energy with additional global mass conservation condition across channel cross section:

$$\frac{\partial u}{\partial x}+\frac{\partial v}{\partial y}=0 \qquad (1)$$

$$\rho\left(u\frac{\partial u}{\partial x}+v\frac{\partial u}{\partial y}\right)=-\frac{\partial P}{\partial x}-\rho g+\mu\left(\frac{\partial^2 u}{\partial y^2}\right) \qquad (2)$$

$$u\frac{\partial T}{\partial x}+v\frac{\partial T}{\partial y}=\frac{\lambda}{\rho c_p}\frac{\partial^2 T}{\partial y^2} \qquad (3)$$

where:

- x,y - axial and transversial coordinates
- u,v - axial and transversial velocities
- ρ - fluid density
- g - gravitational acceleration
- P - absolute pressure
- μ - dynamic viscosity
- T - temperature
- c_p - specific heat of fluid

Absolute pressure in (2) can be substituted with pressure difference p:

$$p=P-(P_0-\rho_\infty gx) \qquad (4)$$

where:

- p - pressure difference within and outside of channel at height x
- P_0 - absolute pressure at height x

pressure and buoyancy terms in (2) then become:

$$-\frac{\partial P}{\partial x}-\rho g=-\frac{\partial p}{\partial x}+g(\rho_{\infty}-\rho) \quad (5)$$

Density difference in (5) can be eliminated via equation of state:

$$\rho_{\infty}-\rho=\beta\rho(T-T_{\infty}) \quad (6)$$

where:

β - thermal expansion coefficient
T_{∞} - temperature of surrounding fluid

After substitution and division with ρ (2) becomes:

$$u\frac{\partial u}{\partial x}+v\frac{\partial u}{\partial y}=-\frac{1}{\rho}\frac{\partial p}{\partial x}+g\beta(T-T_{\infty})+\nu\frac{\partial^2 u}{\partial y^2} \quad (7)$$

Three equations (1),(3) i (7) have four unknowns (u,v,p and T) and additional equation stating global cross-sectional mass conservation is neccesary:

$$G=\int\rho u dy=const \quad (8)$$

where G denotes mass flow.

2.2 Boundary conditions

The above equation system is supplemented with velocity, temperature and pressure boundary conditions.

At entrance crosss section uniform velocity profile and surrounding fluid temperature are accepted:

$$u(x=0)=U_0,\ T(x=0)=T_{\infty} \quad (9)$$

Entrace velocity U_0 is unknown prior to calculation and must be assigned tentative value.

Velocities at wall boundaries are as follows:

$$u(y=0)=0,\ v(y=0)=0,\ u(y=b)=0,\ v(y=b)=0 \quad (10)$$

At exit cross-section velocity and temperature

boundary conditions are not necessary due to the parabolic nature of flow.

Pressure difference p is zero at entrance and exit cross-sections:

$$p(x=0)=p(x=H)=0 \tag{11}$$

Non-homogenous and assymmetric temperature profiles at channel walls in the upper, heat exchange section, can be prescribed in general form:

$$T(La<x<H, y=0)=f_1(x), T(La<x<H, y=b)=f_2(x) \tag{12}$$

At lower, adiabatic section:

$$q(0<x<La, y=0)=0, q(0<x<La, y=b)=0 \tag{13}$$

3. NUMERICAL SOLUTION

For the solution of the equations a Patankar type [6] implicit procedure is used. It is a control volume marching method beginning at the inlet cross section and proceeding upward to the channel exit.
The solution consists of two main levels of iteration:
At the first level the non-linear governing equations are iteratively solved for temperature, velocities and pressures.
The second level encompasses the channel as a whole and is used for the adjustment of the tentative inlet velocity which must be specified at the beginning of the calculation along with:

$$v(x=0)=0, T(x=0)=T_\infty \tag{14}$$

for inlet cross-section.

At each streamwise location equation (3) is solved for temperatures. In this way the influence of local wall temperatures (in the upper section of the channel) is immediately brought in the calculation domain.

In the second step equation (7) with provisional pressure imbalance p is solved for streamwise velocity components. Then the pressure imbalance p is iteratively adjusted with the aid of equation (8).

Cross-stream velocity components v are calculated from continuity equation (1) twice; moving from the left to the right wall and vice versa. Corresponding velocity components are then added and divided by two. Cross-stream velocity componenets obtained in this way are then used for further iterations. The reason for this tehnique is because as one moves from one wall to the other calculation errors of streamwise velocities tend to accumulate in cross-stream velocities. In this way a more eaven error distribution accross channel cross-section is achieved resulting in improved stability and speed of numerical procedure.

After satisfactory convergence movement is made to the next streamwise location and the process is repeated until exit cross-section is reached. The computed pressure at the channel exit will not be zero owing to tentative inlet velocty which will have to be adjusted until the boundary condition p(x=H)=0 is satisfied.

4. RESULTS AND DISCUSSION

As calculation examples a group of seven channels is chosen having common upper section geometry and (for simplicity) symmetric and homogeneous temperature distribution, but variable lower (adiabatic) section length.
Calculation parameters common to all channels are:

- Cooling fluid: Air (Pr=0.7)
- Upper section lenght: Lq = 0.4 m
- Channel width: b = 0.02 m
- Surrounding air temp.: T_3= 20 °C
- Wall temperatures (Lq): T(x,y=0)=T(x,y=b)=40°C

Parameters pertaining to individual channels are:

Ch.#	La,m	H,m	Grid	Unknowns
1	.00	.40	400X50	60400
2	.05	.45	400X50	60400
3	.10	.50	500X50	75500
4	.15	.55	500X50	75500
5	.20	.60	600X50	90600
6	.25	.65	600X50	90600
7	.30	.70	700X50	105700

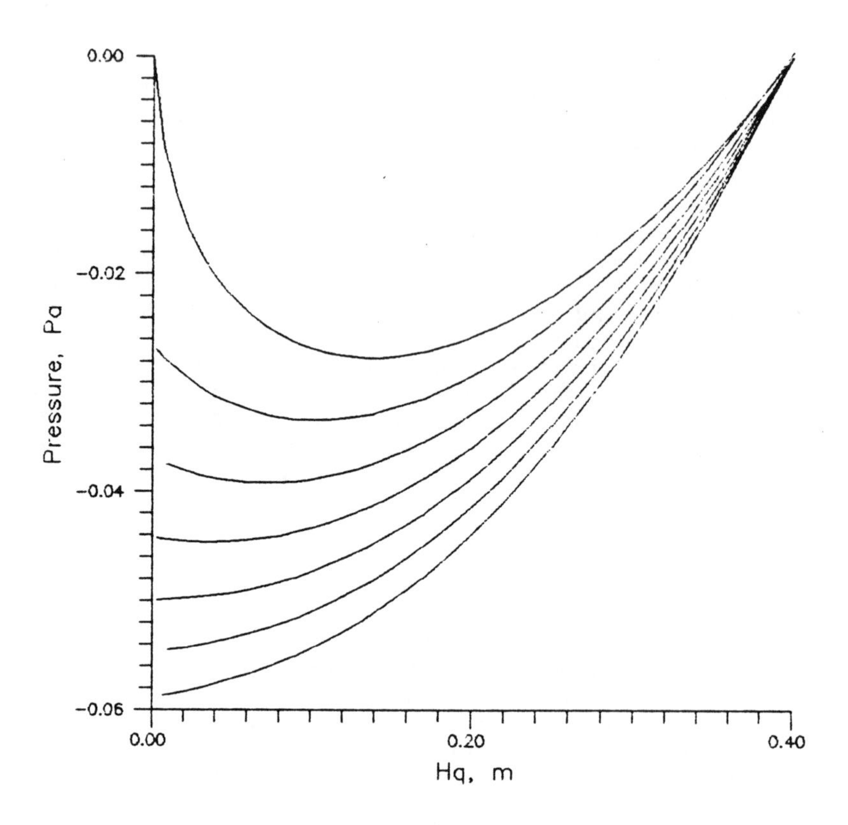

Fig.2: pressure difference

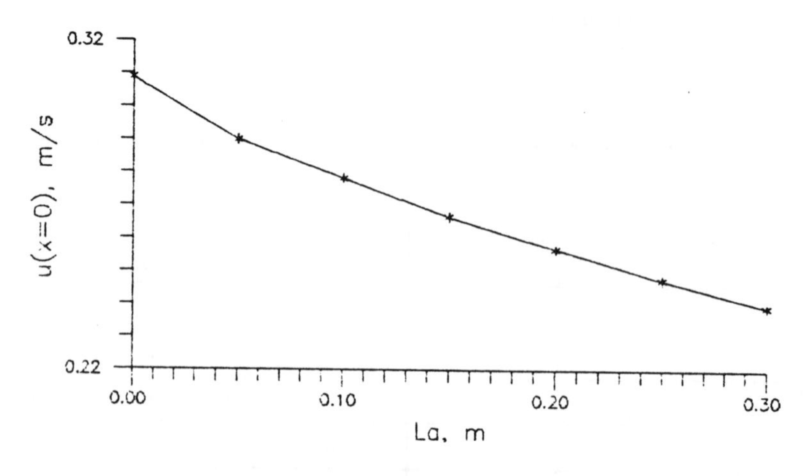

Fig.3: Entrance velocities

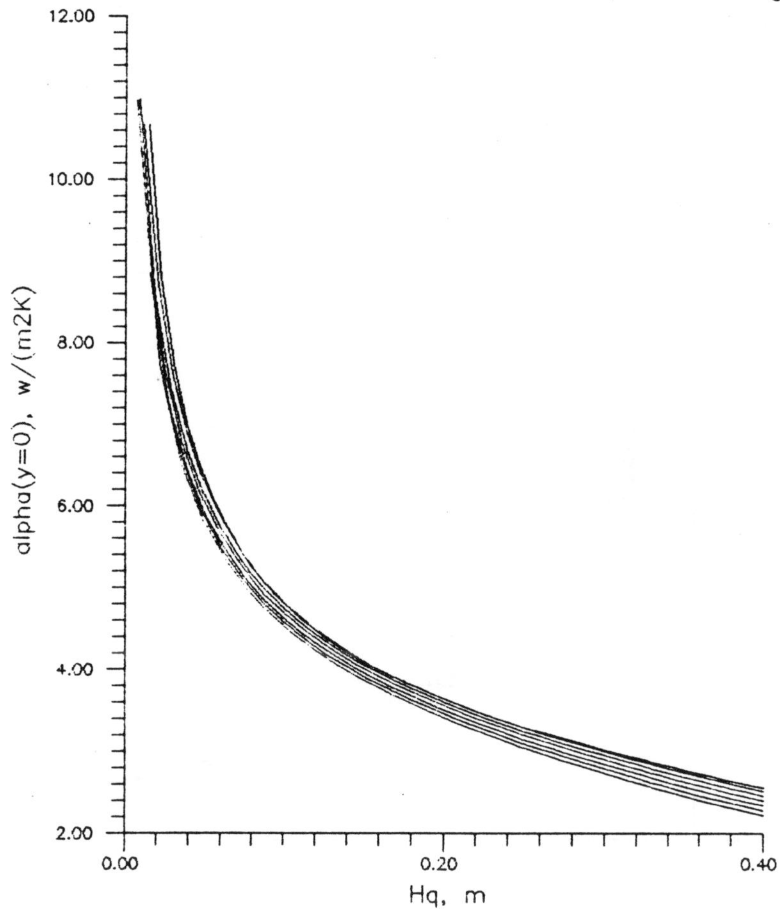

Fig.4: Local heat transfer coeficients

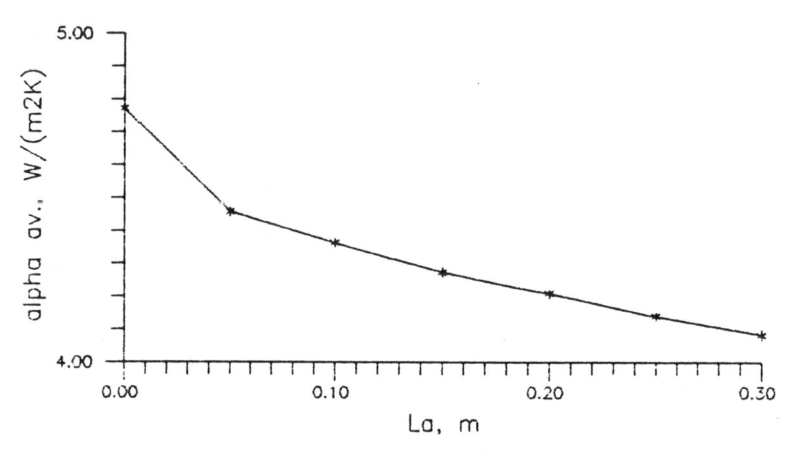

Fig.5: Average heat transfer coefficients

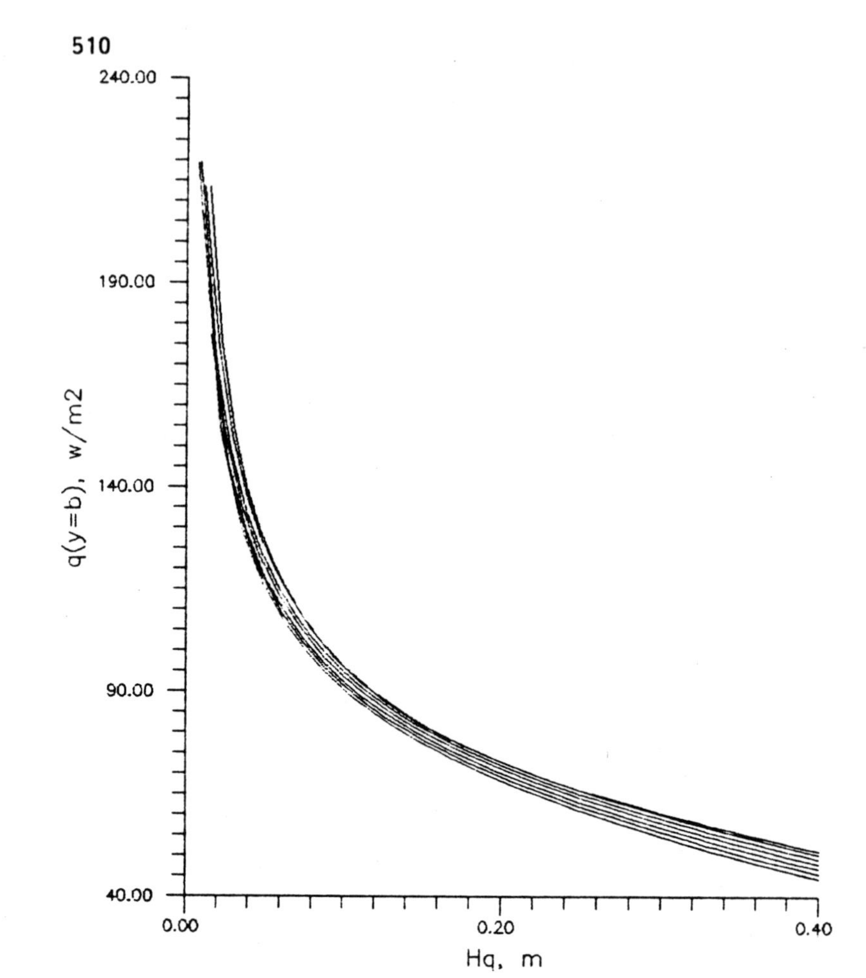

Fig.6: Local wall heat flux

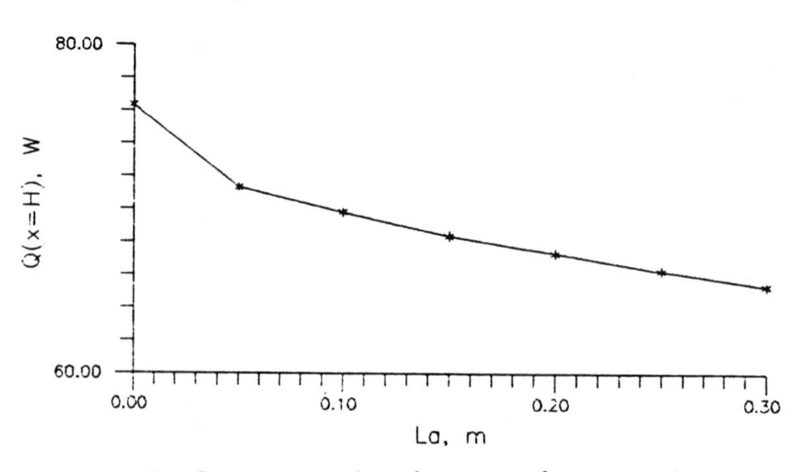

Fig.7: Convective heat exchange rate

Pressure differences within inside and outside of channel along upper section (Lq) are shown in Fig.2 for all seven channels. Our "benchmark" channel #1 with La=0 (top curve) with lowest flow ressistance requires smallest pressure imbalance to drive the flow. Since available buoyant forces are equal for all seven channels longer adiabatic sections cause a decrease of entrance velocities (Fig.3).

Local heat transfer coefficients for left wall (y=0) (Fig.4) are defined as:

$$\alpha(x,y=0)=\frac{q(x,y=0)}{T(x,y=0)-T_{\infty}} \tag{15}$$

Average heat transfer coefficient for each channel is calculated as:

$$\alpha_{av.}=\frac{1}{Lq}\int_{La}^{H}\frac{q(x,y=0)}{T(x,y=0)-T_{\infty}}dx \tag{16}$$

and shown in Fig.5.

q(x,y=0) in (15) (shown in Fig.6) are local heat flux densities:

$$q(x,y=0)=-\lambda\left(\frac{\partial T(x,y)}{\partial y}\right)_{y=0} \tag{17}$$

Convective heat transfer rates (for channels as a whole) shown in Fig.7 are expressed as fluid enthalpy change between entrance and exit cross sections:

$$Q(x=H)=\int_{0}^{H}[(\rho C_p uT)_{x=H}-(\rho C_p uT)_{x=0}]\,dy \tag{18}$$

For group of channels investigated, adiabatic entrance section although not participating in heat transfer, modifies the flow field and thus influences convective heat transfer in the upper section.
Buoyant forces which drive the flow, resulting from wall-ambient temperature difference in the upper part of the channel, must "pull" the flow through entrance section. Depending on it's length the upward flow velocity and overall heat transfer rate are moderately reduced.

5. NOMENCLATURE

b	channel width
c_p	specific heat of fluid
G	mass flow
g	gravitational acceleration
H	channel height
La	channel lower (adiabatic) section length
Lq	channel upper (heated) section length
P	absolute pressure
p	pressure difference, equation (4)
T	temperature
Q	heat flux rate, equation (18)
q	local heat flux rate, equation (17)
U_0	axial velocity at entrance cross-section
u	axial velocity component
v	transversal velocity component
x	axial coordinate
y	transversal coordinate
α	heat ransfer coefficient, equation (15)
α_{av}	average heat transfer coeff., equation (16)
β	- thermal expansion coefficient
μ	- dynamic viscosity
ρ	- fluid density

6. REFERENCES

1. ELENBAAS,W - Heat Dissipation of Paralell Plates by Free Convection, Physica, Vol.9, No.1 , p.75, 1942.
2. BAR-COHEN,A.,ROSENHOW,W.M. - Thermally Optimum Spacing of Vertical Natural Convection Cooled Paralell Plates, ASME Journal of Heat Transfer, Vol.106, p.116, 1984.
3. SPARROW,E.M.,TAO,W.Q. - Buoyancy Driven Fluid Flow and Heat Transfer in a Pair of Interacting Vertical Paralell Channels, Numerical Heat Transfer, Vol.5, p.39, 1982.
4. SPARROW,E.M.,SHAH,S.,PRAKASH,C. - Natural Convection in a Vertical Channel: Interacting Convection and Radiation, The Vertical Plate With and Without Shrouding, Numerical Heat Transfer, Vol.3, p.297, 1980.
5. ZVIZDIC,D.: Calculation of Buoyancy-driven Fluid Flow and Heat Transfer Between Non-isothermal Vertical Walls, Proceedings of Third International Conference for Non Linear Problems, Vol.3, pp.1018-1025, Pineridge Press.Ltd., Swansea England 1986.
6. PATANKAR,S.V. - Numerical Heat Transfer and Fluid Flow, Hemisphere, 1980.

Natural convection during underground coal gasification

R.A. Kuyper, Th.H. van der Meer and C.J. Hoogendoorn
Section Heat Transfer, Faculty of Applied Physics,
Delft University of Technology,
P.O. Box 5046, 2600 GA Delft, The Netherlands

Abstract

A mathematical model for the natural-convection flow arising during the underground gasification of coal layers has been developed. Models for chemical reaction rates, radiative heat transfer and turbulence are included in the simulations. The flow is determined by the combined buoyancy forces due to horizontal temperature and concentration gradients. The relative strength of both forces is depending on the radiative properties of the fluid. For two radiation absorption coefficients the results will be presented. Correlations between the global density differences and the mass transfer will be presented and discussed.

1 Introduction

The coal-reserve in Western Europe is still very large. However, in Western Europe coal layers often appear in thin seams (1-2 meter) and at great depth ($>$ 1 km). For these layers, the conventional technique of digging the coal will be too expensive. The underground gasification of coal is an economically attractive alternative. The process of underground coal gasification is initiated by drilling two holes towards the coal layer and connecting them through this layer. Air is injected in one of the wells and the flow of air through the coal layer will produce combustible gases at the production well (see Figure 1). Since coal is gasified by the air, the width of the channel enlarges. As this channel is expanding, the rock structure above the cavity will collapse, forming two separated channels. Boswinkel [1] developed a model for the gasification of these thin seams, in which an open channel structure is formed. By the gasification of coal, this open channel slowly moves through the coal layer. In order to gasify

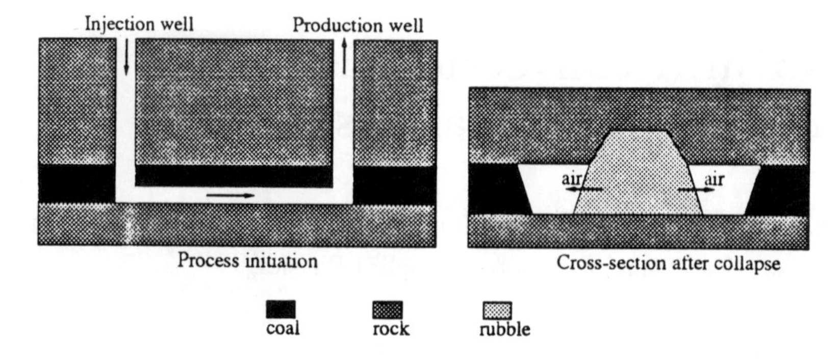

Figure 1: Model of the underground gasification process

a sufficient amount of coal in this seam, it is essential that the mass transfer in the channel is high.

In Figure 1 a cross-section of the resulting configuration is drawn. In this situation, which will be present soon after the start of the gasification process, air will flow through the inert rubble bed towards the open channel. As air reaches the cavity, the oxygen will react with the combustible gases present in the channel (*e.g.* CO, H_2). The combustion products (*e.g.* CO_2, H_2O) can be reduced at the coal face by endothermic reactions. The occurring reactions will produce horizontal temperature and concentration gradients in the channel. At the rubble wall, the gas temperature will be high and the gas molecules will be relatively heavy. At the coal wall, the gas temperature will be low and the molecules will be relatively light. Both the temperature and the gas composition determine the density distribution in the channel and thus the direction of the buoyant forces. The opposing buoyant forces working on the fluid result in an interesting natural convection flow.

If we superpose the forced-convection channel flow from the injection well towards the production well on the natural-convection flow a spiral flow can be expected. The Grashoff number of the natural-convection flow in the process is approximately 10^{10}, while the Reynolds number of the forced convection flow is about 10^4. Comparison of both dimensionless numbers gives $Gr/Re^2 \gg 1$ showing that the first effect is dominant. Obviously, the natural-convection flow determines the transport of the gases to and from the coal wall and thus the quality of the product gas. Since the Grashoff number is high, the natural-convection flow will be turbulent. In this study, the turbulent natural-convection flow due to the

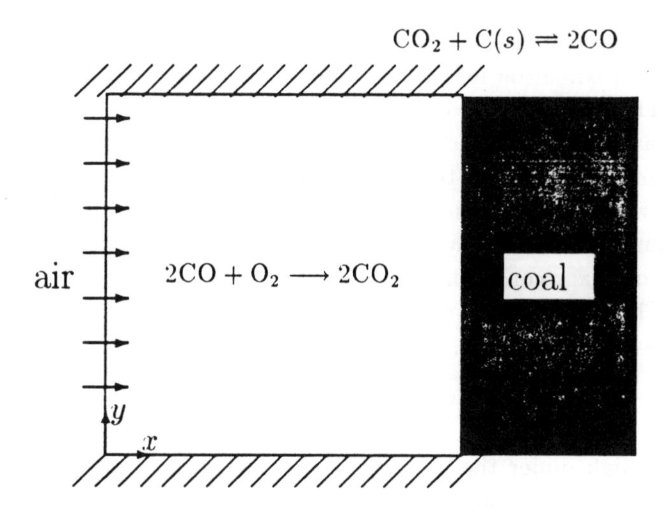

Figure 2: Mathematical model of the gasification channel

combined effects of horizontal temperature and concentration gradients has been investigated. Especially, correlations between the density distribution in the channel and the resulting mass transfer will be discussed.

2 The mathematical model

Since the processes in the gasification channel will probably be more important than the shape of the channel, the geometry is mathematically modelled by a two-dimensional square cavity of dimension 2×2 meter. The process pressure in the cavity is 10 atm. Through the rubble wall, air is injected. At the opposite side, the coal wall is modelled by its chemical property of reducing combustion products. The top and bottom wall consist of rock material and are modelled as impermeable walls. The gases under consideration in this paper are N_2, O_2, CO and CO_2. In the cavity the exothermic reaction of combustible gases with oxygen is modelled. The resulting geometry and reactions are presented in Figure 2.

We will now present the relevant processes described in the model. A more detailed description of the radiation model and the modelling of the coal wall will be given later.

- The conservation of mass and momentum is described by the Navier-Stokes equations. Buoyancy forces are the driving force of the flow.

Because of the large density differences in the cavity, the Boussinesq approximation for natural-convection flows has not been used.

- The temperature distribution is obtained from the conservation law for energy. Radiative heat transfer and reaction energy play an important role in this distribution. The thermal interaction of the process with the surrounding rock and the heat-up of injected air and of the coal layer has been included in the formulation of the boundary conditions.
- The transport of chemical species is modelled from conservation laws for CO, CO_2, N_2 and O_2. The reaction rate of the reaction

$$CO + O_2 \rightarrow CO_2$$

is high under the circumstances present in the cavity. Therefore, the homogeneous combustion of CO in the channel is limited by (turbulent) diffusion and the conversion rate has been modelled by using an eddy break-up model [2].

- Because of the dimensions of the channel, the process pressure and the density differences, the natural-convection flow will be turbulent. In the model, we used a standard $k - \epsilon$ model to describe the turbulence. No wall functions are used in the boundary conditions for the $k - \epsilon$ model. Instead, the Dirichlet boundary conditions $k = 0$ and $\epsilon = \infty$ are used at the cavity walls [3].

We consider a time-dependent flow, therefore the transient differential equations are solved to obtain the fields.

2.1 The radiation model

The underground gasification channel will contain H_2O, CO and CO_2 under a pressure of 10 atm. There will also be dust particles present in the channel. Both effects imply that the gas will act as a participating fluid and a model to describe the radiative emission and absorption by the fluid has to be included. Hottel's Zone Method has been used to describe the radiative heat transfer in the cavity [4]. For a grey gas, a two-dimensional version of this method has been developed. The two-dimensional radiative exchange factors have been calculated by developing a method similar to the three-dimensional calculation method of Siddall [5]. In order to study the effect of the presence of dust particles, the radiative absorption coefficient k is varied. Calculations have been made by considering both a transparent gas ($k = 0$ m^{-1}) and a participating gas ($k = 0.12$ m^{-1}) Radiative heat transfer will decrease the temperature gradients so the concentration gradients will play a larger role in the density distribution.

2.2 The coal wall model

The coal wall properties are used as boundary conditions for the flow field, the temperature distribution and the distribution of chemical species in the channel. In this paper we will not discuss the effect of coal pyrolysis and the coal wall will be modelled by assuming the coal to consist completely out of carbon. Since all oxygen is used to combust the CO present in the channel, the only occurring reaction at the coal wall will be

$$CO_2 + C(s) = 2CO. \tag{1}$$

The amount of CO_2 which has to be reduced at the coal wall per second is small and the reaction is diffusion-limited [6]. The following system of coupled boundary conditions describe the coal wall:

- The relation between CO and CO_2 is described by a pressure and temperature dependent equilibrium relationship [7].
- There is no flux of non-reacting species.
- The energy flux to the coal wall is determined by
  1. the reaction rate of the endothermic reaction 1
  2. the heat-up of the coal layer
  3. the radiative properties of the coal wall.
- The velocity at the coal wall is determined by the net mass flux from this wall.

3 Solution procedure

The equations describing the conservation of mass, momentum, energy and chemical species, combined with the equations for k and ϵ in the turbulence model are solved using a finite-volume method with a pressure-correction method as introduced by Patankar and Spalding [8]. The number of interior grid cells used in the calculations was chosen 50×50. The number of Hottel zones was taken 5×5. By using a sinusoidal distribution of grid points, strong grid refinement was obtained close to the walls.

The flow in the direction of the production well was obtained by assuming a constant velocity profile and solving a global mass balance. The Poisson equation for the pressure correction was solved directly over the full domain.

The boundary conditions for the coal are non-linear and coupled. In order to solve the algebraic equations, a Newton-Raphson iteration method was used. For all considered situations, the calculations converged to time-dependent flows. Since we will only present results for which the time evolution is not important, the time-dependent behaviour of the flow will not be discussed.

4 Results

The results for two radiation absorption coefficients will be presented. In Figure 3 the relevant fields for $k = 0$ m^{-1} are presented for two oxygen-injection rates. Figure 4 show the corresponding fields for $k = 0.12$ m^{-1}. The fields are represented by

1. the velocity field, showing the topology of the flow
2. the mass fraction of CO; plotted are isolines $Y_{CO} = 0(0.03)0.3$
3. the temperature; plotted are isolines $T = 1500(25)2000$ K
4. the distribution of the turbulent viscosity μ_t/μ.

In all plots, the rubble wall (combustion zone) is at the left-hand side of the cavity and the coal wall (reduction zone) at the right-hand side.

4.1 Transparent fluid: $k = 0$ m^{-1}

In the situation $k = 0$ m^{-1}, the temperature gradients are high at the air-inlet because of the combustion of CO. Since the fluid is not emitting radiation, the reaction energy is used to raise the local temperature. However, this reaction also produces heavy CO_2 molecules. In this situation the effect of the temperature raise on the density distribution is higher than the effect of the concentration changes; the fluid moves up. At the coal wall, the fluid is relatively cold because of the endothermic reaction. However, light CO molecules are produced at this wall. For low air-injection rates, the temperature gradients are too small to dominate the concentration gradients; again the fluid moves up. Halfway the top wall, both flows meet and form a free jet into the core region. In this jet turbulence is high and concentration gradients are decreased. For high injection rates, the energy effect of the reactions will be larger on both sides of the cavity, while the concentration effects do not change much. The flow field is driven by the global temperature differences, giving a clockwise flow. At the lower end the coal wall, concentration gradients are strong enough to stop the boundary layer flow and the free jet is formed again.

4.2 Participating fluid: $k = 0.12$ m^{-1}

The flow characteristics change drastically if the fluid is participating in the radiative heat transfer. In Figure 4, we see that only in a small

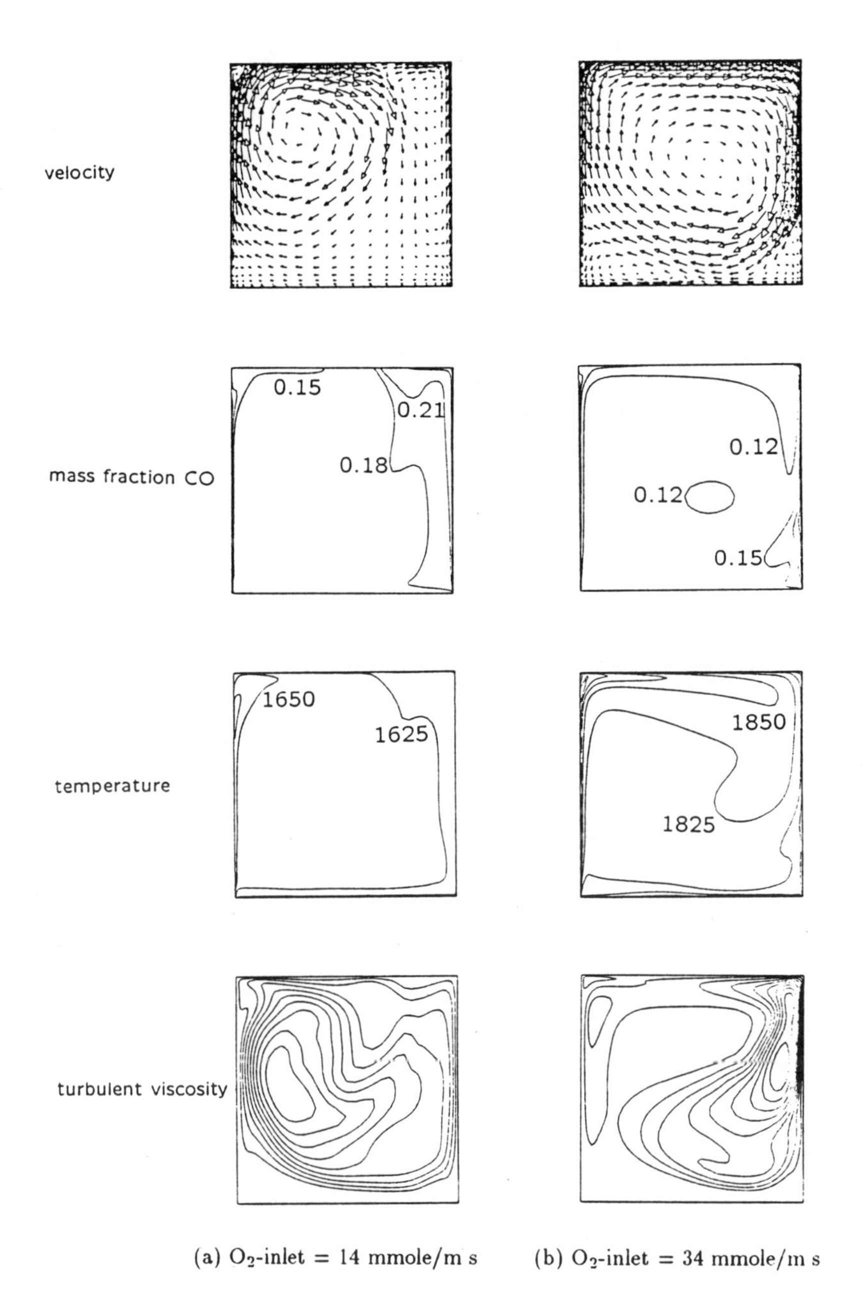

(a) O_2-inlet = 14 mmole/m s (b) O_2-inlet = 34 mmole/m s

Figure 3: Results for $k = 0.0$ m^{-1}

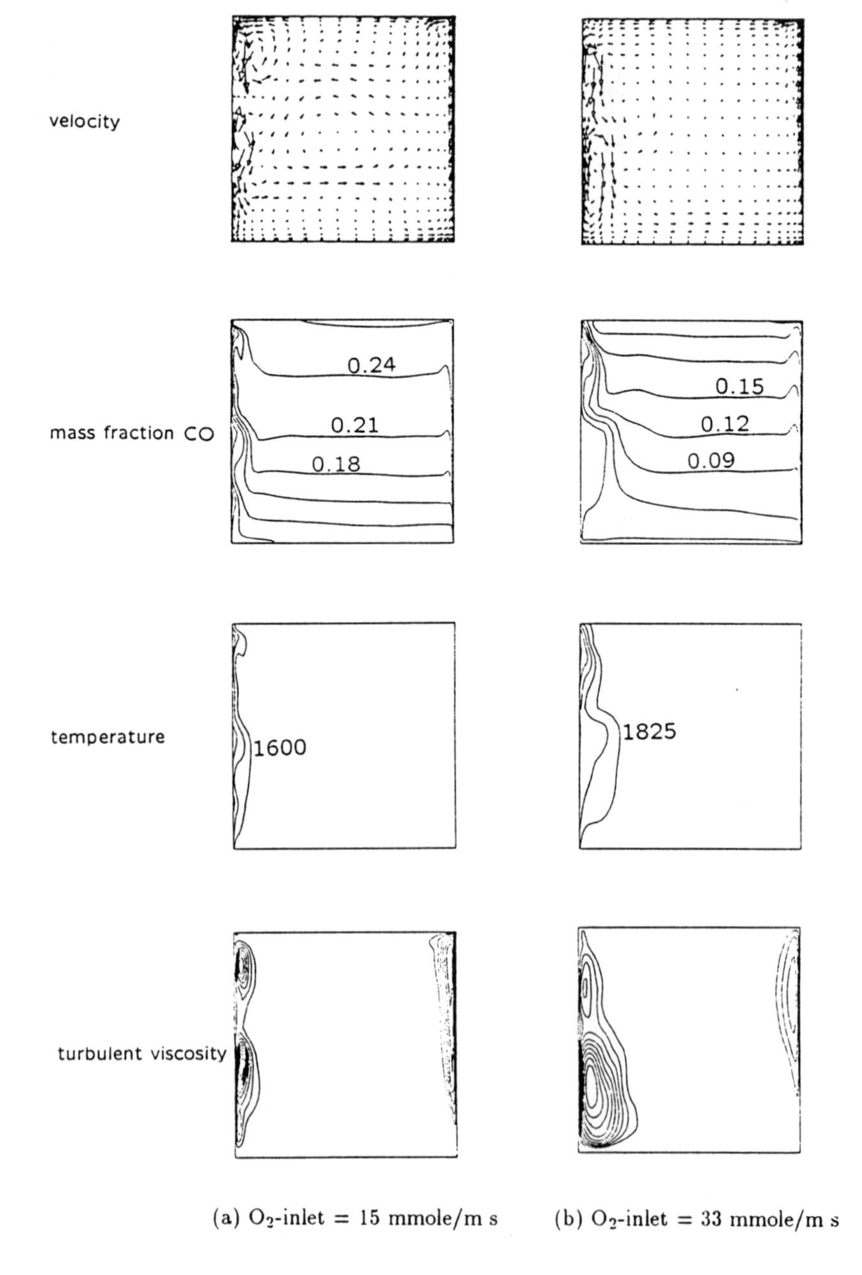

(a) O_2-inlet = 15 mmole/m s (b) O_2-inlet = 33 mmole/m s

Figure 4: Results for $k = 0.12$ m^{-1}

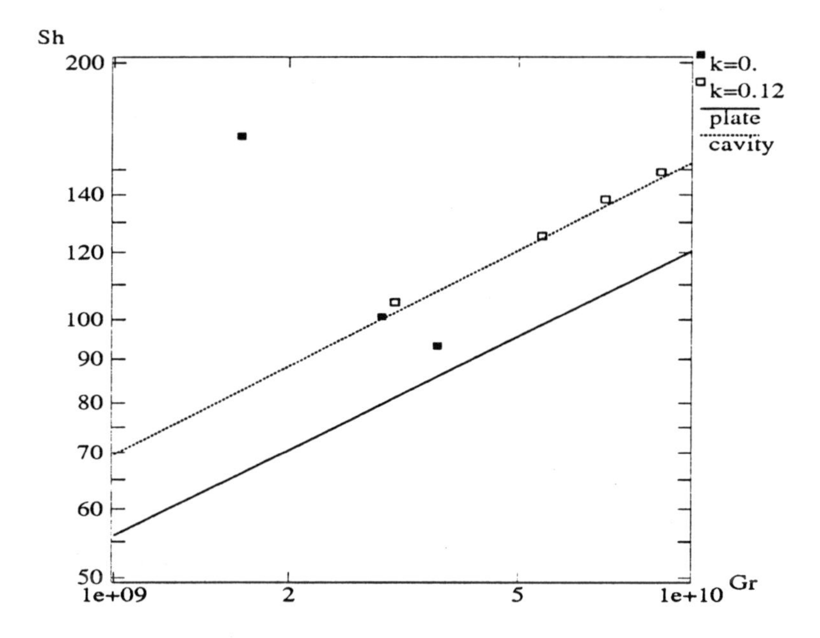

Figure 5: The calculated Sherwood number versus the Grashoff number

part of the cavity temperature gradients exist. In the combustion zone, the temperature gradients can still become high enough to give a local upward flow. After the energy is emitted, the heavy CO_2 molecules will make the fluid move down, producing small cells at the left-hand side of the velocity plots. At the coal wall, temperature gradients are negligible if compared to concentration gradients. Here we find a boundary layer flow induced by only one buoyancy effect. At this side of the cavity turbulence is present at the end of the boundary layer. Although velocities are not horizontal in the core, the concentration distribution is very well stratified. If we look only at the right-hand side of the cavity in Figure 4, the concentration and turbulence distribution resemble the temperature and turbulence distribution along the hot wall of a cavity heated from the side for $Ra=10^{10}$ [3]. For a higher oxygen-injection rate, we see that a larger part of the cavity is filled by the combustion zone. Not too far from the rubble wall, however, most of the reaction energy is emitted to the other parts of the cavity. Even for this situation, the flow along the coal wall is mainly due to concentration gradients.

4.3 Sherwood-Grashoff relations

If we compare the results for $k = 0\ \mathrm{m}^{-1}$ and $k = 0.12\ \mathrm{m}^{-1}$, we see that in the first case the effects of the temperature distribution and the concentration distribution on the buoyancy forces are of equal strength. Local density gradients determine the flow structure and it will be difficult to find an overall relation between density gradients and mass transfer. For $k = 0.12\ \mathrm{m}^{-1}$, both effects are of equal strength only in the combustion zone. In the core region and along the coal wall, the concentration gradients are dominant.

We will consider the relation between density gradients and the mass transfer at the coal wall for both situations. Since the density differences between the rubble wall and the coal wall might equal zero, we will define a Grashoff number based on the density difference between the cavity centre and the coal wall

$$\mathrm{Gr} = \frac{g \dfrac{\rho_{\mathrm{centre}} - \rho_{\mathrm{coal}}}{\rho_0} H^3}{\nu^2}.$$

The mass transfer towards the coal wall is given by the Sherwood number

$$\mathrm{Sh} = \frac{\text{computed mass transfer}}{\text{mass transfer without convection}},$$

in which the mass transfer without convection is defined on half the cavity width. Using the experimental relation for the turbulent natural-convection flow along a heated plate in an isothermal environment [3] in this situation gives

$$\mathrm{Sh} = 0.0595\,(\mathrm{Gr\ Sc})^{\frac{1}{3}} \tag{2}$$

in which Sc is the Schmidt number. Since the core region is not isothermal and since we used a turbulence model, we can also use the correlation obtained previously for a cavity heated from the side [10]

$$\mathrm{Sh} = 0.063\,(\mathrm{Gr\ Sc})^{0.341}\,. \tag{3}$$

In Figure 5 the results for $k = 0\ \mathrm{m}^{-1}$ and $k = 0.12\ \mathrm{m}^{-1}$ are compared to the relations 2 and 3. It can easily be seen that the correlations are applicable for $k = 0.12\ \mathrm{m}^{-1}$. For $k = 0\ \mathrm{m}^{-1}$ local effects determine the flow and no correlation is found between the mass transfer and the global density distribution.

4.4 Product gas composition

The production of CO and CO_2 as a function of the oxygen-injection rate is presented in Figure 6. The predicted production and the production

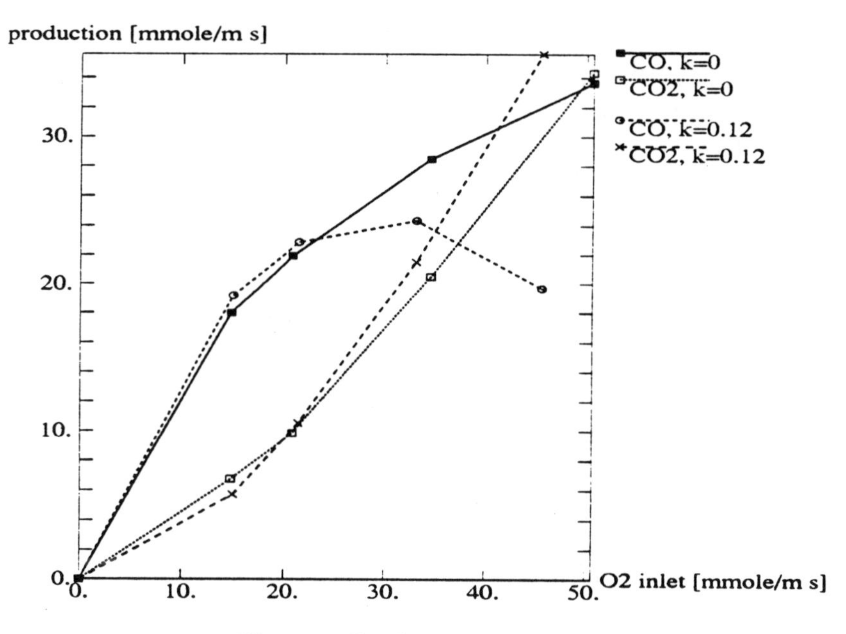

Figure 6: Production of CO and CO_2

obtained during a field test in Pricetown (USA) correspond well [9]. For low oxygen-injection rates the production of CO and CO_2 are the same for $k = 0\ m^{-1}$ and $k = 0.12\ m^{-1}$. If the oxygen-injection rate is high, the boundary layer along the coal wall has to transfer a larger amount of CO_2 towards this wall. For $k = 0\ m^{-1}$, the produced CO_2 is flowing directly from the combustion zone towards the coal wall at high oxygen-injection rates. In this situation the mass transfer is high enough to produce more CO than CO_2. For $k = 0.12\ m^{-1}$, CO_2 is transported slowly through the core region. At a lower oxygen-injection rate than for $k = 0\ m^{-1}$ we will produce more CO_2 than CO.

5 Conclusions

The presented numerical model predicts the heat and mass transfer during underground coal gasification. The flow is determined by opposing buoyancy forces. Radiative heat transfer influences the strength of the buoyancy force due to horizontal temperature gradients. If one of the buoyancy forces is dominant, the flow and mass transfer correspond to the situation with only one buoyancy effect present. However, as soon as the two buoyancy forces become of approximately equal strength, local gradients determine the flow and therefore the heat and mass transfer. No straightforward correlation between mass transfer and global density differences can be found for this situation.

ACKNOWLEDGEMENTS
These investigations are supported by the Netherlands' Foundation for Chemical Research (SON) with financial aid from the Netherlands' Technology Foundation (STW).

References

[1] H. Boswinkel - "Gasifying Dutch coals," in *Proceedings of The Ninth Underground Coal Gasification Symposium, Bloomingdale*, pp. 429–436, 1983.

[2] H. Bockhorn and G. Lutz - "The application of turbulent reaction models to the oxidation of CO in a turbulent flow," in *20th Symposium (International) on Combustion/The Combustion Institute*, pp. 377–386, 1984.

[3] R. A. W. M. Henkes - *Natural-convection boundary layers*. PhD thesis, Delft University of Technology, Delft, The Netherlands, 1990.

[4] R. Siegel and J. R. Howell - *Thermal radiation heat transfer*. Hemisphere Publishing Corporation, 1972.

[5] R. G. Siddall - "Accurate evaluation of radiative direct-exchange areas for rectangular geometries," in *Proceedings of The Eighth International Heat Transfer Conference, San Francisco*, pp. 751–755, 1986.

[6] D. W. van Batenburg - *Heat and mass transfer during underground gasification of thin deep coal seams*. PhD thesis, Delft University of Technology, Delft, The Netherlands, 1992.

[7] A. Belghit and M. Daguenet - "Study of heat and mass transfer in a chemical moving bed reactor for gasification of carbon using an external radiative source," *Int. J. Heat and Mass Transfer*, vol. 32, no. 11, pp. 2015–2025, 1989.

[8] S. V. Patankar - *Numerical heat transfer and fluid flow*. Hemisphere Publishing Corporation, 1980.

[9] H. H. Boswinkel - *De mogelijkheden van in-situ vergassing van steenkool in Nederland*. Energie Studie Centrum 38, 1984.

[10] R. A. Kuyper, T. H. van der Meer, C. J. Hoogendoorn, and R. A. W. M. Henkes - "Numerical study of laminar and turbulent natural convection in an inclined square cavity," *Submitted to Int. J. Heat and Mass Transfer*, 1992.

NUMERICAL STUDY OF FLUID FLOW AND HEAT TRANSFER IN A CIRCULAR DUCT WITH A SUDDEN CONTRACTION

A. Muzzio, P. Parolini
Dipartimento di Energetica, Politecnico di Milano
piazza L. Da Vinci 32, 20133 Milano, Italia

Summary. This paper reports the solution algorithm and the results obtained in the finite-difference analysis of the laminar flow and heat transfer in a circular duct with a sudden contraction. Results are presented with Prandtl number equal to 0.7 and Reynolds number, based on the upstream duct diameter, ranging from 2 to 500. The results include velocity profiles and pressure, temperature and local Nusselt number distributions for contraction ratios ranging from 0.4 to 0.75. The presence of a small recirculation region just upstream of the contraction section is detected and its size has been confronted with the results of other authors confirming the accuracy of the computational scheme adopted. Near the contraction the strong pressure and velocity gradients produce a large increase in the heat transfer rate, shown by a steep increase in the local Nusselt number. Finally downstream, as thermal profiles redevelops along the duct, the Nusselt number decreases asymptotically to the limit foreseen by theory.

1. INTRODUCTION

Fluid flow in a duct with a sudden cross sectional area variation is a situation which occurs in many industrial applications, such as flow meter devices and valves. The main aspect of this situation is the pressure loss caused by flow separation, which is associated by increased heat transfer rate. While a large effort of research has been devoted to the case of the sudden enlargement the situation investigated here has been studied only by few authors. Numerical solutions for the flow field were reported by Greenspan [1], Greppi and Cercignani [2]. More recently Dennis and Smith [3] have studied the similar case of a plane two-dimensional channel with flow field ranging from Stokes flow to Reynolds number of 1000. In their paper a very complete investigation of the size of the recirculation region situated just upstream of the contraction is performed leading the authors to find a minimum of its size for Re = 100, computed with reference to the upstream diameter, and for a contraction ratio of 0.5. Zampaglione [4] and Dobrowolsky [5] have studied the laminar viscous flow through a pipe orifice. More relevant to our work is the paper of Durst and Loy [6], who investigated the laminar flow in a circular pipe with a sudden contraction. In their paper the authors investigated the flow field and the two separation regions in the range $23 \leq Re \leq 1213$. Their numerical results were compared with experimental velocity profiles obtained with a laser-Doppler anemometer.

The geometrical situation of a sudden cross section enlargement has been studied more extensively and has produced a large literature on the subject. We just mention the work of Sparrow [7], who investigated the case of an abrupt asymmetric enlargement in a parallel-plate channel. In this paper, besides the flow field, also heat transfer is investigated and distributions of local Nusselt number at various Reynolds numbers are provided.

In our problem the fluid is steady-state, incompressible with Reynolds number, computed with reference to the upstream duct diameter, ranging from 2 to 500, and with different values of contraction ratio m, defined as the ratio between upstream and downstream diameters, between 0.4 and 0.75. Starting at the inlet section with a fully developed laminar parabolic velocity profile, the flow field is curved towards the entrance of the small diameter duct. As a consequence the flow separates from the wall, yielding a small recirculation region just upstream of the sudden contraction. At the contraction section the velocity profile becomes almost completely uniform, then the flow field separates again from the wall forming the so called vena contracta, which, quite differently from turbolent flow, has a very small reduction in main flow cross sectional area and can be detected with difficulty only at high Reynolds numbers. Afterwards the flow redevelops reaching, at the end of the duct, the shape of a partially developed parabola.

In a similar way the thermal field behaves, as local heat transfer coefficient and local Nusselt number show a sharp variation in the proximity of the contraction. Regarding the heat transfer problem a uniform temperature distribution, different from the inlet fluid temperature, is considered to be on the wall of the domain. The thermo-physical fluid properties are taken into account by Prandtl number, whose value has been set to 0.7 typical for diatomic gases in a wide range of temperature.

The numerical integration of the partial differential equations governing the problem is accomplished by a finite difference scheme similar to the MAC method proposed by Harlow and colleagues [8-9]. In this scheme a derived Poisson equation for pressure is employed, and a stationary problem, as in our specific case, is solved through the time integration of a transient equations system untill steady state is achieved.

2. FORMULATION OF THE PROBLEM AND GOVERNING EQUATIONS

The orizontal circular cross section duct has a diameter D upstream and a diameter mD downstream of the contraction section: the upstream lenght is βD and the downstream lenght is γD as in Fig. 1, where the domain diagram is presented. As mentioned before a fully developed parabolic velocity profile is assumed at the inlet section, with mean axial velocity $\overline{U}$. The duct walls are at a uniform temperature T_w, different from the uniform inlet fluid temperature T_i. This domain geometry and the boundary conditions imposed allow the problem to be two-dimensional axisymmetric.

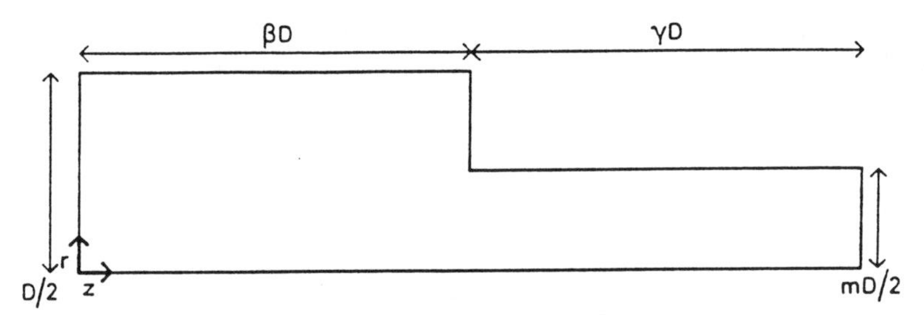

Fig. 1 Domain diagram

As usual in this types of problems viscous dissipation terms are disregarded in the energy equation and the thermo-physical properties are considered to be constant. With these assumptions the governing equations of mass, momentum and energy conservation are written in conservative dimensionless form, cylindrical axisymmetric coordinates in the following way

$$\frac{\partial u}{\partial z}+\frac{1}{r}\frac{\partial (vr)}{\partial r}=0 \tag{1}$$

$$\frac{\partial u}{\partial \tau}+\frac{1}{r}\frac{\partial (uvr)}{\partial r}+\frac{\partial (u^2)}{\partial z}+\frac{\partial p}{\partial z}=\frac{1}{Re}\left(\frac{\partial^2 u}{\partial z^2}+\frac{1}{r}\frac{\partial u}{\partial r}+\frac{\partial^2 u}{\partial r^2}\right) \tag{2}$$

$$\frac{\partial v}{\partial \tau}+\frac{1}{r}\frac{\partial (v^2 r)}{\partial r}+\frac{\partial (uv)}{\partial z}+\frac{\partial p}{\partial r}=\frac{1}{Re}\left(\frac{\partial^2 v}{\partial z^2}+\frac{\partial}{\partial r}(\frac{v}{r})+\frac{\partial^2 v}{\partial r^2}\right) \tag{3}$$

$$\frac{\partial \theta}{\partial \tau}+\frac{1}{r}\frac{\partial (v\theta r)}{\partial r}+\frac{\partial (u\theta)}{\partial z}=\frac{1}{Re\,Pr}\left(\frac{\partial^2 \theta}{\partial z^2}+\frac{1}{r}\frac{\partial \theta}{\partial r}+\frac{\partial^2 \theta}{\partial r^2}\right) \tag{4}$$

In order to write the equations in dimensionless form the contraction upstream diameter is the characteristic lenght and the inlet mean velocity is the characteristic velocity, yielding the following dimensionless variables and parameters

$$z=\frac{Z}{D} \quad r=\frac{R}{D} \quad u=\frac{U}{\overline{U}} \quad v=\frac{V}{\overline{U}} \quad p=\frac{P}{\rho\overline{U}^2} \quad \tau=\frac{t\overline{U}}{D} \quad \theta=\frac{(T-T_i)}{(T_w-T_i)} \tag{5}$$

The boundary conditions of the solution domain are

- at the inlet section, the fully developed laminar parabolic profile and uniform fluid temperature yield

$$u=u(r)=2(1-4r^2) \quad v=0 \quad \theta=0 \tag{6}$$

- on the duct walls a non-slip condition and a uniform wall temperature are assumed

$$u=0 \qquad v=0 \qquad \theta=1 \tag{7}$$

- at the outlet section, conditions of velocity and temperature almost completely developed are assumed

$$\frac{\partial u}{\partial z} = 0 \qquad v = 0 \qquad \frac{\partial \theta}{\partial z} = 0 \tag{8}$$

- on the simmetry axis the usual simmetry boundary conditions are

$$\frac{\partial u}{\partial r} = 0 \qquad v = 0 \qquad \frac{\partial \theta}{\partial r} = 0 \tag{9}$$

Pressure boundary conditions are linked to the particular solution scheme adopted, so they will be defined afterwards.

3. METHOD OF SOLUTION

The partial differential equations system, together with the boundary conditions, is integrated numerically in the domain with a finite-difference scheme similar to the MAC method. One of its main feature is the adoption of the staggered grid, where the velocity components are placed in the middle of the computational cell boundaries, as shown in Fig. 2. Field scalar variables, namely pressure and temperature, are placed in the center of the cells. The position indices i,j are referred to the different variables, not to the geometrical position in the domain. We have, as a consequence, three different grids at a half grid size distance from each one to the other two. This staggered grids arrangement is useful in preventing the insurgence of numerical instability in the solution process [8-12].

We define $\varphi \in (u, p, \theta)$ and $\psi \in (u, \theta)$ as being two dummy variables and r,z the apices which indicate the direction for upwinding, as will be defined later. Accordingly the following discretized differential operators are defined in the computational domain

$$\nabla^2 \varphi_{i,j} = \frac{\varphi_{i+1,j} - 2\varphi_{i,j} + \varphi_{i-1,j}}{\delta z^2} + \frac{\varphi_{i,j+1} - 2\varphi_{i,j} + \varphi_{i,j-1}}{\delta r^2} + \frac{\varphi_{i,j+1} - \varphi_{i,j-1}}{2 r_{j+1/2} \delta r} \tag{10}$$

$$\nabla^2 v_{i,j} = \frac{v_{i+1,j} - 2v_{i,j} + v_{i-1,j}}{\delta z^2} + \frac{v_{i,j+1} - 2v_{i,j} + v_{i,j-1}}{\delta r^2} + \frac{v_{i,j+1} r_{j+1}^{-1} - v_{i,j-1} r_{j-1}^{-1}}{2\delta r} \tag{11}$$

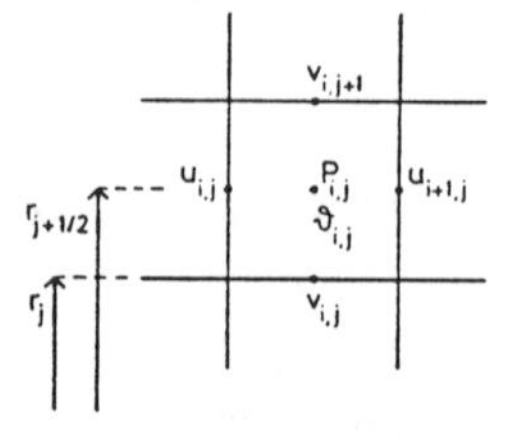

Fig. 2 Computational grid

$$A_{i,j} = \frac{u_{i+1/2,j}\Psi^{z}_{i+1/2,j} - u_{i-1/2,j}\Psi^{z}_{i-1/2,j}}{\delta z} + \frac{v_{i-1/2,j+1}\Psi^{r}_{i,j+1/2}r_{j+1} - v_{i-1/2,j}\Psi^{r}_{i,j-1/2}r_{j}}{r_{j+1/2}\delta r} \quad (12)$$

$$B_{i,j} = \frac{u_{i+1,j-1/2}v^{z}_{i+1/2,j} - u_{i,j-1/2}v^{z}_{i-1/2,j}}{\delta z} + \frac{v_{i,j+1/2}v^{r}_{i,j+1}r_{j+1/2} - v_{i,j-1/2}v^{r}_{i,j-1/2}r_{j-1/2}}{r_{j}\delta r} \quad (13)$$

The conservation equations, discretized in space and time domain, are written at the generic time step $(n+1)\delta\tau$ as follows

$$\frac{u^{n+1}_{i+1,j} - u^{n+1}_{i,j}}{\delta z} + \frac{r_{j+1}v^{n+1}_{i,j+1} - r_{j}v^{n+1}_{i,j}}{r_{j+1/2}\delta r} = 0 \quad (14)$$

$$u^{n+1}_{i,j} = u^{n}_{i,j} + \delta\tau(\mathrm{Re}^{-1}\nabla^{2}u^{n}_{i,j} - A^{n}_{i,j}(u) - \nabla p^{n+1}_{i,j}) \quad (15)$$

$$v^{n+1}_{i,j} = v^{n}_{i,j} + \delta\tau(\mathrm{Re}^{-1}\nabla^{2}v^{n}_{i,j} - B^{n}_{i,j} - \nabla p^{n+1}_{i,j}) \quad (16)$$

$$\theta^{n+1}_{i,j} = \theta^{n}_{i,j} + \delta\tau(\mathrm{Re}^{-1}\,\mathrm{Pr}^{-1}\,\nabla^{2}\theta^{n}_{i,j} - A^{n}_{i,j}(\theta)) \quad (17)$$

By substituting equations (15-16) in the mass conservation equation (14) the following Poisson pressure equation is derived

$$\nabla^{2}p^{n+1} = \left(\frac{u_{i+1,j} - u_{i,j}}{\delta z\delta\tau} + \frac{\nabla^{2}u_{i+1,j} - \nabla^{2}u_{i,j}}{\mathrm{Re}\,\delta z} - \frac{A_{i+1,j}(u) - A_{i,j}(u)}{\delta z}\right.$$

$$\left. + \frac{r_{j+1}v_{i,j+1} - r_{j}v_{i,j}}{r_{j+1/2}\delta z\delta\tau} + \frac{r_{j+1}\nabla^{2}v_{i,j+1} - r_{j}\nabla^{2}v_{i,j}}{\mathrm{Re}\,r_{j+1/2}\delta r} - \frac{r_{j+1}B_{i,j+1} - r_{j}B_{i,j}}{r_{j+1/2}\delta r}\right)^{n} \quad (18)$$

The previously defined differential operators are to be modified in the proximity of the domain boundary, because of the staggered grids, if external points are to be avoided. In this case first and second derivatives are discretized by backward three-point formulas.

The solution time advancement during the false transient from the generic initial time step $n d\tau$ to the next time step $(n+1)d\tau$ is accomplished with the following computational sequence

- Poisson pressure equation (18) is solved iteratively by employing the well known alternating-direction implicit (ADI) method [11]. The differential equation to be solved with ADI method is riarranged to yield two discretized differential equations, each of the two containing unknowns only along one direction. Each of these equations is then solved by making use of the tridiagonal matrix algorithm [10]. The computational domain is swept in two directions alternately untill the following convergence criterium, computed on the pressure field at the iteration (k+1), is satisfied

$$\max_{i,j} \frac{\left|p_{i,j}^{k+1} - p_{i,j}^{k}\right|}{p_{i,j}^{k}} \le \varepsilon_p \tag{19}$$

A typically employed value is $\varepsilon_p = 10^{-3}$

- velocity components fields are computed at the advanced time level $(n+1)d\tau$ solving momentum equations (15–16)
- energy equation (17) is solved
- a convergence test is performed for the four computed variables

$$\max_{\chi}\left(\max_{i,j} \frac{\left|\chi_{i,j}^{n+1} - \chi_{i,j}^{n}\right|}{\left|\chi_{i,j}^{n}\right|}\right) \le \varepsilon \qquad \chi \in (u, v, p, \theta) \tag{20}$$

where ε is assumed equal in value to ε_p.

The correct choice of time step $d\tau$ is dictated by solution numerical stability reasons and a rough estimate is given by the usual limitation based on Courant number [9-11].

3.1 Boundary conditions for pressure

Boundary conditions for Poisson pressure equation are not defined by the problem itself, so that they are to be determined from momentum equations written on the boundaries and then discretized by writing a Taylor series in the velocity component at a wall adjacent grid point (w-1) situated at a mesh size distance from a wall grid point w

- on the cylindrical wall $$\frac{\partial p}{\partial r} = Re^{-1} \frac{2 v_{i,w-1}}{\delta r^2} \tag{21}$$

- on the vertical wall $$\frac{\partial p}{\partial z} = Re^{-1} \frac{2 u_{w-1,j}}{\delta z^2} \tag{22}$$

- on the simmetry axis $$\frac{\partial p}{\partial r} = 0 \tag{23}$$

- at the inlet section $$\frac{\partial p}{\partial z} = Re^{-1} \nabla^2 u_{w,j} \tag{24}$$

- at the outlet section $$p = cost \tag{25}$$

3.2 Convective fluxes and upwinding

The evaluation of convective fluxes by employing a second-order accuracy central difference scheme can lead to numerical instability, depending on Reynolds number. To avoid such inconvenience it is usual to adopt an upwind treatment of the convective terms. This technique , having a first-order accuracy, has the inconvenience of introducing the so called error of false diffusion. In this work the hybrid upwinding technique has been adopted in order to obtain a good trade-off between solution accuracy and numerical stability [12]. Variables ψ^z, ψ^r, v^z, v^r are in the discretized operators of momentum and energy convective fluxes (12-13) and the following scheme is employed for axial hybrid upwinding of axial velocity

$$\text{if } \frac{u_{i,j}\delta z}{\nu} \begin{cases} < -2 & \psi^z_{i,j} = u_{i+1/2,j} \\ > -2 \text{ and } > 2 & \psi^z_{i,j} = u_{i,j} \\ > 2 & \psi^z_{i,j} = u_{i-1/2,j} \end{cases} \tag{26}$$

The tratment is similar for the radial direction. Also for radial velocity and temperature, for which mesh Reynolds number is substituted by Peclet mesh number, a similar procedure is followed.

4. COMPUTATIONAL ASPECTS

Results were obtained with contraction ratios in the range $0.40 \leq m \leq 0.75$ and Reynolds number in the range $2 \leq Re \leq 500$. Prandtl number was assumed equal to 0.7. Most of the results were obtained with dimensionless upstream and downstream lenght $\beta=\gamma=1$. Different uniform meshes were used, though all the results presented here were obtained with a uniform square grid with mesh size equal to D/40 and 1200 computational cells. This grid density has proven to give accurate results and employing not too large computational resources.

Upstream and downstream contraction ducts are to be taken long enough to make the results indipendent of the computational domain size. Since the boundary conditions are imposed at the inlet and outlet sections, this could alter flow and temperature fields obtained in the computations. For this reason the effect of increasing upstream and downstream duct lenght on the results was investigated. For Re=500, m=0.5, increasing upstream lenght ($\beta=1.5, \gamma=1$) produced very small variations on velocity, pressure and temperature fields. A more remarkable effect was produced by increasing downstream lenght ($\beta=1, \gamma=1.5$), particularly near the outlet section. In fact the numerical perturbation didn't propagate upstream and didn't reach the contraction section. As a conclusion dimensionless duct lenghts $\beta=\gamma=1$ have been shown to be sufficient to make the results indipendent of the duct lenght.

5. RESULTS AND DISCUSSION

5.1 Flow field

A complete and general overview of the flow pattern is given by vector velocity field, which, for Re=100 and m=0.5, is displayed in Fig.3. From this figure the flow pattern shows a gradual deviation towards the duct axis approaching the contraction section, with formation, just downstream of it, of an almost uniform axial velocity profile. Subsequently the flow redevelops towards an asymptotic fully developed parabolic profile. This flow pattern is clearly correlated to the pressure field displayed in Fig.4, where dimensionless iso-pressure lines are spaced by 2.03, starting with the outlet section which corrisponds to 0. The dimensionless constant pressure lines converge towards the convex corner , where pressure gradients are very large due to streamlines curvature. Pressure field tends to reach the constant axial gradient, corrisponding to a constant cross section, at a distance of about half diameter from the contraction section.

Downstream of the contraction flow development begins with the rapidity being greater at lower Reynolds numbers, as shown in Fig. 5, for two different contraction ratios. Outlet velocity profiles can be compared with maximum axial velocity corrisponding to a fully developed flow. At high Reynolds number, greater than 100 for a contraction ratio m=0.4, the axial velocity profile reaches its maximum close to the wall at a section just downstream of the contraction section. This trend is much weaker as the contraction ratio increases, as shown in Fig. 5. This upward migration of velocity maximum is explained by momentum conservation, since the fluid layers adjacent to the wall are stopped, due to non-slip boundary conditions, so that momentum is transferred to adjacent more internal layers. These results are in good agreement with numerical and experimental results of Durst and Loy [6].

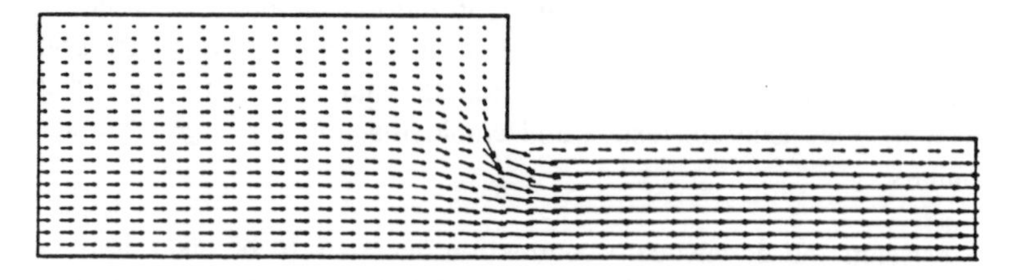

Fig. 3 Vector velocity field (Re=100, m=0.5)

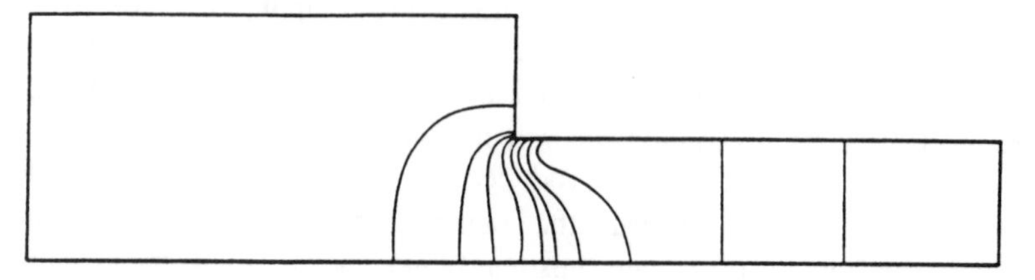

Fig. 4 Pressure field (Re=100, m=0.5)

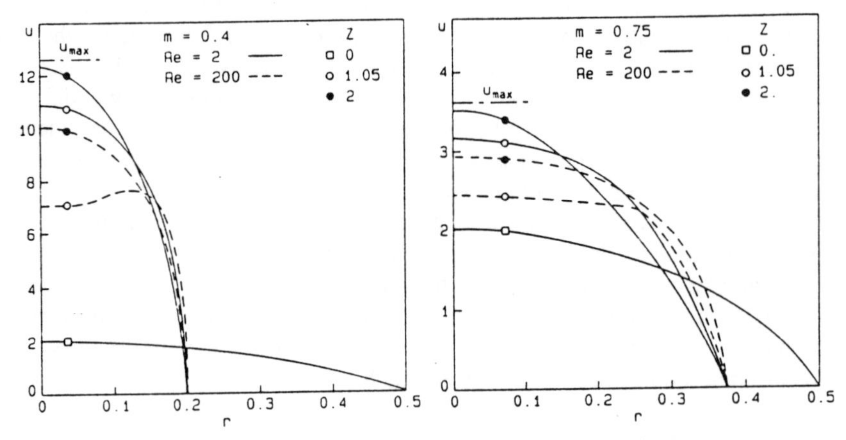

Fig. 5 Axial velocity distributions for m=0.4 and m=0.75

At the flow domain concave corner, just upstream of the contraction section, a small recirculation region is formed, due to flow separation. A complete study about its dimension was carried out by Dennis and Smith [3], Durst and Loy [6]. Their results togheter with those of the present study are plotted in Fig. 6, that shows a good agreement in the determination of the minimum size at Re=100 for m=0.5. Some discrepancies exist about the dimension of the recirculation region, wich is difficult to evaluate depending on accuracy, grid size and discrete velocity interpolation. It appears from Fig. 6 that this minimum is displaced towards lower Reynolds numbers for contraction ratios smaller than 0.5, and towards higher Reynolds number for contraction ratios larger than 0.5. Flow separation region downstream of the contraction section is more difficult to detect. Only for Re=500 this separation region was revealed, but a more refined grid would be needed to be able to define its size with accuracy.

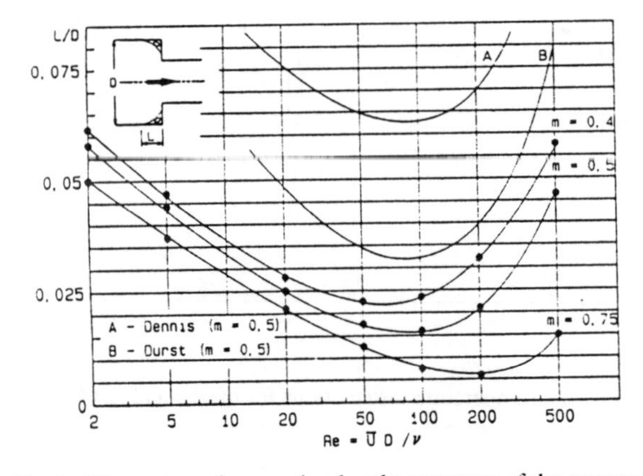

Fig.6 Dimension of separation lenght upstream of the contraction

5.2 Thermal field

A general overwiew of temperature field is represented in Fig.7, where isotherms start at the boundary imposed value of 1.0 followed by the first isotherm with a value of .95 and the others spaced by 0.1. In Fig. 8 radial temperature profiles are plotted at various cross sections for two different contraction ratios. A general overview of this figure shows the thermal boundary layer development after the inlet section due to uniform temperature inlet condition for the fluid. After the contraction section the thermal boundary layer redevelops untill the end of the duct. The rapidity of this development depends, at constant Prandtl number, as in the case of flow field examined before, on the Reynolds numbers, with the rapidity being greater at lower Reynolds numbers. Also contraction ratio plays a role in the thermal boundary layer development in the downstream duct increasing its rapidity at lower contraction ratios, as shown in Fig. 8.

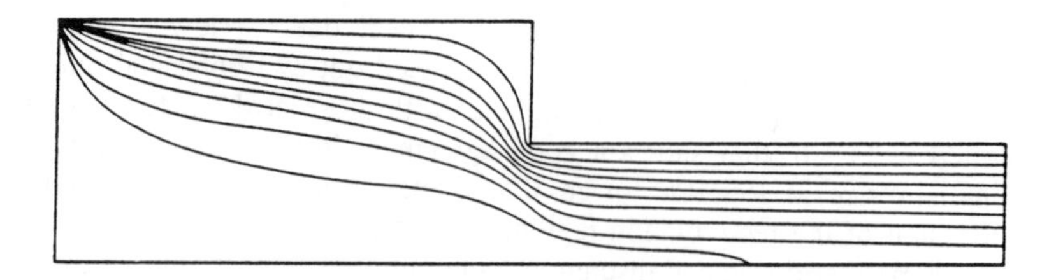

Fig. 7 Temperature field (Re=100, Pr=0.7, m=0.5)

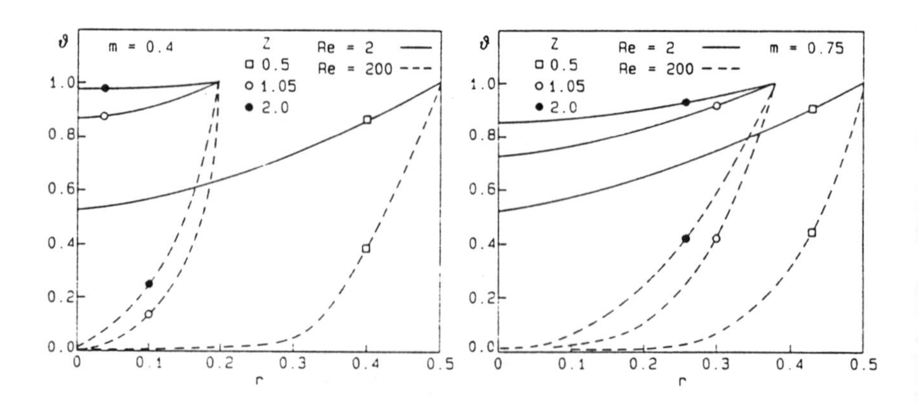

Fig. 8 Temperature distributions for m=0.4 and m= 0.75

From temperature field local Nusselt number profile along the duct was evaluated. From heat transfer coefficient definition, with reference to the contraction upstream duct

$$h(T_w - T_b) = k\left.\frac{\partial T}{\partial R}\right|_{D/2} \tag{27}$$

where T_b is the local bulk temperature. The dimensionless bulk temperature is obtained by integrating temperature and axial velocity distributions across the section, where axial velocity can be positive or negative in the recirculation region. Finally the local Nusselt number is computed

$$\theta_b = \frac{T_b - T_i}{T_w - T_i} = \int_0^{1/2} \theta u r dr \qquad Nu = \frac{hD}{k} = \left.\frac{\partial \theta}{\partial r}\right|_{1/2} \frac{1}{(1-\theta_b)} \tag{28}$$

Axial distributions of local Nusselt number and bulk temperature are plotted in Fig. 9 for two different Reynolds numbers (Re=2,200) and three contraction ratios (m=0.4,0.6,0.75). Upstream and far away from the contraction Nusselt number distribution shows a typical decrease due to thermal profile development in a constant circular cross section duct. Close to the contraction section flow and thermal fields are so distorted that very low values of Nusselt number are observed followed by a sharp increase caused by a steep radial thermal gradient near the contraction section. Then, as velocity and temperature profiles redevelop along the duct, Nusselt number decreases asimptotically along the duct to the limit foreseen by theory of 3.66. Different Reynolds numbers yield different bulk temperature profiles, as the flow and temperature developments are different according to previous considerations. As a consequence, Nusselt number values are larger and bulk temperatures are smaller as the Reynolds number increases, as shown in Fig. 9. Also contraction ratio plays a role, since the lower the contraction ratio is, the larger is the Nusselt number sharp increase at the contraction section.

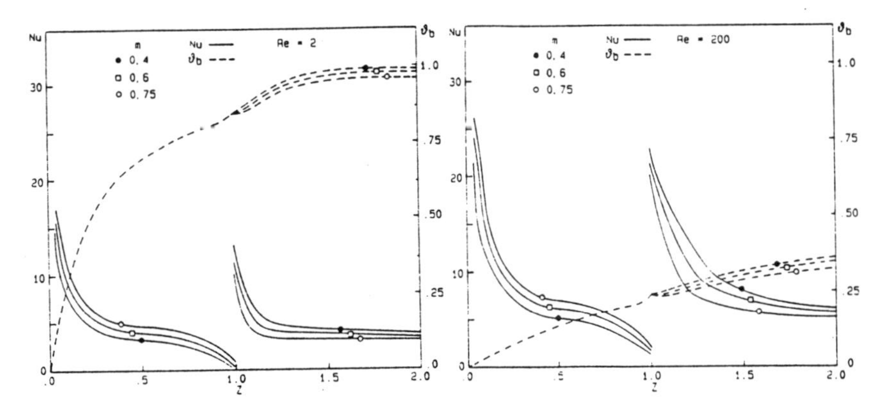

Fig. 9 Local Nusselt number and bulk temperature distributions

REFERENCES

1. GREENSPAN, D. - Numerical Studies of Steady, Viscous, Incompressible Flow in a Channel with a Step. Journal of Engineering Mathematics, Vol. 3, No. 1, 1969.
2. GREPPI, M., CERCIGNANI, C., - Computer Simulation of a Viscous Channel Flow. Meccanica, No. 9, 1971.
3. DENNIS, S., SMITH, F. - Steady Flow Through a Channel with a Symmetrical Constriction in the Form of a Step, Proceedings R. Soc. London, A372, 1980.
4. ZAMPAGLIONE, D., GREPPI, M. - Numerical Study of a Viscous Flow through a Pipe Orifice. Meccanica, No. 9,1972.
5. DOBROWOLSKY, B., KABZA, Z. - Numerical Analysis of a Laminar Viscous Flow through a Pipe Orifice, Studia Geotechnica et Mechanica, Vol IV, No. 1-2, 1982.
6. DURST,F., LOY,T. - Investigations of Laminar Flow in a Pipe with Sudden Contraction of Cross Sectional Area. Computers & Fluids, Vol. 13, No. 1, pp. 15-36, 1985.
7. SPARROW, E., CHUCK, W. - PC Solutions for Heat Transfer and Fluid Flow Downstream of an Abrupt, Asymmetric Enlargement in a Channel. Numerical Heat Transfer, Vol. 12, 1987.
8. HARLOW, F., WELCH, J. - Numerical Calculation of Time-dipendent Viscous Incompressible Flow of Fluid with Free Surface. Phis. Fluids, No.8, 1965.
9. HARLOW, F., AMSDEN, A. - A Numerical Fluid Dynamics Calculation Method for all Flow Speed. Jou. of Computational Physics, No. 8, 1971.
10. ROACHE, P. - Computational Fluid Dynamics, Hermosa Ed., Albuquerque, New Mexico, 1972.
11. PEYRET, R., TAYLOR, T. - Computational Methods for Fluid Flow, Springer Verlag, New York, 1982.
12. SPALDING, D. - Basic Equations of Fluid Mechanics and Heat and mass Transfer and Procedures for their Solution, Imperial College, Mech. Eng. Dept., London, 1980.

NOMENCLATURE

D	upstream diameter
h	heat transfer coefficient
k	thermal conductivity
m	contraction ratio
Nu	local Nusselt number
P,p	pressure, dimensionless pressure
Pr	Prandtl number $[= \nu / \alpha]$
R,r	radial coordinate,dimensionless radial coordinate
Re	Reynolds number $[= \overline{U}D / \nu]$
t,τ	time,dimensionless time
T,θ	temperature,dimensionless temperature
U,u	axial velocity, dimensionless axial velocity
$\overline{U}$	mean axial velocity at the inlet section
V,v	radial velocity, dimensionless radial velocity
Z,z	axial coordinate, dimensionless axial coordinate
α	thermal diffusion coefficient
β	upstream duct dimensionless lenght
γ	downstream duct dimensionless lenght
ε	convergence tolerance
ν	kinematic viscosity
ρ	density

LAMINAR MIXED CONVECTION IN HORIZONTAL FLAT DUCTS WITH HEAT SOURCES ON THE BOTTOM WALL

C. Nonino† and S. Del Giudice‡

SUMMARY

Laminar mixed convection in horizontal flat ducts heated by arrays of square heat sources on the bottom wall is analized. A finite element procedure based on a strategy similar to the Patankar-Spalding's SIMPLE algorithm is employed for the numerical simulations. Two typical arrangements of the heat sources are considered for different values of the Prandtl, Grashof and Reynolds numbers.

1. INTRODUCTION

In laminar internal flows, buoyancy effects can significantly enhance heat transfer with respect to pure forced convection. Although the topic of laminar mixed convection has been frequently addressed in the literature, only little has been done to consider the effects of discontinuous heating from discrete sources. This problem is of significant interest because of its practical applications in the field of cooling of electronic equipment [1]. In fact, in many cases the geometry of a multichip electronic module can be modelled as an array of discrete heat sources mounted to one wall of a duct.

This paper concerns the laminar mixed convection in horizontal flat ducts heated by arrays of square heat sources located on the bottom wall. A finite-element procedure, based on the Boussinesq approximation for taking into account the buoyancy effects [2], is employed in the analysis. Velocity and temperature distributions over the cross section of the duct are computed employing a step-by-step procedure in which momentum and energy equations are solved in sequence, and the transverse velocity field is estimated according to a strategy similar to the Patankar-Spalding's SIMPLE algorithm [2,3].

Numerical results, concerning two typical arrangements of the heat sources, are obtained for different values of the Prandtl, Grashof and Reynolds numbers. Calculated Nusselt number distributions over the heat sources are also reported and discussed.

† Associate Professor, ‡ Professor
Istituto di Fisica Tecnica e di Tecnologie Industriali, Università di Udine, Viale Ungheria 41, 33100 Udine, ITALY

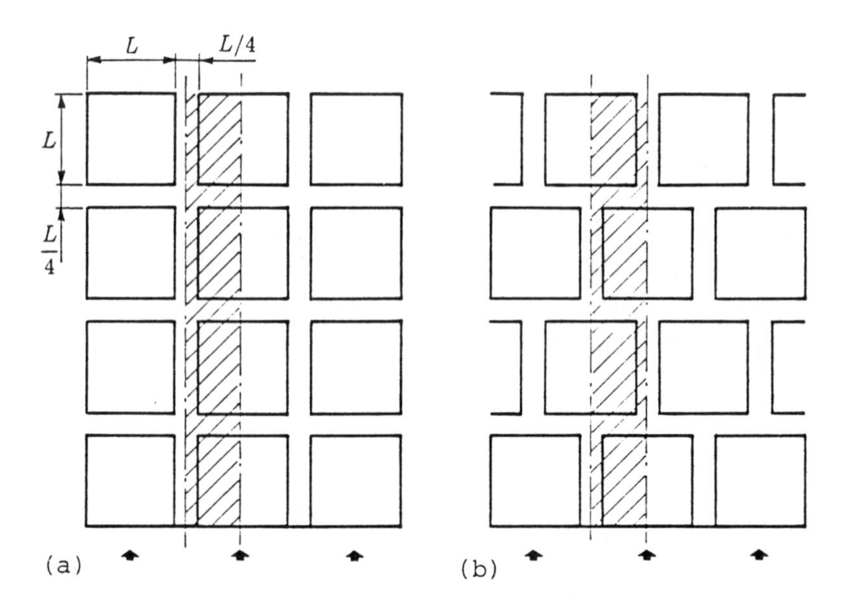

Fig. 1 - Heat source layouts on the bottom wall of flat ducts: (a) aligned and (b) staggered heaters.

2. MATHEMATICAL MODEL AND SOLUTION STRATEGY

The systems investigated are horizontal flat ducts with adiabatic walls and with arrays of discrete heat sources flush-mounted to the bottom wall. These systems are representative of multichip electronic modules with different layouts. Two arrangements are considered in this paper, namely, four rows of aligned and staggered square heat sources. As shown in Fig. 1, the side length $<L>$ of the square sources is equal to the duct height, and the sources are $<L>/4$ apart from each other. All heat sources produce the same constant and uniform heat flux $<q_s''>$. A uniform temperature and a fully developed velocity profile are assumed upstream of the first row of heat sources. Due to the existing symmetries, the analysis can be limited to the parts of the ducts corresponding to dashed areas in Fig. 1. Two fluids are considered as coolants: air, with a Prandtl number $Pr = 0.7$, and a dielectric liquid with Pr=7 [1]. For each fluid, different flow rates through the duct and different heat fluxes from the discrete sources can be modelled by varying the Reynolds number and the Rayleigh number, respectively.

The main flow is in the positive axial direction x, while y and z are the horizontal and vertical cross-sectional coordinates, respectively. Since the Reynolds numbers considered in this study are always larger than 100, axial diffusion can be neglected [2-4]. In the absence of recirculation in the axial direction, the coordinate x becomes a "one-way" coordinate in the upstream to downstream march. A flow of this kind is called "parabolic" [4,5], and can be solved in the x direction by means of a marching procedure, which consists in the step-by-step solution of parabolized equations over the two-dimensional domain represented by the cross-section of the duct.

With the above assumptions, the steady laminar mixed convection in the horizontal duct considered is governed by the continuity equation

$$D = \frac{\partial u}{\partial x} + \frac{\partial v}{\partial y} + \frac{\partial w}{\partial z} = 0 \tag{1}$$

by the momentum equations, written here in the following conservative form in which the divergence D of the velocity field is retained

$$u\,\frac{\partial u}{\partial x} = \frac{1}{Re}\left(\frac{\partial^2 u}{\partial y^2} + \frac{\partial^2 u}{\partial z^2}\right) - \left(v\,\frac{\partial u}{\partial y} + w\,\frac{\partial u}{\partial z} + D\,u\right) - \frac{d\overline{p}}{dx} \tag{2}$$

$$u\,\frac{\partial v}{\partial x} = \frac{1}{Re}\left(\frac{\partial^2 v}{\partial y^2} + \frac{\partial^2 v}{\partial z^2}\right) - \left(v\,\frac{\partial v}{\partial y} + w\,\frac{\partial v}{\partial z} + D\,v\right) - \frac{\partial p}{\partial y} \tag{3}$$

$$\begin{aligned} u\,\frac{\partial w}{\partial x} = {} & \frac{1}{Re}\left(\frac{\partial^2 w}{\partial y^2} + \frac{\partial^2 w}{\partial z^2}\right) - \left(v\,\frac{\partial w}{\partial y} + w\,\frac{\partial w}{\partial z} + D\,w\right) - \frac{\partial p}{\partial z} \\ & + \frac{Gr}{Re^2}\,(T - T_r) \end{aligned} \tag{4}$$

and by the conservative energy equation

$$u\,\frac{\partial T}{\partial x} = \frac{1}{Pe}\left(\frac{\partial^2 T}{\partial y^2} + \frac{\partial^2 T}{\partial z^2}\right) - \left(v\,\frac{\partial T}{\partial y} + w\,\frac{\partial T}{\partial z} + D\,T\right) \tag{5}$$

In Eqs. (1) to (5), $u = <u> / <\overline{u}>$, $v = <v> / <\overline{u}>$ and $w = <w> / <\overline{u}>$ are dimensionless velocity components, $x = <x> / <L>$, $y = <y> / <L>$ and $z = <z> / <L>$ are dimensionless coordinates, $p = <p> / (<\varrho><\overline{u}>^2)$ is the dimensionless pressure, $T = <k>\,(<T> - <T_e>)/(<q_s''><L>)$ is the dimensionless temperature, $\overline{p}$ is the average value over the cross section of the dimensionless pressure p, T_r is a dimensionless reference temperature to account for local density variations, which can be assumed as the mean value of T over the cross section [1]. The Reynolds and the Grashof numbers are defined as $Re = <\varrho><\overline{u}><L> / <\mu>$ and $Gr = <g><\beta><q_s''><\varrho>^2<L>^4 / (<k><\mu>^2)$, respectively, while $Pe = Re\,Pr$ is the Péclet number, being Pr the Prandtl number $Pr = <c_p><\mu> / <k>$. In the above expressions, $<\overline{u}>$ is the mean axial velocity, $<\varrho>$ is the density, $<k>$ is the thermal conductivity, $<\mu>$ is the viscosity, $<g>$ is the gravitatational acceleration, $<\beta>$ is the thermal expansion coefficient, and $<c_p>$ is the specific heat at constant pressure.

In the cases studied here, the computational domains, represented by the cross-sections of the ducts, are surrounded by upper and lower rigid walls and by lateral symmetry boundaries. At rigid walls, no-slip boundary conditions $u = w = w = 0$ apply to the momentum equations, while the second kind (or prescribed flux) boundary condition is appropriate for the energy equation. In particular the condition

$$\frac{\partial T}{\partial z} = -1 \tag{6}$$

applies on the parts of the lower wall corresponding to the heat sources, and the condition

$$\frac{\partial T}{\partial z} = 0 \tag{7}$$

applies on the unheated parts of the rigid walls. Therefore, in the marching procedure, conditions (6) and (7) are alternatively imposed to the same part of the lower boundary of the cross-section to accout for finite size of the discrete heat sources in the axial direction x. On vertical boundaries, instead, the symmetry boundary conditions

$$\frac{\partial u}{\partial y} = v = \frac{\partial w}{\partial y} = \frac{\partial T}{\partial y} = 0 \tag{8}$$

are appropriate.

Finally, inflow conditions, both for velocity and temperature, correspond to the initial conditions for the numerical simulation. As already stated, a fully developed velocity distribution and a uniform temperature $T_e = 0$ are assumed at the entrance. In addition, a pressure field accomodating the inlet velocity field must be assumed. In this case, a uniform pressure field is the appropriate initial choice.

In the numerical procedure, momentum and energy equations are solved in sequence. In three-dimensional parabolic flows, two distinct velocity-pressure couplings result: one in the main parabolic (axial) direction and the other in the cross-flow plane [2-4]. The first one is used to determine the value of $d\bar{p}/dx$ that gives the correct mass flow in the $x-$direction, while the second one corrects cross-stream velocities to enforce local continuity, according to Eq. (1) [2,3]. Equation (2) is first solved with the term $d\bar{p}/dx$ evaluated from a global momentum balance [2,3]. Then, on the assumption that the shape of the velocity profile remains unchanged, a correction is applied to the estimated axial velocity field which, in general, does not satisfy the integral mass flow constraint. In order to compute the cross-stream pressures and velocities, a SIMPLE-like strategy is used [2-6]. The important operations, in order of execution, are:

1. Estimate the transverse pressure field and compute the buoyancy terms from the previous axial step.
2. Using the estimated pressure, solve the momentum equations to obtain approximate cross-flow velocity components.
3. Calculate the pressure corrections that enforce continuity.
4. Calculate the corresponding velocity corrections.
5. Treat the corrected pressure as the estimated pressure for the new axial step.

Finally the temperature distribution on the cross section can be computed through the solution of the energy equation (5).

Once the temperature distribution has been computed, the spanwise averaged Nusselt number can be calculated as

$$Nu = \frac{\langle q_s'' \rangle \langle L \rangle}{\langle k \rangle \left(\langle \overline{T}_s \rangle - \langle T_b \rangle \right)} = \frac{1}{\overline{T}_s - T_b} \tag{9}$$

where $\overline{T}_s$ is the spanwise average dimensionless temperature over the heated perimeter of the cross-section and T_b is the dimensionless bulk temperature.

The procedure outlined above has been implemented with reference to the finite element method. A Galerkin approach is used to obtain space-discretized equations which are solved in the x-direction by means of a marching technique based on finite difference approximations. A detailed description of the solution strategy and of the finite element formulation is reported in Reference 2.

3. NUMERICAL RESULTS

The finite element procedure described in the previous section has been validated elsewhere with reference to mixed convection in the entrance region of horizontal rectangular ducts [2]. Results of further computations [7], concerning mixed convection heat transfer from an array of discrete heat sources in horizontal rectangular ducts, compare successfully with those obtained by Mahaney, *et al.* [1].

In this study, the laminar mixed convection in flat ducts heated from below by two arrays of discrete sources, schematically represented in Fig. 1, is analyzed with reference to the same range of dimensionless parameters considered by Mahaney, *et al.* [1]. As already stated, two values of the Prandtl number, $Pr = 0.7$ and $Pr = 7$, are assumed. For both fluids, the same values of the Grashof number are chosen, corresponding to Rayleigh numbers $Ra = Gr\,Pr$ equal to 1×10^6, 2.5×10^6, and 5×10^6 for $Pr = 0.7$, and to 1×10^7, 2.5×10^7, and 5×10^7 for $Pr = 7$. In addition, the corresponding cases of pure forced convection are also considered for comparison. The selected values of the Reynolds number Re_D based on the hydraulic diameter ($Re_D = 2\,Re$) are equal to 500, 625, 750, 1000, 1250, 1750, and 2500.

The computational domain, reduced to a rectangle of height L and width $W = L/2 + L/8 = 0.625\,L$ because of the existing vertical synmmetries, is discretized using a nonuniform mesh consisting of 140 parabolic elements and 469 nodes. Since buoyancy-induced secondary flows necessitate the same resolution across the entire cross section, a uniform mesh is employed over most part of the domain. However, in order to improve the accuracy of the solution, smaller elements are used in the region near the rigid walls, where steeper gradients of the variables are expected. This mesh was chosen since several tests performed for selected values of Re_D and Ra with a finer mesh (280 parabolic elements and 1777 nodes) did not show significant differences in the computed velocity and temperature fields. Over each heater, the longitudinal step size is allowed to linearly increase with increasing axial coordinate x, from a minimum value $\Delta x = 0.005$ to a maximum value $\Delta x = 0.02$, reached in 80 axial steps. In the nonheated zones beyond each heater, the step increases with x at the same rate, starting again from the value $\Delta x = 0.005$.

Numerical simulations confirm that, as already found by Mahaney, *et al.* with the same values of the dimensionless parameters, velocity and temperature fields are strongly affected by the buoyancy forces originating from bottom heating. In fact, a secondary flow develops which forms multiple recirculation cells increasing in size in the direction of the main flow. Of course, for a given set of dimensionless parameters, different recirculation patterns and different axial velocity distributions are found, depending on the heat source layout considered. As an example, two vector plots of the transverse velocity field are reported in Fig. 2 together with the corresponding axial velocity contours. All plots refer to the case $Pr = 7$, $Re_D = 500$, and $Ra = 2.5 \times 10^7$, and show velocity distributions at the end of the fourth row of heat sources ($x = 4.75$).

The influence of the Rayleigh number on the Nusselt number is shown in Figs. 3 to 6 with reference to $Re_D = 625$. As a general remark, it must be pointed out that the values of the Nusselt number obtained with $Pr = 7$ (Figs. 3 and 4) are much higher than those obtained, for the same Grashof number ($Gr = Ra/Pr$), with $Pr = 0.7$ (Figs. 5 and 6). In fact, at the same axial location, the dimensionless axial coordinate $x^+ = x/(D_h\,Re_D\,Pr)$ for $Pr = 7$ is one-tenth of the corresponding x^+ for $Pr = 0.7$. The heat transfer enhancement due to the buoyancy-induced secondary flow is evident from

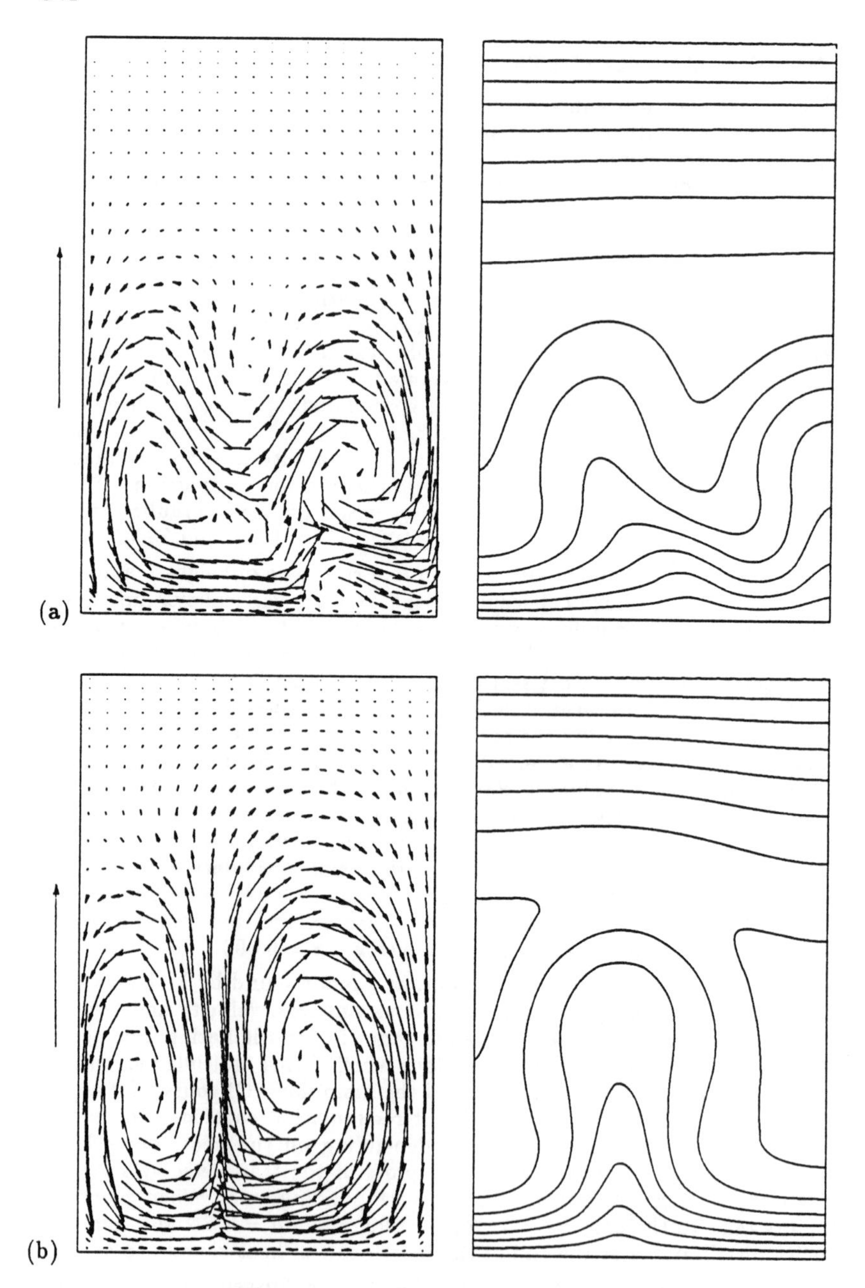

Fig. 2 - Transverse velocity vectors and axial velocity contours at the end of the fourth row of heat sources, $x = 4.75$, for $Pr = 7$, $Re_D = 500$, and $Ra = 2.5 \times 10^7$: (a) aligned and (b) staggered heaters. The length of the arrows on the left of the vector plots correspond to a dimensionless velocity equal to 0.5, while dimensionless axial velocity contours range from 0.2 to 1.4, with a step equal to 0.2.

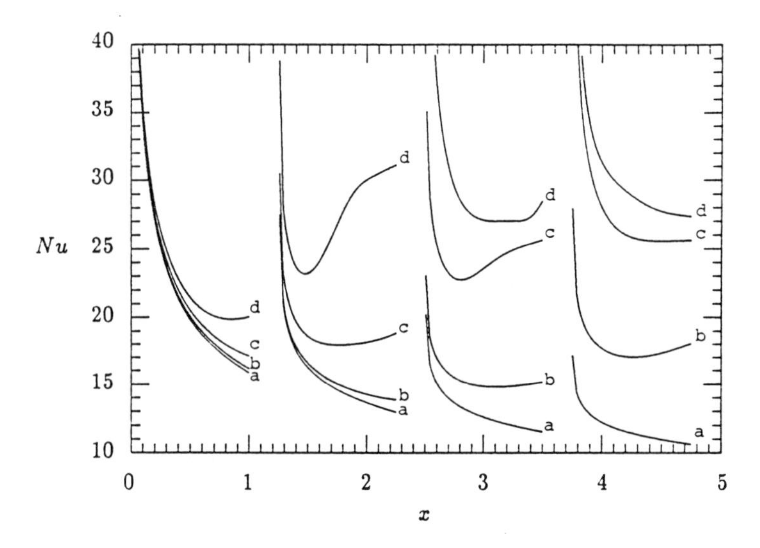

Fig. 3 - Longitudinal distributions of the spanwise average Nusselt number Nu over an array of aligned square heaters for $Pr = 7$, $Re_D = 625$, and different values of the Rayleigh number Ra: (a) pure forced convection ($Ra = 0$), (b) $Ra = 1 \times 10^7$, (c) $Ra = 2.5 \times 10^7$, (d) $Ra = 5 \times 10^7$.

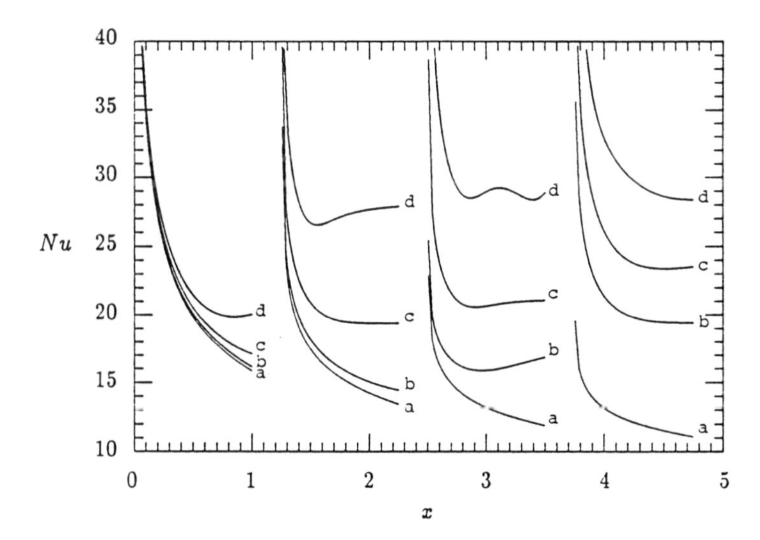

Fig. 4 - Longitudinal distributions of the spanwise average Nusselt number Nu over an array of staggered square heaters for $Pr = 7$, $Re_D = 625$, and different values of the Rayleigh number Ra: (a) pure forced convection ($Ra = 0$), (b) $Ra = 1 \times 10^7$, (c) $Ra = 2.5 \times 10^7$, (d) $Ra = 5 \times 10^7$.

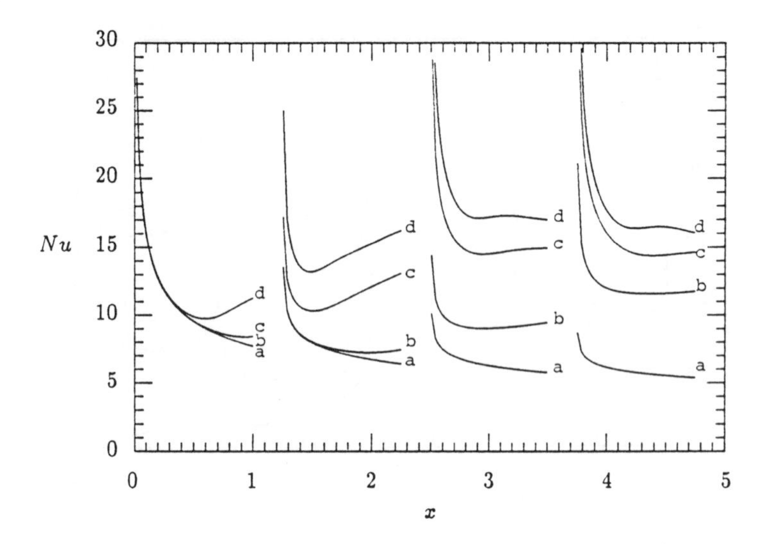

Fig. 5 - Longitudinal distributions of the spanwise average Nusselt number Nu over an array of aligned square heaters for $Pr = 0.7$, $Re_D = 625$, and different values of the Rayleigh number Ra: (a) pure forced convection ($Ra = 0$), (b) $Ra = 1 \times 10^6$, (c) $Ra = 2.5 \times 10^6$, (d) $Ra = 5 \times 10^6$.

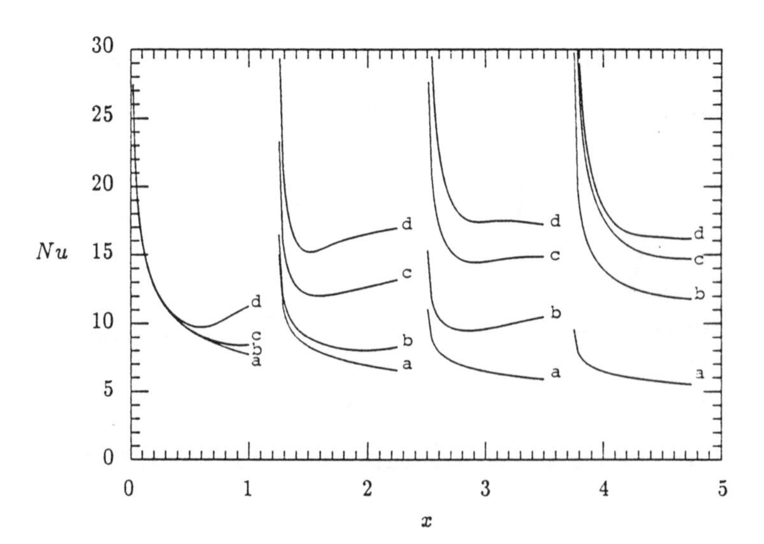

Fig. 6 - Longitudinal distributions of the spanwise average Nusselt number Nu over an array of staggered square heaters for $Pr = 0.7$, $Re_D = 625$, and different values of the Rayleigh number Ra: (a) pure forced convection ($Ra = 0$), (b) $Ra = 1 \times 10^6$, (c) $Ra = 2.5 \times 10^6$, (d) $Ra = 5 \times 10^6$.

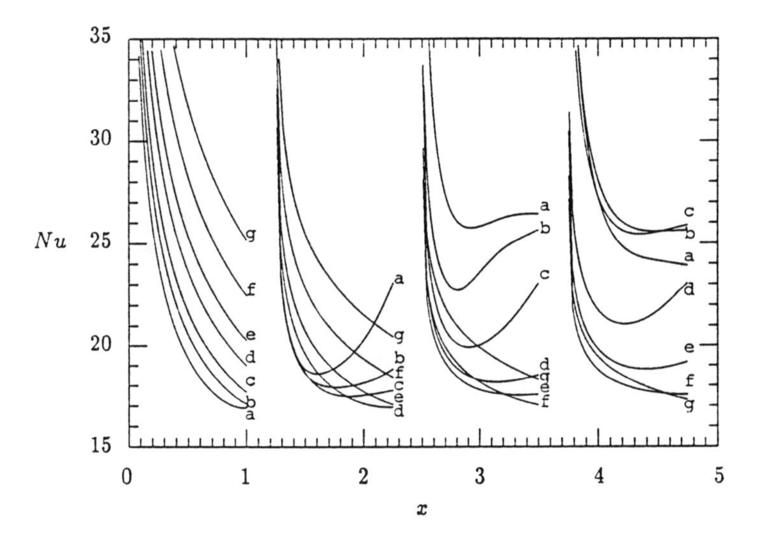

Fig. 7 - Longitudinal distributions of the spanwise average Nusselt number Nu over an array of aligned square heaters for $Pr = 7$, $Ra = 2.5 \times 10^7$, and different values of the Reynolds number Re_D: (a) $Re_D = 500$, (b) $Re_D = 625$, (c) $Re_D = 750$, (d) $Re_D = 1000$, (e) $Re_D = 1250$, (f) $Re_D = 1750$, (g) $Re_D = 2500$.

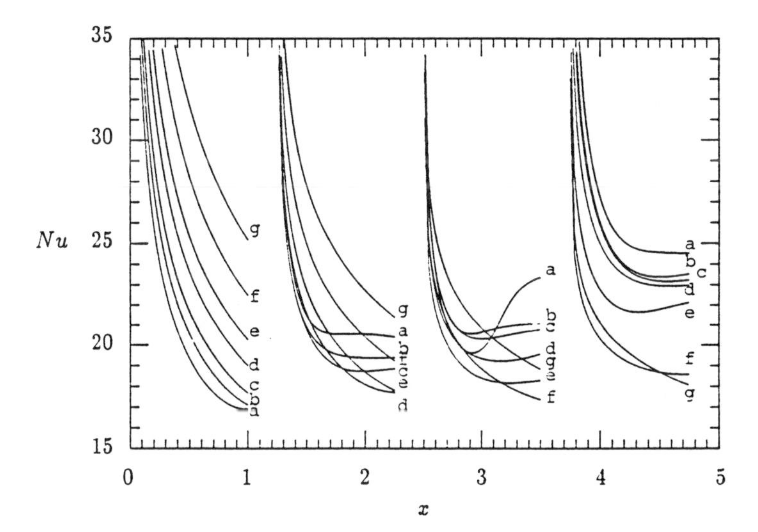

Fig. 8 - Longitudinal distributions of the spanwise average Nusselt number Nu over an array of staggered square heaters for $Pr = 7$, $Ra = 2.5 \times 10^7$, and different values of the Reynolds number Re_D: (a) $Re_D = 500$, (b) $Re_D = 625$, (c) $Re_D = 750$, (d) $Re_D = 1000$, (e) $Re_D = 1250$, (f) $Re_D = 1750$, (g) $Re_D = 2500$.

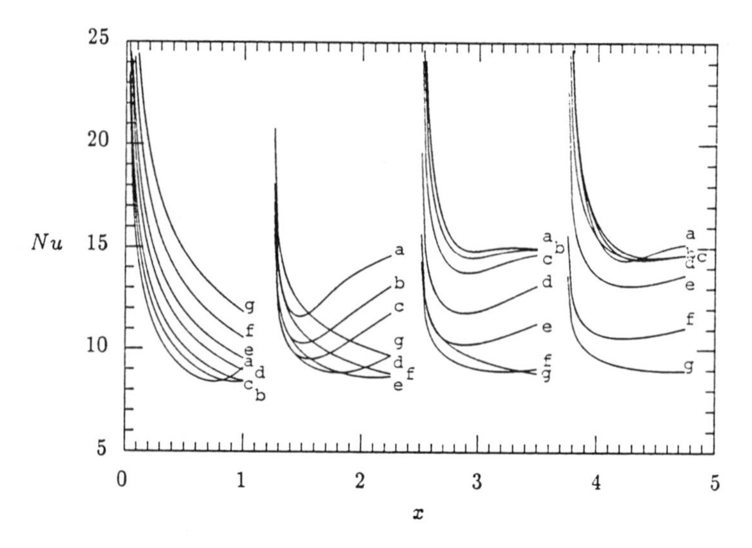

Fig. 9 - Longitudinal distributions of the spanwise average Nusselt number Nu over an array of aligned square heaters for $Pr = 0.7$, $Ra = 2.5 \times 10^6$, and different values of the Reynolds number Re_D: (a) $Re_D = 500$, (b) $Re_D = 625$, (c) $Re_D = 750$, (d) $Re_D = 1000$, (e) $Re_D = 1250$, (f) $Re_D = 1750$, (g) $Re_D = 2500$.

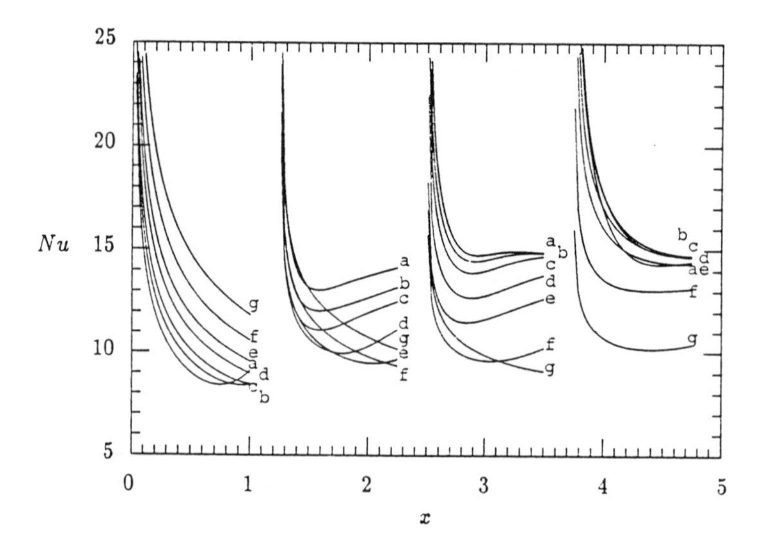

Fig. 10 - Longitudinal distributions of the spanwise average Nusselt number Nu over an array of staggered square heaters for $Pr = 0.7$, $Ra = 2.5 \times 10^6$, and different values of the Reynolds number Re_D: (a) $Re_D = 500$, (b) $Re_D = 625$, (c) $Re_D = 750$, (d) $Re_D = 1000$, (e) $Re_D = 1250$, (f) $Re_D = 1750$, (g) $Re_D = 2500$.

Table 1 - Average value of the Nusselt number $\overline{Nu}$ over each heater row as a function of Re_D for $Pr = 7$ and $Ra = 2.5 \times 10^7$. Heat source layout: (a) aligned, (s) staggered.

Re_D	1st row	2nd row		3rd row		4th row	
	(a)≡(s)	(a)	(s)	(a)	(s)	(a)	(s)
500	22.7	20.3	21.8	27.7	22.0	27.1	27.3
625	24.1	18.9	20.8	24.6	21.7	28.1	25.8
750	25.4	18.6	20.4	21.4	21.4	27.5	25.5
1000	27.7	18.8	20.5	19.1	20.3	22.3	24.7
1250	29.7	19.7	21.3	18.5	19.3	19.7	22.9
1750	33.2	21.6	23.4	18.8	19.7	18.5	19.9
2500	37.5	24.1	26.0	20.6	21.7	18.9	20.3

all figures, leading to Nusselt numbers two to three times larger than those obtained for pure forced convection. Even though the average values of the Nusselt number over each heater do not significantly change when different heat source layouts are considered, Nusselt number profiles are different, especially for the fluid with the larger Prandtl number.

The same considerations apply to Figs. 7 to 10, which show the influence of the Reynolds number on the Nusselt number. Results for $Pr = 7$ refer to $Ra = 2.5 \times 10^7$ and those for $Pr = 0.7$ to $Ra = 2.5 \times 10^6$, corresponding to the same Grashof number ($Gr = Ra/Pr$). On the first row of heaters, where forced convection is predominant, the Nusselt number always increases with increasing Reynolds numbers. Instead a more complicated behaviour is observed on the second and third rows, where the effects of the secondary flow become stronger. Finally, higher Nusselt numbers for decreasing Reynolds numbers are generally observed on the fourth row, where natural convection is dominating. A quantitative comparison of heat transfer performaces of the different layouts considered can be carried out on the basis of the average values of the Nusselt number over each heater row

$$\overline{Nu} = \frac{1}{L} \int_L Nu \; dx \tag{10}$$

reported in Tables 1 and 2 for $Pr = 7$ and $Pr = 0.7$, respectively. Of course, on the first row of heaters the values of $\overline{Nu}$ are coincident for both layouts, and increase with increasing Re_D. On the second and the third rows, instead, a minimum value of $\overline{Nu}$ is in general observed for an intermediate value of Re_D, while on the last row $\overline{Nu}$ decreases with increasing Re_D. As a consequence, the rows where the minimum and the maximum values of $\overline{Nu}$ are reached depend on Re_D. In particular, higher values of $\overline{Nu}$ are found on the last row for lower values of Re_D and on the first row for higher values of Re_D. By comparing Table 1 with Table 2, we can observe that the position of the minimum $\overline{Nu}$ moves upstream when Pr is changed fron 7 to 0.7. As a final remark it can be observed that, for a given Re_D, differences between maximum and minimum values of $\overline{Nu}$ are, in general, larger for aligned than for staggered heat sources.

Table 2 - Average value of the Nusselt number $\overline{Nu}$ over each heater row as a function of Re_D for $Pr = 0.7$ and $Ra = 2.5 \times 10^6$. Heat source layout: (a) aligned, (s) staggered.

Re_D	1st row	2nd row		3rd row		4th row	
	(a)≡(s)	(a)	(s)	(a)	(s)	(a)	(s)
500	10.8	13.2	14.1	16.2	15.9	16.1	16.8
625	11.3	11.6	12.9	15.6	15.5	16.1	17.3
750	11.8	10.5	12.0	14.7	14.8	16.1	17.2
1000	12.9	9.5	10.9	12.6	13.5	15.5	16.6
1250	13.8	9.5	10.6	10.8	12.2	13.8	15.8
1750	15.4	10.3	11.2	9.5	10.2	11.0	13.7
2500	17.3	11.4	12.3	9.9	10.3	9.5	10.7

4. CONCLUSIONS

Mixed convection in flat ducts, caused by bottom heating from arrays of discrete square sources, has been investigated with reference to two different layouts, namely, aligned and staggered heaters. Even if both layouts do not determine significantly different behaviours, the duct with staggered heat sources performs slightly better if uniform heat transfer coefficients are required over the whole array of heaters.

Acknowledgements: The support to this research by M.U.R.S.T. and C.N.R. is gratefully acknowledged.

REFERENCES

1. MAHANEY, H.V., RAMADHYANI, S., and INCROPERA, F.P. - Numerical Simulation of Three-Dimensional Mixed Convection Heat Transfer from an Array of Discrete Heat Sources in an Horizontal Rectangular Duct, *Numerical Heat Transfer*, Part A, Vol. 16, 267-286, 1989.
2. NONINO, C. and DEL GIUDICE, S. - Laminar Mixed Convection in the Entrance Region of Horizontal Rectangular Ducts, *Int. j. numer. methods fluids*, Vol. 13, 33-48, 1991.
3. NONINO, C., DEL GIUDICE, S., and COMINI, G. - Laminar Forced Convection in Three-Dimensional Duct Flows', *Numer. Heat Transfer*, Vol. 13, 451-466, 1988.
4. COMINI, G. and DEL GIUDICE, S. - Parabolic Systems: Finite-Element Method, *Handbook of Numerical Heat Transfer*, Ed. W.J. Minkowycz, *et al.*, Chapter 4, 155-181, Wiley, New York, 1988.
5. PATANKAR, S.V. and SPALDING, D.B. - Calculation Procedure for Heat, Mass and Momemtum Transfer in Three-Dimensional Parabolic Flows', *Int. J. Heat Mass Transfer*, Vol. 15, 1787-1806, 1972.
6. PATANKAR, S.V. - *Numerical Heat Transfer and Fluid Flow*, Hemisphere, Washington (D.C)., 1980.
7. NONINO, C. and DEL GIUDICE, S. - unpublished results, 1992.

NUMERICAL INVESTIGATION OF LAMINAR HEAT TRANSFER AND FLUID FLOW IN SOME CORRUGATED DUCTS

Bijan Farhanieh* and Bengt Sundén**

*Sharif University. Department of Mechanical Engineering, P.O. Box 11365-9567, Tehran, I.R. Iran

**Lund Institute of Technology, Division of Heat Transfer, Box 118, 221 00 Lund, Sweden

SUMMARY

Numerical investigations of laminar fully developed periodic heat transfer and fluid flow in some corrugated ducts are presented. The governing equations are solved numerically by a finite-volume method for elliptic flows in arbitrary geometries using collocated variables and Cartesian velocity components. Results were obtained for some ducts at uniform wall temperatures and Reynolds numbers in the laminar range. Streamline plots show complex flow patterns which give an insight of the governing physical phenomena. Enhancement of the overall heat transfer coefficient was found but the accompanied pressure drop was also increased.

1. INTRODUCTION

Knowledge of laminar forced convective heat transfer in ducts of various geometries plays an important rule in the design and performance calculations of compact heat exchangers. Demands for reduction of the heat exchanger size (volume), weight and costs are present in several applications. Analytical and experimental laminar flow and heat transfer solutions for a variety of ducts are available in [1].

To approach the above-mentioned demands, the utilization of corrugated ducts for still higher achievements in convective heat transfer has been taken seriously into account. Some experimental and numerical results have been presented in [2]. The numerical approach for solution of such problems is usually based upon the assumption that the duct is long and consists of many identical modules in the streamwise direction where the fully developed velocity and temperature fields repeat themselves in a cyclic manner. This assumption enables the ignorance of the entrance and exit regions and confine the calculation domain to just cover one such module. The approach was introduced by Patankar et al. [3] and has since then been widely adopted, see e.g. [4-15].

The present paper concerns numerical calculations of fully developed periodic forced laminar flow and heat transfer in ducts having streamwise-periodic corrugations. The duct walls are heated isothermally. The numerical procedure is based on a finite-volume method for elliptic flows in arbitrary coordinates which utilizes collocated variables and Cartesian velocity components. Details of the procedure have been presented thoroughly in [16] and briefly in e.g. [17]. The calculations are performed in the Reynolds number range corresponding to laminar flow and for Prandtl number Pr = 0.72 (air).

2. DUCTS UNDER CONSIDERATION

The ducts considered are presented schematically in Fig. 1. In Fig. 1a, the cross-sectional area is varying in the flow direction and thus converging-diverging effects occur besides the corrugation effects. Fig. 1b shows a duct with constant cross-sectional area and only influence of the corrugation is present. The oblique walls are positioned at an angle $\phi = 45^o$ to the main flow direction. The geometry of these ducts is specified by the periodic axial length L, the height H and inclination angle ϕ. Fig. 1c presents a grooved duct. The groove height is h and the ratio of duct height to groove height is 2:1 resulting in a hydraulic diameter of 6h. The walls of the ducts in Figs. 1a and 1b are kept at a uniform temperature. The upper wall in Fig. 1c is hot while the lower wall is cold. The main flow is in the x-direction.

3. BASIC EQUATIONS

The basic equations are the continuity, momentum and energy equations. Steady state, constant fluid properties, negligible viscous dissipation and no natural convection are assumed. For the ducts in Figs. 1a and 1b periodic situations are assumed and the pressure p is then expressed as

$$p(x,y) = -\beta x + p^*(x,y) \tag{1}$$

where p^* behaves periodically from cycle to cycle. In βx, β represents the non-periodic pressure gradient in the flow direction.

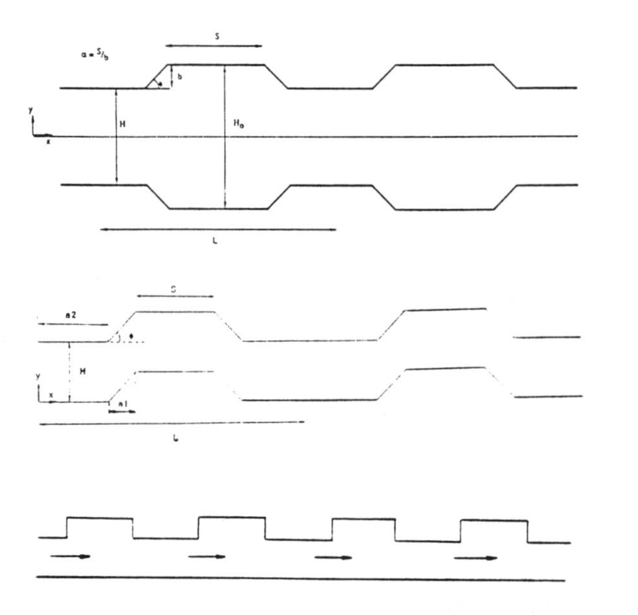

Figure 1. Ducts under consideration.

The following dimensionless variables are introduced:

$$X_i = x_i / L, \quad U_i = u_i L / \upsilon, \quad P = \frac{p^*}{\rho\left(\frac{\upsilon}{L}\right)^2}, \quad B = \frac{\beta L}{\rho\left(\frac{\upsilon}{L}\right)^2}, \quad \theta = \frac{t - t_w}{t_b - t_w}$$

In Cartesian coordinates the basic equations take the following non-dimensional form:

$$\frac{\partial U}{\partial X} + \frac{\partial V}{\partial Y} = 0 \qquad (2)$$

$$U\frac{\partial U}{\partial X} + V\frac{\partial U}{\partial Y} = -\frac{\partial P}{\partial X} + B + \nabla^2 U \qquad (3)$$

$$U\frac{\partial V}{\partial X} + V\frac{\partial V}{\partial Y} = -\frac{\partial P}{\partial Y} + \nabla^2 V \qquad (4)$$

$$U\frac{\partial \theta}{\partial X} + V\frac{\partial \theta}{\partial Y} - \frac{1}{\mathrm{Pr}}\nabla^2\theta = \frac{\sigma}{\mathrm{Pr}} \qquad (5)$$

where

$$\sigma = \lambda\left(2\frac{\partial \theta}{\partial X} - \mathrm{Pr}\, U\theta\right) + \theta\left(\lambda^2 + \frac{\partial \lambda}{\partial X}\right) \qquad (6)$$

and

$$\lambda = \frac{\partial(t_b - t_w)}{\partial X} / (t_b - t_w) \tag{7}$$

σ and λ are periodic parameters for the case of a constant wall temperature.

The shape of the non-dimensional temperature profile θ(x,y) repeats itself in the fully developed periodic area.

The boundary conditions are:

$Walls: U = V = \theta = 0$

Inlet and outlet (periodicity):

$\phi(x,y) = \phi(x+L,y), \quad \phi = U,V,P,\theta,\lambda$

Since the equation of the temperature field contains two unknowns, θ(x,y) and λ(x), an additional condition is needed to close the problem. This condition is obtained from the definition of the bulk temperature. We then have:

$$\int |U| \theta dA_c = \int |U| dA_c \tag{8}$$

For the duct in Fig. 1c periodicity is not assumed. The dimensionless variables are instead introduced as

$$X_i = \frac{x_i}{D_h}, \quad U_i = \frac{u_i}{u_m}, \quad P = \frac{p}{\rho u_m^2}, \quad \theta = \frac{t - t_{in}}{t_h - t_{in}}$$

where u_m is the average velocity, t_{in} the inlet temperature (equal to the temperature of the cold wall), t_h the temperature of the hot wall and D_h the hydraulic diameter. In this case the governing equations read:

$$\frac{\partial U}{\partial X} + \frac{\partial V}{\partial Y} = 0 \tag{9}$$

$$U\frac{\partial U}{\partial X} + V\frac{\partial U}{\partial Y} = -\frac{\partial P}{\partial X} + \frac{1}{\text{Re}}\nabla^2 U \tag{10}$$

$$U\frac{\partial V}{\partial X} + V\frac{\partial V}{\partial Y} = -\frac{\partial P}{\partial Y} + \frac{1}{\text{Re}}\nabla^2 V \tag{11}$$

$$U\frac{\partial \theta}{\partial X} + V\frac{\partial \theta}{\partial Y} = \frac{1}{\text{Re Pr}}\nabla^2 \theta \tag{12}$$

4. NUMERICAL SOLUTION PROCEDURE

A general-purpose finite-volume method using boundary fitted coordinates was employed.

The complex flow domain in the physical space is mapped to a rectangular domain in the computational space by using a curvilinear coordinate transformation. The Cartesian coordinate system in the physical space is replaced by a general non-orthogonal coordinate system. The procedure is similar to that of Burns and Wilkes [18].

The momentum equations are solved for the velocity components U and V in the fixed Cartesian directions on a non-staggered grid. The velocity components at the control volume faces are calculated by the Rhie-Chow interpolation method, [19], and the pressure-velocity coupling is handled by the SIMPLEC-method. The convective terms are treated by the hybrid upwind/central difference scheme whereas the diffusive terms are treated by a central difference scheme. TDMA-based algorithms are applied for solving the algebraic equations. Further details are provided in [16, 17].

4.1 Pressure drop

The pressure drop Δp is compared by the corresponding pressure drop Δp_0 of a straight and smooth duct. The ratio $\Delta p/\Delta p_0$ vs Reynolds number is presented.

4.2 Heat transfer coefficients

In this paper, only distributions of the local heat transfer coefficients are provided. The local heat flux, q_w, is defined as

$$q_w = -k\left(\frac{\partial t}{\partial \eta}\right)_w = h(t_w - t_b) \tag{13}$$

where $(\partial t/\partial \eta)_w$ is the temperature gradient normal to the wall at any cross-sectional position, k the thermal conductivity of the fluid and h the heat transfer coefficient. The heat transfer coefficients are presented in dimensionless form by the Nusselt number Nu defined as

$$Nu = \frac{hD_h}{k} \tag{14}$$

where D_h is the hydraulic diameter.

4.3 Computational details

The Prandtl number was set equal to 0.72 and the Reynolds number was in the laminar regime. The solution procedure is iterative for all variables and the computations were terminated when the sum of absolute residuals normalized by the inflow fluxes were below 10^{-4} for all the variables. To achieve convergence in the solution, an under-relaxation factor of 0.5 was chosen for all the variables.

The grid points are generally distributed in a non-uniform manner with a higher concentration of grid points close to the walls. Each control volume contains one node at its centre but the boundary adjacent volumes contain two nodes. Accuracy tests on various aspects of the numerical procedure have been performed carefully and is reported in [9, 10, 14, 15, 17].

All calculations were carried out on a DEC 3100 work-station.

5. RESULTS AND DISCUSSION

Figure 2 shows the fully developed percycle pressure drop ratio vs Reynolds number for the duct in Fig. 1b. The highest pressure drop is obtained for H/L = 1/5 and $\phi = 45^{o}$. For an inclination angle of 45°, the pressure drop increases by decreasing the module aspect ratio whereas for an inclination angle of 15° the module aspect ratio has only a small effect on the pressure drop. Irrespective of the inclination angles, the pressure drop increases with increasing Reynolds number. However, this increase is very moderate in the case of $\phi = 15^{o}$. For $\phi = 15^{o}$, the curves show a cross-over at Re ≈ 900.

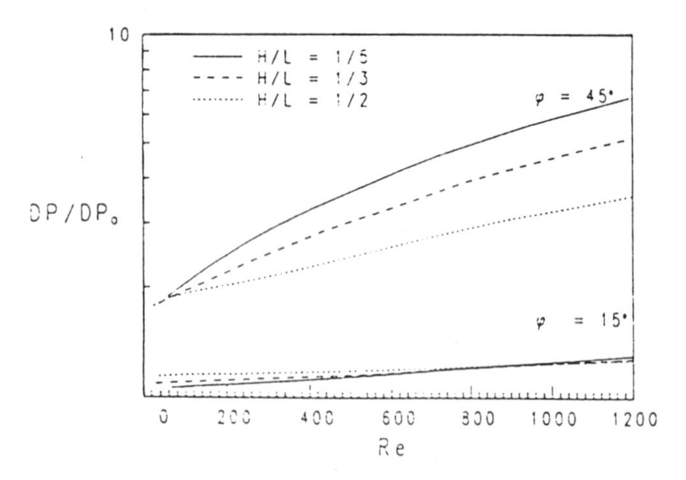

Figure 2. Pressure drop ratio vs Reynolds number for the duct in Fig. 1b.

Distributions of local Nusselt numbers (duct in Fig. 1b) are provided in Figs. 3 and 4. In Fig. 3, Nu distributions for Re = 250 and H/L = 1/5 are plotted for two different inclination angles. Fig. 4 shows local Nu distributions for $\phi = 45^{o}$ and H/L = 1/5 for three different Reynolds numbers.

As is shown in these Figs., the local Nusselt number at the top and bottom walls have a similar pattern of development. The duct can be divided into five different sections. Section one is the upstream horizontal section, section two is the upstream oblique section, section three is the middle horizontal section, section four is the downstream oblique section and section five is the downstream horizontal section. At the top wall of the first section, the local Nusselt number increases until we enter the second section. In this section at the oblique wall of the top wall, the Nusselt number decreases sharply until we approach section three. This latter section by itself consists of two parts. The length of the first part is 1/3 of this section and the remaining length is the second part. In the first part the local Nusselt number increases. This increase is stronger for higher Reynolds number (Fig. 4) and for higher inclination angle (Fig. 3). In the second part of the third section the local Nusselt number decreases until we enter the fourth section. At the downstream oblique wall the local Nu increases sharply to its maximum value which lies at the interface of the fourth and fifth sections. Right at the beginning of the fifth section the local Nu falls sharply and continues to decrease at a moderate rate. In this section, contrary to the first and third section, the value of the local Nu is lower for higher Reynolds number and higher inclination angle and for smaller module aspect ratio.

Let us now consider the bottom wall. In the first section the local Nusselt number decreases moderately until we enter the second section. Here the local Nu increases very sharply to its maximum value and falls very rapidly again in the beginning of the third section. In the second part of the third section the local Nu starts to increase to another maximum value. At the oblique wall of the fourth section, Nu decreases sharply until we approach the fifth section where Nu increases over the separated region and starts to decrease thereafter.

The results can be interpreted physically as follows. In the separated flow regions, the recirculating fluid behaves as a conveyor carrying heat from the core to the wall. The recirculating fluid itself should be divided into two parts, a downstream part and an upstream part. In the downstream part of the recirculating fluid, the heat is conveyed from the separated layer to the wall causing a high temperature difference at the wall and thus an increase in the local Nusselt number. The temperature gradient between the wall and the upstream part of the recirculation becomes smaller and therefore the heat transfer between the wall and the fluid is reduced which causes reduction in the local Nusselt number. In reattached regions the heat is directly convected to the wall by the main flow.

A double-peak can be observed in the Nu distributions at the upstream corner of the top wall and downstream corner of the bottom wall. This appearance can not be judged physically but tests have shown that the double-peaks are not generated numerically.

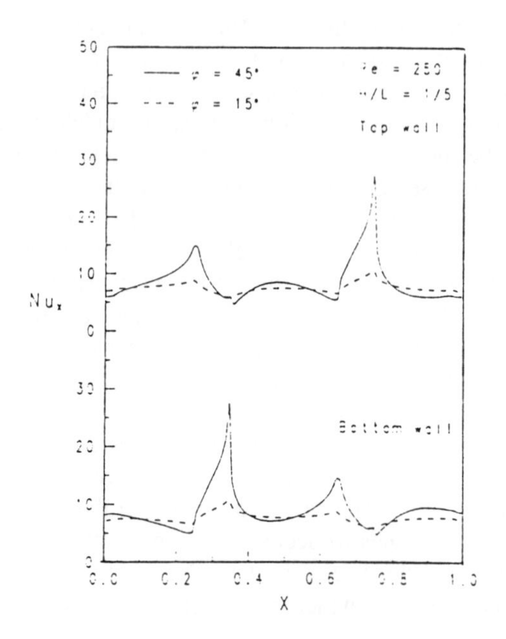

Figure 3. Local Nusselt number distributions for H/L = 1/5 and Re = 250.

Figure 5 shows a comparison of the computed isotherms with experimental ones for two Reynolds numbers for the duct in Fig. 1c. Decent agreement is achieved.

Figure 6 presents distributions of experimental and numerical local Nusselt numbers (Nu_X) along the heated upper wall. The agreement between computations and experiments is quite good. Immediately at the beginning of the first groove the local Nusselt number falls sharply. This is due to the low flow velocities in the recirculating zone. Within the groove it increases gradually and reaches a maximum value at the last quarter section of the groove. After this peak the value of the local Nu falls sharply near the rear wall of the groove. Due to the low convective velocities inside the grooves, the values of the local Nusselt number in the grooves are lower than in the main duct. Immediately downstream of the groove, Nu_X reaches a sharp maximum. This peak is followed by a gradual decrease along the straight section towards the next groove. This behaviour corresponds to re-establishment of thermal boundary layers.

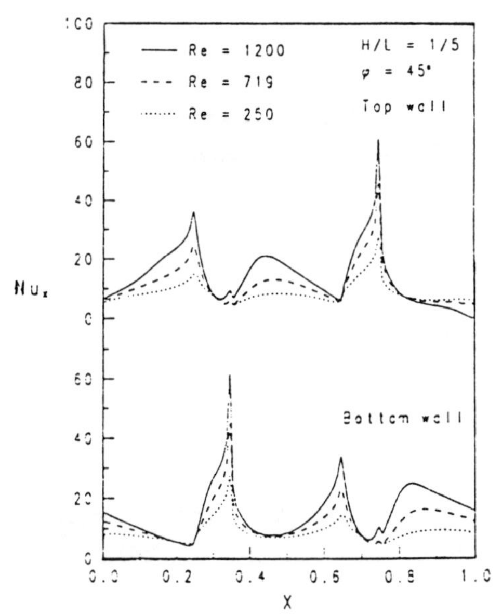

Figure 4. Local Nusselt number distributions for H/L = 1/5 and $\phi = 45^{o}$.

6. CONCLUSION

A numerical method has been applied for analysis of convective heat transfer in some corrugated ducts. Enhancement of the heat transfer compared to a smooth duct was found. For the duct where experimental data was available, the agreement between computed and experimental results was good.

7. ACKNOWLEDGEMENTS

The major part of the work was carried out as the authors were employed at Chalmers University of Technology, Göteborg, Sweden. Financial support was obtained from the former National Swedish Board for Technical Development (STU).

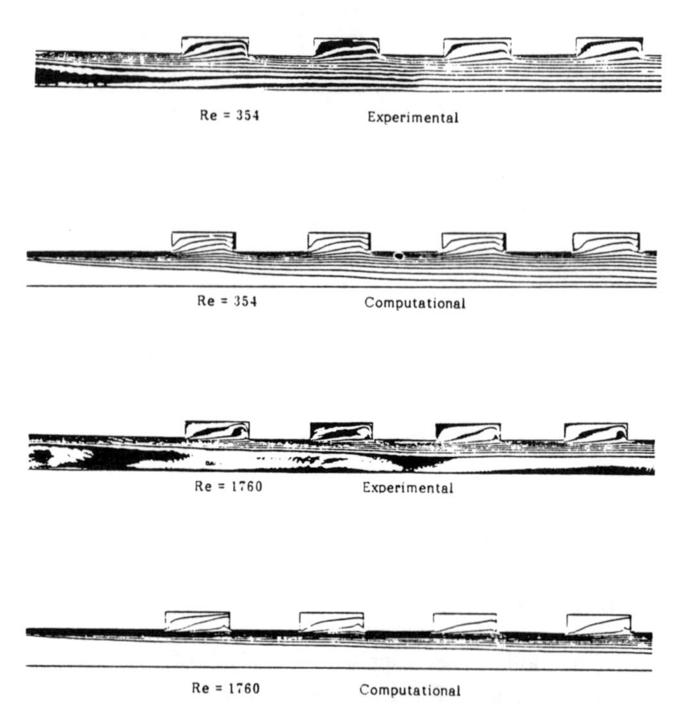

Figure 5. Computed and experimental isotherms. Duct in Fig. 1c.

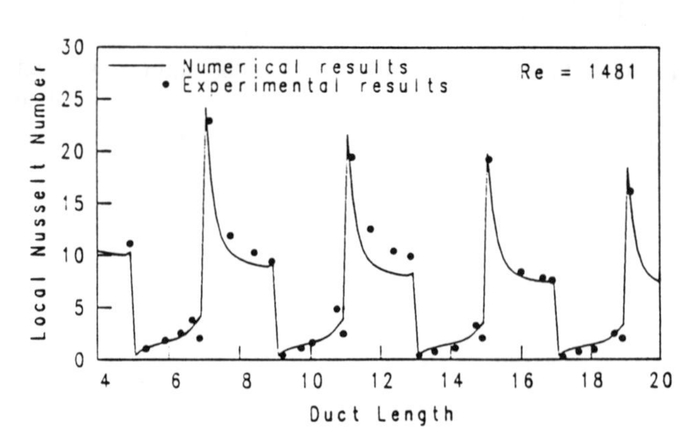

Figure 6. Computed and experimental local Nusselt numbers for the duct in Fig. 1c.

8. REFERENCES

1. SHAH, R.K. and BHATTI, M.S. - Laminar Convective Heat Transfer in Ducts, in Handbook of Singie-Phase Convective Heat Transfer, Eds. Kakac, S., Shah, R.K. and Aung, W., J. Wiley-Interscience Publ., 1987.

2. WEBB, R.L. - Enhancement of Single-Phase Heat Transfer, in Handbook of Single-Phase Convective Heat Transfer, Eds. Kakac, S., Shah, R.K. and Aung, W., J. Wiley-Interscience, 1987.

3. PATANKAR, S.V., LIU, C.H. and SPARROW, E.M. - Fully Developed Flow and Heat Transfer in Ducts having Streamwise-Periodic Variation of Cross-Sectional Area. J. Heat Transfer, Vol. 99, 180-186, 1977.

4. SPARROW, E.M. and PRATA, A.T. - Numerical Solution for Laminar Flow and Heat Transfer in a Periodically Converging-Diverging Tube with Experimental Confirmation. Numer. Heat Transfer, Vol. 6, 441-461, 1983.

5. ROWLEY, G.L. and PATANKAR, S.V. - Analysis of Laminar Flow and Heat Transfer in Tubes with Internal Circumferential Fins. Int. J. Heat Mass Transfer, Vol. 27, no 4, 553-560, 1984.

6. ASAKO, Y. and FAGHRI, M. - Heat Transfer and Fluid Flow Analysis for an Array of Interrupted Plates, Positioned Obliquely to the Flow Direction, in Proc. 8th Int. Heat Transfer Conference 1986, Vol. 2, 421-427, Hemisphere Publ. Corp., 1986.

7. WEBB, B.W. and RAMADHYANI, S. - Conjugate Heat Transfer in a Channel with Staggered Ribs. Int. J. Heat Mass Transfer, Vol. 28, 1679-1687, 1985.

8. SUNDÉN, B. and TROLLHEDEN, S. - Periodic Laminar Flow and Heat Transfer in a Corrugated Two-Dimensional Channel. Int. Comm. Heat Mass Transfer, Vol. 16, 215-225, 1989.

9. FARHANIEH, B. and SUNDÉN, B. - Numerical Investigation of Periodic Laminar Heat Transfer and Fluid Flow Characteristics in Parallel Plate Ducts with Streamwise-Periodic Cavities. Int. J. Num. Meth. Heat Fluid Flow, Vol. 1, no 2, 143-157, 1991.

10. FARHANIEH, B. and SUNDÉN, B. - Fully Developed Laminar Fluid Flow and Heat Transfer in a Streamwise-Periodic Corrugated Duct with Constant Cross-Sectional Area. Int. J. Num. Heat Fluid Flow, Vol. 2, 379-390, 1992.

11. RUSTUM, I.M. and SOLIMAN, H.M. - Numerical Analysis of Laminar Forced Convection in the Entrance Region of Tubes with Longitudinal Internal Fins. J. Heat Transfer, Vol. 110, 310-313, 1988.

12. SOH, W.Y. and BERGER, S.A. - Fully Developed Flow for a Curved Pipe of Arbitrary Curvature Ratio. Int. J. Num. Meth. Fluids, Vol. 7, 733-755, 1987.

13. HWANG, G.J. and CHUNG-HSING, C. - Forced Laminar Convection in a Curved Isothermal Square Duct. J. Heat Transfer, Vol. 113, 48-55, 1991.

14. FARHANIEH, B. and SUNDÉN, B. - Laminar Heat Transfer and Fluid Flow in Streamwise-Periodic Corrugated Square Ducts for Compact Heat Exchangers, ASME HTD - Vol. 201, Eds. Shah, R.K., Rudy, T.M., Robertson, J.M. and Hostetler, K.M., 37-49, 1992.

15. FARHANIEH, B. and SUNDÉN, B. - Numerical Investigation of Turbulent Fluid Flow and Heat Transfer Characteristics in a Streamwise-Periodic Corrugated Duct with Constant Cross-Sectional Area. Int. J. Num. Meth. Heat Fluid Flow, Vol. 3, 1993. (in press)

16. DAVIDSON, L. and FARHANIEH, B. - CALC-BFC: A Finite-Volume Code Employing Collocated Variable Arrangement and Cartesian Velocity Components for Computation of Fluid Flow and Heat Transfer in Complex Three-Dimensional Geometries, Publ. no 91/14, Department of Thermo and Fluid Dynamics, Chalmers University of Technology, Göteborg, Sweden, 1991.

17. FARHANIEH, B. and SUNDÉN, B. - Three Dimensional Laminar Flow and Heat Transfer in the Entrance Region of Trapezoidal Ducts. Int. J. Num. Meth. Fluids, Vol. 13, 537-556, 1991.

18. BURNS, A.D. and WILKES, N.S. - A finite difference method for the computation of fluid flows in complex three dimensional geometries, AERE R 12342, Harwell Laboratory, Oxfordshire, UK, 1987.

19. RHIE, C.M. and CHOW, W.L. - Numerical Study of the Turbulent Flow Past an Airfoil with Trailing Edge Separation. AIAA J., Vol. 21, 1527-1532, 1983.

THREE-DIMENSIONAL NUMERICAL STUDY OF MIXED CONVECTION IN CYLINDRICAL CAVITIES

A. Ivančić, A. Oliva, C.D. Pérez Segarra, H. Schweiger

Laboratori de Termotècnia i Energètica
Dept. Màquines i Motors Tèrmics. Universitat Politècnica de Catalunya
Colom 9, 08222 Terrassa, Barcelona (Spain)

SUMMARY

Three-dimensional and transient finite difference calculations have been done in order to investigate fluid flow and heat transfer in vertical cavities of cylindrical geometry with inlet and outlet ports. The model takes into account both forced and natural convection. The influence of inlet and outlet position on Nusselt number has been analysed for aspect ratio A=2; the range of calculation has been $10 \leq Re \leq 2000$ and $5000 \leq Ra \leq 64000$ assuming laminar flow. The effects of solid structures inside the cavity have also been studied with the aim of simulating the existence of a heat exchanger.The coil heat exchanger is represented as a solid annulus. The influence of the coil position on the outlet temperature has been observed. Flow patterns and temperature distributions are presented. In all calculations water has been the working fluid (Pr=6.97).

1. INTRODUCTION

The problem of natural convection in cylinders has been studied in many papers. Crespo del Arco and Bontoux [1] have numerically analysed flows produced by a slightly supercritical Rayleigh number in a vertical cylinder of aspect ratio A=4. Schneider and Straub [2] have studied natural convection inside inclinated cylinders of aspect ratio $1 \leq A \leq 4$ and inclination angle $0 \leq \alpha \leq 180°$; they found that the maximum heat transfer, for A=1, can be achieved for angles between 40° and 60°. These results have been confirmed experimentally. A vertical cylinder with isothermal bottom and top, and an asymmetrically heated lateral wall have been investigated by Pulicani et al.[3].

Several numerical works deal with mixed convection in enclosures, the majority of them consider a two dimensional problem in a rectangular cavity. The case of laminar convection has previously been studied by Simoneau et al.[4]. They describe two configurations related to the effect of forced convection on natural convection (buoyancy flow), and define them as adding and opposing flows. The authors find a few characteristic zones of mixed convection.To and Humprey [5] have analysed a similar geometry for turbulent flows solving the problem with a k-ε turbulence model.

Transient, two dimensional laminar flow has been examined by Chan et al.[6] in order to simulate the behaviour of thermal storage tanks. The optimal inflow and outflow position was pointed out. Guo and Wu [7] have solved the two dimensional transient problem, but with two inflow and two outflow ports. It was found that for $Gr/Re^2 >> 1$ buoyancy force controls the process within the cavity, and for $Gr/Re^2 << 1$ the process is dominated by forced convection.

Ozoe et al.[8] developed a three-dimensional simulation for a cubic air-filled cavity. Both laminar and turbulent situation have been considered; k-ε turbulence model was used. The conclusion was that the flow is really three dimensional.

Mixed convection in a cylindrical geometry often occurs in technical applications and deserves special attention. With certain boundary conditions it permits axial symmetry considerations and two dimensional treatment; but in general, with non-axisymmetric inlet and outlet port or strong natural convection, the problem loses axial symmetry and requires three-dimensional calculation. So the aim of our study is to investigate sensible heat storage devices in three dimensions.

2. MATHEMATICAL FORMULATION AND NUMERICAL METHOD

The phenomena of a laminar viscous fluid flow in a cylindrical cavity are described by the Navier-Stokes equations:

$$\nabla V = 0 \qquad (1)$$

$$\rho \frac{DV}{Dt} = -\nabla P + \mu \nabla^2 V + \rho g \qquad (2)$$

$$\rho\, c_p \frac{DT}{Dt} = k\, \nabla^2 T \qquad (3)$$

the scalar components in a cylindrical coordinate system can be written in compact form:

$$\frac{\partial(\rho\phi)}{\partial t} + \frac{1}{r}\left[\frac{\partial(\rho r u \phi)}{\partial r} + \frac{\partial(\rho v \phi)}{\partial \theta} + \frac{\partial(\rho r w \phi)}{\partial z} \right] =$$

$$\frac{\Gamma}{r}\left[\frac{\partial}{\partial r}\left(r \frac{\partial \phi}{\partial r} \right) + \frac{\partial}{\partial \theta}\left(\frac{1}{r}\frac{\partial \phi}{\partial \theta} \right) + \frac{\partial}{\partial z}\left(r \frac{\partial \phi}{\partial z} \right)\right] + S \qquad (4)$$

where the dependent variable ϕ, the diffusion coefficient Γ and the source term S are given in table 1.

All fluid properties are taken as constant except the density in the buoyancy term of the momentum equations (Boussinesq approximation). Thus, the momentum equations become temperature dependent and so a simultaneous calculation of velocities and temperature is required. The calculation process is based on the SIMPLEC method [9]. The equations are discretized by the Power Law scheme [10] and solved by the modified strongly implicit procedure [11]. Scalar values are calculated in central nodes and for velocities staggered grids are applied.

Even if the goal of our work has been to obtain a permanent solution the transient approach has been used. At every instant of time upon mass balance verification has been insisted. The process terminates when the Nusselt number change in time is smaller than a prescribed value.

equation	ϕ	Γ	S
mass	1	0	0
r-momentum	v_r	μ	$\rho \frac{v_\theta^2}{r} - \frac{\partial P}{\partial r} - \mu \frac{2}{r^2}\frac{\partial v_\theta}{\partial \theta} - \mu \frac{v_r}{r^2} + \beta\, g_r\, (T-T_0)$
θ-momentum	v_θ	μ	$-\rho \frac{v_\theta v_r}{r} - \frac{1}{r}\frac{\partial P}{\partial \theta} + \mu \frac{2}{r^2}\frac{\partial v_r}{\partial \theta} - \mu \frac{v_\theta}{r^2} + \beta\, g_\theta\, (T-T_0)$
z-momentum	v_z	μ	$-\frac{\partial P}{\partial z} + \beta\, g_z\, (T-T_0)$
energy	T	$\frac{k}{c_p}$	0

Table 1 Dependent variables, diffusion coefficients and source terms

3. RESULTS

Different initial conditions have been imposed: -temperature and velocities equal to zero in entire calculation domain, -pure conduction solution and velocities equal to zero, -natural convection solution (Re=0), -solution obtained with a previous calculation for different conditions (Re or Ra numbers), -solution of coarser grid. No difference in the final solution has been observed, however they have some influence on CPU time; in general, the solution of the previous calculation has been used as an initial condition.

Due to the lack of data about mixed convection in a cylindrical enclosure, the numerical code was compared with a series of numerical and experimental studies of forced and pure natural convection available from literature. In general, close agreement was achieved. These calculations were used to find an appropriate grid. The CPU time increases significantly with the number of grid nodes; a grid of 10*20*20 nodes has usually been selected. This grid gives good results (flow pattern, velocity and Nu), for pure natural convection and forced convection cases, in the range of parameters used in our study.

Two different situations have been studied: situation 1 examines the case of a cylinder with two jets, while situation 2 refers to a cylinder with a coil heat exchanger inside.

3.1 Situation 1

The geometry analysed consists of a cylinder of aspect ratio A=2 with adiabatic lateral wall and isothermal horizontal walls (T_h on the bottom wall, T_c on the top wall, $T_h > T_c$). Inlet and outlet port are situated as shown in figure 1, at the distance $l = H/10$ from the nearest cylinder edge. The section that contains center of inlet and outlet port is the principal plane ($\theta=0$ and $\theta=\pi$)

The fluid entering is always at temperature T_C, while for the one leaving the condition $\partial T/\partial x=0$ is applied (in cases A and B: x=z; and in cases C, D, E and F: x=r). On the ports an uniform velocity profile is employed. The ports cross sections are adapted to the existing grid.

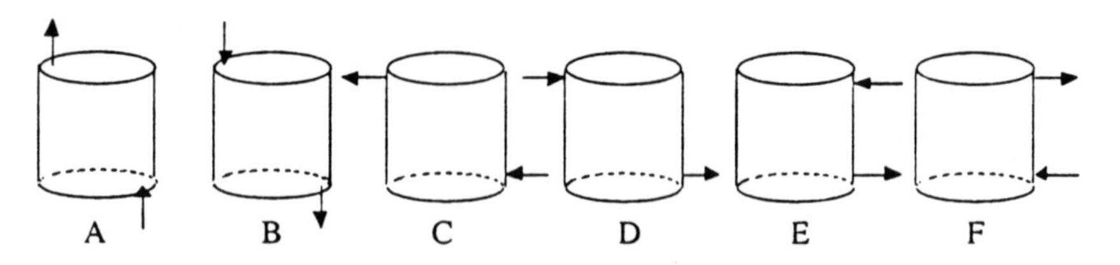

Figure 1 Inlet and outlet port position

The characteristic non-dimensional numbers are defined as:

$$Re = \frac{VR}{\nu} \; ; \; Ra = \frac{\beta g \Delta T H^3}{\nu k} \; ; \; \overline{Nu} = \int_{S_a} \left(V_x T - \frac{\partial T}{\partial x} \right) dS_a$$

The range of our study is $10 \leq Re \leq 2000$ and $5000 \leq Ra \leq 64000$; pure natural convection is also calculated as a limiting case (Re=0).

The figure 2 presents the variation of the average Nusselt number on the inlet velocity (Re) for various Rayleigh numbers for cases A and B (see figure 1). In the case A the Nusselt number increases monotonously with increasing Reynolds number.

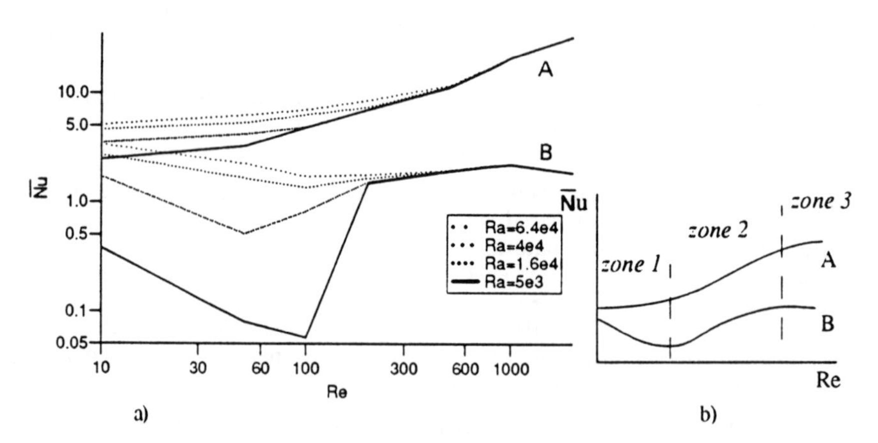

Figure 2 Nusselt number vs. Reynolds number for cases A and B

Three characteristic zones can be observed in figure 2a (schematically represented in figure 2b). In *zone 1* a weak vortex exists near the inlet port in the principal plane; the enhanced Rayleigh number induces more recirculation, the center of the vortex moves toward the outlet port (figure 4a). The isotherms are horizontal for low Ra (figures 3 a1 and 3 b1), similar to natural convection. In case B, as a forced fluid motion opposes the buoyancy force, an increasing Reynolds number diminishes heat transfer. With the rise of the Rayleigh number *zone 1* increments.

In *zone 2* an axial vortex grows breaking stratified isotherm disposition (figures 3 a2 and 3 b2); the growing vortex increases the heat transfer on the cold wall. With the rise of the Rayleigh number *zone 2* diminishes.

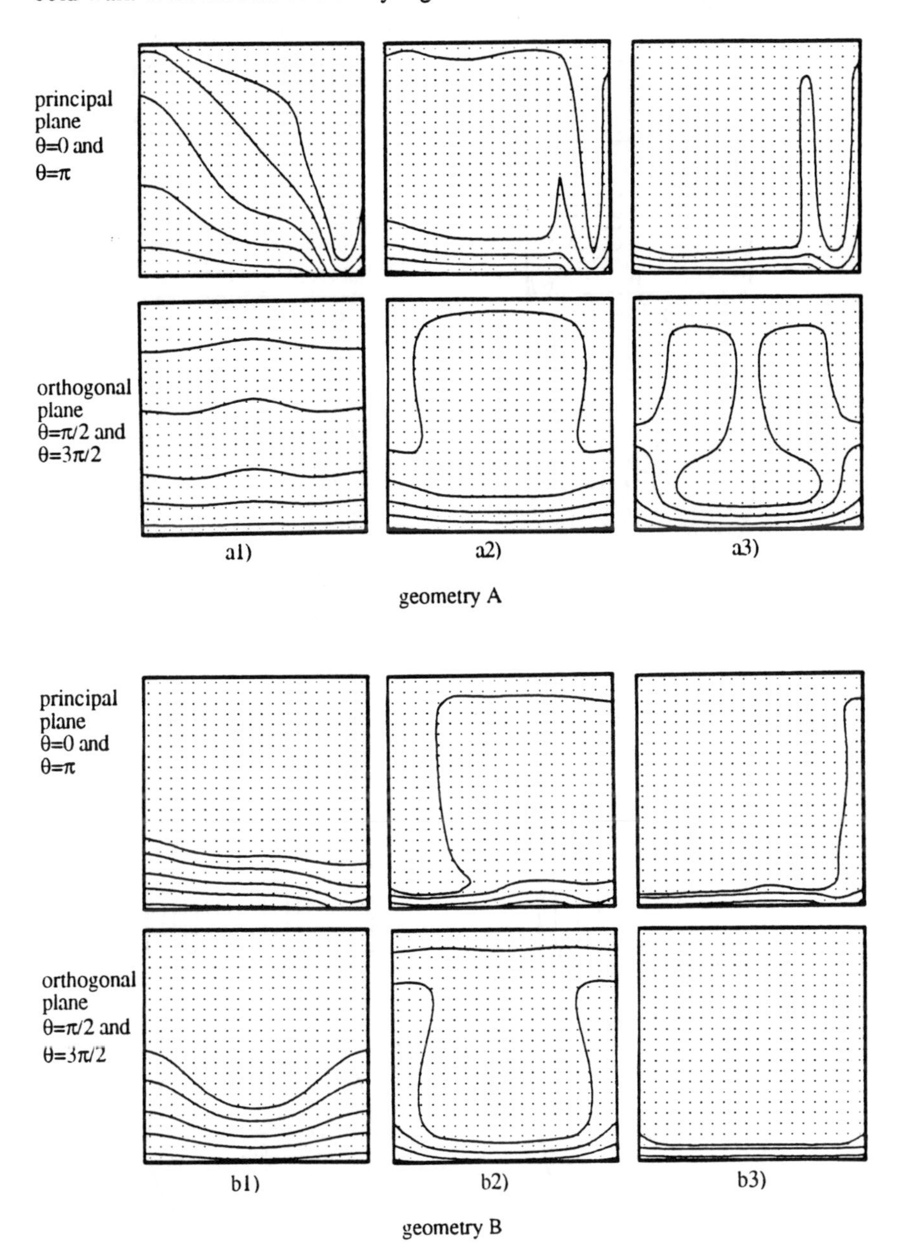

Figure 3 Isotherms for Ra = 5000; Geometry A: a1) Re=50; a2) Re=200; a3) Re=1000; Geometry B: b1) Re=50; b2) Re=200; b3) Re=1000

In *zone 3* the axial vortex in the principal plane vanishes (figure 6a) and vortices in the transversal plane (figure 6d) appear; the influence of the Rayleigh number loses importance, the Nusselt number depends on forced convection only.

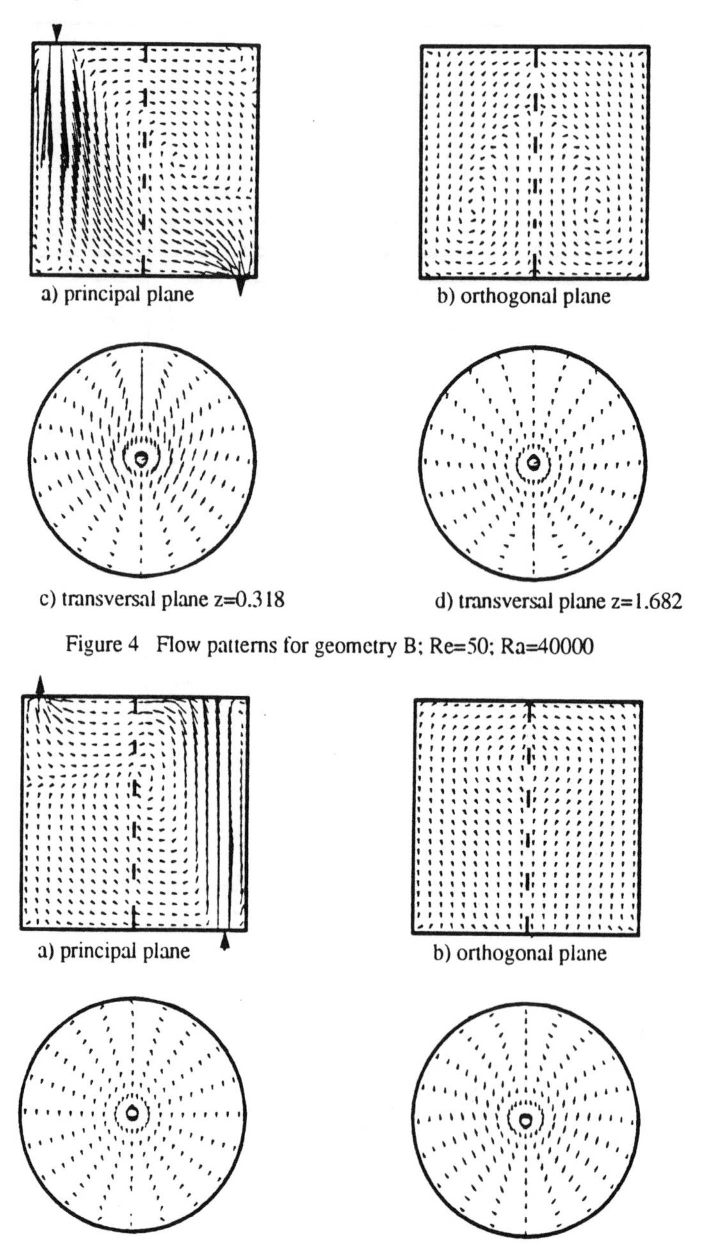

Figure 4 Flow patterns for geometry B; Re=50; Ra=40000

Figure 5 Flow patterns for geometry A; Re=200; Ra=5000

For the cases C, D, E and F three corresponding zones exist; an axial vortex in the principal plane occurs near the lateral wall, and two symmetric transversal recirculations develop near the inlet port. The maximum heat transfer in *zone 1* and *zone 2* is achieved with port position C and in *zone 3* with port position E, while the minimum is reached by geometry B.

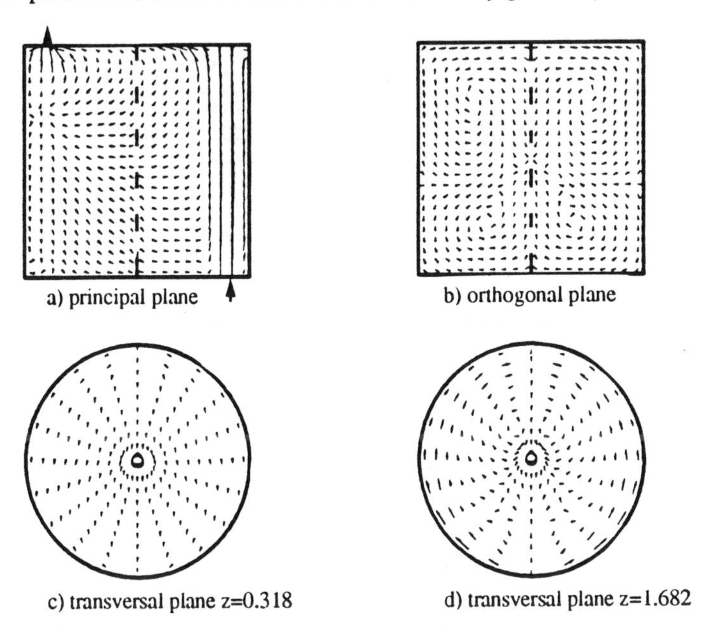

Figure 6 Flow patterns for geometry A; Re=1000; Ra=5000

The effect of increasing Reynolds number on the Nusselt number for various port position is shown in figures 7 and 8.

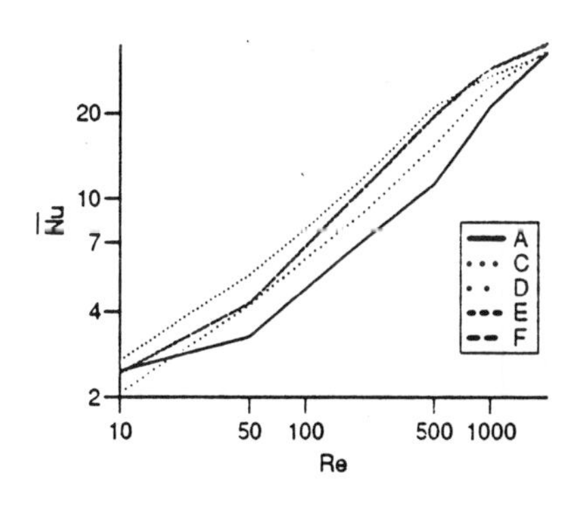

Figure 7 Nusselt number vs. Reynolds number, for Ra=5000

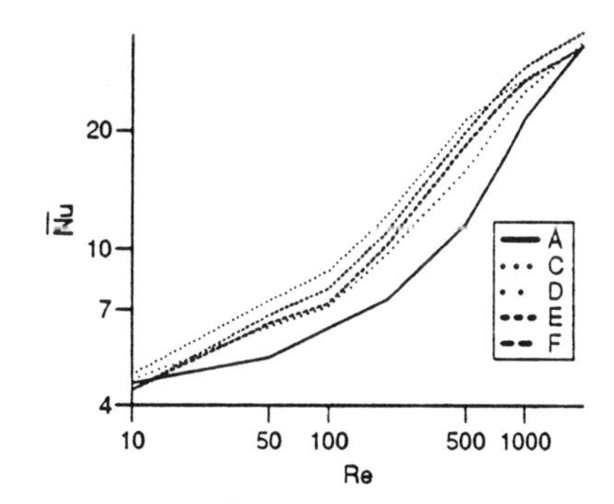

Figure 8 Nusselt number vs. Reynolds number, for Ra=40000

The difference in heat transfer at the hot wall and cold wall that occurs in the cases C, D, E and F is indicated in figure 9 for case C; this difference is equal to the heat the fluid receives. For geometries A and B, as inlet and outlet ports are situated on the cold and hot wall, the Nusselt number is equal on the isothermal walls.

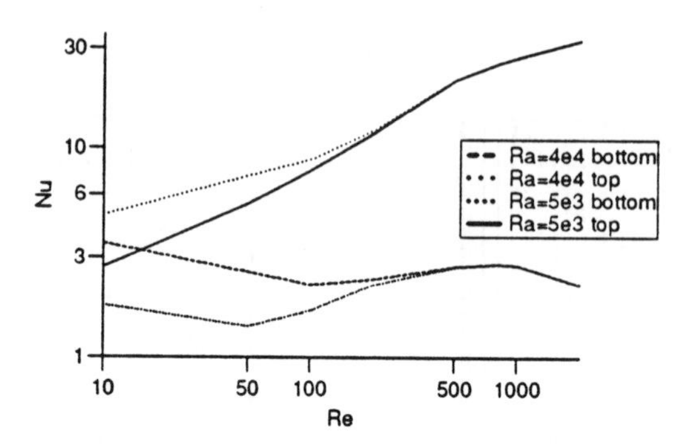

Figure 9 Nusselt number vs. Reynolds number on bottom and top wall, case C

3.2 Situation 2

The other situation studied in our work concerns to heat storage tanks with a heat exchanger inside. The problem is composed of an insulated cylinder with inlet and outlet ports situated as in the figure 10, and a single loop coil heat exchanger presented as an annulus. The dependence of the outlet temperature on the annulus position has been observed in the range of $10^3 \leq Ra \leq 10^4$ and $100 \leq Re \leq 500$. The upper limit of the Reynolds number is established to avoid unstable regime, known as the Karman vortex street, behind the annulus. The same boundary conditions as in the previous case are applied on the inlet and outlet fluid (case C from figure 1). The annulus is defined by the radius R_{an}=0,35R and thickness D_a=0,1R. Its cross section is a rectangular, adapted to the actual grid. The annulus position in the cavity H_a has been varied; its surface is assumed to be isothermal at temperature T_h ($T_h > T_c$). The Rayleigh number is based on the annulus thickness D_a as a characteristic length:

$$Ra = \frac{\beta g \Delta T D_a^3}{\nu k}$$

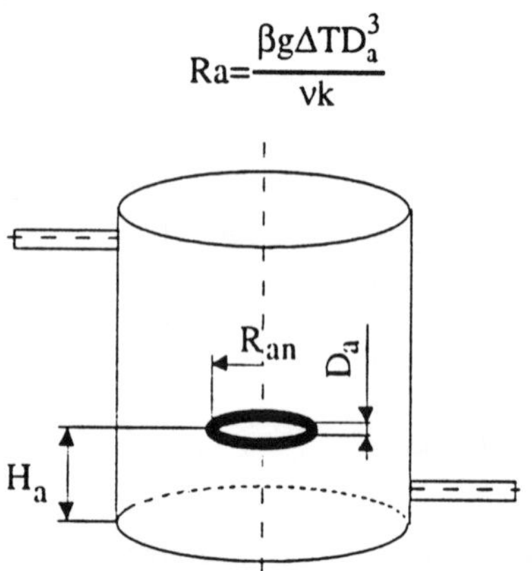

Figure 10 Situation 2

The changes in the flow patterns provoked by the annulus have been observed in three cases: forced convection calculation without annulus, the same conditions with annulus, and mixed convection with annulus. The forced flow without annulus is distinguished by a vortex which occupies half the principal plane (figure 11a) and two vortices near the inlet port in the transversal plane. If the annulus is included, in the principal plane the vortex is displaced toward the lateral wall (figure 11b). When natural convection is taken into consideration the recirculations in the transversal plane fade; as a consequence of the buoyancy force the axial velocities above the annulus are more intensive, the vortex in the principal plane is smaller but another one is formed (figure 11c). In the orthogonal plane two symmetric vortices appear above the annulus.

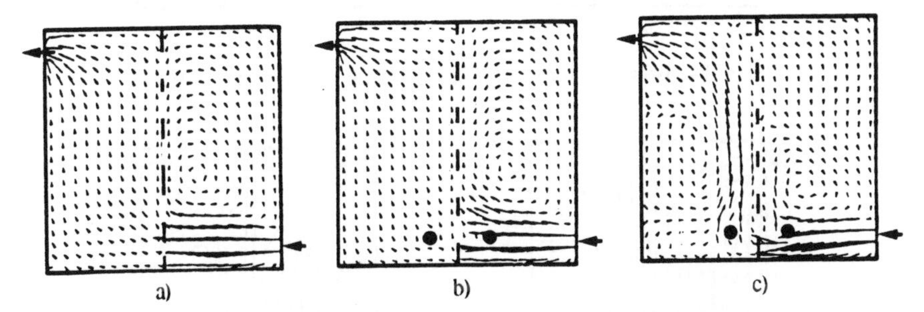

Figure 11 Flow patterns in orthogonal plane for: a) forced convection without annulus; b) forced convection with annulus, H_a=0.238; c) mixed convection with annulus, H_a=0.238

The influence of the annulus position on the outflow temperature for various Reynolds numbers is presented in figure 12. The outflow temperature is shown in adimensional form $T^*=(T-T_c)/(T_h-T_c)$. For higher Reynolds numbers the dependence of the outlet temperature on annulus position is stronger. The maximum heat transfer is obtained with the lowest analysed annulus position for all values of the Reynolds number. The isotherms for Re=100 and Re=500 are shown in figure 14.

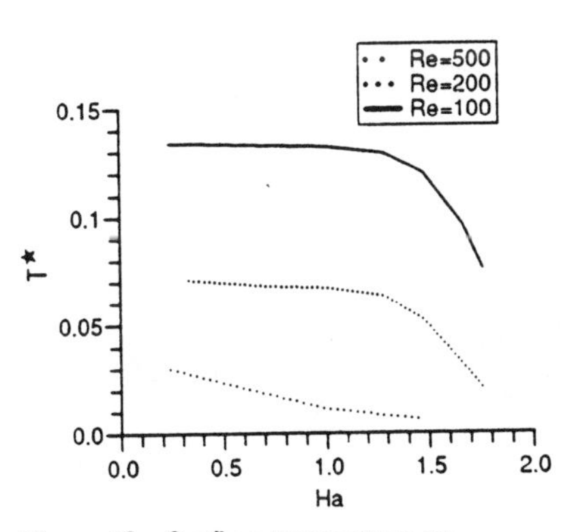

Figure 12 Outflow temperature vs. annulus position for Ra=1000

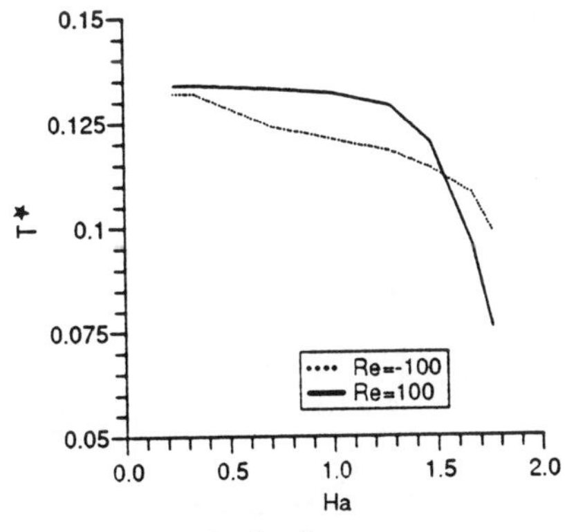

Figure 13 Outflow temperature vs. annulus position for Re=100 and Re=-100 (inverse velocity), Ra=1000

For the case of inverse velocity (such as case D in figure 1), buoyancy force and forced fluid motion act in opposite directions, the maximum is also achieved for the lowest analysed position, as shown in the

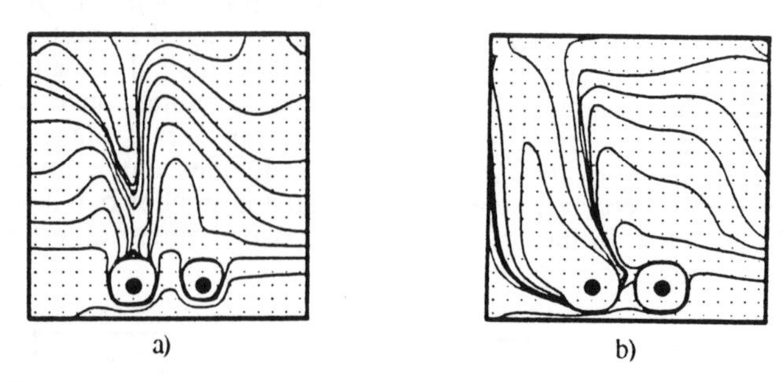

Figure 14 Isotherms for Ra=1000; H_a=0.238; a) Re=100; b) Re=500

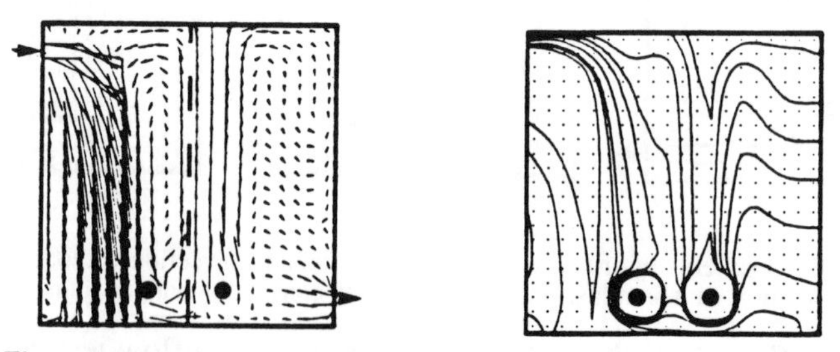

Figure 15 Flow pattern and isotherms for the inverse velocity case; principal plane; Re=100; Ra=1000; H_a=0.238

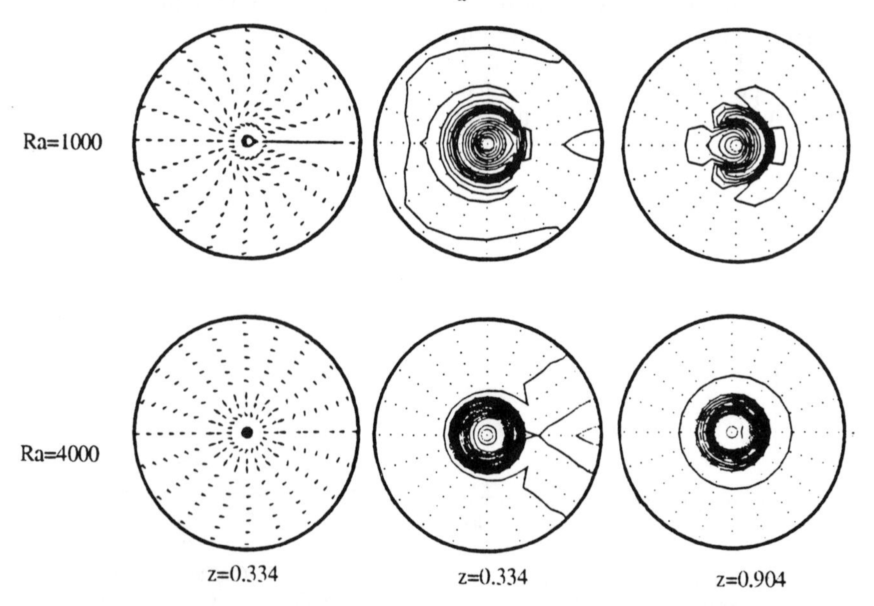

Figure 16 Flow pattern and isotherms in transversal plane H_a=0.238; Re=100;

figure 13, but the outlet temperature difference caused by the annulus position change is smaller and almost linear. The flow pattern and isotherms for the inverse velocity case are presented in figure 15. The radial velocity component of the cold entering fluid diminishes rapidly and the axial one rises, as a consequence of natural convection.

The enhancement of the Rayleigh number yields falling of the radial and azimuthal velocity components, and rising of the axial one, above the annulus (figure 16).

When Ra number is increased the maximum outlet temperature is achieved for slightly higher annulus position and for $Ra=10^4$ the maximum is obtained for $H_a=H/3$ (figure 17)

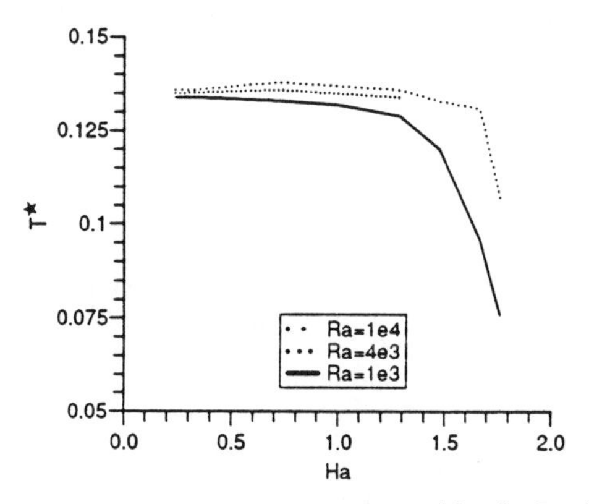

Figure 17 Outflow temperature vs. annulus position for Re=100

4. CONCLUDING REMARKS

A three-dimensional, time dependent, mixed convection model in a cylindrical coordinate system has been developed. A range of Reynolds and Rayleigh numbers has been examined considering laminar flow regime. The analysed geometry has been cylinder with one inlet and outlet port. The effects of the port position on Nusselt has been observed. Three characteristic zones are identified. The port position C (for the zones 1 and 2) and E (for the zone 3) gives the maximum Nusselt number, while with the position B the minimum is achieved.

The influence of the hot annulus inside the cylinder has also been analysed. The maximum outlet temperature is provided for the lowest annulus position, for Ra=1000; increasing the Rayleigh number the maximum is obtained for slightly higher annulus position.

ACKNOWLEDGEMENTS

This study has been supported by the Comisión Interministerial de Ciencia y Tecnología, Spain (ref. no. PTR89-0188) and the Dirección General de Investigación Científica y Técnica, Spain (ref. no. PB90-0606).

NOMENCLATURE

Symbol	Meaning
A	aspect ratio (H/R)
c_p	specific heat at constant pressure
D_a	annulus thickness
g	gravitational acceleration
Gr	Grashof number
H	cylinder height
H_a	annulus height
l	port distance from the nearest edge
k	thermal conductivity
$\overline{Nu}$	average Nusselt number
P	pressure
Pr	Prandtl number
R	cylinder radius
r	radial coordinate
Ra	Rayleigh number
R_{an}	annulus radius
Re	Reynolds number
S	source term
S_a	surface
T	temperature
t	time
T^*	adimensional temperature
T_0	reference temperature
ΔT	temperature difference (T_h-T_c)
V	velocity
x	general coordinate
z	axial coordinate

Greek:

Symbol	Meaning
β	thermal expansion coefficient
Γ	diffusion coefficient
μ	dynamic viscosity
ν	kinematic viscosity
ρ	density
θ	azimuthal coordinate

Subscripts:

Symbol	Meaning
c	cold
h	hot
r	radial component
z	axial component
θ	azimuthal component

REFERENCES

1. CRESPO DEL ARCO, E., BONTOUX, P. - Numerical solution and analysis of asymmetric convection in a vertical cylinder: An effect of Prandtl number. Phys. Fluids A, Vol. 1, No. 8, pp. 1348-1359, 1989.
2. SCHNEIDER, S., STRAUB, J. - Laminar natural convection in a cylindrical enclosure with different end temperatures. Int. J. Heat Mass Transfer, Vol 35, No 2, pp. 545- 557, 1992.
3. PULICANI, J.P., KRUKOWSKI, S., ALEXANDER, J.I.D., OUAZZANI, J., ROSENBERGER, F. - Convection in an asymmetrically heated cylinder. Int. J. Heat Mass Transfer, Vol 35, No 9, pp. 2119- 2130, 1992.
4. SIMONEAU, J.P., INARD, C., ALLARD, F. - Numerical approach of interaction between an injection and laminar natural convection in a thermally driven cavity. Natural Convection in Enclosures, The winter annual meeting of ASME, HTD-Vol. 99, pp. 45-51, 1988.
5. TO, W.M., HUMPHREY, J.A.C. - Numerical simulation of buoyant, turbulent flow. Int. J. Heat Mass Transfer, Vol 29, No 4, pp. 573-610, 1986.
6. CHAN, A.M., SMEREKA, P.S., GIUSTI, D. - A numerical study of transient mixed convection flows in a thermal storage tank. ASME J. of Solar Energy Engineering, Vol. 105, No. 3, pp. 246-253, 1983.
7. GOU, K.L., WU, S.T. - Numerical study of flow and temperature stratifications in a liquid thermal storage tank. ASME J. of Solar Energy Engineering, Vol. 107, No. 1, pp. 15-20, 1985.
8. OZOE, H., MIYACHO, H., HIRAMITSU, M.,MATSUI, T. - Numerical computation of natural convection in a cubical enclosure with ventilation for both a laminar and a two-equation turbulent model. Proceedings of the 8th int. heat transfer conference, San Francisco , Vol. 3, pp. 1489-1494, August 1986.
9. VAN DOORMAAL, J.P., RAITHBY, G.D. - Enhancements of the Simple Method for Predicting Incompressible fluid flow. Numerical Heat Transfer, Vol. 7, pp. 147-163, 1984.
10. PATANKAR, S.V. - Numerical Heat Transfer and Fluid Flow, Hemisphere, Washington, D.C.,1980.
11. SCHNEIDER, G.E., ZEDAN, M. - A modified strongly implicit procedure for the numerical solution of field problems. Numerical Heat Transfer, Vol. 4, No. 1, pp. 1-19, 1981.

DOUBLE DIFFUSIVE CONVECTION IN AN ENCLOSURE WITH A TRAPEZOIDAL CROSS-SECTION

Z. F. Dong and M. A. Ebadian

Department of Mechanical Engineering
Florida International University
Miami, FL 33199 USA

SUMMARY

A numerical study of double diffusive convection has been conducted in a trapezoidal enclosure with aspect ratios of Ar=1 and 2. The solutions were obtained by solving the conservation equations of mass, momentum, energy, and concentration. The buoyancy forces were approximated by Boussinesq assumption. The Prandtl number of Pr=7 and Lewis number of Le=100 were used in the computation to simulate the double diffusive convection of salt water. It was found that the layered cell flow structure is formed on the cross section of the trapezoidal enclosure when the thermal buoyancy is compatible with solutal buoyancy. The unicell flow structure is seen when thermal convection or solutal convection dominates. It must be noted that no layered flow structure can be seen in the assisting flow in the trapezoidal enclosure when Ar=1. The mean Nusselt and Sherwood numbers decrease as the buoyancy ratio increases for assisting flow in a trapezoidal enclosure.

1. INTRODUCTION

It is well known that double diffusive convection commonly refers to the natural convection driven by combined buoyancy forces of heat and mass transfer. Double diffusive convection occurs in various fields, such as oceanography, astrophysics, geology, biology, chemical processes, the materials solidification process, crystal growth, etc. Studies of double diffusive convection were first conducted by oceanography and emphasis was given to the development of the linear stability theory for a simple salt-stratified fluid heated from below [1]. When there exist considerable differences in diffusivities between heat and salt, the double diffusive convection phenomena, such as salt fingers and sharp diffusive interfaces, is observed. Double diffusive convection by lateral heating in a stable, salt- stratified flow have been studied since the 1970's, the conspicuous feature of which is the layered flow structure. Several reviews on double diffusive convection have been found in the literature by [2,3,4]. Inspection of the literature reveals that most of the research emphasized double diffusive convection in the rectangular enclosure or square cavity. Lee's group [5,6,7] has conducted research on double diffusive convection in a rectangular enclosure with an aspect ratio equal to 2. The transient process of natural convection due to thermal and solutal gradients has been investigated experimentally or numerically by [7,8,9], and included the assisting flow, as well as the opposing flow in their investigations. The layered flow structure has been confirmed under appropriate conditions in the case of lateral heating. In [10,11], double diffusive natural convection in a vertical rectangular enclosure with an aspect ratio reaching 4 has been studied experimentally and numerically. The layered multicell flow structure in the enclosure for both assisting and opposing flows has been visualized by the Schlieren method and simulated by the numerical method. The double diffusive phenomena in a square cavity was investigated in [12]. In addition to the above cited survey, double diffusive convection in the solidification and melting process was investigated by [13,14] with phase changes and a non-stationary irregular interface. Based on the authors' knowledge, double diffusive convection in irregular geometries has not been investigated experimentally or numerically; therefore, basic understanding of double diffusive convection in irregular geometry, such as in a trapezoidal cavity, is absent from the open literature. In continuous casted alloys, the liquid region in the casting can be approximated as trapezoidal geometry. Double diffusive convection in the liquid region has a significant effect on the quality of alloys. Consequently, investigation of double diffusive convection in the liquid region is essential to improve the quality of the product. In view of that, the study of double diffusive convection in a trapezoidal enclosure with both lateral thermal and solutal gradients is imposed on the side wall of the trapezoidal wall in this paper. As to the natural convection in a trapezoidal cavity, in [15], Rayleigh numbers of 10^2-10^5 and Prandtl numbers of 0.001 ~ 100 are considered and the geometry effects on thermal natural convection were studied as well.

The purpose of the present study is to conduct a numerical simulation of double diffusive convection in a trapezoidal enclosure. The flow is induced by the horizontal temperature and concentration gradients. The steady state of double diffusive convection is considered in this analysis, since the unsteady state requires extensive CPU time on a super computer. In the following sections, the mathematical formulation for the problem is given first, then the numerical scheme is employed, and the grid system and solution procedures are described next. Finally, results from the numerical computations are discussed in detail and the conclusion follows.

2. MATHEMATICAL FORMULATION

A schematic of the present problem is shown in Fig. 1. The two-dimensional trapezoidal enclosure has a width, b. of the bottom wall, the height of which is h. The sidewall angle, ψ, is 75°, and aspect ratios are defined as h/b. The flow in the enclosure is considered to be two-dimensional and steady laminar flow. The fluid is assumed to be incompressible and Newtonian in behavior with negligible viscous dissipation. The thermal properties of the fluid are treated as constants except for the density in the buoyancy forces term, which is approximated by Boussinesq assumption. Depending on the directions of the buoyancy forces, flow can be either an assisting or opposing flow. In this investigation, only assisting flow is taken into account due to page limitation. By employing the above assumptions, the conservation equations of mass, momentum, energy, and heavier species, are obtained as follows:

$$\frac{\partial U}{\partial X} + \frac{\partial V}{\partial Y} = 0 \tag{1}$$

$$U\frac{\partial U}{\partial X} + V\frac{\partial U}{\partial Y} = -\frac{\partial P}{\partial X} + Pr\left[\frac{\partial^2 U}{\partial X^2} + \frac{\partial^2 U}{\partial Y^2}\right] \tag{2}$$

$$U\frac{\partial V}{\partial X} + V\frac{\partial V}{\partial Y} = -\frac{\partial P}{\partial Y} + Pr\left[\frac{\partial^2 V}{\partial X^2} + \frac{\partial^2 V}{\partial Y}\right] + PrRa_t(\theta - SN) \tag{3}$$

$$U\frac{\partial \theta}{\partial X} + V\frac{\partial \theta}{\partial Y} = \frac{\partial^2 \theta}{\partial X^2} + \frac{\partial^2 \theta}{\partial Y^2} \tag{4}$$

$$U\frac{\partial S}{\partial X} + V\frac{\partial S}{\partial Y} = \frac{1}{Le}\left[\frac{\partial^2 S}{\partial X^2} + \frac{\partial^2 S}{\partial Y^2}\right] \quad . \tag{5}$$

The above equations have been dimensionalized by using the following parameters:

$$\begin{aligned} &U = [u/(\alpha_T/D_h)] \ , \ V = [v/(\alpha_T/D_h)] \ , \ X = x/D_h \ , \\ &Y = y/D_h \ , \ P = p/(\rho\alpha_T^2/D_h^2) \ , \ \Delta T = T_h - T_\ell \ , \ \Delta c = c_h - c_\ell \ , \\ &\theta = (T - T_\ell)/\Delta T \ , \ S = (c - c_\ell)/\Delta c \ , \ Pr = \nu/\alpha_T \ , \ Le = \alpha_T/D \ , \\ &Ar = h/b \ , \ Ra_t = g\beta_t \Delta T D_h^3/\alpha_T\nu \ , \ Ra_s = g\beta_s \Delta c D_h^3/\alpha_T\nu \ , \\ &N = (\beta_s\Delta c)/(\beta_t\Delta T) = Ra_s/Ra_t \ . \end{aligned} \tag{6}$$

In the above dimensionless quantities, the relevant fluid properties are kinematic viscosity, ν; thermal diffusivity, α_T; solutal diffusivity, D; and the coefficients of volumetric expansion with temperature and solute, β_t and β_s, respectively. The hydraulic diameter, D_h, is simply used as a scale length. The Prandtl number, Pr, the Lewis number, Le, the thermal Rayleigh number, Ra_t, and the solutal Rayleigh number, Ra_s, are all included in the governing equations.

The associated boundary conditions considered in the present study are:

$$\theta = 1, \qquad S = 0 \qquad \text{on the left sidewall} \tag{7}$$

$$\theta = 0, \qquad S = 1 \qquad \text{on the right sidewall} \tag{8}$$

$$\frac{\partial \theta}{\partial Y} = 0, \qquad \frac{\partial S}{\partial Y} = 0 \qquad \text{on the bottom and top walls} \tag{9}$$

A nonslip boundary condition has been applied on all walls, which is:

$$U = V = 0 \quad . \tag{10}$$

3. NUMERICAL SOLUTIONS

Because of the irregularity of geometry of the trapezoidal enclosure, the boundary fitted coordinate system is applied. Thus, the trapezoidal duct in the physical domain can be transferred into the computation domain, and therefore, the governing equations are solved in the computational domain instead of in the physical domain. This makes the governing equations more complicated, as one can observe later. In the present study, the grids are generated by solving two coupled differential equations with two sets of coordinates (X,Y) and (ξ,η) as suggested by [16 and 17]. Nonuniform grid spacing is used in the X direction. The grid spacing is minimum near the left and right sidewalls and is increased exponentially away from the walls to the center. Grid spacing in the Y direction is uniformly maintained. The resulting grid configurations for Ar=1 in the physical domain are shown in Fig. 2.

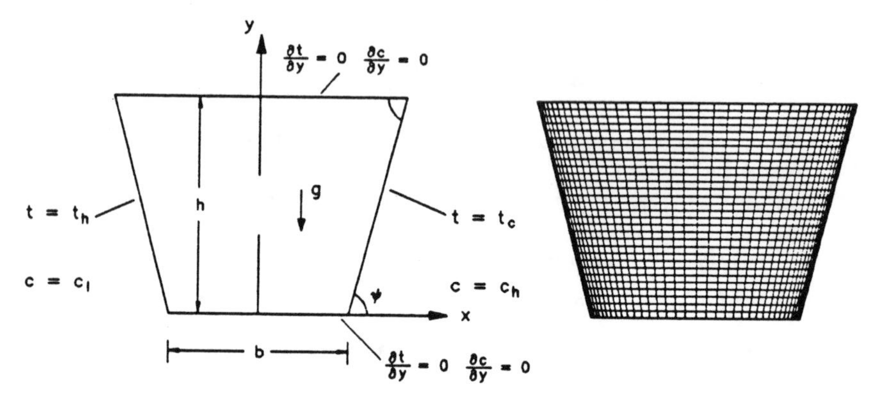

Fig. 1 Schematic of enclousure and coordinate system.

Fig. 2 Configuration of grid.

The dimensionless governing equations, Eqs. (1) to (5), can be transformed into the boundary coordinate system:

$$\frac{\partial \bar{U}}{\partial \xi} + \frac{\partial \bar{V}}{\partial \eta} = 0 \tag{11}$$

$$\frac{\partial}{\partial \eta}(\bar{U}U) + \frac{\partial}{\partial \eta}(\bar{V}U) = \frac{\partial}{\partial \xi}\left[\frac{Pr}{J}\left(\alpha \frac{\partial U}{\partial \xi} - \beta \frac{\partial U}{\partial \eta}\right)\right] \tag{12}$$
$$+ \frac{\partial}{\partial \eta}\left[\frac{Pr}{J}\left(-\beta \frac{\partial U}{\partial \xi} + \gamma \frac{\partial U}{\partial \eta}\right)\right] - \left(\frac{\partial P}{\partial \xi}\frac{\partial Y}{\partial \eta} - \frac{\partial P}{\partial \eta}\frac{\partial Y}{\partial \xi}\right)$$

$$\frac{\partial}{\partial \eta}(\bar{U}V) + \frac{\partial}{\partial \eta}(\bar{V}V) = \frac{\partial}{\partial \xi}\left[\frac{Pr}{J}\left(\alpha \frac{\partial V}{\partial \xi} - \beta \frac{\partial V}{\partial \eta}\right)\right]$$
$$+ \frac{\partial}{\partial \eta}\left[\frac{Pr}{J}\left(-\beta \frac{\partial V}{\partial \xi} + \gamma \frac{\partial V}{\partial \eta}\right)\right] + J Pr Ra_t (\theta - SN) \tag{13}$$
$$- \left(\frac{\partial P}{\partial \eta}\frac{\partial X}{\partial \xi} - \frac{\partial P}{\partial \xi}\frac{\partial X}{\partial \eta}\right)$$

$$\frac{\partial}{\partial \eta}(\bar{U}\theta) + \frac{\partial}{\partial \eta}(\bar{V}\theta) = \frac{\partial}{\partial \eta}\left[\frac{1}{J}\left(\alpha \frac{\partial \theta}{\partial \eta} - \beta \frac{\partial \theta}{\partial \eta}\right)\right] \tag{14}$$
$$+ \frac{\partial}{\partial \eta}\left[\frac{1}{J}\left(-\beta \frac{\partial \theta}{\partial \xi} + \gamma \frac{\partial \theta}{\partial \eta}\right)\right]$$

$$\frac{\partial}{\partial \xi}(\bar{U}S) + \frac{\partial}{\partial \eta}(\bar{V}S) = \frac{\partial}{\partial \xi}\left[\frac{1}{J Le}\frac{\partial}{\partial \xi}\left(\alpha \frac{\partial S}{\partial J} - \beta \frac{\partial S}{\partial \eta}\right)\right] \tag{15}$$
$$+ \frac{\partial}{\partial \eta}\left[\frac{1}{J Le}\frac{\partial}{\partial \eta}\left(-\beta \frac{\partial S}{\partial \eta} + \gamma \frac{\partial S}{\partial \eta}\right)\right],$$

where $\bar{U}$ and $\bar{V}$ are defined as:

$$\bar{U} = U \frac{\partial Y}{\partial \eta} - V \frac{\partial X}{\partial \eta}, \quad \bar{V} = V \frac{\partial X}{\partial \xi} - U \frac{\partial Y}{\partial \eta}, \tag{16}$$

and

$$\alpha = \left(\frac{\partial X}{\partial \eta}\right)^2 + \left(\frac{\partial Y}{\partial \eta}\right)^2 \tag{17}$$

$$\beta = \frac{\partial X}{\partial \xi}\frac{\partial X}{\partial \eta} + \frac{\partial Y}{\partial \xi}\frac{\partial Y}{\partial \eta} \tag{18}$$

$$\gamma = \left(\frac{\partial X}{\partial \xi}\right)^2 + \left(\frac{\partial Y}{\partial \eta}\right)^2 \tag{19}$$

$$J = \frac{\partial X}{\partial \xi}\frac{\partial Y}{\partial \eta} - \frac{\partial X}{\partial \eta}\frac{\partial Y}{\partial \xi} \tag{20}$$

The governing equations (11-15) are approximated with finite difference equations by a control volume based finite difference method in the boundary fitted coordinate system. A power law scheme is adopted for the convection-diffusion formulation. A staggered grid is employed for the velocity components, U and V, as well as $\bar{U}$ and $\bar{V}$. The coefficients such as α, β, γ, J are calculated by second order central difference approximation. The coupling between the nonlinear algebraic equations is controlled with the SIMPLE algorithm [18]. The solutions are obtained on a line-by-line scheme. Underrelaxation is required to ensure the convergence of the iterative procedure. The range of the underrelaxation factor is 0.2 ~ 0.6 for velocities, pressure, temperature, and concentration. Also, a block correction scheme is incorporated to accelerate the convergence rate. The fully converged solution is reached when the following criterion is satisfied for all nodes:

$$\frac{||\phi_{ij}^{k+1} - \phi_{ij}^{k}||_\infty}{||\phi_{ij}^{k+1}||_\infty} \leq 10^{-5} , \qquad (21)$$

where ϕ refers to U, V θ, and S. Subscript, ij, represents the node number in the boundary fitted coordinate system. Superscript, k, refers to the kth iteration, and $||\cdot||_\infty$ is the infinite norm.

RESULTS AND DISCUSSION

For the purpose of validating the present numerical code, the natural convection of air in a square enclosure has been numerically analyzed. The results for $Ra=10^3$-10^6, Pr=0.71 of the present investigation have been compared with the benchmark results in [19 and 20]. Excellent agreement between those results have been achieved.

On the other hand, the comparison of the presented results for double diffusive convection in a square enclosure with those from [12] have been made for assisting flow with parameters of Le = 1, Pr = 0.71, $Ra_t = Ra_s = 10^4$. The local Nusselt and Sherwood numbers along the left vertical wall of the square cavity are matched with those in Table 2 [12]. No significant difference can be found between these two results.

As is seen from the governing equations, Eqs. (1)-(5), the principal non-dimensional parameters are the Prandtl number, Pr; the Lewis number, Le; the thermal Rayleigh number, Ra_t; the solutal Rayleigh number, Ra_s and the aspect ratio of the trapezoidal enclosure, Ar. The effects of the buoyancy forces ratio, Ra_s/Ra_t, and the aspect ratio on the double diffusive flow in the enclosure are emphasized in the present study. Therefore, the Prandtl number, Pr, and the Lewis number, Le, are set to 7.0 and 100.0, respectively, to simulate the values of salt water, which is often used for the simulation of binary alloy solidification. Aspect ratios equal to 1 and 2 are considered, and the solutal Rayleigh number, Ra_s, is set to be 1.0×10^6. The buoyancy forces ratio is varied to explore the various characteristics of double diffusive flow.

Numerical experiments have been conducted to choose the independent grid size. Three grids of 31x21, 41x31, and 51x41, for an aspect ratio of 1, are used to calculate the mean Nusselt and Sherwood numbers. To obtain a convergent solution at a finer grid, extensive CPU time is needed. Fortunately, the deviation of the mean Nusselt and Sherwood numbers from the finer grid of 51x41 with the grid of 41x31 is only less than 1.5 percent. Therefore, the 41x31 grid is employed for Ar=1 as a compromise between accuracy and CPU time. For Ar=2, the 41x61 grid is applied. All of the computations are performed on the CRAY-YMP.

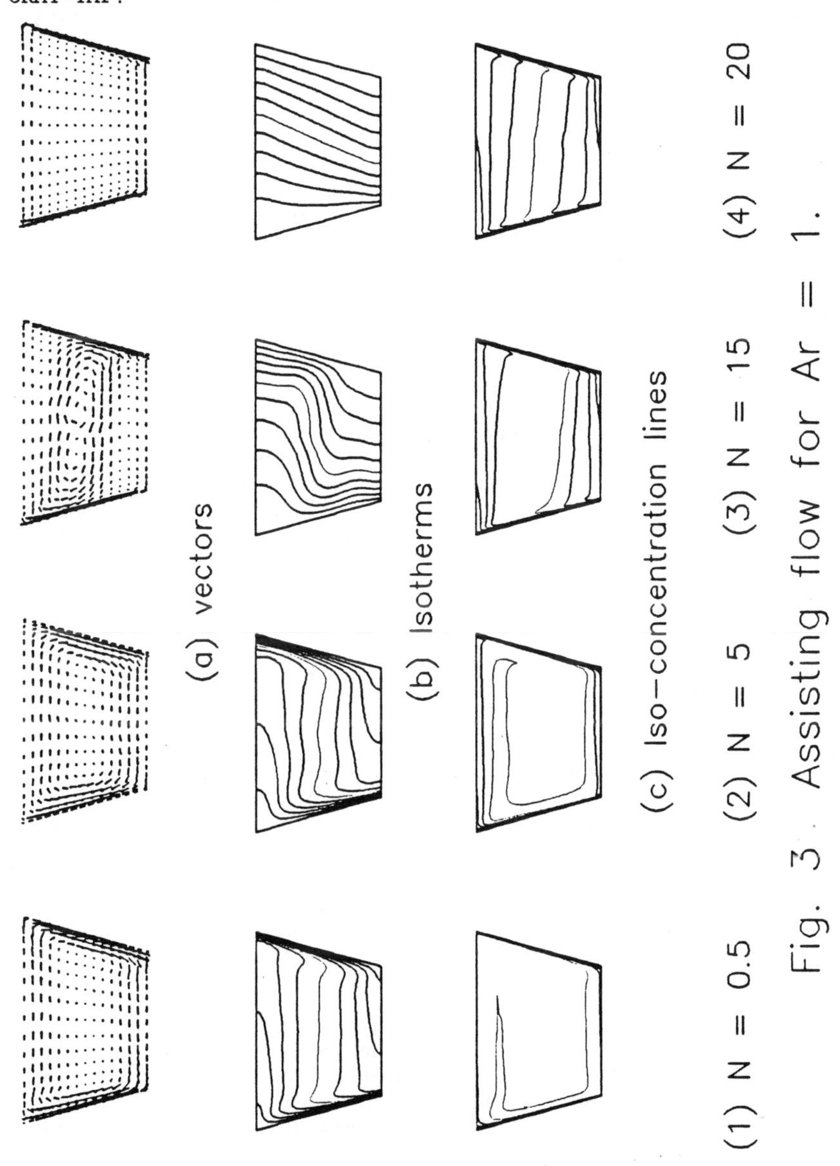

Fig. 3 Assisting flow for Ar = 1.

In Fig. 3, the typical features of assisting double diffusive flow are plotted for Ar=1; Figs. 3a, 3b, and 3c represent the vectors, isotherms and iso-concentration lines, respectively. It is seen from Fig. 3a that fluid in the enclosure flows in a clockwise manner. When N = 0.5, the fluid near the left hot and low concentration wall flows up to the top and the fluid near the right cold and high concentration wall moves toward the bottom, leading to a unicell flow in the enclosure. The center region of the cell is located in the middle level of the enclosure. When the buoyancy forces ratio increases, the flow is gradually changed. Inspection of this figure reveals that two small cells may be formed in the center region at N=5. However, when the buoyancy forces ratio is increased to N=15, two small cells can be detected. Furthermore, for N=15, the flow in the boundary layer region as well as the cell flow are much stronger. The flow occurs only in the boundary region of concentration when the buoyancy forces ratio equals 20. The stagnant fluid can be seen in a greater portion of the enclosure. Figure 3b shows the temperature distribution in the enclosure. At N=0.5, more isotherms appear near the left and right walls, and the deformation of isotherms is observed near the boundary with the temperature stratification in the center region. When the buoyancy forces ratio increases to 5, the deformation of isotherms is enlarged from the boundary region to the whole cross section and the temperature gradient on the left and right walls is decreased. At N=15, the deformation of isotherms occurs in the center region of the cross section, which indicates stronger flow there. The temperature distribution resembles conduction at N=20. The temperature distribution variation, when N increases from 0.5 to 20, reflects the domination of flow from the thermal convection to the solutal convection. From the distribution of concentration, Fig. 3c, it is apparent that the solutal convection is gradually enhanced as the buoyancy ratio increases. Because of the larger Lewis number, the boundary layer of concentration is much thinner than that of temperature. When N equals 0.5, the concentration is changed mainly in the boundary layer. Only in the case of a greater value of N do the stratified iso-concentration lines take up the whole cross section. When the iso-concentration line forms in the cross section at N = 20, the flow induced by solutal force makes the iso-concentration lines deform only in the boundary layer region. It follows from Fig. 3 that the layer-cell of the flow structure is not seen for the assisting flow in a trapezoidal enclosure when Ar=1.

The characteristics of assisting flow in the trapezoidal enclosure when Ar=2 are shown in Fig. 4a. Thermal convection dominates and the unicell flow can be seen when N=0.5. When the buoyancy forces ratio increases, the unicell is separated into two small cells in the center region when N=5, with a two-layered cell flow at N=10, and one strong flow cell in the middle level of the enclosure when N=20. Figure 4b displays the temperature distribution variation with the buoyancy forces ratio. When N is smaller than N=5, the isotherms shape deforms in the boundary layer region, and the temperature of the fluid in the core region is more stratified. The greater variation takes place when the buoyancy forces ratio is large, and as a result, it generates a two-layered cell flow and the zig-zag shape of isotherms appears. When solutal convection dominates the flow, the temperature distribution is one of conduction, where heat transfer is mainly dominated by conduction. Furthermore, when thermal convection is dominant the wall temperature gradient is greater for the small buoyancy forces ratio, which leads to a larger Nusselt number. The iso-concentration lines are shown in Fig. 4c. It is seen that when solutal convection is not strong (N=0.5, N=5), the concentration varies only in its thin boundary layer. However, when the two-layered cell flow structure forms, the concentration distributions are more or less stratified horizontally (N=10). Finally, the stratified concentration distribution will be seen in the whole cross section, except at the boundary layer (N=20 or larger), where thermal convection is very weak.

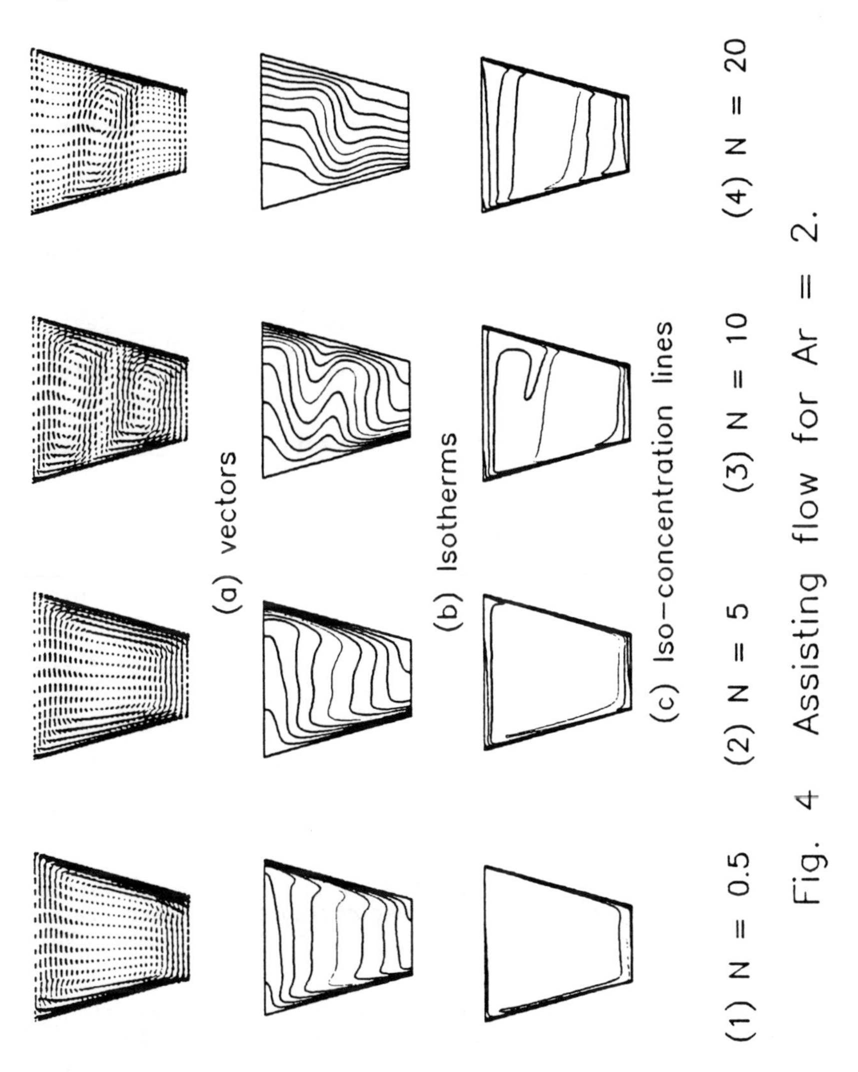

Fig. 4 Assisting flow for Ar = 2.

CONCLUSION

Double diffusive convection in a trapezoidal enclosure has been numerically investigated. The two aspect ratios of Ar=1 and Ar=2 are considered. The boundary conditions imposed on the trapezoidal enclosure walls are lateral temperature and concentration gradients, which lead to assisting and opposing flows. Based on the above results and discussion, it is found that the layered cell flow structure can be formed when both buoyancy due to temperature and concentration are comparable. Otherwise, there exists unicell flow in the cross section of the trapezoidal enclosure. In the trapezoidal enclosure with assisting flow and Ar=1, the unicell flow is dominant at all buoyancy forces ratios. However, when Ar=2, two-layered cell flow is seen when the buoyancy forces flow ratio is around 15.

ACKNOWLEDGMENT

The results presented in this paper were obtained in the course of research sponsored by the National Science Foundation under Grant No. HRD-9250087.

REFERENCES

1. TURNER, J.S. - *Buoyancy Effects in Fluids*, Cambridge University Press, London, 1979.

2. TURNER, J.S. - Double Diffusion Phenomena. Annual Reviews of Fluid Mechanics, Vol. 6, pp. 37-56, Eds., M. Van Dyke, W. G. Vincent, and J. V. Wehausen, Annual Reviews Inc., Polo Alto, California, 1974.

3. TURNER, J.S. - Multicomponent Convection. Ann. Rev. Fluid Mech., Vol. 17, pp. 11-44, 1985.

4. VISKANTA, R., BERGAN, T.L., AND INCROPERA, F.P. - Double Diffusive Natural Convection in Natural Convection. Fundamentals and Applications, Ed., Kakac, S. Aung, W, and Viskanta, R., Hemisphere Publishing Corporation, Washington, pp. 1075-1099, 1985.

5. LEE, J.W. and HYUN, J.M. - Time Dependent Double Diffusive in a Stable Stratified Fluid Under Lateral Heating. *Int. J. Heat Mass Transfer*, Vol. 34, No. 9, pp. 2409-2421, 1991.

6. LEE, J.W. and HYUN, J.M. - Double Diffusive Convection in a Cavity Under a Vertical Solutal Gradient and a Horizontal Temperature Gradient. *Int. J. Heat Mass Transfer*, Vol. 34, No. 9, pp. 2403-2427, 1991.

7. LEE, J.W., HYUN, J.M., and KIM, K.W. - Natural Convection in Confined Fluids with Combined Horizontal Temperature and Concentration Gradients. *Int. J. Heat Mass Transfer*, Vol. 31, pp. 1969-1977, 1988.

8. HYUN. J.M. and LEE, J.W. - Double Diffusive Convection in a Rectangle with Cooperating Horizontal Gradients of Temperature and the Concentration Gradients. *Int. J. Heat*

Mass Transfer, Vol. 33, pp. 1605-1716, 1990.

9. LEE, J.W. and HYUN, J.M. - Double Diffusive Convection in a Rectangle with Opposing Horizontal Temperature and the Concentration Gradients. *Int. J. Heat Mass Transfer*, Vol. 33, pp. 1619-1622, 1990.

10. HAN, H. and KUEHN, T.H. - Double Diffusive Natural Convec tion in a Vertical Rectangular Enclosure--I: Experimental Study. *Int. J. Heat Mass Transfer*, Vol. 34, pp. 449-459, 1991a.

11. HAN, H. and KUEHN, T.H. - Double Diffusive Natural Convection in a Vertical Rectangular Enclosure--II: Numerical Study. *Int. J. Heat Mass Transfer*, Vol. 34, pp. 461-471, 1991.

12 BEGHEIN C., HAGHIGHAT, F., and ALLARD, F. - Numerical Study of Double Diffusive Natural Convection in a Square Cavity. *Int. J. Heat Mass Transfer*, Vol. 35, pp. 833-846, 1992.

13. BECKERMAN, C. and VISKANTA, R. - Double Diffusive Convection Due to Melting. *Int. J. Heat Mass Transfer*, Vol. 31, No. 10, pp. 2077-2089, 1989.

14. BECKERMAN, C. and VISKANTA, R. - Double Diffusive Convection During Dendritic Solidification of a Binary Mixture. *PCH Physicochemical Hydrodynamics*, Vol. 10, No. 2, pp. 195-213, 1988.

15. LEE, L.S. - Numerical Experiment with Fluid Convection in Tilted Nonrectangular Enclosures. *Numerical Heat Transfer*, Part A., Vol. 19, pp. 487-499, 1991.

16. THOMPSON, J.F., THEMES, F., and MASTIN, C. - Automatic Numerical Generation of the Body-Fitted Curvilinear Coordinate System for a Field Containing any Number of Arbitrary Two-Dimensional Bodies. *J. Comp. Phys.*, Vol. 24, pp. 299-319.

17. THOMAS, P.D. and MIDDLECOFF, J.F. - Direct Control of Grid Point Distribution in Meshes Generated by Elliptic Equation. *AIAA J.*, Vol. 18, pp. 652-656.

18. PATANKAR, S.V. - *Numerical Heat Transfer and Fluid Flow*, Hemisphere, Washington, DC, 1980.

19. DE VAHL DAVIS, G. - Natural Convection of Air in a Square Cavity, A Bench Mark Numerical Solution. *Int. J. Numer. Math. Fluids*, Vol. 3, pp. 249-264, 1983.

20. LIN, T.F., HUANG, C.C., and CHANG, T.S. - Transient Binary Mixture Natural Convection in Square Enclosures. *Int. J. Heat Mass Transfer*, Vol. 33, No. 2, pp. 287-299, 1990.

NOMENCLATURE

Ar aspect ratio, h/b

b	width of bottom wall of trapezoidal
c	dimensional concentration
D	mass diffusivity [m^2 s^{-1}]
D_h	hydraulic diameter
g	gravity [m s^{-2}]
h	height of the cavity [m]
J	coefficient, Eq. (12)
Le	Lewis number, α/D
N	buoyancy ratio, $\beta_s \Delta C/\beta_t \Delta T$
P	dimensionless pressure
Pr	Prandtl number, ν/α
p	dimensional pressure
Ra_s	solutal Rayleigh number, $g\beta_s \Delta C D_h^3/\alpha\nu$
Ra_t	thermal Rayleigh number, $g\beta_t \Delta t D_h^3/\alpha\nu$
S	dimensionless concentration
T	dimensional temperature [K]
$\underline{U},\underline{V}$	dimensionless velocities
$\bar{U},\bar{V}$	defined in Eq. (16)
u,v	dimensional velocities [m s^{-1}]
X,Y	dimensionless coordinates
x,y	dimensional coordinates

<u>Greek Symbols</u>

α	coefficient, Eq. (17)
α_T	thermal diffusivity [m^2 s^{-1}]
β	coefficient, Eq. (18)
β_s	coefficient of volumetric expansion with concentration
β_t	coefficient of volumetric expansion with temperature
γ	coefficient, Eq. (19)
Δc	concentration gradient, C_h - C_ℓ
ΔT	temperature gradient, T_h - T_L
θ	dimensionless temperature
ν	kinematic viscosity [m^2 s^{-1}]
ρ	density [K g m^{-3}]

<u>Subscripts</u>

h	higher value
ℓ	lower value

EFFECT ON VELOCITY OF COMBINED CONVECTION HEAT TRANSFER IN INCLINED DUCTS

R. Smyth

Department of Mechanical Engineering, University of Sheffield.

ABSTRACT

Heat Transfer by combined free and forced convection in laminar air flow in the entrance region of inclined rectangular cross-sectioned ducts of aspect ratio 5:1 has been studied numerically to examine the effect on fluid velocity and average Nusselt numbers. The investigation was made with one or both of the upper and lower walls of the duct heated with a constant heat flux and the vertical side walls unheated. The three heated situations studied were, (a) one pair of inclined parallel sides of the duct heated, (b) one upper side heated only, and (c) one lower side heated only. Velocity and temperature profiles have revealed that secondary flows created by natural convection have a significant effect on the flow behaviour and the heat transfer process at the lower plate.

1. INTRODUCTION

Heat transfer in inclined rectangular cross-sectioned ducts can take place partly due to the movement of fluid provided by some external mechanism and partly due to buoyancy forces. Combined (or mixed) convection commences when both modes of convection (natural and forced) are present simultaneously and when natural convection effects are of the same order of magnitude as the forced effects. Thus in applications in which the flow velocity is sufficiently low or if large temperature differences are encountered, the natural convection may be important. For simultaneously developing air flow, the hydrodynamic and thermal boundary layers develop at approximately same rate from the duct entrance and for the situation of constant wall heat flux, a fluid to wall temperature difference exists throughout the duct entrance, with possibly significant natural convection prevalent.

Mixed or combined convection heat transfer inside a duct is either assisting or aiding when the forced convection flow direction is upward and assisted by the fluid buoyancy or opposing when the forced convection flow direction is downward and is opposed by the fluid buoyancy. From experimental observation by other workers it has been found that in vertical tubes the laminar forced convection heat transfer coefficient in the aiding situation can be as much as 2.5 times that for pure forced convection.

A number of theoretical studies by other workers have been made of convection in rectangular cross-sectioned channels. Cheng and Hwang [1] have analysed combined natural and forced convection heat transfer for steady fully developed laminar flow in a horizontal rectangular channel with constant wall heat flux and at two Prandtl numbers Pr. Abou-Ellail and Morcos [2] have investigated numerically the buoyancy effect on laminar forced convection heat transfer in the entrance region of uniformly heated horizontal rectangular ducts having aspect ratio of 1 and 4 for different values of Pr. An analysis of fully-developed combined forced and free convection in vertical rectangular channels has been reported by Agrawal [3]. Combined free and forced laminar convection has been investigated in non-circular vertical ducts by Igbal et al [4], comprising rectangular, elliptical and rhombic cross-sections with uniform axial heat input and either uniform peripheral heat flux or uniform peripheral temperature distribution. Chow et al [5] have studied numerically the effects of free convection and axial conduction on forced convection heat transfer to fully developed flow in a vertical channel at low Peclet numbers Pe.

2. THEORETICAL ANALYSIS

The numerical analysis scheme of Patankar [6], which has been found to be applicable to a wide range of practical engineering problems and has been encapsulated in commercially available fluid dynamic simulation computer codes, was used to solve the governing differential equations of continuity, momentum and energy for steady flow in three dimensions:

$$\frac{\partial}{\partial x_j}\left(\rho u_j\right) = 0$$

$$\frac{\partial}{\partial x_j}\left(\rho u_j \psi\right) = \frac{\partial}{\partial x_j}\left(\Gamma \frac{\partial \psi}{\partial x_j}\right) + S$$

with $\psi = u, v, w$ (the co-ordinate fluid velocities) for the momentum equations (with Γ the viscosity) and $\psi = T$ (the fluid temperature) for the energy equation (with Γ the thermal conductivity). The source term S in the momentum equations contains the buoyancy body forces.

The boundary conditions for the solution of these equations were (1) uniform axial velocity and temperature at the duct inlet, (2) prescribed constant heat flux at one or both of the wider sides of the duct, and (3) insulated narrow sides of the duct.

3. RESULTS

Secondary flow patterns across half the duct width are represented by a set of vectors in Fig. 1 to Fig. 4, at a dimensionless axial position $z^* = 0.615$ for five cases, $\phi = 0^o, 30^o, 45^o, 60^o$ and 90°, each having $Gr_{Dq} = 0.15 \times 10^8$, $Re_D = 801$ and $Ri = 4.3$.

For both sides heated, Fig. 1 shows a clear existence of 4 vortices in the section and reveals that the effects of the secondary flow is delayed due to the higher value of Reynolds number. In the vertical duct a diverging of the flow towards the left and right plate is indicated in order to accelerate the flow at both plates. In Fig. 2, for a lower Reynolds number, the vortices dominate the flow field for all the angles of inclination. For $\varphi = 0^o$, the vector scale reveals the strong acceleration of the flow, driven by the secondary flow. The secondary flow pattern, represented by the number of vortices and their position in the section, seems to vary along the duct and there appears to be 2 major and 4 minor vortices in the whole section. This secondary flow, at the upper plate, always drives the flow from the duct centre region toward the duct upper corners and the unheated vertical wall. It then appears to decelerate and be diverted horizontally to create a long vortex adjacent to the top plate causing an increase in its surface temperature.

At the lower plate, the secondary flows are strong, reducing its temperature in general and creating large temperature fluctuations in the traverse direction. Therefore, the temperature difference between the upper and lower plates is greatly affected by the Re_D and GR_{Dq} values. Fig. 1 and Fig. 2 indicates that increasing the heat flux or reducing the flow rate will lead to more isolation of the upper plate, improving the upper plate secondary flow, raising its temperature and protecting it from the lower plate secondary flow penetration. Meanwhile the lower plate secondary flow increases, which creates in general a lower surface temperature with more traverse temperature fluctuations. For $\varphi = 30^o$, 45° and 60°, the strong secondary flow at the upper plate is gradually lifted and the lower plate secondary flow seems to be able to penetrate the upper plate boundary layer and the vortices existence is extended to include $\varphi = 60^o$. The secondary flow effect is to accelerate the flow at the heated plate.

One side heated only cases are shown in Fig. 3 and Fig. 4. The upper plate heated only results Fig. 3 display the existence of a very wide vortex adjacent to the upper plate with its velocity maximum at the duct corners, leaving the duct centre and the region adjacent to lower plate unaffected. The effect of the secondary flow is gradually reduced and limited to a small region at the upper corners as φ moves toward vertical, with a complete one dimensional flow field displayed for $\varphi = 90^o$. For the lower plate heated only, Fig. 4, at $\varphi = 0^o$, there are 4 very well established vortices at the lower plate edged, leaving the duct centre undisturbed. The secondary flow displays the ability

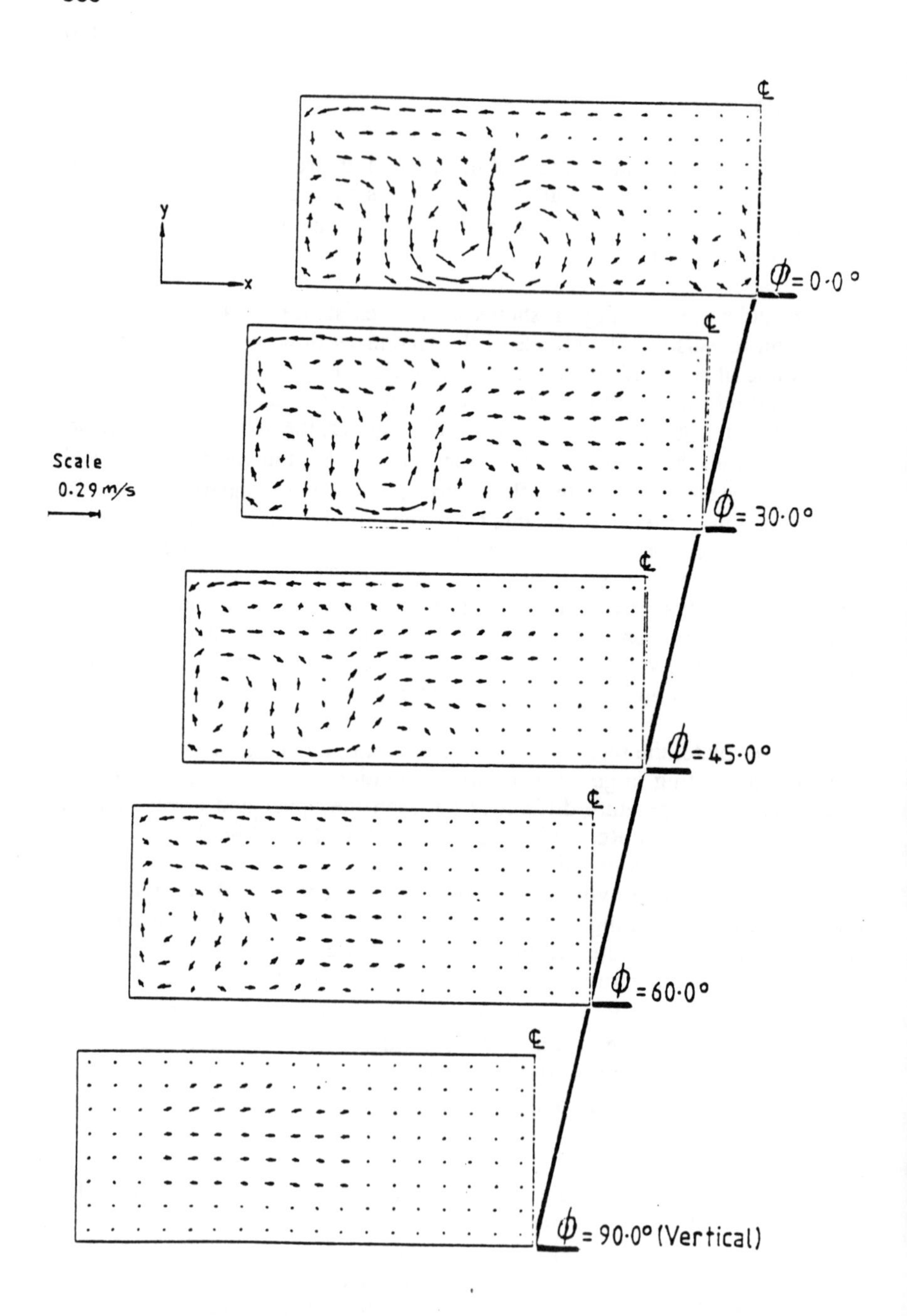

Fig. 1. Velocity vector in x-y plane at (z/L = 0.615, for different duct angles of inclination, both sides heated $Gr_{Dq} = 1.518 \times 10^7$, $Re_D = 1803$, $z^* = 0.00733$

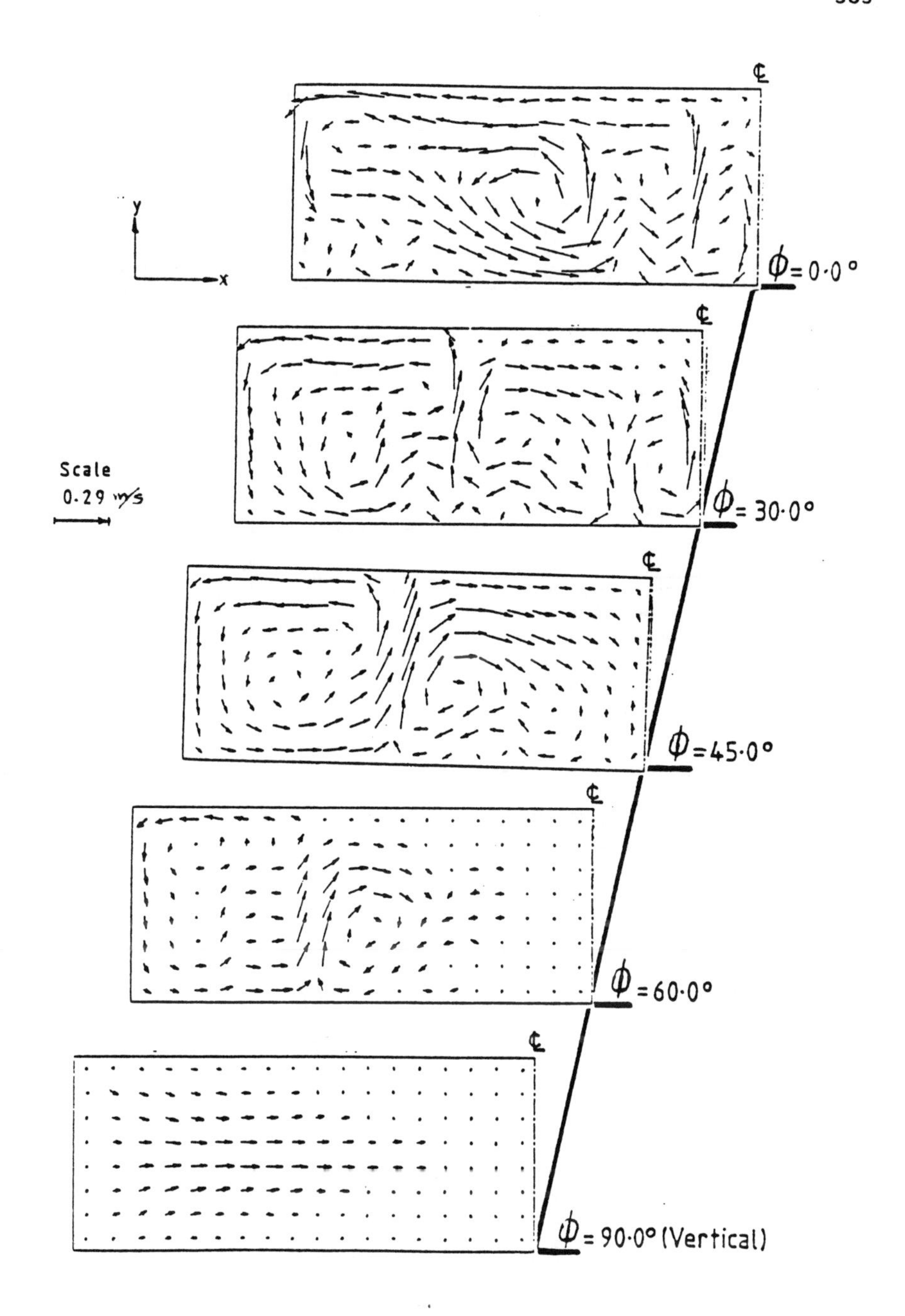

Fig. 2. Velocity vector in x-y plane at (z/L) = 0.615 for different duct angles of inclination, both sides heated $Gr_{Dq} = 1.518 \times 10^7$, $Re_D = 801$, $z^* = 0.01650$

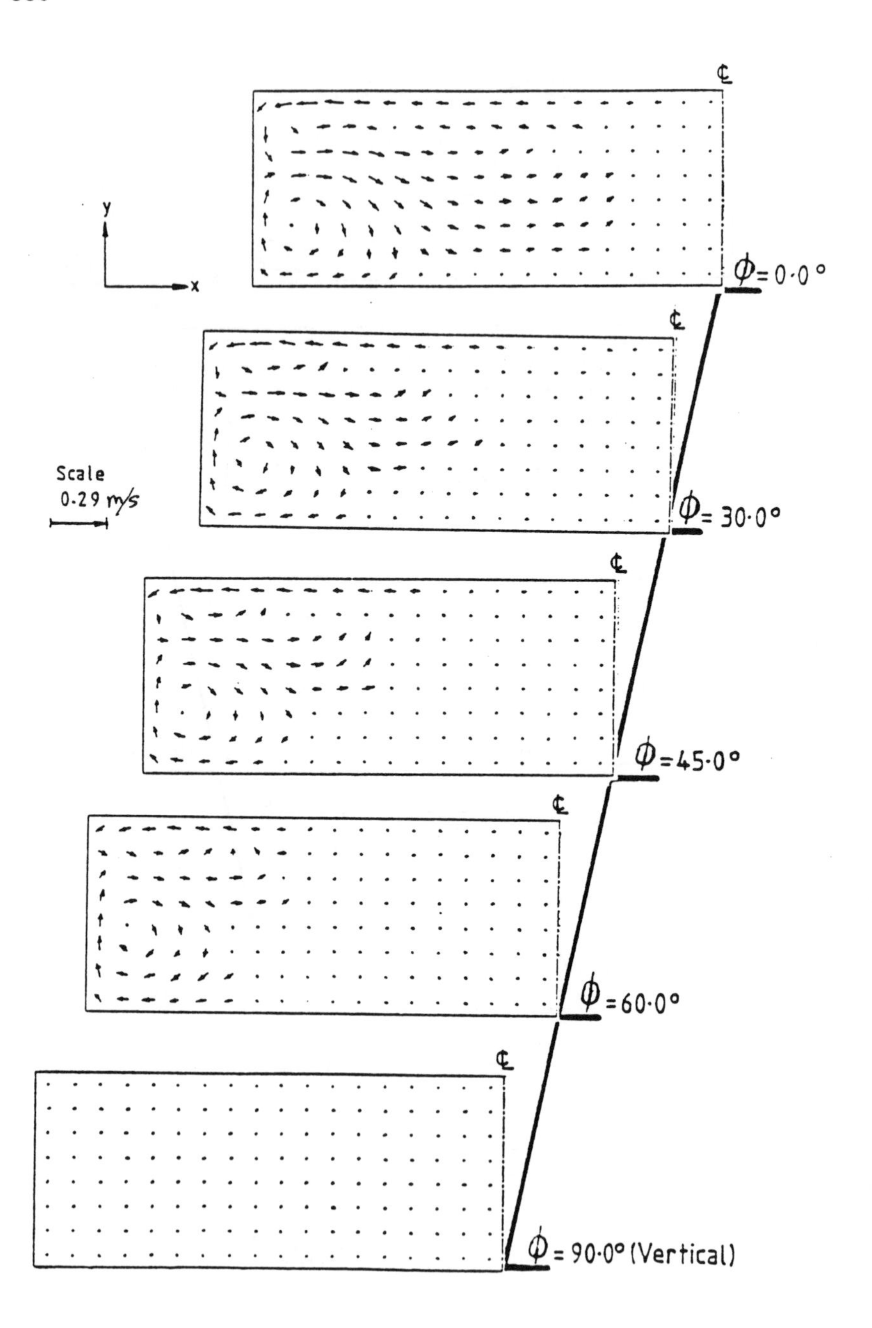

Fig. 3. Velocity vector in x-y plane at (z/L) = 0.615 for different duct angles of inclination, upper plate heated $Gr_{Dq} = 1.518 \times 10^7$, $Re_D = 1803$, $z^* = 0.0733$

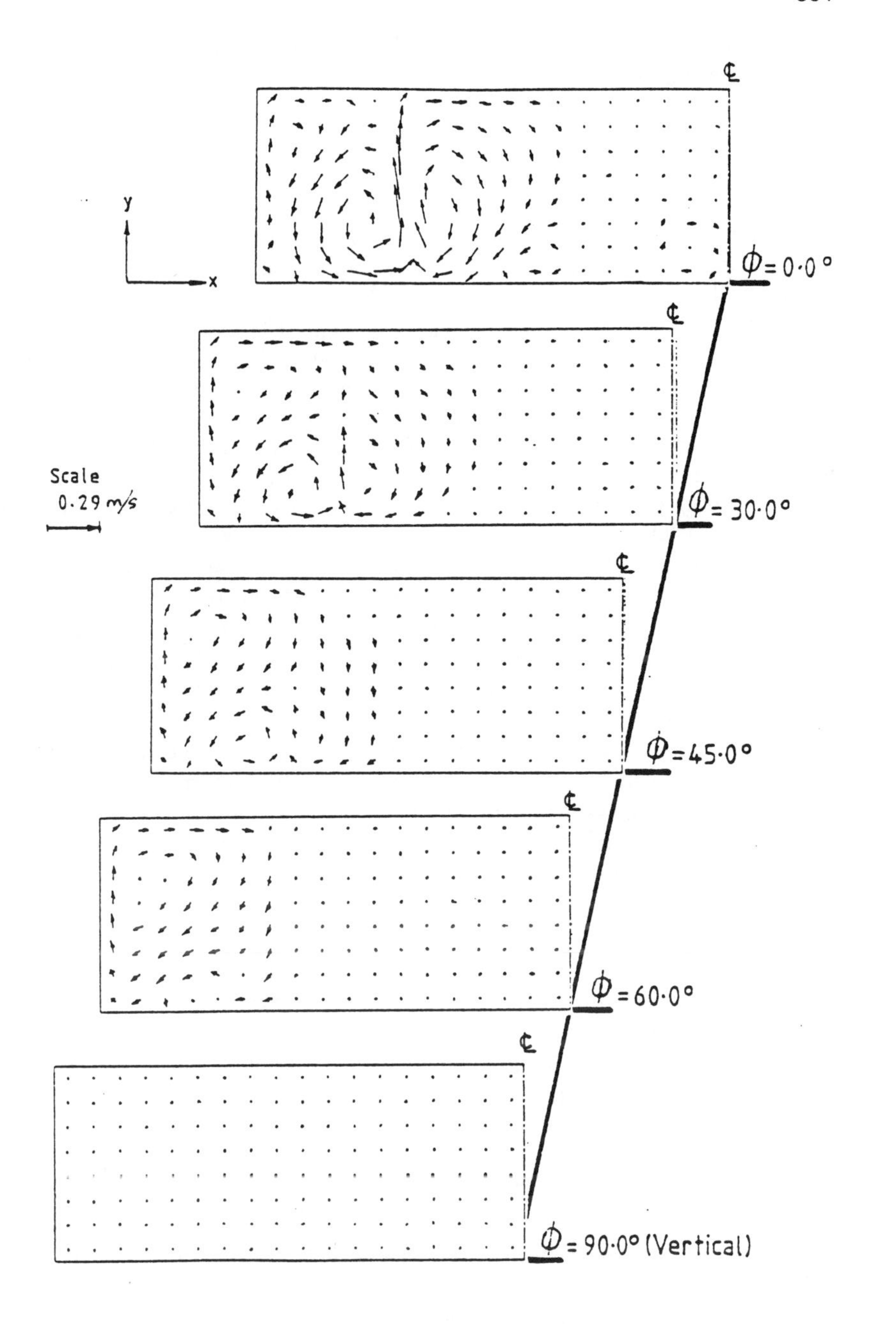

Fig. 4. Velocity vector in x-y plane at (z/L) = 0.615 for different duct angles of inclination, lower plate heated Gr_{Dq} = 1.518 x 10^7, Re_D = 1803, z* = 0.0733

to reach the upper plate. For $\varphi = 30^{o}$ the velocity of the secondary flow is reduced and it appears only to be able to penetrate up to a limited height and be not capable of reaching the upper plate. The secondary velocity continues to reduce as φ moves to the vertical and leads to a one-dimensional flow for $\varphi = 90^{o}$.

The contours of isotherms in the x-y plane, for the same cases shown in Fig. 1, Fig. 3 and Fig. 4 are displayed in Fig. 5, Fig. 6 and Fig. 7. In these figures is shown the average temperature of the horizontal surfaces, the field isotherms in ^{o}C and the bulk air temperature T_B. The both sides heated case with a relatively high Reynolds number is shown in Fig. 5. For $\varphi = 0^{o}$, there is a thin thermal boundary layer, with a sharp increase in air temperature, adjacent to the upper plate in comparison with the lower plate thermal boundary layer. As was indicated in Fig. 1, the appearance of the vortices accelerate the development of temperature distribution. The isotherms reveal the irregularity of the air temperature in the transverse direction. A high temperature difference between the average upper and lower plate temperature is shown here in comparison with the results for other angles of inclination and is equal to approximately 8°C. For $\varphi = 30^{o}$, the results show a great improvement in the upper plate heat transfer as the average plate temperature reduces by approximately 9°C. The lower plate boundary layer becomes more regular especially at the plate centre while at the plate edge there is still shown a transverse temperature irregularity due to upward secondary flow depicted in Fig. 1. For $\varphi = 45^{o}$, 60° and 90°, the upper plate results show a small drop in average plate temperature while the lower plate show no variation in its temperature. The boundary layer temperature variation rate, for the upper plate, reduces as φ moves to the vertical while the lower plate temperature variation becomes more regular. For $\varphi = 90^{o}$, complete similarity is found on both left and right surfaces.

Fig. 6 and Fig. 7 show the contours of isotherms in the x-y plane for the upper plate heated only and the lower plate heated only respectively. For the upper plate heated only, Fig. 6 shows approximately similar temperature distributions for all φ, but the gradient of temperature is larger for $\varphi = 0^{o}$ than for $\varphi = 90^{o}$. Average surface temperatures show a large drop between $\varphi = 0^{o}$ to 30°. The lower plate results, as shown in Fig. 7, reveals a small irregularity in the temperature distribution along the plate, especially at the edges of the heated plate as the results of the secondary flow vortices which were shown in Fig. 4 to exist at the plate edges. This disturbance in flow temperature distribution is reduced as φ moves towards the vertical while the average surface temperature shows a small reduction. In the results for $\varphi = 90^{o}$, the left side result is exactly similar to the right side.

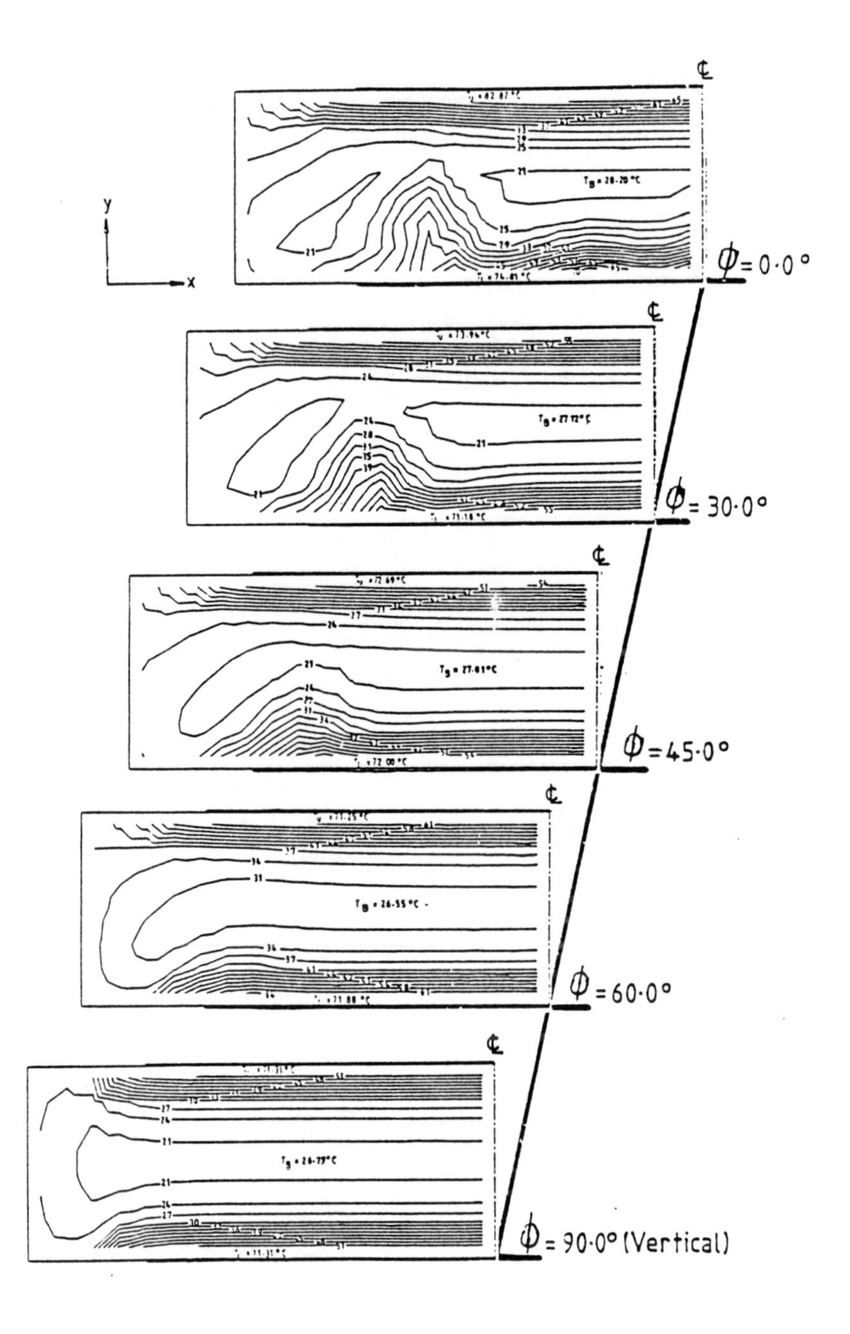

Fig. 5. Contours of isotherms in x-y plane at (z/L) = 0.615 for different duct angles of inclination, both sides heated $Gr_{Dq} = 1.518 \times 10^7$, $Re_D = 1803$, $z^* = 0.0733$

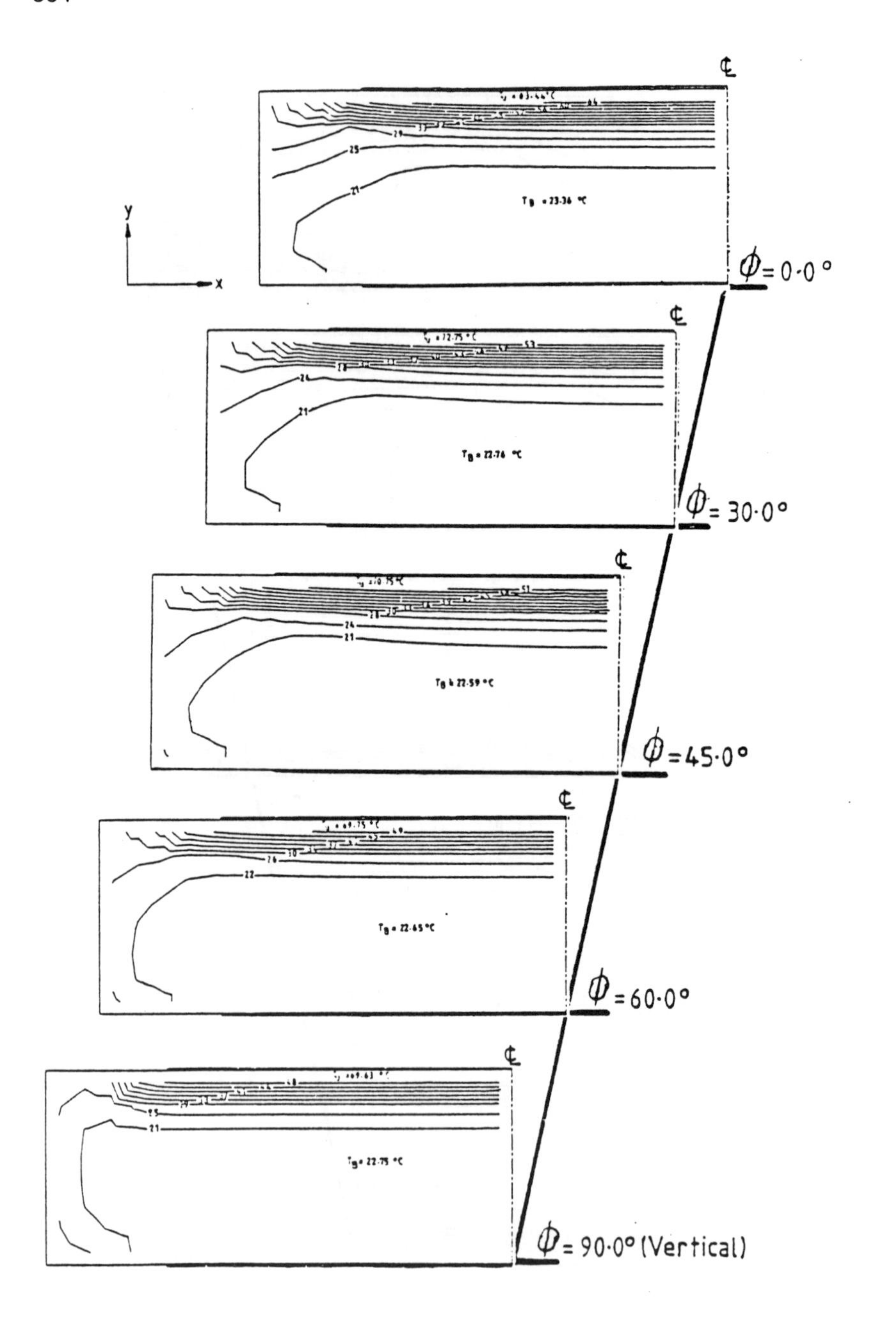

Fig. 6. Contours of isotherms in x-y plane at (z/L) = 0.615 for different duct angles of inclination, lower side heated Gr_{Dq} = 1.518×10^7, Re_D = 1803, z* = 0.0733

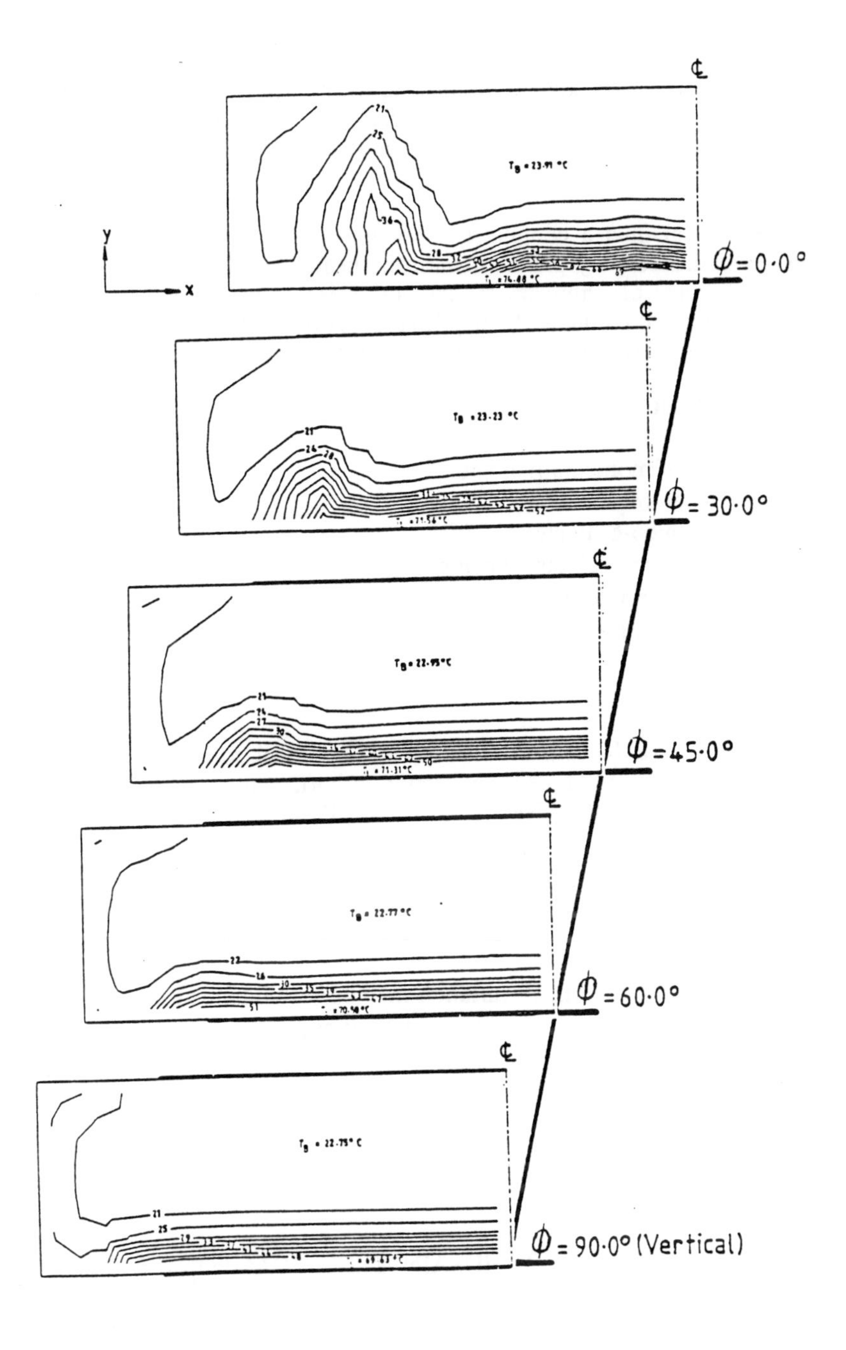

Fig. 7. Contours of isotherms in x-y plane at (z/L) = 0.615 for different duct angles of inclination, lower side heated Gr_{Dq} = 1.518×10^7, Re_D = 1803, z* = 0.0733

4. CONCLUSIONS

For the inclined ducts studied the biasing of the velocity profile towards the heated surface increases as the duct angle of inclination moves towards the vertical, increasing the flow at the duct centre and reducing it at the duct sides and corners.

For the vertical duct a completely one-dimensional flow with a very clear skewing of the velocity profile toward the heated surface is obtained with a very high degree of symmetry for the both sides heated case. Increasing the plate heat flux and reducing Re may lead to a reversed flow at the duct core as the above two factors accelerate the flow at the heated surfaces.

5. NOMENCLATURE

S source of momentum or energy
T temperature
u_j cartesian co-ordinate velocities
x_j cartesian co-ordinate velocities
ψ general variables (u_j *or* T)
ϕ duct angle of inclination

6. REFERENCES

1. K. C. CHENG and G-J HWANG, Numerical solution for combined free and forced laminar convection in horizontal rectangular channels, Trans. ASME, J. of Heat Transfer, pp. 53-66, Feb. (1969).

2. M. M. M. ABOU-ELLAIL and S. M. MORCOS, Buoyancy effects in the entrance region of horizontal rectangular channels, ASME paper 80-HT-139, (1980).

3. H. C. AGRAWAL, A variation method for combined free and forced convection in channels, Int. J. of Heat and Mass Transfer, vol. 5, pp. 439-444 (1962).

4. M. IGBAL, A. K. KHATRY and B. D. AGGARWAL, On the second fundamental problem of combined free and forced convection through vertical non-circular ducts, App. Sci. Res., 26, pp. 183-208, June (1972).

5. L. C. CHOW, S. R. HUSAIN and A. CAMPO, Effects of free convection and axial conduction on forced convection heat transfer inside a vertical channel at low Peclet number, Trans. ASME, J. of Heat Transfer, vol. 106, pp. 297-303, May (1984).

6. S. V. PATANKAR, Numerical Heat Transfer and Fluid Flow, Hemisphere Publishing Corp., McGraw-Hill Book Co., 1980.

COMBINED FORCED AND FREE LAMINAR CONVECTIVE HEAT TRANSFER FROM A VERTICAL PLATE WITH COUPLING OF DISCONTINUOUS SURFACE HEATING

Koki KISHINAMI, Hakaru SAITO and Jun SUZUKI
Department of Mechanical Engineering, Muroran Institute of Technology
27-1 Mizumoto-cho, Muroran, 050, Japan

ABSTRACT

Combined free and forced laminar air convective heat transfer from a vertical composite plate with isolated discontinuous surface heating elements has been studied numerically and experimentally, simplified by neglecting heat conduction in unheated elements of the plate to accomplish better understanding of the complicated combined/complicated convection problem. In this study, it is most important in explaining the heat transfer behavior to clarify the interactions between buoyancy and inertia forces in the convective field and also the coupling effects of unheated elements upon the combined flow fields. Therefore, the temperature distributions of the wall surface and local Nusselt number, obtained by numerical calculations and experiments, have been discussed based on the various parameters associated with the present convection problem, i.e., Grashof number Gr_L, Reynolds number Re_L, geometry factor D/L and stage number N. Heat transfer characteristics $Nu_t/Re_L^{1/2}$ of this combined and coupled convection of air are presented as a function of a generalized coupling dimensionless number Gr_L/Re_L^2, and stage number N for certain values of the geometry factor of D/L.

1 INTRODUCTION

Combined/coupled convective heat transfer has been becoming important in recent years, both in academic and in practical fields. A problem considered here has become a subject of considerable interest in several technological applications, particularly in electronic semi-conductor devices, any industrial manufacturing systems and heat transfer elements in heat exchangers appearing in engineering. The convection problem considered in this paper should be treated exactly as a combined problem of convection due to buoyancy and inertia forces, and also coupled with thermal conduction in unheated elements though it is very difficult to solve the combined /coupled free and forced convection because of the many parameters including.

Problems with flow situations similar to those considered here have been studied by many investigators [1],[2]. For example, an effect of buoyancy on forced convection flow and heat transfer is reported by Mori [2], and combined laminar convection in horizontal rectangular channels by Cheng, et al. [3]. Most of the papers presented up to now have treated heat transfer by considering the heated surface to be isothermal and have paid little attention to the case of a composite plate with discontinuous surface heating in which peculiar heat transfer may appear [4],[5].

In this study, combined forced and free laminar convective heat transfer of air from a composite plate with isolated discontinuous surface heating has been studied, where buoyancy and inertia forces are in the same direction and the induced convective flow is affected by the existence of unheated elements. Numerical calculations are carried out for laminar, forced and free mixed convection air flow along a vertical composite wall with isolated heating elements. Numerical results such as Nusselt number and velocity/temperature distributions in boundary layers are presented for the purpose of illustrating the effects of these parameters on the convective heat transfer. Experimental corroboration is provided by measuring the temperatures of the plate surface and boundary layers with ϕ 0.1 mm C-C thermocouple under the conditions of temperature difference between the heating element and the surroundings $\Delta\theta$=15~45 ℃, free stream velocity U_∞=0.2, 0.5 and 1.0 m/s, geometry factor D/L=0.5, 1.0 and 1.3 for the stage number N=4, on the basis of the characteristic length of L=100mm. Heat transfer behavior predicted by the numerical calculations and observed in the experiments are discussed based on the dimensionless number for coupling forced and free convection Gr_L/Re_L^2, stage number N, and the geometry factor D/L.

The results are summarized to propose the generalized expressions of $Nu_t/Re_L^{1/2}$ for predicting the combined and coupled convective heat transfer by using these parameters of $Gr_L/Re_L^{1/2}$, D/L and N.

2 PHYSICAL MODEL AND ANALYSIS

The schematic diagram for the physical situation of forced-free combined air convection under consideration is shown in Fig. 1. The composite vertical plate, located in the opposite direction to the gravity, is set up in the parallel flow of the free stream velocity U_∞, in which x-axis originated from the leading edge and y-axis measured from the surface of the plate are taken as the longitudinal and transverse length coordinates. Isolated heating elements with the length of L mounted on an adiabatic surface are separately located by a prescribed distance D, i.e., unheated element length D. The vertical composite plate is composed of N stages from the leading edge, counting a pair of combinations with one heated element and one adiabatic element as one stage.

The combined/coupled convection heat transfer is treated as a steady, laminar boundary layer flow developed by the cooperation of the buoyancy and inertia forces over the vertical plate. In treating the multi-stage heated elements, the following assumptions are made; the heating elements with the length of L are kept at the identical isothermal heating temperature θ_w, and unheated elements with the length of D are regarded to be adiabatic .

2.1 Governing Equation

The following dimensionless variables, commonly used for the forced con-

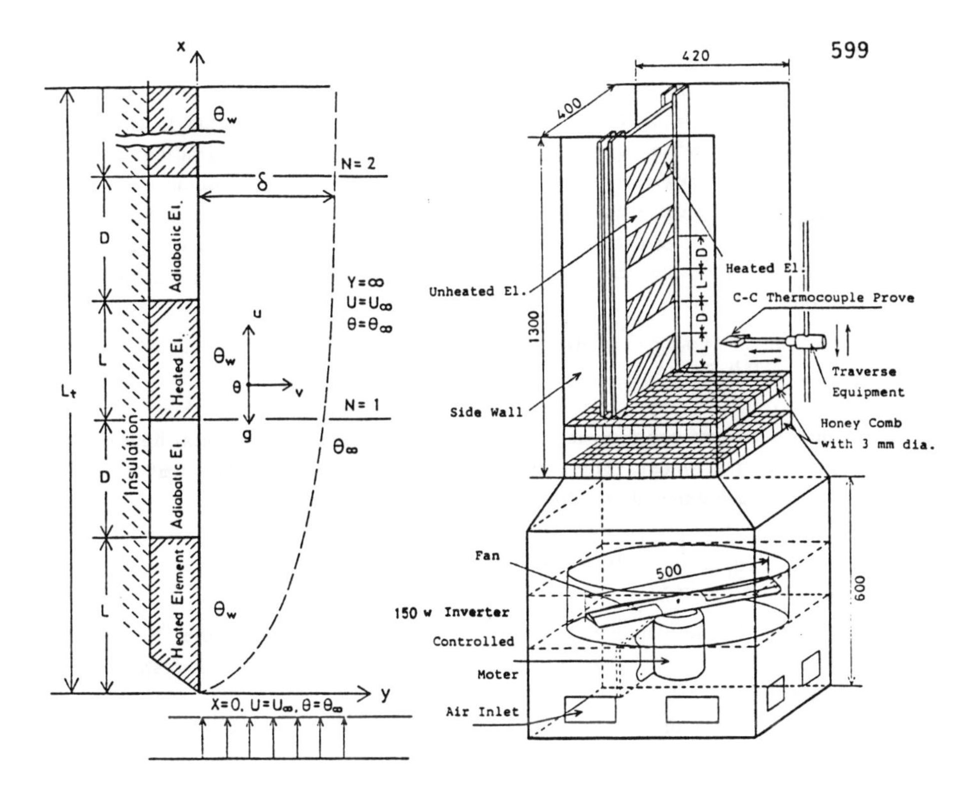

Fig. 1 Physical Model

Fig. 2 Experimental Apparatus

vection on the multi-stage heating plate [5], can be defined, by taking the heating element length L and free stream velocity U_∞ as the characteristic length and velocity.

$$X=x/L,\ Y=y/L,\ U=u/U_\infty,\ V=v/U_\infty,\ \Theta=(\theta-\theta_\infty)/(\theta_w-\theta_\infty)$$

$$Re_L=U_\infty L/\nu,\ Gr_L=g\beta L^3(\theta_w-\theta_\infty)/\nu^2,\ Pr=\nu/\alpha \tag{1}$$

The governing equations for the laminar boundary layer are given in the following dimensionless form by applying the variables of Eq. (1).

Continuity equation
$$\frac{\partial U}{\partial X}+\frac{\partial V}{\partial Y}=0 \tag{2}$$

X-momentum and energy equations

$$U\frac{\partial U}{\partial X}+V\frac{\partial U}{\partial Y}=\frac{Gr_L}{Re_L^2}\cdot\Theta+\frac{1}{Re_L}\cdot\frac{\partial^2 U}{\partial Y^2} \tag{3}$$

$$U\frac{\partial \Theta}{\partial X}+V\frac{\partial \Theta}{\partial Y}=\frac{1}{Pr\cdot Re_L}\left(\frac{\partial^2\Theta}{\partial X^2}+\frac{\partial^2\Theta}{\partial Y^2}\right) \tag{4}$$

where the standard boundary layer assumptions for the forced convection have been employed, except that heat conduction in the x-direction in the fluid is included in order to accommodate possible large temperature gradient appearing near the contact regions between the heated and unheated elements.

Based on the difference in temperature between the heated surface and surroundings ($\theta_w - \theta_\infty$), the heat transfer coefficient on the surface, regardless of heating or unheating, can be expressed as follows, by using the dimensionless variables of Eq.(1).

$$\mathrm{Nu} = \frac{h_x L}{\lambda_f} = -\left.\frac{\partial \Theta}{\partial Y}\right|_0 \tag{5}$$

In the present problem, it is most important to estimate the total heat delivery from the entire system. Then, the average/accumulated Nusselt number, Nu_t, on every stage are defined as follows,

$$\mathrm{Nu}_t = \frac{1}{L}\int_0^x \mathrm{Nu}\,\mathrm{d}x = \int_0^X \mathrm{Nu}\,\mathrm{d}X \tag{6}$$

where the average/accumulated Nusselt number, Nu_t, on the N stages is interpreted as an integration of the local Nusselt number Nu with respect to the dimensionless length of X.

Only the boundary conditions on the surfaces of heating and unheating elements are expressed as follows since the rest parts of the boundary conditions are the same as in the case of forced convection.

For the heating elements: $Y=0;\ \Theta=1$ (7)

For the adiabatic elements: $Y=0;\ \dfrac{\partial \Theta}{\partial Y} = 0$ (8)

2.2 Outline of Numerical Solution Technique

In the numerical calculation for a multi-stage combined/coupled convection flow, the governing equations (2), (3) and (4) for convection in the boundary layer are written in the form of finite difference, by taking an up-wind difference for a non-linear terms and a central difference form for other derivatives, based on the control volume technique. Hence, non-uniform node spacings are used in both X- and Y-directions in order to guarantee the accuracy of the numerical results, especially fine node spacings in the regions near the surface of the wall and the leading edge. The difference forms of the governing equations are numerically solved by the iterative technique of the SOR method. The numerical results obtained are estimated to involve the relative error less than 3 % for Nusselt number on the 1st stage.

3 EXPERIMENTAL APPARATUS AND PROCEDURE

Figure 2 shows an outline of the experimental apparatus with a composite vertical plate set up in the opposite direction to the gravity. The heat transfer composite vertical plate was sharpened at the lower leading edge and comprised of N=4 stages with isothermal heating elements in length L=100 mm and adiabatic elements in length D=50 or 100 or 130 mm. Heating elements were made of aluminum plate of 3 mm thick and warmed from the rear by an electrical heater (Nichrome wire of ϕ 0.26 mm) set in the wooden plate with the thickness of 10 mm, and controlled at a heating temperature between 35~65 ℃ within the accuracy of 0.2 ℃. Unheated elements were made from a balsa plate with thermal conductivity λ_b =0.55 W/(mK) assumed to be nearly an adiabatic. Thin aluminum foil with emissivity of 0.09 and 15 μmm in thickness was bonded tightly to the surface of the unheated balsa elements in order to minimize the thermal radiation loss. A low velocity wind tunnel (U_∞=0.05~4.0 m/s), with a 500 mm dia. propeller driven by a 150 W inverter controlled motor, was installed at the bottom of the composite vertical plate to regulate the free stream velocity U_∞ from 0.2 m/s to 1.0 m/s through the 2 stages honeycomb with dimension 400x420x30 mm(hole dia. ϕ 3 mm). The side walls, made of plywood with 1300 mm in height and 420 mm in width, were installed on the both sides of the convection plate to prevent external disturbance. The heated and adiabatic elements could be replaced to change the number of the stage N and the geometry factor of D/L. The temperature profiles in the boundary layer and surface temperature distribution on the plate were measured by a ϕ 0.1 mm C-C thermocouple prove which was possible to traverse precisely by a remote-controller, and the free stream velocity by thermistor anemometer. Experimental conditions were as follows; the difference in temperatures between the heated surface and the surroundings $\Delta\theta$ =15, 30 and 45 ℃, the free stream velocity U_∞=0.2, 0.5 and 1.0 m/s and the geometry factor D/L=0.5, 1.0 and 1.3. A steady state condition was attained at about 1 hour after the starting of the run. Then, temperatures of the boundary layer were measured at 52~64 points along the surface by remote-controlling C-C thermo-couple probe. The data obtained directly linked to a personal computer to estimate the local Nusselt number and draft figure at real time.

4 RESULT AND DISCUSSION

The basic behavior of combined free and forced convective heat transfer in this model is considerably different from that of an ordinary single isothermal plate because of the existence of the intermittent adiabatic elements.

4.1 Vectorial Dimensional and Numerical Analyses

By performing a vectorial dimensional analysis for the mixed forced-free convection heat transfer, the following expressions are obtained for the wall temperature and heat transfer coefficient when the heat conduction in unheated element is neglected.

$$Nu/Re_L^{1/2} \text{ or } \Theta_w = f(X,\ Pr,\ Gr_L/Re_L^2,\ D/L) \tag{9}$$

where X=x/L=f(N).

The main parameter (Gr_L/Re_L^2) explaining this combined/coupled problem can be provided by other aspect of the dimensionless transformation of the governing equations applying new dimensionless variables as follows,

$$X = \frac{x}{L} \quad Y = \frac{y}{L}Re_L^{1/2} \quad U = \frac{u}{U_\infty} \quad V = \frac{v}{U_\infty}Re_L^{1/2} \tag{10}$$

where the above variables are based on the forced convection boundary layer assumption while the dimensionless temperature Θ, Reynolds number Re_L, Grashof number Gr_L are the same forms defined by Eq. (1).

The dimensionless governing equations and local Nusselt number are given as follows,

$$\frac{\partial U}{\partial X}+\frac{\partial V}{\partial Y}=0 \ , \qquad U\frac{\partial U}{\partial X}+V\frac{\partial U}{\partial Y}=\frac{Gr_L}{Re_L^2}\cdot\Theta+\frac{\partial^2 U}{\partial Y^2} \qquad (11),\ (12)$$

$$U\frac{\partial \Theta}{\partial X}+V\frac{\partial \Theta}{\partial Y}=\frac{1}{Pr}\frac{\partial^2\Theta}{\partial Y^2} \ , \qquad \frac{Nu}{Re_L^{1/2}}=-\left.\frac{\partial\Theta}{\partial Y}\right|_{y=0} \qquad (13),\ (14)$$

where the heat conduction in X-direction in energy equation is neglected because of little effect on the velocity and temperature fields.

These relations by Eqs. (11), (12), (13) and (14) indicate that the results obtained by numerical calculation should be generalized solution for the every parameter $Gr_L/Re_L^{1/2}$, regardless of the values of Gr_L and Re_L. Consequently, it is concluded that the obtained results of U, Θ_w and $Nu/Re_L^{1/2}$ should be treated as the coupling parameter Gr_L/Re_L^2, geometry factor D/L and stage number N from both analyses.

4.2 Surface Temperature and Local Nusselt Number

The distributions of the surface temperature on the plate with N=4 stages predicted by the numerical calculations are shown in Fig. 3, for the geometry factor D/L=0.5, 1.0 and 1.3, indicated by a broken line, a solid line, and a chained line, respectively. Also in the figure indicated are the free stream velocity U_∞ =0.2 m/s, and temperature difference $\Delta\theta$ =30 ℃, i.e., Gr_L/Re_L^2 =2.325. The numerical results are compared with the experimental data of corresponding

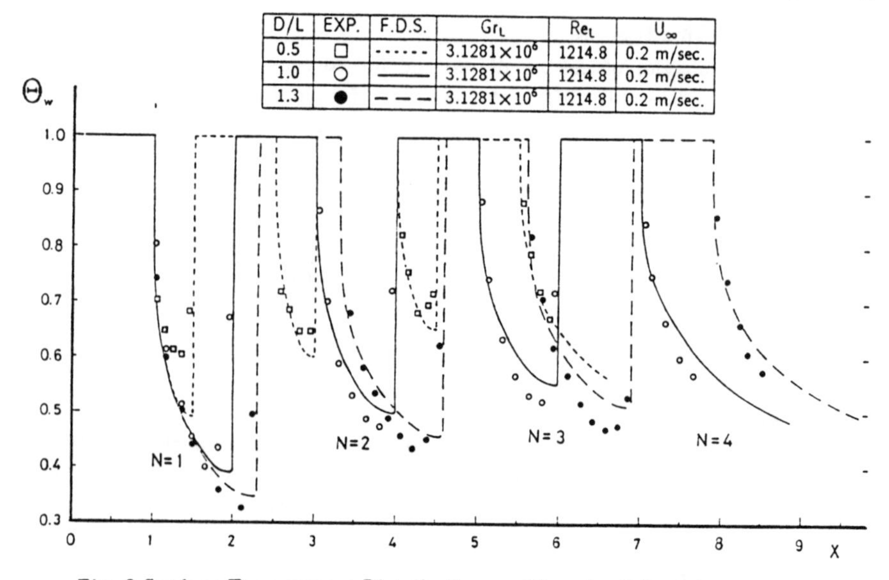

Fig. 3 Surface Temperature Distribution on Plate for D/L= 0.5, 1.0, 1.3

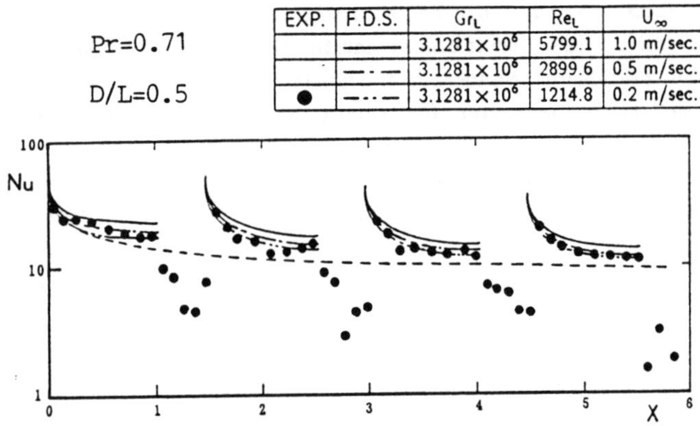

Pr=0.71

D/L=0.5

EXP.	F.D.S.	Gr_L	Re_L	U_∞
	———	3.1281×10^6	5799.1	1.0 m/sec.
	—·—	3.1281×10^6	2899.6	0.5 m/sec.
●	—··—	3.1281×10^6	1214.8	0.2 m/sec.

Fig. 4 Local Nusselt Number for D/L=0.5

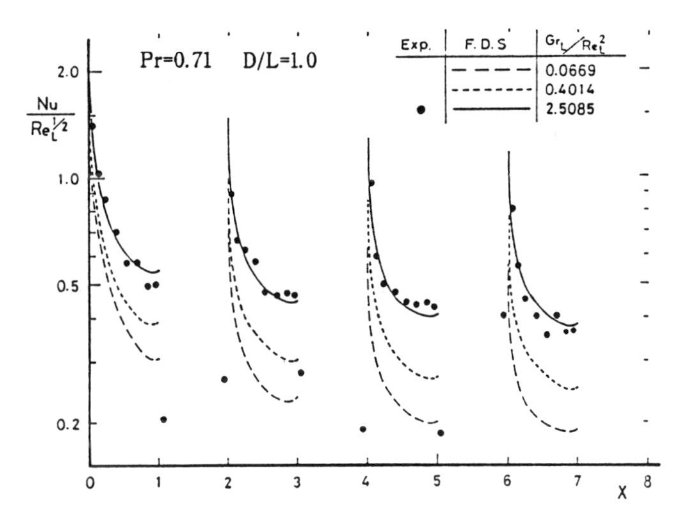

Fig. 5 Local Heat Transfer Characteristics $Nu/Re_L^{1/2}$ for D/L=1.0

conditions indicated as symbols of □, ○ and ● , respectively. The surface temperature distributions on the unheated elements are found to sharply fall near the lower contact point and then gradually decrease downstream. Also the surface temperature of the unheated elements increases in accordance with the stage number N. From the figure, the predicted numerical results are found to be in good agreement with the corresponding experimental data except the contact points region. Paying attention to the effect of the geometry factor D/L on the distribution, the surface temperature for small value of D/L=0.5 is found to be considerably higher than that for a large value of D/L while the distributions are very similar with each other. The small effect of heat conduction in unheated elements (balsa wood) are also recognized on their distribution near the contact region between the heated and unheated elements in the range of small D/L.

Figure 4 shows the local Nusselt number on every heated element under the same condition as above for D/L=0.5, with slightly changed free stream velocity U_∞ =0.2, 0.5 and 1.0 m/s for the constant Gr_L=3.125x10^6($\Delta\theta$=30 ℃). Figure 5 indicates the relation of the local heat transfer characteristics $Nu/Re_L^{1/2}$ and X with Gr_L/Re_L^2 as the parameter for D/L=1.0 under the same condition as above. In these figures, the numerical corresponding results are expressed by a solid line, a dotted chained line and a double-dotted chained line in Fig. 4, and a broken line, a dotted line and a solid line in Fig. 5, comparing with the experimental data for the U_∞=0.2 m/s expressed by a symbol ●, respectively. According to the results in these figures, Nusselt number on the first heated element changes in the same manner as an ordinary isothermal single plate , and Nusselt numbers on the following heated elements abruptly increases at every of the lower contact point due to the edge effect while the Nusselt number becomes larger with the increase in the free stream velocity U_∞. The effect of the free stream velocity changed from U_∞=0.2 m/s to 1.0 m/s at constant Gr_L is found to take place to 15～30 % promotion of Nu on the N=1 stage and 10～20 % on the following stages. The numerical results of Nusselt number are recognized to be consistent with the corresponding experimental data in Figs. 4 (D/L=0.5) and 5 (D/L=1.0), but are not always in agreement with the data in the region of unheated element due to the heat conduction effect from the upper and lower contact regions, especially in the range of small D/L. From the Fig. 5, small range of Gr_L/Re_L^2such as 0.067 is interpreted to be small effect of the buoyancy, i.e., pure forced convection flow. So, it is concluded that the buoyancy effect on Nu at constant U_∞ plays an important role of heat transfer promotion.

The heat transfer enhancement due to the edge effect is clearly recognized occurring on every heated element and trending toward weakening its effect in the region of small D/L. For reference, a pure natural convection heat transfer for a single isothermal plate at the same condition are compared with the present results expressed by a broken line.

4.3 Further Consideration for Various Thermal Conditions

Figure 6 shows the temperature distributions along the surface of the wall and local Nusselt number Nu estimated through the numerical calculations for various parameters and experiments under the specified conditions of $\Delta\theta$=30 ℃, U_∞=0.2 m/s, i.e., the coupling parameter Gr_L/Re_L^2=2.325 for the D/L =1.0. The results illustrated in the figure includes the some cases taking the combination effect of the buoyancy on the inertia forces and the existence of the unheated elements into consideration; one is the result obtained by considering the effect of heat conduction in unheated balsa wood elements (solid line), the others are by neglecting the thermal conduction in the unheated elements as adiabatic (double-dotted line), and the others by only the inertia forces, i.e., pure forced convection (broken line). It seems that there is little difference in both numerical results indicated by a solid line and a double-dotted chained line. This is due to the insulation effect of the balsa wood and the results indicated by the solid line are found to be in better agreement with the experimental data than the results by the double-dotted chained line, especially in the vicinity of the contact region of the unheated elements. Therefore, it can be said that the simplification employed in this calculation (the unheated elements as adiabatic) is

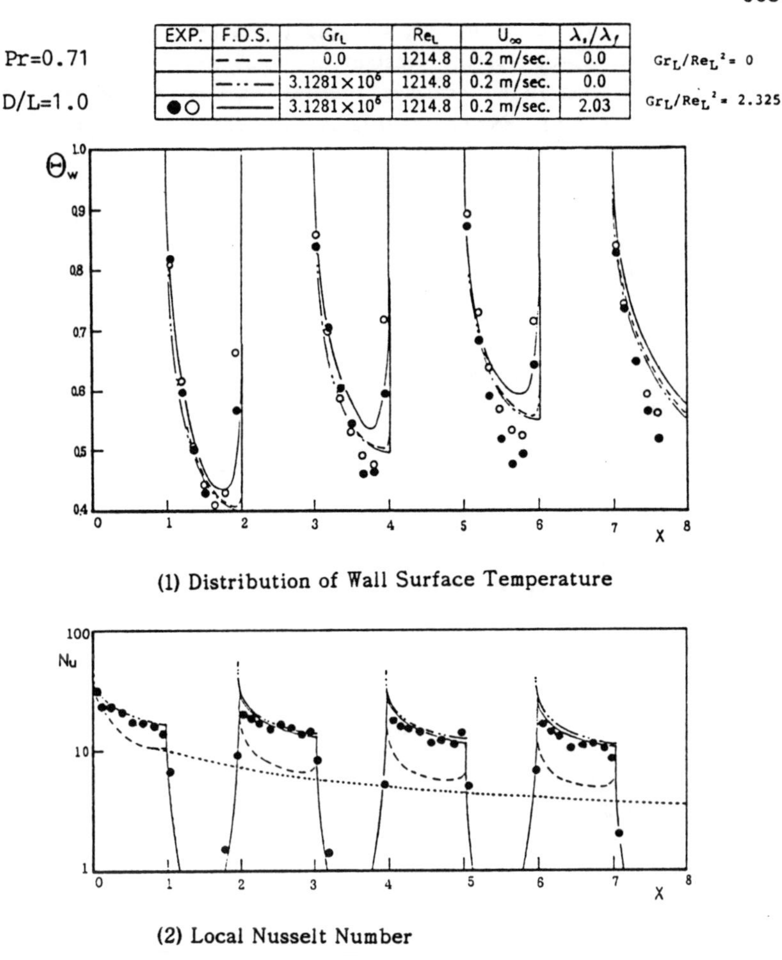

EXP.	F.D.S.	Gr_L	Re_L	U_∞	λ_s/λ_f
	— — —	0.0	1214.8	0.2 m/sec.	0.0
	—··—	3.1281×10^6	1214.8	0.2 m/sec.	0.0
●○	———	3.1281×10^6	1214.8	0.2 m/sec.	2.03

(1) Distribution of Wall Surface Temperature

(2) Local Nusselt Number

Fig. 6 Results for Various Thermal Conditions

reasonably applicable to the present problem for the case of large D/L. In the figure, the broken line corresponds to the case of pure forced convection by neglecting the buoyancy term in the momentum equation. Accordingly, the local Nusselt number indicated by the broken line is found to be reduced by about 2/3 ~1/2 in its value in accordance with the stage number N, comparing with the solid line for the combined free-forced convection. Contrary to this, little difference in the surface temperatures can be seen between both results. The fine dotted line, i.e., local Nusselt number for the case of the pure forced convection on a continuous isothermal heating plate, indicates that Nu is considerably lower than the solid line for the combined convection. Hence, it is expected that the strong effects of the buoyancy to the boundary layer exists in the vicinity of the wall surface.

Figures 7 and 8 show the temperature and velocity profiles of the boundary layers in the transition region from N=1 to N=2 stages for the conditions of

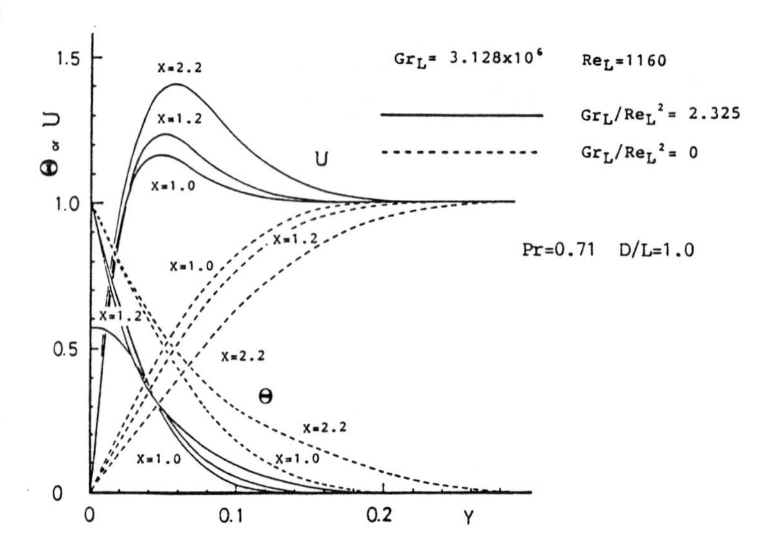

Fig. 7 Behavior of Boundary Layer Affected by Buoyancy for Gr_L/Re_L^2=2.325

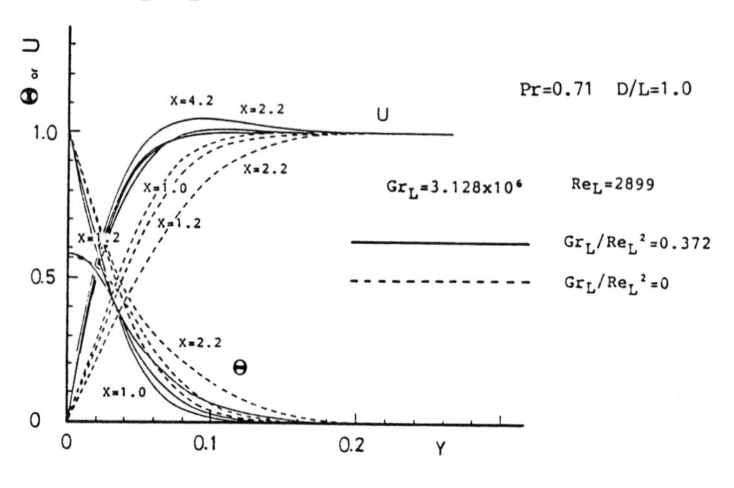

Fig. 8 Behavior of Boundary Layer Affected by Buoyancy for Gr_L/Re_L^2=0.372

Gr_L/Re_L^2=2.325 ($\Delta\theta$=30 ℃, U_∞=0.2 m/s) and Gr_L/Re_L^2 =0.372 ($\Delta\theta$=30 ℃, U_∞=0.5 m/s). Also, the solid line and dotted line correspond to the cases of the combined free-forced convection (buoyancy and inertia forces being considered) and the pure forced convection (buoyancy force being neglected), respectively. The results represented by the solid line for the Gr_L/Re_L^2=2.325 reveal that the strong effects of the buoyancy on the velocity and temperature distributions exist in the vicinity of the wall surface and become more evident with an increase in the X-location, in comparison with the case of the dotted line. Subsequently, the local Nusselt number of the combined free-forced convection is promoted by 3

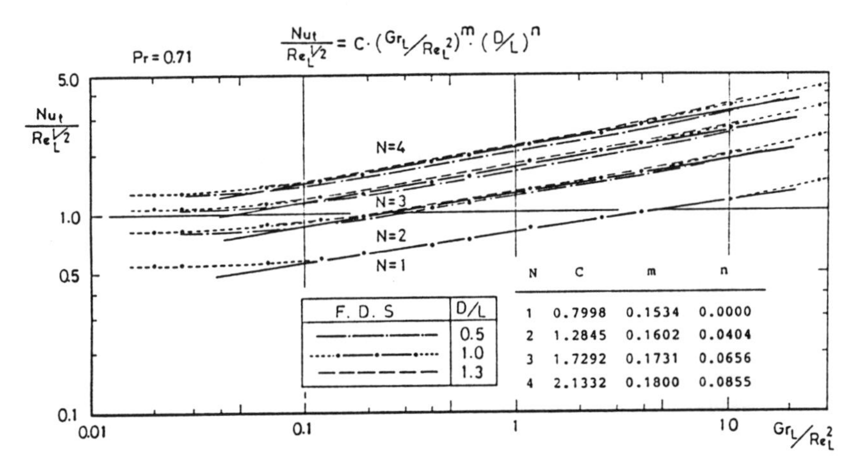

Fig. 9 Relationship between Mean Heat Transfer Characteristics $Nu_t/Re_L^{1/2}$ and Generalized Coupling Parameter Gr_L/Re_L^2 with D/L as Parameter

~5 times in comparison with that of the pure forced convection as shown in Fig. 6. On the other, for the case of $Gr_L/Re_L^2=0.372$, the effects of the buoyancy on the velocity and temperature boundary layers are found to be not so strong as that of $Gr_L/Re_L^2=2.325$, even though the velocity and temperature distributions near the wall surface are considerably affected by the buoyancy.

An observation of the flow patterns and measurement of the temperature reveal that the stable laminar flow covers the entire surface of the plate. It is also recognized that the combined free and forced convection flow considered here is more stable than the pure natural convection flow.

4.4 Heat Transfer Characteristics, $Nu_t/Re_L^{1/2}$

Figure 9 shows the relationship between the heat transfer characteristics $Nu_t/Re_L^{1/2}$ at every heating element (N=1~4) and the generalized coupling parameter Gr_L/Re_L^2 of the mixed convection based on the dimensional and numerical analyses for the case of D/L=0.5,1.0 and 1.3. It can be concluded that the heat transfer characteristics $Nu_t/Re_L^{1/2}$ are completely dependent on the generalized coupling parameter Gr_L/Re_L^2 and geometry factor D/L in the following relation:

$$Nu_t/Re_L^{1/2}=C\,(Gr_L/Re_L^2)^m(D/L)^n \quad (15)$$

where C is a constant and an exponent m is the slope of the relation at every stage. For convenience, the constant C and the exponents m,n determined by the least square method for multi variables are tabulated in the figure for the range of $0.07< Gr_L/Re_L^2<7$. Eq. (10) is interpreted as the generalized relation of the heat transfer characteristics of the present combined/coupled free and forced convection and within the 3 % error in the prescribed range.

For the range of $Gr_L/Re_L^2<0.07$, the exponent m asymptotically approaches zero, i.e., Nu_t is proportional to $Re_L^{1/2}$. On the other, for the range of

$Gr_L/Re_L^2>7$, the constant m gradually approaches 0.25, i.e., Nu_t is proportional to $Gr_L^{1/4}$. Consequently, forced convection heat transfer predominates for the range of <0.03 and natural convection heat transfer for the range of $Gr_L/Re_L^2>7$, even though the transition range of Gr_L/Re_L^2 seems to become narrow in accordance with the stage number N as shown dotted line in Fig. 9.

5 CONCLUSION

In this paper, a combined and coupled convective heat transfer from the composite plate with an isolated heating surface was studied by numerical analysis and experiments for the wide range of thermal conditions. It is clear that the heat transfer characteristics $Nu_t/Re_L^{1/2}$ at every stage are strongly dependent on the existence of the unheated section, i.e., D/L, and on the generalized coupling parameter, Gr_L/Re_L^2.

NOMENCLATURE

D : length of unheated element
L : length of heated element as characteristic length
Gr_L : Grashof number based on the temperature difference between the heated element and surroundings, ($\theta_w - \theta_\infty$), and L
Re_L : Reynolds number based on the free velocity U_∞ and L
h : Heat transfer coefficient based on ($\theta_w - \theta_\infty$)
N : stage number from the leading edge
Nu : local Nusselt number defined by Eq. (5)
Nu_t : mean Nusselt number defined by Eq. (6)
Pr : Prandtl number
Θ : dimensionless temperature defined by Eq. (1)
U,V : dimensionless vertical and transverse velocities defined by Eq. (1)
U_∞ : free stream velocity
X,Y : dimensionless vertical and transverse coordinates

REFERENCES

1. Mabuti. I. -Heat Transfer from a Heated Vertical Plate in Flow Field of Low Reynolds Number. Trans. of JSME, Vol. 27, No. 180, pp.1299-1305, 1961.
2. Mori. Y. -Bouyancy Effects in Forced Laminar Convection Flow over a Horizontal Flat Plate. Trans. of ASME, J. of Heat Transf., Vol.83, pp.479-482, 1961.
3. Cheng. K. C. -Numerical Solution for Combined Free and Forced Laminar Convection in Horizontal Rectangular Channels. Trans. of ASME, J. of Heat Transf., Vol. 91, pp.59-66, 1969.
4. Kishinami. K. -A Fundamental Investigation of Laminar Natural Convective Heat Transfer from a Vertical Plate with Discontinuous Surface Heating. 29th Series of Archival Publishings of the ICHMT, pp.139-153, 1991.
5. Kishinami. K. -Natural Convective Heat Transfer on a Vertical Plate with Discontinuous Surface-Heating. Proceedings of 1987 ASME-JSME Thermal Engn. Joint Conf., pp.61-68, 1987.

FINITE ELEMENT SIMULATION OF COMBINED BUOYANCY AND THERMOCAPILLARY FREE BOUNDARIES GOVERNED BY THE NAVIER-STOKES EQUATIONS

Bala Ramaswamy
Department of Mechanical Engineering and Materials Science
Rice University, Houston, Texas 77251-1892, U.S.A.

Hwar-Ching Ku
Johns Hopkins University Applied Physics Laboratory
Johns Hopkins Road, Laurel, MD 20723, U.S.A.

SUMMARY

Buoyancy and thermocapillary driven convection is studied using Finite Element Method. The moving free surface is handled through Arbitrary Lagrangian Eulerian (ALE) description of flow. The Navier-Stokes equations are discretized in time using the Fractional Step method and pressure is calculated from the Poisson equation. The free surface is assumed to be resting on vertical spines and its evolution in time is determined from the kinematic equation for the free surface. The physical domain is an open cavity with an upper free surface and a hot left wall and a cold right wall. The influence of various non-dimensional numbers such as Marangoni number, Grashof number, Ohnesorge number, Bond number and Prandtl number on the flow field and heat transfer is investigated.

1. INTRODUCTION

Temperature gradients in the presence of a body force such as gravity induce bulk fluid flow. The heat transfer and fluid flow associated with the above phenomena are called natural convection or Grashof convection. Presence of free surface can cause convective motions in the fluid as well. Convective motion induced by surface tension variations resulting from temperature gradients is termed thermocapillary convection or Marangoni convection. In most situations thermocapillary convection is quite small compared to natural convection and is often neglected. Thermocapillary convection becomes important in situations where gravitational force is small such as in microgravity environments and in cases where the fluid dimensions are very small such as in thin fluid layers. Combined buoyancy-induced and thermocapillary-driven convection phenomena is important in crystal-growth processes under microgravity environments[1, 2, 3, 4] and in laser surface melting, surface coating and welding [5].

The governing equations are the Navier-Stokes equations and the energy equation. Analytical solution of Navier-Stokes is very difficult due to the presence of the non-linear advection terms. In addition to this we have additional non-linearity in the form of free surface whose location is not known a *priori* and needs to be determined as part of the solution. Careful examination of the problem would reveal that flow is influenced by viscous forces, inertial forces, buoyancy forces and surface tension forces. All these phenomena interact in a complex way giving rise to a variety flow and heat transfer phenomena. Most of the theoretical studies to date have been under simplifying assumptions with the aim of isolating a single feature and studying it in detail. Analytical solutions for surface tension driven convection in infinitely long thin liquid layers neglecting inertial forces has been given by Levich etal[6] and Pimputkar etal[7].

With the advent of high speed computing devices numerical studies have gained prominence in the last few years. Bergman etal[8] assumed the free surface to be flat and investigated combined buoyancy and thermocapillary driven convection using finite difference method. Boundary-fitted coordinate transformations have been used in finite differences and control volume formulations to overcome the problem posed by irregular free surface [9, 10, 11, 12]. Finite element method can handle irregular geometries more easily and has been used by various researchers for the study of combined convection due to buoyancy and thermocapillary[13, 14].

Since the free surface location is not known a *priori*, it needs to be determined either through transient analysis or an iterative procedure. At the free surface three boundary conditions are to be imposed, namely, normal stress continuity, tangential stress continuity and kinematic equation. But only two boundary conditions can be imposed in the Navier-Stokes solution. Thus two of the three boundary conditions are imposed during the solution of Navier-Stokes equations and the the remaining boundary condition is used to establish an iteration procedure to determine the steady state free surface location. There are many ways this can be done, with the most popular being normal stress iteration[9, 13].

In this paper the steady state free surface is obtained through transient calculations of Navier-Stokes equations and energy equation. Navier-Stokes equations in primitive variables are solved. Galerkin Finite Element method is used for spatial discretization and the flow field is updated in time using Fractional Step Method. Arbitrary Lagrangian Eulerian description of flow is employed and the free surface is considered to be resting on vertical spines. The changing shape of the free surface is determined from the kinematic equation for the free surface.

2. GOVERNING EQUATIONS

The physical problem is a open cavity of length L. The left, right and bottom surfaces of the cavity are rigid and impermeable walls. The top surface of the cavity is the free surface and changes with time. The physical description of the problem with the boundary conditions and other relevant information pertaining to the geometry of the problem is shown in Fig. 1. The origin of the coordinate frame of reference is at the lower left corner of the cavity. $h(x,t)$ is the height of the free surface from the bottom wall. The velocity field is described by the vector $\mathbf{u} = (u, v)$ and the temperature field by the scalar T.

The governing equations are the conservation laws for mass, momentum and energy and the kinematic equation for free surface. Some of the important assumptions made in simplifying these equations are:

1. fluid flow is Newtonian, laminar and incompressible;

2. dynamic viscosity (μ), thermal diffusivity (α) and specific heat at constant pressure (c_p) are constant (i.e., not dependent on temperature or pressure);

3. gravity is the only external body force and there are no internal heat sources;

4. Boussinesq approximation for density(ρ) is valid. Thus the variations in density are neglected everywhere except in the gravity term.

5. surface tension(σ) is a linear function of temperature. Moreover, we assume the surface tension to be constant everywhere except in the boundary condition for shear stress continuity at the free surface.

The dependence of density and surface tension on temperature can be expressed as:

$$\rho = \rho_0(1 - \beta(T - T_0)) \tag{1}$$

$$\sigma = \sigma_0(1 - \gamma(T - T_0)) \tag{2}$$

The subscript 0 refers to a reference state. $\beta = -\frac{1}{\rho_0}\frac{d\rho}{dT}|_{T_0}$ is the volume expansion coefficient, and $\gamma = -\frac{1}{\sigma_0}\frac{d\sigma}{dT}|_{T_0}$ is the temperature coefficient of the surface tension. γ could be positive or negative but for most of the liquids and crystals it is positive.

Non-dimensionalization not only makes the equations independent of units, but also helps gain insight into the various physical phenomena involved. Let L be the reference length and U be the reference velocity. The non-dimensional parameters are:
$x^* = x/L$, $y^* = y/L$, $t^* = tU/L$;
$u^* = u/U$, $v^* = v/U$, $p^* = p/(\rho_0 U^2)$, $h^* = h/L$. $\theta = (T - T_0)/\Delta T$;
where $\Delta T = T_h - T_c$. g is the gravitational acceleration, ν is the kinematic viscosity and $T_0 = (T_h + T_c)/2$ is the reference temperature.

The non-dimensional governing equations in ALE formulation can then be written as (dropping the superscript * for convenience):

$$u_{,x} + v_{,y} = 0 \tag{3}$$

$$u_{,t} + (u - w_x)u_{,x} + (v - w_y)u_{,y} = -p_{,x} + \frac{1}{Re}\nabla^2 u \tag{4}$$

$$v_{,t} + (u - w_x)v_{,x} + (v - w_y)v_{,y} = -p_{,y} + \frac{1}{Re}\nabla^2 v + \frac{Gr}{Re^2}\theta - \frac{Bo}{(OhRe)^2} \tag{5}$$

$$\theta_{,t} + (u - w_x)\theta_{,x} + (v - w_y)\theta_{,y} = \frac{1}{RePr}\nabla^2\theta \tag{6}$$

$$h_{,t} + uh_{,x} = v \tag{7}$$

w_x and w_y are the velocities of the grid point in x and y directions respectively. Above equations may be interpreted as material conservation laws with respect to arbitrary moving grid points. In the event that a grid point may coincide with a material point, the relative velocity term $(u_i - w_i)$ becomes zero, resulting in the vanishing of the convective terms, and consequently the set of equations become Lagrangian. Similarly, a pure Eulerian description is obtained by simply setting $w_i = 0$.

The hydrodynamic and thermal boundary conditions are as follows:

$$u = 0, \quad v = 0, \quad \theta = 0.5, \quad \text{at } x = 0. \tag{8}$$

$$u = 0, \quad v = 0, \quad \theta = -0.5, \text{at } x = 1. \tag{9}$$

$$u = 0, \quad v = 0, \quad \theta = 0.5 - x, \quad \text{at } y = 0. \tag{10}$$

The boundary conditions at the free surface are obtained from the traction conditions for normal and tangential stresses. Let n and τ denote the normal and tangential directions on the free surface boundary $y = h(x, t)$.

$$-p + \frac{2}{Re}\frac{\partial u_n}{\partial n} = \frac{1}{(OhRe)^2}\frac{1}{R} - p_a, \quad \text{at } y = h(x, t). \tag{11}$$

$$\frac{1}{Re}\left(\frac{\partial u_n}{\partial \tau} + \frac{\partial u_\tau}{\partial n}\right) = \frac{Ma}{Re^2 Pr}\frac{\partial \theta}{\theta \tau}, \quad \text{at } y = h(x, t). \tag{12}$$

where $R = h_{,xx}/(1 + h_{,x}^2)^{3/2}$ is the radius of curvature. The free surface is assumed to be insulated.

$$\frac{\partial \theta}{\partial n} = 0, \quad \text{at } y = h(x, t). \tag{13}$$

The boundary conditions for $h(x, t)$ are the specified contact angles.

$$h_{,x}(0, t) = tan(\phi - \pi/2), \qquad h_{,x}(1, t) = -tan(\phi - \pi/2) \tag{14}$$

Since the liquid volume remains constant for any instant, it means

$$\int_0^1 h(x, t)dx = V_0 \tag{15}$$

where V_0 is the dimensionless initial volume. Equations 3-7 are to be solved for the five unknowns u, v, p, θ, h subject to the boundary conditions given by equations 8-14. The non-dimensionalization procedure has resulted in the following non-dimensional numbers:

- $Re = \frac{UL}{\nu}$, Reynolds number
- $Gr = \frac{g\beta\Delta T L^3}{\nu^2}$, Grashof number
- $Ma = \frac{\gamma\Delta T L}{\mu\alpha}$, Marangoni number
- $Oh = \frac{\mu}{(\rho_0\sigma_0 L)^{1/2}}$, Ohnesorge number
- $Pr = \frac{\nu}{\alpha}$, Prandtl number
- $Bo = \frac{g\rho_0 L^2}{\sigma_0}$, Bond number

4. NUMERICAL PROCEDURE

The absence of pressure in the continuity equation of the incompressible Navier-Stokes equations makes straight forward time discretization of the equations difficult. Penalty formulation, stream function- vorticity formulation and primitive variable formulation are some of the most widely used procedures. The boundary conditions at the free surface are in terms of pressure and are easier to impose if pressure is also a primary variable. Due to this reason we chose to use primitive variable formulation. The continuity equation is replaced with an equivalent pressure Poisson equation. The philosophy of the Fractional Step Method is to first determine approximate velocities satisfying the boundary conditions but which are not necessarily divergence free. These approximate velocities are then projected onto divergence free spaces resulting in final divergence free velocities. This method of discretizing Navier-Stokes equations in time is widely used[15, 16, 17]. The procedure by which the $(n+1)$th time step velocities are calculated from nth time step values can be divided into three steps.

Step 1: The intermediate velocities $\tilde{u}^{n+1}, \tilde{v}^{n+1}$ are calculated subject to the velocity and stress boundary conditions described in the previous section. Viscous terms are treated implicitly and advection terms are treated explicitly using Euler Forward scheme.

$$\frac{\tilde{u}^{n+1} - u^n}{\Delta t} - \frac{1}{Re}\nabla^2 \tilde{u}^{n+1} = -p^n_{,x} - (u^n - w_x)u^n_{,x} - (v^n - w_y)u^n_{,y} \tag{16}$$

$$\frac{\tilde{v}^{n+1} - v^n}{\Delta t} - \frac{1}{Re}\nabla^2 \tilde{v}^{n+1} = -p^n_{,y} - (u^n - w_x)v^n_{,x} - (v^n - w_y)v^n_{,y} + \frac{Gr}{Re^2}\theta^n - \frac{Bo}{(OhRe)^2} \tag{17}$$

Step 2: The pressures on free surface boundary are calculated from the normal stress balance equation (11) and imposed as essential boundary conditions in the pressure Poisson equation.

$$\nabla^2 \hat{p} = \frac{1}{\Delta t}(\tilde{u}^{n+1}_{,x} + \tilde{v}^{n+1}_{,y}) \tag{18}$$

$$p^{n+1} = p^n + \hat{p} \tag{19}$$

Step 3: The final divergence free velocities u^{n+1}, v^{n+1} are calculated from:

$$\frac{u^{n+1} - \tilde{u}^{n+1}}{\Delta t} = -\hat{p}_{,x} \tag{20}$$

$$\frac{v^{n+1} - \tilde{v}^{n+1}}{\Delta t} = -\hat{p}_{,y} \tag{21}$$

In this step the mass matrix obtained from Galerkin formulation is lumped.

After the velocities are calculated temperatures are determined from the energy equation. The convection terms are treated explicitly using Euler Forward scheme and the conduction terms are treated implicitly.

$$\frac{\theta^{n+1} - \theta^n}{\Delta t} - \frac{1}{RePr}\nabla^2 \theta^{n+1} = -(u^n - w_x)\theta^n_{,x} - (v^n - w_y)\theta^n_{,y} \tag{22}$$

A method to update the free surface with time is required. There are three hydrodynamic boundary conditions that need to be satisfied at the free surface, namely, normal stress balance, shear stress balance and kinematic equation for free surface. However during the solution of Navier-Stokes equations only two boundary conditions can be imposed. Thus there are different approaches possible depending upon which boundary condition is chosen to update the free surface position. Most of the algorithms use the shear stress as natural boundary condition and kinematic equation as the essential boundary condition for steady state formulation, then apply the normal stress equation to correct the free surface. This is the so called normal stress iteration[9, 13]. One alternative is to substitute the normal stress and shear stress boundary conditions into the momentum equations as natural boundary conditions to calculate the domain variables, then apply the kinematic equation to calculate the new free surface. This corresponds to the kinematic iteration. We used the kinematic equation to determine the evolution of free surface in time. Thus the position of the free surface is calculated from the following equation:

$$\frac{h^{n+1} - h^n}{\Delta t} + u^n h^n_{,x} = v^n \tag{23}$$

The grid points are moved in the vertical direction while keeping them stationary in the horizontal direction which implies $w_x = 0$ where as w_y is non-zero. This is the same as mesh points lying on vertical spines and has already been used by many authors (e.g. Saito[18], FIDAP[19]) to describe free surface flow. Once the free surface location is known the mesh points are redistributed along the spines in a predetermined way. Knowing the new location of the grid points the mesh point velocity can be calculated as:

$$w_x = 0, \qquad w_y = \frac{y^{n+1} - y^n}{\Delta t} \tag{24}$$

The above equations are discretized in space using 3-noded non-staggered triangular elements. Thus the unknowns u, v, p, θ are at the three vertices of the triangle. The kinematic free surface equation is evaluated using explicit first order upwind finite difference method. No upwinding or artificial viscosity was used for the momentum and energy equations. A typical finite element mesh is shown in Fig. 2. Most of the calculations have been done using 21 grid points in x and y directions giving rise to 800 triangular elements and 441 grid nodes. The vertical spines are equispaced in x direction. Once the new free surface location is known the grid points are redistributed along the vertical spines such that they are equispaced in y direction.

Since there is no fluid entering or leaving the flow domain the total volume of the fluid should not change with time. This fact was verified by comparing the final steady state volume to the initial volume and the error was found to be less than 2% in all our simulations. The maximum possible time step such that the system of equations are stable is chosen as the time step. The system of equations is solved using banded Cholesky factorization and all the simulations have been run on CRAY-YMP at NCSC. It took less than 50 CPU seconds to obtain the final steady state answers.

5. RESULTS AND DISCUSSION

The fluid is assumed to be at rest initially and the non-dimensional volume is taken to be one. The initial free surface is taken to be flat and the temperatures to vary linearly from the hot wall to the cold wall. The physical problem considered including the boundary conditions is the same as that investigated by Cuvelier etal[13]. The Reynolds number Re is taken to be one in all our simulations. This corresponds to assuming the reference velocity U to equal ν/L. We investigated the effects of buoyancy, surface tension, combined effects of buoyancy and surface tension and Prandtl number. The isotherms and velocity vectors are plotted on the same picture. For comparison purpose the maximum total velocity $|\mathbf{u}|_{max}$, free surface height at the hot wall $h(0)$ and free surface height at the cold wall $h(1)$ are given in tabular form below the pictures. The values given in brackets are those obtained by Cuvelier[13].

The two driving forces are the buoyancy and thermocapillary phenomena. $Ma = 0$ corresponds to the case where there is no thermocapillary driven convection which means that the surface tension is independent of temperature.

However, surface tension itself is non-zero and Ohnesorge number represents the ratio viscous forces to capillary forces. Grashof number expresses the ratio of buoyancy forces to viscous forces. Increased Gr leads to greater flow velocities thus increasing the convective transport. Moreover increased Gr leads to increased surface deformation as shown in Fig. (a,b). Positive Grashof number implies the volumetric expansion coefficient β is positive and the fluid becomes lighter on being heated. Thus the fluid tends to rise at the hot wall and drop at the cold wall. Grashof convection is a bulk flow phenomenon and generates fluid circulation throughout the fluid domain. Positive Grashof numbers in our problem generate clockwise circulating vortex.

Unlike Grashof convection Marangoni convection is a surface phenomena. The shear stress caused by thermocapillary phenomena is transmitted into the interior of the fluid due to the viscous action of the fluid. The direction of the circulating flow generated by positive Marangoni convection is the same as that of positive Grashof convection. However, unlike in Grashof convection the free surface drops at the hot wall and rises at the cold wall. Increased Marangoni number leads to increased flow and increased free surface deformation as shown in Fig. (c,d).

Ohnesorge number represents the ratio of viscous forces to capillary forces. Decreased Ohnesorge number implies increased surface tension and this leads to reduced free surface deformation. However, its influence on the flow and temperature distribution is not much as can be seen from Fig. 4. This is true for both buoyancy driven flow and thermocapillary driven flow.

Bond number is the ratio of gravitation forces to capillary forces. Increasing the Bond number flattens the free surface but its effect on flow and temperature distribution is not significant as can be seen clearly from Fig. 5.

Prandtl number represents the ratio of momentum diffusion to thermal diffusion. Low Prandtl number implies that the medium has high thermal conductivity and consequently thermal conduction is large compared to thermal convection. High Prandtl number implies that thermal conduction is small and most of the heat transfer is through convection. The influence of Prandtl number on buoyancy flow doesn't appear to be much as is evident from Fig. 6(a,b). However, in the case of Marangoni convection Prandtl number seems to have a significant effect on the flow and temperature field and free surface deformation. Marangoni convection is entirely dependent on thermal gradients. In the case of high Prandtl numbers heat transfer is mostly by convection and hence large thermal gradients exist only in the thermal boundary layer adjacent to the wall and the rest of the domain sees only small thermal gradients.

Finally we studied the combined effects of buoyancy and thermocapillary driven flow which is shown in Fig. 7. The two phenomena may be augmenting each other or counteracting each other depending on whether Gr and Ma have the same sign or not. Both Gr and Ma when positive generate clockwise circulating vortices and vice versa when negative. However, the free surface behaves differently when both Gr and Ma have the same sign. For positive Gr the free surface rises at the free surface whereas for positive Ma the free surface drops at the hot wall. When Gr and Ma have opposite signs we could have two vortices circulating in opposite directions generated in the fluid as shown in Fig. 7(d). The vortex near the free surface is due to the Marangoni convection and the one in the interior of the fluid is due to the Grashof convection.

6. CONCLUDING REMARKS

A projection type finite element algorithm for non-stationary buoyant and thermocapillary free boundaries in crystal growth is developed in this research.

Kinemtaic equation is used for determining the free surface and the steady state flow and temperature distribution is obtained through transient calculations. The results show good agreement with Cuvelier's results.

Gr and *Ma* effect both the flow and temperature field and free surface deformation. *Oh* and *Bo* seem to have less of an effect on the flow and thermal field but have significant effect on the free surface shape. Reduced *Oh* leads to reduced free surface deformation, whereas increased Bond number leads to flattening of the free surface.

Acknowledgements
This research was partially supported by National Science Foundation under grant DMS-9112847 and Texaco Corporation, U.S.A. The computations were carried out using CRAY-YMP at NCSC, U.S.A.

References

[1] D. T. J. Hurle and E. Jakeman, 'Introduction to the Techniques of Crystal Growth,' *PhysicoChemical Hydrodynamics*, **2**, pp.237-244, (1981).

[2] D. Schwabe, 'Marangoni effects in crystal growth melts,' *PhysicoChemical Hydrodynamics*, **2**, pp.263-280, (1981).

[3] S. Ostrach, 'Low-gravity fluid flows', *Ann. Rev. Fluid Mech.*, **14**, 313-345 (1982).

[4] S. Ostrach, 'Fluid Mechanics in Crystal Growth - The 1982 Freeman Scholar Lecture,' ASME *J. Fluids Engg.*, **105**, pp.5-20, (1983).

[5] J. Srinivasan and B. Basu, 'A numerical study of thermocapillary flow in a rectangular cavity during laser melting,' *Int. J. Heat Mass Transfer*, **29**, pp.563-572, (1986).

[6] V. G. Levich and V. S. Krylov, 'Surface-tension-driven phenomena', *Ann. Rev. Fluid Mech.*, **1**, 293-316 (1969).

[7] S. M. Pimputkar and S. Ostrach, 'Transient thermocapillary flow in thin fluids,' *Phys. Fluids*, **23**, pp.1281-1285, (1980).

[8] T. L. Bergman and S. Ramadyani, 'Combined buoyancy - and thermocapillary - driven convection in open square cavities', *Num. Heat Transfer*, **9**, pp.441-451, (1986).

[9] G. Ryskin and L. G. Leal, 'Numerical solution of free-boundary problems in fluid mechanics', *J. Fluid Mech.*, **148**, 1-47 (1984).

[10] W. Shyy and M.-H. Chen, 'A study of transport process of buoyancy- induced and thermocapillary flow of molten alloy,' *AIAA-90-0255*, pp.1-21, (1990).

[11] J. C. Chen, J. C. Sheu and S. S. Jwu, 'Numerical computation of thermocapillary convection in a rectangular cavity', *Numer. Heat Transfer Part A*, **17**, 287-308 (1990).

[12] H. B. Hadid and B. Roux, 'Buoyancy- and thermocapillary-driven flows in differentially heated cavities for low-Prandtl-number fluids,' *J. Fluid Mech.*, pp.1-36, (1992).

[13] C. Cuvelier and J. M. Driessen, 'Thermocapillary free boundary in crystal growth', *J. Fluid Mech., 169*, 1-26 (1986).

[14] A. I. Van De Vooren and H. A. Dukstran, 'A finite element stability analysis for the Marangoni problem in a rectangular container with rigid sidewalls.' *Comput. Fluids*, **17**, pp.467-485, (1989).

[15] A. J. Chorin, 'A numerical method for solving viscous incompressible flow problems,' *J. Comput. Phys.*, **2**, pp.12-26, (1967).

[16] B. Ramaswamy, T. C. Jue and J. E. Akin, 'Semi-implicit and explicit finite element schemes for coupled fluid/thermal problems', *Int. j. numer. meth. eng.*, **34**, 675-696 (1992).

[17] B. Ramaswamy and T. C. Jue, 'Some recent trends and developments in finite element analysis for incompressible thermal flows', *Int. j. numer. meth. eng.*, **35**, pp.671-708, (1992).

[18] H. Saito and L. E. Scriven, 'Study of the coating flow by the finite element method,' *J. Comput. Phys*, **42**, pp.53-76, (1981).

[19] Fluid Dynamics International, FIDAP 5.1 Users Manuals (1990).

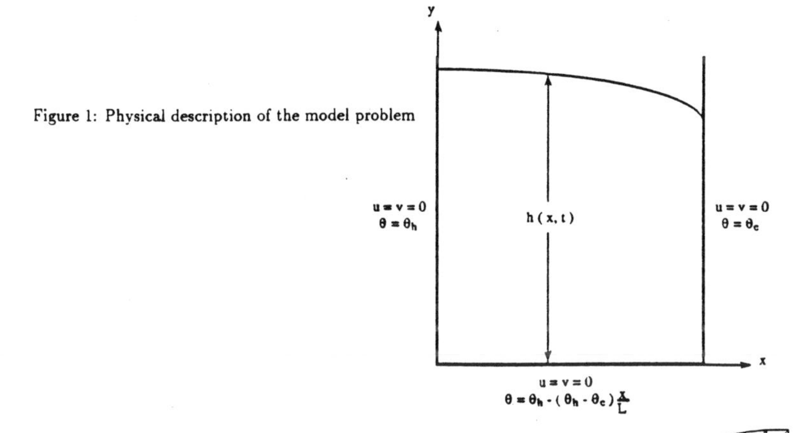

Figure 1: Physical description of the model problem

Figure 2: Typical Finite Element mesh

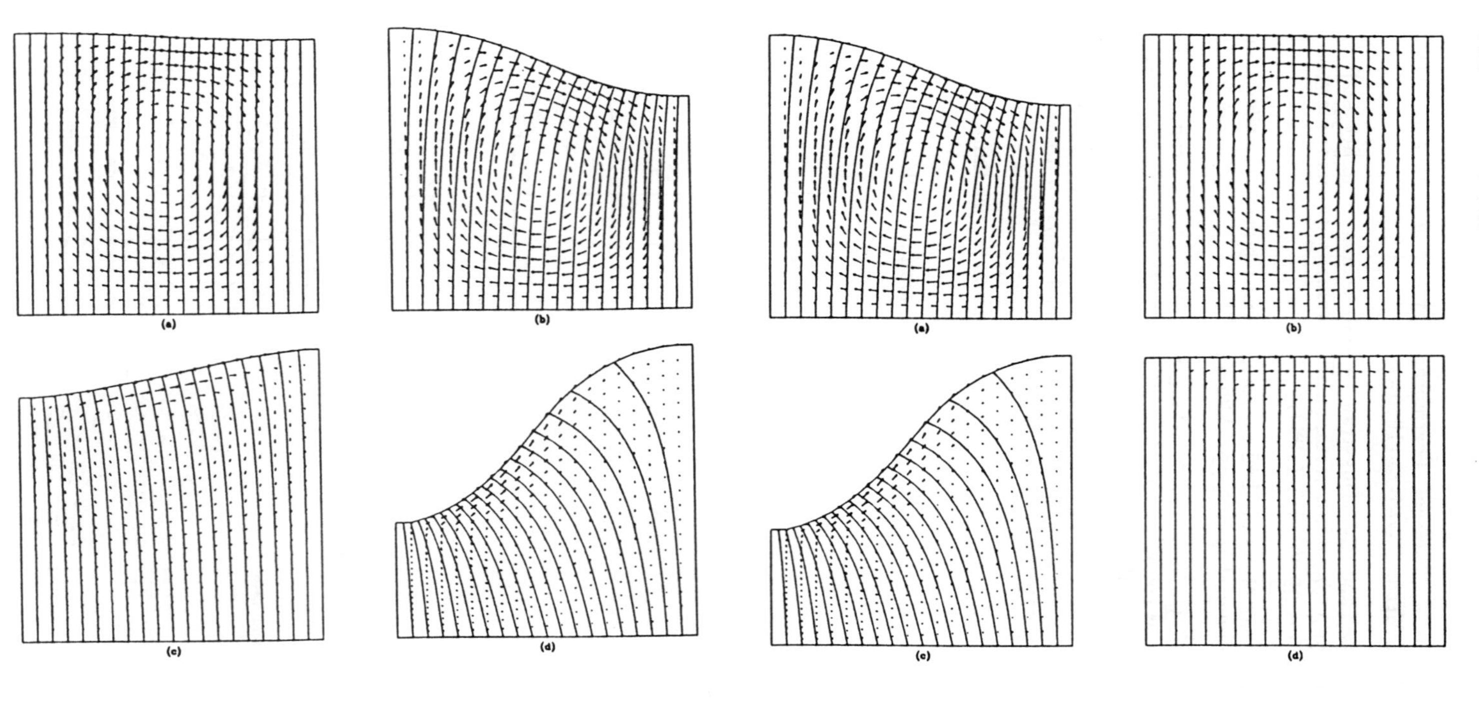

Fig.	Gr	Ma	Pr	Oh	Bo	$\mid u \mid_{max}$	h(0)	h(1)
a	2.0	0.0	0.73	1.0	0.0	0.0119(0.0125)	1.016(1.021)	0.984(0.978)
b	14.0	0.0	0.73	1.0	0.0	0.0818(0.0785)	1.143(1.204)	0.862(0.773)
c	0.0	0.5	0.73	1.0	0.0	0.098(0.095)	0.915(0.944)	1.085(1.063)
d	0.0	2.5	0.73	1.0	0.0	0.486(0.418)	0.554(0.567)	1.401(1.488)

Figure 3: Effect of Grashof number and Marangoni number.

Fig.	Gr	Ma	Pr	Oh	Bo	$\mid u \mid_{max}$	h(0)	h(1)
a	14.0	0.0	0.73	1.0	0.0	0.0818(0.0785)	1.143(1.204)	0.862(0.773)
b	14.0	0.0	0.73	0.1	0.0	0.087	1.001	0.998
c	0.0	2.5	0.73	1.0	0.0	0.486(0.418)	0.554(0.567)	1.401(1.488)
d	0.0	2.5	0.73	0.1	0.0	0.476	0.995	1.004

Figure 4: Effect of Ohnesorge number.

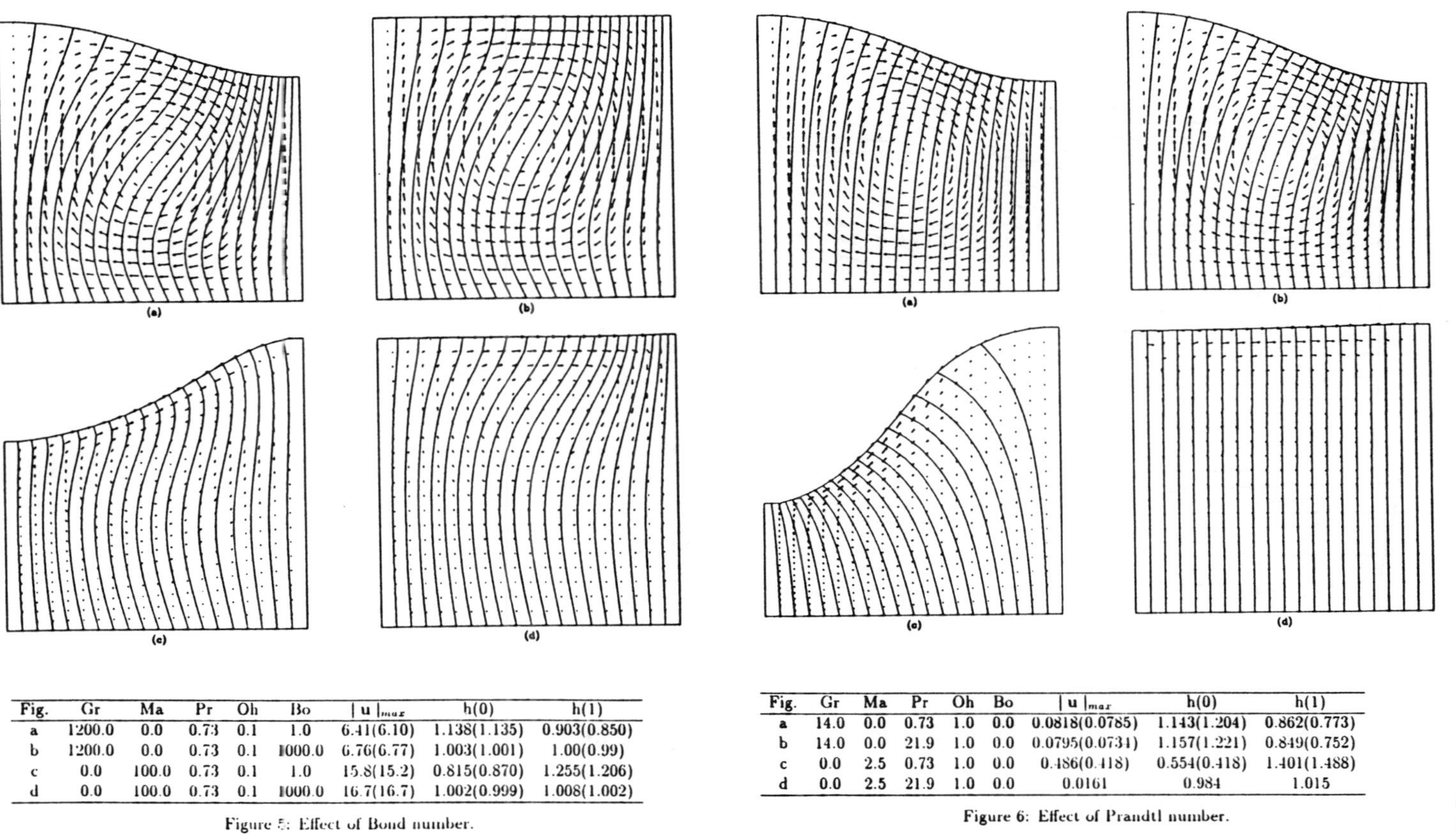

Fig.	Gr	Ma	Pr	Oh	Bo	$\|u\|_{max}$	h(0)	h(1)
a	1200.0	0.0	0.73	0.1	1.0	6.41(6.10)	1.138(1.135)	0.903(0.850)
b	1200.0	0.0	0.73	0.1	1000.0	6.76(6.77)	1.003(1.001)	1.00(0.99)
c	0.0	100.0	0.73	0.1	1.0	15.8(15.2)	0.815(0.870)	1.255(1.206)
d	0.0	100.0	0.73	0.1	1000.0	16.7(16.7)	1.002(0.999)	1.008(1.002)

Figure 5: Effect of Bond number.

Fig.	Gr	Ma	Pr	Oh	Bo	$\|u\|_{max}$	h(0)	h(1)
a	14.0	0.0	0.73	1.0	0.0	0.0818(0.0785)	1.143(1.204)	0.862(0.773)
b	14.0	0.0	21.9	1.0	0.0	0.0795(0.0734)	1.157(1.221)	0.849(0.752)
c	0.0	2.5	0.73	1.0	0.0	0.486(0.418)	0.554(0.418)	1.401(1.488)
d	0.0	2.5	21.9	1.0	0.0	0.0161	0.984	1.015

Figure 6: Effect of Prandtl number.

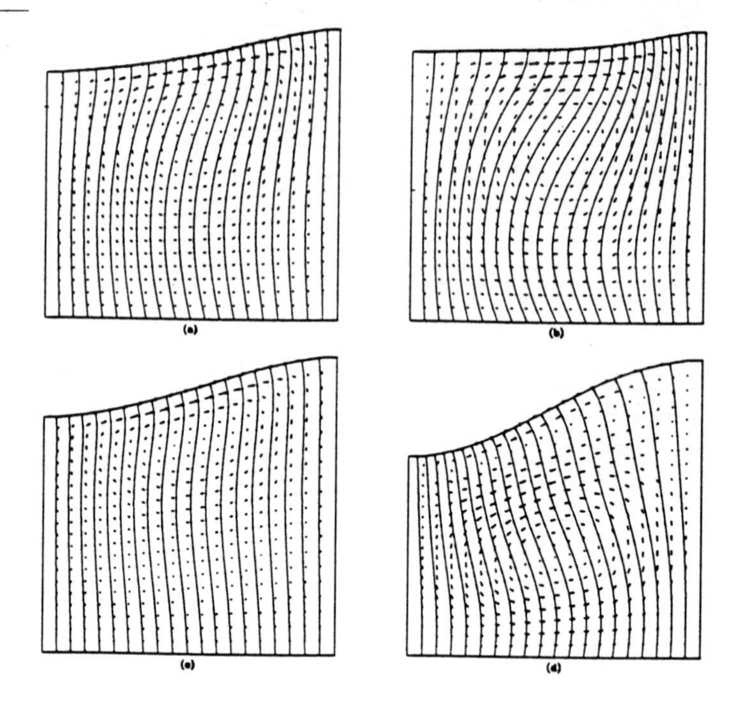

Fig.	Gr	Ma	Pr	Oh	Bo	$\mid u \mid_{max}$	h(0)	h(1)
a	200.0	50.0	0.73	0.1	0.0	10.0	0.926	1.095
b	800.0	50.0	0.73	0.1	0.0	12.7	0.982	1.059
c	-200.0	50.0	0.73	0.1	0.0	7.80(7.76)	0.884(0.919)	1.125(1.097)
d	-800.0	50.0	0.73	0.1	0.0	5.07(4.99)	0.799(0.827)	1.203(1.177)

Figure 7: Combined buoyancy and thermocapillary driven flow.

Transient Simulation of Non-Newtonian Non-isothermal Flows in Two and Three-dimensions

D. Ding[I], P. Townsend[II] and M.F. Webster[III]

ABSTRACT

In this article we report on progress in the development of software tools for fluid flow prediction in the polymer processing industry. This involves state-of-the-art numerical techniques and the study of a number of non-trivial model problems, in an effort to investigate realistic transient problems relevant to industrial processes. Here we study particularly the effects of variations in non-Newtonian and heat transfer properties of the flowing materials, both throughout the transient development period and at steady-state.

1 INTRODUCTION

Over a period of time work has been underway at Swansea to develop a computer code to simulate the types of flow which are important in the polymer processing industry. In particular finite element based facilities have been implemented for dealing with flows of non-Newtonian materials under transient and non-isothermal conditions. The algorithms used in the code are novel, state-of-the-art and incorporate high accuracy time-integration of the governing differential equations. Full details of the mathematics have already been published extensively in the literature [1-4].

The main problem facing the developers of such a computer code is to establish confidence in its performance, both in terms of accuracy, and flexibility to deal with a wide range of possible flow conditions and also in terms of its efficiency to compute results without wasting expensive computing resources. In order to consider these issues with respect to the code referred to above, we have chosen to solve a set of non-trivial model problems. This data has been provided by our industrial collaborators. Such problems either have an analytical solution or represent some flow conditions for which experimental data is available. These model problems allow us to investigate the effects on specific flows of varying the non-Newtonian properties and the heat transfer properties of the flowing material. Also the transient capabilities of the computer code allow investigation of the build-up of flows to a steady-state when, for example, temperature boundary conditions are impulsively changed.

In the literature, analytical solutions for steady flows of non-isothermal non-Newtonian fluids under isothermal boundary conditions have been given by various authors. Turian [5] presented a solution for drag flows using a

---

I: Senior Research Assistant, II: Professor, III: Senior Lecturer.
University of Wales Institute of non-Newtonian Fluid Mechanics,
Department of Computer Science, University of Wales,
Swansea, SA2 8PP, United Kingdom

perturbation method. Martin [6] obtained a solutions for pressure-driven flows. Lindt [7] recently gave a solution for pure drag flow without pressure gradients. Existing analytical solutions provide a fundamental way to validate our code by comparing steady state results with the above analytical solutions. However, they do not cover the transient development of flows.

Numerical simulations of non-isothermal and non-Newtonian polymer flows under steady conditions have been reported widely. Winter [8] investigated the temperature effect of thermal power-law fluids in two dimensional extruder dies under various boundary conditions. The momentum and the energy equations were solved based on a decoupled finite difference method. It was shown that viscous heating created an increase of the temperature in the die. These results agree with our two-dimensional studies very well. Mitsoulis et al. [9] presented a high speed wire-coating problem for polyethylene with shear rate and temperature-dependent viscosity under complex boundary conditions. A finite element method was employed to solve the momentum and energy equations using a decoupled scheme. Effects of viscous dissipation were also found in the temperature field and the non-isothermal predictions gave better predictions than the isothermal analysis. Karagiannis et al. [10] studied non-isothermal Newtonian flow through a straight three-dimensional die with a square cross section, again employing a finite element approach. Isothermal boundary conditions were applied to the die walls and viscous heating was observed. This agrees with our solutions for three-dimensional flows. The above investigations also offer comparisons with our results at a converged steady state. Again, they are not available for the transient analysis.

In this paper the results of simulations for a number of different flow conditions and geometries are presented. Velocity and temperature profiles are given, both for the final steady-state and for stages throughout the development of the flow. The influence on such profiles of the non-Newtonian power-law index m on the viscous heating generated is considered as well as the effect of variation in the material constant β, a thermal exponential factor. It is shown that the computer code performs well under a variety of conditions, giving confidence in results predicted for other flows for which no analytical solution is available, or for which it is not possible to obtain experimental measurements. The range of problems discussed are of varying complexity. These include problems where a transient phase is significant to those where only the steady-state is relevant, and cover various instances of Couette and Poiseuille flows with a variety of thermal boundary and initial conditions imposed. Reference is made to an earlier study [3] as a basis from which to develop the work presented here.

Other areas of current interest which are being addressed with the same computer code include different material, e.g., glass flow, fibre suspensions flow, visco-elastic flow etc., and also various industrial interests, for example capillary flow, multi-layer injection moulding involving particle tracking and moving fronts, etc. In a separate project the simulation software has been interfaced to a commercially available CAD/CAM package and a visualisation package, and it is envisaged that this will provide ultimately a general purpose software tool of wide ranging application in computational fluid dynamics but particularly for polymer processing.

2 DESCRIPTION OF MODEL PROBLEMS

In reference [3] four model problems were addressed, three Couette flows and one Poiseuille flow. First for Couette flows, the effects of a temperature dependent viscosity were considered and compared against steady-state analytic solutions. Then a power-law shear-thinning functionality was incorporated into the viscosity to extend consideration to generalised Newtonian fluids. Again analytic comparisons were to hand and excellent agreement was achieved with the numerical solutions obtained for power-law index $0.2 \leq m \leq 1.0$. Subsequently two transient problems were solved, one of a Couette and the other of a Poiseuille flow. Some interesting transient phenomena were reported in the temperature fields with variation in m and Péclet number. At the chosen level of $\beta=1$ and for a fixed value of m, there was little noticable effect on the velocity fields from the corresponding isothermal forms. It is therefore to this issue that present attention is drawn by altering β, as well as additional variation with m and the imposition of realistic initial and boundary conditions to the problems studied. Here two Poiseuille flows are studied which incorporate a wide variation in shear-rate and are capable therefore of manifesting significant field effects as desired.

The particular geometry under study is that of a two-dimensional plane channel of one unit width and ten units length. This domain is discretised uniformly according to reference [3], using some 200 triangular finite elements where a pair of such elements subdivide a rectangle of side 0.1*1.0 units. The interpolation functions used are piecewise continuous quadratics for velocities and temperature, and linears for pressure. This implies a total number of nodal unknowns of 441 for each velocity component and temperature, and 121 for pressure. The numerical parameters chosen follow reference [3] with typically, time-step of $\Delta t=10^{-2}$, time-stepping convergence tolerance of 10^{-4}, and three Jacobi iterations per fractional solution stage.

The first transient problem studied follows problem 4 of reference [3] a thermal Poiseuille flow problem with constant temperature boundary conditions and a unit flowrate that develops under viscous heating. In the present context this problem is referred to as Poiseuille problem A. The effects of an increased thermal factor from $\beta=1$ to $\beta=5$ are considered. The initial conditions are taken as the Newtonian isothermal equivalent conditions, and to achieve a physically realistic transient inlet boundary condition some attempt is made to recycle the exit flow conditions back to the inlet at certain well-chosen times in the development of the flow. This has the effect of continually looking down a longer and longer geometry and effectively simulates the imposition of periodic boundary conditions. A final test to this procedure is to take the ultimate solution and rerun such a problem with the entry condition fixed to confirm the correctness of this solution, which has been demonstrated in all instances tested.

A second transient problem considered here is that of a thermal Poiseuille flow problem of unit flowrate that develops under viscous heating when there is a sudden increase in temperature boundary condition to the second half only of the channel walls. This problem is referred to as Poiseuille problem B and is shown schematically in Figure 1. The temperature boundary condition applied initially obeys a step function and the initial conditions are those adopted from problem A above, being both generalised Newtonian and non-isothermal as appropriate. This is a new problem to this investigation and one that closely approximates industrially realisable conditions, perhaps only to be superceded by adopting a more gradually changing function as the wall boundary condition on temperature. Consideration is given again here to the influence of variation in the power-law index and change in β from 1 to 5.

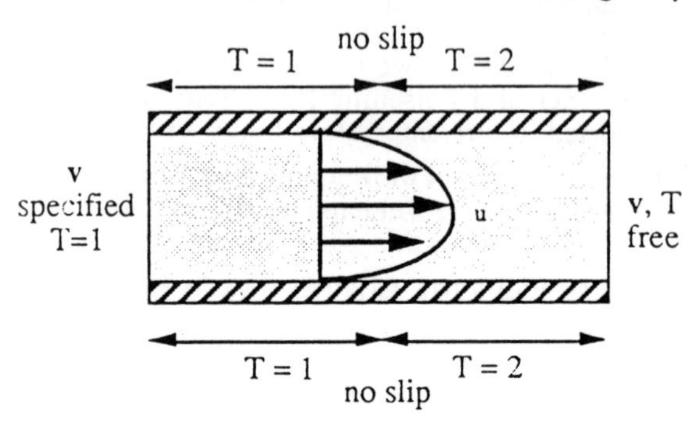

Figure 1 Schematic diagram for Poiseuille problem B

A third transient problem concerned is fully three-dimensional in geometry. The problem has been investigated for isothermal flows in an early publication [11]. Here, we study the non-isothermal and non-Newtonian effects for the same geometry. Figure 2 shows the plan view of the problem, where a channel with square cross section is curved through 90° and a straight extension is involved on both inlet and outlet sections. Only the upper half of the channel is employed for the simulation because of the symmetry. A fully developed axial velocity profile of the straight pipe with rectangular cross section is applied at the inlet, and the outlet is set to be free. Temperature is assumed to be fixed as unitary on the wall and the inlet, and free elsewhere. A thermal Carreau model is employed to simulate the behaviour of the viscosity.

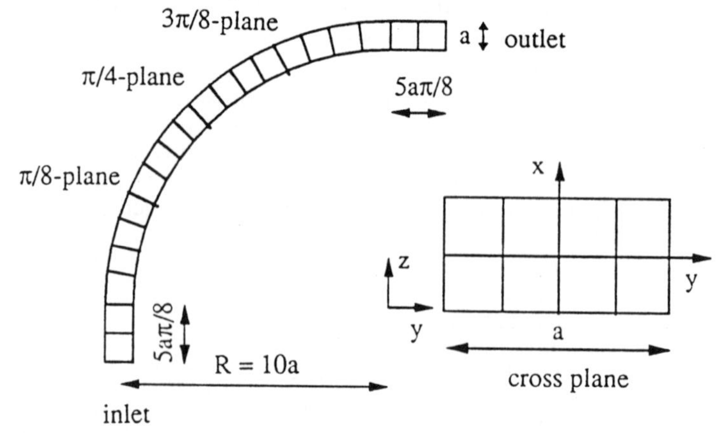

Figure 2 Plan view of 3-D problem

3 THEORETICAL CONSIDERATIONS

The present version of the algorithm follows closely that already appearing in the literature cf. [1-4]. The purpose here is to direct specific attention to those key areas where changes arise. For example, here there is a specific requirement to accommodate a generalised Newtonian non-isothermal fluid prescription and as such the viscosity function is assumed to behave according to the relationship

$$\mu = \mu_0 \gamma^{m-1} e^{-\beta(T-T_0)}$$

where μ_0 is a fixed viscosity value, $\gamma=0.5(I_2)^{1/2}$ for the second invariant I_2 (see [6] for definition), m is the power-law index, β is a material constant. T_0 is a reference temperature and T is the temperature of the fluid.

Treatment of the generalised momentum equations involves consideration of the diffusion in terms of the rate-of-deformation tensor D (see references [1] for more detail). Under the semi-implicit Taylor-Galerkin formulation of [4] this leads to a full (as opposed to a diagonal) subsystem matrix for the momentum component equations. With the Crank-Nicolson treatment for the diffusion terms as before, typically the corresponding new subsystem matrix in fully discrete weak variational form yields the new additional terms as follows:

$$(S_{lm})_{ij} = [\int_\Omega \mu\{\chi_{lk} \frac{\partial \phi_i}{\partial x_k} \frac{\partial \phi_j}{\partial x_k}\} d\Omega] \qquad \text{if } l = m$$

$$(S_{lm})_{ij} = [\int_\Omega \mu\{\frac{\partial \phi_i}{\partial x_m} \frac{\partial \phi_j}{\partial x_l}\} d\Omega] \qquad \text{if } l \neq m$$

$$\chi_{lm} = \begin{cases} 2 & \text{if } l = k \\ 1 & \text{if } l \neq k \end{cases}$$

for each component momentum equation *l* and each velocity component *m*, where repeated indicial notation implies summation. In addition μ is evaluated at an appropriate time level using the most recent known velocity and temperature solution components, and i, j range over all velocity nodal indices. It follows directly that the new momentum subsystem matrix is symmetric positive definite. This can be observed from the fact that for any nonzero vector U, diffusion subsystem matrix [S] and viscous dissipation function Φ_v (see [3] for definition)

$$U^T [\, S \,] U = \int_\Omega \mu \{ \Phi_v \} d\Omega$$

which is non-negative as $\Phi_v \geq 0$ by definition and μ is a positive function. This property underpins the iterative solution procedure that is implemented at stages 1 and 3 of the fractional stage scheme reported earlier. The Jacobi nature of this iteration with its diagonalised preconditioning matrix and the element-wise construction for matrix-vector products in the right-hand side vectors do not alter from those cited in references [3,4]. However, the velocity solution components are now coupled together in the right-hand side products, and quadrature is necessary to accurately capture the spatial dependency of the variable viscosity function and the diagonalised preconditioner is now itself time-dependent.

It is worth pointing out that a variation to the above scheme with the diffusion subsystem matrix split into diagonal and offdiagonal component sections has also been considered. The diagonal section was treated in a Crank-Nicolson implicit manner and the offdiagonal section dealt with via the explicit Taylor-Galerkin predictor-corrector approach using a half-time -step. Such a partially implicit treatment of the diffusion terms was found to yield as expected a less stable scheme which became apparent when studying a Couette flow problem due to Turian and Bird [12] with a temperature dependent viscosity. Such an alternative, less costly scheme is certainly not to be advocated here therefore.

4 SIMULATION RESULTS

4.1 Poiseuille problem A

Many aspects of this problem were discussed in a previous article [3], where an attempt was made to predict velocity and temperature development for a range of material parameters. In the work described here we investigate further the effect of temperature variations on the flow field. The temperature within the fluid changes with time as a result of viscous heating, and provided one has sufficiently sensitive material parameter settings, this change in temperature feeds back to modify the velocity profile. The latter, of course, also changes as a result of shear-thinning. One might argue that the boundary conditions used in reference [3] were somewhat artificial and it is certainly true that they did not represent any truly physical transient behaviour. However one can obtain a more realistic steady-state by integrating the above problem to convergence and then feeding back the converged outlet profiles for velocity and temperature as new inlet conditions. This procedure is adopted in this paper but in practice, apart from the inlet conditions themselves, this proves to have little or no effect on the remainder of the flow region. Thus it is shown that the true Poiseuille non-Newtonian solution had developed very rapidly only a short distance downstream of the imposed isothermal Newtonian inlet flow.

In Figure 3, converged steady-state (css) velocity profiles are given for different values of m and β, where the annotation follows that used in Figure 5. Outlet profiles only are displayed as there are no significant changes observed over the channel length. If we first consider the effect of shear-thinning as illustrated by curves (a) and (b), it can be seen that for values of m less than unity a significantly flatter profile is obtained with a boundary layer behaviour near the walls where the viscosity of the sheared fluid is greatly reduced. If we compare curves (a) and (c), where in both cases $m=1$, i.e. no shear-thinning, we can isolate the effect of the change in the thermal parameter β. For $\beta=1$ there is little or no feed-back of viscous heating into the velocity profile which remains essentially the same as a Newtonian parabolic profile. When β is increased to 5, however, a significant change takes place. The small temperature rise in the fluid lying midway between the bounding channel walls is sufficient to reduce the viscosity to such an extent that the fluid in this region moves more rapidly and one sees developing a characteristic bell-shaped velocity profile. If one combines both shear-thinning and thermal effects, a compromise profile, as shown in curve (d), develops. Figure 4 shows the corresponding temperature profiles for the four different cases. A comparison of curve (a) with (b) and (c) shows that the reduction in viscosity due to either shear-thinning or increased temperature feed-back has the effect of reducing the viscous heating in the fluid. From curve (d) it is seen that when both influences are taken into account, however, they do not reinforce one another, but rather an intermediate profile is achieved. Figures 5 and 6 summarise the transient development to a steady-state of the velocity and temperature channel exit profiles respectively. For the velocity there is a gradual change from the Newtonian profile, whilst for the temperature one sees a characteristic increase in temperature in the high shear regions near the walls which eventually diffuses into the remainder of the fluid to give a flat-topped steady-state temperature profile.

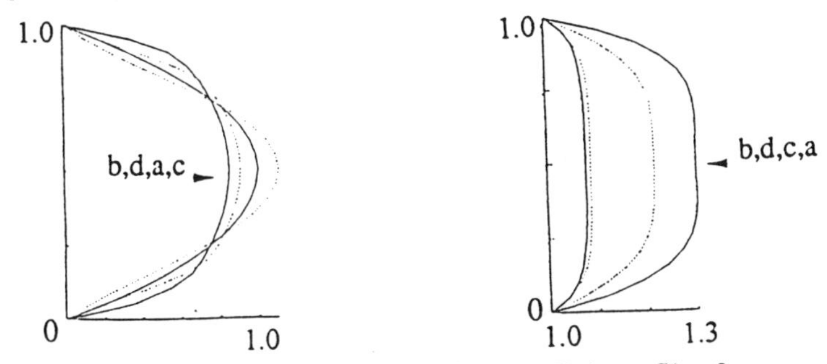

Figure 3 Exit profiles for velocity

Figure 4 Exit profiles for temperature

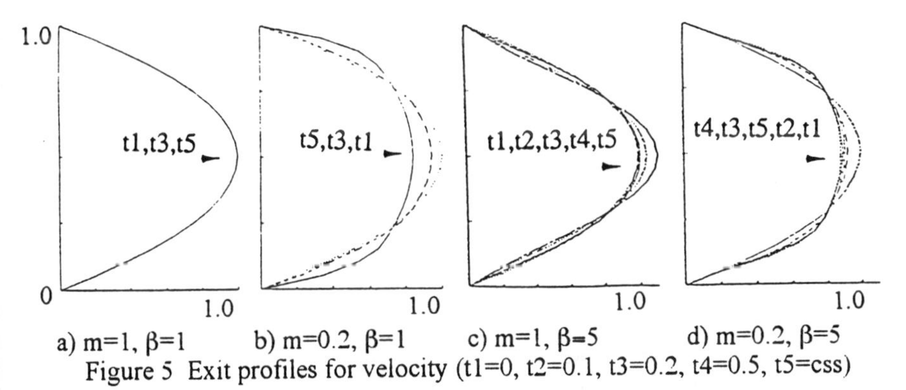

a) $m=1$, $\beta=1$ b) $m=0.2$, $\beta=1$ c) $m=1$, $\beta=5$ d) $m=0.2$, $\beta=5$

Figure 5 Exit profiles for velocity (t1=0, t2=0.1, t3=0.2, t4=0.5, t5=css)

The temperature profiles of Winter [8], as they develop down the die for a thermal power-law fluid under isothermal boundary conditions, have the same characteristics as our results shown in Figures 6b and 6c. It is found that the temperature at the wall region increases more than that at the centre of the channel due to viscous dissipation, and radial heat conduction balances the dissipation. This leads to a fully developed temperature field at the flow outlet.

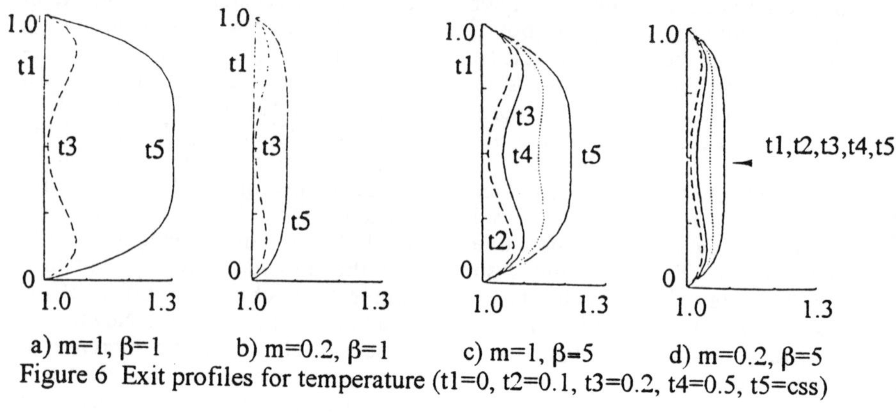

Figure 6 Exit profiles for temperature (t1=0, t2=0.1, t3=0.2, t4=0.5, t5=css)

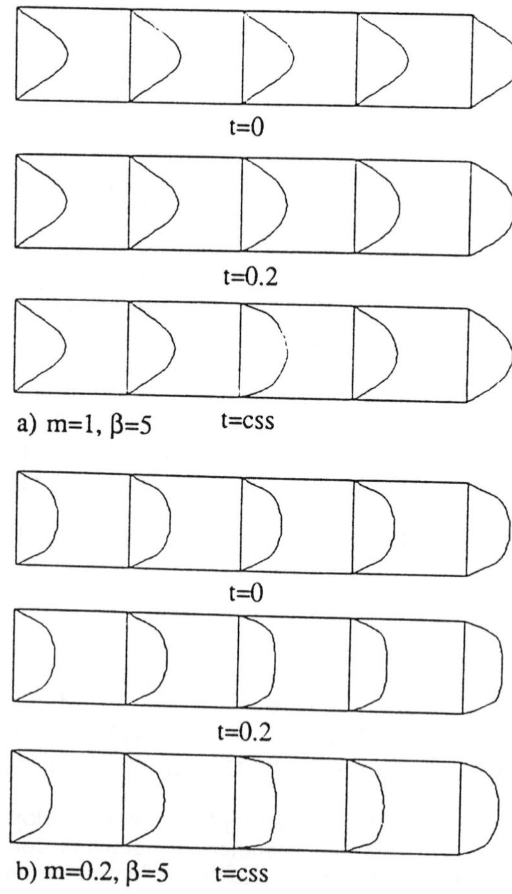

Figure 7 Velocity profiles along the mesh

4.2 Poiseuille problem B

For reasons outlined earlier, the Poiseuille problem A represented somewhat artificial flow conditions. In an attempt to simulate a truly realisable transient flow, a similar problem has been solved with a set of initial and boundary conditions as outlined in section 2, and is designated Poiseuille problem B. For this problem the converged steady-state developed in Poiseuille problem A is taken as the starting field for both velocity and temperature, and also an instantaneous step change in the temperature is made to the downstream halves of the bounding channel walls (see figure 1).

Figures 7 and 8 show the velocity and temperature development in space and time for various material parameter settings. For $\beta=1$ the feed-back of temperature rise into the momentum equations is of such little consequence that no discernible change from the inlet velocity profile is apparent. For this reason these plots have been omitted. For $\beta=5$, however, the feed-back of temperature rise into the momentum equations does effect velocity profiles.

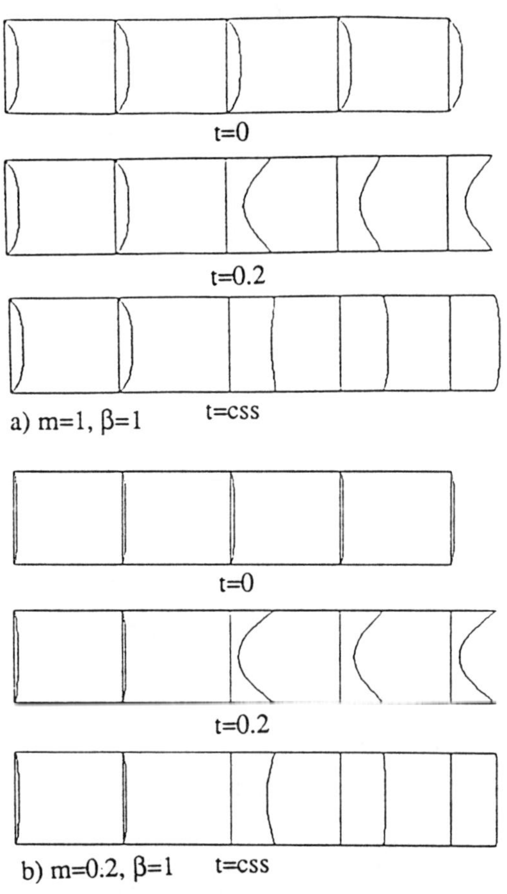

Figure 8 Temperature profiles along the mesh

Figures 7(a) and 7(b) show the velocity development for m=1 and m=0.2 respectively. Velocity profiles flatten compared with $\beta=1$, as larger β leads to less viscous heating for a fixed value of m. Shear-thinning again causes flatter velocity profiles. Figure 8(a) shows the temperature development for m=1 and $\beta=1$. The instantaneous increase in temperature at the walls in the right half of the flow region is seen to gradually diffuse in time until one obtains the same profile as at the inlet but offset by the step change in temperature. Figure 8(b) shows the effect of shear-thinning, where the pattern of development is essentially the same as for figure 8(a). The same behaviour is observed in Figures 8(c) and 8(d) for $\beta=5$, but the thermal viscous contribution is reduced from that in Figures 8(a) and 8(b).

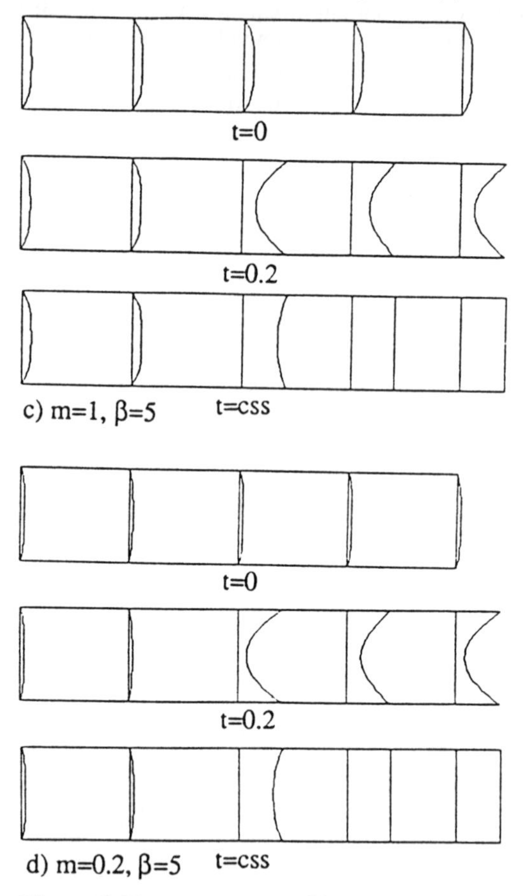

Figure 8 Temperature profiles along the mesh

4.3 Thermal 3-D problem

A three-dimensional curved channel flow under non-isothermal conditions with a Carreau model has been studied for Re=1 and 100. Figure 9 shows converged dimensionless axial velocity profiles for Re=1 and 100 on the horizontal symmetry plane of the bend, and the cross velocity vector plots for Re=100 along the bend. In the axial velocity profile plots, the zero position on the horizontal axis indicates the outer wall of the bend and unity indicates the inner wall of the bend. It is observed that the fluid near the outer wall moves faster than that near the inner wall for the Re=100 case and the secondary flow occurs at the cross-plane of the bend. This is not observed for the lower value of inertia of Re=1 where the fluid moves more slowly.

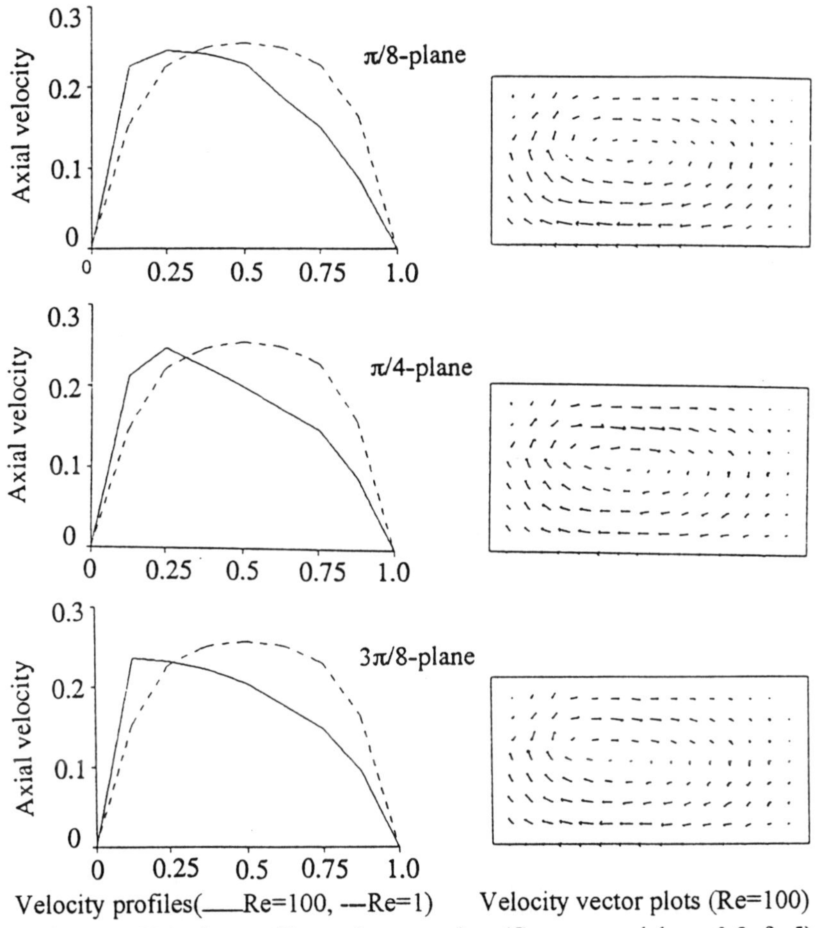

Velocity profiles(——Re=100, ---Re=1) Velocity vector plots (Re=100)

Figure 9 Velocity profiles and vector plots (Carreau model, m=0.2, β=5)

Figure 10 shows converged dimensionless temperature profiles along the bend for Re=1 and 100, which are equivalent to thermal Péclet number Pé=1 and 100 as the value of Re is adjusted by the fluid density. It is found that: first, the internal fluid temperature rises due to viscous heating; second, the temperature increase in the flow domain for Re=100 is much less than that for Re=1, as higher level of Pé reduces heat dissipation; third, for Re=100, the fluid near the wall is hotter than that in the core flow since high Pé effects the level of heat diffusion, therefore heat stays where it is produced by viscous heating; and fourth, for Re=100, the fluid in the outer wall vicinity is hotter than that near the inner wall, because larger shear rates are experienced in the outer wall region (cf. Figure 9).

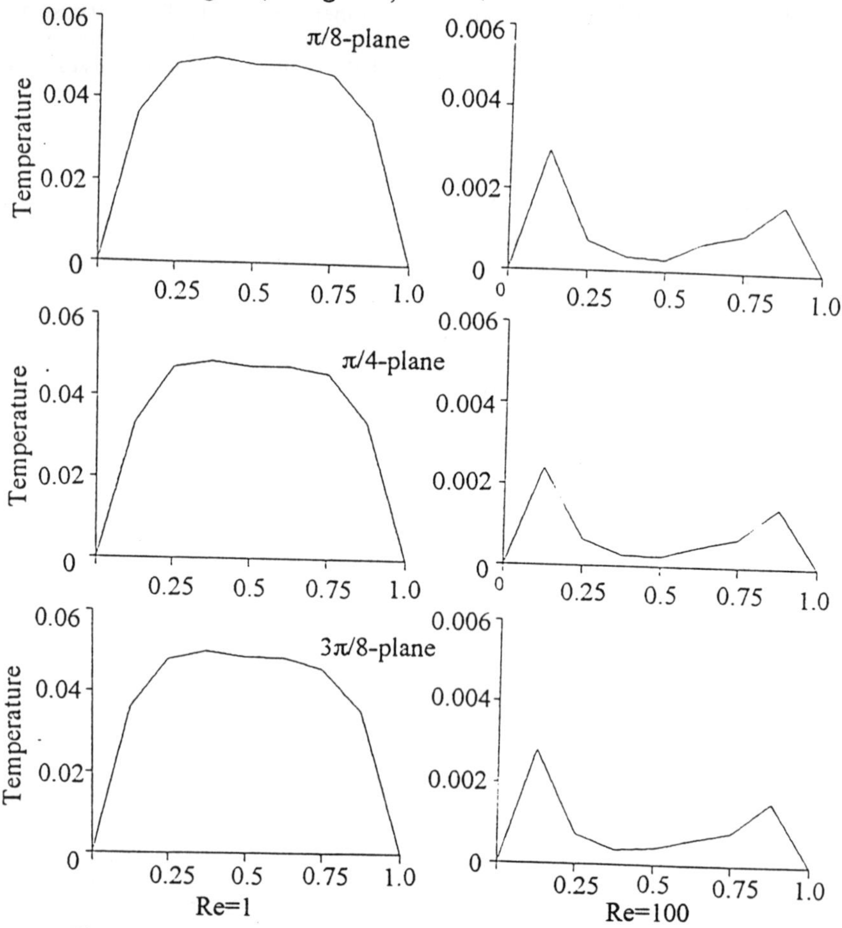

Figure 10 Temperature profiles (Carreau model, m=0.2, β=5)

Karagiannis et al. [10] studied a similar problem of the temperature development in a non-isothermal pressure-driven three-dimensional flow of a Newtonian fluid down a straight pipe of square cross-section under viscous dissipation, isothermal boundary conditions and for Re of the order of 10^{-6}. Their thermal Newtonian steady solution is in broad terms in agreement with our own results, taking into account the arguments of section 4.1, where we observe the reduction in the level of viscous heating due to reduction in power-law index. For the higher level of inertial of Re=100, due to asymmetry generated in the axial velocity profiles caused by the bending in the geometry (cf. Figure 9), we observe asymmetry in the cross-flow temperature profiles, peak temperatures near the walls with maximum near the outer wall, and cooler core flow than in the wall vicinities.

5 CONCLUSIONS

In this paper progress in the development of software tools for fluid flow prediction in the polymer processing industry has been reported. In particular the nonlinear effect of viscous heating coupled with variations in viscosity of the flowing materials due to shear and increased temperature has been studied, both throughout the transient development period and at steady-state. The work presented forms part of a phased evaluation of the software tools for model problems, although the capabilities of the software are quite general, as illustrated by the inclusion of a complex three-dimensional example. The results reported have demonstrated significant effects within the flow fields for the chosen problems at the material parameter settings selected. The influence of shear and temperature change on the fluid viscosity may produce either contra or like effects, although when both are present an average of the individual effects takes place. For Poiseille problem B a sudden change in wall temperature brought about significant changes to the velocity profile upstream of the hotter portion of the channel walls.

6 REFERENCES

1. TOWNSEND, P. and WEBSTER, M.F. - An algorithm for the three-dimensional transient simulation of non-Newtonian fluid flow. Transient/Dynamic Analysis and Constitutive Laws for Engineering Materials, Eds. Pandé, G.N. and Middleton, J., Nijhoff, 1987.
2. WEBSTER, M.F. and TOWNSEND, P. - Development of a transient approach to simulate Newtonian and non-Newtonian flow. Numerical Methods in Engineering: Theory and Applications, Ed. Pandé, G.N. and Middleton, J., Elsevier Applied Science, 1990.
3. DING, D., TOWNSEND, P. and WEBSTER, M.F. - The development of flow simulation software for polymer processing applications. Rev.Port. Hemorreologia, Vol. 4, Suppl.1/Pt A, pp. 31-38, 1990.
4. HAWKEN, D.M., TAMADDON-JAHROMI, H.R., TOWNSEND, P. and WEBSTER, M.F.- A Taylor-Galerkin based algorithm for viscous incompressible flow. Int. J. Num. Meth. Fluids, Vol. 10, No. 3, pp. 327-351, 1990.
5. TURIAN, R.M.- Viscous heating in the cone-and-plate viscometer-III. Non-Newtonian fluids with temperature-dependent viscosity and thermal conductivity. Chem.Eng. Sci., Vol. 20, pp. 771-781, 1965.
6. MARTIN, B. - Some analytical solutions for viscometric flows of power-law fluids with heat generation and temperature dependent viscosity. Int.J. Non-linear Mechanics, Vol. 2, No. 4, pp. 285-301, 1967.
7. LINDT, J.T.- Flow of a temperature dependent power-law model fluid between parallel plates: an approximation for flow in a screw extruder. Polym. Eng. Sci., Vol. 29, No. 7, pp. 471-478, 1989.
8. WINTER, H.H. - Temperature fields in extruder dies with circular annular or slit cross-section. Polm. Eng. Sci., Vol. 15, No. 2, pp. 84-89, 1975.
9. MITSOULIS, E. WAGNER, R. and HENG, F.L. - Numerical simulation of wire-coating low-density polyethylene: theory and experiments. Polym. Eng. Sci., Vol. 28, No. 5, pp. 291-310, 1988.
10. KARAGIANNIS, A., HRYMAK A.N. and VLACHOPOULOS, J.- Three-dimensional extrusion flows. Rheologica Acta, Vol. 28, No. 2, pp. 121-133, 1989.
11. HASSAGER, O., HENRIKSEN, P., TOWNSEND, P., WEBSTER, M.F. and DING, D. - The quarterbend: a three-dimensional benchmark problem. J. Comput. Fluids, Vol. 20, No. 4, pp. 373-386, 1991.
12. TURIAN, R.M. and BIRD, R.B. - Viscous heating in the cone-and-plate viscometer-II. Newtonian fluids with temperature-dependent viscosity and thermal conductivity. Chem.Eng. Sci., Vol. 18, pp. 689-696, 1963.

SECTION 5

TURBULENT FLOW

AN IMPROVED MODEL TO NUMERICALLY PREDICT HEAT TRANSFER IN TURBULENT FLUID FLOW

JP Cilliers, JA Visser and A Bekker

Department of Mechanical Engineering, University of Pretoria, RSA.

ABSTRACT

This paper outlines a method that improves the accuracy of simulated heat transfer to turbulent fluid flow by introducing an effective turbulent conductivity. The effective turbulent conductivity compensates for mixing in turbulent fluid flow not accounted for by the ordinary turbulence closure models, like the k-ϵ model, often used for such applications. The mathematical model and boundary conditions used in conjunction with the effective turbulent conductivity model are also included in the paper. The effective turbulent conductivity method was evaluated in several two-dimensional finite difference simulations of fluid flowing through a smooth pipe while being heated. Simulation and verification were performed over a range of Reynolds numbers between 2580 and 10315. Verification of the model is accomplished by firstly comparing the numerically predicted heat flux to the fluid with an analytical solution for the same problem. Secondly the predicted temperature distribution in the flow field is compared to measured temperatures in a similar pipe. With the modified effective turbulent conductivity vast improvements on the accuracy of simulating a temperature distribution in the fluid flow was achieved.

1. INTRODUCTION

Several large industrial installations rely on effective heat transfer processes to optimize their efficiency while minimizing the use of energy resources. This requires detailed knowledge of the flow and temperature distribution in the fluids taking part in the heat transfer process. One-dimensional heat transfer calculations have therefore become absolute insufficient for the design of complex modern heat exchangers. During the development phase, more experiments are thus required to reach specified standards which increases the development cost and time of new heat exchangers.

An attractive alternative to this problem is the use of numerical methods to simulate the fluid flow and heat transfer in heat exchangers numerically. Through the years a fair amount of work has been done by several authors on the prediction of fluid flow in heat exchangers. Devalba and Rispoli [1] developed a numerical model to solve the oscillating fluid flow in the tubes. Boundary layer effects and abrupt cross sectional variations were found to influence the temperature distribution in the tube material. Extensive work has been done on the numerical prediction of heat transfer in compact heat exchangers coupled with experimental verification. Sparrow and Ohadi [2] investigated turbulent heat transfer in a tube and verified the numerical model experimentally, for Reynolds numbers between 5400 and 83500. Results deviated from the measurements by a maximum of 6%. Faghri and Rao [3] used a numerical model to solve heat transfer and fluid flow in in-line finned and plain tube banks. Their experimental measurements corresponded well with the results of the numerical model.

In several previous simulations the k-e turbulence model was used to predict heat transfer in turbulent fluid flow and inaccurate results were obtained [4-5]. The inaccurate simulated temperatures were reported to have been caused by the turbulence model. Few methods have, however, been developed that can be used to improve results in turbulent heat transfer using the k-ϵ turbulence model. One such a model was introduced by Kadle and Sparrow [6] in which they investigated turbulent heat transfer in longitudinal fin arrays and used an effective turbulent conductivity (k_t) of the fluid in their numerical model. The results agreed well with both experimental measurements and results predicted by standard empirical formulae.

In this paper a technique is investigated to accurately simulate heat transfer to fluid flowing through a pipe and the temperature distribution in the flow with the view to optimize the design of heat exchangers. To evaluate the model water flowing through a smooth tube was simulated and compared to measurements for the same configuration. The model will be outlined in the following paragraphs.

2. NUMERICAL MODEL

2.1 The governing equations

The partial differential equations for the conservation of momentum, mass and energy in a fluid are widely published [7-8]. These equations can be written in the following general form:

Momentum Equation

$$\frac{\partial}{\partial t}(\rho\phi) + \nabla\cdot\rho v\phi = \nabla\cdot\tau + S_\phi \tag{1}$$

Continuity Equation

$$\frac{\partial\rho}{\partial t} + \nabla\bullet\rho\upsilon = 0 \tag{2}$$

Energy Equation

$$\rho\frac{DH}{Dt} = -[\nabla\bullet q] - [\tau:\nabla v] + \frac{Dp}{Dt} \tag{3}$$

Turbulence Equations

$$\frac{\partial\epsilon}{\partial t} + (\upsilon\bullet\nabla\epsilon) = \frac{1}{\rho}[\nabla\bullet(\frac{\mu_t}{\sigma_\epsilon}\nabla\epsilon)] + \frac{C_1\mu_t\epsilon}{\rho k}\Gamma - C_2\frac{\epsilon^2}{k} \tag{4}$$

$$\frac{\partial k}{\partial t} + (\upsilon\bullet\nabla k) = \frac{1}{\rho}[\nabla\bullet(\frac{\mu_t}{\sigma_k}\nabla k)] + \frac{\mu_t}{\rho}\Gamma - \epsilon \tag{5}$$

The finite difference equations used in the numerical model are obtained by integrating the partial differential equations over control volumes surrounding grid points in the flow field [8]. The general finite difference equation can be written in the following form:

$$a_p\phi_p = \Sigma\, a_{nb}\phi_{np} + b \tag{6}$$

where a_p is the coefficient at the central point and a_{nb} coefficients at neighbouring grid points in the flow field.

2.2 Boundary conditions

a. Momentum equations

At solid boundaries of the tube walls velocity components are set to zero. Near these walls the local Reynolds numbers are low and do not fall in the fully turbulent region. The k-ϵ turbulence model used in this study is not applicable close to the wall since the assumption of fully turbulent flow is not valid here. The turbulent viscosity model is thus incorrect close to a wall and must be altered by introducing an extra shear stress term next to the solid boundary in the momentum equations [10-11].

The extra shear stress term required close to solid boundaries in the momentum equations depends upon the local Reynolds number at each point along the wall. This shear stress is calculated in the turbulence model with special wall functions. These wall functions are only valid in the viscous sub-layer and transition zones up to a local Reynolds number of 11.5 based on the distance from the wall of the tube. For a local Reynolds number greater than 11.5 the logarithmic velocity profile is used to derive the wall function, while a linear velocity profile is used if the Reynolds number is below 11.5. The local Reynolds number is defined by the following equation:

$$y^{+} = C_{\mu}^{0.25} \rho k^{0.5} dw/\mu \tag{7}$$

A zero gradient boundary condition is used along the centre line for the axial velocity of the flow in the tube. This implies that no flow enters or leaves the flow field in the radial direction. A zero velocity boundary condition is therefore used for the radial velocity of the flow. At the outlet of the tube a zero gradient velocity profile is used for all velocity components. A zero gradient temperature boundary condition is used along the centre line of the fluid flowing in the tube. At the inlet and exit a zero gradient temperature boundary condition was implemented.

b. Energy equation

The energy equation is solved over the complete grid which includes the fluid and solid tube. In the solid tube material, the convection terms become zero and diffusion dominates the heat transfer process. As an inlet boundary condition a constant temperature equal to the measured temperature was used. At the outflow boundary and along the centre of the combustor a zero gradient temperature boundary condition is used [12].

On the interface between the tube and fluid no boundary condition is required. The heat transfer coefficient on the outer surface of the tube is the sum of the radiation and convection coefficients and is calculated from the following equation [13-14]:

$$h_o = 0.68\ Re_o^{0.464}\ Pr_o^{1/3}\ \frac{k_o}{d_o} + \frac{\sigma(T_1^2 + T_2^2)(T_1 + T_2)}{1/\epsilon_1 + (A_1/A_2)(1/\epsilon_2 - 1)} \tag{8}$$

c. Turbulence equations

The general boundary conditions for the turbulent transport equations differ greatly from those for the momentum equations. The inlet values for (k) and (ϵ) in these simulations are taken as 0.25 J/kg and 50 W/kg respectively [12]. Zero gradient boundaries for both (k) and (ϵ) are used along the centre line and outflow boundary. Near the solid boundaries the local Reynolds numbers are low, resulting in transitional and laminar flow. The extra wall stress term introduced in the momentum boundaries give a better description of the shear stresses near the walls resulting in more accurate values for the turbulence kinetic energy (k). At solid boundaries of the tube walls (ϵ) reaches its highest value. A fixed value for (ϵ) is adopted at the walls. The value of (ϵ) next to a solid boundary is found from the approximate equilibrium equation between generation and dissipation of turbulence [10-11]. These assumptions lead to the following equation for (ϵ) near walls:

$$\epsilon = \frac{C_\mu^{0.75} k^{1.5}}{\kappa d w} \tag{9}$$

3. EXPERIMENTAL WORK

Experiments were carried out to calibrate the external heat transfer coefficient as well as the numerical predicted results. The experimental work consisted of measuring the heat transfer to water flowing in a smooth tube. The tube was made of Inconel 625 with an outer diameter of 4.5 mm and an inner diameter of 3.4 mm. The tube was placed in a continuous diesel combustion chamber operating at atmospheric pressure. Temperatures in the combustion chamber were measured with a Rhenium-Tungsten (B-type) thermocouple for a constant preset mass flow rate of air and diesel into the combustion chamber.

Water was pumped through the pipe at different flow rates resulting in a range of Reynolds numbers from 2580 to 10315. Chromel-Alumel (K-type) thermocouple were placed in the tube and sealed off with a non-conducting material. Water temperatures were only measured over a flow range in which accurate readings could be obtained and at a distance of at least 40 tube diameters from the inlet so that entrance effects did not influence the readings. Figure 1 illustrates the positions of the thermocouple in the experimental model. The flow rate was measured with a calibrated rotameter. The experiment was repeated many times to ensure a good statistical sample of the readings.

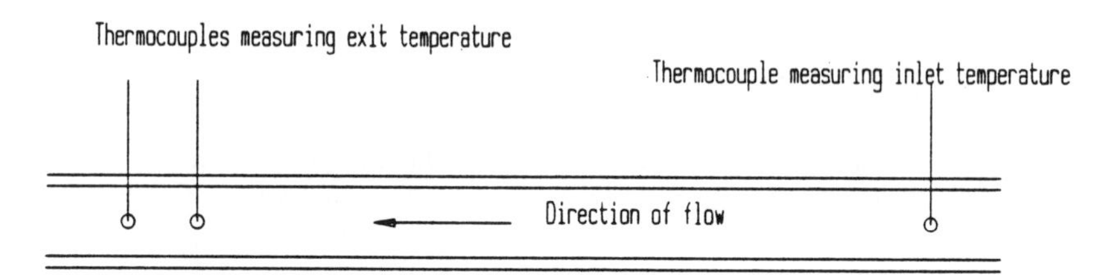

Figure 1 Experimental Configuration

4. THE TURBULENT CONDUCTIVITY MODEL

Simulations were first carried out without any model to compensate for the small scale turbulent mixture of the fluid. The initial numerical simulations of the experiment predicted a fluid temperature of 202°C to 140 °C, depending on the flow rate, at the tube wall while the inlet temperature of 16°C was still predicted along the centre line. The predicted temperature distribution in the flow is shown in figure 2. These results did not agree with the measurements. It was evident that the rate of diffusion of heat from the tube wall through the fluid was not being modelled accurately.

Convection and diffusion are the mechanisms by which heat is transferred in turbulent flow. A numerical simulation does, however, not compensate for mixing of turbulent fluid on a small scale. The turbulence model only compensates for the effect of the turbulence on the fluid flow. When modelling turbulent fluid flow numerically a technique is therefore required to artificially increase the rate of heat transfer due to small scale mixing in turbulent flow. This can be achieved by increasing either the convection or the diffusion term in the energy equation.

The respective effects of convection and diffusion on the total heat transfer is given by the Peclet number and is defined by the following equation [9]:

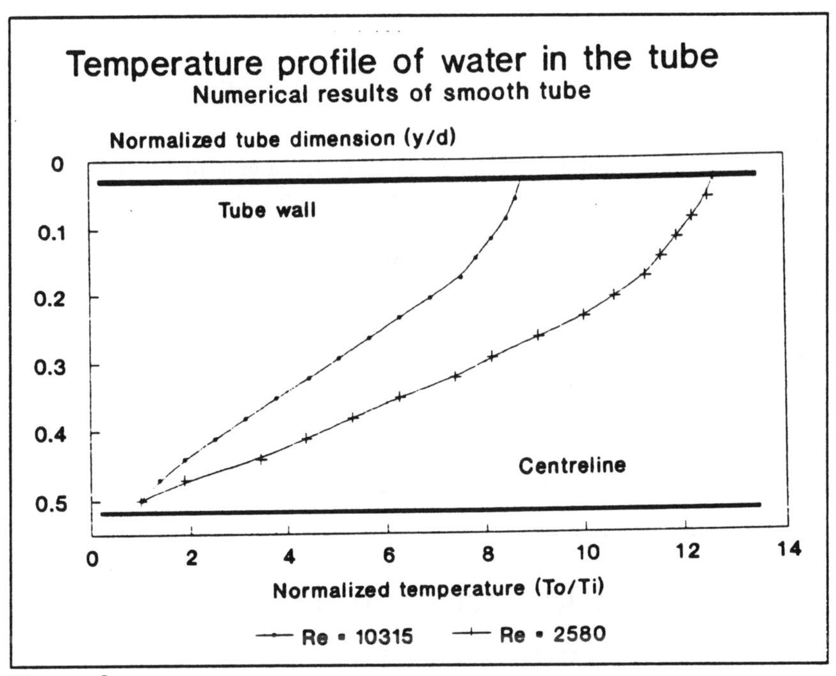

Figure 2

$$Pe = \frac{d_i \ \nu \ \rho \ C_p}{k_i} \tag{10}$$

At small Peclet numbers diffusion dominates the heat transfer process. The influence of convection increases as the Peclet number becomes larger. Similar to μ_t in fluid flow an effective turbulent conductivity (k_t) appears to be required to solve the heat transfer problem accurately.

Kadle and Sparrow [6] recognised this problem and used an effective turbulent conductivity to increase the diffusion term in their numerical simulations of a compact heat exchanger. They used the function outlined in equation (11) to calculate the turbulent conductivity which was verified experimentally for Reynolds numbers from 5000 to 35000. The model is, however, not a function of the Reynolds number and will not change as the parameters included in the Reynolds number changes. The model also lacks a numerical constant that can be altered for fine tuning to specific applications.

$$k_t = (\frac{C_p}{Pr_t})\ \mu_t \qquad (11)$$

In view of these restraints in the existing model, a modified function for calculating effective turbulent viscosity is introduced in this paper. The function is defined in terms of the turbulent kinetic energy (k), the rate of dissipation (ϵ) and the Reynolds number and is calculated from the following equation:

$$k_t = (C\ Re_i^n\ (\frac{k}{\epsilon}) + 1)\ k_L \qquad (12)$$

From expeiments conducted the values of the constants C and n were determined as 713.4402 x 10^{-6} and 1.968 respectively.

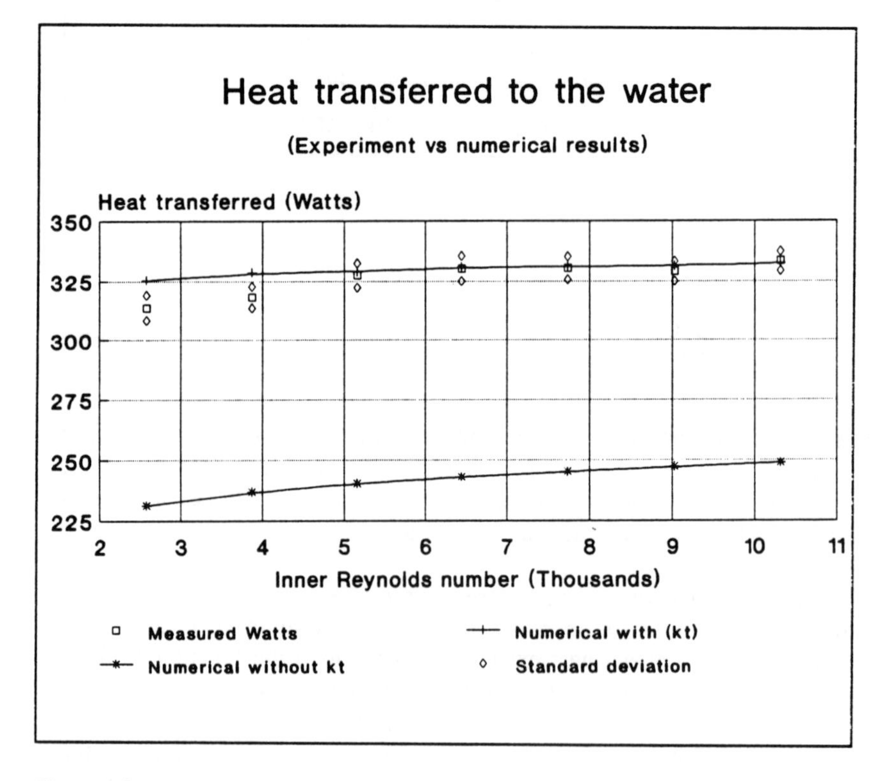

Figure 3

5. RESULTS

The effect of the turbulent conductivity in the numerical model is illustrated in figure 3. The heat transfer predicted without (k_t) is considerably lower than the measured results. This is mainly due to the numerical model's inability to compensate for the physical mixing of the fluid due to turbulence. The turbulence model in the numerical simulations only increases the shear stresses which are not effective to compensate for increased heat transfer due to turbulence.

With the inclusion of the effective turbulent conductivity numerically predicted results compare well with measurements as shown in figure 3. At the lower Reynolds numbers the flow is not completely turbulent and poor results can be expected if the k-e turbulence model is used. With the introduction of the effective turbulent viscosity term numerically predicted results compare well with measurements at the low Reynolds numbers. At the higher Reynolds numbers the correlation is very good since the current model also compensates for increases in the Reynolds number.

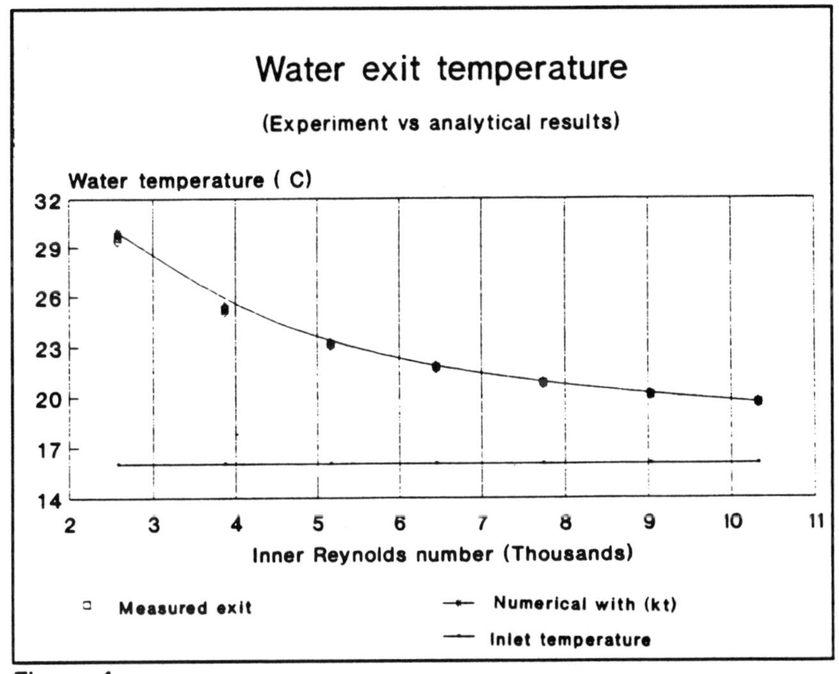

Figure 4

Temperatures at the centre of the tube were also recorded at various flow rates and compared to numerically predicted temperatures shown in figure 4. From these results it can be concluded that the introduced model for effective turbulent conductivity, not only improved the accuracy of predicting heat transferred to the fluid, but also enables the accurate predicting of the temperatures in heated fluids.

6. CONCLUSION

In this study an effective turbulent conductivity is introduced to a finite difference numerical model to accurately simulate the heat transfer rate to heated turbulent fluid flow in a tube. With this model the numerical simulations can now also be used to simulate the temperature distribution in the turbulent fluid as it passes through the tube. When calibrated, a numerical model thus provides a simple and cost effective method for evaluating the performance of compact heat exchangers. The advantage of the new model lies in the fact that it compensates for changes in Reynolds number. More experiments will therefore have to be carried out to evaluate the accuracy of the model at higher Reynolds numbers.

NOMENCLATURE

A	Surface area	[m^2]
C_p	Specific heat	[kJ/kg°C]
C_μ	Turbulence constant	
d	Tube diameter	[m]
h	Heat transfer coefficient	[W/m^2°C]
k	Turbulent kinetic energy	[J]
k_{subs}	Thermal conductivity of a fluid/material	[W/m°C]
L	Tube length	[m]
Pe	Peclet number	
Pr	Prandtl number	
Q	Heat flow	[W]
Re	Reynolds number based on tube diameter	
r	Tube radius	[m]
S_ϕ	Source term	
T	Absolute temperature	[K]
t	Time	[s]
U	Overall heat transfer coefficient	[W/m^2°C]
v	Velocity	[m/s]
w	Molecular weight of species	[g/mol]
y^+	Local Reynolds number	

Greek letters

ϵ	Rate of dissipation	$[s^{-1}]$
ϕ	General dependent variable	
ρ	Density	$[kg/m^3]$
σ	Stefan-Boltzmann constant	$[W/m^2K^4]$
τ	Laminar shear stress	[Pa]
μ	Viscosity	[kg/ms]

Subscripts

i	inner
L	laminar
m	material
o	outer
t	turbulent

REFERENCES

1. DEVALBA, M. and RISPOLI, F., 'The oscillating fluid flow effects on the Stirling engine heat exchanger design,', ISEC 5th International Stirling Engine Conference, pp. 195-210, 8-10 May, 1991, Dubrovnik, Yugoslavia.
2. SPARROW, E.M. and OHADI, M.M., 'Numerical and experimental studies of turbulent heat transfer in a tube,', Numerical Heat Transfer, Vol. 11, pp. 461-476, 1987.
3. FAGHRI, M. and RAO, N., 'Numerical computation of flow and heat transfer in finned and unfinned tube banks,',International Journal of Heat and Mass Transfer, Vol. 30, pp. 363-372, 1987.
4. WORMECK, J.J; Computation of turbulent reacting flows; Dynamics Technology Inc; California; 90505; September 1983.
5. HABIB, M.A. and WHITELAW, J.H.; Velocity Charactiristics of Confined Coaxial Jets with and without Swirl; Journal of Fluids Engineering; Vol 102; pp 47-53; 1980.
6. Kadle, D.S., and Sparrow, E.M., 'Numerical and experimental study of turbulent heat transfer and fluid flow in longitudinal fin arrays,', pp. 16 to 23, Journal of Heat Transfer - Transactions of the ASME, Vol. 108, 1986.
7. ANDERSON, D.A., TANNEHILL, J.C. and PLETCHER, R.H.; Computational Fluid Mechanics and Heat Transfer; Hemisphere Publishing Corporation; USA, 1984.
8. THOMPSON, J.F., WARSI, Z.U.A. and MASTIN, C.W.; Numerical Grid Generation Foundations and Applications; Elserivier Science Publishing Co. Inc.; New York, 1985.

9. PATANKAR, S.V., Numerical Heat Transfer and Fluid Flow, Hemisphere Publishing Corporation, New York, 1980.

10. LAUNDER, B.E. and SPALDING, D.B.; The Numerical Computation of Turbulent Flows; Computer Methods in Applied Mechanics and Engineering; pp 269-289; 1974.

11. GOSMAN, A.D. and IDERIAH, F.J.K.; A General Computer Program for Two-Dimentional Turbulent, Recirculating Flows, Imperial College, London, 1976.

12. MARLIN, M.R., TATCHELL, D.G. and YING-SHI, Calculations of steady three-dimensional turbulent reacting flow in a combustion chamber, CHAM Paper, PDR/CHAM UK/19, October 1982.

13. VISSER, J.A., The Numerical Prediction of Temperature Distribution in and Around a Steel Bar during Hot Rolling, Communications in Applied Numerical Methods, Vol. 4, pp 657 to 664, 1988.

14. HOLMAN, J.P. Heat Transfer, McGraw-Hill, 1989.

Modelling the Effect of Turbulence on Subcooled Void with the ASSERT Subchannel Code

G.R. Dimmick and M.B. Carver

A.E.C.L. Chalk River laboratories,
Ontario, Canada.

ABSTRACT

AECL is currently constructing a 10 MW(t) pool type research and isotope producing reactor, designated MAPLE-X10, at its Chalk River Laboratories. An important parameter in the thermalhydraulic design of this type of reactor is subcooled boiling, which may be influenced by fuel grid spacers that promote inter-subchannel mixing and also affect the subcooled void bubble size. A set of subcooled void data was available for an 18-pin bundle from thermalhydraulic tests done to support the reactor design. The data gave the bundle cross-sectional average void at various axial locations in the bundle and showed that there was a significant reduction in subcooled void immediately downstream of grid spacers.

The matching of the experimental data with the ASSERT code is described in this paper. It is concluded that the use of a bubble breakup parameter based on the grid spacer pressure loss coefficient is a viable way to describe the experimentally observed subcooled void variations.

1 INTRODUCTION

AECL is currently constructing a 10 MW research reactor, designated MAPLE-X10, at its Chalk River Laboratories. The reactor concept [1], Figure 1, is a 10 MW(t) pool type, light-water cooled and moderated, with a heavy-water reflector. The reactor core is about 0.4 m in diameter and 0.6 m high, and consists of 19 fuel assemblies in a close-packed hexagonal array. A mix of fuel assemblies is used: a hexagonal cluster of 36 pins, a circular cluster of 18 pins, and an annular cluster of 12 pins. The fuel is cooled by 35°C water pumped to the bottom of the core. The water exits the core into an open chimney, from which it is pulled off to the suction side of the primary pump.

Reactor design and safety analysis uses sophisticated computer codes supported by a number of experimental measurements to validate the codes for the particular application. The primary thermalhydraulic codes used in the MAPLE design are ASSERT [2], a subchannel code used to describe the detailed core thermalhydraulic behaviour, and CATHENA [3],

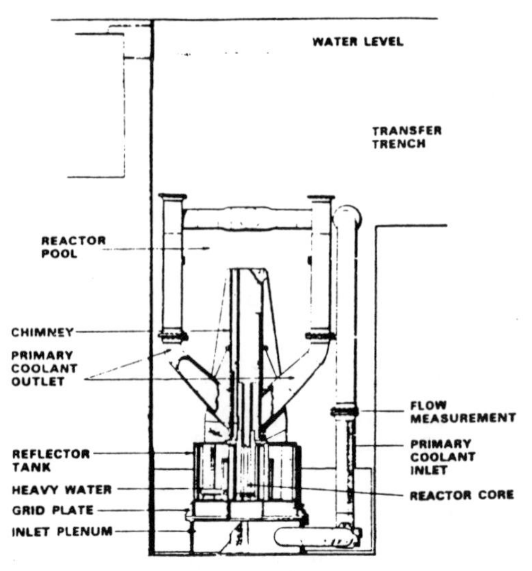

Figure 1. **MAPLE X10 REACTOR STRUCTURE**

a one-dimensional transient code used to describe the overall primary circuit thermalhydraulics. Various models and empirical correlations specific to the MAPLE fuel geometry that describe, for example, critical heat flux and single- and two-phase heat transfer, were derived from experimental tests with electrically heated single pins in an annulus, or with clusters of pins in a bundle geometry. The work reported here describes the modification of one aspect of the ASSERT code using void data taken with an 18-pin electrically heated half segment of a 36-pin hexagonal cluster.

Subcooled boiling can be an important phenomena in the design of nuclear reactors, because of its strong influence on the neutron kinetics and the possiblity of void-induced thermalhydraulic instabilities. It is especially important for the design of low-pressure pool-type reactors where, because of the low steam density, a significant void volume will contain only a small amount of thermal energy.

A large number of correlations and models exist for predicting void. Most of these were developed for high-pressure conditions and are generally inapplicable to the low pressures that exist in pool-type reactors. For low-pressure applications, the most successful type of model appears to be that based on the balance between simultaneous void generation and condensation. An early version of this type of model [4] showed good agreement with high-pressure experimental data. Subsequently, Dix [5], using Freon 114, showed that the condensation of the bubbles in the liquid core was a controlling factor in the void equation. More recently, Dimmick and Selander [6], and Chatoorgoon et al. [7], using experimentally measured condensation rates, have developed void generation/condensation balance models suitable for use at low pressures.

These models are generally derived by assuming a physically plausible condensation model and fitting it to experimentally derived condensation rate data. Experimental measurements in a heated tube, where both generation and condensation are occurring simultaneously, are then used along with the previously derived condensation model to extract a void generation model. Dimmick and Selander [6] found that a condensation rate proportional to the interfacial area best fitted the then available experimental data, whereas Chatoorgoon et al. [7] took the condensation rate to be proportional to the interfacial heat transfer coefficient and the interfacial area.

All of the condensation rate data available have been taken in either tubes or annuli, where the flow is essentially smooth and generally fully developed. Intuitively, it can be seen that the condensation rate will depend on the surface area, and thus the bubble size distribution, and that it should also depend on the liquid micro-turbulence around the bubble, as this will determine the heat transfer rate from the bubble to the bulk liquid. In our experimental tests with heated bundles we have visually observed that there is a reduction of void immediately downstream of an obstruction or flow disturbance, even though there is no significant change in any other parameter.

A set of subcooled void data was available for an 18-pin bundle from thermalhydraulic tests done to support the design of the MAPLE-X10 reactor. These data gave the bundle cross-sectional average void at various axial locations. The bundle had three grid spacers at the 1/4 axial positions, as well as inlet and outlet end plates. The experimental measurements showed that there was a significant reduction in the amount of subcooled void present immediately downstream of the grid spacers.

It is probable that the turbulence caused by the obstruction not only increases the condensation rate but would retard void generation. The retardation is anticipated to be a much smaller effect, as the void generation is controlled primarily by conditions in the boundary layer, rather than in the bulk fluid. In the model proposed here, the void reduction is assumed to be caused solely by the reduction in bubble diameter (and hence increase in interfacial area) that is caused by the obstruction, which in turn increases the condensation rate.

A number of papers exist on the inverse problem (i.e., that of droplet breakup around grid spacers during reflood). Sugimoto and Murao [8] showed that the droplet diameter was smaller just downstream of the spacer compared to upstream, and that the heat transfer was 20 to 50% better just downstream. Their data was used to develop a model to describe the effects of grid spacers on reflood thermalhydraulics [9] which was implemented in RELAP5. This showed that correct modelling of the grid thermalhydraulics was important in the prediction of the peak clad temperatures.

In our study, the ASSERT code was initially used to model the pressure and flow distributions within the bundle, and the code predictions were compared to single-phase low-power experimental measurements. This enabled the correct overall pressure loss factors to be determined for the various components. These overall loss factors were then split up and ascribed to the individual subchannels according to the relative blockage that the components gave in each subchannel. ASSERT was then used to model the axial void distribution in the bundle. To match the experimentally observed axial profile, it was necessary to introduce a bubble breakup parameter that depended on the local pressure-loss coefficient. This parameter effectively decreased the bubble diameter, and thus increased the interfacial area, promoting increased condensation. By introducing this parameter at the grid spacers, and thus perturbing the bubble size at that axial location only, the void was allowed to recover to its unperturbed value further downstream.

2 EXPERIMENTAL

2.1 Test Desctiption

The subcooled void measurements were made in an electrically heated bundle in a low-pressure, low-temperature water loop. The major components of the loop are a pump, preheater, the MAPLE test station, a plate-type heat exchanger for heat rejection, a header tank and a water reservoir, along with the associated instrumentation and control systems, as shown in Figure 2.

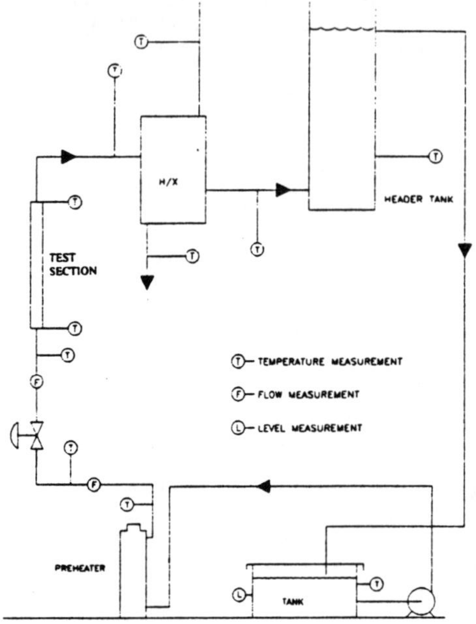

Figure 2. TEST LOOP SCHEMATIC

The test bundles used a simulated half section of a MAPLE 36-pin hexagonal fuel bundle. Each of the 18 heaters consisted of a stainless steel heater tube (the wall thickness of which varied along its length, to provide the required axial heat flux profile), which was coated with a thin layer of alumina followed by finned aluminum cladding. A schematic of a heater and

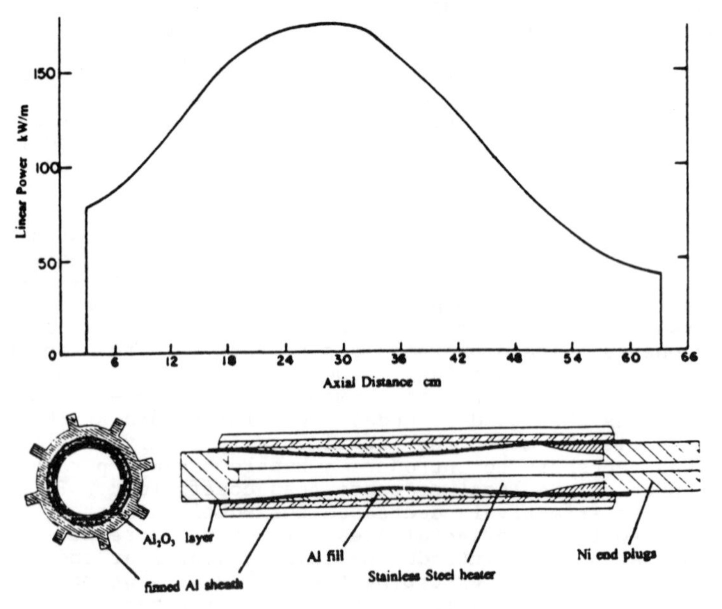

Figure 3 HEATER DESIGN AND AXIAL POWER PROFILE

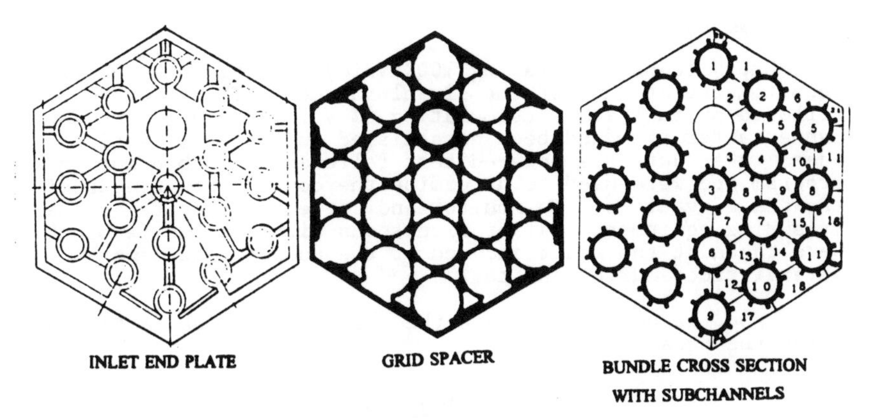

Figure 4. BUNDLE, SPACER AND END PLATE, CROSS SECTIONS

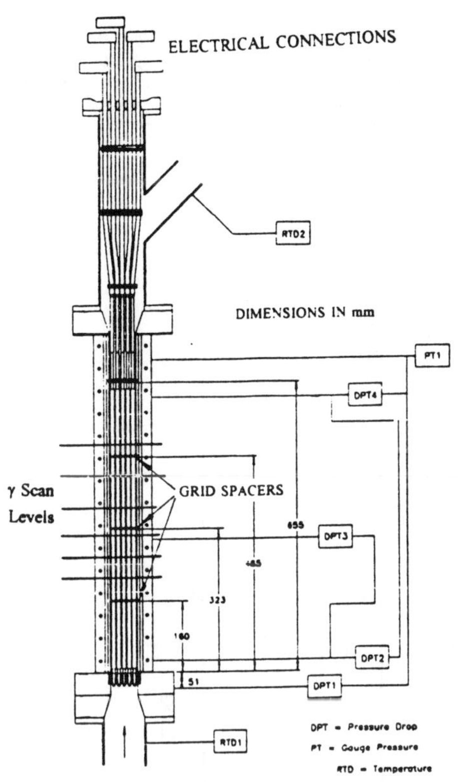

Figure 5 TEST SECTION DETAILS

the axial heat flux profile is shown in Figure 3. The cluster of heaters was contained within a thick-walled polycarbonate flow tube, which allowed visual observation of the bundle during a test. A cross section of the test bundle is shown in Figure 4. All rods were heated, except for the central tie rod and rod #2 which was electrically disconnected prior to the test. The overall arrangement of the test section is shown in Figure 5.

The void present at various axial locations in the test station was measured using a gamma densitometer system. The detector was shielded with lead, so that only gamma rays from a fan shaped beam passing through a 10 mm high slot (slightly wider than the inside of the test section) were accepted.

The source/detector assembly was mounted on two roller slides, which in turn were mounted on a vertical motion table. The roller slides permitted about 15 cm of horizontal movement, which was enough to move the gamma beam clear of the test section and into a lucite block that acted as a calibration standard. A scan program that controls the table motion provided count rate data at six elevations, shown in Figure 5, which allowed the axial variation in the void along the bundle to be determined.

2.2 Void Results

Empty scans (simulating 100% void) were taken with the densitometer at each of the six elevations. These were followed by full scans (simulating 0% void) taken in the same manner after flow had been established in the test section. Scans taken during a powered test followed the table motion program and were used to determine the count rate at each of the six elevations. To monitor and correct for drift in the measuring equipment, calibration scans with the lucite block were taken at the start and end of the bundle test, and at several times during the test.

From the count rate data the void fraction was obtained using the equation

$$\alpha = \frac{\ln I_m - \ln I_l}{\ln I_g - \ln I_l} \tag{1}$$

where I is the count rate and subscript l refers to the count rate with all liquid present (0% void), subscript g refers to all steam (gas) present (100% void) and subscript m refers to the measured rate with the unknown amount of void present.

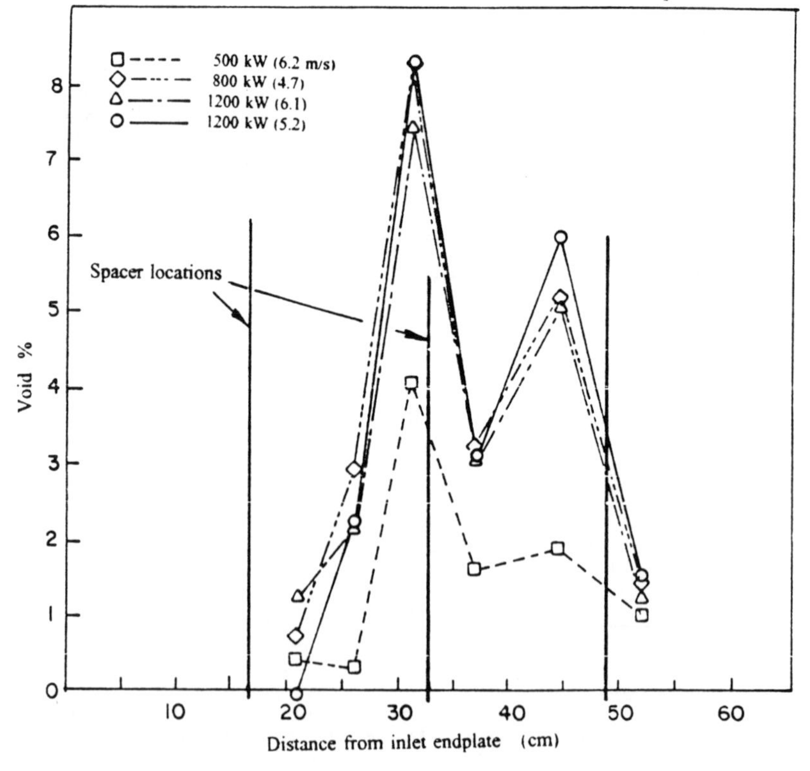

Figure 6 EXPERIMENTALLY MEASURED AXIAL VOID PROFILES

The void fractions measured during the experiment are plotted in Figure 6 for each power and flow condition, as a function of elevation. In this test series the onset of nucleate boiling occurred at 419 kW, and scans taken at greater than that power indicated void.

Little void is seen near the inlet of the bundle, due to the relatively low heat flux and the high subcooling. The void rises rapidly to a peak at the 30 cm measurement location, which corresponds to the peak heat flux (see Figure 3). It then experiences a dramatic fall downstream of a spacer grid, followed by a rise between the 40 and 45 cm level prior to a sharp fall downstream of the last grid spacer. In the 40 to 45 cm region, the heat flux is falling, which would tend to reduce void generation, but this is apparently more than offset by the reduction in subcooling, which would increase the generation and/or reduce the condensation.

Because the bulk coolant was highly subcooled, the mixing induced by the spacers would promote rapid void collapse immediately downstream of the spacers. In other tests, it has been visually observed that subcooled void collapse occurs downstream of turbulence promoters, although the effect has not been quantified previously.

3 ASSERT CODE

3.1 General Description

The ASSERT code [2] is an advanced subchannel code developed to model single- and two-phase flow and heat transfer in vertical and horizontal rod bundles. The thermal nonequilibrium flow model used in ASSERT not only requires that the heat generated in subcooled boiling be partitioned between void generation and further heating of the subcooled liquid, but also that these components enter the individual energy equations. Full details of the methodology used in ASSERT are discussed in the manual [2]; however, the methodology used in subcooled boiling is independently modular to the subchannel analysis and can be discussed separately.

3.2 Void Model

Separate energy equations are written for each of the liquid, gaseous and mixture enthalpies. During boiling, the total heat released from the wall to the fluid mixture can be partitioned into components that, respectively, heat the liquid and generate void:

$$Q_W = Q_{WL} + Q_{GV} \tag{2}$$

This partition and the associated heat transfer coefficients are computed by the widely accepted Chen correlation [10]. During the generation process, the liquid also receives heat from condensing void, Q_C. Thus, the net heat transfer to the liquid is:

$$Q_L = Q_{WL} + Q_C \tag{3}$$

and the effective weight quality, expressed as usual in the terms of mixture, liquid and latent enthalpies, is:

$$\frac{h_m - h_l}{h_{fg}} = f \frac{Q_W - Q_L}{h_{fg}} = f \frac{Q_{GV} - Q_C}{h_{fg}} \tag{4}$$

The actual values of h_m and h_l are obtained in ASSERT from the indvidual energy equations. Unfortunately, no comprehensive set of correlations for interfacial area and heat transfer is available. What does exist in other computer codes is based on simplified ideas, and is not entirely satisfactory for ASSERT. The primary difficulty is discontinuities in heat transfer at flow regime boundaries, and the interfacial heat transfer does not necessarily go to zero at $\alpha = 0$. Such behaviour is necessary for the implicit numerical methods used in ASSERT. The approach used in ASSERT assumes that the interfacial area can be defined as a continuous function of void fraction, and that it goes to zero at $\alpha = 0$. In the work reported here, the void fraction was less than 0.2, which is defined as the bubbly to slug transition in ASSERT. In the bubbly region the bubble diameter is defined as being linear with void fraction as:

$$d = d_{min} + \frac{d_{max} - d_{min}}{0.2}\alpha \quad ; \quad d < d_{max} \tag{5}$$

$$or \qquad d = d_{max} \quad ; \quad otherwise \tag{6}$$

where d_{min} = 0.5 mm and d_{max} = subchannel hydraulic diameter. The interfacial heat transfer coefficient is taken from the work of Moalem and Sideman [11].

4 ASSERT/EXPERIMENTAL COMPARISON

4.1 Pressure Drop

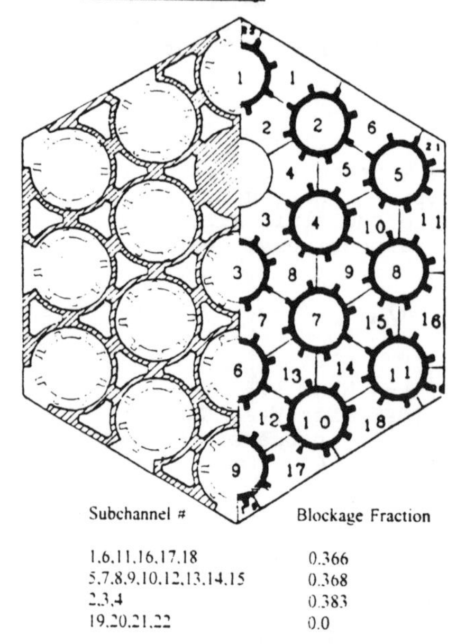

Subchannel #	Blockage Fraction
1,6,11,16,17,18	0.366
5,7,8,9,10,12,13,14,15	0.368
2,3,4	0.383
19,20,21,22	0.0

Figure 7. ASSERT NODALIZATION AND GRID BLOCKAGE

Previous measurements, where the axial pressure profiles along a bundle had been measured, had shown that the Colebrook correlation as implemented in ASSERT adequately described the frictional pressure gradient. In the experiments reported here, the pressure drops across the various sections of the test bundle were measured, and by using ASSERT to calculate the frictional gradient for a low-power test where no two-phase was present, the irrecoverable pressure losses could be extracted from the experimental measurements. These losses were used to calculate the cross-sectional average K factor for the two bundle end plates and the three grid spacers.

However, ASSERT requires an obstruction coefficient (K factor) for each subchannel, rather than an overall cross-sectional average value. Figure 7 indicates the blockage fraction for the grid spacers in individual subchannels and also gives the ASSERT nodalization. It shows that the blockage varied for all the subchannels from 36 to 38%, except for a few small subchannels with zero blockage. The overall average flow blockage was 35%. Based on the above experimental measurements the cross sectional average K factor for the grid spacer was 1.0. Obviously, the local K factor in a subchannel with zero blockage should be 0.0. It is also assumed that a fully blocked subchannel would effectively deflect the flow into the other subchannels in a zigzag motion, and this would add an additional velocity head loss (i.e. would raise the K factor to 2.0 for a fully blocked subchannel).

With these two assumptions, the subchannel grid spacer loss coefficient for a partially blocked individual subchannel can be described by:

$$K_{local} = 2.0 * (blockage\ fraction)^{0.66} \tag{7}$$

A similar approach was used to obtain the individual subchannel K factors for the inlet and outlet end plates. Comparison of the ASSERT pressure drop predictions with both the single- and two-phase experiments when using these individual subchannel blockage factors gave excellent agreement, as shown in Figure 8.

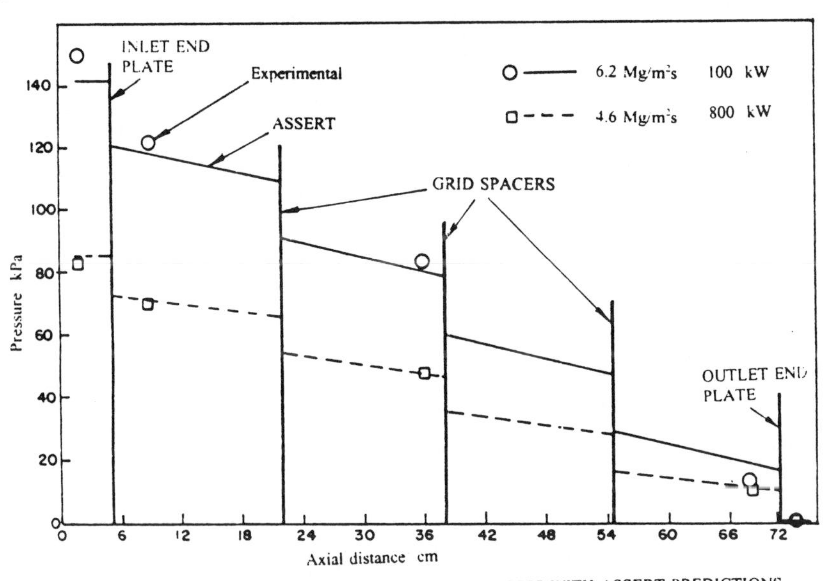

Figure 8 COMPARISON OF MEASURED PRESSURE LOSSES WITH ASSERT PREDICTIONS

4.2 Void Prediction Without Turbulent Condensation

The bundle average void profile for run 305 using the ASSERT program without any increase in condensation due to turbulence is given in Figure 9. It follows a smooth curve with no discontinuities or significant changes in slope at the grid spacers, even though there is a significant pressure change from upstream to downstream of the grids, as shown in Figure 8. This indicates that local pressure variations do not significantly affect the void.

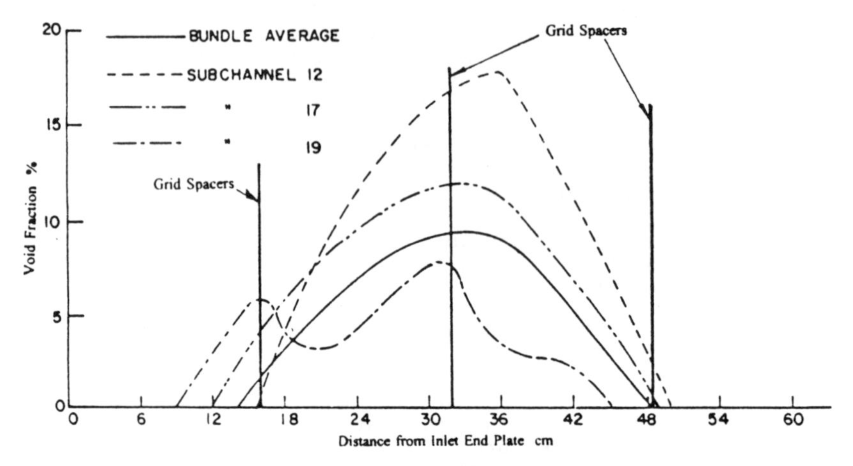

Figure 9 EFFECT OF FLOW DIVERSION ON SUBCHANNEL VOID

Investigation of the void in the individual subchannels for this case, Figure 9, showed that there was a local drop in subcooled void at the grid spacer location in the small unblocked subchannel #19 (see Figure 7 for the subchannel designations). The adjacent large subchannels #17 and #12, which were partly obstructed, showed no such drop. The drop in void in #19 was caused by divertion of lower enthalpy coolant from the partially obstructed subchannel #17, which both increased the local mass flux and reduced the local enthalpy in #19 compared to the undiverted case. However, other than reducing the void locally in the small unobstructed subchannel, the spacer had no effect on the overall cross-sectional average void profile.

4.3 With Turbulent Condensation

As shown in Figure 6, the experimental measurements gave a significant drop in void immediately downstream of a grid spacer on a cross-sectional average basis. This was not captured by ASSERT in its existing form, as was demonstrated in section 4.2. Drawing on the parallel with droplet breakup around grid spacers [8], it was decided to introduce an empirical bubble breakup parameter into ASSERT. Because of the lack of detailed subchannel experimental measurements, this parameter was applied equally to all subchannels. It was implemented as a multiplier on the average bubble diameter at the node where the obstruction exists as

$$D_{bub} = D_{bub} * \exp(-0.693 * K) \qquad (8)$$

where D_{bub} is the bubble diameter and K the local obstruction coefficient. This form has the correct asymptotic trend of: as

K goes to zero, exp(-0.693*K) goes to 1.0; i.e., there is no change in bubble diameter. The constant 0.693 was determined from the experimental data.

The results of the comparison between ASSERT, with the modified bubble diameter, and the experimental data is shown in Figure 10.

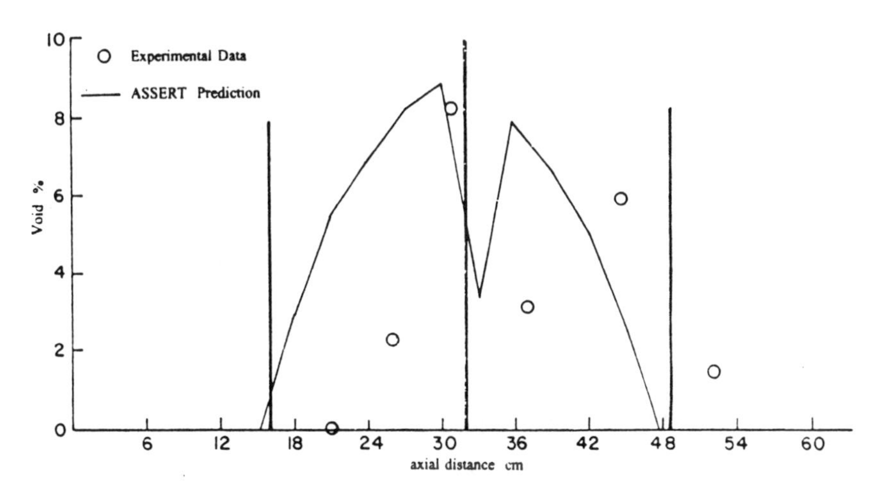

Figure 10 COMPARISON OF ASSERT PLUS TURBULENCE MODEL WITH EXPERIMENTAL DATA

The ASSERT cross-sectional average void shows a significant drop immediately downstream of the grid spacer, followed by a recovery as the bubbles again grow in size. The overall magnitudes of the voids are predicted very well, but it appears that both the growth and the collapse of the voids is too rapid. In previous work with ASSERT, Chatoorgoon et al.[7] concluded that the condensation appeared to be too rapid. As the overall void values are predicted fairly well, this means that the generation rate must also be too rapid to maintain the correct balance. Figure 10 appears to confirm this.

5 CONCLUSIONS.

The ASSERT subchannel code has been compared to some experimentally obtained bundle pressure drop and subcooled void data. These data were taken at low pressures typical of pool-type nuclear reactors, a region where it is known to be difficult to model subcooled void.

The initial comparisons were reasonably good, except that there was no mechanism for modelling the void collapse that occurs downstream of a grid spacer or channel obstruction. To account for this, a bubble breakup parameter which was a function of the obstruction coefficient was introduced into ASSERT. With this parameter in place, the phenomenon was captured and the numerical agreement between the experimental results and the predictions was adequate.

6 REFERENCES

1 Lidstone, R.F. "MAPLE: A Multipurpose Reactor for the

Nineties", The Seventh Pacific Basin Nuclear Conference in San Diego, California, 1990 March.

2 Carver, M.B., Tahir, A., Rowe, D.S., Banas, A.O., and Midvidy, W.I. "Simulation of the Distribution of Flow and Phases in Vertical and Horizontal Bundles using the ASSERT Subchannel Code", Nucl. Eng. Design, Vol. 122, p 413, 1990.

3 Richards, D.J., Hanna, B.N., Hobson, N. and Ardron, K.H. "ATHENA. A Two-Fluid Code for CANDU LOCA Analysis", 3rd. Intl. Topical Meeting on Reactor Thermalhydraulics, Newport, Rhode Island, USA. 1985 October 15-18.

4 Lavigne, P. "Modele D'evolution du titre et du taux de vide en ebullition locale et zone de transition", CEA report 2365, 1963.

5 Dix, G.E. "Vapour Void Fractions for Forced Convection with Subcooled Boiling at Low Flow Rates", PhD Dissertation, University of California, Berkeley, 1971.

6 Dimmick, G.R. and Selander, W.N. "A Dynamic Model for Predicting Subcooled Void: Experimental Results and Model Development", AECL Report, AECL-10304, 1990.

7 Chatoorgoon, V., Dimmick, G.R., Carver, M.B. and Selander, W.N. "Application of Generation and Condensation Models to Predict Subcooled Boiling Void at Low Pressures", Nuc. Technology, Vol. 98, 1992 June, pp 366 to 378.

8 Sugimoto, J. and Murao, Y. "Effect of Grid Spacers on Reflood Heat Transfer in PWR-LOCA", J. of Nuclear Science and Technology, Vol 21, no. 2, 1984 February.

9 Nithianandan, C.K., Lowe, R.J. and Biller, J.R. "Thermalhydraulic Effects of Grid Spacers and Cladding Rupture During Reflood", Transactions of ANS meeting on Transient Thermalhydraulic Analysis and Simulation, Chicago, USA, 1992.

10 Chen, J.C. "A Correlation for Boiling Heat Transfer to Saturated Fluids in Convective Flow" ASME 63-HT-34, American Society of Mechanical Engineers, 1963.

11 Moalem, D. and Sideman, S. "The Efect of Motion on Bubble Collapse", Int. J. Heat and Mass Transfer, Vol. 16, p 2321-2329, 1973.

INFLUENCE OF PRESSURE PULSATION AMPLITUDE ON HEAT TRANSFER IN FULLY DEVELOPED TURBULENT FLOW

*Artur S. Bartosik and Adam J. Wanik**
Kielce University of Technology
Al. 1000–Lecia P.P. 7
25–314 Kielce, Poland

** Technical University of Wrocław*
Wyb. Wyspianskiego 27
50–370 Wroclaw, Poland

Abstract

The paper deals with thermally fully developed turbulent flow of a Newtonian fluid in a straight pipe supplied with a constant heat flux per axial length unit. It is assumed that the pressure gradient in the main flow direction is a harmonic function of time imposed on a constant pressure gradient. The mathematical model is based on Reynolds equations and the energy equation and a convection term is determined from the energy balance. The problem of closure is solved using a low–Reynolds number turbulence model (k–ε). It has been found that in some flow situations the heat transfer coefficient has two maxima which appear in decelerating phase of pulsation. In addition it has been found that imposing the pulsation upon steady flow changes the flow structure causing an increase in generation and intensity of turbulence [15].

1. INTRODUCTION

Pressure pulsations influencing heat transfer both in laminar and turbulent flow have been intensively studied for several years. The problem is important in a number of devices, especially in heat exchangers. Some of the earliest experimental works concerning pulsating flow with heat transfer is presented by Martinelli [1],[2], West and Taylor [3], Havemann and Rao [4], Lemlich and Armour [5] and recently Ludlow et al. [6] and Fallen [7],[8]. Among theoretical works it is appropriate to mention Barnett and Vachon [9], Niida et al. [10], Faghri et al. [11]. The phenomenon of laminar pulsating flow is quite well known in the literature [12]. Turbulent pulsating flow with heat transfer is more complex and still is not fully understood. Some of experiments show an increase or a decrease of heat transfer coefficient which is usually due to the range of chosen parameters. At the present time it appears appropriate to select key parameters and examine their qualitative influence on the heat transfer.

The main aim of this paper is creating a possibly simple mathematical model and examine the qualitative influence of pressure pulsation amplitude on heat transfer coefficient.

The viscous region in the turbulent flow is the dominant parameter affecting efficiency of heat transfer. Introducing disturbances into this region plays a significant role in the intensification of the heat transfer between the fluid and the wall. The efficiency of heat transfer is strongly dependent upon the choice of pulsating flow parameters, particularly upon the amplitude of the pressure gradient shown later.

2. THE MATHEMATICAL MODEL

This study deals with the fully developed pulsating turbulent flow of a Newtonian fluid in a straight pipe supplied with a constant heat flux per axial length unit. The flow is axially symmetrical without a mean circumferential velocity component. It is assumed that the pressure gradient in the main flow direction is a harmonic function of time imposed on a constant pressure gradient (2). The mathematical model is based on Reynolds equations and the energy equation in which the convection term is determined from the energy balance. The problem of closure is solved using a low-Reynolds number turbulence model (k-ε).

The final form of the governing equations is presented in cylindrical coordinates with the x-axis lying on the symmetry axis of the pipe and "r" being the radial distance. Including aforementioned assumptions the final form of the time averaged momentum equation in x-direction is,

$$\bar{\rho}\frac{\partial \bar{U}}{\partial t} = \frac{1}{r}\frac{\partial}{\partial r}\left[r\left(\mu + \mu_t\right)\frac{\partial \bar{U}}{\partial r}\right] - \frac{\partial \bar{P}}{\partial x} \tag{1}$$

$$\frac{\partial \bar{P}}{\partial x} = A_M + A_P \sin \Theta \tag{2}$$

and the temperature equation,

$$\rho\frac{\partial \bar{T}}{\partial x} + \bar{\rho}\bar{U}\frac{\partial \bar{T}}{\partial x} = \frac{1}{r}\frac{\partial}{\partial r}\left[r\left(\frac{\mu}{Pr} + \frac{\mu_t}{\sigma_T}\right)\frac{\partial \bar{T}}{\partial r}\right] \tag{3}$$

where axial temperature gradient for thermally developed flow was determined from the energy balance as follows,

$$\frac{\partial \bar{T}}{\partial x} = \frac{2q}{\bar{\rho}_b \bar{U}_b (c)_b \, R^2} \tag{4}$$

The turbulent viscosity appears in (1) and (3) is the result of approximation of turbulent stress component as,

$$-\bar{\rho}\overline{u'v'} = \mu_t \frac{\partial \bar{U}}{\partial r} \tag{5}$$

and

$$-\bar{\rho}\overline{v't'} = \frac{\mu_t}{\sigma_T}\frac{\partial \bar{T}}{\partial r} \tag{6}$$

To determinate turbulent viscosity, we used k-ε turbulence model of Jones and Launder [13],

$$\mu_t = C_\mu f_\mu \bar{\rho}\frac{k^2}{\varepsilon} \tag{7}$$

$$f_\mu = C_\mu \, EXP\left(\frac{-2.5}{1 + R_*/50}\right) \tag{8}$$

and

$$R_* = \frac{\bar{\rho}\, k^2}{\mu\, \varepsilon}$$

The equation of k and ε are presented as (10) and (11).

$$\bar{\rho}\,\frac{\partial k}{\partial t} = \frac{1}{r}\,\frac{\partial}{\partial r}\left[r\left(\mu + \frac{\mu_t}{\sigma_k}\right)\frac{\partial k}{\partial r}\right] + \mu_t\left(\frac{\partial \bar{U}}{\partial r}\right)^2 - \rho\varepsilon - 2\mu\left(\frac{\partial \sqrt{k}}{\partial r}\right)^2 \qquad (10)$$

$$\bar{\rho}\,\frac{\partial \varepsilon}{\partial t} = \frac{1}{r}\,\frac{\partial}{\partial r}\left[r\left(\mu + \frac{\mu_t}{\sigma_\varepsilon}\right)\frac{\partial \varepsilon}{\partial r}\right] + C_1\frac{\varepsilon}{k}\,\mu_t\left(\frac{\partial \bar{U}}{\partial r}\right)^2 - C_2 f_\varepsilon \bar{\rho}\,\frac{\varepsilon^2}{k} - 2\,\frac{\mu}{\bar{\rho}}\,\mu_t\left(\frac{\partial^2 \bar{U}}{\partial r^2}\right)^2 \qquad (11)$$

where

$$f_\varepsilon = 1 - 0.3\ \mathrm{EXP}\left(-\ R_*^2\right) \qquad (12)$$

The turbulence model constants are following: $C_\mu=0.09$; $C_1=1.55$; $C_2=2.0$; $\sigma_k=1.0$; $\sigma_\varepsilon=1.30$; The mathematical model contains four partial differential parabolic equations, namely (1), (3), (10) and (11) together with complementary relations (2), (4), (7) -(9) and (12). After determining the initial and boundary conditions, and defining a criterion of convergence (13), the problem is solved numerically using an implicit scheme with iterations on each time level.

$$\max_{r=0}\left|\frac{\Phi(r,t) - \Phi(r,t-T^*)}{\Phi(r,t)}\right| < 10^{-3} \qquad (13)$$

The criterion (13) defines the maximum relative difference obtained in time interval T^*.

The pulsation period T is divided into 96 time steps and the pipe radius "R" into 60 intervals but the majority of nodal points are distributed at the pipe wall.

3. NUMERICAL PREDICTIONS

The pressure gradient phase shift with respect to dependent variable h, requires definition of initial conditions in the proper order. The calculations where carried out in two stages:

(a) For a constant value of pressure gradient, i.e. $-\partial\bar{P}/\partial x=A_M$, the following assumptions where made:

$$\bar{T}(r,t) = \bar{T}(r,0) = \bar{T}_W\ ,$$
$$\bar{U}(r,t) = \bar{U}(r,0) = \bar{U}_0\left(\frac{R-r}{R}\right)^{1/7},$$

"k" is defined using turbulence intensity, i.e.

$$k(r,0) = \bar{U}^2(r,0)\ I/2\ ,$$

"ε" is defined using following relation,

$$\varepsilon(r,0) = C_\mu^{3/4} \frac{k^{3/2}(r,0)}{L}$$

The results of the above calculations is a fully thermally developed turbulent flow. The following profiles are obtained:
$\bar{U}_D(r)$, $\bar{T}_D(r)$, $k_D(r)$, and $\varepsilon_D(r)$, where "D" means thermally developed flow.

(b) For designated values of A_M, A_P and f, the set of equations is being solved by several pulsation periods until repeatability in two successive cycles is attended (13). The following profiles are obtained: $\bar{U}_D(r,\Theta)$, $\bar{T}_D(r,\Theta)$, $k_D(r,\Theta)$, $\varepsilon_D(r,\Theta)$ and following dependent variables: $Nu(\Theta)$, and/or $h(\Theta)$.

The quantitative changes of predicted Nusselt number had been compared with the experimental data of Fallen [8], for steady fully developed turbulent flow of water ($A_P = 0$). The results indicate that for flow without pulsation, the maximum relative error is less than 5% in the range of Prandtl number $Pr_D = 2.8\text{-}4.7$, [14].

Generally, for fully developed turbulent pulsating flow with heat transfer, general dependent variable Φ depends on r, Θ, $(K_R)_t$, Re_M, Pr_M, where $(K_R)_t$ is dimensionless frequency parameter [15].

$$\bar{U}_P(r,\Theta) = \bar{U}(r,\Theta) - \bar{U}_M(r) \tag{14}$$

$$\bar{T}_P(r,\Theta) = \bar{T}(r,\Theta) - \bar{T}_M(r) \tag{15}$$

The qualitative character of changing of heat transfer coefficient is compared with the results of numerical calculations (carried out for steady flow) in which the value of the pressure gradient is equal to the mean value (A_M) in pulsating flow. For turbulent flow it has been proved that in some determined ranges of dimensionless pulsation frequency, the pulsating component of velocity (14) and temperature (15), have parabolic, ring or double ring profiles even though no double ring flow has been noted in the literature so far [15]. The amplitude of the pressure gradient seems to be one of the critical parameters influencing the heat transfer.

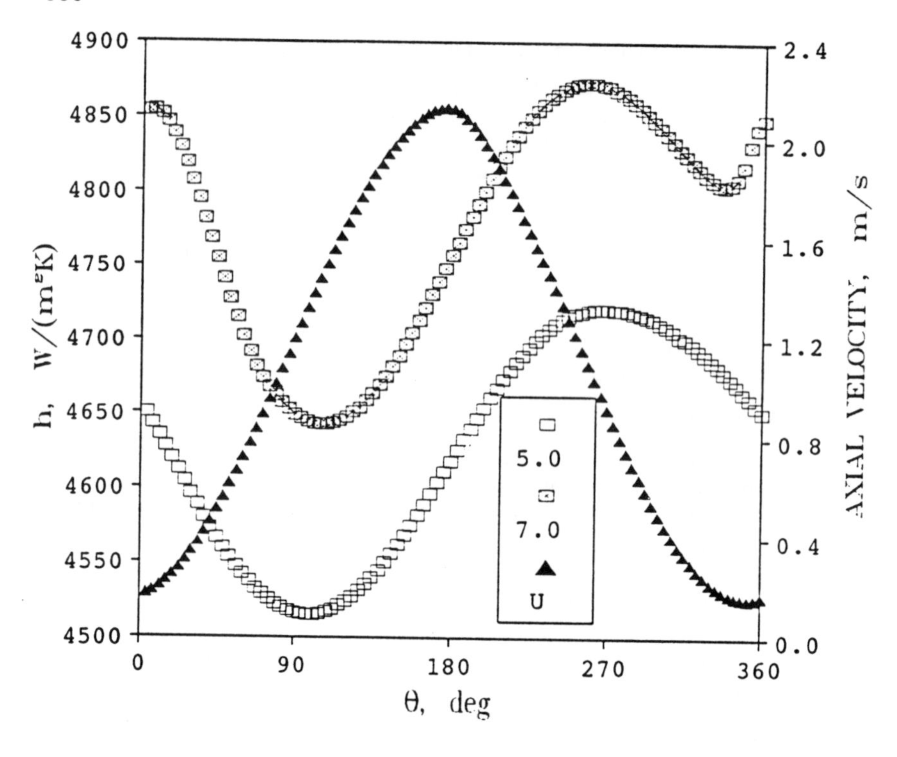

Fig.1 Phasewise variation of the heat transfer coefficient and axial velocity at f=0.4; A_P/A_M = 5 and 7; U_o at A_P/A_M=7.

In pulsating flow the heat transfer coefficient varies periodically in time. However, the shape of these changes is not sinusoidal for higher values of A_P/A_M, as shown in Fig.1. At f=0.4 and A_P/A_M=5.0 the shape of h is similar to sinusoidal while at A_P/A_M=7.0 it has two maxima. It is proper to note that when flow is decelerating ($\Theta \cong 180^0 - 360^0$), the heat transfer coefficient is higher than in the case of accelerating flow. It seems to be strictly connected with the level of turbulence generation and intensity which is higher during deceleration than acceleration.

4. CONCLUSIONS

For a given pulsation frequency there is a critical value of A_P/A_M. When A_P/A_M is higher than the critical value, the shape of h is not sinusoidal and has two maxima during pulsation period.

When A_P/A_M is higher than critical one, the value of h_M is higher than "h" for steady turbulent flow in which Reynolds number is equal to the Re_M in pulsating flow.

NOMENCLATURE

A - pressure gradient value, N/m^3
C_i - turbulence model constant ($i=1,2,\mu$)
c - unit heat capacity of fluid, W/(kg K)
D - pipe internal diameter, m
f - pulsation frequency, 1/s
f_i - damping function in turbulence model, ($i=\mu,\varepsilon$)
h - heat transfer coefficient, $W/(m^2K)$
I - intensity of turbulence,
k - kinetic energy of turbulence, m^2/s^2
$(K_R)_t$ - dimensionless frequency, $(K_R)_t = \sqrt{\frac{\omega}{[(\nu_t)_b]_M}}\,R$
L - Nikuradse length, m
Nu - Nusselt number,
P - static pressure, N/m^2
Pr - Prandtl number,
q - heat flux, W/m
r - distance from symmetry axis, m
R - pipe radius
R_* - turbulent Reynolds number
t - time, s
t' - fluctuating component of temperature, K
T - temperature, K
T^* - pulsation period, s
u', v' - fluctuating components of velocity in direction x and r respectively, m/s
U - velocity component in direction x, m/s
x - axial coordinate, m
ε - rate of dissipation of turbulence kinetic energy, m^2/s^3
Φ - general dependent variable ($\Phi = \bar{U}, \bar{T}, k, \varepsilon$ or Nu, h),
μ - dynamic viscosity of slurry, Ns/m^2
ρ - density of slurry, kg/m^3
σ_i - effective Prandtl-Schmidt number ($i=k,\varepsilon$)
ω - pulsation frequency, rad/s

Subscripts

b - bulk
M - averaged after pulsation period T
o - symmetry axis
t - turbulent
w - wall
$\underline{\Theta}$ - phase angle, $\Theta=\omega t$
‾ - time averaged value

LITERATURE

1. MARTINELLI,R.C.,1941,Analysis of mechanism of heat transfer trough a solid-fluid interface with application to fluids flowing in pipes-Part III: Periodic Flow, Ph.D., University of California, Berkeley, CA.
2. MARTINELLI,R.C.,BOELTER,L.M.K.,WEINBERG,E.B.,YAKAHI,S.,1943, Heat transfer to a fluid flowing periodically at low frequencies in a vertical tube, Trans.ASME, vol.65,789-798.
3. WEST,F.B. and TAYLOR,A.T.,1952, The effect of pulsations on heat transfer, Chem. Engng. Progress, vol.48, No.1, 39-43.
4. HAVEMANN,H.A. and RAO,N.N.,1954, Heat transfer in pulsating flow, Nature, vol.174, No.4418, 41.
5. LEMLICH,R. and ARMOUR,J.C., 1963, Forced convective heat transfer to a pulsed liquid, Chem. Engng. Progress Symp., vol.61, No.57, 83-88.
6. LUDLOW,J.C.,KIRWAN,D.J.,GAINER,J.L., 1980, Heat transfer with pulsating flow, Chem. Engng. Commun., vol.7, 211-218.
7. FALLEN,M., 1982, Heat transfer in a pipe with superimposed pulsating flow, Kaiserslautern Univ., Warme und Stoffubertragung, vol.16, No.2, 89-99.
8. FALLEN,M.,1981, Warme bergang bei laminarer und turbulenter Srómung im starren glatten Rohr mit berlagerter Strómungs-pulsation, Diss. Universitat Kaiserslautern.
9. BARNETT,D.O. and VACHON R.I., 1970, An analysis of convective heat transfer for pulsating flow in a tube, Fourth Int. Heat Transfer Conference, Elsevier Publishing Company, Amsterdam.
10. NIIDA,T., YOSHIDA,T., YAMASHITA,R., NAKAYAMA,S., 1974, The influence of pulsation on laminar heat transfer in pipes, Heat Transfer-Japenese Res., vol.3, No.3, 19-28.
11. FAGHRI,M. JAVDANI,K., FAGHRI,A., 1979, Heat transfer with laminar pulsating flow in a pipe, Letters in Heat and Mass Transfer, vol.6, 259-270.

12. CREF,A. and ANDRE,P., 1986, Computational techniques for fluid flow, Ed. Taylor C.,Johnson J.A.,Smith W.R.,Pineridge Press.

13. JONES,W.P. and LAUNDER,B.E., 1973, The Calculation of Low-Reynolds-Number Phenomena with a Two Equation Model of Turbulence, Int. J. Heat and Mass Transfer, vol.16.

14. BARTOSIK,A.S., SOBOCINSKI,R.A., WANIK,A.J., 1987, Numerical Prediction of heat transfer in fully developed pulsating turbulent flow, Proc. 5th Int.Conf. on Numerical Methods in Thermal Problems, Editors Lewis,R.W.,Morgan,K.,Habashi,W.G., Pineridge Press, Swansea, vol.5, Part 1, 632-643.

15. BARTOSIK,A.S., 1989, Numerical modelling of fully developed pulsating flow with heat transfer,Ph.D., Kielce University of Technology, Poland.

MODELLING OF THE TURBULENT THERMAL BOUNDARY LAYER IMPLEMENTED IN CFD CODE N3S AND APPLICATION TO FLOWS IN HEATED ROOMS

B. DELENNE(1) , G. POT(2)

ELECTRICITE DE FRANCE

SUMMARY

In this paper we present an efficient finite element method for solving the unsteady Navier-Stokes equations for turbulent incompressible flow coupled with thermal problems. This method has been implemented in the N3S code, developed at Electricité de France. The time discretization is first described. We precise then the "projected Uzawa" algorithm used for the Stokes problem. After that we present the modelling of the thermal boundary layer used to simulate walls with fixed temperature in turbulent flows, which has been implemented in the code. We give then some recent applications using this modelling and compared with experimental data, and we conclude with the future development.

1 INTRODUCTION

The "Direction des Etudes et Recherches" of EDF has been working since 1982 on N3S, a 3D finite element (F.E.) code for simulating turbulent incompressible flows [1] and developed under quality assurance procedures. This code is used for many industrial applications nowadays (internal flows [1], thermal problems [2], turbomachinery [3]). Some applications, involving temperature in turbulent flows (as natural and forced convection, or heat transfer) need a very accurate description of the thermal boundary layer at the walls and of its effects on the flow, so an important work has been done to improve this modelling in the code N3S.

2 PRESENTATION OF N3S

2.1 Presentation of the problem

The equations governing the fluid motion in a regular open bounded subset Ω of $\mathbf{R}^N$ ($N = 2$ or 3) and over a time interval $[0,T]$ are the Navier-Stokes equations for velocity **v** and pressure p. For some flows of the thermal convection type, we assume that density variations with temperature T are small enough to be taken into account with the Boussinesq

---

(1) Department TTA, EDF/DER, 6 quai Watier, 78400 CHATOU CEDEX, FRANCE

(2) Department LNH, EDF/DER, 6 quai Watier, 78400 CHATOU CEDEX, FRANCE

approximation. The energy equation gives the evolution of the temperature T which satisfies a convection diffusion equation.

Industrial flows computed with N3S code are generally turbulent and characterized by very high Reynolds numbers. To simulate such complex non linear flows, we consider the average value of physical quantities (velocity, pressure and temperature if necessary) which we calculate by means of a model for correlations between fluctuating velocities and between velocity and temperature fluctuations. The model used or k-ε model is made up of two equations in which k denotes the turbulent kinetic energy and ε the turbulent dissipation rate.

2.2 Boundary conditions

Boundary conditions depend on the type of boundary which is to be dealt with. For the inlet Γ_{in} of the Ω fluid domain, forced constrained conditions (Dirichlet) are used on all the variables. For the outlet Γ_{out} of the Ω fluid domain, vanishing normal stress for the velocity and vanishing flux conditions (homogeneous Neumann) for scalar quantities are used. For walls Γ_w a modelling based on the generalisation of the analysis of the boundary layer on a flat plate is used. For velocity, the normal component satisfies an impermeability condition ($\mathbf{v}.\mathbf{n} = 0$, where $\mathbf{n}$ is the normal exterior to the wall). This condition is completed by a friction condition on the tangential stress. We furthermore assume that near the wall, there is an equilibrium between turbulent production and dissipation which enables to express k and ε at the wall. Flux conditions can be imposed for temperature T : the modelling of the thermal boundary layer is detailed in the next chapter.

2.3 Time discretization

Time discretization of scalar equations of convection diffusion as well as Navier-Stokes equations is realized thanks to a fractional step scheme. The convection step is processed by a characteristics method and the diffusion or Stokes steps thanks to an implicit Euler scheme.

Convection step : The kth-order characteristics scheme consists in computing an approximation at time t^{n+1} in $[0,T]$ of the total derivative of the quantity C (or the velocity $\mathbf{v}$) with the help of a kth-order backward differentiation scheme integrated along the characteristics curve defined on the time interval $[t^{n-k+1}, t^{n+1}]$.

Diffusion or Stokes step : We can now compute C^{n+1} ($\mathbf{v}^{n+1}$ and p^{n+1}) which denotes an approximation at time t^{n+1} by solving a diffusion problem (or a Stokes problem for velocity $\mathbf{v}$ and pressure p) by a classical k^{th}-order backward differentiation scheme.

A theoretical analysis of the whole scheme has been done in [4]. In N3S code, the 1st-order and 2nd-order schemes have been implemented.

2.4 Space discretization

The Stokes problem is discretized in space thanks to a finite element method. The unstructured meshes use triangles or tetrahedra with a mixed formulation for the velocity and the pressure so as to get the Stokes problem well posed. The elements available in the N3S code are P1-P2 or P1-isoP2 elements. In most industrial cases, we use the P1-isoP2 element. The velocity matrix can be mass-lumped without diminishing the global spatial precision.

2.5 Projected gradient Uzawa algorithm

$\Gamma_{in,h}, \Gamma_{out,h}, \Gamma_{w,h}$ are the portions of the boundary of the calculation domain Ω_h associated to $\Gamma_{in}, \Gamma_{out}, \Gamma_w$ respectively. We consider in this section the P1-P2 triangle. We introduce discrete spaces :

$X_h = \{\varphi \in C^0(\Omega_h) \,/\, \varphi$ is a 2nd-degree polynomial on each triangle$\}^N$

$M_h = \{q \in C^0(\Omega_h) \,/\, q$ is a 1st-degree polynomial on each triangle$\} \cap L^2(\Omega_h)$

$V_{0,h} = \{\mathbf{w} \in X_h \,/\, \forall\, a \in \Gamma_{in,h}, \mathbf{w}(a) = 0$ and $\forall\, a \in \Gamma_{w,h}, \mathbf{w}(a)\,\mathbf{n}_h = 0\}$

where $\mathbf{n}_h$ is a discret normal vector defined in [5].

The discretized Stokes problem according to boundary conditions, as defined at section 2.2 naturally uses test functions of $V_{0,h}$. Another way can be to solve the Stokes problem in a larger space (X_h) than $V_{0,h}$ and apply the boundary conditions thanks to a projection operator P_h from X_h onto $V_{0,h}$.We introduce the continuous bilinear form $a_h : X_h \times X_h \to \mathbf{R}$ related to the "mass+diffusion" discretized operator and $b_h : X_h \times M_h \to \mathbf{R}$ related to the "divergence/gradient" discretized operator $l_h : X_h \to \mathbf{R}$ and $g_h : M_h \to \mathbf{R}$ are linear forms. They come from source terms and non-homogeneous Dirichlet conditions. The discretized Stokes problem is as follows :

Find $\mathbf{u}_h \in X_h$, $p_h \in M_h$ solutions of

$$\begin{cases} a_h(P_h\mathbf{u}_h,\ P_h\mathbf{w}) + b_h(P_h\mathbf{w},\ p_h) = l_h(P_h\mathbf{w}), & \forall\, \mathbf{w} \in X_h \\ b_h(P_h\mathbf{u}_h, q) = g_h(q), & \forall\, q \in M_h \\ P_h\mathbf{u}_h = \mathbf{u}_h & \end{cases}$$

P_h is symmetric. This leads to the matrical system (A is associated to a_h, B to b_h, L to l_h and G to g_h) :

Find $\mathbf{u} \in X_h$, $p \in M_h$ solutions of

$$\begin{cases} P\,A\,P\,\mathbf{u} + P\,B^t\,p = P\,L \\ B\,P\,\mathbf{u} = G \\ P\,\mathbf{u} = \mathbf{u} \end{cases} \qquad (1)$$

The projected gradient Uzawa algorithm applied to the Stokes problem with boundary conditions consists in applying the Uzawa algorithm to system (1) and using a conjugated gradient algorithm on the matrix PAP in the velocity system. In this case, the most adapted preconditioner is the diagonal one which can be easily computed thanks to the symmetry property of matrix P.

2.6 Numerical methods

The analysis of CPU time and memory requirement on a standard 3D industrial application reveals that one important part of the time is needed by F.E. calculation, that is assembling of the different matrices (varying in time with the eddy viscosity) and r.h.s ; another important part of the CPU time is also used to solve big linear systems **A**x=b (see 1)), where **A** is a well-conditioned mass plus diffusion matrix (diffusion of all the variables such as velocity, temperature, k, ε), or where **A** is the Laplacian-like preconditioning matrix of the pressure system, equal to $\mathbf{BB}^T$. To solve these systems, we use an iterative Preconditioned Conjugate Gradient algorithm. Details concerning the optimisation are given in references [6] and [7].

3 MODELLING OF THE THERMAL BOUNDARY LAYER

3.1 The boundary conditions

The flows studied with thermohydraulic codes are usually bounded by walls. The modelling of the turbulence and the heat transfer in the vicinity of these walls is quite difficult. Many solutions are suggested in the literature.

The turbulent effects in the core of the flow are taken into account in N3S by a k-ε model and the different solutions to simulate the turbulent boundary layer can be divided into two groups. First, the most sophisticated methods, like the low-Reynolds models, compute all the gradients in the boundary layer considering the physically correct no slip condition. These methods are the most accurate but require very thin meshes in the wall area and the number of nodes is far too high for industrial calculations. The other classical ways to simulate the boundary layers are the *wall laws*, that is, rather than apply the no slip condition at the wall, the modelling of the turbulent boundary layer supply the shear stress τw at a given distance from the wall (that is small compared to the characteristic size of the mesh at the wall). This method does not require any great refinement in the wall area.

On a thermal point of view, different kinds of boundary conditions could be used in N3S, such as Dirichlet conditions in inlet, Neumann conditions on heating or cooling walls. As an analogy with the momentum transfer, walls that are kept at fixed temperature raise many problems, and one more time, instead of applying numerically the continuity of the temperature profile at the wall, the modelling of the thermal boundary layer evaluates the heat flux from wall to fluid. Many models can be found, the simplest ones, based on Prandtl-Taylor analogy, consider that viscous and thermal sublayers are of the same order of magnitude. The main disadvantage of these proceedings is that the very small, or, on the opposite, very high Prandtl number (Pr) fluids cannot be taken into account. The formulation that was chosen is the one suggested by Arpaci and Larsen [8].

3.2 The modelling of the thermal boundary layer for Pr << 1

Considering a steady flow on a plate kept at fixed temperature Tw, the heat flux (q_w) from the plate can be written :

$$q_w = \rho C_p (a + a_t) \frac{dT}{dy} \tag{2}$$

on a dimensionless form :

$$1 = a_e^+ \frac{dT^+}{dy^+} \tag{3}$$

The purpose of the model is to describe with precision the diffusivity parameter a_e^+. On a first approach, the boundary layer can be divided into two parts, one near the wall dominated by the molecular diffusion, the other by the turbulent diffusion. In the first region, the turbulent terms can be neglected and the equivalent diffusivity is reduced to :

$$a_e^+ = \frac{a}{\nu} = \frac{1}{Pr} \tag{4}$$

On the opposite, in the second region, the heat transfer is mainly due to turbulent diffusion. The parameter a_t is simply related to the eddy viscosity ν_t through the turbulent Prandtl number concept. The turbulent Prandtl number (Pr_t) is supposed to be constant. The expression of the eddy viscosity is given by the mixing length model applied in the core of the boundary layer where the logarithmic velocity distribution is valid :

$$\nu_t = (l_m^+)^2 \left| \frac{\partial U^+}{\partial y^+} \right| = \nu \kappa y^+ \quad \Rightarrow \quad a_e^+ = \frac{\kappa y^+}{\mathrm{Pr}_t} \tag{5}$$

Those two regions in the boundary layer are not necessary separated. Considering fluids for which the conductivity is high enough, the molecular diffusion contributes to the heat transfer in the turbulent region, that is, the conduction sublayer is thicker than the viscous one. For this kind of fluids, clearly small Prandtl number fluids, the equations (4) and (5) intersect and the temperature distribution for Pr << 1 can be described by :

$$\begin{cases} y^+ \le \dfrac{Pr_t}{k\,Pr} = y_0{}^+ & : \quad a_e^+ = \dfrac{1}{Pr} \quad \Rightarrow \quad T^+ = Pr\,y^+ \\ y^+ > \dfrac{Pr_t}{\kappa Pr} = y_0{}^+ & : \quad a_e^+ = \dfrac{\kappa y^+}{Pr_t} \quad \Rightarrow \quad T^+ = \dfrac{Pr_t}{\kappa}\left(Ln\left(\dfrac{\kappa Pr}{Pr_t} y^+\right) + 1\right) \end{cases} \tag{6}$$

The dimensionless thickness of the conduction layer : $y_0{}^+$ and the integration constant are determined by assuming that the equivalent diffusivity and the temperature are continuous.

3.3 The modelling of the thermal boundary layer for Pr ≥1

A modelling of the thermal boundary layer that consider only very small Prandtl number fluids is not satisfying. That would mean that only liquid metal flows can be simulated by the code, and that the model cannot be applied to oil (Pr>100), water (Pr≈10) or even gases (Pr=0,7 for the air). For those fluids, the near wall thermal sublayer is immersed in the viscous sublayer. This suggests to introduce a third intermediate layer between the molecular and the logarithmic regions in which the heat transfer is determined not only by conduction but also by turbulent diffusion. The mixing length theory cannot be applied crudely because the flow in this part of the boundary layer is not fully turbulent. Arpaci and Larsen retain Levitch suggestion who considers : $\nu_t \approx (y^+)^3$. This formulation is in good agreement with Van Driest [9] who developed an expression for the mixing length all over the boundary layer. So, in this third layer, the equivalent diffusivity is :

$$a_e^+ = \frac{1}{\mathrm{Pr}} + \frac{a_1}{\mathrm{Pr}_t (y^+)^3} \tag{7}$$

Kader and Yaglom [10] expected from experimental results the relation $a_1/Pr_t = 10^{-3}$. According to dimensional arguments, the thickness of the thermal conduction sublayer $y_1{}^+$ can be determined by equating the molecular terms to the turbulent ones. The continuity of the temperature all over the boundary layer and the continuity of the equivalent diffusivity at the boundary between the intermediate and logarithmic layers supply the thickness $y_2{}^+$ of the intermediate zone and the two integration constants c_1 and c_2. The temperature profile for Pr ≥ 1 is given by :

$$0 < y^+ \le y_1^+ \quad : \quad a_e^+ = \frac{1}{Pr} \quad \Rightarrow \quad T^+ = Pr\ y^+$$

$$y_1^+ < y^+ \le y_2^+ \quad : \quad a_e^+ = \frac{1}{Pr} + \frac{a_1}{Pr_t (y^+)^3} \quad \Rightarrow$$

$$T^+ = \frac{1}{3} a b^{-\frac{2}{3}} \left[Ln\left(y^+ + b^{\frac{1}{3}} \right) - \frac{1}{2} Ln\left(y^{+2} - b^{\frac{1}{3}} y^+ + b^{\frac{2}{3}} \right) + \sqrt{3} Arctg\left(\frac{2y^+}{\sqrt{3} b^{\frac{1}{3}}} - \frac{1}{\sqrt{3}} \right) \right] + c_1 \quad (8)$$

$\underline{y_2^+ < y^+}$: $a_e^+ = \dfrac{1}{Pr} + \dfrac{\kappa y^+}{Pr_t} \Rightarrow T^+ = \dfrac{Pr_t}{\kappa} Ln\left(\dfrac{Pr_t}{Pr\,\kappa} + y^+ \right) + c_2$

with $a = \dfrac{Pr_t}{a_1}$ and $b = \dfrac{a}{Pr}$

It must be emphasized that the dimensionless temperature profile depends only on the property of the fluid (i.e. the Prandtl number) and on the dimensionless distance to the wall. These equations are a bit different from those derived by Arpaci and Larsen who neglected the conduction term in the intermediate region. The present model is supposed to be more precise in the description of the temperature profiles especially for high Prandtl number fluids. It has been validated by comparisons with thermal measurements from the literature, one application is reported in reference [2].

4 APPLICATION TO FLOWS IN HEATED ROOMS

The comfort notion in habitation is related to fluid dynamic parameters. The temperature gradients have to be as small as possible, the air currents indiscernible but strong enough to evacuate the pollutants. All these conditions are challenges to the conception of heating apparatus. The flows generated by heating or cooling systems are generally very complex and rise many difficulties to be modelled. Numerical methods seem to be useful to study the movements of air in habitations. Different complementary approaches are developed by EDF. For the first ones, the global models, each room is cut in few areas, the solver evaluates the heat and mass transfer from one area to another. An other approach consists in evaluating the ability of thermohydraulic codes like N3S to predict the thermal and dynamic characteristics of this kind of flows. Different heating systems have been simulated with N3S. The configurations correspond to an experimental setup that was elaborated by EDF and the LET[(3)] in order to create flows similar to those in real habitations.

4.1 The experimental set-up

The set-up consists in a square cavity, in which small gaps have been arranged in walls to inject hot or cold air. The test cell is 300 mm wide, 1040 mm long and 1040 mm high. In order to generate two dimensional flows, there are two other identical cavities on each side of the test cell. The heating systems that have been simulated are the heating floor and the convector. For each configuration, an external flow enters the enclosure through an opening (18 mm) in the top of the left vertical wall and exits from another opening (24 mm) located in the bottom of the right vertical wall. The

---

[(3)] Laboratoire d'Etudes Thermiques de Poitiers
40, avenue du Docteur Pineau 86022 Poitiers cedex

convector is simply represented by an injection of hot air on the left end of the floor through a 18 mm gap (figure 1). The mixed convection in the test cell was studied considering isothermal boundary conditions. All the walls are kept at fixed cold temperature, generally the same temperature as the aeration except the floor in the heating floor configuration, the thermal source being then the low horizontal wall. The velocities can be chosen between 0. m/s and 1. m/s. The temperature of the air varies between 10°C and 40°C for the aeration and between 20°C and 70°C for the convector. The temperature inside the cavity is evaluated from readings of thermocouples regularly arranged in the cell. Velocity measurements are made by Laser-Doppler Velocimetry. Prior to the measurements and to the calculations, a succession of visualizations was made in different configurations. These experiments revealed that the flows are extremely sensitive to the boundary conditions. A small alteration on the incoming velocity or temperature can provide two absolutely different behaviours.

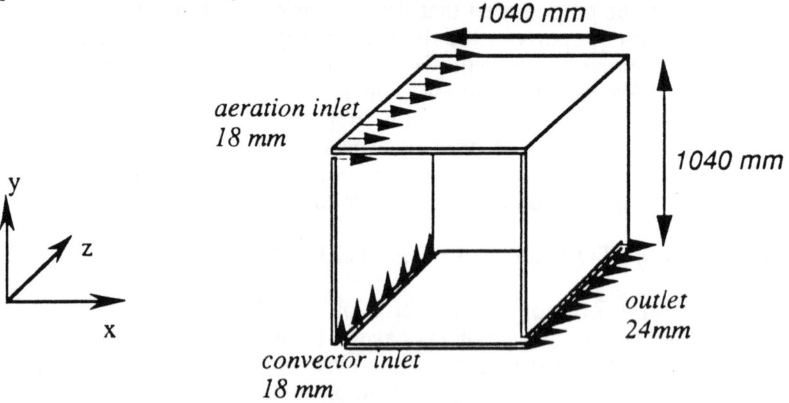

figure 1 : Experimental setup

4.2 Experimental results

Isothermal experiments lead to a great vortex all over the cavity area turning clockwise : the jet from the aeration gap hits the right wall and falls. The same structure was found with thermal sources either in the heating floor or in the convector configuration ;however, if there is an important temperature difference and small velocities, the buoyancy terms can deviate the aeration jet and make it fall along the left wall as it comes into the cavity, generating an anticlockwise vortex. Clearly these flows are ruled by inertia or buoyancy effects and mainly two parameters arise in this kind of problem : the Reynolds number and the Froude (Fr) number.

In the heating floor configuration, there is only one incoming flow and the Froude number (or the Richardson number $Ri = 1/Fr^2$) is the only criterion. The Froude number is defined as

$$Fr = \frac{Va}{\sqrt{ge\beta\Delta T}}$$

where e is the width of the opening, ΔT the temperature difference between the aeration and the heating floor, g the magnitude of gravitational acceleration and b the coefficient of thermal expansion. It was observed that for Fr greater than 3.67 the flow is dominated by the inertia effects and turns clockwise and for Fr lower than 1,83 the flow is dominated by buoyancy effects and turns in the opposite direction. An intermediate flow corresponding to a critical value of the Froude number was obtained for Fr ≈ 3,23, in which case the incoming flow is not strong enough to reach the right vertical wall and is deviated in the middle of the ceiling (figure 2).

In the case of the convector, the thermal source is a hot jet that gives rise to perturbations that have to be taken into account. The same structures have been noticed, as far as the big vortices are concerned, when the flow is fully dominated either by buoyancy or inertia, but they cannot be classified in respect of only one criterion. Four cases are detailed in table 1. In the critical flow configuration, the cold air from aeration falls along the left wall and get mixed in the middle of the cavity with the up-coming hot air from the convector (figure 3).

V_a	V_c	T_a	T_c	T_w	Fr_a	Fr_c	flow dominated by
0,3	0,3	18,5	23,5	18,5	11,6	5,8	inertia
0,3	0,3	13	63	40	2,5	2,3	inertia
0,18	0,10	11	28	20	2,3	1,9	buoyancy
0,15	0,15	13	63	40	1,2	1,1	inertia

table 1 : The dominating phenomena in the convector configuration for various parameters (Fra and Frc represent respectively the Froude numbers in the aeration and convector areas).

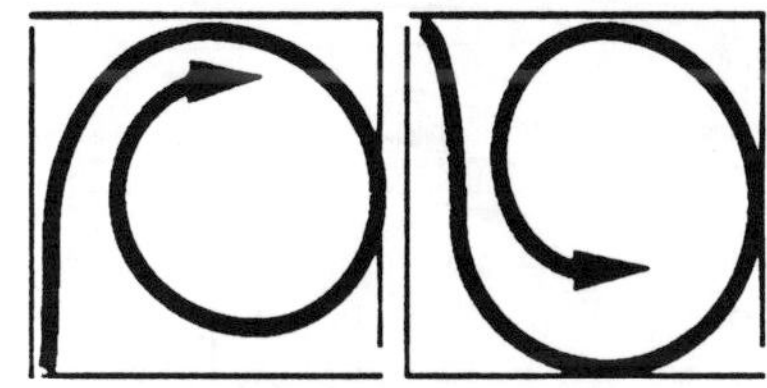

flow dominated respectively by inertia terms and buoyancy terms both in convector or heating floor configuration.

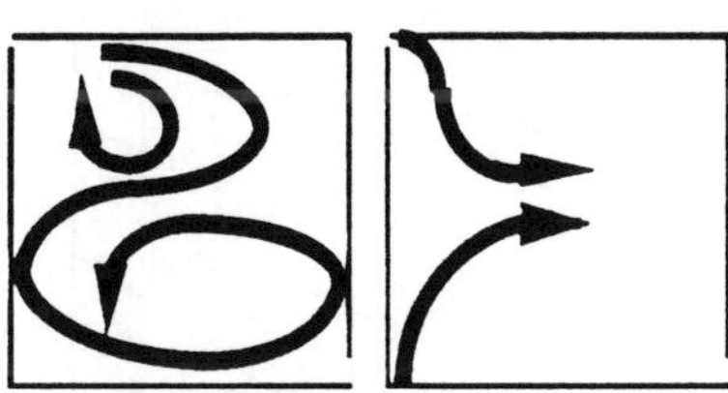

critical flows respectively for the heating floor and for the convector

figure 2 : flow structures

4.3 <u>Calculations</u>

All the flows generated are turbulent. Numerically, the turbulence effects are represented with the k-ε model and the heat transfer on walls is simulated by the modelling presented above. It can be expected that a pure natural convection problem requires a low-Reynolds model, but in each configuration studied here, the velocities of the air injected seem to be high enough to consider that the heat transfer is due to mixed convection. In order to take into account the buoyancy effect, the Navier Stokes equations are coupled to the energy equation through the Boussinesq approximation. One configuration of each heating apparatus has been computed.

The data for the heating floor are Va = 0.6 m/s, Ta = Tw = 15°C, Tf = 35°C ; this leads to Fr = 5.19. Actually, the flow generated experimentally is very stable, and the clockwise vortex occupies the whole cavity. As a result, the temperature field is very homogeneous between 19°C and 20°C in the middle of the cell. For the numerical study, the grid used is made of 687 elements and 1526 nodes ; it is reported on figure 3. The elements are triangle P1-isoP2. The time step applied for this calculation is dt = 0.02 s. 7500 time steps have been computed during 1h20 CPU time on Cray YMP (that corresponds to 0.43 s for 1000 nodes per time step).

Velocities and turbulent parameters fields converge in less than 5000 iterations but the convergence of the temperature field is harder to reach, specially in the centre of the cell where all the velocities are very low.

The data for the convector are Va = 0.4 m/s, Vc = 0.2 m/s, Ta = Tw = 20°C, Tc = 40°C. There is actually a big vortex occupying most of the cell, but there is also a smaller one turning in the opposite direction just under the aeration opening. In this case, both buoyancy and inertia act on the flow structure, which corresponds to a critical flow. Near the aeration gap, the buoyancy effects are dominating and the cold air falls along the left wall. Then, lower, this air is submitted to the thrust of the up-coming hot jet and goes up again, creating the smaller vortex and deviating the hot jet to the right. Apart from this secondary vortex region and the regions close to the walls, the temperature in the main part of the cavity is one more time very homogeneous between 25°C and 26°C. For the numerical study, a different mesh was used, made of 881 elements (triangle P1-isoP2) and 1952 nodes (figure 3). The time step applied for this calculation is dt = 0.05 s. 6000 time steps have been computed during 1h20 CPU time on Cray YMP (that corresponds to 0.41 s for 1000 nodes per time step). The first iterations were computed neglecting the buoyancy terms in order to create a proper initial state before taking into account the buoyancy terms through the Boussinesq approximation.

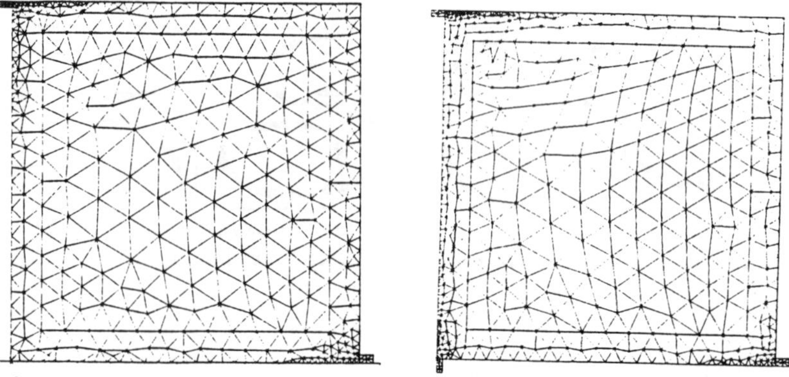

figure 3 grids used for the heating floor and for the convector simulations

4.4 <u>Numerical results</u>

In the heating floor case, the numerical results are in good agreement with the experimentation. N3S simulates a big vortex occupying all the cavity. The calculated velocities near the walls are a bit more important than the measured ones as can be seen on velocities distributions figure 4. The temperature in the middle of the cell is too small for about one Kelvin compared to the measurements. Many other iterations would reduce this discrepancy. The temperature still progresses after 7500 time steps although the velocity field is absolutely converged. The last 2500 iterations raised the temperature in the middle of the cell by 0.3 K only, without any consequence on the dynamic parameters. A difference of 0.5 K between the measured and the calculated temperature appears on the wall areas, this difference does not decrease when increasing the iterations, so it can be expected that the inaccuracy in the middle would be reduced from 1 K to 0.5 K in the best case. The comparisons with the measurements of the temperature profiles for the heating floor are plotted on figure 5.

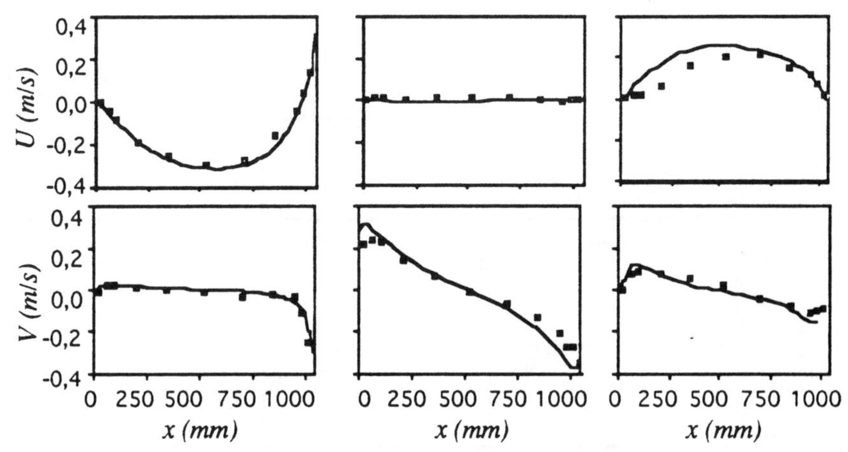

figure 4 : horizontal (U) and vertical (V) velocities profiles in the heating floor configuration for y = 25 mm ; y = 500 mm ; y = 920 mm
▫ *measurement,* —— *calculation*

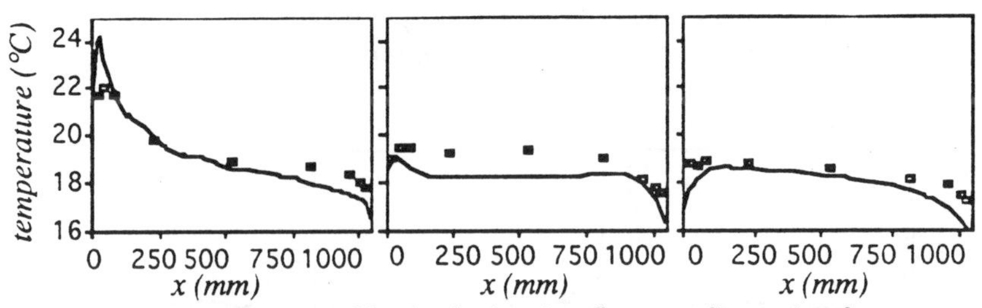

figure 5 : temperature profiles in the heating floor configuration for y = 30 mm, y = 520 mm ; y = 960 mm ▪ *measurement,* —— *calculation*

The simulation of the convector gave rise to a lonely vortex turning clockwise and occupying all the cavity. According to the numerical results, the buoyancy effects are not strong enough to generate the secondary vortex under the aeration opening. The cold air entering the cell keeps an horizontal direction. This inaccuracy generates many other discrepancies that appear on the velocities profiles (figure 6). The hot jet is not slowed down any more by the down-coming cold air. N3S simulates the jet all along the left wall although the experiment founds negative velocities in the top of this wall. Entering the cavity, the cold air is mixed to the vortex ; this generates all along the edges of the cell too large velocities.
The temperature found by N3S in the core of the cavity is higher than the ones revealed by the measurements by a bit less than 1 K (figure 7).

These two calculations reveal that the numerical solution does not detect the transition. However, in a stable configuration, the calculations provide very satisfying results although some discrepancies could be observed near the walls. This phenomenon appeared for the two simulations so the code seems to underestimate the shear stress. This can be related to the fact that for this kind of flows, the viscous sublayer might be very thick, and the boundary layer modelling was designed for fully turbulent flows where the molecular viscosity interferes on very small length scales.

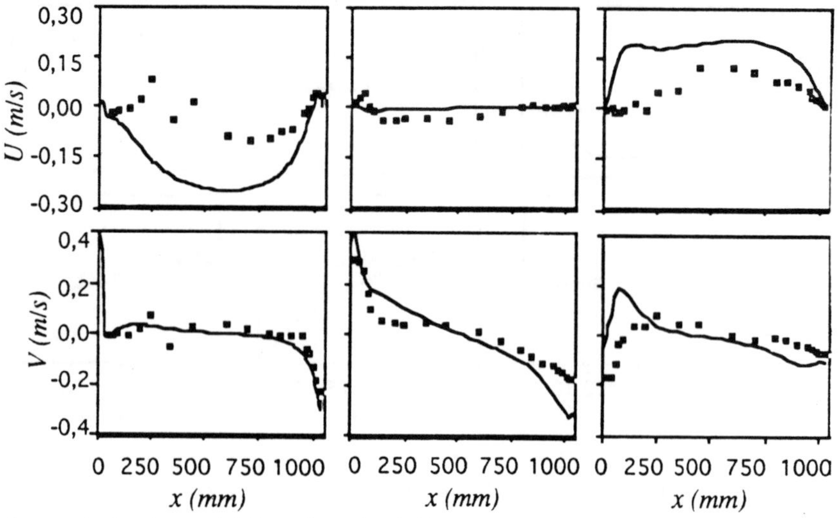

figure 6 : horizontal (U) and vertical (V) velocities profiles in the convector configuration for y = 40 mm ; y = 500 mm ; y = 920 mm
▫ *measurement,*—— *calculation*

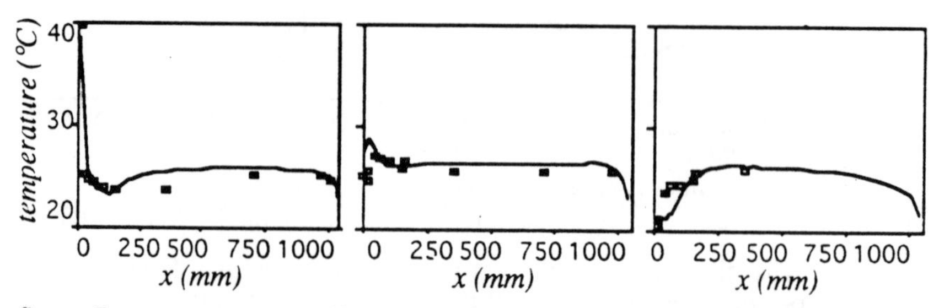

figure 7 : temperature profiles in the convector configuration for y = 50 mm, y = 620 mm ; y = 1000 mm ▪ *measurement,* —— *calculation*

CONCLUSION

Flows generated by heating apparatus have been simulated with a finite element code. On a qualitative point of view the numerical results are in good agreement with the measurements although some differences could be pointed out in unstable configurations. A wall law model for the thermal boundary layer, described above, seems to be able to simulate the heat transfer in such mixed convection flows in enclosures.

In order to compute pure natural convection problems in realistic geometries, more work is under way : it consists of implementing in the code a low-Reynolds model and taking into account the variations of density with temperature. Some other works in progress, specially the implementation of a R_{ij}-ε model, more realistic than the k-ε model, or the development of adaptative meshing procedures may also improve the accuracy of the solution.

NOMENCLATURE

a, at, $a_e^+=(a+at)/\nu$	molecular, turbulent and equivalent diffusivity
Cp	specific heat at constant pressure
lm	mixing length
q_w	heat flux from wall to fluid
$T^+=(T-Tw)/T^*$	dimensionless temperature
$T^*=q_w/\rho Cpu^*$	
Ta, Tc	air temperatures from aeration and from convector
Tf, Tw	temperature of the floor and of the other walls
u*	friction velocity
Va ,Vc	the velocity of the air from aeration andconvector
$y^+=yu^*/\nu$	dimensionless distance to the wall
ν, ν_t	kinetic molecular and turbulent viscosity
ρ	density

REFERENCES :

[1] CHABARD J.-P., METIVET B., POT G., THOMAS B. :
An efficient finite element method for the computation of 3D turbulent incompressible flows, Finite Elements in Fluids, Vol. 8, 1992.

[2] CHABARD J.P., DELENNE B., POT G., RAZAFINDRAKOTO E.:
Application of N3S code to the computation of an axisymetric jet impinging a flat plate. 15th Meeting of the A.I.H.R. Working Group on Refined Modelling, Lyon, France, October 1991.

[3] COMBES J.-F., GRIMBERT I., RIEUTORD E. :
An analysis of the viscous flow in a centrifugal pump with a finite element code, Proceedings of Fluid Machinery Forum, Portland, Oregon, USA, June 1991.

[4] BOUKIR K., MADAY Y., METIVET B. :
A high order characteristics method for the incompressible Navier-Stokes equations, Proceedings of International Conference on Spectral and High Order Methods, Montpellier, France, June 1992

[5] ENGELMAN M.S., SANI R.L., GRESHO P.M. :
The implementation of normal and tangential velocity boundary conditions in finite element codes for incompressible fluid flow, Int. J. Num. Math. Fluids, 1981.

[6] POT G., GREGOIRE J.P., NITROSSO B. :
Conjugate gradient's performances enhanced in the C.F.D. code N3S by using a better vectorization of matrix vector product, Proceedings of IMAC'S 91, Dublin, Ireland, July 1991.

[7] POT G., BONNIN O., MOULIN V., THOMAS B. :
Improvement of finite element algorithms implemented in CFD code N3S for industrial turbulent applications, Proceedings of Numerical Methods for Laminar and Turbulent Flow, Swansea, U.K., July 1993.

[8] ARPACI V., LARSEN P.
Convective Heat Transfer, Prentice Hall, 1984

[9] VAN DRIEST E.R.
On turbulent flow near a wall, J. Aero. Sci., Vol. 23, pp. 1007, 1956

[10] KADER B. A., YAGLOM. M.
Heat and Mass transfer laws for fully turbulent walls flows, J. of Heat and Mass Transfer, vol. 15, pp. 2329, 1972

NUMERICAL SIMULATION OF TURBULENT FLOW AND FIRE PROPAGATION IN 3D MOUNTAIN RIDGES

A.M.G. Lopes*, A.C.M. Sousa and D.X. Viegas***

* *Grupo de Mecânica dos Fluidos, Universidade de Coimbra, Portugal*
** *Dept. of Mechanical Engg., University of New Brunswick, Fredericton, N.B., Canada*

ABSTRACT

A numerical model for the simultaneous calculation of velocity and temperature fields, and fire propagation in mountain ridges is presented. Turbulent fluid flow calculations are performed using the SIMPLEC procedure applied to a boundary fitted coordinate system, while the fire rate of spread is computed using a combination of Rothermel's fire spread model, a two semi-ellipse formulation for fire shape, and the Dijkstra dynamic programming algorithm for fire growth simulation. Computations are made for two ridge configurations, with different angles of intersection between the two slopes, using two types of fuel. Results show a much higher rate of spread for the ridge with a lower slope intersection angle, confirming the danger due to unusual propagation rates of fires in these topographies.

INTRODUCTION

Forest fires occurring in wildland terrain very often take dramatic proportions, which may result in high material costs, not to mention tragic loss in human lives. Among large fires that occur world-wide, a few of them have been singled out due to their dangerous behaviour. The aim of the present paper is to provide a better understanding of the high propagation rates of fires occurring in mountain ridges, a topography, which, in the past, has been related to large killing fires, and that can be visualised as the intersection of two inclined slopes with the ground. It is widely recognised that the ability to predict the rate of spread of a fire is of outmost importance for its efficient attack and rapid suppression. Among the several models that have been proposed to describe the rate of spread of wind driven fires over diverse topography, Rothermel's model [1], with its application in the Behave system [2], has been one of the most popular and widely used fire spread models. The previous knowledge of the midflame wind speed difficults and limits the use of the Behave system in complex terrain topography, where wind speed is extremely difficult to estimate. The present approach is designed to bridge this gap by proposing a computer procedure that integrates both a numerical algorithm for turbulent 3D flow calculations over general topography, with a fire propagation model based on Rothermel's work for the calculation of fire spread.

GRID GENERATION

For appropriate discretization of the equations in the selected domain, a boundary fitted co-ordinate system is employed. The accuracy of the solution obtained and its numerical stability are highly dependent on the grid used, which should be smooth, refined in locations where high gradients are expected to occur, and with a low degree of skewness. To meet these requirements, the physical locations of each grid node were determined by solving a set of elliptic differential equations. Grid generation using a Laplace equation was first proposed by Thompson [3], and further improved through the introduction of control functions, which allow proper adjustment of the clustering and orthogonality of grid lines near the boundaries. The transformation from the physical space (x, y, z) to the computational (deformed) space (ξ,η,ζ) is done by solving a set of three partial differential equations. This method has proved to be quite robust and of easy adaptation to new geometries [4].

FLUID FLOW AND TEMPERATURE CALCULATIONS

Transport Equations

The air velocity at mid flame height for the input of the fire propagation model is obtained through the solution of the Navier-Stokes (NS) equations discretized in a boundary fitted coordinate system. The original NS equations are transformed from the physical space to a computational domain by substituting the dependent Cartesian variables by the deformed grid variables. In the present approach, the Cartesian components of the velocity are kept as the dependent variables. The resulting set of equations written in terms of a generic velocity component u_i, is as follows :

momentum equations :

$$J\frac{\partial \rho u_i}{\partial t}+\frac{\partial}{\partial \xi_j}\left(J\rho U_j u_i \right) = -\frac{\partial}{\partial x_i}\frac{\partial P}{\partial x_j} + \frac{\partial}{\partial \xi_m}\mu_{eff} J \times \tag{1}$$

$$\left[g^{mn}\frac{\partial u_i}{\partial \xi_n}+\frac{\partial \xi_m}{\partial x_j}\frac{\partial \xi_n}{\partial x_i}\frac{\partial u_j}{\partial \xi_n}-\frac{2}{3}\frac{\partial \xi_m}{\partial x_i}\frac{\partial \xi_n}{\partial x_j}\frac{\partial u_j}{\partial \xi_n}\right] -\frac{2}{3}J\left[\frac{\partial \xi_j}{\partial x_i}\frac{\partial(\rho k)}{\partial \xi_j}\right] - Jg_i(\rho-\rho_0)$$

continuity equation :

$$\frac{\partial \rho}{\partial t}+\frac{\partial}{\partial \xi_i}\left(\rho J U_i\right) = 0 \tag{2}$$

In the above equations, U_j represent the contravariant components of the velocity vector and J is the transformation Jacobian.

The turbulent viscosity is computed with a low Reynolds number k-ε model, in a formulation similar to the one proposed by Zhang and Sousa [5], here extended to generalized coordinates in 3 dimensions. For simplicity, these equations are here presented in their Cartesian form :

$$\frac{D}{Dt}(\rho k)=\frac{\partial}{\partial x_i}\left[\left(\mu+\frac{\mu_t}{\sigma_k}\right)\frac{\partial k}{\partial x_i}\right]+P_1+G-\rho\tilde{\varepsilon}+D_1 \quad \text{where} \quad P_1=-\rho\overline{u_i' u_j'}\frac{\partial u_i}{\partial x_j}$$

$$\tilde{\varepsilon}=\varepsilon+D_1 \quad \tilde{\varepsilon}=f_\mu\frac{C_\mu \rho k^2}{\mu_t} \quad G=-\beta g\frac{\mu_t}{\sigma_t}\frac{\partial T}{\partial z} \quad D_1=-2\mu\left(\frac{\partial\sqrt{k}}{\partial n}\right)^2 \quad Re_t=\frac{\rho k^2}{\mu\tilde{\varepsilon}} \tag{3}$$

$$\frac{D}{Dt}(\rho\tilde{\varepsilon})=\frac{\partial}{\partial x_i}\left(\frac{\partial\tilde{\varepsilon}}{\partial x_i}\right)+\frac{\tilde{\varepsilon}}{k}\left(C_1(P_1+C_3 G)-C_2 f_1\rho\tilde{\varepsilon}\right)+E \;;\; f_\mu=\left[1-\exp\left(-\frac{Re_{kn}}{26.5}\right)\right]^2$$

$$f_1=1-0.3\exp\left(-Re_t^2\right) \;;\; E=\frac{\mu\mu_t\left(1-f_\mu\right)^2}{\rho} \;;\; Re_{kn}=\frac{\rho\sqrt{k}n}{\mu} \;;\; C_3=\tanh\left|\frac{w}{u}\right|$$

The variables n and u_p, in the original formulation [5], were taken as the normal distance to the closest wall and the velocity component parallel to the wall, respectively. In the present case, due to the geometry, and also for the sake of generality, n was replaced by an 'equivalent distance' to the wall, computed on the basis of a potential like function. For a location faraway from the wall, the 'equivalent distance' will be equal to the actual distance. Similarly, u_p is the velocity component parallel to surfaces of constant 'equivalent distance'.

The transformed energy equation is obtained following the procedure already presented for the NS equations :

$$J\frac{\partial \rho T}{\partial t}+\frac{\partial}{\partial \xi_i}(J\rho U_i T)=\frac{\partial}{\partial \xi_i}\left[J\left(\frac{\mu}{\sigma_l}+\frac{\mu_t}{\sigma_t}\right)\left(g^{ij}\frac{\partial T}{\partial \xi_j}\right)\right]+S_c \tag{4}$$

where g^{ij} is a generic contravariant metric.

Numerical Algorithm

The control volume formulation is used for the discretization and integration of the transport equations in each computational cell, which for convenience is cubic with a unit mesh spacing. A staggered grid approach [6] is followed for the location of the three components of the velocity vector, which are thus located at the centre of each face of the control volume (contravariant positions). Other scalars, such as temperature and turbulence quantities are positioned at the geometric centre of the control volumes.

Continuity and momentum equations are linked through pressure following the SIMPLEC procedure [7]. Transport equations are discretized using linear profiles for the viscous derivatives, while the convection terms are treated using a hybrid formulation [6]. These discretized equations, including the pressure correction equation, are all cast in a single general form:

$$a_P\phi_P \Sigma a_{nb}\phi_{nb} + Sc \quad ; \quad \Sigma a_{nb}\phi_{nb} = a_E\phi_E + a_W\phi_W + a_N\phi_N + a_S\phi_S + a_T\phi_T + a_B\phi_B \quad (5)$$

E - East W - West N - North S - South T - Top B - Bottom

where ϕ is a generic variable and Sc is the source term. These equations are solved using the TDMA algorithm with sweeps in the three computational directions. Exception is made for the pressure correction equation for which, due to its low rate of convergence mainly in grids with a large aspect ratio, a three level multigrid technique is used in conjunction with the TDMA algorithm [8].

Boundary Conditions

Velocity, temperature and turbulence quantities are specified at the inlet of the domain. These were obtained as solution of the respective transport equations when the gradients in the two directions perpendicular to the flow vanish, a situation which corresponds to a developed flow with an imposed boundary layer height. At the other boundaries, except to the ground, the second derivatives of scalars are made equal to zero, and for the momentum equations, a global mass balance is imposed. At the bottom boundary (the ground), an adiabatic condition is imposed for the energy equation, while for momentum, a no-slip impervious boundary is considered.

THE FIRE SPREAD MODEL

The model for fire propagation used in the present study can be decomposed in three main parts:

- Fire rate of spread in the direction of maximum spread is computed using the equations proposed by Rothermel [1].
- A model of double ellipse is used to compute fire spread in any arbitrary direction [9].
- The fire propagation from cell to cell is computed using the Dijkstra's dynamic programming algorithm [10].

Each of these parts is next presented separately.

Rothermel's model for fire spread calculation

Rothermel's model is an empirical fire spread model, developed essentially from laboratory experiments (as opposed to field experiments), using three different artificial fuels (random excelsior fuel beds, regular geometry fuel beds made of wood sticks arranged in two configurations, and grass like natural fuel beds). The model was developed for a stationary fire spread in a statistically homogeneous fuel bed (at least in what concerns relevant parameters), and is not suited to be applied to high intensity fires, crown fires, or fires where spotting plays a relevant role in fire spread. A recent review of Rothermel's model is presented in [11].

The keystone of the model is equation (6), which expresses an energy balance within a unit volume of the fuel ahead of the flame. It illustrates the concept that the rate of spread, R, is just a ratio between the rate of heating of the fuel, and the energy required to bring that same fuel to ignition.

$$R = \frac{I_R \pi (1 + \phi_w + \phi_s)}{\rho_b \omega Q_i} \tag{6}$$

I_R - Reaction Intensity - Heat release per unit area of the flame front [J/m²s]
π - Propagating Flux Ratio - Fraction of heat release that is responsible for fuel heating and consequent ignition.
φ_w - Wind Factor
φ_s - Slope Factor
ρ_b - Bulk Density - Mass of fuel per unit volume [Kg/m³].
ε - Effective Heating Number - ratio between the bulk density and the mass of fuel involved in the ignition process
Q_i - Heat of Pre-Ignition - the heat required to bring a unit weight of fuel to ignition [J/Kg].

All the quantities in the above equation are calculated using the fuel and environmental characteristics supplied by the user, which are :

- ovendry fuel loading [Kg/m2]
- fuel depth [m]
- fuel particle surface-area-to-volume- ratio [m^{-1}]
- ovendry particle density [Kg/m^3]
- fuel particle moisture content
- fuel particle mineral content
- wind speed at midflame height [m/s]
- terrain topography (slope and direction of maximum slope)

A detailed explanation of this spread model and respective equations is given in [1].

The effect of wind and slope is contained in the two coefficients ϕ in Eq. (6). In the original Behave system, the direction of maximum spread is computed by summing vectorially the fire spread obtained in the wind direction, assuming a no-slope situation, with the spread obtained in the direction of maximum slope in a no-wind condition. This procedure seems to be physically unrealistic, since situations may arise where a zero rate of spread may occur (if wind and maximum slope directions are 180º apart). Furthermore, in this formulation, the no-wind, no-slope contribution for the total fire spread is taken into account twice. The approach adopted in the present work is to sum up as vectorial quantities, not the computed rates of spread in the two mentioned situations, but the wind and slope coefficients φ_w and φ_s.

The double ellipse model

Rothermel's model provides as output the spread rate in the maximum spread direction. Anderson [9], using data from 198 fire experiments, proposed a two semi-ellipse fire shape model, whose equations shown bellow, are used in the present work to compute fire size and shape :

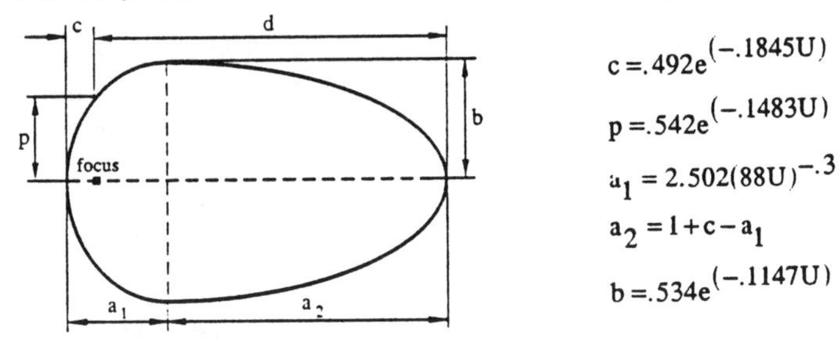

$$
\begin{aligned}
c &= .492e^{(-.1845U)} \\
p &= .542e^{(-.1483U)} \\
a_1 &= 2.502(88U)^{-.3} \\
a_2 &= 1+c-a_1 \\
b &= .534e^{(-.1147U)}
\end{aligned}
\tag{7}
$$

Dijkstra's Algorithm

Simulation of fire growth in the presence of spatial non-homogeneities of the diverse parameters governing fire spread requires a sub-division of the domain in cells, in which the input conditions are assumed as locally homogeneous. In the present case, the fuel is considered as homogeneous, and the spatially changing conditions are: slope, wind velocity and wind direction. Starting with a certain number of ignited cells, the fire spread calculation is a matter of finding the next cell to ignite, which may not be the closest one. Dijkstra's dynamic programming algorithm is thus a procedure designed to find the shortest path between a specified pair of nodes. For each burning cell, the time the fire takes to propagate to all its neighbours is computed with Rothermel's model and the two semi-ellipse model. The closest instant of time each non-burning cell may be ignited is computed, and the next cell to ignite will be the one with the lowest assigned value.

THE OVERALL MODEL

Interaction between the fire spread model and flow field calculations is processed in the following way :

Grids

For computational expediency, and within the objectives and expected accuracy of the model used in the present study, the calculations are carried out in two different grids : a coarser one, three-dimensional, where temperature and velocity fields are computed ; and a finer grid, quasi-two dimensional, mapping the ridge soil, where fire size and shape are calculated. Information between these two grids is interchanged through linear 3D interpolation.

Interaction Fire - Velocity Field

The heat released by the fire is responsible both for ignition of the fuel ahead of the flame front, due to heat transfer to the fuel, as well as for heating of the air and soil. Rothermel's model provides a means of quantifying both the total heat release and the part of this heat absorbed by the fuel, by defining the variable π (propagating flux ratio). For the remaining energy, based on some experimental evidence, it was considered that 30% was absorbed by the soil, and 70% was responsible for air heating. Treatment of the fuel bed immediately after the passage of the flame front was done by assigning an exponential decay of heat release, using a physically realistic relaxation time chosen to match field data.

Effects of fire on the velocity and temperature field are introduced in the numerical model by imposing a volumetric heat release through the first row of control volumes. The amount of heat release is dependent on the corresponding percentage of ground area that is burning.

Numerical procedure

The solution is carried out iteratively, starting with a nearly punctual fire. Taking as initial condition the velocity field computed for an isothermal situation, a few newly ignited cells are computed in the 'fine grid', and the instant of time at which each cell ignites is registered. The computation is then transferred to the 'coarse grid', where the heat release rate trough each control volume is calculated as a linear function of the percentage burning area of the control volume. Velocity, temperature and turbulent quantities are solved, and after convergence of the iterative method, the new velocity field is interpolated to the fine grid and a few new cells are ignited. The process is continued until the burning/burned area reaches a specified limit.

The advance in time of the process is thus controlled by the combustion model. The number of cells ignited within each cycle is made sufficiently small, typically around 10% of the already ignited cells, so that the changes in the velocity field are not too large.

GEOMETRY DEFINITION AND RESULTS

Simulations were made using two different types of fuel, characterised by distinct reaction intensities and rates of spread. The characteristics of the fuels were taken from data published by Rothermel [1], and they correspond to a tall grass field, and to debris of light logging slash, hereafter designated as fuel models a and b, respectively (fuel models 2 and 9, respectively, in [1]). Characteristic fire spread rate and reaction intensity for these two fuels, for a no wind condition, and for an incident wind speed of 5 m/s, both in flat terrain, are:

	Reaction Intensity [KW/m^2]	Rate of Spread [m/min]	Rate of Spread [m/min]
Fuel a	501	1.2	66
Fuel b	465	.174	3.54

For comparison purposes, two different canyon geometries were considered :

- Geometry A is represented in Fig. 1, where the grid used for flow field calculations can observed. The two slopes are inclined at an angle of 30º to the ground, and intercept one another with an angle of 90º.
- Geometry B is shown in Fig. 2. The slopes intercept now at an angle of 180º, and are inclined at an angle of 22.2º.

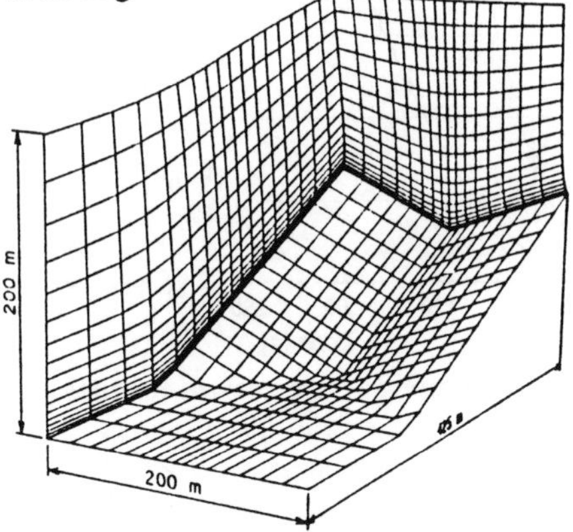

Figure 1 - Geometry of Ridge A.

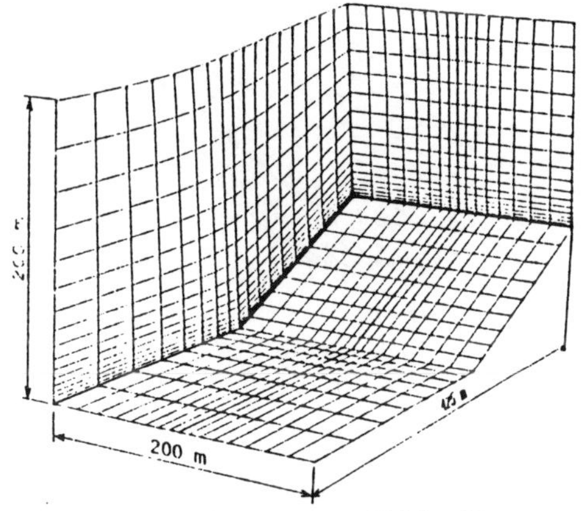

Figure 2 - Geometry of Ridge B

For all the cases considered, the fire was ignited at the point of intersection of the slopes with the ground. The incident wind velocity is 7 m/s. Figure 3 represents the fire growth, for fuel a, in terms of burning and burned areas as well as the distance of head fire from the ignition point, as a function of the time elapsed since fire start. Figure 4 depicts the fire shape for fuel a in geometry A (G.A) and geometry B (G.B), along with some selected velocity vectors. The shaded areas are the fire contours at the times represented in the tables (t = 0, time at which fire is just ignited). In each figure, the velocity vectors correspond to the latest instant of time. The difference in plume shape for these two situations is clearly noticeable.

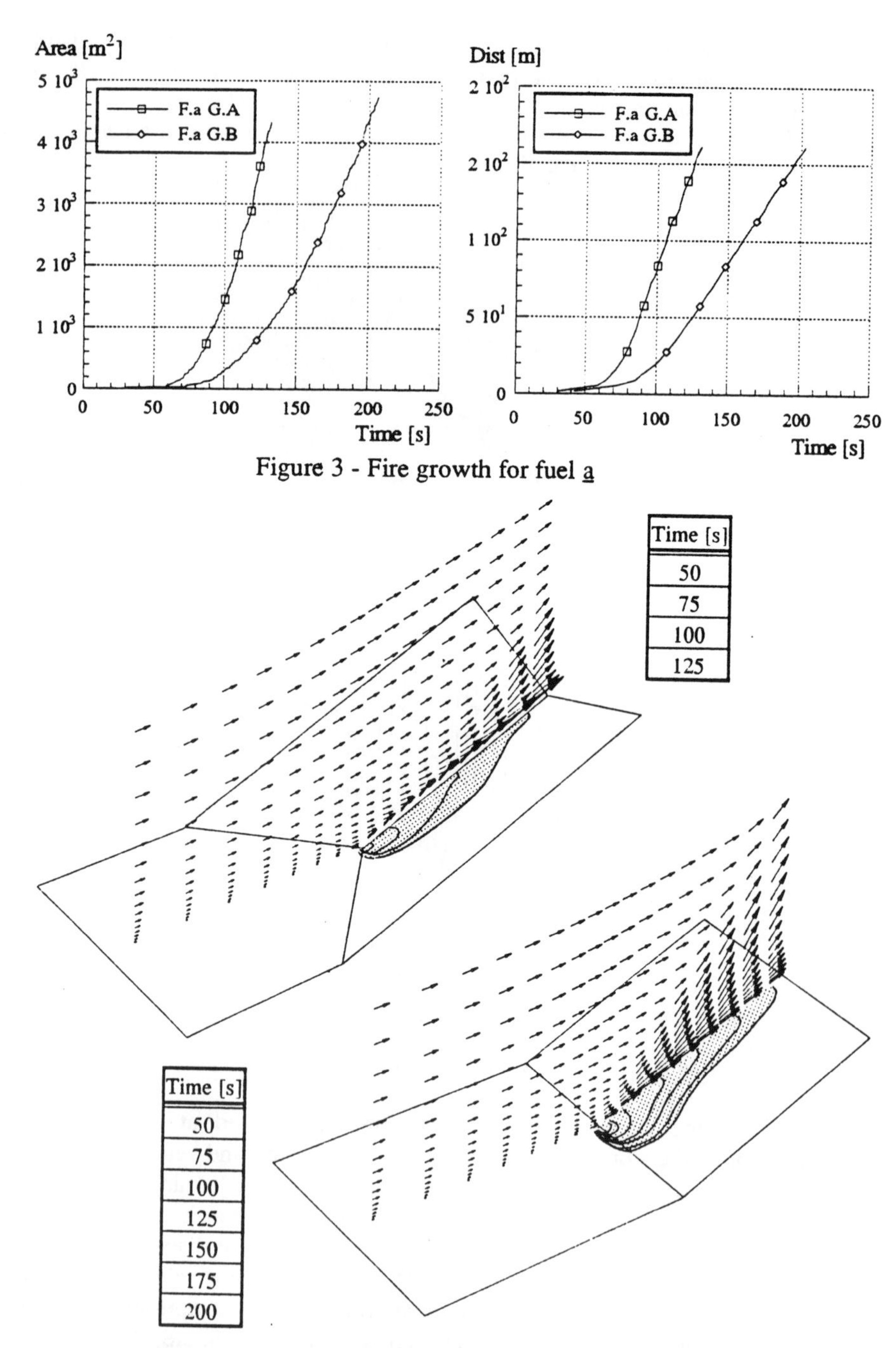

Figure 3 - Fire growth for fuel a

Figure 4 - Fire shape and velocity vectors for fuel a.

Figures 5 and 6 correspond to fuel b , for the conditions already described for Figures 3 and 4.

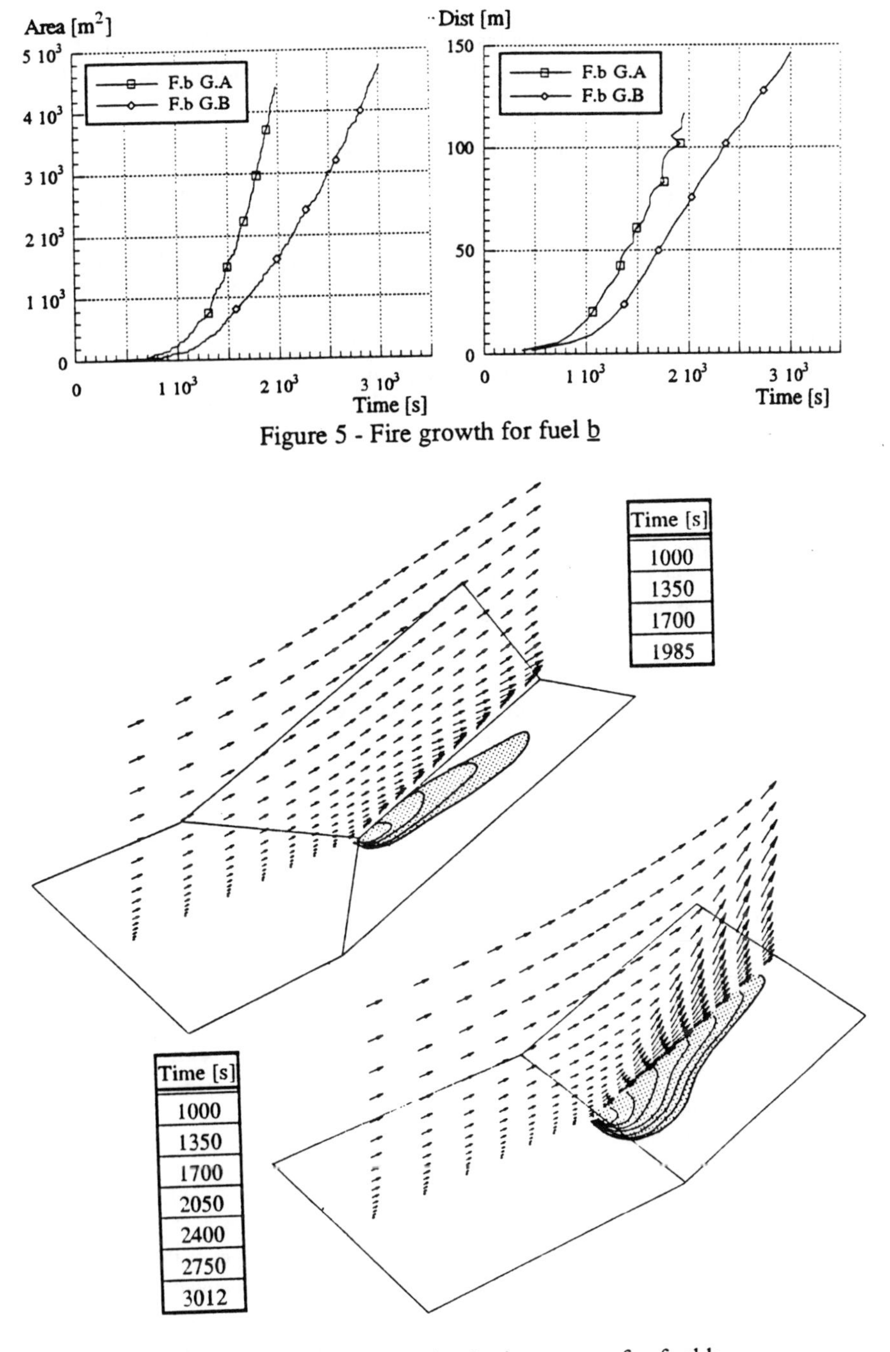

Figure 5 - Fire growth for fuel b

Figure 6 - Fire shape and velocity vectors for fuel b.

An interesting feature arising from this situation is the change in fire shape, for the geometry A. This is a result of the lower sensitivity of fuel b to wind speed, which, in turn, yields an increased relative influence of terrain slope. Close examination of the velocity field shows the formation of a small roll vortex for this case.

DISCUSSION AND CONCLUDING REMARKS

The goal of the proposed study was to demonstrate how the ridge geometry can significantly modify fire behaviour relatively to other terrain topographies, with consequent increased danger and risk to the personnel directly involved in the fire. The results for fire spread and growth in geometries A and B put in evidence the remarkable role that topography has in fire spread; the terrain slope is an important factor, but the overall geometry may be even more important. In the present case, the canyon proved to have an important role in flow acceleration, with evident consequences in fire rate of spread. This effect is more pronounced for fuels with a characteristic rate of spread more sensitive to wind speed. One must be aware that the procedure for the calculation of rate of fire spread here presented, most likely, does not give accurate quantitative results. The authors consider, nevertheless, that it takes into account in a more realistic way the dynamic interaction between fire spread and consequent changing conditions in the case of complex geometries, than a stationary approach.

ACKNOWLEDGEMENTS

The authors want to express their gratitude to Engº Jorge C. S. André for his valuable suggestions and interest in this project. This work was conducted under the auspices of a research program funded by Junta Nacional de Investigação Científica e Tecnológica, Portugal, which is acknowledged.

NOMENCLATURE

- $a_E, a_W,..$ - coefficients of the discretized equations
- a_1, a_2 - variables in the ellipse definition
- c,p,b - variables in the ellipse definition
- C_1, C_2 - turbulence model constants
- C_3, C_μ - turbulence model constants
- $\rho\ t_1, t_\mu$ - damping functions in the turbulence model
- D_1,E - turbulence model constants
- G - buoyancy term
- g_i - gravity acceleration in direction i
- g^{ij} - 'equivalent distance'
- J - Jacobian of the transformation
- P - pressure
- P_1 - production of turbulent kinetic energy
- Q_i - heat of pre-ignition
- R - rate of spread
- Re_t - turbulence Reynolds number $= \rho k^2 / (\mu \tilde{\varepsilon})$
- Re_{kn} - local turbulence Reynolds number $= \rho\sqrt{kn}/\mu$
- S_c - generic source term
- t - time
- T - temperature
- u,v,w - Cartesian velocity components
- u_i - $u_1 = u$; $u_2 = v$; $u_3 = w$
- U_i - i component of the contravariant velocity component
- u_p - velocity component parallel to lines of constant equivalent distance
- x,y,z - axes of Cartesian coordinate system

$\tilde{\varepsilon}$	- isotropic dissipation of turbulent kinetic energy	π	- propagating flux ratio
ε	- total dissipation of turbulent kinetic energy	σ_l	- laminar Prandtl number
ϕ	- generic variable	$\sigma_t, \sigma_k, \sigma_\varepsilon$	- turbulent Prandtl for T, k and ε
ϕ_w	- wind factor	ρ	- density
ϕ_s	- slope factor	ρ_0	- reference density
μ, μ_t	- dynamic viscosity (laminar and turbulent)	ρ_b	- bulk density
μ_{eff}	- effective dynamic viscosity = $\mu + \mu_t$	ω	- effective heating number
		ξ, η, ζ	- axes of curvilinear coordinate system
		ξ_i	- $\xi_1 = \xi$; $\xi_2 = \eta$; $\xi_3 = \zeta$

REFERENCES

[1] - ROTHERMEL, R.C.,"A Mathematical Model For Predicting Fire Spread In Wildland Fuels", USDA F.S. Research Paper, INT-115, 1972

[2] - ROTHERMEL, R.C.,"How To Predict the Spread and Intensity of Forest and Range Fires", Gen. Tech. Rep. INT-143, Ogden, UT, USDA, 1983

[3] - THOMPSON, J.F., THAMES, F.C. and MASTIN, C.W., "Automatic Numerical Generation of Body-Fitted Curvilinear Coordinate System for Field Containing Any Number of Arbritary Two-Dimensional Bodies", J. Computational Physics, Vol. 15, pp. 299-319, 1974

[4] - LOPES, A.M.G., SOUSA, A.C.M. and VIEGAS, D.X., "Numerical Generation of Three-Dimensional Grids Using Control Functions", submitted for publication in J. of Computational Physics, 1993

[5] - ZHANG, C. and SOUSA, A.C.M., "Numerical Simulation of Turbulent Shear Flow in an Isothermal Heat Exchanger Model", Journal of Fluids Engineering, Vol. 112, pp.48-55, 1990.

[6] - PATANKAR, S.V., "Numerical Heat Transfer and Fluid Flow", Hemisphere Publishing Corporation, Washington, D.C., 1980.

[7] - VAN DOORMAAL, J.P. and RAITHBY, G.D., "Enhancements of The Simple Method For Predicting Incompressible Fluid Flows", Numerical Heat Transfer, Vol.7, pp.147-163, 1984.

[8] - BRAATEN, M.E. AND SHYY, W., "Study of Pressure Correction Methods With Multigrid for Viscous Flow Calculation in Non-Orthogonal Curvilinear Coordinates", Numerical Heat Transfer, Vol. 11, pp.417-442, 1987.

[9] - ANDERSON, E.A., "Predicting Wind-Driven Wild Land Fire Size and Shape", USDA F.S. Research Paper, INT-305, 1983

[10]- FEUNEKES, U., "Error Analysis in Fire Simulation Models", M. Sci. Thesis, Dept. of Forest Res., Univ. of New Brunswick, Canada, 1991

[11]- ANDRÉ, C.S.A., LOPES, A.M.G. and VIEGAS, D.X., "A Broad Synthesis of Research on Physical Aspects of Forest Fires", Cad. Cient. Inc. Forestais, N. 3, Coimbra, 1992

Numerical Analysis of Coupling Relation Between Near and Far Field Regions of a Horizontal Buoyant Jet into Crossflow

Hin-Fatt Cheong[1], Jothi Shankar N[2], Wen Lei[3]

ABSTRACT

A submerged buoyant jet discharging normally into a fully developed open channel flow has been numerically analysed by using the three dimensional Navier-Stokes equations, the energy equation and the standard k-ε model. The calculation procedure is based on the SIMPLE algorithm. The flow patterns and temperatures at different sections near the jet outlet are presented. The near field of the jet outlet is characterised by steep temperature gradients before tapering off as the far field is approached. The extent of the near field region may be determined by the gradient of the temperature along the centre line of the jet. The numerical results show that the extent of the near field region is related to the jet flow at the outlet and the upstream flow conditions.

1. INTRODUCTION

There are many field problems which involve the turbulent transport of pollutants such as treated waste water discharges or heated effluents in riverine and coastal regions. These are of great concern to the environmental engineers as the discharges can inflict serious damages to the delicate ecological balance in the coastal waters.

The horizontal side discharge at a certain depth below the water surface is a common method of disposing heated effluent from a power plant into a river or an estuary. It is generally believed that a submerged horizontal discharge into crossflowing current gives better initial mixing. This kind of flow is equivalent to a horizontal thermal jet discharging into a crossflow. It is usual to consider the problem in two parts in which different mixing processes are known to predominate. In the vicinity of the jet outlet, the submerged jet is deflected by the crossflow and there is strong mixing effected by the buoyancy of the heated effluent and the exchange of momentum fluxes with the ambient current stream. The flow is fully three-dimensional and this region is commonly called the initial mixing zone or near field. Mixing with consequent

---

[1]Associate Professor , Dept. of Civil Engineering , National University of Singapore
[2]Associate Professor , Dept. of Civil Engineering , National University of Singapore
[3]Research Scholar , Dept. of Civil Engineering , National University of Singapore

dilution due to ambient turbulence predominates in the region further downstream which is often referred to as the far field.

It is often of practical importance to estimate the downstream distance of the initial mixing zone (near field region) beyond which depth integrated two-dimensional models may be used. Only limited investigations have contributed to the coupling relation between the near and far field region. Rodi et al [1], in their study on side discharges into a river, mentioned the existence of the near field and the far field but did not provide any information on the coupling relationship between them. Kaufman et al [2] gave an empirical relation for the estimation of the extent of the near field area in the study of side discharge of heated water into the sea. Using dimensional analysis and field and experimental data, Lee et al [3] gave empirical formulae to estimate the degree of dilution for side discharges of thermal effluents. Again, through dimensional analysis and laboratory experimentation for the case where waste effluent is introduced from the bottom into a uniform stream, Roberts et al [4] showed that a simple relationship exists between the downstream extent of the initial mixing zone in terms of the mean flow velocity and an ambient stratification factor.

The present study is an attempt to investigate the coupling relationship between the near field and the far field of a situation in which a submerged buoyant jet issues laterally into a uniform open channel flow. In the case of a thermal jet issuing into crossflow, the flow in the near field is characterized by the large pressure gradients and buoyancy due to temperature difference. The three-dimensional Navier-Stokes equations, the energy equation and the k-ε turbulence model are used in the analysis of the coupling relation.

2. MATHEMATICAL MODEL AND NUMERICAL PROCEDURE

Figure 1 shows the schematic diagram of the computational domain with the heated jet discharging normally into crossflowing current at 2/3 water depth from the free surface.

Under the Boussinesq assumption, the time-averaged partial differential equations for the three-dimensional flow can be written as

$$\frac{\partial \Phi}{\partial t} + \frac{\partial U_j \Phi}{\partial x_j} = \frac{\partial}{\partial x_j} \left(\Gamma_\Phi \frac{\partial \Phi}{\partial x_j} \right) + S_\Phi \qquad (1)$$

and

$$\rho = \rho(T) \qquad (2)$$

where Φ ,Γ_Φ and S_Φ are as defined in Table 1. In the Table 1,

$$P = \upsilon_t \left(\frac{\partial U_i}{\partial x_j} + \frac{\partial U_j}{\partial x_i} \right) \frac{\partial U_i}{\partial x_j} \qquad (3)$$

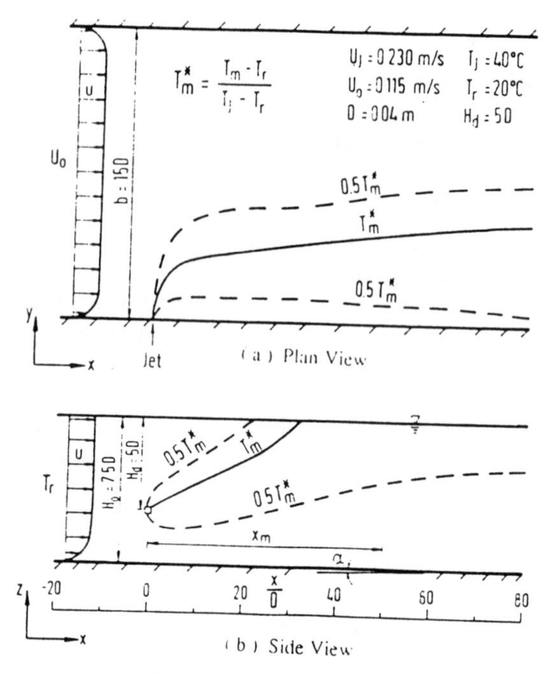

Fig. 1 Flow Configuration

Table 1 Summary of governing equations

Equation	Φ	Γ_Φ	S_Φ
Continuity	1	0	0
X-momentum	U	υ_e	$-\frac{1}{\rho}\frac{\partial p}{\partial x}+\frac{\partial}{\partial x}(\upsilon_e\frac{\partial U}{\partial x})+\frac{\partial}{\partial y}(\upsilon_e\frac{\partial V}{\partial x})+\frac{\partial}{\partial z}(\upsilon_e\frac{\partial W}{\partial x})+gSin\alpha$
Y-momentum	V	υ_e	$-\frac{1}{\rho}\frac{\partial p}{\partial y}+\frac{\partial}{\partial x}(\upsilon_e\frac{\partial U}{\partial y})+\frac{\partial}{\partial y}(\upsilon_e\frac{\partial V}{\partial y})+\frac{\partial}{\partial z}(\upsilon_e\frac{\partial W}{\partial y})$
Z-momentum	W	υ_e	$-\frac{1}{\rho}\frac{\partial p}{\partial z}+\frac{\partial}{\partial x}(\upsilon_e\frac{\partial U}{\partial z})+\frac{\partial}{\partial y}(\upsilon_e\frac{\partial V}{\partial z})+\frac{\partial}{\partial z}(\upsilon_e\frac{\partial W}{\partial z})+(\frac{\rho_r}{\rho}-Cos\alpha)g$
Energy	T	$\frac{\upsilon}{P_r}+\frac{\upsilon_t}{\sigma_t}$	0
Turbulent kinetic energy	k	$\upsilon+\frac{\upsilon_t}{\sigma_k}$	$P + G - \varepsilon$
Kinetic energy dissipation rate	ε	$\upsilon+\frac{\upsilon_t}{\sigma_\varepsilon}$	$c_{1\varepsilon}\frac{\varepsilon}{k}(P + c_{3\varepsilon}G) - c_{2\varepsilon}\frac{\varepsilon^2}{k}$

$$G = -\beta g \frac{\upsilon_e}{\sigma_t} \frac{\partial T}{\partial z} \qquad (4)$$

$$\upsilon_e = \upsilon + \upsilon_t \quad , \quad \upsilon_t = c_\mu \frac{k^2}{\varepsilon} \qquad (5)$$

where, P_r is the Prandtl number. All the symbols and variables used above are consistent with those adopted by the ASCE Task Committee [5]. The empirical constants in the governing equations are given in Table 2.

Table 2. Constants in governing equations

$c_{1\varepsilon}$	$c_{2\varepsilon}$	$c_{3\varepsilon}$	σ_k	σ_ε	c_μ	σ_t	P_r
1.44	1.92	0.2	1.0	1.3	0.09	0.7	0.8

For the elliptic problem defined by equation (1), the boundary conditions must be specified at the upstream , the downstream, the free surface and the jet outlet as well as at the wall. The computational domain here has six boundaries and the jet outlet.

Upstream Condition

The upstream conditions are described by the dependent variables taken for a fully developed channel flow. The fully developed open channel conditions are obtained from the preliminary run where the initial condition of a uniform velocity profile is imposed and the computations are allowed to proceed until fully developed conditions are achieved downstream. The conditions for the preliminary runs are given by Equation (6)

$$U = U_0 \ , V = W = 0 \ , T = T_r \ , \ k_0 = 0.04 U_0^2 \ , \ \varepsilon_0 = \frac{c_\mu^{3/4} k_0^{3/2}}{0.06R} \qquad (6)$$

where R is hydraulic radius of the cross section. In the preliminary run, the jet was absent.

Downstream Condition

The downstream conditions are described by a zero-gradient condition for the dependent variables, i.e. $\partial\Phi/\partial x = 0$.

Free Surface

The rigid lid approximation is used in the current model for the free surface condition.

$$p = 0 \ , \ W = 0 \ , \ \frac{\partial U}{\partial z} = \frac{\partial V}{\partial z} = \frac{\partial T}{\partial z} = \frac{\partial k}{\partial z} = \frac{\partial \varepsilon}{\partial z} \qquad (7)$$

Jet Outlet

The jet condition at the outlet is described by the following equations with D as the size of the outlet,

$$V=U_J\ ,\ U=W=0\ ,\ T=T_J\ ,\ k_J=0.04U_J^{\ 2}\ ,\ \varepsilon_J=\frac{c_\mu^{\ 3/4}k_J^{\ 3/2}}{0.5D} \tag{8}$$

Wall Function

As for the wall boundaries, in the near wall region, the wall function, proposed by Launder and Spalding [6], is commonly used. But it has certain limitations, particularly in the low velocity region. In the present model, the velocity profile, proposed by Dou [7], is used near the wall as follows

$$\frac{U_{res}}{U_*}=2.5\ln(1+\frac{U_*z_w}{5\upsilon})+7.05\left\{\frac{U_*z_w/(5\upsilon)}{1+U_*z_w/(5\upsilon)}\right\}^2+2.5\left\{\frac{U_*z_w/(5\upsilon)}{1+U_*z_w/(5\upsilon)}\right\} \tag{9}$$

where

$$k_w=\frac{U_*^{\ 2}}{\sqrt{c_\mu}} \tag{10}$$

$$\varepsilon_w=\frac{\alpha U_*^{\ 3}}{0.4z_w} \tag{11}$$

$$\alpha=\frac{2}{1+5\upsilon/U_*z_w}+\frac{0.008(23.2-U_*z_w/\upsilon)(U_*z_w/\upsilon)^2}{(0.2U_*z_w/\upsilon+1)^3} \tag{12}$$

where z_w is the distance to the wall, U_* is the friction velocity on the wall. The velocity profile is supposedly universal, covering the laminar, transition and turbulent zones and is a function of the frictional Reynolds number (U_*z_w/υ). U_{res} is the resultant velocity computed from the momentum equations near the wall and U_* is estimated from the Eq. (9).

The governing equations are discretized into finite difference forms with a hybrid central/upstream scheme. The solution code is based on the SIMPLE algorithm of Patankar and Spalding [8], as described by Patankar [9]. The algorithm makes use of a nonuniform staggered grid arrangement. The convergence criterion is imposed by the requirement to satisfy the continuity equation and the allowable maximum deviation from the continuity equation is set at 0.01. Further, the relative difference of the accumulated discharge on the downstream section is also set at less than 5% of the total discharge.

3. NUMERICAL RESULTS AND DISCUSSION

The numerical simulations were conducted for different mean flow velocities U_0 in the open channel ranging from 0.06 m/s (Fr_0=0.05) to 0.46 m/s (Fr_0=0.38) with the ambient temperature T_r set at 20°C and the Froude number defined as $Fr_0=U_0/\sqrt{gR}$. The densimetric Froude number at the jet outlet, which is defined as $Fr_J=U_J/\sqrt{g'D}$, where $g'=g(\rho_r-\rho)/\rho_r$ varies from 1.2 to 10.6 (corres-ponding to jet velocities from 0.1 m/s to 0.35 m/s with the jet temperature set at 30°C to 60°C and the characteristic length of the outlet dimension D set at 0.04 m). Figures 2, 3 and 4 show velocity fields at the vertical longitudinal section corresponding to $y/D = 0.075$, the horizontal plan section corres-ponding to $z/D = 3$ just at the jet outlet and the vertical transverse section corresponding to x/D = 0, 3.95 & 21.75 respectively. In contrast to the case of a two-dimensional jet into a crossflow, there is no significant longitudinal recirculation zone. Figures 4(a,b,c) suggest a helical flow structure accom-panying the buoyant plume.

The centre line of the buoyant plume may be represented by the locus describing the maximum temperature (T_m) in the plume as shown schematically in Fig. 1. Isothermal tubes define the spread of the plume. Figures 5, 6 and 7 show the projection of the isothermal tubes onto the horizontal plane, the vertical longitudinal plane and the vertical transverse plane respectively. These figures demonstrate the strong three-dimensional characteristics of the buoyant plume near the outlet showing steep temperature gradients.

The numerical solutions permit the determination of the line defining T_m and Fig. 8 shows the variation of the temperature along the lines as a function of the dimensionless longitudinal distance (x/D) from the jet outlet for a certain simulation run (Fr_0=0.1, Fr_J=4.56). It is evident that very steep temperature gradients prevail in the vicinity of the jet outlet. The rate of change in the temperature diminishes rapidly with distance from the outlet.

A criterion for defining the extent of the near field (x_m/D) may be defined to be the longitudinal distance from the jet outlet to the point where the centre line temperature gradient T_m' falls below 0.001 (Fig. 8), for which the mixing can be assumed to be complete.

From a series of 14 simulation runs the values of x_m/D for which T_m'=0.001 are obtained and depicted in Figure 9. In these simulation runs, the jet outlet temperature is set at 40 °C and the jet outlet velocity is chosen to be 0.23 m/s (Fr_J=4.56). From the Figure 9, strong mixing by the entrainment occurred near the outlet is evident and the extent of the mixing zones increases with the velocity in the open. The same observation can be made from Fig. 10. Figure 10 shows the plot of x_m/D against the jet densimetric Froude number for two open channel flow conditions (Fr_0 = 0.1 & Fr_0 = 0.2). It is noted that the curves in both figures 'flatten' off with increasing Fr_0 and Fr_J. The extent of the mixing zone is expected to increase with the increase in Fr_0 or Fr_J. A possible explanation for the flattening of the curves as depicted in Figs. 9 and 10 could be due to the x_m-projection of the actual T_m locus is plotted and not the actual distance measured along the locus. The trajectory of the buoyant plume is dictated to a large extent by the relative strength of the jet and the open channel flow.

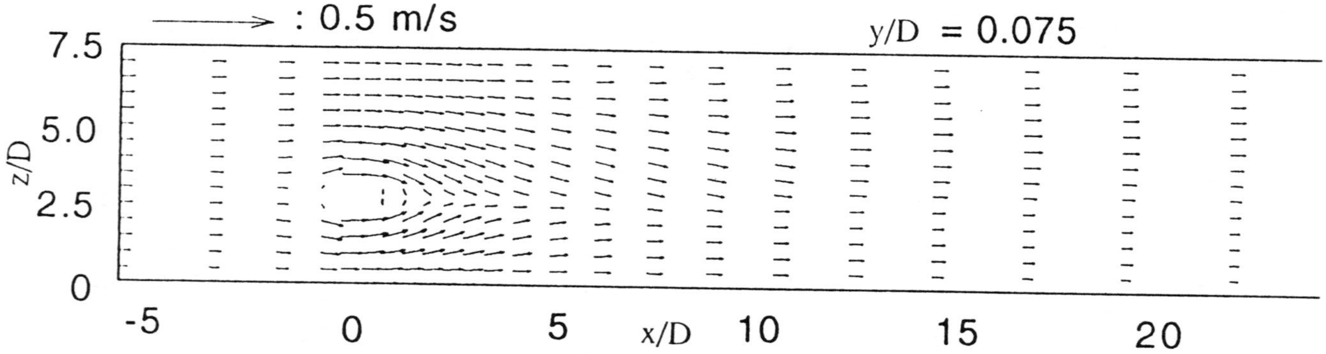

: 0.5 m/s
y/D = 0.075
z/D
7.5
5.0
2.5
0
-5
0
5
x/D
10
15
20

Fig. 2 Flow Pattern Near Side Wall

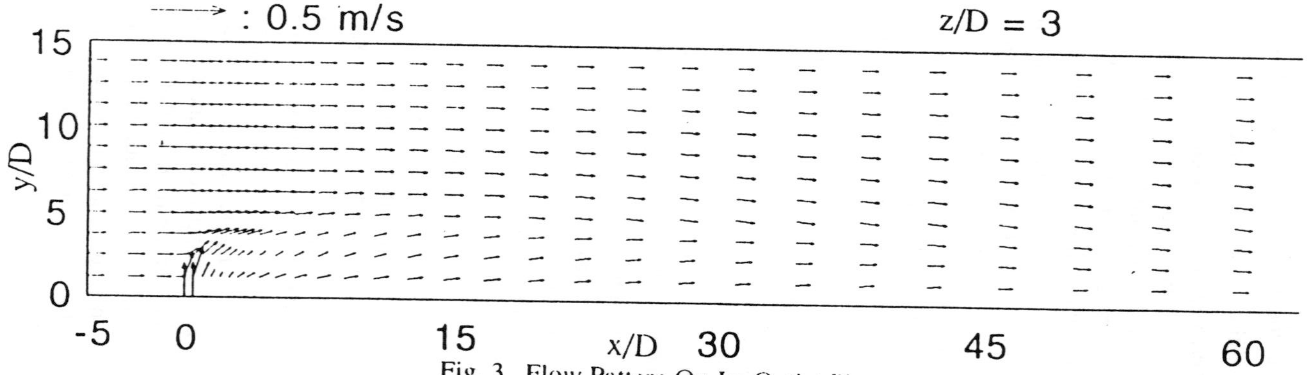

: 0.5 m/s
z/D = 3
y/D
15
10
5
0
-5
0
15
x/D
30
45
60

Fig. 3 Flow Pattern On Jet Outlet Plan

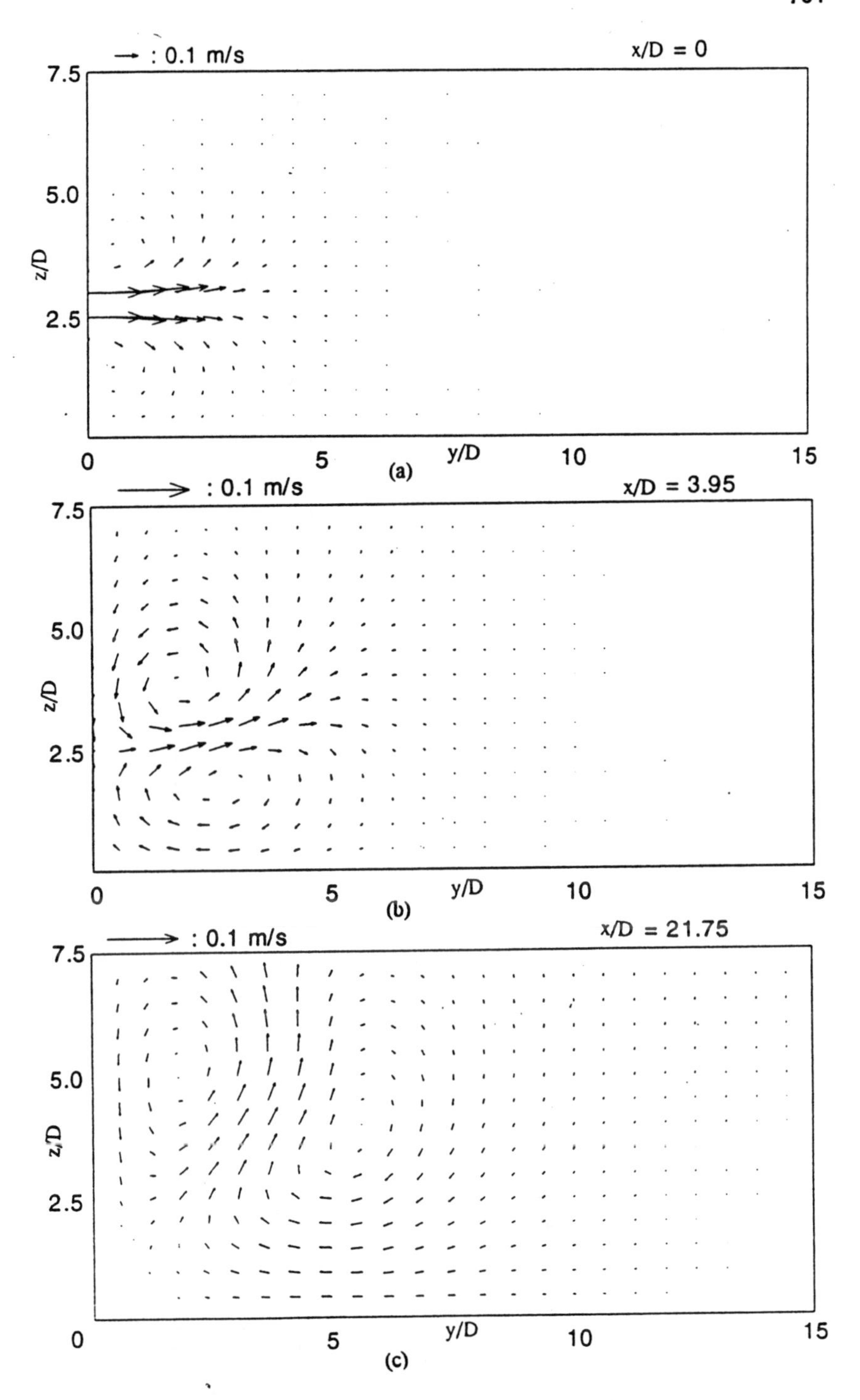

Fig. 4 Flow Pattern On Transverse Section

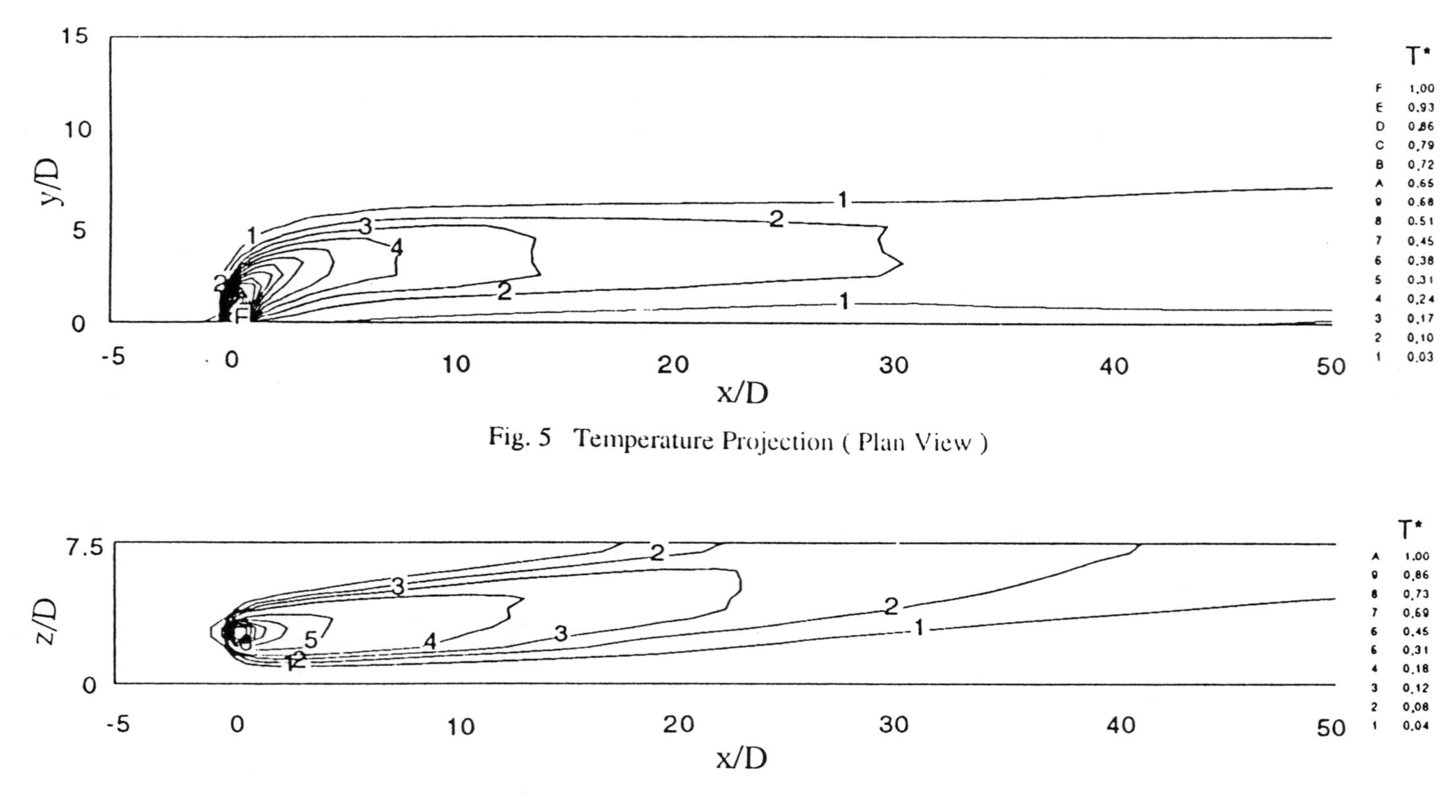

Fig. 5 Temperature Projection (Plan View)

Fig. 6 Temperature Projection (Side View)

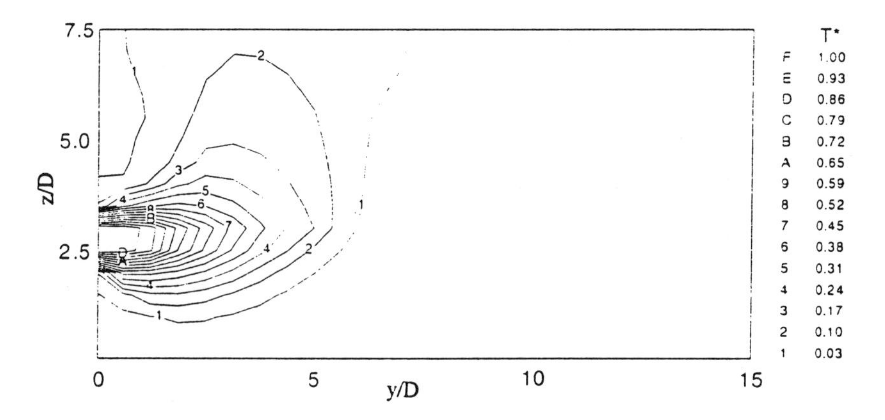

Fig. 7 Temperature Projection (Transverse Section)

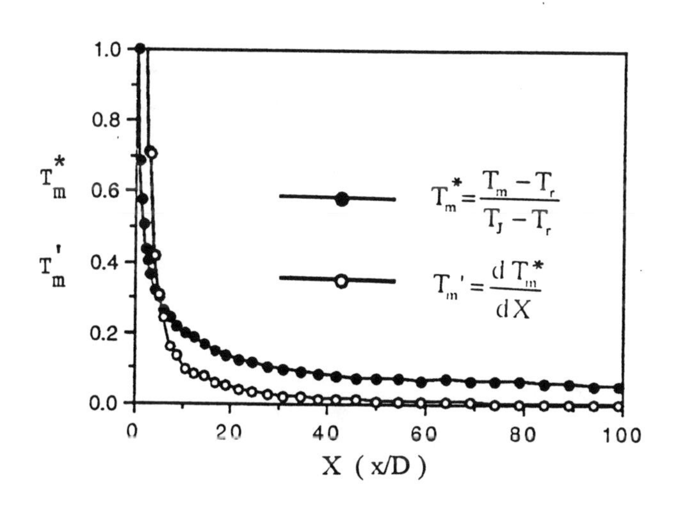

Fig. 8 Centerline Temperature Deficiency Decay

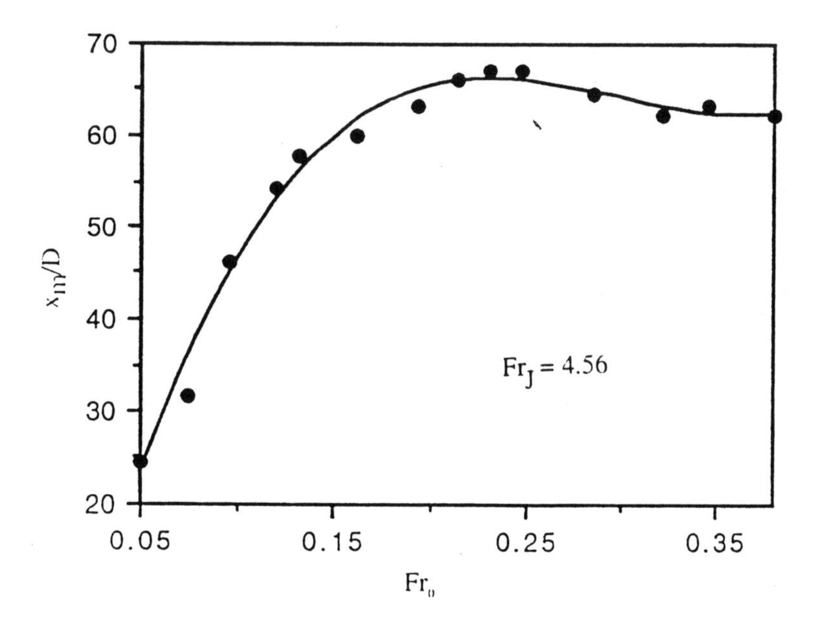

Fig. 9 Variation Of x_m With Fr_o

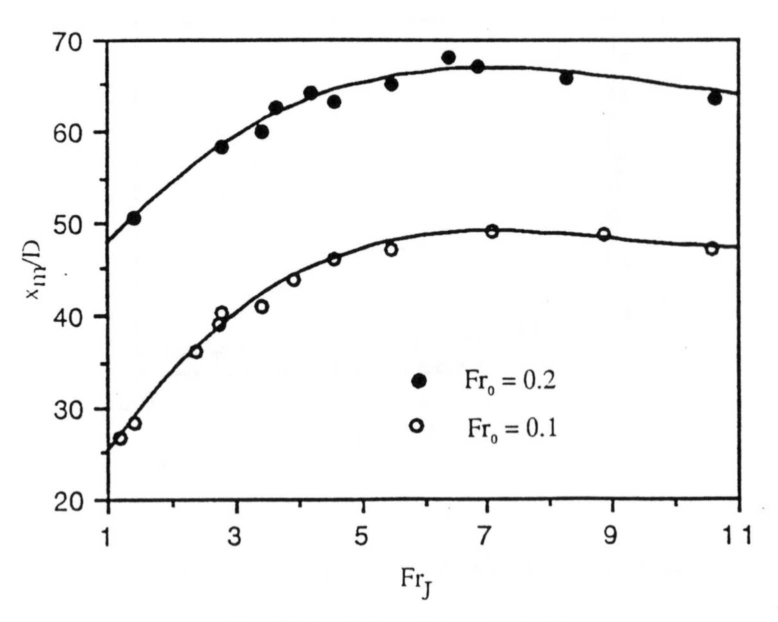

Fig. 10 Variation Of x_m With Fr_J

The initiation of the possibility of surface undulation or waves resulting from high velocity of the channel flow and the jet cannot be overemphasised as these will mean that the analysis will have to be reformulated.

4. CONCLUSION

(1) A study of the numerical simulation of the three-dimensional behaviour of a thermal jet discharging horizontally into a crossflow by using the Navier-Stokes equation, the energy equation and the k-ε turbulence closure model was initiated. A wall function, based on the universal velocity profile proposed by Dou [7], was used in the present investigation.

(2) Velocity patterns and temperature contours were obtained to show the three-dimensional characteristics of the thermal plume in the crossflow. The numerical prediction revealed helical flows accompanying the three-dimensional jet.

(3) The numerical results show that the longitudinal extent of the initial mixing zone or near field is related to the jet densimetric Froude number (Fr_j) and the Froude number of the channel flow (Fr_o).

5. REFERENCES

1. RODI, W., and SRIVATSA, S. - Prediction of flow and pollution spreading rivers, in : Transport models for inland and coastal waters,ed. FISCHER, H. B., Academic Press, New York, pp. 63 - 111, 1981
2. KAUFMAN, and J.T., ADAMS, E. E. - Coupled near and far field thermal plume analysis using finite element techniques, Energy laboratory report No. MIT-EL 81-036, M.I.T., pp.35-41, 1981
3. LEE, J. H. W. and NEVILLE-JONES, P. - Initial dilution of horizontal jet in crossflow, J. Hydr. Eng., ASCE, 113 (5), pp. 615-629, 1987
4. ROBERTS, P. J. W., SNYDER, W.H., and BAUMGARTNER, D.J. - Ocean outfall II : Spatial evolution of submerged wastefield, J. Hydr. Eng., ASCE, 115 (5), pp. 26-48, 1989
5. ASCE TASK COMMITTEE - Turbulent modelling of surface water flow and transport, J. Hydr. Eng., ASCE, 114 (9), pp. 970-1073, 1988
6. LAUNDER, B. E. and SPALDING, D. B. - The numerical computation of turbulent flow, Comp. Math. in Appl. Mech. and Eng., 3, pp.269, 1974
7. DOU, G.R. - Basic laws in mechanics of tubulent flows , Int. Res. Train. Centre Eros. Sedim. (IRTCES), Beijing, pp. 24, 1990
8. PATANKAR, S. V. and SPALDING, D. B. - A calculation procedure for heat, mass and momentum transfer in three-dimensional parabolic flow, Int. J. Heat Mass Transfer, 15, pp. 1787, 1972
9. PATANKAR, S.V. - Numerical heat transfer and fluid flow, Hemisphere Publishing Corporation, Washington, 1980

7. LIST OF NOTATIONS

D	dimension of outlet
F_{ro}	Froude no. of the channel flow
g	gravitational constant
G	buoyancy production of k
k	turbulent kinetic energy
ϵ	kinetic energy dissipation rate
p	pressure scalar
P	stress production of k
P_r	Prandtl number
R	hydraulic radius of the channel
t	time
T	temperature
T_r	upstream ambient temperature
x	downstream space coordinate (x = 0 at the jet outlet)
y	transverse space coordinate (y = 0 at the jet outlet)
z	vertical space coordinate
U, V, W	temporal means of the velocity in the x, y, z direction
U_o	uniform flow velocity for preliminary run
U_J	velocity at jet outlet
β	volumetric expansion coefficient
ρ	mass density
σ_t	turbulent Prandtl/Schmidt number
$\sigma_{k,\epsilon}$	constants in k-ϵ model
υ_t	turbulent viscosity

Computational Modelling of Flow in IC Engine Intake Systems

N.I.A. HAIDAR+, and A.R. PLEWS *

+Lecturer, *Research Student,
Department of Engineering, University of Leicester,
University Road, Leicester LE1 7RH, UK.

SUMMARY

A computational study of three-dimensional, steady air flow in a hot internal combustion engine intake duct has been conducted. A fully-elliptic, finite volume computational model using the SIMPLE algorithm has been adopted. For the main part of the flow the k-ε turbulence model is adopted, while wall functions are employed across the low Reynolds number sublayer. The computational results for air mass flow rates at inlet temperatures ranging between 5 to 65°C in 10°C increments are presented. The predictions are in good agreement with available experimental data, thus demonstrating the accuracy and generality of the present method. Analysis of the results reveals that the mass flow rate of air with an inlet temperature of 5°C is about 9% higher than that for 65°C. The computational findings indicate the desirability of using this technique for improving the volumetric efficiency of IC engines.

1. INTRODUCTION

One of the many aims of the Internal Combustion (IC) engine designer is to increase the amount of air drawn into the cylinder during the intake stroke, so that more fuel can be burnt efficiently per cycle of the engine. In order to improve this volumetric efficiency, the induction system design must be optimised to maximise the mass flow rate of air through it. Traditionally new designs were tested experimentally using bench flow tests which resulted in a cut and try process. Today new designs can be drawn and flow simulations performed using Computational Fluid Dynamics (CFD), which results in a quicker and cheaper design cycle. Aspects of engine design are summarised in Barnes-Moss [1].

In the past, CFD investigations concerning IC engine intake systems have predominantly concentrated on the simulation of steady cold air flow through the intake system, such as those reported in Balasubramanian et al [2] and Seppen and Visser [3], while Kuo and Chang [4] considered a uniform air temperature of 47°C. On the other hand, the numerical investigation of Benson [5] has considered the heat transfer from the inlet valve only. Therefore, the true effect of heat transfer from the hot walls of the intake duct due to conduction of combustion temperatures has not been accounted for in the above publications.

This paper deals primarily with the modelling of steady air flow through a "hot" IC engine intake system, with a fully open valve. The temperature of air at entry ranged between 5 to 65°C. From the knowledge of the computed dependant variables and geometry, the mass flow rates are predicted. Computational results for an inlet air temperature of 20°C are compared with the findings of Sugiura et al [6]. A series of similar experimental investigations are reported in Tindal et al [7], Yianneskis et al [8] and Annand [9].

2. EXPERIMENTAL STUDY

The present numerical study is based on the "combined" experimental investigations of Sugiura et al [6], Brown and Ladommatos [10] and Heywood [11]. Details of the valve, port and cylinder geometry are given in Fig.1.

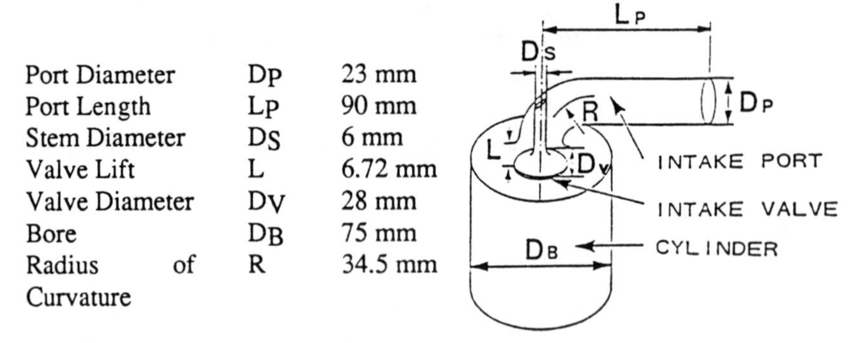

Port Diameter	D_P	23 mm
Port Length	L_P	90 mm
Stem Diameter	D_S	6 mm
Valve Lift	L	6.72 mm
Valve Diameter	D_V	28 mm
Bore	D_B	75 mm
Radius of Curvature	R	34.5 mm

Fig.1 Valve, port and cylinder geometry (from Sugiura et al [6])

The wall and valve temperature distributions, which are equivalent to those experienced under running conditions, are presented in Fig.2.

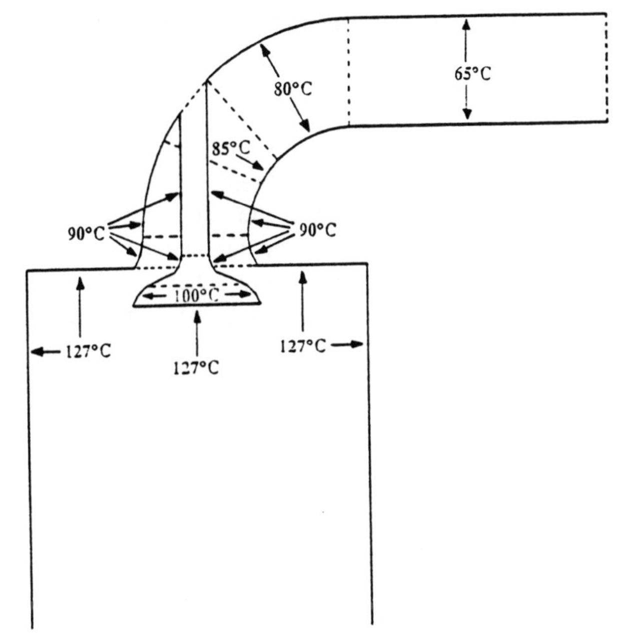

Fig.2 Intake duct and valve wall temperature distributions

The pressure difference between inlet and outlet of the intake system when the valve was fully open was 6.67 KPa. For an inlet air temperature of 25°C, this drop in pressure is equivalent to an average flow speed of about 75 m/s, at inlet.

3. NUMERICAL AND PHYSICAL MODEL

All notations appearing in this section are defined separately in the Nomenclature Section, available at the end of the text.

3.1 Governing Equations

The basic differential mass conservation and momentum conservation equations governing the physical situation under consideration can be written as follows, in Cartesian tensor notations, and using the repeated-suffix summation convention:

- Mass conservation:

$$\frac{\partial \rho}{\partial t}+\frac{\partial}{\partial x_j}(\rho\, u_j)=0 \tag{1}$$

- Momentum conservation:

$$\frac{\partial \rho u_i}{\partial t}+\frac{\partial \rho u_i u_j}{\partial x_j}=-\frac{\partial p}{\partial x_i}+\frac{\partial \sigma_{ij}}{\partial x_j}+F_i \tag{2}$$

The equations are solved by using the fully-elliptic procedure.

3.2 The Turbulence Model

In the present study, a two-equation turbulence model has been used in the high Reynolds number region. Therefore, two additional differential equations are solved for the transport of the kinetic energy of turbulence, k, and its rate of dissipation, ε. Details of the turbulence model are explained in Launder and Spalding [12,13] and many other publications since. The governing differential equations for k and ε, after the neglect of terms of small order of magnitude are given as:

$$\frac{\partial k}{\partial t}+u_j\frac{\partial k}{\partial x_j}=\frac{\partial}{\partial x_j}\left(\frac{\nu_t}{\sigma_k}\frac{\partial k}{\partial x_j}\right)+P_k-\varepsilon \tag{3}$$

and

$$\frac{\partial \varepsilon}{\partial t}+u_j\frac{\partial \varepsilon}{\partial x_j}=\frac{\partial}{\partial x_j}\left(\frac{\nu_t}{\sigma_\varepsilon}\frac{\partial \varepsilon}{\partial x_j}\right)+\frac{\varepsilon}{k}(C_{\varepsilon 1}P_k-C_{\varepsilon 2}\varepsilon) \tag{4}$$

The turbulent viscosity, ν_t, is obtained by assuming that it is proportional to the product of the turbulent velocity scale and length scale. The turbulent viscosity (Prandtl-Kolmogorov) relationship is defined as:

$$\nu_t=C_\mu\frac{k^2}{\varepsilon} \tag{5}$$

The empirical parameters in the above turbulence model are assigned the following values [13,14]:

$$C_\mu=0.09;\ C_{\varepsilon 1}=1.44;\ C_{\varepsilon 2}=1.92;\ \sigma_k=1.0;\ \sigma_\varepsilon=1.3$$

3.3 The Wall-Function Method

Across the low Reynolds number sublayer wall-functions have been used. The code used employs the Log-law of the wall to compute the wall shear stress, τ_w, [13] for turbulent flows, when $y^+ > 11.63$. The equation is defined as:

$$\tau_w=\frac{\rho C_\mu^{1/4}k^{1/2}\kappa u}{\ln\left(Ey^+\right)} \tag{6}$$

where y^+ is evaluated from $y^+ = y\sqrt{\tau_w/\rho}/\nu$

When y^+ falls below 11.63, the wall shear stress is calculated by using the relationship for laminar flow:

$$\tau_w = \left(\mu \frac{\partial u}{\partial y}\right)_{y=0} \tag{7}$$

While the near-wall value of the turbulence kinetic energy, k_w, is computed via the solution of the full transport equation (3), the near-wall value for ε_w, is computed assuming equilibrium in the turbulent boundary layer [13] as follows:

$$\varepsilon_w = \frac{C_\mu^{3/4} k_w^{3/2}}{\kappa \partial y} \tag{8}$$

3.4 Solution Procedure

The basic numerical scheme has been described in Patankar [15] and several other earlier publications and thus does not require a detailed presentation here. It provides a finite volume solution of the three-dimensional momentum and continuity equations in Cartesian co-ordinates. The present method is based on the non-staggered control volume technique and the use of the SIMPLE algorithm for correcting the pressure field. This algorithm resolves the coupling between pressure and velocity in order to obtain an equation for pressure. The one-dimensional discretised momentum equation is given as:

$$a_p u_p = \sum a_{nb} u_{nb} + (p_e - p_w) Ap + S^* \tag{9}$$

where a_p and a_{nb} are the coefficients containing both convection and diffusion contributions. The a-coefficients link the node p to its two neighbour points, while S* accounts for linkage to other nodes as well as for sources and sinks. The fully-elliptic scheme, in which all dependant variables are stored three-dimensionally, and the power law interpolation scheme have both been adopted in the present study.

4. DISCUSSION OF RESULTS

The predictions for mass flow rates of air through an IC engine intake system, at inlet temperatures of 5, 15, 25, 35, 45, 55 and 65°C are presented in this section. The velocity and temperature fields for the 25°C test case are also included. Computational results were obtained using the k-ε turbulence model in conjunction with wall functions. Along most of the intake duct and engine cylinder, the grid distributions were fairly uniform. A large concentration of nodes was used in the vicinity of the valve seat. A typical three-dimensional grid system is shown in Fig.3. The grid systems employed represent the exact physical dimensions of the intake system shown in Fig.1.

All computational mass flow rate results included in this paper were obtained using a coarse mesh of 13 x 10 x 29 nodes for the intake and 22 x 22 x 16 nodes for the cylinder, and a fine mesh of 18 x 13 x 43 nodes for the intake and 31 x 31 x 21 nodes for the cylinder.

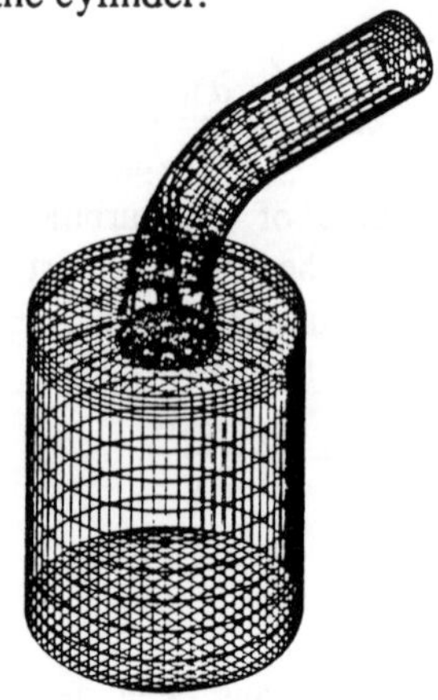

Fig.3 Three-dimensional grid distribution

A two-dimensional grid system, taken along the plane of symmetry, and the corresponding velocity field for the 25°C test case are depicted in Fig.4. It is apparent from Fig.4b that the flow accelerates as it enters the cylinder and forms two small and two large, almost symmetrical, vortices.

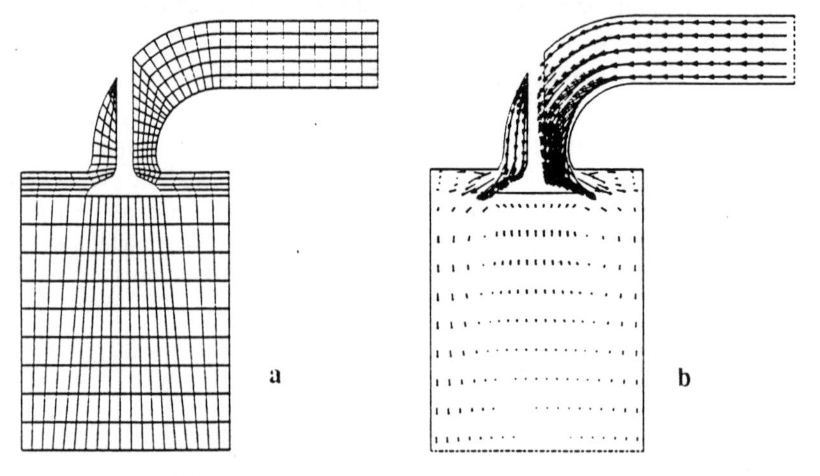

Fig.4 Grid distribution and velocity field along the plane of symmetry

In Fig.5, a "cross-sectional" grid system, taken at the exit plane of the intake duct, and the corresponding secondary velocity field are presented. In Fig.5b, it is evident that the secondary motion is symmetrical. Considering the top half portion, the secondary flow motion is characterised by two opposing vortices: the first is a "weak" clockwise vortex with its centre situated near the top-right wall, and the second is a "strong" counter clockwise vortex with its centre located on the left hand side of the valve, near the symmetry plane.

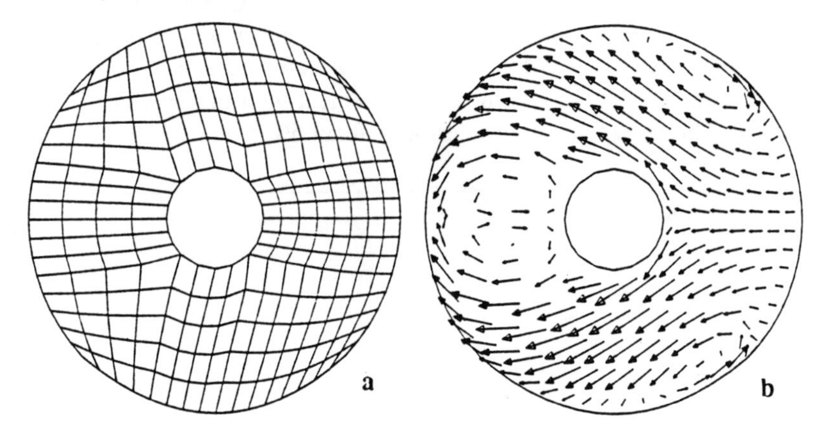

Fig.5 Grid distribution and secondary velocity field across the intake duct at the exit plane

The temperature fields for the 25°C test case, corresponding to the grid distributions shown in Figs.4a and 5a, are given in Fig.6. It is apparent that the bulk of the flow is hardly affected by the intake duct and valve wall temperature distributions. In other words, the heat transfer from the hot walls is only dominant in the near wall regions. In Fig.6a, only the 25°C (inner) temperature contours are identified. The outer wall temperatures are the same as those depicted in Fig.2.

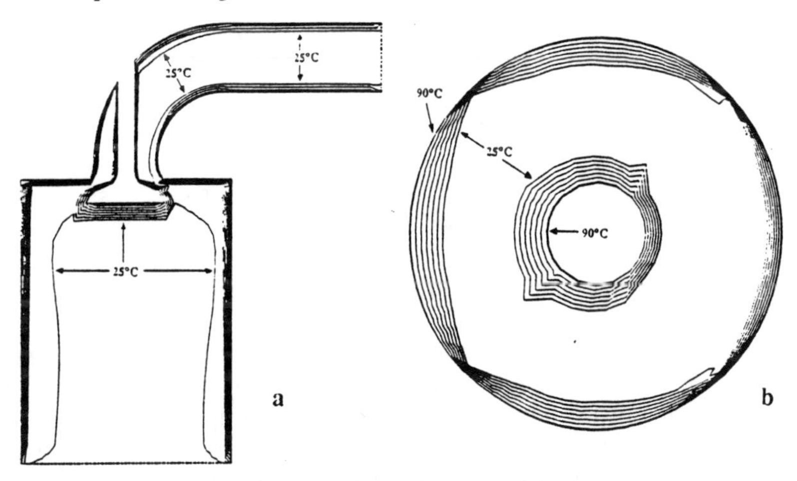

Fig.6 Temperature fields for the 25°C test case

The variation of the predicted air mass flow rates with the inlet air temperatures is shown in Fig.7. As expected, the continuous cooling of the air flow, entering the intake duct, brings about a linear increase in the mass flow rate. For example, by cooling the air from 65 to 5°C, the mass flow rate through the intake system can be increased by about 9%. The successive refinement of the grid was found to bring the computed mass flow rate for the 20°C test case (36.36 g/s) in close agreement with that measured (38.68 g/s) in [6]. This is equivalent to an error of about 6%.

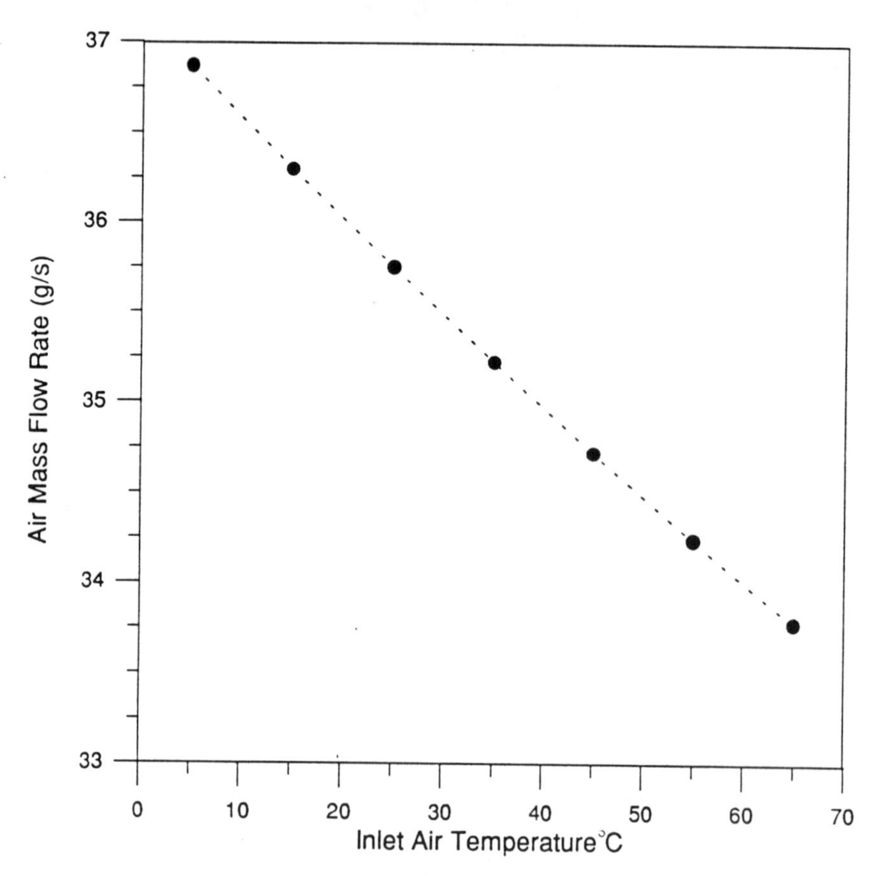

Fig.7 Variation of mass flow rate with inlet air temperature

Rozas et al [16] state that low air temperatures can be tolerated in engines without affecting the fuel evaporation in the system. Additionally, the selective heating of the intake duct is preferable to evaporate fuel films that accumulate on the intake and valve walls. The results of a similar computational investigation employing a time-dependent, three-dimensional computer program using either the Algebraic Stress Model, reported in

Haidar, Iacovides and Launder [17], or the Reynolds Stress Model is due to be published in Haidar and Plews [18].

5. CONCLUDING REMARKS

A qualitative type of approach to the problem of predicting the steady air mass flow rates through a hot intake system with a fully open valve has been presented. A three-dimensional, fully-elliptic, finite volume procedure using a k-ε turbulence model in conjunction with wall functions has been employed. The geometry of the intake system and wall temperature distributions considered are similar to those present in real life applications.

Predictions confirm that the uniform cooling of the air flow in an IC engine intake system lead to improved air mass flow rates, without upsetting any parameters affecting the fuel evaporation from the walls. Cooling the air flow from 65 to 5°C resulted in approximately a 9% increase in the mass flow rate. Comparison between computational results and available experimental data yielded good agreement.

This type of investigation is of practical importance as it can provide a better understanding of the air flow behaviour occurring in hot IC engine intake systems and can assist in the design of more efficient engines.

NOMENCLATURE

a	coefficient in finite-volume equation
A	area of control volume
$C_{\varepsilon 1}$, $C_{\varepsilon 2}$, C_μ	constants in turbulence model
E	roughness parameter
F_i	external body force (x_i-component)
k	kinetic energy of turbulence
p	static pressure
P_k	generation function of kinetic energy of turbulence
S^*	source term
t	time
u	instantaneous fluid velocity
x,y	co-ordinate directions
ε	dissipation of turbulence kinetic energy
κ	von Karman constant
μ	dynamic viscosity
ν	kinematic viscosity
ρ	fluid density
σ	Prandtl-Schmidt number
τ	stresses in fluid

Subscripts and Superscripts

i,j	directional indices in tensorial notation
k	property used in equation for kinetic energy of turbulence
nb	refers to neighbour points (e.g. E,W in 1D)
(P, N, S, E, W, U, D)	nodes forming 3D computational star around central node P
t	turbulent
w	wall
ε	property used is dissipation equation
+	denotes quantity non-dimensionalised

REFERENCES

1. BARNES-MOSS, H.W. - A Designer's Viewpoint, Proceedings of the IMechE. Conference Publication 19 , Vol. C343, pp. 133-147, 1973.

2. BALASUBRAMANIAN, B., SVOBODA, M., and BAUER, W. - Numerical Simulation in Passenger Car Engine Design and Development, Proceedings of the IMechE. Computers in Engine Technology, Vol. C430, No. 068, pp. 235-241, 1991.

3. SEPPEN, J.J., and VISSER, A.H. - Intake and Exhaust Processes in Combustion Engines, Development of Siflex. 1st International Phoenics Users Conference, Ch. 37, pp. 73-83, 1985.

4. KUO, T.W., and CHANG, S., - 3-Dimensional Computations of Flow and Fuel-Injection in an Engine Intake Port, Journal of Engineering for Gas Turbines and Power-Transactions of the ASME, Vol. 113, pp. 427-432, 1991.

5. BENSON, R.S., - Steady and Non-Steady Flow Through an I.C. Engine Inlet Valve with Heat Transfer, International Journal of Mechanical Science, Vol. 19, pp. 673-692. Pergamon Press, 1977.

6. SUGIURA, S., SAITO, Y., YAMADA, T., and SATOFUKA, N., - Evaluation of New Three-Dimensional Codes for Flow in an Induction System, Proceedings of the IMechE. Computers in Engine Technology, Vol. C430, No. 053, pp. 39-45, 1991.

7. TINDAL, M.J., CHEUNG, R.S., and YIANNESKIS, M. - Velocity Characteristics of Steady Flows Through Engine Inlet Ports and Cylinders, International Congress and Exposition, No. SAE 880383, 1988.

8. YIANNESKIS, M., CHEUNG, R.S., and TINDAL, M.J. - A Method of Investigating Flows in Inlet Ports of Complex Shape, Proceedings of the IMechE Conference, Combustion in Engines, Vol. C62, No. 88, pp. 51-58, 1988.

9. ANNAND, W.J.D. - Experiments on a Model Simulating Heat Transfer Between the Inlet Valve of a Reciprocating Engine and the Entering Stream, Proceedings of the IMechE, Vol. 182, Pt. 3H, pp. 37-41, 1967.

10. BROWN, C.N., and LADOMMATOS, N. - A Numerical Study of Fuel Evaporation and Transportation in the Intake Manifold of a Port-Injected Spark-Ignition Engine, Proceedings of the IMechE, Vol. 205, pp. 161-175, 1991.

11. HEYWOOD, J.B. - Internal Combustion Engine Fundamentals, McGraw-Hill International Editions, Automotive Technology Series, 1989.

12. LAUNDER, B.E., and SPALDING, D.B. - Lectures in Mathematical Models of Turbulence, Academic Press, 1972.

13. LAUNDER, B.E., and SPALDING, D.B. - The Numerical Computation of Turbulent Flows, Computer Methods in Applied Mechanics and Engineering, Vol. 3, pp. 269-289, 1974.

14. KOLMAGOROV, A.N. - Equations of Turbulent Motion of an Incompressible Fluid, Izv. Akad. Nauk SSSR Ser. Phys, Vol. VI, No. 1-2, 1942.

15. PATANKAR, S.V. - Numerical Heat Transfer and Fluid Flow, Hemisphere Publishing Corp, 1980.

16. ROZAS, T., THOMAS, U., and MEYERDIERKS, D. - Optimisation of Intake Manifold Heating for Internal Combustion Engines, International Symp on Automotive Technology and Automation, Vol. 1, Ch. 30, 1985.

17. HAIDAR, N.I.A., IACOVIDES, H., and LAUNDER, B.E. - Computational Modelling of Turbulent Flow in S-Bends, CFD Techniques for Propulsion Applications, AGARD-CP-510, No. 34, pp. 1-16, 1991.

18. HAIDAR, N.I.A., and PLEWS, A.R. - The Numerical Computation of Pulsating Flow in IC Engine Intake Systems, In preparation, 1994.

COMPUTER SIMULATION OF FIRE-INDUCED AIR FLOW IN A FULL SCALE VENTILATED TUNNEL

H. Xue[1], T. C. Chew[2], H. F. Cheong[3]
National University of Singapore
10 Kent Ridge Crescent
Singapore 0511

ABSTRACT

A modified three-dimensional version of $k-\varepsilon$ turbulence model is applied to simulate the fire-induced air flow and smoke movement in a full-scale ventilated tunnel. The model adopts general curvilinear coordinate system to keep the real geometry of the tunnel. Both effects of buoyancy and mean streamline curvature on turbulence are accounted for in the model. The smoke concentration is expressed in terms of smoke obscuration and calculated by its conservative equation. The predictions are shown to be in reasonable agreement with the available experiment data. The results have confirmed the temperature stratification of the heated air flow in a full scale tunnel which can be evaluated by a mixed convection parameter $Gr^*/Re_H^{5/2}$. The leading edges of the heated air flow and the heated reverse air flow are predicted.

1. INTRODUCTION

With the significant progress made in tunneling and tunnel ventilation system, more and more long road and railway tunnels have been built. The problem of the fire-induced air flow in tunnels therefore has attracted considerable attention in the last decade. A good physical insight into fire dynamics is desirable for systematic planning of fire safety in design and maintenance. A quantitative knowledge about the detailed behavior of the fire-induced air flow in a ventilated tunnel is required for fire detection, control, extinguishment, personal evacuation and hazard evaluation.

It is known that the complete similarity between full and scale models of fire cannot be preserved. Some compromises are always required for relaxing complete scaling rigor. Recently, the so-called field model

---

[1]Research Scientist, Department of Civil Engineering
[2]Senior Lecturer, Department of Mechanical & Production Engineering
[3]Associate Professor, Department of Civil Engineering

approach, which is to formulate a mathematical model of the tunnel fire system and obtain solutions using numerical procedures over the field of interest, has achieved encouraging results for a number of fire problems. In our previous studies[1],[2], a three-dimensional turbulence $k-\varepsilon$ model to simulate the heated air flow in a ventilated tunnel has been developed and validated by laboratory scale experiments. The main interest in the present study is to extend the model to simulate a full-scale tunnel fire situation and compare the numerical results with some available experiment data. The motivation is twofold. Being able to correlate laboratory scale experiments with full-scale experiments is desirable from a cost standpoint. More importantly, however, from a life safety and operational standpoint, the ability to make accurate predictions of the spread of fire-induced air flow opens up many possibilities for combating these problems, as well as taking effective preventive measures.

2. VALIDATION EXPERIMENT

The date for validation are based on the experiment reported by Iida et al[3]. The experiment was carried out in 1987 in Seikan Tunnel, the longest undersea railway tunnel connecting the main island and Hokkaido of Japan. The total length of Seikan Tunnel is 53.85 km. The experiment was conducted in part of the tunnel where 1.2% gradient exists. Longitudinal ventilation system provides a mean air flow at 0.61 m/s. The ventilation air flows at the declining direction of the tunnel slope. The combustion test involved fires of methanol (equivalent to the heat release of 1.34×10^{10} J) in twelve 1 m^2 trays put in central part of the tunnel floor. The total combustion area was 17.82 m^2 and combustion period was estimated to be 25 minutes. Diesel oil was also burnt at the same time in two 0.5 m^2 trays for observation of smoke movement.

The above full scale tunnel fire configuration is simplified for the present numerical investigation as shown in Fig. 1. The model has a same geometric cross section and size as the experimental part of Seikan Tunnel,

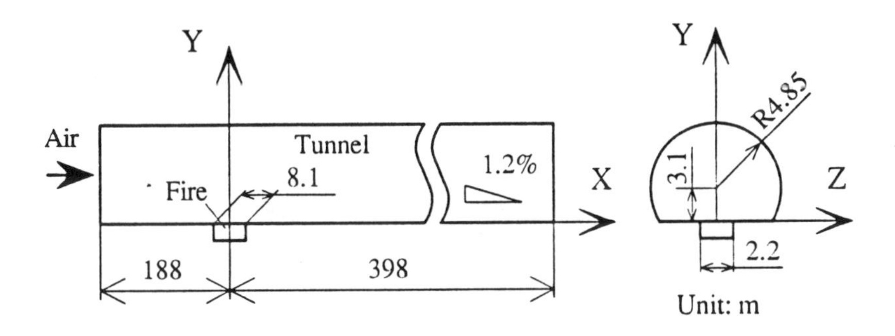

Fig. 1 Sketch of a model tunnel

but the length is largely shortened. The tunnel consists of a local fire source on the floor and longitudinal ventilation. Same values as the experiment have been specified for the calculation. The combustion model of the fire is not made. The fire and smoke sources are simplified as local heat and smoke release rates given as boundary conditions in the numerical model.

3. THE MATHEMATICAL MODEL

3.1 Governing Equations and Turbulence model

The mathematical model employs the time-averaged balance equations for mass, momentum, energy and smoke concentration. The smoke obscuration, which is expressed in terms of optical density (OD) per meter, is chosen for the evaluation of smoke concentration, because it is directly related to the field of human vision which dominates the evacuation speed of people in the smoke. The Boussinesq approximation is not made. The unsteady term in the mass conservation equation is retained. The equations are written in a Cartesian tensor form

$$\frac{\partial \rho}{\partial t}+\frac{\partial(\rho u_i)}{\partial x_i}=0, \tag{1}$$

$$\frac{\partial(\rho u_i)}{\partial t}+\frac{\partial(\rho u_i u_j)}{\partial x_j}=-\frac{\partial P}{\partial x_i}+\frac{\partial}{\partial x_j}(\mu\frac{\partial u_i}{\partial x_j}-\rho\overline{u_i' u_j'})+g_i(\rho-\rho_0), \tag{2}$$

$$\frac{\partial(\rho\phi)}{\partial t}+\frac{\partial(\rho u_j\phi)}{\partial x_j}=\frac{\partial}{\partial x_i}(\Gamma\frac{\partial\phi}{\partial x_i}-\rho\overline{u_i'\phi'})+S_\phi, \tag{3}$$

where ϕ represents h, c, k or ε. The term S_ϕ means an appropriate source or sink of the variable ϕ concerned.

The fluctuation of the density is ignored so that the density is only related with the temperature of the air through the equation of state

$$\rho=\rho(T). \tag{4}$$

For the closure of the governing equations, the two-equation $k-\varepsilon$ model of turbulence[4] is used. In order to take into account the effects of both buoyancy and mean streamline curvature on turbulence, the model is modified through the buoyancy production term G_B and flux Richardson number R_f. The modified G_B and R_f are given as follows

$$G_B=-\beta g\frac{\mu_t}{\sigma_T}\frac{\partial T}{\partial y}-\rho\overline{v_r' v_\theta'}r\frac{\partial(v_\theta/r)}{\partial r}-\rho\overline{v_s' v_\gamma'}s\frac{\partial(v_\gamma/s)}{\partial s}, \tag{5}$$

$$R_f = -\frac{G_B - 2\rho\overline{v_r' v_\theta'}\, v_\theta / r - 2\rho\overline{v_s' v_\gamma'}\, v_\gamma / s}{G_k + G_B}, \tag{6}$$

where

$$\rho\overline{v_r' v_\theta'} = -\mu_t r \frac{\partial (v_\theta / r)}{\partial r}, \tag{7}$$

$$\rho\overline{v_s' v_\gamma'} = -\mu_t s \frac{\partial (v_\gamma / s)}{\partial s}. \tag{8}$$

In above equation, v_θ is the swirl velocity component and $\rho\overline{v_r' v_\theta'}$ is the tangential shear stress in $x-y$ plane, v_γ is the swirl velocity component and $\rho\overline{v_s' v_\gamma'}$ is the tangential shear stress in $y-z$ plane.

The physical explanation of the modification has been described in the previous work[2]. The same turbulence model and constants have been used for the present simulation. The Schmidt number and turbulent Schmidt number in smoke obscuration equation are given by 0.7.

3.2 Initial and Boundary Conditions

Two types of boundaries are considered in the present investigation: solid surfaces and free surfaces at inlet and outlet. On solid wall, the non-slip condition for the velocity components is employed. And the wall is considered as a thermally insulated surface. Wall functions for velocity components, temperature as well as k and ε are applied to a near-wall point. Zero-gradient condition is prescribed for smoke concentration.

At the inlet, variables are given by the initial solution of a previous computation. At the outlet, for all dependent variables, zero-gradient conditions are prescribed.

The rate of fire spread depends on many factors: material properties, radiative and convective heat transfer to the flammable surface from the fire, conduction heat transfer within the surface material, the material pyrolysis chemistry, the composition of the pyrolysis gases, the chemical kinetics of the postpyrolysis surface material and the availability of oxygen in the vicinity of the surface and so on. No attempt has been made to simulate the details of ignition and the immediate postignition flame spread of the experiments. Instead, a fire strength of 5.24×10^5 W/m^2 is assumed and given as a constant heat flux. The fire source covers an area of 17.82 m^2, located at 188 m from the inlet of computation and in the central floor of the cross section. The smoke was generated by burning diesel oil in a 1 m^2 area in the experiment. By assuming that the burning rate is 1.5×10^{-5} m^3/s, the density of the diesel oil is 800 kg/m^2, and the coefficient of the smoke obscuration is 820 $1/\text{m} \cdot \text{m}^3/\text{kg}$, then the equivalent smoke release rate is

$$C_g = 800 \times 1.5 \times 10^{-5} \times 820 = 9.84 \quad 1/\mathrm{m} \cdot \mathrm{m}^3/\mathrm{s} \tag{9}$$

The heat and smoke release rates are given at the boundary. They do not appear in the governing equations.

The program commences with the computations for the ventilation air flow to its steady state before the introduction of the heat and smoke sources at a time set as zero. A uniform temperature (set at 19° C) and a zero smoke obscuration distribution are also assumed as an initial condition.

4. NUMERICAL SOLUTION TECHNIQUE

Since general Cartesian or cylindrical polar coordinate system suffers severely from geometric limitations, the application to curved surfaces, as most tunnel cross sections have, must involve interpolation between grid points not coincident with the boundaries. This may adversely affect the accuracy of the solution. In the present study, the geometric limitation is removed by adopting a general curvilinear coordinate system.

4.1 Numerical Grid Generation

The grid generation scheme of an elliptic system suggested by Thompson et al.[4] is adopted. The curvilinear coordinates are generated by solving the Poisson equations with a control function fashioned to control the spacing and orientation of the coordinates lines.

The generated grids for the tunnel symmetrical plane and cross section are shown in Fig. 2. It can be seen that the grid distribution of surfaces is smooth with concentration in regions of strong solution variation. The present computation is performed on $46 \times 14 \times 21$ grids.

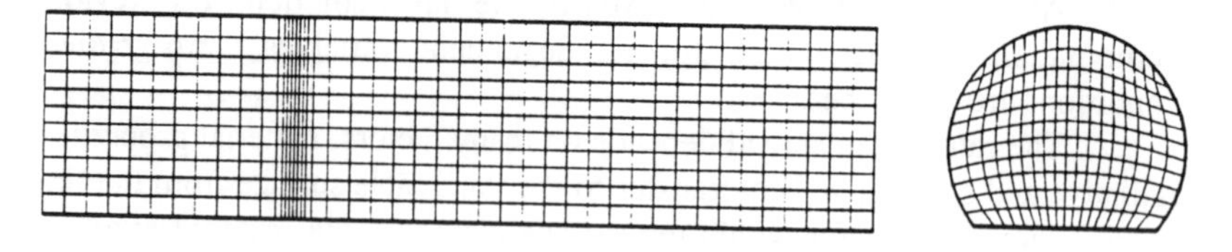

Fig. 2 Grid detail of the symmetrical plane and cross section

4.2 Transformation of the Governing Equations

By adopting the general curvilinear coordinate system, the set of conservation equations in Cartesian coordinates (x, y, z) can be transformed into curvilinear coordinates (ξ, η, ζ) as

$$\frac{J\partial\rho}{\partial t} + \frac{\partial(\rho U)}{\partial\xi} + \frac{\partial(\rho V)}{\partial\eta} + \frac{\partial(\rho W)}{\partial\zeta} = 0, \tag{10}$$

$$\frac{J\partial(\rho\Phi)}{\partial t} + \frac{\partial(\rho U\Phi)}{\partial\xi} + \frac{\partial(\rho V\Phi)}{\partial\eta} + \frac{\partial(\rho W\Phi)}{\partial\zeta}$$

$$=\frac{\partial}{\partial\xi}\left[\Gamma J\left(q_{11}\Phi_{\xi}+q_{12}\Phi_{\eta}+q_{13}\Phi_{\zeta}\right)\right]$$

$$+\frac{\partial}{\partial\eta}\left[\Gamma J\left(q_{21}\Phi_{\xi}+q_{22}\Phi_{\eta}+q_{23}\Phi_{\zeta}\right)\right]$$

$$+\frac{\partial}{\partial\zeta}\left[\Gamma J\left(q_{31}\Phi_{\xi}+q_{32}\Phi_{\eta}+q_{33}\Phi_{\zeta}\right)\right]+JS_{\Phi}, \tag{11}$$

Where q_{ij} is the metric of transformation. The contravariant velocity components are defined as

$$U=(u\partial\xi/\partial x+v\partial\xi/\partial y+w\partial\xi/\partial z)J, \tag{12}$$

$$V=(u\partial\eta/\partial x+v\partial\eta/\partial y+w\partial\eta/\partial z)J, \tag{13}$$

$$W=(u\partial\zeta/\partial x\pm v\partial\zeta/\partial y+w\partial\zeta/\partial z)J, \tag{14}$$

and, J represents Jacobian written as

$$J=\begin{pmatrix}\partial x/\partial\xi & \partial y/\partial\xi & \partial z/\partial\xi\\ \partial x/\partial\eta & \partial y/\partial\eta & \partial z/\partial\eta\\ \partial x/\partial\zeta & \partial y/\partial\zeta & \partial z/\partial\zeta\end{pmatrix}. \tag{15}$$

4.3 Method of Solution

The governing equations along with their boundary conditions are approximated with finite difference equations by a control volume based method. The convection-diffusion terms are discretized by a hybrid scheme. Since the common staggered grid system introduces problems, especially when the system is extended to curvilinear nonorthogonal coordinates, a collocated grid system is employed for the computation. A specific scheme developed in the previous work[2] is used to suppress the pressure oscillations which is a possible consequence of the collocated grid system. The velocity components and solutions are obtained by an iterative scheme(SIMPLER) suggested by Patankar[6].

The present solution method employs the efficient line-by-line solver based on the Tri-Diagonal Matrix Algorithm. All the dependent variables are updated at each iteration. At the same iteration level, the energy equation is solved first. Then, the density is computed from the equation of state to solve the momentum equations, followed by the transport equations for c, k, ε. Although the solution sequence is arbitrary in iterative method, the present scheme represents a realistic mechanism of the evolution of buoyancy induced flows due to the density variation in the field. The computation are carried out on the Digital Alpha 4000 computer at the Computer Center of National University of Singapore.

5. SIMULATION RESULTS AND DISCUSSION

Predictions have been compared with measurement[3] at three cross sections, namely -100m ($x/H = -12.6$), 100m ($x/H = 12.6$) and 300m ($x/H = 37.3$) from the center of the fire source. The comparisons of temperature and smoke concentration profile are shown in Fig. 3. Symbols represent the experiment values and lines represent computation results.

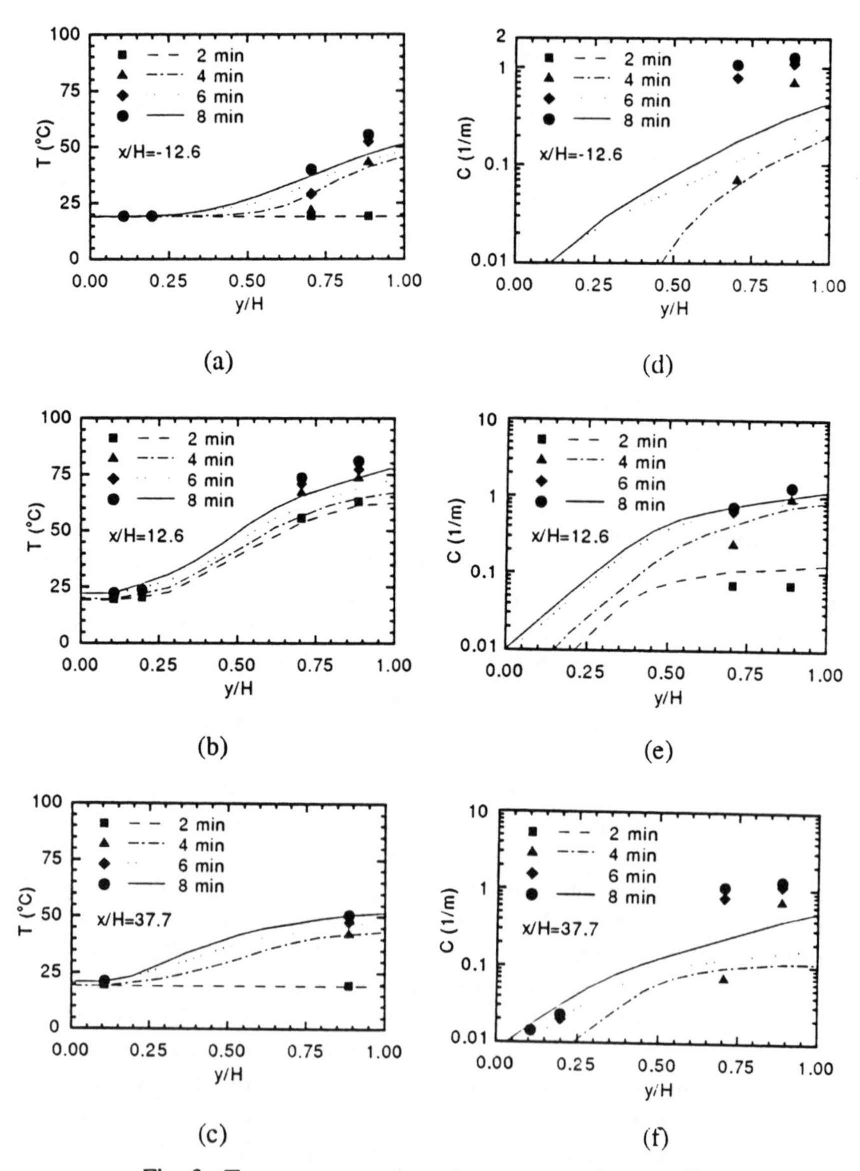

Fig. 3 Temperature and smoke concentration profiles

Figure 3 (a)~(c) show the temperature profiles. At the cross section remote from the fire source, the agreement between the predictions and measurements is quite remarkable. Close to the fire, however, thermocouple temperatures are considered to be dominated by radiative heat transfer which is not explicitly included in the prediction and thus discrepancies are apparent. In most of the regions, the predicted temperature is lower than the measured one. This suggests that the effect of the radiative heat transfer in full scale tunnel fire is much more significant than that in the laboratory scale experiment. Smoke concentration profiles can be seen in (d)~(f). Although the overall agreement is not perfect, reasonable agreement is obtained at early time stage from the ignition and at the near floor area. It is noted that the smoke obscuration less than 0.1 is not plotted in the figures. Basically the smoke concentration profile is analogous to that of temperature when Lewis number is close to 1. Some measurement points have shown different behavior. It is possible that the smoke movement may be affected by the ventilation system in the experimental tunnel.

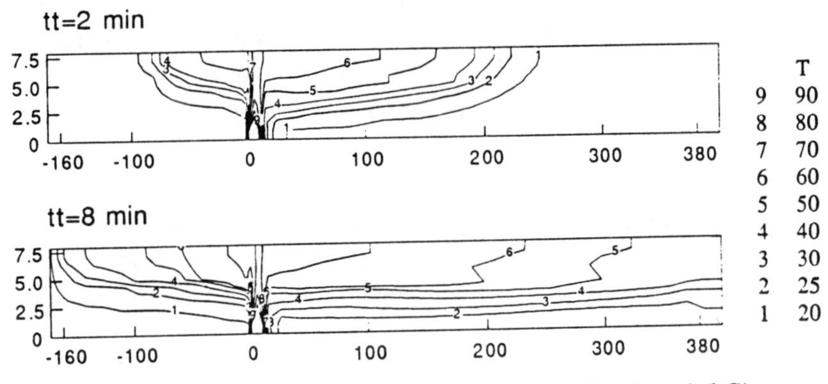

Fig. 4 Calculated isotherms ($x-y$ plane, $z/H=0$, unit:$^{\circ}$ C)

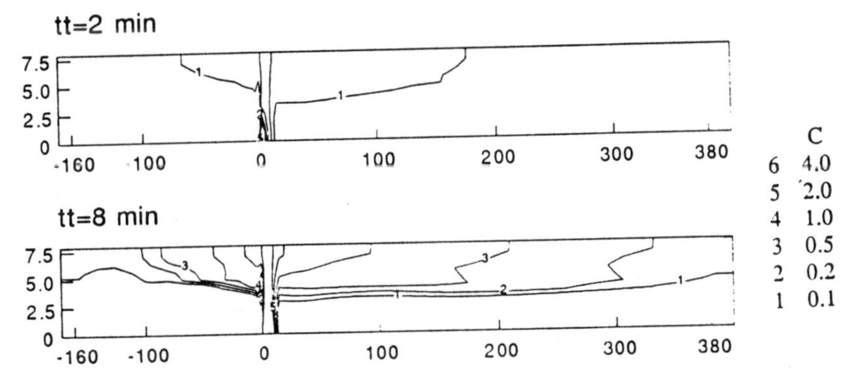

Fig. 5 Calculated smoke concentration contour ($x-y$ plane, $z/H=0$, unit: 1/m)

Figures 4 and 5 show the development with time of temperature and smoke concentration contours in the symmetry plane of the tunnel. These contours exhibit expected forms and are biased in the direction of forced ventilation flow even though it is in the declining direction of tunnel gradient. The report on temperature stratification in scale fire tunnel models[7] introduced a temperature stratification parameter expressed as

$$S = \frac{1}{\Delta T_{avg}} \frac{\Delta T_{cf}}{\Delta Y}, \tag{16}$$

where $\Delta Y = \Delta y / H$, $\Delta y = y_c - y_f$. y_c and y_f are the locations near the ceiling and floor area respectively. For the above three cross sections, the values of the stratification parameter S are all larger than 1.5, which is considered as a critical value for a highly stratified region. This indicates the temperature and smoke concentration are well stratified. The report also showed that the stratification behavior of the heated air flow is governed by a mixed convection parameter $\mathrm{Gr}^*/\mathrm{Re}_H^{5/2}$ representing the overall effect of natural and forced convection at different ventilation velocities and heat release rate. It was concluded that the heated air flow is always highly stratified even at the region close to the fire source, whenever $\mathrm{Gr}^*/\mathrm{Re}_H^{5/2} > 2\times10^2$. The parameter $\mathrm{Gr}^*/\mathrm{Re}_H^{5/2}$ in the present calculation is 3.17×10^2. The result obtained in the full-scale tunnel has confirmed the above conclusion.

Figures 6 and 7 show the development with time of temperature and smoke concentration contours in four cross sections of the tunnel. At the $x/H = 0$ cross section, two low temperature and low smoke concentration regions exist at both left and right floor corners even at 8 minutes from the ignition. This implies that a strong secondary flow has been formed in the

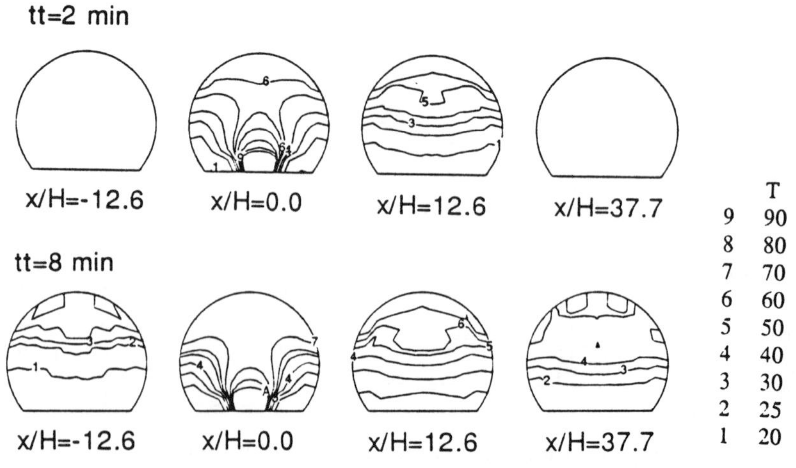

Fig. 6 Calculated isotherms ($z-y$ plane, unit: $^\circ$C)

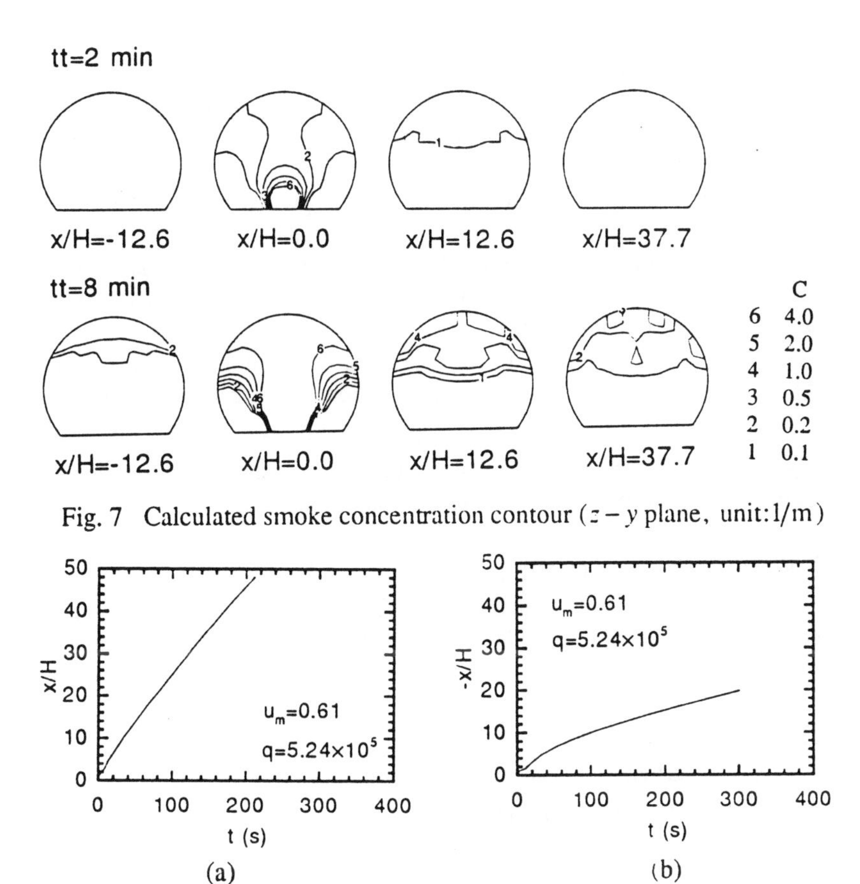

Fig. 7 Calculated smoke concentration contour ($z-y$ plane, unit:1/m)

Fig. 8 Leading edges of the heated air flow and heated reverse air flow

area close to the fire source. The convective effect of the large swirls continuously introduces the fresh air into this region and therefore keeps the temperature and smoke concentration at a relatively low level. But it should be pointed out that these are the results obtained when the radiative heat transfer is not included and the flame spread is not considered. These two factors are reported in some papers[8], [9] as a main cause in bringing about the disastrous accidents.

The leading edge of the heated air flow x/H (or the heated reverse air flow $-x/H$), which is defined as a position where the fluid temperature is increased by $1^{\circ}C$, is a function of burning time. They are considered as an important parameter for fire hazard prevention. The leading edge of the heated air flow and heated reverse air flow are shown in Fig. 8 (a) and (b) respectively. No direct experimental data is available for comparison. The relationship between the leading edge of the heated air flow and burning time is linear, but a polynomial functional relationship is shown in Fig 8 (b).

6. CONCLUSIONS

The predictions of heated air flow and smoke in a full-scale tunnel have provided a direct vision of real tunnel fire situation. The present model performed well to predict the temperature distribution at the region far from the fire source. Smoke concentration is also predicted reasonably well except for some data points near the ceiling of the tunnel. Disagreement with experimental data in the near field may be attributed to exclusion of the effect of radiative heat transfer and the coarse grid spacing near the fire source in the numerical model.

The temperature and smoke concentration contour illustrate a high degree of stratification in the present calculation. It has been confirmed that the mixed convection parameter $Gr^*/Re_H^{5/2}$ obtained in the scale model study can also be used for the full scale tunnel fire situation. The parameter is directly relevant to the safety consideration in tunnel fires.

ACKNOWLEDGMENTS

The authors wish to thank Prof. T. Saito and E. Hihara, Dept. of Mechanical Engineering, the University of Tokyo, Japan, for their helpful advice and encouragement.

NOMENCLATURE

g	gravitational acceleration
G_k	shear production term
Gr^*	modified Grashof number, $g\beta qH^4/\nu^2\lambda$
H	height of the tunnel
h	specific enthalpy
k	turbulence kinetic energy
p	mean pressure
q	heat release rate of the fire source
r, s	spatial coordinates in cylindrical system
Re_H	Reynolds number, HU_m/ν
T	mean temperature
ΔT_{avg}	average temperature rise in a cross section, $T_{avg} - T_0$
ΔT_{cf}	temperature difference between near ceiling and floor area
t	time
tt	time from ignition
U_m	mean velocity of the cross section
u,v,w	mean velocity components
u',v'	fluctuating velocity components

Greek symbols

β	thermal expansion coefficient
Γ	exchange coefficient
γ,θ	angular coordinates in cylindrical system
ε	turbulence energy dissipation
λ	thermal conductivity
ν	kinematic viscosity
ν_t	turbulent viscosity
ρ	air density
σ_t	turbulent Prandtl number
Φ	general dependent variable
ϕ	general scalar quantity

REFERENCES

1. XUE, H., HIHARA, E. and SAITO, T. - Analysis and Computation of Three-dimensional Transient Flow in a Fire Tunnel with Ventilation, Proceeding of 7th International Conference on Numerical Methods for Thermal Problems, pp.1127-1137, 1991.
2. XUE, H., HIHARA, E. and SAITO, T. - Turbulence Model of Fire-induced Air Flow in a Ventilated Tunnel, Int. J. Heat Mass Transfer, (in print).
3. IIDA, T., HARAKAWA, M. TERAGUCHI, H., SHIGENAGA, T. and KUWATA, H. - Test on the Smoke Exhausting System of the Seikan Tunnel, Air-conditioning and Hygienic Engineering, Vol. 63, No. 3, pp11-22, 1989. (in Japanese)
4. RODI, W. - Turbulence Models and Their Application in Hydraulics ---- A State of the Art Review, Book Publication of International Association for Hydraulic Research, Delft, Netherlands, 1980.
5. THOMPSON, J. F., WARSI, Z. U. A. and WAYNE MATIN, C. - Numerical Grid Generation, North-Holland, New York, 1985.
6. PATANKAR, S. V. - Numerical Heat Transfer and Fluid Flow, McGraw-Hill, New York, 1980.
7. XUE, H., HIHARA, E. and SAITO, T. - Temperature Stratification of Heated Air Flow in a Fire Tunnel, JSME Int. J., (in print).
8. FUSEGI, F. and FAROUK, B. - Numerical Study on Interactions of Turbulent Convection and Radiation in Compartment Fires, Fire Science & Technology, Vol. 8, No. 1, pp15-28, 1988.
9. MALALASEKERA, W. M. G and LOCKWOOD, F. - Computer Simulation of the King's Cross Fire: Effect of Radiative Heat Transfer on Fire Spread, Proc. Instn. Mech. Engrs., Vol. 25, pp201-208, 1991.

SECTION 6

RADIATIVE HEAT TRANSFER

NUMERICAL SOLUTION OF THE RADIATION INDUCED IGNITION OF SOLID FUELS

Jennifer R. Koski [1]
Georgia Institute of Technology, Atlanta, GA 30332-0405 U.S.A.
Pandeli Durbetaki [2]
Georgia Institute of Technology, Atlanta, GA 30332-0405 U.S.A.

SUMMARY

Depending on the radiation intensity and oxygen concentration, ignition occurs either on the surface or in the boundary layer gas mixture produced through gasification of the solid fuel. The development of the temperature, fuel, and oxygen concentrations that lead to ignition depends on the competing processes between the gas phase and the solid surface. It is this interdependence of the processes in the gas phase and on the surface that causes difficulty in the numerical analysis of the problem. This research uses a unique approach developed in previous studies for numerically solving the solid and gas phase domains simultaneously by using the Patankar-Spalding algorithm. This paper focuses on the Patankar-Spalding method in producing accurate results for incident radiation on solid PMMA (Polymethylmethacrylate) with forced convection.

1. INTRODUCTION

A wall suddenly becomes engulfed in flames as a fire sweeps through a building destroying everything in its path. The characteristics of such a fire have been under investigation by many researchers [1-4]. Fire is still not fully understood and warrants more research, especially in the prediction of when a fire may occur. Increasing the time to ignition by just a few seconds can result in a significant reduction in losses of life and property. Therefore, the conditions that result in a surface ignition need to be studied. Researchers in the School of Mechanical Engineering at the Georgia Institute of Technology have been investigating methods to predict the

1. Graduate Research Assistant, The George W. Woodruff School of Mechanical Engineering
2. Professor, The George W. Woodruff School of Mechanical Engineering

radiation induced ignition of solid fuels [5-9].

As a solid is subjected to a uniform radiant flux, the surface temperature increases. This increase in surface temperature causes a corresponding increase in the temperature of the air near the surface. The amount of forced convection along the surface will influence the time to ignition. A convective thermal boundary layer forms along the surface due to the interaction of the fluid with the heated surface. Eventually, pyrolysis begins in the solid and the surface temperature essentially remains constant [1, 2]. Through a gasification process, gaseous fuel from the surface is generated and enters the boundary layer, thus creating a combustible mixture. The gas begins to absorb the radiation leading to a further increase of the mixture temperature. This absorbed radiant energy is a critical factor in any ignition which would occur in the gas phase. Park and Tien [2] reported that even though natural convection flows dilute the gas mixture and increase heat losses, the boundary layer gases must be taken into account in the ignition process. Thus, depending on the radiation intensity and oxygen concentration, ignition occurs either on the surface or in the boundary layer gas mixture. Once ignition begins, a thermal runaway condition exists and the solid and gas are engulfed in fire.

The Patankar-Spalding finite control volume method can deal with the coupling effect resulting from the gas-solid interactions by forming a tridiagonal matrix of transformed governing equations that can solve the transient, multi-phase problem as a whole, rather than matching the solutions at the interface as it was done in the past [6]. Li [6] introduced a unique approach for solving the solid-gas interface which involves solving for the solid and gas phases simultaneously by using the control volume method. Using a grid fitted to the boundary layer and variable time step sizes, the current study employs the Patankar-Spalding method with Li's approach to numerically solve the radiation induced ignition problem in the presence of forced convection along a horizontally positioned semi-infinite solid piece of PMMA.

The method begins by examining the velocity field in the boundary layer. Then, the steady energy equation for the boundary layer is solved. Through a transient conjugate formulation, the transient conduction equation for the solid and the boundary layer equations are solved as a whole. Based on results of Li and Durbetaki [6], the Patankar-Spalding method provides good agreement with experimental analysis.

2. THEORETICAL MODEL

The Patankar-Spalding method will be employed in investigating ignition with forced convection along a horizontally positioned semi-infinite solid piece of PMMA (Polymethylmethacrylate). PMMA is chosen as the solid fuel due to its uniform properties and non charring characteristics. The model assumes: (i) two-dimensional, laminar flow, (ii) Fourier's Law of conduction and Fick's Law of diffusion apply, (iii) constant specific heat across the boundary layer and among the gases, (iv) viscous dissipation is negligible, (v) ideal gases, (vi) properties of the solid are constant, (vii) oxidation occurs at the solid surface in a single step, irreversible first order Arrhenius type reaction with no particulates produced, (viii) the gas phase chemical reaction is a single step, irreversible second order Arrhenius type reaction, (ix) Sorret and Defour effects are negligible, (x) pressure diffusion is negligible, and (xi) for the solid phase, incident radiation is absorbed at the surface. Note that the gas phase and the solid phase must be treated seperately, except at the interface. Thus, the equations are presented in a way that they are easily grouped for properties and solution.

1.1 Gas Phase

(a) Conservation of Mass

$$\frac{\partial \rho}{\partial t} + \frac{\partial(\rho u)}{\partial x} + \frac{\partial(\rho v)}{\partial y} = 0 \qquad (1)$$

(b) Species Conservation

$$\rho\frac{\partial Y_i}{\partial t} + \rho u\frac{\partial Y_i}{\partial x} + \rho v\frac{\partial Y_i}{\partial y} = \omega_i + \frac{\partial}{\partial y}(\rho D\frac{\partial Y_i}{\partial y}) \quad \text{for } i = O,F,N \qquad (2)$$

$$Y_P + Y_O + Y_F + Y_N = 1 \qquad (3)$$

(c) Conservation of Momentum in the x direction

$$\rho\frac{\partial u}{\partial t} + \rho u\frac{\partial u}{\partial x} + \rho v\frac{\partial u}{\partial y} = \frac{\partial}{\partial y}(\mu\frac{\partial u}{\partial y}) - \frac{dP}{dx} \qquad (4)$$

(d) Conservation of Energy

$$\rho c_p\frac{\partial T}{\partial t} + \rho c_p u\frac{\partial T}{\partial x} + \rho c_p v\frac{\partial T}{\partial y} = \frac{\partial}{\partial y}(\lambda\frac{\partial T}{\partial y}) - \sum_{i=1}^{n} h_i^o\omega_i - \frac{\partial q_r}{\partial y} \qquad (5)$$

1.2 Solid Phase

Conservation of Energy

$$\rho_s c_s \frac{\partial T}{\partial t} = \lambda_s \frac{\partial^2 T_s}{\partial y^2} \tag{6}$$

1.3 Gas Phase Chemical Kinetics

The chemical reaction in the gas phase prior to ignition is assumed to follow a second order, one-step, irreversible Arrhenius equation. The resulting mass consumption rates are

$$\omega_i = -\frac{A}{M_F M_O} \rho^2 \upsilon_i M_i Y_F Y_O e^{-E/RT} \quad \text{for } i = O, F, P \tag{7a}$$

$$\omega_N = 0 \tag{7b}$$

1.4 Solid Phase Chemical Kinetics

The oxidation reaction on the solid surface is assumed to follow a first order Arrhenius equation. Thus, the net energy effect is shown below.

$$q_{sR} = A_{sR} Q_{sR} \rho_w Y_{Ow} e^{-E_{sR}/RT_{sw}} \tag{8}$$

Pyrolysis at the surface creates a mass flux that follows the equation below.

$$\dot{m}_o'' = \rho_s \frac{1.5 \times 10^{11} \exp(-21651.6/T)}{1 + 5.6 \times 10^{11} \exp(-16112.8/T)} \tag{9}$$

1.5 Radiation Equations

Siegel and Howell [10] derived the radiation equations for a scattering, absorbing, and emitting medium. In particular, they reported that

$$\frac{di_\lambda'}{dk_\lambda} + i_\lambda'(k_\lambda) = I_\lambda'(k_\lambda, \omega) \tag{10}$$

where the directional radiant intensity, i_λ', can be integrated over the two hemispheres formed by the energy going into and out of the boundary layer to obtain the radiant flux.

$$\frac{\partial E_+}{\partial y} + \alpha E_+ = \alpha\sigma T^4 \qquad ; \qquad -\frac{\partial E_-}{\partial y} + \alpha E_- = \alpha\sigma T^4 \tag{11}$$

Thus, the radiant flux in the surface normal direction is described as

$$q_r = E_+ + E_- \tag{12}$$

The radiant term of the energy equation becomes

$$-\frac{\partial q_r}{\partial y} = -\alpha(2\sigma T^4 - E_+ - E_-) \tag{13}$$

where the absorption coefficient, α, is the hemispherical total absorption coefficient for the boundary layer gas.

Figure 1 shows the schematic diagram of the ignition problem along with the radiation terms. The E_+ and E_- indicate the two beam concept discussed above. The forced convection is applied by the gas stream made up of oxygen and nitrogen. The velocity boundary layer formed over the solid surface due to the gas stream is shown in the diagram, also.

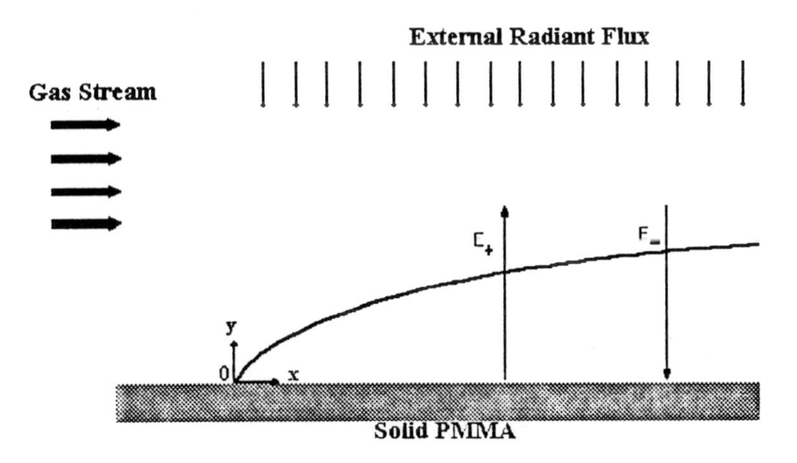

Figure 1. Schematic diagram of the ignition problem

The previous equations are simplified according to the assumptions before they are discretized. The quantities of interest are the velocity, species concentrations, and temperature for the gas phase and temperature for the solid phase. Thus, the temperature through the solid and the gas is of particular interest in the ignition problem. The results discussed later in this paper indicate that the method handles the interface appropriately.

1.6 Initial Conditions

The conservation equations must be solved subject to a set of initial and boundary conditions.

$$\text{at } t = 0\_ \text{ and } x = 0: \quad u = u_\infty \; ; \; v = 0 \quad Y_F = 0 \quad Y_O = Y_{O\infty} \quad Y_N = Y_{N\infty}$$
$$T = T_\infty = 300K \qquad T_s = T_\infty \tag{14}$$

1.7 Boundary Conditions

(a) Momentum condition

$$y = 0: \quad u = 0 \text{ and } \rho v = (\rho v)_w = \dot{m}''_o \quad ; \quad y \to \infty: \; u = u_\infty \tag{15}$$

(b) Fuel species

$$y = 0: \quad (\rho v)_w = (\rho v)_w Y_{Fw} - \rho D \frac{\partial Y_F}{\partial y})_0 \; ; \quad y \to \infty: \; Y_F = 0 \tag{16}$$

(c) Oxygen species

$$y = 0: \quad (\rho v)_w Y_{Ow} = \rho D \frac{\partial Y_O}{\partial y})_0 - \rho_w Y_{Ow} A_{sR} e^{-E_{sR}/RT_{sw}} \quad ;$$
$$y \to \infty: \; Y_O = Y_{O\infty} \tag{17}$$

(d) Inert species

$$y = 0: \quad (\rho v)_w Y_{Nw} = \rho D \frac{\partial Y_N}{\partial y})_0 \quad ; \quad y \to \infty: \; Y_N = Y_{N\infty} \tag{18}$$

(e) No slip condition

$$y = 0: \qquad u = 0 \tag{19}$$

(f) Free stream condition

$$y \to \infty: \qquad u = u_\infty \tag{20}$$

(g) Temperature conditions

$$y = 0: \qquad T = T_{sw} \tag{21}$$

$$y \to \infty: \qquad T = T_\infty \tag{22}$$

$$y \to -\infty: \qquad T_s = T_\infty \tag{23}$$

(h) Energy balance at the interface

$$-\lambda_s \frac{\partial T_s}{\partial y})_0 = -\lambda \frac{\partial T}{\partial y})_0 + E_+(0) - E_-(0) - \rho_w Y_{Ow} A_{sR} Q_{sR} e^{-E_{sR}/RT_{sw}} + \overset{o}{m}{}_0'' Q_{sp} \tag{24}$$

(i) Radiative transfer equations

$$E_+(0) = rE_-(0) + \varepsilon\sigma T^4{}_{sw} \quad ; \qquad E_-(\infty) = I_o + \sigma T_\infty^4 \tag{25}$$

3. NUMERICAL PROCEDURE

Patankar [11] and Spalding [12] recognized that the all the conservation equations, except for the overall mass conservation equation and the radiation transfer equations follow the general structure listed below.

$$\frac{\partial}{\partial t}(\rho\phi) + \frac{\partial}{\partial x_j}(\rho u_j \phi) = \frac{\partial}{\partial x_j}(\Gamma \frac{\partial \phi}{\partial x_j}) + S \tag{26}$$

where ϕ = generic dependent variable, Γ = generic diffusion coefficient, and S = the source term.

The overall mass conservation equation is included in the analysis through the stream function.

$$\frac{\partial \psi}{\partial x} = -\rho v \quad ; \qquad \frac{\partial \psi}{\partial y} = \rho u \tag{27}$$

The radiative transfer equations are handled by noting that these equations do not have to be evaluated simultaneously with the other conservation equations. The source term, however, must be evaluated with the updated values of the unknowns T, Y_F, Y_O, etc. as they are computed. Note that the solid energy equation fits the general structure without any modifications. The Patankar-Spalding method derives the discretization equations by integrating the transformed equations along the surface of the control volume described below around each node point. Since the principles of the equations are maintained through the transformation process, the control volumes can vary in size. The resulting discretization equations form a band matrix that is solved through a simple tridiagonal matrix algorithm for each node point.

The Patankar-Spalding method is a marching scheme which successively solves the downstream locations using the known values for the upstream locations, solving only one downstream location at a time. The method begins by solving the velocity field in the boundary layer using a computer program based on the structure used by Spalding in his computer program, GENMIX [12]. Then, the steady energy equation for the boundary layer is solved. Temperature and concentration profiles are generated and stored for later use. The transient conduction equation for the solid and the boundary layer equations are solved as a whole by sweeping along the entire surface. Thus, different solid conduction equations are solved at different x locations [6].

In order to ensure an adequate amount of grid points to be available for the front edge of the boundary layer and not too many points at large x, define a new coordinate variable, ϖ, where I and E are the internal and external surfaces of the boundary layer, respectively.

$$\varpi = \frac{\psi - \psi_I}{\psi_E - \psi_I} \qquad (28)$$

The finite control volume used in this procedure is shown on the next page in Figure 2. In general, the known values of ϕ at the upstream location x_u are used to find the values of ϕ at the downstream location x_d, which is a step size of Δx away. Thus, the values of ϕ at the points NU, PU, and SU are known, and those at N, P, and S are unknown [11]. n and s indicate the lateral faces of the control volume. The ϕ values at NU, PU, and SU are assumed to dominate over the respective control volume faces, while the downstream ϕ values will be used to find the fluxes across the lateral faces.

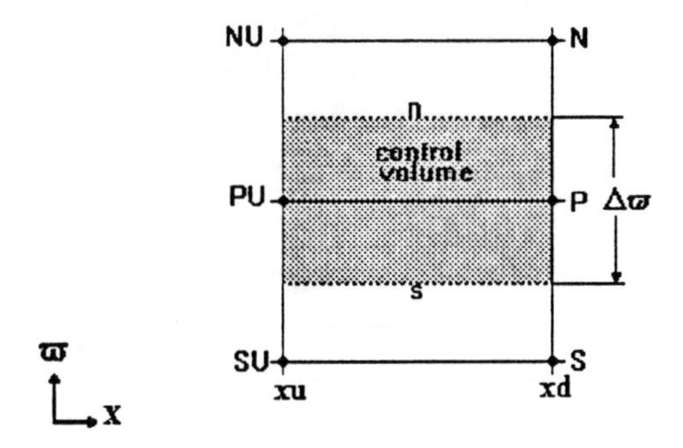

Figure 2. Control volume layout

A unique method to solve the solid-gas interface was introduced by Li [6], which involves solving for the solid and gas phase domains simultaneously. The same approach is used in the current study to find the temperatures for the interfacial nodes. In general, the interface becomes a control volume face where radiation exchanges take place as the two phases are compined. Then, an energy balance can be conducted on the interface to find the temperatures. The marching procedure continues until all the temperatures are known.

4. RESULTS

Currently, the computer program being developed to solve the radiation induced ignition with forced convection has been able to successfully model solid absorption of the radiation in the presence of forced convection along a piece of solid PMMA. Figure 3 shows the temperature profile through the solid and the gas phases for solid absorption, but no gas absoption or solid pyrolysis, with an incident radiant flux of 22 W/cm^2 initially with both the solid and gas at the dimensionless temperature of 1. The thermal wave through the solid can be seen, as well as the expected temperature profile for the gas phase with a heated surface in the presence of forced convection.

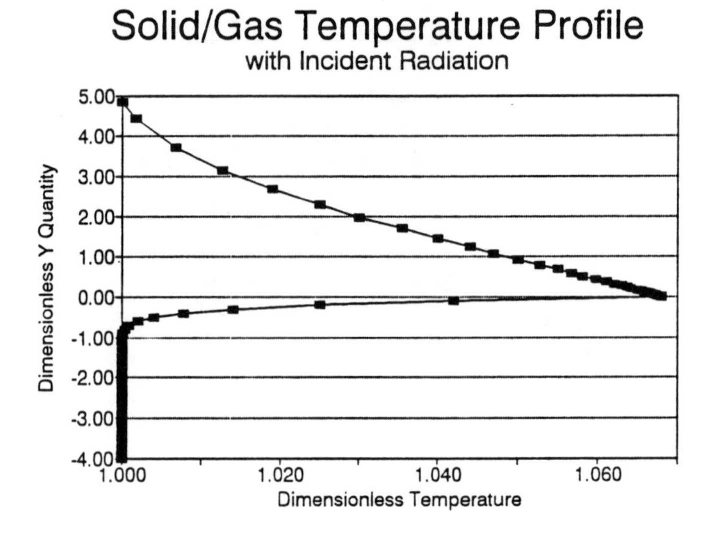

Figure 3. Temperature profile through the solid and gas phases

Figure 4 shows the temperature profile at a later time. Note that the temperature at the interface has increased along with the corresponding temperatures near the surface.

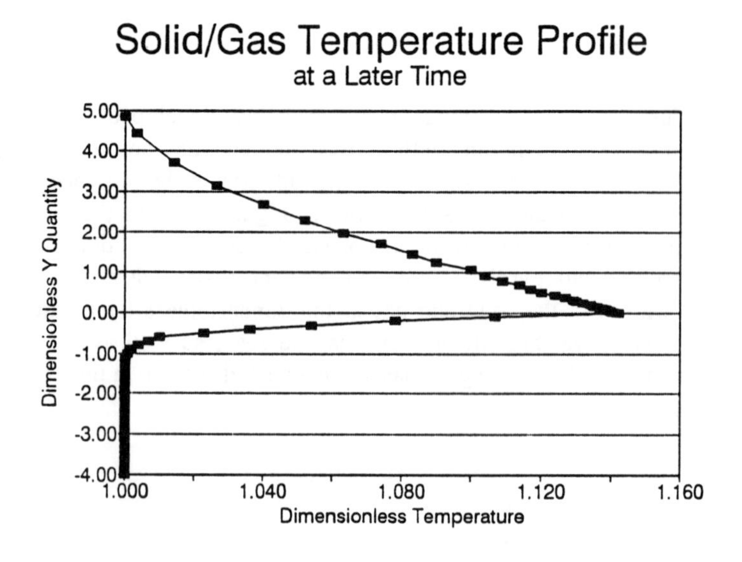

Figure 4. Temperature profile of the solid and gas phases at a later time

5. NOMENCLATURE

A	pre-exponential factor	c	specific heat
D	diffusion coefficient	E	activation energy
E_+	hemispherical total emissive power in the outward normal direction	E_-	hemispherical total emissive power coming into the layer normal to the surface
I_o	initial external radiant intensity	i_λ'	directional radiant intensity
h_i^o	standard heat of formation	k_λ	optimal thickness
M	molecular weight	$\dot{m}_o$	mass flux at the surface
P	pressure	Q	energy effect of chemical reaction
q_r	radiant heat flux	R	universal gas constant
r	surface reflectivity	T	temperature
t	time	u,v	velocity in the x, y direction
Y_i	species mass fraction	α, σ	absorption coefficient, scattering coefficient
ε	surface emissivity	λ	thermal conductivity
μ	dynamic viscosity	υ	stoichiometric coefficient
ρ	density	ψ	stream function
ω	solid angle	ω_i	rate of species i production

Subscripts

g	gas	s	solid
F, N	fuel, inert species	O, P	oxygen, product species
sp	surface pyrolysis,	sR	surface reaction
w	interface	∞	free stream

6. REFERENCES

1. BAEK, SEUNG WOOK and KIM, JEONG SOO. - Ignition of a pyrolyzing solid with radiative active fuel vapor. Combustion Science and Technology, Vol. 75, Gordon and Breach Science Publishers S.A., United Kingdom. pp. 89-102, 1991.
2. PARK, S. H. and TIEN, C. L. - Radiation induced ignition of solid fuels. International Journal of Heat and Mass Transfer, Vol. 33, No. 7, Pergamon Press, Great Britain. pp. 1511-1520, 1990.
3. KASHIWAGI, T. - Experimental observations of radiative ignition mechanisms. Combustion and Flame, Vol. 34, pp.231-244, 1979.
4. DI BLASI, COLOMBA. - Ignition and flame spread across solid fuels. Numerical Approaches to Combustion Modeling, Vol. 135, American Institute of Aeronautics and Astronautics, Inc., Washington, D. C. pp. 643-671, 1991.

5. LI, XIANMING. - The effect of gas-surface interactions on radiative ignition of PMMA, Ph.D. Dissertation, School of Mechanical Engineering, Georgia Institute of Technology, 1990.
6. LI, XIANMING and DURBETAKI, PANDELI. - The conjugate formulation of a radiation induced transient natural convection boundary layer. International Journal for Numerical Methods in Engineering, Vol. 35, John Wiley & Sons, Ltd. pp. 853-870, 1992.
7. PHUOC, T. X. - Ignition of polymeric material under radiative and convective exposure, Ph.D. Dissertation, School of Mechanical Engineering, Georgia Institute of Technology, 1985.
8. DURBETAKI, P. and PHUOC, T. X. - Ignition phase transition of a polymer: convective exposure. International Journal for Numerical Methods in Engineering, Vol. 25, John Wiley & Sons, Ltd. pp. 373-386, 1988.
9. DURBETAKI, P. and PHUOC, T. X. - Radiative ignition of fabrics and assemblies: experiments and modelling. International Journal for Numerical Methods in Engineering, Vol. 30, John Wiley & Sons, Ltd. pp. 859-873, 1990.
10. SIEGEL, R. and HOWELL, J. R. - Thermal Radiation Heat Transfer (Second Edition), Hemisphere Publishing Corp., Washington, 1981.
11. PATANKAR, S. V. - Parabolic systems: Finite-difference Method I. Handbook of Numerical Heat Transfer, Minkowycz, Sparrow, Schneider, and Pletcher, John Wiley and Sons, New York. pp. 89-115, 1988.
12. SPALDING, D. B. - GENMIX: A general computer program for two-dimensional parabolic phenomena, Pergamon Press, Oxford, 1977.

LOW COST ALGORITHMS FOR 2-D AND 3-D RADIATION VIEW FACTORS COMPUTATIONS BY FINITE ELEMENTS

M. HOGGE[+] , P. MAGERMANS[++] , Ph. MICHEL[++]

[+]LTAS-Thermomécanique, University of Liège
Rue E. Solvay, 21, B-4000 LIEGE, BELGIUM

[++]SAMTECH s.a., Bld. F. Orban, 25, B-4000 LIEGE, BELGIUM

ABSTRACT

Radiation heat transfer modelling faces at least two difficulties in a FEM (Finite Element Method) context : first, natural sets for radiation computations are surface variables whereas nodal unknowns are the standard variables in FEM ; second, defining the visibility of these surfaces (or part of surfaces) is an essential task to be carried out at the lowest possible cost, generally before the effective heat transfer computations take place.

We present in this paper a FEM methodology which is well suited to meet such requirements :
- first, special surface elements are developped, which enable automatic transfer between the natural surface quantities appearing in radiation heat transfer (the radiosities) and the nodal variables associated with the surfaces (or part of them) [1]. These elements are provided with appropriate equivalent conductivities depending on the associate surface emissivities, and exhibit new auxiliary nodal unknowns linked with the effective surface radiosities.
- second, these auxiliary nodes are linked together by special 1-D elements with thermal resistance functions of the radiation view factor between the associate surfaces.

Special attention is thus devoted to economical computations of these view factors in a FEM context. 2-D situations are dealt with simply via the geometric Hottel rule [2] with special care for cavities and obstacles. For 3-D situations, several classifying algorithms [3,4] are presented and compared : the basic idea is to definie global entities comprising individual surfaces. Computation time is reduced to a minimum by searching directly for the visibility of these entities one with respect to the other, instead of doing it at the elementar surface levels. Model problems show that up to 90 % of computer cost can be gained by this method with respect to a complete search.

1. FINITE ELEMENT MODELLING FOR RADIATION HEAT TRANSFER

1.1. General

Standard heat conduction problems are easily treated via FEM since precise representation of the conducting bodies is dealt with through geometric solid modelling. On the contrary, radiation inputs are barely treated in a comprehensive way in finite element computer programs. The common difficulty for radiant diffuse interchange between finite gray surfaces limiting conducting bodies is that angle (or view) factors are defined in terms of surfaces whereas finite element models treat nodal unknowns which are not directly representative of those surfaces as it is the case in finite difference lumped methods : the nodes in FEM are located at the boundary of the radiant areas. Specialized surface finite elements have thus been developed to relate these surface nodes with auxiliary nodes associated with the radiosity of each acting surface. These auxiliary nodes are finally linked together with linear elements exhibiting heat conductance proportional to their relative angle factors. The procedure is shown to be a generalization of the classical lunped method based on electric circuit analogy. The interesting feature of the proposed method is that it enables an automatic translation of radiative surface loads into nodal inputs in the FEM context and that it does not require separate or additional work for conductive links : the conduction is easily taken into account by the volumic FE for which the surface elements are the boundaries.

1.2. The radiosity equations [4]

Engineering calculations of radiant interchange among diffusely emitting and reflecting surfaces limiting opaque bodies are frequently performed under the assumption that the participating surfaces are non specular and gray and that

$$\varepsilon_i = \alpha_i \tag{1}$$

where ε_i and α_i are respectively the emittance and the absorptance of the ith surface (Kirchhoff law). Such an assumption is reasonable when both the emitted and the incident radiation are confined to the same wave-length range and the spectral emittance is relatively constant within that range.

Consideration is given to a set of surfaces (e.g. an enclosure) composed of N finite surfaces such as that pictured in Fig. 1. Each surface of the enclosure is assumed gray, isothermal, emitting or reflecting radiant energy in a perfectly diffuse manner.

The rate at which radiant energy streams away from surface i per unit area is termed the radiosity of the surface and is denoted by the symbol $R_i [W/m^2]$; on the other hand, the incident radiant energy arriving on that surface per unit time and unit area is denoted by the symbol H_i and it follows that

$$R_i = \varepsilon_i\,\sigma\,\bar{T}_i^4 + \rho_i\,H_i = \varepsilon_i\,\sigma\,\bar{T}_i^4 + (1-\varepsilon_i)H_i \tag{2}$$

the radiosity being the sum of the emitted and reflected radiation of the surface.

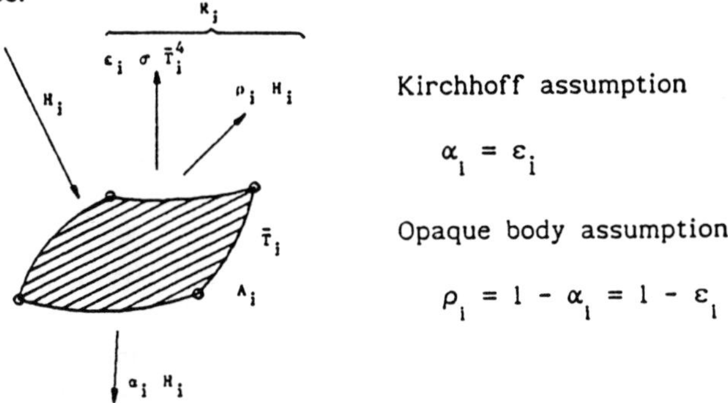

Figure 1. Radiant thermal balance for gray diffuse surface i

The net rate of heat loss Q_i of the typical surface of area A_i is the difference between the emitted radiation and the absorbed portion of the incident radiation :

$$Q_i/A_i = \varepsilon_i\,\sigma\,\bar{T}_i^4 - \alpha_i\,H_i = \varepsilon_i(\sigma\,\bar{T}_i^4 - H_i) \tag{3}$$

or, equivalently, between the energy leaving and arriving at that surface

$$Q_i/A_i = R_i - H_i \tag{4}$$

Note that the gray body condition Eq. (1) and the assumption of opaque materials have been introduced in both Eqs. (2) and (3). The next step is to find the incident radiant flux H_i : it comes from the other surfaces ($j \neq i$) of the enclosure through the angle factors F_{ij}. In order that these be independent of the magnitude and surface distribution of the associate energy flux, we have to require that the radiosity of any surface is constant along that surface. From the definition Eq. (2), it is clear that if R_i is to be uniform along an isothermal surface at $\bar{T}_i$, it is necessary that the incident radiant flux H_i be also uniform. Though it is generally unlikely that this latter condition will be fully satisfied in practice, we thus assume that

$$H_i = \left(\sum_{j=1}^{N} R_j\,A_j\,F_{ji}\right)/A_i = \sum_{j=1}^{N} R_j\,F_{ij} \tag{5}$$

where use has been made of the reciprocity rule for angle factors.

The radiant energy conservation principle for an enclosure yields in addition

$$\sum_{j=1}^{N} F_{ij} = 1 \qquad (i = 1,2, \ldots N) \tag{6}$$

which leads to the identity

$$R_i = \sum_{j=1}^{N} R_i F_{ij} \tag{7}$$

By combining Eqs. (2-3-4-5-7), we finally arrive to the set of N equations

$$Q_i = \frac{A_i \varepsilon_i}{1-\varepsilon}(\sigma \bar{T}_i^4 - R_i) = A_i \sum_{j=i}^{N} F_{ij}(R_i - R_j) \quad (i=1,\ldots,N) \tag{8}$$

termed the radiosity equations : it is a linear systems of N inhomogeneous algebraic equations for the N unknown radiosities R_i, that can be written in matrix form as

$$[B_{ij}] \left\{R_j\right\} = \left\{G_i\right\} \tag{9}$$

in which the elements of the array $[\underset{\sim}{B}]$ and of the vector $\left\{\underset{\sim}{G}\right\}$ are known values, either

$$B_{ij} = \frac{\varepsilon_i \delta_{ij} - (1 - \varepsilon_i)F_{ij}}{\varepsilon_i}, \quad G_i = \sigma \bar{T}_i^4 \tag{10}$$

for a surface with prescribed temperature $\bar{T}_i$, or

$$B_{ij} = \delta_{ij} - F_{ij} \quad , \quad G_i = Q_i/A_i \tag{11}$$

for a surface with prescribed heat flux Q_i, δ_{ij} denoting the Kronecker delta. This system can be built and inverted once for ever if the emittances ε_i may be regarded as constants. When the radiosities have been obtained for each surface, Eq. (8) yields the associated heat flux (for a surface with prescribed temperature) or the fourth power of the surface temperature (for a surface with prescribed heat flux).

1.3. Finite Element implementation

Radiant surfaces whose mutual heat exchanges are governed by Eq.(9) are generally boundaries of heat conducting bodies for which a finite element model has been established. This model deal with nodal unknowns (corner or mid-side nodes) located at the boundary of the radiant areas (Fig. 2). Hence these unknowns are not directely representative of the radiant surfaces as it is the case in Finite Difference (F.D.) lumped methods. We are thus faced with the key problem for finite element modelling of radiant heat transfer, i.e. converting radiation surface heat inputs into nodal heat inputs associated with the nodal unknowns. Subsidiarily, we shall have to link the radiative surfaces together for mutual exchange according to Eq. (5).

One could first think of defining auxiliary radiative "nodal" areas (Fig. 2.A) to circumvent this problem, in such as way that F.E. nodes be centered on new exchange areas different from the element radiative areas. This F.D. type of solution is discarded because of the amount of additional work required to define these surfaces and due to the lack of universality for nodes located at the intersection of two or more planar surfaces.

The present solution is in fact to define specialized surface elements (Fig. 2.B) relating the boundary nodes of a F.E. radiant surface with an auxiliary node associated with the radiosity of that surface.

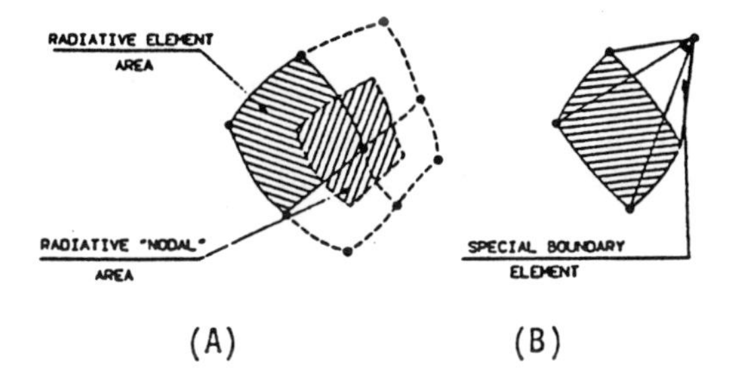

Figure 2. Radiative surfaces and nodal unknowns.

By doing so, the user benefits of some obvious advantages with respect to a classical F.D. nodal discretization :

- the radiative surfaces coincide with the surfaces naturally generated by the F.E. spatial discretization for conduction ;

- he is freed from the "quasi-isothermal" radiant surface concept since radiative inputs are automatically transfered to the F.E. nodal points. $\bar{T}_i$ in section 2.2 may thus be interpreted as the radiant surface mean temperature ;

- no additional work is required at a node for internal link by conduction with its neighbours since this task will be naturally taken up by the F.E. conduction model.

Finally, note that radiant mutual exchange between surfaces is only through the radiosity nodes as in a F.D. procedure, and consequently computation of view factors can be performed separetely (see section 3).

1.4. Solution procedure

Once the global F.E. model has been built, we are faced with two set of simultaneous equations, a first set yielded by Eq. (9) for the N radiosity nodes

$$[\underset{\sim}{B}] \left\{\underset{\sim}{R}\right\} = \left\{\underset{\sim}{G}\right\} \quad \text{(dimension N)} \qquad (12)$$

and a second set yielded by the conductive/ diffusive links between the M finite element nodal unknowns,

$$[\underset{\sim}{K}] \left\{\underset{\sim}{T}\right\} + [\underset{\sim}{C}] \left\{\underset{\sim}{\dot{T}}\right\} = \left\{\underset{\sim}{G}_t\right\} \quad \text{(dimension M >> N)} \qquad (13)$$

$[\underset{\sim}{K}]$ being the array of conduction influence coefficients between nodal unknowns, $[\underset{\sim}{C}]$ the one of thermal capacitances (in case of transient situation), $\{\underset{\sim}{T}\}$ the vector collecting the nodal unknowns, $\{\underset{\sim}{\dot{T}}\}$ the one for their time rate of change and $\{\underset{\sim}{G}_t\}$ the one for the total heat inputs on these nodes, part of which being given by Eq. (12). In steady state situations, Eq. (13) reduces to

$$[\underset{\sim}{K}] \{\underset{\sim}{T}\} = \{\underset{\sim}{G}_t\} \quad \text{(dimension M)} \qquad (14)$$

One could think of solving systems (12-13) iteratively in a two-pass sweeping procedure :

(i) assume or compute temperatures (by Eqs. (13) or (14)) for the problem in steady-state or at a given time station (in transient situations) with global thermal loads as if there was no radiation exchange ;

(ii) solve Eq. (12) separetely for radiosities and deduce by Eq. (8) new thermal loads to be added to the global problem ;

(iii) repeat step (i-ii) until thermal equilibrium is reached.

To avoid definition of subsystem (12), we prefer in fact solving problems (12-13 or 12-14) in one-pass by rewritting the whole problem in terms of unknown temperatures only : this is by means of auxiliary temperatures associated with each radiosity node and such that

$$R_i = \sigma(T_i^{aux})^4 \qquad (15)$$

Eq. (8) leads thus to the definition of two types of radiant conductances (Fig. 3) : one for the specialized surface elements, linking the radiant surface nodes to the radiosity nodes, and given by the first part of that equation, i.e.

$$Q_i = \frac{A_i \ \varepsilon_i}{1-\varepsilon_i} \sigma[\bar{T}_i^4 - (T_i^{aux})^4] = (B.C.)_i(\bar{T}_i - T_i^{aux}) \qquad (16)$$

in which appears the temperature-dependent boundary conductance

$$(B.C.)_i = \frac{A_i \ \varepsilon_i}{1-\varepsilon_i} \sigma[\bar{T}_i^2 + (T_i^{aux})^2](\bar{T}_i + T_i^{aux}) \qquad (17)$$

The second is for the N(N-1)/2 mutual exchange elements, coupling all the radiosity nodes together via one-to-one links, and yielded by the second part of Eq. (8)

$$Q_{ij} = A_i \ F_{ij} \ \sigma[(T_i^{aux})^4 - (T_j^{aux})^4] = (L.C.)_{ij} \ (T_i^{aux} - T_j^{aux}) \qquad (18)$$

in which appears the temperature-dependent linking conductance

$$(L.C.)_{ij} = A_i \ F_{ij} \ \sigma[(T_i^{aux})^2 + (T_j^{aux})^2] \quad (T_i^{aux} + T_j^{aux}) \qquad (19)$$

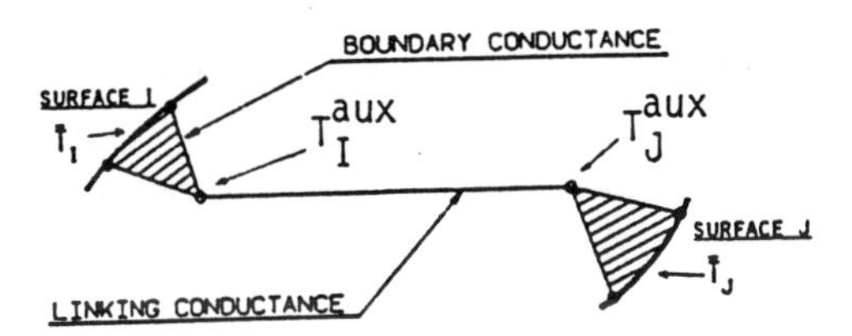

Figure 3. Boundary and linking conductances for radiant exchange

Note that frequent particular cases arising in practical computations are easily taken into account by the present formulation, for instance :

- a black radiant surface ($\varepsilon_i = 1$) corresponds to an infinite (at least very large) B.C., in other words $\bar{T}_i = T_i^{aux}$;

- a radiant source with prescribed temperature T_s (e.g. outer space or sun considered as ponctual radiant sources) corresponds to a black radiant surface with $\bar{T}_i = T_i^{aux} = T_s$;

- an axis of symetry (from both the geometrical and thermal points of view) corresponds to an adiabatic surface which is modelled by a black surface with $Q_i = 0$;

- the case of temperature-dependent emittances in Eq. (17) is straightforward.

Finally, let us stress the fact that the geometric positions of the auxiliary radiosity nodes are quite free ; each of them is however usually associated with the center of the corresponding radiant surface for easy geometrical visualization of the mutual interchanges.

2. VIEW FACTORS COMPUTATION [3-5]

2.1. General

View factors computation is an essential task to be performed in the preceding formulation, though completely separate from the actual thermal computations. We start from the general formula for diffuse gray surfaces obeying to Lambert's law :

$$F_{12} = \frac{1}{\pi A_1} \int_{A_1} \int_{A_2} \frac{\cos\theta_1 \cos\theta_2}{||\overrightarrow{r_{12}}||^2} dA_1 \, dA_2 \qquad (20)$$

in which A_1, A_2 are the areas of the emitting and receiving surfaces respectively,

θ_1, θ_2 the angles between the local normals $\vec{n}_1, \vec{n}_2$ to surfaces 1 and 2, and the ray traced from one surface to the other,

$||\overrightarrow{r_{12}}||$ the length of this ray (fig. 4).

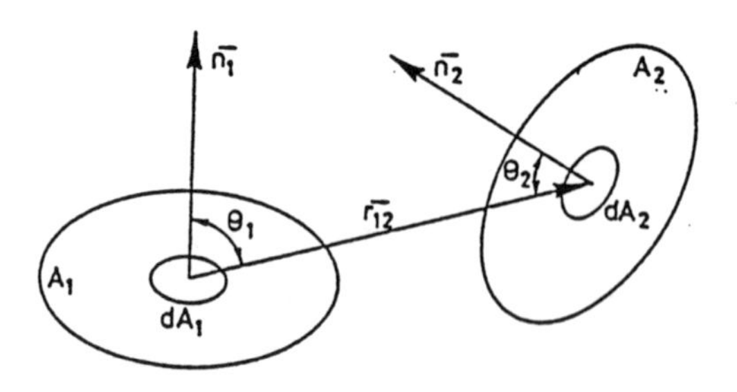

Fig. 4. View factor Data

Many numerical algorithms may be used to perform efficiently and automatically these view factors. We have retained four specific methods in the frame of our developments, each one applying to definite situations. A major criterion for selecting these methods has been the case of implementing them in a FEM context (excluding for instance Monte Carlo ray tracing methods [6] at this stage).

2.2. Gaussian Integration Algorithm

Depending on the required accuracy (or on the affordable computational cost), a number of sampling points are chosen on both surfaces according to Gauss rules, say P_{ij} for A_1 and P_{kl} for A_2. The effective surfaces (maybe with curved boundaries) are in addition transformed in reference unit surfaces with straight edges (triangles or quadrilaterals), on which numerical integration is performed (fig. 5) :

$$A_1F_{12} = \frac{1}{\pi} \sum_{i=1}^{NGX1} \sum_{j=1}^{NGY1} \sum_{k=1}^{NGX2} \sum_{l=1}^{NGY2} W_i\, W_j\, W_k\, W_l\, J_1\, J_2\, \frac{\cos\theta_{ij}\,\cos\theta_{kl}}{r^2_{ijkl}} \qquad (21)$$

in which NGX1,NGX2 are the number of Gauss points in the x direction for reference surface 1 and 2 respectively

NGY1,NGY2 the same for the y direction

W_m the associate weighting functions

J_1 J_2 the appropriate Jacobians for isoparametric transformations

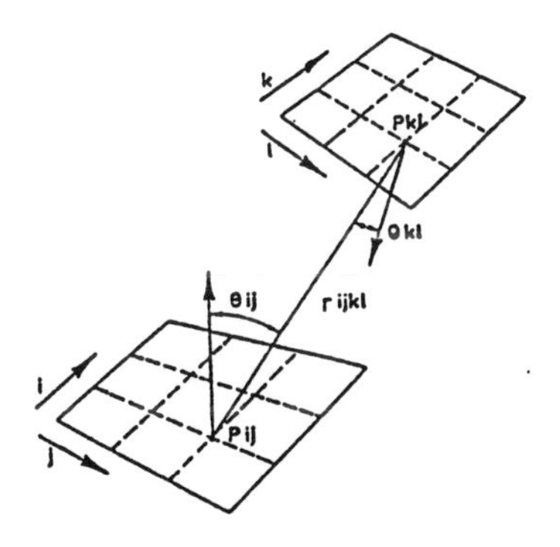

Fig. 5. Gaussian integration rule

The former rule is very effective with regard to computational cost since Gauss points are known to be sampled in that sake. However varying weights make it sensible to the presence of an obstacle between the surfaces and it is obviously not well taylored for close surfaces.

2.3. Double Surface Integration Algorithm

Here numerical integration of (20) is also performed in a similar way to (21), but with regularly spaced integration points, as in a

Newton-Cotes procedure. In addition associated integration weights are constant in order to overcome the problem of screening sensibility. Close surfaces are still difficult to treat.

2.4. Contour integration and Reduced contours integration

Applying stoke's theorem to eq. (20), we get

$$A_1F_{12} = \frac{1}{2\pi} \oint_{A_1} \oint_{A_2} \ln ||\vec{r}|| \, d\vec{s}_1 \cdot d\vec{s}_2 \tag{22}$$

where $d\vec{s}_1$ and $d\vec{s}_2$ are contour segments of A_1 and A_2 respectively. Eq. (22) is actually performed by numerical integration using Gauss points on the two boundaries of the surfaces. The procedure is very effective and works for close surfaces.
If the two surfaces have an edge in common (joining surfaces), a special procedure is required, known as reduced contour integration allowing for a definite kernel in (22) even if $||\vec{r}||$ tends to zero [3,4].

2.5. Selection of a method

The performances of the preceding algorithms are readily very much dependent upon the distance between the interacting surfaces. We state thus a selection criterion based on a reduced distance, i.e.

$$\rho = \frac{\text{min } dist}{\text{max } edge} \tag{23}$$

in which min *dist* denotes the smallest distance in 3-D between the surfaces

and min *edge* denotes the largest edge of the receiving surface.

Very often surfaces are attached to separated parts of structures or components : we denote these individual parts as entities, and each of these entities is enclosed in a definite surrouding box. Criterion (23) is then applied between the emitting box and the receiving one, which results in large computer savings [3,4].

Table 1 summarizes the rules for algorithm selection with respect to the selection criterion values and the presence of screens or not. Standard choices for the number of integration points are also displayed.

Fig. 6 and 7 give the comparison in performances for two benchmark cases with the preceeding algorithms, in the case of close surfaces and for surfaces exhibiting a common edge, respectively.

ρ criterion	Screens Presence	Center-Center ray captation	Partial visibility	Method	Integration points number
$\rho \le 0$	Yes	Yes	Yes/No	-	$F_{ij} = 0$
		No	Yes	Reduced boundary	$N_G = 4$/edge
	No	-	Yes	Double discretization	$N_G = 36$/surface
			No	Reduced boundary	$N_G = 4$/edge
$0 < \rho < 0.8$	Yes	Yes	Yes/No	-	$F_{ij} = 0$
		No	Yes	Double discretization	$N_G = 25$/surface
			No	Boundary	$N_G = 4$/edge
	No	-	Yes	Double discretization	$N_G = 25$/surface
			No	Boundary	$0 \le \rho \le 0.4$ $N_G = 4$/edge $0.4 \le \rho \le 0.8$ $N_G = 3$/edge
$\rho \ge 0.8$	Yes	-	Yes/No	Double discretization	$\rho \le 1.2$ $N_G = 16$/surface
					$\rho \le 5.0$ $N_G = 9$/surface
	No		Yes		$\rho > 5.0$ $N_G = 4$/surface
	No	-	No	Gauss points	$\rho \le 1.2$ $N_G = 9$/surface
					$\rho \le 5.0$ $N_G = 4$/surface
					$\rho > 5.0$ $N_G = 1$/surface

Table 1. Method Selection

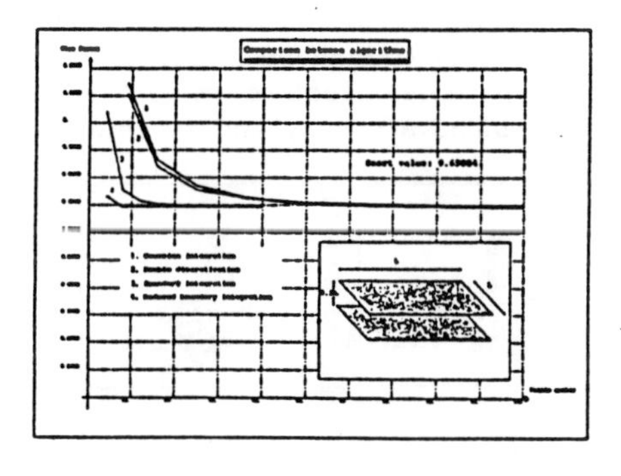

Fig. 6. Two parallel planes

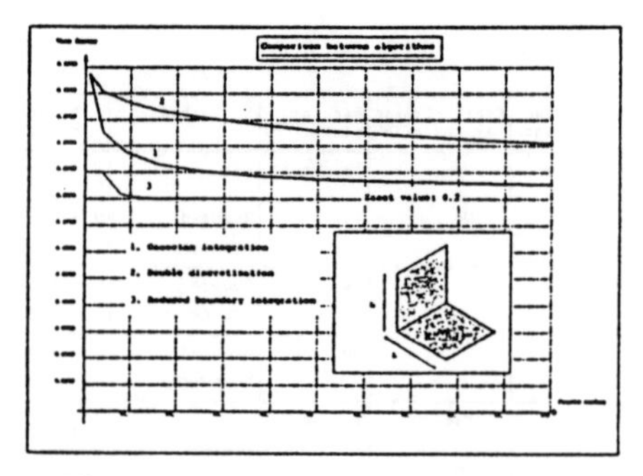

Fig. 7. Two orthogonal planes

3. AN EXAMPLE

As a final example, we treat the case of radiative interaction between a cylinder and a plate (fig. 8)

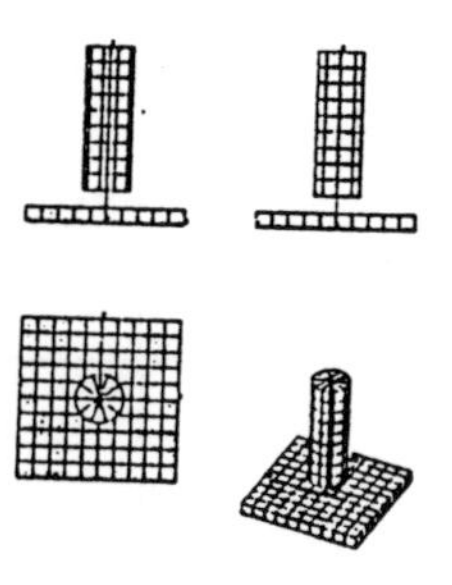

Fig. 8. Cylinder and plate geometrical set up

The discretization parameter is N the number of grids elements taken for the plate (uniform NxN grid) and the cylinder (N axis stations and N angular sectors).

Fig. 9 and 10 display the surface grids, computer cost and radiative links for respectively N = 4, 6, 10 and 20. Computer cost explodes in an exponential manner for fine grids (fig. 11), which is needless to say : view factors computations are definitely a compromise between desired accuracy and acceptable computer cost.

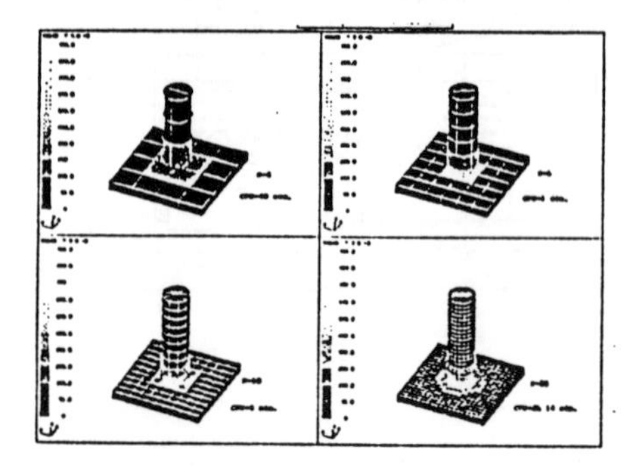

Fig. 9. Cylinder and plate. Grids and computer cost

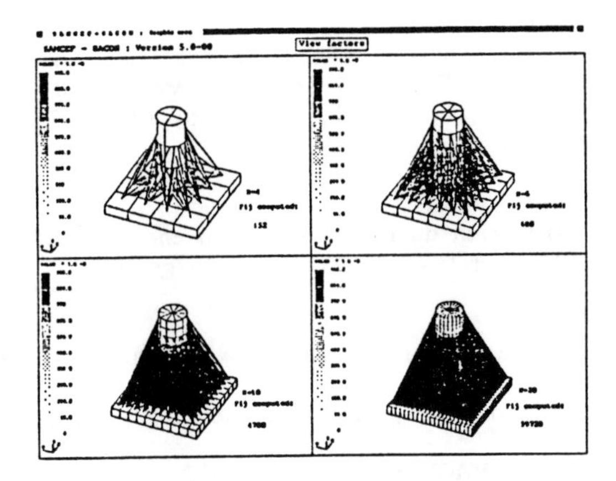

Fig. 10. Cylinder and plate. Radiative links

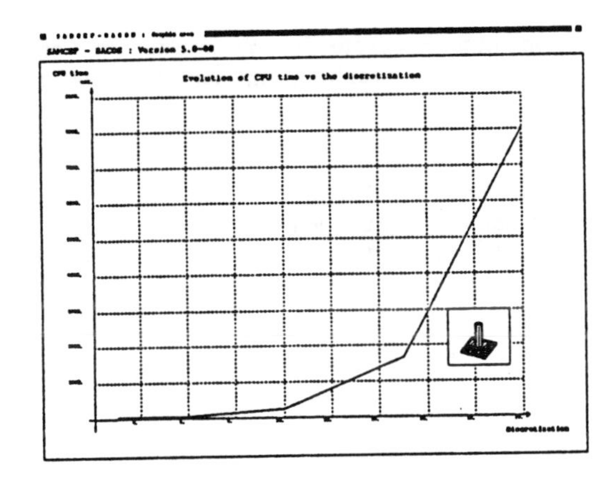

Fig. 11. Cylinder and plate. Evolution of computer cost

4. FUTURE WORK

Intensive developments are currently carried on to reduce the overall cost of view factors computation in large radiation heat transfer models. This is done on the basis of entity selection and interaction, each individual surface pertaining to one entity only.

Since entity classification and visibility comprises the indivual surface ones, effective time savings are easily achieved.

5. REFERENCES

[1] M. HOGGE et al
E.R.M. Thermal Analysis : A Heat Radiation Case Study via the FEM, Proc. 2nd ESA Workshop on Mechanical Technology for Antennas, ESTEC, Noordwijk, 20-22 May 1986, ESA SP-261 (August 1986), 227-235.

[2] P. MAGERMANS et al
Notice d'utilisation du rayonnement en enceintes dans THERNL. Report LTAS N° NT-7, University of Liège, 1988.

[3] Ph. MICHEL
Algorithmes de calcul de facteurs de vue 3-D pour échanges radiatifs.
Travail de fin d'études, Université de Liège, 1990.

[4] Ph. MICHEL, P. MAGERMANS, M. HOGGE
Calcul des facteurs de vue 3-D pour échanges radiatifs.
SAMTECH Report 33, March 1991.

[5] A.F. EMERY
View : A Computer Program for View Factors Computation, Dept. of Mech. Eng. Univ. of Washington, Seattle, Now. 1988.

[6] C. KOEK
Improved Ray Tracing Technique for Radiative Heat Transfer Modelling, Proc. 3rd European Symposium on Space Thermal Control and Life Support Systems, ESTEC, Noordwijk, 3-6 Oct. 1988.

Numerical Solution of Three-Dimensional Radiation Heat Transfer by Flux and Finite Element Methods

A. Bermúdez[I] J.L. Ferrín[I] O. López[I]
E. Fernández[II]

Abstract

In this paper we solve, in the context of modelling pulverized-coal furnaces, the three-dimensional radiation heat transfer equations for a gray medium using the classical Chu and Churchill's six-flux method to discretize the directions of propagation and a finite element method of degree one associated with a tetrahedral mesh of the domain for space discretization. We consider general boundary conditions which lead us to a system similar to that encountered in elasticity and give numerical results obtained in the particular case of diffuse reflection.
Subject Classifications: M.R.: 65N35, 80A20.

1 INTRODUCTION

One of the factors that accounts for the importance of thermal radiation in some applications is the manner in which radiant emission depends on temperature. While for conduction and convection the transfer of energy between two locations depends on the temperature difference of the locations to approximately the first power, the transfer of energy by thermal radiation between two bodies depends on the differences of the individual temperatures of the bodies each raised to a power in the range of about 4 or 5. From this basic difference between radiation and the convection and conduction energy-exchange mechanisms, it is evident

[I]Department of Applied Mathematics. University of Santiago de Compostela. A Coruña. SPAIN.
[II]ENDESA (Spanish National Electric Company). Central of As Pontes. A Coruña. SPAIN.

that the importance of radiation becomes intensified at high temperature levels, which is the case, for instance, of coal-fired boilers of power plants (see [3]).

On the other hand, in the case of combustion chambers we must add to that fact the highly emitting and absorbing nature of components of pulverized-coal flame, as well as the low velocity of gases and small temperature differences at the boundary between the flame and the fouled surfaces of waterwall tubes, which makes the convective component of heat transfer to be comparatively small and usually ignored in engineering calculations (see [2], [4]). Therefore the problem of solving the radiation heat transfer equations for an absorbing, emitting and scattering medium unavoidably arise in this context.

Besides, even when the absorption and scattering coefficients for a gas usually vary strongly with the wavelength, this dependence is week in our case because of the presence of soot particles which enhance and make homogeneous these coefficients over all the spectrum (see [3]). For that reason we will consider a medium where absorption (a) and scattering (σ_s) coefficients are independent of wavelength, what is called a gray medium.

2 EQUATIONS IN Ω

The total energy equation must be coupled to that modelling the evolution of the *spectral radiation intensity* $i(x, \lambda, \omega)$ which is given, for each direction ω in the unit sphere S^2 (see [3]), by

$$\omega \cdot \nabla_x i + (a + \sigma_s) i - \frac{\sigma_s}{4\pi} \int_{S^2} \phi(\omega^*, \omega) i(x, \lambda, \omega^*) \, d\omega^* = a i_b, \qquad (2.1)$$

where i_b is the spectral intensity of radiation from a blackbody.

This equation tells us that variation of i in the direction ω ($\omega \cdot \nabla_x i$) is due to the loss by absorption (ai) and scattering ($\sigma_s i$), and the gain by emission (ai_b) and scattering into ω direction (the integral term).

The scattering coefficient σ_s is a measure of the inverse of the mean free path that a photon will travel before undergoing scattering and $a + \sigma_s$ is the inverse of the radiation mean penetration distance (both strictly true only when neither a nor σ_s vary along the considered path).

The quantity $[\sigma_s/(4\pi)]\phi(\omega^*, \omega)$ expresses the fraction of radiant intensity which comes from the direction ω^* and is scattered into the direction ω. So the equation

$$\frac{\sigma_s}{4\pi} \int_{S^2} \phi(\omega, \omega^*) \, d\omega^* = \sigma_s \quad \forall\, \omega \in S^2 \qquad (2.2)$$

must be verified, and consequently we have

$$\frac{1}{4\pi} \int_{S^2} \phi(\omega, \omega^*) \, d\omega^* = 1 \quad \forall\, \omega \in S^2. \qquad (2.3)$$

Naturally, for isotropic scattering $\phi = 1$

As a and σ_s do not depend on wavelength λ, we can get from (2.1), by integrating over all the spectrum, the equation for the *total radiation intensity* $I(x,\omega)$:

$$\omega \cdot \nabla_x I + (a+\sigma_s)I - \frac{\sigma_s}{4\pi}\int_{S^2}\phi(\omega^*,\omega)I(x,\omega^*)\,d\omega^* = aI_b, \qquad (2.4)$$

where the total intensities I and I_b are defined as

$$I = I(x,\omega) = \int_0^\infty i(x,\lambda,\omega)\,d\lambda, \quad I_b = I_b(x) = \int_0^\infty i_b(x,\lambda)\,d\lambda. \qquad (2.5)$$

We have already said that equation (2.4) must be solved together with the energy equation. In the stationary case considering neither convection nor conduction that equation is

$$\nabla \cdot q_r = f_c, \qquad (2.6)$$

where q_r is the radiant flux vector and f_c the internal heat source.

Coupling between (2.4) and (2.6) lies on the relations

$$I_b = \frac{\sigma T^4}{\pi} \qquad (2.7)$$

$$q_r = \int_{S^2} I\omega\,d\omega, \qquad (2.8)$$

where σ is the Stefan-Boltzmann constant. The first one, which is strictly true only in the vacuum, is known as the Stefan-Boltzmann's radiation law and the second one means that the component i of q_r is $q_{ri} = \int_{S^2} I\omega_i\,d\omega$.

Thanks to (2.8), we have

$$\nabla \cdot q_r = \int_{S^2}\nabla_x\cdot(I\omega)\,d\omega = \int_{S^2}\omega\cdot\nabla_x I\,d\omega = \int_{S^2} a(I_b - I)\,d\omega, \qquad (2.9)$$

where in the last equality we have made use of the fact that I verifies (2.4). Thus we may write the energy equation (2.6) in the following way:

$$4\pi a I_b - a\int_{S^2} I\,d\omega = f_c. \qquad (2.10)$$

From now on, our efforts will be directed towards solving the system (2.4), (2.10) together with certain boundary conditions.

3 BOUNDARY CONDITIONS

In this section we suppose that the emission from the walls takes place in a diffuse fashion, i. e. the emission does not depend on the direction ω. We are going to pose the equations that must be verified on the boundary

Γ in two important cases of reflecting surfaces: specular, in which case reflection depends on the direction ω, and diffuse, where that dependence does not exist.

The specular reflectivity used is assumed to be independent of the incident angle of radiation; that is, the same fraction of incident energy is reflected, regardless of the angle of incidence of energy. In addition, all the surfaces are assumed to have gray properties; that is, radiative properties do not depend on wavelength.

3.1 Specularly reflecting surfaces

In this case, the radiant intensity in each direction ω such that $\omega \cdot n < 0$ (where n is the outward unit normal to the boundary walls) is the sum of the fraction that comes from the incident direction ω_{inc} and is reflected into direction ω plus the emission of the wall:

$$I(x,\omega) = (1-\varepsilon_w)I(x,\omega_{inc}) + \varepsilon_w \frac{\sigma T_w^4}{\pi} \quad \text{on } \Gamma. \tag{3.1}$$

The symbol ε_w stands for the emissivity of the walls, which is a measure of how well they can radiate energy as compared with a blackbody. It is a number between 0 and 1 and is dimensionless. The quantity $1-\varepsilon_w$ is the reflectivity, and expresses the fraction of incident radiation that is reflected. At last, σ is the Stefan-Boltzmann constant and T_w the temperature of the walls.

The incident and reflected directions are related through the following formula:

$$\omega_{inc} = \omega - 2(\omega \cdot n)n. \tag{3.2}$$

3.2 Diffusely reflecting surfaces

In the case of diffusely reflecting surfaces, the reflection does not depend on the direction ω and so $I(x,\omega) = I(x)$ if $\omega \cdot n < 0$. The total amount of energy which comes from the directions ω such that $\omega \cdot n < 0$ is due to the fraction of the total incident energy which is reflected added to the total emission of the walls. This is what the following equation represents:

$$\begin{aligned} I(x)\left|\int_{\omega\cdot n<0} \omega \cdot n \, d\omega\right| &= (1-\varepsilon_w)\int_{\omega\cdot n>0} I(x,\omega)\, \omega\cdot n \, d\omega \\ &+ \varepsilon_w \frac{\sigma T_w^4}{\pi}\left|\int_{\omega\cdot n<0} \omega\cdot n\, d\omega\right| \end{aligned} \tag{3.3}$$

Using the equalities

$$\int_{\omega\cdot n<0} \omega\cdot n \, d\omega = -\pi \quad , \quad \int_{\omega\cdot n>0} \omega\cdot n\, d\omega = \pi, \tag{3.4}$$

we have

$$\pi I(x) = (1-\varepsilon_w)\int_{\omega\cdot n>0} I(x,\omega)\,\omega\cdot n\,d\omega + \varepsilon_w \sigma T_w^4. \tag{3.5}$$

Thanks to the expession (2.8) for q_r we can write

$$q_r\cdot n = \int_{S^2} I(x,\omega)\,\omega\cdot n\,d\omega = I(x)\int_{\omega\cdot n<0}\omega\cdot n\,d\omega + \int_{\omega\cdot n>0} I(x,\omega)\,\omega\cdot n\,d\omega, \tag{3.6}$$

and consequently

$$\int_{\omega\cdot n>0} I(x,\omega)\,\omega\cdot n\,d\omega = q_r\cdot n + \pi I(x). \tag{3.7}$$

Finally, we get from (3.5) that the total intensity $I(x)$ in any direction ω such that $\omega\cdot n<0$ verifies

$$I(x) = \frac{1-\varepsilon_w}{\pi\varepsilon_w}\,q_r\cdot n + \frac{\sigma T_w^4}{\pi} \quad \text{on } \Gamma. \tag{3.8}$$

3.3 The linking between I and T_w on the boundary

Since T_w is unknown, another boundary condition is needed, in any of the previous cases, to relate the total radiant intensity I to the temperature T_w on the boundary. Such an equation comes from the fact that the radiant energy that impinges on the tubes which constitute the walls is absorbed by the water inside them. If T_a is the known temperature of the water, we will take the following law to represent that phenomenon:

$$q_r\cdot n = h(T_w - T_a) \quad \text{on } \Gamma, \tag{3.9}$$

where h is the conduction heat transfer coefficient between the wall tubes and the water inside them.

Thus we will consider boundary conditions (3.1), (3.9) for specular reflection and (3.8), (3.9) for diffuse reflection.

4 SEMIDISCRETIZATION IN ω

For each $x\in\Omega$, we consider an approximation $\hat{I}(x,\cdot)$ of $I(x,\cdot)$ in a six-dimensional space of functions defined on S^2:

$$\hat{I}(x,\cdot)\ \in\ W = \langle w_1,\ldots,w_6\rangle. \tag{4.1}$$

Let $\omega^1=(1,0,0)$, $\omega^2=(-1,0,0)$, $\omega^3=(0,1,0)$, $\omega^4=(0,-1,0)$, $\omega^5=(0,0,1)$ and $\omega^6=(0,0,-1)$. We have chosen the following w_j:

$$w_j(\omega) = \begin{cases} \cos^2\theta_j(\omega) & \text{if } 0\le\theta_j(\omega)\le\pi/2 \\ 0 & \text{if } \pi/2\le\theta_j(\omega)\le\pi \end{cases}, 1\le j\le 6, \tag{4.2}$$

where $\theta_j(\omega)$ is the angle between ω^j and ω.

If we use the notation

$$\begin{array}{lll} I_1^+(x) = \hat{I}(x,\omega^1) & , & I_1^-(x) = \hat{I}(x,\omega^2) \\ I_2^+(x) = \hat{I}(x,\omega^3) & , & I_2^-(x) = \hat{I}(x,\omega^4) \\ I_3^+(x) = \hat{I}(x,\omega^5) & , & I_3^-(x) = \hat{I}(x,\omega^6), \end{array} \tag{4.3}$$

we have

$$\hat{I}(x,\omega) = I_1^+(x)w_1(\omega) + I_1^-(x)w_2(\omega) + \cdots + I_3^+(x)w_5(\omega) + I_3^-(x)w_6(\omega), \tag{4.4}$$

because $w_j(\omega^i) = \delta_{ij}$, $1 \leq i,j \leq 6$.

If we substitute (4.4) in equation (2.4) for the six directions ω^i (i. e. we make **collocation** in ω^i) we get a system consisting of six equations with first-order partial derivatives. The coupling of this system comes from the scattering represented by the integral term which is also the responsible for the presence of the thirty six coefficients

$$d_{ij} = \frac{1}{4\pi}\int_{S^2} \phi(\omega^*,\omega^i)\; w_j(\omega^*)\; d\omega^*. \tag{4.5}$$

Under the assumption that the value of $\phi(\omega^*,\omega)$ depends only upon the angle between ω^* and ω those coefficients are actually three. We use for them the classical notation f (forward), b (backward) and s (sidewise). Explicitly,

$$f = \frac{1}{2}\int_0^{\frac{\pi}{2}} \phi(\theta)\cos^2\theta \sin\theta \; d\theta \tag{4.6}$$

$$b = \frac{1}{2}\int_{\frac{\pi}{2}}^{\pi} \phi(\theta)\cos^2\theta \sin\theta \; d\theta \tag{4.7}$$

$$s = \frac{1}{8}\int_0^{\pi} \phi(\theta)\sin^3\theta \; d\theta, \tag{4.8}$$

and so we have the following property:

$$f + b + 4s = 1, \tag{4.9}$$

which is the discrete version of (2.3). For isotropic scattering, $f = b = s = 1/6$.

The mentioned system of six equations with the coefficients f, b and s is just the one proposed by Chu and Churchill in [1]. The interpretation that we have given here provides us with a good understanding of that method as well as a way for making improvements and generalizations of it.

By adding and substracting two by two those six equations, we can obtain a system with three equations and second-order partial derivatives. With the notations

$$F_i = I_i^+ + I_i^- \;,\; q_i = I_i^+ - I_i^- \; (1 \leq i \leq 3), \tag{4.10}$$

we have

$$q_i = -\beta \frac{\partial F_i}{\partial x_i} \quad (1 \leq i \leq 3), \tag{4.11}$$

where

$$\beta = [a + \sigma_s(1 - f + b)]^{-1}. \tag{4.12}$$

In that way we obtain the system

$$-\frac{\partial}{\partial x_i}\left(\beta \frac{\partial F_i}{\partial x_i}\right) + KF_i - 2\sigma_s s \sum_{\substack{j=1 \\ j \neq i}}^{3} F_j = \frac{2a\sigma T^4}{\pi}, \quad 1 \leq i \leq 3, \tag{4.13}$$

where

$$K = a + \sigma_s(1 - f - b). \tag{4.14}$$

Finally, from the energy equation (2.10), we may conclude that

$$T^4 = \frac{1}{2\sigma}\left(\frac{f_c}{2a} + \frac{\pi}{3}\sum_{i=1}^{3} F_i\right), \tag{4.15}$$

and so remove T from (4.13) to get

$$-\frac{\partial}{\partial x_i}\left(\beta \frac{\partial F_i}{\partial x_i}\right) + KF_i - 2\sigma_s s \sum_{\substack{j=1 \\ j \neq i}}^{3} F_j - \frac{a}{3}\sum_{j=1}^{3} F_j = \frac{1}{2\pi} f_c, \quad 1 \leq i \leq 3, \tag{4.16}$$

a system in second-order partial derivatives with three equations and three unknowns ($F_i(x)$, $1 \leq i \leq 3$). The problem will be closed with the semidiscretized boundary conditions.

Remark. *If we substitute $\hat{I}$ in the expression (2.8) for q_r and then calculate $\nabla \cdot q_r$ the system changes slightly, although it seems to be as good approximation as the one presented here.*

Remark. *If the energy equation is not so simple (for instance, if convection or conduction are not negligible) we cannot obtain T as a function of the fluxes F_i and we must solve simultaneously the energy equation together with (4.13).*

5 SEMIDISCRETIZED BOUNDARY CONDITIONS

At each point on Γ we will have information about the heat exchange between the wall and the water through the expression (3.9). If we substitute $\hat{I}$ in the expression (2.8) for q_r and then operate we obtain that

$$q_r = \frac{\pi}{2}(q_1, q_2, q_3). \tag{5.1}$$

Then we have

$$q_r \cdot n = -\frac{\pi}{2}\beta \sum_{i=1}^{3} \frac{\partial F_i}{\partial x_i} n_i, \tag{5.2}$$

thanks to (**4.11**), and (**3.9**) turns into

$$-\frac{\pi}{2}\beta\sum_{i=1}^{3}\frac{\partial F_i}{\partial x_i}n_i = h(T_w - T_a) \text{ on } \Gamma. \tag{5.3}$$

On the other hand, we will also have information about the total intensity in the directions ω^i $(1 \leq i \leq 6)$ such that $\omega^i \cdot n < 0$. Those directions are one, two or three depending on how many of the components of n are not zero. In general, we will have information about $I_i^{-sg(i)}$ if $n_i \neq 0$, where

$$sg(i) = \begin{cases} + & \text{if } n_i > 0 \\ - & \text{if } n_i < 0 \end{cases}, 1 \leq i \leq 3. \tag{5.4}$$

From (**4.10**) and (**4.11**) we get

$$I_i^+ = \frac{1}{2}\left(F_i - \beta\frac{\partial F_i}{\partial x_i}\right) \quad , \quad I_i^- = \frac{1}{2}\left(F_i + \beta\frac{\partial F_i}{\partial x_i}\right), \tag{5.5}$$

and so

$$I_i^{sg(i)} = \frac{1}{2}\left(F_i - \frac{n_i}{|n_i|}\beta\frac{\partial F_i}{\partial x_i}\right). \tag{5.6}$$

We will only show the case of diffusely reflecting surfaces.

For those $i \in \{1,2,3\}$ such that $n_i \neq 0$, continuous boundary condition (**3.8**) becomes, using (**5.2**) and (**5.3**),

$$I_i^{-sg(i)} = \frac{1-\varepsilon_w}{\pi\varepsilon_w}\, h(T_w - T_a) + \frac{\sigma T_w^4}{\pi}, \tag{5.7}$$

or, in terms of F_i,

$$F_i + \frac{n_i}{|n_i|}\beta\frac{\partial F_i}{\partial x_i} = 2\,\frac{1-\varepsilon_w}{\pi\varepsilon_w}\, h(T_w - T_a) + 2\,\frac{\sigma T_w^4}{\pi}. \tag{5.8}$$

For diffuse boundaries, the system (**4.16**) is solved together with the boundary conditions (**5.3**) and (**5.8**).

6 SOLVING THE SEMIDISCRETIZED PROBLEM

In order to solve the semidiscretized problem (**4.16**) together with the boundary conditions (**5.3**) and (**5.8**) we first make a finite element discretization and then propose an iterative algorithm.

6.1 Finite element discretization

In a classic way, multiplying each equation of system (4.16) by a test function v_i and using the Green's formula to take into account the boundary condition (5.8) we get the following Galerkin formulation of the problem: " Find F_i, $1 \leq i \leq 3$, such that

$$\int_\Omega \beta \sum_{i=1}^{3} \frac{\partial F_i}{\partial x_i}\frac{\partial v_i}{\partial x_i}\, dx + \int_\Omega [K \sum_{i=1}^{3} F_i v_i - 2\sigma_s s \sum_{i=1}^{3}(\sum_{\substack{j=1\\ j\neq i}}^{3} F_j) v_i$$

$$-\frac{a}{3}\sum_{i=1}^{3} F_i \sum_{i=1}^{3} v_i] dx + \int_\Gamma \sum_{i=1}^{3} F_i v_i |n_i|\, d\Gamma = \frac{1}{2\pi}\int_\Omega f_c \sum_{i=1}^{3} v_i\, dx \tag{6.1}$$

$$+\int_\Gamma [2\,\frac{1-\varepsilon_w}{\pi\varepsilon_w}\, h(T_w - T_a) + 2\,\frac{\sigma T_w^4}{\pi}] \sum_{i=1}^{3} v_i |n_i|\, d\Gamma \text{ ''},$$

which can be solved if we know T_w. To approximate (6.1) we use lagrangian finite elements of degree one associated with a tetrahedral mesh of the domain.

Since T_w is unknown, we need another equation to close the system. This can be obtained thanks to (5.3) and (5.8), from which we deduce that T_w verifies on Γ the *nonlinear* equation

$$\frac{\pi}{2}\sum_{i=1}^{3} F_i|n_i| = \sigma T_w^4 \sum_{i=1}^{3} |n_i| + h(T_w - T_a)\left(\frac{1-\varepsilon_w}{\varepsilon_w}\sum_{i=1}^{3}|n_i| + 1\right). \tag{6.2}$$

6.2 Iterative algorithm

We have solved the problem (6.1), (6.2) using the following iterative algorithm:

Step 0.- The problem (6.1) is solved for a given initial value of T_w, T_w^0. In this way we obtain F_i^0, $1 \leq i \leq 3$.

Step n.- We know F_i^{n-1}, $1 \leq i \leq 3$ and calculate T_w^n via (6.2), which is solved by Newton's method. Then we obtain F_i^n, $1 \leq i \leq 3$ from (6.1).

The problem with specular reflection can be solved in an analogous way.

7 NUMERICAL RESULTS

We present here numerical results in the case of a furnace whose shape is the one that can be seen in the figures and for the following data:

a=0.01	σ_s=0.02	σ=5.6696d-8
f=0.25	b=0.25	s=0.125
ε_w=0.7	h=260	T_a=573.15

$$f_c = \sum_{j=1}^{5} f_{cj}$$

where

$$f_{cj}(x) = \frac{A_j}{(4\pi r_j)^{1.5}} e^{-\frac{|x-p_j|^2}{4r_j}}$$

and

A_1=5.098d5 r_1=3.5 p_1=(.7845, .7858, 1.38885)
A_2=5.091d5 r_2=3.5 p_2=(.7845, .7858, 1.57135)
A_3=5.010d5 r_3=3.5 p_3=(.7845, .7858, 1.90635)
A_4=4.000d5 r_4=3.5 p_4=(.7845, .7858, 2.06385)
A_5=4.000d5 r_5=3.5 p_5=(.7845, .7858, 2.39485)

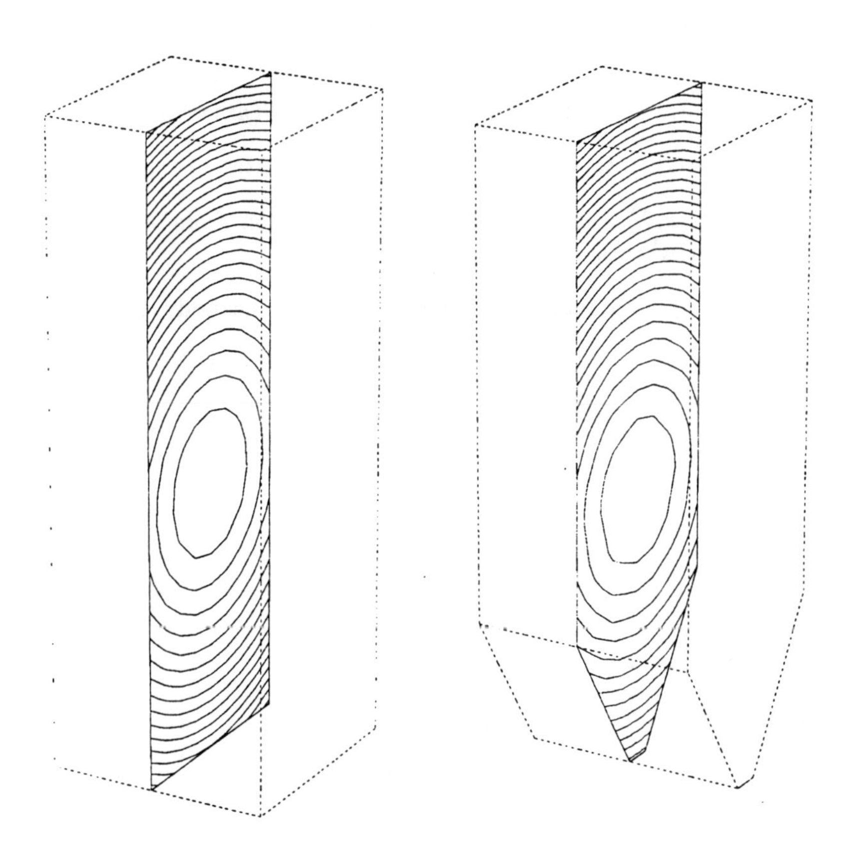

Figure 1: isothermal lines.

Figure 2: isothermal lines.

The domain Ω in the first example is a rectangular prism of $1.5690 \times 1.5716 \times 4.3485$.

We have used a finite element mesh with 8448 tetrahedrons and 1863 vertices for the first example, and 5400 tetrahedrons and 1178 vertices for the second one. Figures 1 and 2 show the corresponding isothermal lines.

In both cases, convergence was achieved after 8 iterations. The computing times were 555s and 325s, respectively, on a IBM RISC 6000/320.

Results for the first example are in good agreement with those obtained in [5] by using an "alternate directions" method. However the computing time for the latter was 23 seconds only.

8 NOMENCLATURE LIST

- $a = a(x)$: absorption coefficient $[m^{-1}]$
- $f_c = f_c(x)$: internal heat source $[W/m^3]$
- $h = h(x)$: conduction heat transfer coefficient between the wall tubes and the water inside them $[W/(m^2\,K)]$
- $i = i(x, \lambda, \omega)$: spectral radiation intensity $[W/(\mu m\, m^2\, str)]$
- $I = I(x, \omega)$: total radiation intensity $[W/(m^2\, str)]$
- $i_b = i_b(x, \lambda)$: spectral intensity of radiation from a blackbody $[W/(\mu m\, m^2\, str)]$
- $I_b = I_b(x)$: total intensity of radiation from a blackbody $[W/(m^2\, str)]$
- $n = n(x) = (n_1, n_2, n_3)$: outward unit normal to the boundary walls
- $q_r = q_r(x)$: radiant flux vector $[W/m^2]$
- $S^2 = \{\omega \in \mathbb{R}^3 : \|\omega\|_2 = 1\}$: unit sphere
- $T = T(x)$: temperature inside the domain $[K]$
- $T_a = T_a(x)$: temperature of the water circulating inside the wall tubes $[K]$
- $T_w = T_w(x)$: temperature of the walls $[K]$
- $\Gamma = \partial\Omega$: boundary of the domain
- $\varepsilon_w = \varepsilon_w(x)$: radiant emissivity of the walls
- λ: wavelength $[\mu m]$
- σ: Stefan-Boltzmann constant $[5.6696 \times 10^{-8}\ W/(m^2\, K^4)]$

- $\sigma_s = \sigma_s(x)$: scattering coefficient $[m^{-1}]$
- $\phi = \phi(\omega^*, \omega)$: phase function for scattering
- Ω: spatial domain (an open and connected set)

References

[1] CHU, C.M., CHURCHILL, S.W. - *Numerical Solution of Problems in Multiple Scattering of Electromagnetic Radiation.* J. Physical Chemistry 59, pp. 855-863. 1955.

[2] SMOOT, L.D., PRATT, D.T. -*Pulverized-Coal Combustion and Gasification*, Plenum Press, New York, 1979.

[3] SIEGEL, R., HOWELL, J.R. - *Thermal Radiation Heat Transfer*, Hemisphere Publishing Corporation, New York, 1981.

[4] BLOKH, A.G. - *Heat Transfer in Steam Boiler Furnaces*, Hemisphere Publishing Corporation, New York, 1988.

[5] BERMÚDEZ, A., LÓPEZ, O. -*Análisis de un método de flujos para el cálculo de la radiación térmica*, Actas del XII Cedya, pp. 381-386, Oviedo, 1991.

RADIATION TRANSFER IN A TWO-LAYER PLANNAR SLAB WITH FRESNEL'S BOUNDARY

J.S.Chiou* and J.S.Liu**

*Associate professor **Graduate student

Department of Mechanical Engineering
National Cheng Kung University Taiwan

SUMMARY

This study investigates the steady state radiative transfer in an azimuthally symmetric two-layer plannar slab with Fresnel's boundaries. The radiative transfer equation is solved by the discrete ordinate method developed by Fiveland [1]. In applying that method the S-N ordinate sets were extended and modified to solve the problems that the refractive index of both layers is different from unity or one layer is different from the other.

The result indicates that the key parameter to influence the hemispherical transmissivity is the optical thickness for isotropic scattering material and may be the refractive index for anisotropic scattering material.

1. INTRODUCTION

Each kind of semitransparent material has its own physical strength and optical characteristics such as refractive index, absorbing and scattering coefficients etc. In certain circumstance, it is necessary to combine two or more layers of different materials to form a composite that can poccess the desired overall properties. In order to take full advantage of all attractive features of composite material, it is important to have a good understanding of their energy transfer characteristics.

Many studies were conducted on the calculation of hemispherical transmissivity and reflectirity of a one-dimensional two-layer plannar slab. Özisik and Shouman [2] applied the source function expansion technique to solve the radiative heat transfer in a two-layer slab with isotropic scattering and specularly reflecting boundaries. In their analysis the refractive indexes of both layers are assumed equal to one. Clements and Özisik [3] investigated the effect due to the variation of single scattering albedo for the similar problem. Swathi et el. [4] analyizied a two-layer composite porous media by the P-11 approximation. They employed the linear-anisotropic scattering phase function to show that the effect due to anisotropic scattering on the hemispherical reflectance was not negligible. Again, the refractive index was considered as unity and the slab had specularly and diffusely reflecting boundaries. The study considering the directional dependent reflectivity at the boundary was performed by Cengel and Özisik [5] for a single layer slab.

In this study, a one-dimensional two-layer slab with Fresnel's boundaries is investigated. The refractive index of first layer may be equal to, greater than, or smaller than that of second layer.

2. ANALYSIS

The radiative transfer propagates across a two-layer plannar slab is scketched in Fig. 1.

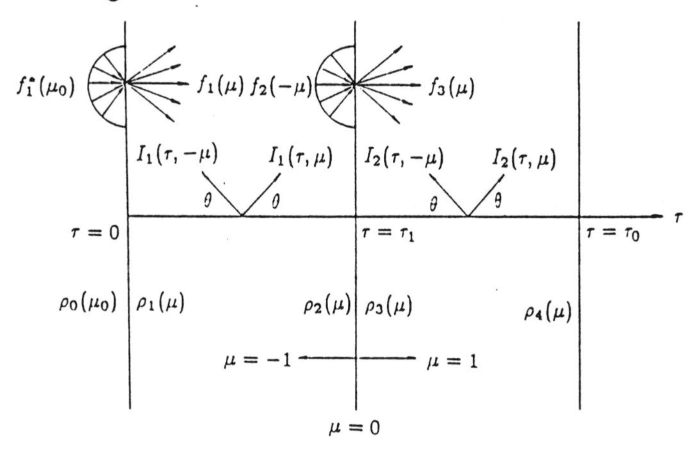

Fig 1 The diagram of a two-layer slab

The incident radiation on the left boundary is assumed diffuse, and the medium is considered as a grey body which absorbs, scatters isotropically or anisotropically but doesn't emitt radiation. The physical properties of the medium are assumed constant and homogeneous, and the interfaces are assumed to reflect according to Fresnel's law. Based on these assumptions, the radiative transfer equation can be written as

$$\mu_1 \frac{\partial I_1(\tau,\mu_1)}{\partial \tau} + I_1(\tau,\mu_1) = \frac{\omega_1}{2}\int_{-1}^{1} p(\mu_1,\mu^a) I_1(\tau,\mu^a) d\mu^a \qquad 0 \le \tau \le \tau_1 \ldots\ldots\ldots (1)$$

$$\mu_2 \frac{\partial I_2(\tau,\mu_2)}{\partial \tau} + I_2(\tau,\mu_2) = \frac{\omega_2}{2}\int_{-1}^{1} P(\mu_2,\mu^a) I_2(\tau,\mu^a) d\mu^a \qquad \tau_1 \le \tau \le \tau_0 \ldots\ldots\ldots (2)$$

where I is the radiation intensity, ω is the single-scattering albedo, μ is the cosine of the angle between the radiant direction and the direction normal to the planner surface. μ^a is only a dummy variables. τ is the optical variable, $p(\mu, \mu^a)$ is the scattering phase function, and the subscripts 1 and 2 represent layer 1 and layer 2 respectively.

The boundary conditions that allow for Fresnel's reflection and their corresponding interface transmission are taken as

$$I_1(0,\mu_1) = f_1(\mu_1) + \rho_1(\mu_1)I_1(0,-\mu_1) \qquad \text{for } n_1 < n_2$$
$$f_1(\mu_1) = (\frac{n_1}{1})^2[1-\rho_0(\mu_0)]f_1^*(u_0) \quad \text{(3a)}$$

$$I_1(0,\mu_1) = f_1(\mu_1) + \rho_1(\mu_1)I_1(0,-\mu_1) \qquad \text{for } n_1 \geq n_2$$
$$f_1(\mu_1) = (\frac{n_1}{1})^2[1-\rho_0(\mu_0)]f_1^*(\mu_0) \quad \text{(3b)}$$

$$I_1(\tau_1 - \mu_1) = f_2(-\mu_1') + \rho_2(\mu_1)I_1(\tau_1,\mu_1) \qquad \text{for } n_1 < n_2$$
$$f_2(-\mu_1') = (\frac{n_1}{n_2})^2[1-\rho_3(\mu_2)]I_2(\tau_1,-\mu_2) \quad \text{(3c)}$$

$$I_1(\tau_1,-\mu_1) = f_2(-\mu_1) + \rho_2(\mu_1)I_1(\tau_1,\mu_1) \qquad \text{for } n_1 \geq n_2$$
$$f_2(-\mu_1) = (\frac{n_1}{n_2})^2[1-\rho_3(\mu_2')]I_2(\tau_1,-\mu_2) \quad \text{(3d)}$$

$$I_2(\tau_1,\mu_2) = f_3(\mu_2) + \rho_3(\mu_2)I_2(\tau_1,-\mu_2) \qquad \text{for } n_1 < n_2$$
$$f_3(\mu_2) = (\frac{n_2}{n_1})^2[1-\rho_2(\mu_1')]I_1(\tau_1,\mu_1') \quad \text{(3e)}$$

$$I_2(\tau_1,\mu_2) = f_3(\mu_2') + \rho_3(\mu_2)I_2(\tau_1,-\mu_2) \qquad \text{for } n_1 \geq n_2$$
$$f_3(\mu_2') = (\frac{n_2}{n_1})^2[1-\rho_2(\mu_1)]I_1(\tau_1,\mu_1) \quad \text{(3f)}$$

$$I_2(\tau_0,-\mu_2) = \rho_4(\mu_2)I_2(\tau_0,\mu_2) \qquad \text{for } n_1 < n_2 \quad \text{(3g)}$$
$$I_2(\tau_0,-\mu_2) = \rho_4(\mu_2)I_2(\tau_0,\mu_2) \qquad \text{for } n_1 \geq n_2 \quad \text{(3h)}$$

where n_1 and n_2 are the absolute refractive indexes of the first and second layers. If $n_1 = n_2$, then f_2 and f_3 vanish automatically. The reflectivity $\rho_i(\mu)$ is calculated by the Fresnel's law, which defines n_i as a relatire index, when $n_i \geq 1$

$$S = \sqrt{n_i^2 - (1-\mu^2)}$$

$$\rho_i = \frac{1}{2}\left[\left(\frac{s-\mu}{s+\mu}\right)^2 + \left(\frac{s-\mu n_i^2}{s+\mu n_i^2}\right)^2\right] \quad \text{(4a)}$$

and when $n_i < 1$, the cosine of critical incident angle $\mu_{cr} = \sqrt{1-n_i^2}$, and

$$\begin{cases} \rho_i = \frac{1}{2}\left[\left(\frac{s-\mu}{s+\mu}\right)^2 + \left(\frac{s-\mu n_i^2}{s+\mu n_i^2}\right)^2\right] & \text{if } \mu > \mu_{cr} \\ \rho_i = 0 \quad \text{if} \quad \mu \leq \mu_{cr} \end{cases} \quad \text{(4b)}$$

n_0 and n_4 denote the environmental refractive indexes outside the two-layer slab, their values are less than n_1 and n_2. In calculating ρ_0, ρ_1, ρ_2, ρ_3, and ρ_4, the n_i are set to be $\frac{n_1}{n_0}$, $\frac{n_0}{n_1}$, $\frac{n_2}{n_1}$, $\frac{n_1}{n_2}$, and $\frac{n_4}{n_2}$ respectively.

The direction of incident radiation after refraction is determined by Snell's law

$$\frac{\sin\theta_i}{\sin\theta_r} = \frac{\sqrt{1-u_i^2}}{\sqrt{1-\mu_r^2}} = \frac{n_r}{n_i} \quad \text{(5)}$$

where the subscripts i and r respectively represent the incident and refraction radiations. n_r is the absolute refractive index of the medium that contains the refracted beam and n_i is that contains the incident beam.

The hemispherical reflectivity, R_1, and transmissivity, T, of the two-layer slab medium are defined as

$$R = \frac{q_{reflected}}{q_{incident}} \quad \text{and} \quad T = \frac{q_{transmitted}}{q_{incident}} \quad \text{(6)}$$

$$q_{incident} = 2\pi \int_0^1 f_1^*(\mu_0)\mu_0 d\mu_0 \quad f_1^*(\mu_0) = 1 \quad \text{(7)}$$

$$q_{ref} = 2\pi \left\{ \int_0^1 \rho_0(\mu_0) f_1^*(\mu_0)\mu_0 d\mu_0 + \int_0^1 [1-\rho_1(\mu)] I_1(0,-\mu)\mu d\mu \right\} \quad \text{(8)}$$

$$q_{trans} = 2\pi \int_0^1 [1-\rho_4(\mu)] I_2(\tau_0,\mu)\mu d\mu \quad \text{(9)}$$

3. NEMERICAL METHOD

The governing equations (1), (2) are solved by the discrete ordinate method developed by Fiveland [1], called S-N method. Eqs (1), (2) can be expressed as

$$\mu_m \frac{I_{1,j+1}^m - I_{1,j}^m}{\Delta\tau} + \frac{1}{2}(I_{1,j+1}^m + I_{1,j}^m) = \frac{\omega_1}{2}\int_{-1}^1 p(\mu_1,\mu^a) I_1(\tau,\mu^a) d\mu^a \qquad j = 1,2,...,N_{s1-1} \quad \text{(10)}$$

$$\mu_m \frac{I_{2,j+1}^m - I_{2,j}^m}{\Delta\tau} + \frac{1}{2}(I_{2,j+1}^m + I_{2,j}^m) = \frac{\omega_2}{2}\int_{-1}^1 p(\mu_2,\mu^a) I_2(\tau,\mu_a) d\mu^a \qquad j = 1,2,...,N_{s2-1} \quad \text{(11)}$$

For forward propagation ($\mu > 0$)

$$I_{i,j+1}^m = \frac{(\frac{\mu_m}{\Delta\tau} - \frac{1}{2}) I_{i,j}^m + \frac{\omega_i}{2}\sum_{k=1}^{NQ} P_{mk} W_k (\frac{I_{i,j+1}^k + I_{i,j}^k}{2})}{(\frac{\mu_m}{\Delta\tau} + \frac{1}{2})} \quad \text{(12)}$$

$$i = 1 \text{ or } 2 \qquad m = N1, N1+1, ..., NQ$$

For backward propagation ($\mu < 0$)

$$I_{i,j}^m = \frac{(\frac{\mu_m}{\Delta\tau} + \frac{1}{2}) I_{i,j+1}^m - \frac{\omega_i}{2}\sum_{k=1}^{NQ} P_{mk} W_k (\frac{I_{i,j+1}^k + I_{i,j}^k}{2})}{(\frac{\mu_m}{\Delta\tau} - \frac{1}{2})} \quad \text{(13)}$$

$$i = 1,2 \qquad m = 1,2,...,N1-1$$

where NQ is the total number of direction μ $(-1 \leq \mu \leq 1)$ and $N1 = NQ/2 + 1$. N_{S1} and N_{S2} are the number of spacial grid in the first and second layers. The anisotropic scattering phase function is expressed as

$$P(\mu, \mu^a) = \sum_{n=0}^{N} a_n P_n(\mu) P_n(\mu^a) \quad \text{(14)}$$

where $P_n(\mu)$ and $P(\mu^a)$ are the Legendre polynomials of order n, and a_n are constant coefficients, as shown in Table 1, which are obtained from reference [6].

n	Forward scattering a_n	Backward scattering a_n
0	1.0	1.0
1	1.98398	- 0.56524
2	1.50823	0.29783
3	0.70075	0.08571
4	0.23489	0.01003
5	0.05133	0.00063
6	0.00760	0.0
7	0.00048	
8	0.0	

Table 1 The coefficients of anisotropic scattering phase function.

3.1 Determination of ordinate direction μ

In Five land's paper [1], the S-N method was developed for a single medium. The accuracy of integral approximation could be satisfied even the moments of intensity were less than 12. Therefore, the weighting factors and point values of S-N method listed in his paper were up to S-12.

In this study, the radiative transfer propagates through mediums with different refrative indexes. S-12 is obviously not good anymore. For example, if the refractive indexes of both layers are 1.5, the cosine of critical angle for the layers is $\mu_{cr} = \sqrt{1 - \frac{1}{(1.5)^2}} = 0.74536$. From Table 2, it shows there is only one direction corrdinate $(\mu = 0.933123)$ greater than μ_{cr} if S-12 is used, the approximation of positive radiation by a single moment is definitely very poor. To overcome this drawback, we increase the total number of direction up to 18, the weighting factors and point values of S-18 are also listed in Table 2.

The obtained results by the use of S-18 for a special case $(\omega_1 = \omega_2 = 1.0)$ were still not satisfied, see Table 3, in which the sum of reflectivity R and transmissivity T should be 1.0, since the mediums are nonabsorbing.

S-N	Weighting factor	$\pm\mu$	S-N	Weighting factor	$\pm\mu$
S-12	1/6	0.066877	S-18	1/9	0.044205
	1/6	0.288732		1/9	0.199491
	1/6	0.366693		1/9	0.235619
	1/6	0.633307		1/9	0.416047
	1/6	0.711267		1/9	0.500000
	1/6	0.933123		1/9	0.583953
				1/9	0.764381
				1/9	0.800509
				1/9	0.955795

Table 2 Weight and point values for S-N method

τ_0	R	T	$R+T$
0.1	0.22628	0.81221	1.03848
1.0	0.44345	0.59504	1.03848
5.0	0.71809	0.32039	1.03848

Table 3 The results before making corrdinate transformation

Fortunately, the accuracy can be greatly improved after we performed the coordinate transformation. Fig. 2 shows the original S-12 spreads in the range of $-1 \leq \mu \leq 1$ was transformed into four new ranges, namely $-1 \leq \mu < -\mu_{cr}$, $\mu_{cr} \leq \mu < 0$, $0 \leq \mu < \mu_{cr}$, and $\mu_{cr} < \mu \leq 1$. In the new domain, each range has 12 monents of intensity, so the total direction ordinate is 48.

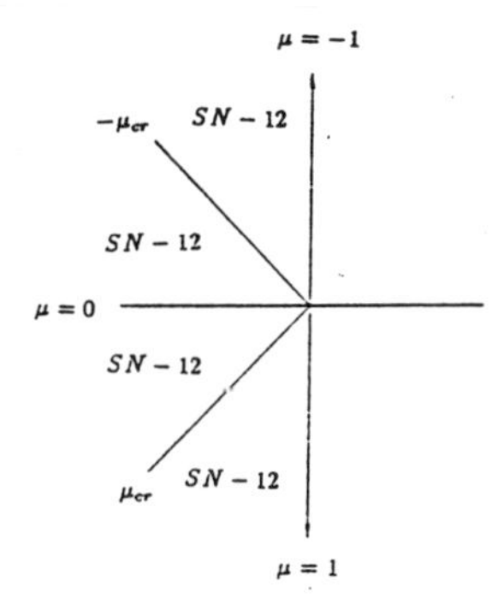

Fig 2 The coordinate transformation of S-N method

3.2 The procedure of coordinate transformation

Let $x = a + bt$ then $dx = bdt$, the function in the new coordinate can be expressed as

$$\int_A^B f(x)dx = b\int_{-1}^1 f(a+bt)dt \quad \text{......} (15)$$

where t is the original S-N qvadrature points as listed in Table 1, where $-1 \le t \le 1$. From (15), we can obtain

$$\begin{aligned}\int_0^{u_{cr}} f(x)dx &= \frac{\mu_{cr}}{2}\int_{-1}^1 f(\frac{\mu_{cr}}{2}+\frac{\mu_{cr}}{2}t)dt \\ &= \frac{\mu_{cr}}{2}\sum_{i=1}^N w_i \cdot f(\frac{\mu_{cr}}{2}+\frac{\mu_{cr}}{2}t)\end{aligned} \quad \text{......} (16)$$

$$\begin{aligned}\int_{\mu_{cr}}^1 f(x)dx &= \frac{1-\mu_{cr}}{2}\int_{-1}^1 f(\frac{1+\mu_{cr}}{2}+\frac{1-\mu_{cr}}{2}t)dt \\ &= \frac{1-\mu_{cr}}{2}\sum_{i=1}^N w_i \cdot f(\frac{1+\mu_{cr}}{2}+\frac{1-\mu_{cr}}{2}t)\end{aligned} \quad \text{......} (17)$$

where μ_{cr} is the cosine of the critical angle of incidence

N is the total number of S-N points from -1 to 1

w_i is the weighting factor for SN points

For the case of $n_1 > n_2$, $\mu_{cr1} = \sqrt{1-(n_0/n_1)^2}$ then

$$\mu_1 = \frac{\mu_{cr1}}{2} + \frac{\mu_{cr1}}{2}t \quad 0 \le \mu_1 < \mu_{cr1} \quad \text{......} (18)$$

$$\mu_1 = \frac{1+\mu_{cr1}}{2} + \frac{1-\mu_{cr1}}{2}t \quad \mu_{cr1} < \mu_1 \le 1 \quad \text{......} (19)$$

Eqs (18) and (19) are applied for the first layer, the same formula are good for the second layer if μ_1 and μ_{cr1} are replaced by μ_2 and μ_{cr12}, $\mu_{cr12} = \sqrt{1-(n_1/n_2)^2}$.

For the case of $n_1 > n_2$, both μ_{cr1} and μ_{cr12} exist in the first layer. We simply set $\mu_2 = \mu_1$. In the case of $n_1 = n_2$, μ_{cr12} does not exist. μ_2 is thus equals to μ_1.

The transformation for x from -1 to 0 is similar to equations (18) and (19). After the coordinate transformation, the calculated R and T compare very well with that obtained by Cengel [5] for the isotropic scattering case with $n_1 = n_2 = 1.5$, see Table 4.

τ	ω_1	ω_2	R	T	R^*	T^*
$\tau_1 = 0.05$	.0	.0	.141684	.752503	.14137	.75208
$\tau_2 = 0.05$	1.0	1.0	.195031	.805828	.19466	.80535
$\tau_1 = 0.5$	.0	.0	.097579	.270671	.09756	.27060
$\tau_2 = 0.5$	1.0	1.0	.420749	.580109	.42033	.57967
$\tau_1 = 2.5$	.0	.0	.091779	.003188	.09178	.00319
$\tau_1 = 2.5$	1.0	1.0	.687895	.312963	.68723	.31277

* the solution of cengel [5]

Table 4 The results after making coordinates transformation

3.3 Calculation for f_1, f_2, and f_3

The incident radiation to the left boundary from outside, $f_1^*(\mu)$, is assumed diffuse. As presented by Congel [5], the relation between $f_1(\mu)$ and $f_1^*(\mu)$ can be expressed by

$$f_1(\mu) = (\frac{n_1}{1})^2[1-\rho_0(\mu_0)]f_1^*(\mu_0) \qquad \text{for } \mu > \mu_{cr1} \quad \text{(3a, 3b)}$$
$$f_1(\mu) = 0 \qquad \text{for } \mu \le \mu_{cr1}$$

where $u_{cr1} = \sqrt{1-(\frac{1}{n_1})^2}$, $f_1^*(\mu_0) = 1.0$ for all μ_0 (diffuse assumption).

In the calculation of $f_3(\mu)$, two cases are considered.Firstly, $n_1 < n_2$ (the medium of second layer is denser than that of first layer), so there exists a critical angle in the denser medium, $\mu_{cr12} = \sqrt{1-(n_1/n_2)^2}$. it is easy for the case of $\mu \le \mu_{cr12}$, since $f_3(\mu) = 0$ due to the total reflection. when $\mu > \mu_{cr12}$, it seems straight forward to obtain $f_3(\mu)$ from Eq. (3e)

$$f_3(\mu_2) = (\frac{n_2}{n_1})^2[1-\rho_2(\mu_1^{'})]I_1(\tau_1,\mu_1^{'}) \quad \text{(3e)}$$

Because $\rho_2(\mu_1^{'})$ can be calculated from Fresnel's law, Eq. (4b). However, $\mu_1^{'}$ calculated by Snell's law, $\mu_1^{'} = \sqrt{1-(1-\mu^2)(n_2/n_1)^2}$ is not located at the coordinate we had set for μ_1 in the calculation of $f_1(\mu_1)$. There fore, $I_1(\tau_1,\mu^1)$ can not be directly obtained from the S-N method. In this study, the value of $I_1(\tau_1,\mu^{'})$ is thus obtained from the cubic spline interpolation on $I_1(\tau_1,\mu)$.

In the other case, when $n_1 > n_2$, the medium of first layer is denser than that of second layer. There exists another critical angle (beside μ_{cr1}) in the first medium, $\mu_{cr12} = \sqrt{1-(n_1/n_1)^2}$, in this case, the incident radiation from the first medium can penetrate into the second medium only when $\mu^{'} > \mu_{cr12}$, and

$$f_3(\mu^{'}) = (\frac{n_2}{n_1})^2[1-\rho_2(\mu)]I_1(\tau_1,\mu) \quad \text{(3f)}$$

now, $I_1(\tau_1,\mu)$ can be directly obtained from the S-N method, but the calculated $f_3(\mu^{'})$ is not located at the coordinate we had set before, again, the cubic spline interpolation is employed to get the value of $f_3(\mu)$.

The calculation of f_2 is similar to that of f_3. when $n_1 < n_2$, f_2 is obtained from the same procedure as that for f_3 in the case of $n_1 \ge n_2$, see Eq. (3c). when $n_1 \ge n_2$, f_2 is calculated following the same procedure as that for f_3 in the case of $n_1 < n_2$, see Eq. (3d).

The sequence of problem calculation starts by guessing the radiative intensities at the boundaries, substitute them into Eq. (12) to get the all forward direction intensities, and then obtain the all backward direction intensities from Eq. (13). The incident radiation at every grid point can then be calculated by

$$G_i(\tau) = 2\pi\int_{-1}^{1} I_i(\tau,\mu)d\mu \qquad i = 1,2 \qquad i = 1,2 \quad \text{(20)}$$

repeat the above calculations until the solution is converge, the converge criterion is set as $| G_{new} - G_{old} | < 10^{-6}$. Finally, the hemispherical reflectivity and transmissivity of the slab are calculated by Eqs. (6)-(9).

All the results presented in the next section used the transformed S-18 method with the total direction ordinates of 72. The numerical grid number in the medium was tested by sensitivity study and determined to be 20 nodes for $\tau = 0.1$, 50 nodes for $\tau = 1.0$, and 100 nodes for $\tau = 5.0$.

RESULTS AND DISCUSSIONS

The calculations were performed for those cases including the following parameter variations:
the refractive indexes, n_1 and n_2, equal 1.335 or 1.5. the single scattering albedo, ω_1 and ω_2, range from 0.0 to 1.0. the total optical thickness, τ_0, covers 0.1 (thin), 1.0 (medium) and 5.0 (thick). the individual optical thicknesses, τ_1 and τ_2, ratio as

$\tau_1 = 0.2\tau_0$ and $\tau_2 = 0.8\tau_0$,
$\tau_1 = \tau_2 = 0.5\tau_0$,
$\tau_1 = 0.8\tau_0$ and $\tau_2 = 0.2\tau_2$

In addition, both isotropic scattering and anisotropic scattering are examined in this study. Only selected cases are presented in the following.

Table 5 illustrates that R and T increase with the increase of ω regardless the change of τ_0. In the case of completely absorbing ($\omega_1 = \omega_2 = 0.0$), R and T decrease as τ_0 increases, which means thicker medium can absorb more energy. On the other hand, the completely scattering mediums ($\omega_1 = \omega_2 = 1.0$), thinner medium allows more penetration energy (T is the highest when $\tau_0 = 0.1$), and thicker medium reflects more energy (R is the highest when $\tau_0 = 5.0$).

τ_0		0.1		1.0		5.0	
τ_1	τ_2	0.05	0.05	0.5	0.5	2.5	2.5
ω_1	ω_2	R	T	R	T	R	T
.0	.0	.127432	.757265	.073469	.268374	.066733	.002988
.0	.5	.132686	.761641	.082632	.283056	.066855	.003939
.0	1.0	.144716	.771688	.112147	.331532	.067647	.016344
.5	.0	.135055	.762905	.110962	.287495	.126724	.004260
.5	.5	.142000	.768615	.126115	.308503	.127120	.005943
.5	1.0	.159702	.783405	.179821	.384401	.129874	.030129
1.0	.0	.148516	.772767	.206063	.336491	.509592	.017568
1.0	.5	.159922	.782266	.244660	.378527	.524378	.028509
1.0	1.0	.194296	.810977	.423436	.576093	.704275	.297092

Table 5 Isotropic scattering. $n_1 = 1.335, n_2 = 1.5$

Table 6 shows the results with the same parameters as that of Table 5 except the radiative transfer is considered as anisotropic here. when $\omega_1 = \omega_2 = 0.0$, the results of Table 5 and Table 6 are identical. Because the radiation is not influenced by the scattering phase function for the completely absorbing case. When $\omega_1 = \omega_2 = 1.0$ the values of T in Table 6 is about 0.8 for all three different optical thickness. While the results of Table 5 shows the value of T is about 0.8 for $\tau_0 = 0.1$, about 0.6 for

$\tau_0 = 1.0$, and about 0.3 for $\tau_0 = 5.0$. The discrepency of the above results shows the effect of anisotropic scattering. The reason that transmissirity is hardly influenced by the optical thickness in the anisotropic scattering case comes from the characteristics of scattering phase function, see Table 1. The expansion coefficient (a_n) for forward scattering is much larger than that of backward scattering. This will lead to a transfer behavior with strong forward tendency. This means that most of the radiative energy tend to propagate in a forward direction.

τ_0		0.1		1.0		5.0	
τ_1	τ_2	0.05	0.05	0.5	0.5	2.5	2.5
ω_1	ω_2	R	T	R	T	R	T
.0	.0	.127432	.757265	.073469	.268374	.066733	.002988
.0	.5	.129844	.767870	.075995	.307266	.066745	.006103
.0	1.0	.135859	.783908	.089344	.393075	.067006	.037809
.5	.0	.130629	.770145	.079031	.314421	.073312	.006777
.5	.5	.133960	.782445	.084130	.367873	.073377	.015176
.5	1.0	.143104	.804142	.111095	.504532	.074606	.111825
1.0	.0	.136735	.786817	.102010	.391848	.150883	.031047
1.0	.5	.142575	.803792	.115671	.479832	.152257	.078778
1.0	1.0	.161772	.843241	.200837	.797640	.178819	.822404

Table 6 Anisotropic scattering, $n_1 = 1.335, n_2 = 1.5$

Figures 3a and 3b show the transmissivity distributions for isotropic and anisotropic scattering respectively. Both cases are calculated with the same other parameters, in which ω_1 is fixed at 0.5, and $\tau_1 = 0.2$, $\tau_2 = 0.8$. The maximum variation of T is about 0.2 in Fig. 3a and about 0.4 in Fig. 3b. In both figures, the case of $n_1 = n_2 = 1.335$ has the highest T, and the case of $n_1 = n_2 = 1.5$ has the lowest T. In general, the results of $n_1 = 1.335$ and $n_2 = 1.5$ is higher than that of $n_1 = 1.5$ and $n_2 = 1.335$. However, both are closer to the results of $n_1 = n_2 = 1.5$ rather than to that of $n_1 = n_2 = 1.335$.

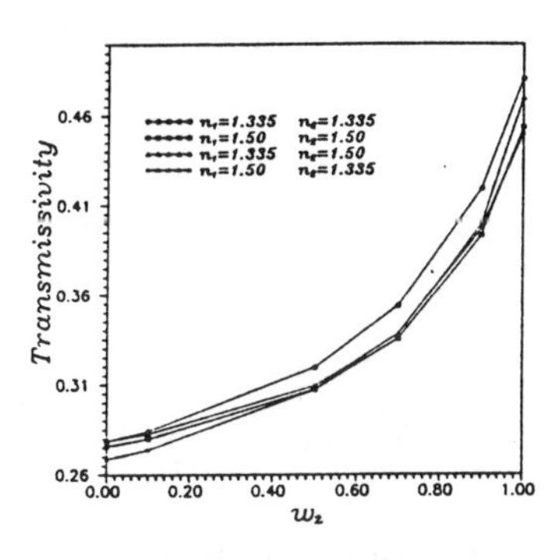

(a) isotropic scattering

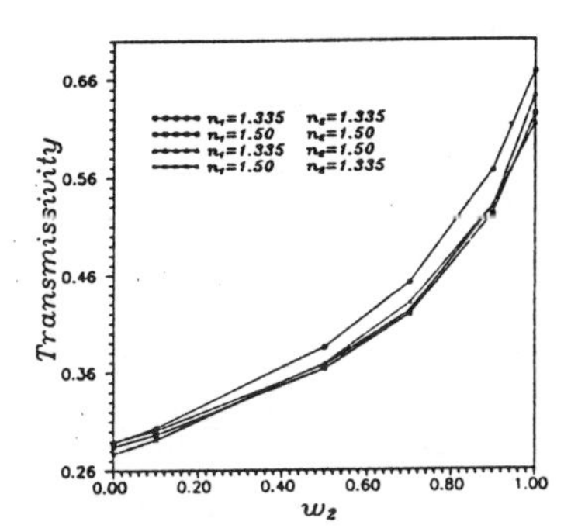

(b) anisotropic scattering

Fig 3 Transmissivity distribution ($\tau_1 = 0.2, \tau_2 = 0.8$)

Figures 4a and 4b are respectively similar to figures 3a and 3b except the optical thicknesses are put in reverse, i.e. $\tau_1 = 0.8$ and $\tau_2 = 0.2$. The absolute variation of T is less than that found in Fig. 3, about 0.034 for isotropic case and 0.076 for anisotropic case. However, the trend that T with mixed indexes tends to closer to the T with $n_1 = n_2 = 1.5$ becomes more obvious, and T with $n_1 = 1.335$ and $n_2 = 1.5$ is generally larger than T with $n_1 = 1.5$ and $n_2 = 1.335$.

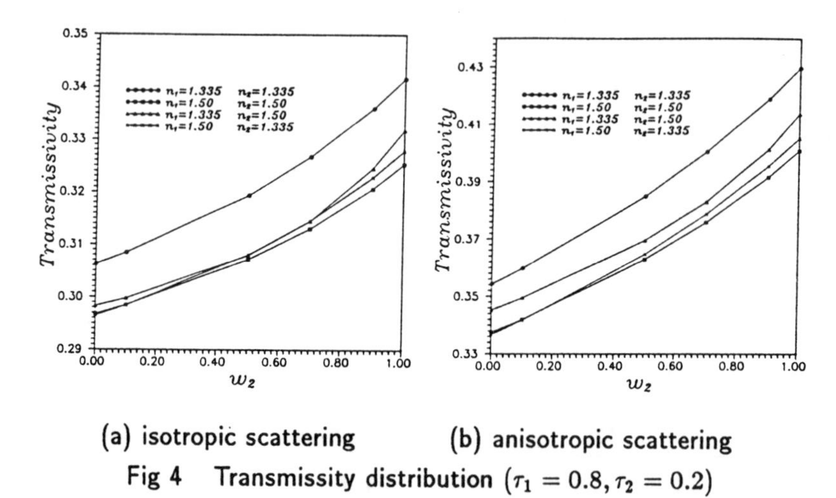

(a) isotropic scattering (b) anisotropic scattering

Fig 4 Transmissity distribution ($\tau_1 = 0.8, \tau_2 = 0.2$)

5. CONCLUSION

The hemispherical transmissivity and reflectivity are generally known to be function of single scattering albedo and optical thickness. For a two-layers slab with grey body medium, the transmissivity, however, may not obiously vary with optical thickness due to the forwarding transfer characteristics existing in the anistropic scattering medium. The key parameter to determine the transmissivity of those anisotropic scattering mediums is the refractive index. In practical application, the optical charateristics may be substantially altered by even applying a thin coating with a refrative index larger the original one.

REFERENCES

1. FIVELAND, W. A. — Discrete Ordinate Methods for Radiative Heat Transfer in Isotropically and Anisotropically Scattering Media, ASME J. of Heat Transfer, Vol.109, p.809 (1987).
2. ÖZISIK, M. N. and SHOUMAN, S. M. — Source Function Expansion Method for Radiative Transfer in a Two-Layer Slab, J. Quant. Spectrosc. Rodiat. Transfer, Vol.24, pp.441~449 (1980).
3. CLEMENTS, T. B. and ÖZISIK, M. N. — Effects of Stepwise variation of Albedo on Reflecivity and Transmissivity of an Isotropically Scattering Slab, Int. J. Heat Mass Transfer, Vol.20, No.10, pp.1419-1426. (1983).

4. SWATHI, P. S., TONG, T. W. AND CUNNINGTON, JR, G. R. — "Reflectance of Two-Layer Composite Porous media with Linear-Anisotropic Scattering," J. Ouant. Spectrose. Radiat. Transfer, Vol.38, No.4, pp.273-279, (1987).

5. CENGEL, Y. A. and ÖZISIK, M. N. — Radiation Transfer in an Anisotropically Scattering Slab With Directional Dependent Reflectivities, ASME 86-HT-28, June, (1986).

6. ÖZISKIK, M. N., — Radiative Transfer and Interaction with Conduction and Convection, Wiley, New York (1973).

IGNITION OF A VERTICALLY ORIENTED SOLID FUEL UNDER RADIATIVE HEATING

Xianming Li
Fluent Inc., 10 Cavendish Court, Lebanon, NH 03766 U.S.A.

Pandeli Durbetaki[1]
Georgia Institute of Technology, Atlanta, GA 30332-0405 U.S.A

SUMMARY

Ignition of a semi-infinite, vertically oriented solid polymethylmethacrylate (PMMA) under radiative heating is an idealized phenomenon that has great implications in fire prevention and hazard assessment studies. This paper establishes a mathematical model of the ignition phenomenon by taking the heat transfer, mass transfer and chemical processes into account. The Patankar-Spalding finite control volume method was adapted to deal with the transient, two-phase, and chemically reacting systems. Adaptive grid refinement and variable time step sizes were employed to assure convergence and expedite computations. This method of solution was first validated using nonreactive flow systems and the results were presented earlier. The current paper emphasizes the numerical treatment extended to include the aspects of reactive flow and radiation absorption. The results indicate that the ignition occurs on the surface at low radiation intensities. Past a critical radiation intensity, the ignition site moves into the gas mixture and the ignition time shortens. Across the critical intensity, the ignition time changes discontinuously. These results agree with experimental observations.

1 INTRODUCTION

Ignition is the initiation of a flame and a part of the flame propagation. In large scale fires, radiant heat exchange controls the rate of flame propagation [1], thus the study of radiative ignition becomes essential to understanding of fires of large dimensions. In the study of the radiative ignition phenomenon for solid fuels, the role of radiation absorption by the pre-ignition gas mixture was recognized and was confirmed [2-6]. The

[1]Professor. The George W. Woodruff School of Mechanical Engineering, Atlanta, GA 30332-0405 U.S.A. Address all correspondence.

same researchers also found that the radiation ignition phenomenon possessed a phase transition from the heterogeneous ignition on the surface at low radiation intensities to the homogeneous gas phase ignition at higher fluxes. Across the transition, the ignition delay time changed abruptly. Phuoc and Durbetaki [7] was the first to confirm analytically that the heterogeneous and homogeneous characteristics of ignition are two facets of the same problem, the particular occurrence of which depends on the intensity of the applied flux. Later, Amos and Fernandez-Pello [8] conducted a similar study in which pyrolysis and chemical reactions in the solid phase were excluded. With their own absorption data for PMMA pyrolysis gases [9], Tien and coworkers [10] constructed a one-dimensional ignition model which further confirmed the importance of gas phase absorption in the radiative ignition phenomenon.

In this study, the gas-surface radiant exchange is incorporated in the Phuoc-Durbetaki ignition model with the experimental data of Tien et al. [9]. The partial differential equations describing the reactive flow and the solid are solved numerically. The finite control volume method developed originally by Patankar [11] and Spalding [12] was adapted in this research to handle the transient, multi-phase and chemically reactive system seamlessly.

2 METHODS

2.1 Governing Equations

A thick slab of solid with height L is vertically placed and exposed to a uniform radiative heat flux. The origin starts at the bottom of the slab. The x axis runs upward along the surface, and the y axis is normal to the surface. Gravity acts in the negative x-direction. Two sites of major activity exist in this system: the chemically active, natural convection boundary layer and the chemically active gas-solid interface. It is assumed that the boundary layer approximations are valid and the flow is laminar. Furthermore, the mixture in the boundary layer obeys the ideal gas law with equal binary diffusion coefficients, and the thermal properties of the solid are constant. The gas phase chemical reaction is assumed to be one-step, irreversible, and of second order. The pyrolysis and oxidation of the solid are assumed to take place at the surface. The rate of oxidation is assumed to be governed by a single-step, irreversible first order reaction. Heating of the solid is assumed to be by surface absorption and convection from the gas boundary layer. The effect of in-depth absorption by the solid is neglected. The conservation principles of mass, momentum and energy can be stated mathematically [13]

$$\frac{\partial \rho}{\partial t} + \frac{\partial(\rho u)}{\partial x} + \frac{\partial(\rho v)}{\partial y} = 0 \tag{1}$$

$$\rho\frac{\partial Y_i}{\partial t} + \rho u\frac{\partial Y_i}{\partial x} + \rho v\frac{\partial Y_i}{\partial y} = \omega_i + \frac{\partial}{\partial y}\left(\rho D\frac{\partial Y_i}{\partial y}\right), \qquad i = O, F, N \tag{2}$$

$$\rho\frac{\partial u}{\partial t} + \rho u\frac{\partial u}{\partial x} + \rho v\frac{\partial u}{\partial y} = -\frac{dP}{dx} + \frac{\partial}{\partial y}\left(\mu\frac{\partial u}{\partial y}\right) - \rho g \tag{3}$$

$$\rho c_p \frac{\partial T}{\partial t} + \rho c_p u \frac{\partial T}{\partial x} + \rho c_p v \frac{\partial T}{\partial y} = \frac{\partial}{\partial y}\left(\lambda \frac{\partial T}{\partial y}\right) - \sum_{j=1}^{n} h_j^\circ \omega_j - \frac{\partial q_r}{\partial y} \tag{4}$$

$$\rho_s c_s \frac{\partial T_s}{\partial t} = \lambda_s \frac{\partial^2 T_s}{\partial y^2} \tag{5}$$

2.2 Constitutive Relations

Pre-ignition chemical activity in the gas phase may be represented as

$$\nu_F M_F + \nu_O M_O + \nu_N M_N \longrightarrow \nu_P M_P + \nu_N M_N \tag{6}$$

with the following reaction rate expressions

$$-\omega_F = -\frac{\omega_O}{r_s} = \frac{\nu_F M_F}{\nu_P M_P}\omega_P = \hat{A}\frac{\rho}{\rho_\infty}\rho Y_F Y_O e^{-E/RT} \tag{7}$$

where $r_s = \nu_O M_O / \nu_F M_F$ is the stoichiometric oxidizer-fuel mass ratio, and $\hat{A} = A\rho_\infty \nu_F / M_O$ is the modified frequency factor.

The first-order surface oxidation reaction results in a net energy effect

$$q_{sR} = A_{sR}\, Q_{sR}\, \rho_w\, Y_{Ow.}\, e^{-E_{sR}/RT_{sw}} \equiv Q_{sR}\, \omega_{sR} \tag{8}$$

where ρ_w is the gas density at the interface. Pyrolysis of the solid follows a linear rate [14]

$$k = \frac{1.5 \times 10^{11} \exp(-21651.6/T)}{1 + 5.6 \times 10^{11} \exp(-16112.8/T)} \quad [\mathrm{m/s}] \tag{9}$$

where T is temperature of the fuel in degrees Kelvin. Thus the mass flux at the surface due to pyrolysis is $\dot{m}_0'' = \rho_s k$.

The pyrolysis gases of PMMA attenuate external radiation mainly by absorption and further participate in the radiant exchange through emission. Let E_+ be the hemispherical total emissive power in the positive y-direction, and E_- be the counterpart in the negative y-direction. Then, the process of absorption and augmentation along the beam path can be described as

$$\frac{\partial E_+}{\partial y} + \alpha E_+ = \alpha\sigma T^4, \qquad -\frac{\partial E_-}{\partial y} + \alpha E_- = \alpha\sigma T^4 \tag{10}$$

The radiant flux in the surface normal direction is $q_r = E_+ - E_-$. Consequently the radiation source term in the energy equation (4) becomes $-\partial q_r/\partial y = -\alpha(2\sigma T^4 - E_+ - E_-)$. The absorption coefficient α becomes

$$\alpha = \kappa_p\, x_F\, P, \quad x_F = (Y_F/M_F)\,/\,(Y_F/M_F + Y_O/M_O + Y_P/M_P) \tag{11}$$

where x_F is the mole fraction of fuel, and P is the total pressure. The pressure-based absorption coefficient κ_p is evaluated according to Reference [9].

2.3 Auxiliary Conditions

Initially, the system is quiescent and in thermal equilibrium. Thus

$$t = 0^+ : \quad u = v = Y_F = 0, \quad Y_O = Y_{O\infty}, \quad Y_N = Y_{N\infty}, \quad T = T_s = T_\infty \tag{12}$$

These conditions must also be satisfied at $x = 0$ at all times. At the solid-gas interface ($y = 0$)

$$u = 0, \quad \rho v = \dot{m}_0'', \quad \dot{m}_0'' Y_{Ow} = \rho D \left.\frac{\partial Y_O}{\partial y}\right)_0 - \omega_{sR} \tag{13}$$

$$\dot{m}_0''(Y_{Fw} - 1) = \rho D \left.\frac{\partial Y_F}{\partial y}\right)_0, \quad \dot{m}_0'' Y_{Nw} = \rho D \left.\frac{\partial Y_N}{\partial y}\right)_0 \tag{14}$$

$$T(0) = T_s(0) \equiv T_{sw}, \quad E_+(0) = r\, E_-(0) + \epsilon\sigma T_{sw}^4 \tag{15}$$

$$-\lambda_s \left.\frac{\partial T_s}{\partial y}\right)_0 = -\lambda \left.\frac{\partial T}{\partial y}\right)_0 + E_+(0) - E_-(0) + \omega_{sR} Q_{sR} + \dot{m}_0'' Q_{sp} \tag{16}$$

At the free stream

$$y = \infty : \quad u = Y_F = 0, \quad Y_O = Y_{O\infty}, \quad Y_O = Y_{O\infty} \tag{17}$$

$$T = T_s(-\infty) = T_\infty, \quad E_-(\infty) = I_o + \sigma T_\infty^4 \tag{18}$$

2.4 Material Properties and Baseline Parameters

The thermophysical and chemical kinetics data of PMMA are listed in Table 1. A baseline case was defined in which the solid fuel, $L = 0.07$ m long, is placed in room air at 300 K and exposed to a radiant flux I_o at 22 W/cm^2. Air viscosity and diffusivity are used with temperature dependence [15].

$$\frac{\mu}{\mu_\infty} = 1.00 + 2.91\beta - 1.71\beta^2 + 0.73\beta^3, \quad \frac{D}{D_\infty} = 1.39 + 11.07\beta + 8.30\beta^2 \tag{19}$$

where $\beta = (\theta - 1)/4$, and $\theta = T/T_\infty$.

2.5 Numerical Technique and Validation

Equations (1)–(5) are coupled and nonlinear, and together with the auxiliary conditions described in Section 2.3 form a two-point boundary value problem. In addition, this set of equations is for transient behavior and involves both the solid phase and the gas phase boundary layer. The finite control volume method developed by Patankar [11] and Spalding [12] with its strictly conservative numerical properties appeared to be the most promising in handling the thermochemical system under consideration. The structure of all equations describing the thermochemical phenomena is recognized to be

$$\rho\frac{\partial \phi}{\partial t} + \rho u \frac{\partial \phi}{\partial x} + \rho v \frac{\partial \phi}{\partial y} = \frac{\partial}{\partial y}\left(\Gamma \frac{\partial \phi}{\partial y}\right) + S \tag{20}$$

Table 1: Thermophysical and Chemical Data for PMMA

Description	Value	Units	Reference
Thermophysical data			
ρ_s	1200	kg/m^3	[16]
λ_s	0.1875	W/m/K	[16]
c_s	1464	J/kg/K	[16]
r	0.05		[5]
ϵ	0.9		[5]
Gas chemical reaction			
A	5.27×10^7	1/s	[17]
E	1.2549×10^8	J/kmole	[17]
Q (exo-)	2.6353×10^7	J/kg	[17]
Solid pyrolysis and in-depth absorption			
α	100	cm^{-1}	[5]
Q_{sp} (endo-)	1.610×10^6	J/kg	[5]
Surface reaction			
A_{sR}	26.183	m/s	[18]
E_{sR}	5×10^7	J/kmole	[19]
Q_{sR} (exo-)	2.6353×10^7	J/kg	[18]

where ϕ is a generic dependent variable, Γ is a generic diffusivity, and S is the source term which can also lump anything unaccountable by other terms. The parameters ϕ, Γ and S together completely specify one equation. The details of the discretization procedure and the treatment of the gas-solid interface for the energy equations have been reported [13, 20]. Particular aspects for the radiation and the chemically reactive flow are described below.

As constitutive relations for the gas phase energy equation, the radiative transfer equations appear in the source term of the gas phase energy equation. For numerical stability, the source term in the gas phase energy equation (4) must be delayed one time step and is evaluated with the most recently available values of the unknowns (T, Y_F, Y_O, etc.) Such considerations lead to a separate or "off-line" solution of the radiative transfer equations.

In flows whose driving force comes from nonlinear source terms in the governing equation, such as the body force term in the momentum equation and the chemical and radiation source terms in the energy equation, the solution hinges on the ability to capture quickly the gradient of the driving potential. A Richardson type error estimator of the local truncation error in the solution was employed to detect the need for grid adaptation [21], and a weighting scheme similar to the Galerkin weighted residual method was used to modify the grid placement. With grid adaptation according to the activities in the flow field, uncertainty levels otherwise too costly to achieve were obtainable. A variable time step strategy similar to grid adaptation was also developed.

A systematic study was carried out to validate the numerical procedure in nonreactive flows [20]. For chemically reactive flows in the current setting, the numerical precision was established through grid-dependence studies. It is estimated that with the solution adaptive grid refinement and the adaptive time step sizes the numerical uncertainty of the full ignition problem with 70 cross-stream nodes is within 3.0%.

2.6 Ignition Criterion

An ignition criterion quantifies the thermal runaway condition associated with the first flame spot. In a radiantly heated solid fuel, the chemical heating rates are negligible initially. As the external heating continues, chemical reactions become more vigorous and their associated energy release becomes more important. Eventually, the induced (or secondary) heating rates reach a comparable level to that of the external (or primary) heating source. Such comparable magnitudes mark the ability for the secondary heating source to maintain continued chemical reactions even if the primary heating source is removed. Consequently, the ignition event may be defined as the equality of the secondary heating rate to the primary heating rate. This requirement gives rise to

$$\Gamma_g = \frac{\omega_F Q / c_p}{\alpha(2\sigma T^4 - E_+ - E_-)}, \quad \Gamma_s = \frac{\dot{m}_0'' Q_{sp}}{-\lambda \left.\frac{\partial T}{\partial y}\right)_0 + E_+(0) - E_-(0) + \omega_{sR} Q_{sR}} \tag{21}$$

Ignition occurs in the gas phase if $\Gamma_g = 1$ or on the interface if $\Gamma_s = 1$. It must be emphasized that although one ignition criterion is designated for each phase, the two criteria come from the same reasoning and describe the same problem, only for different facets.

3 RESULTS

3.1 Baseline Case

Results for the baseline case indicate that the ignition occurs in the gas phase with a delay time of 1.01 second. The homogeneous ignition character was confirmed experimentally by Kashiwagi [4] and the predicted ignition delay time is within 16% of Kashiwagi's measurements. Due to thermal expansion and the blowing velocity of the pyrolysis gases, the hydrodynamic boundary is twice as large as the nonreacting case with identical heating conditions, and the maximum velocity location is 6.5 times farther away into the boundary layer.

The amount of absorption of the external energy by the boundary layer gas mixture at various locations along the fuel surface is described in Figure 1. The ordinate variable is the ratio of radiant flux at the surface to that of the boundary layer outer edge, or the energy transmission across

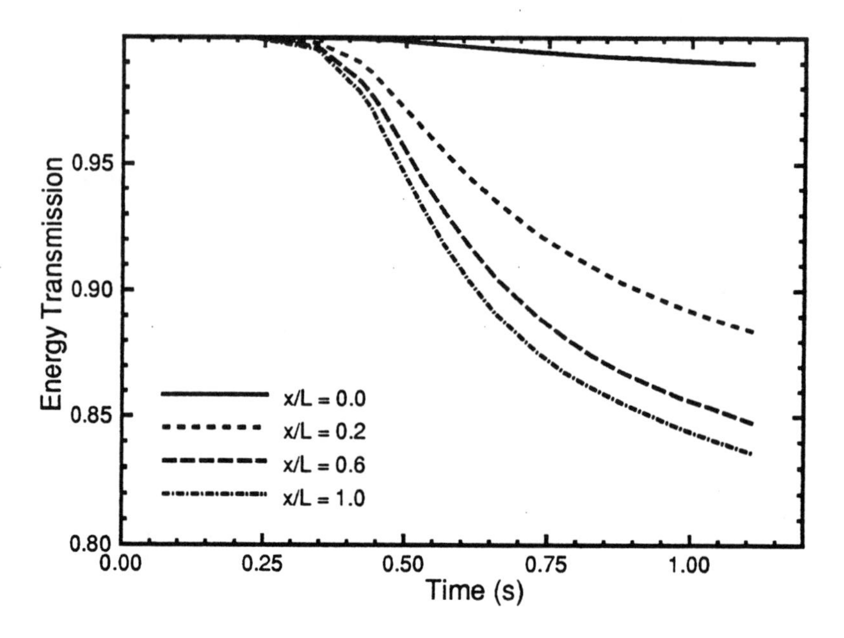

Figure 1: Radiation transmission through the gas boundary layer

the boundary layer:

$$\frac{q_{r,w}}{q_{r,\infty}} = \frac{E_+(0) - E_-(0)}{E_+(\infty) - E_-(\infty)} \tag{22}$$

As seen from the figure, up to 17% of the energy is absorbed by the gas mixture across the boundary layer. The absorption is the largest towards the trailing edge.

The most favorable site of gas phase ignition is near the trailing edge because the optical thickness of the mixture is the largest there. For the same reason, the leading edge is the most probable site of surface ignition. Shown in Figure 2 are values of the trailing edge gas phase ignition criterion and the leading edge surface ignition criterion, both of which represent the maximum in their respective category. Once the pyrolysis process is under way, the energy release due to gas phase chemical reaction takes off and the gas phase ignition criterion at the trailing edge reaches unity exponentially. In the solid phase, chemical activities are not fast enough to compete with the gas phase, and the solid ignition criterion is outpaced by its counterpart in the boundary layer. Consequently, ignition occurs in the gas phase in the current case.

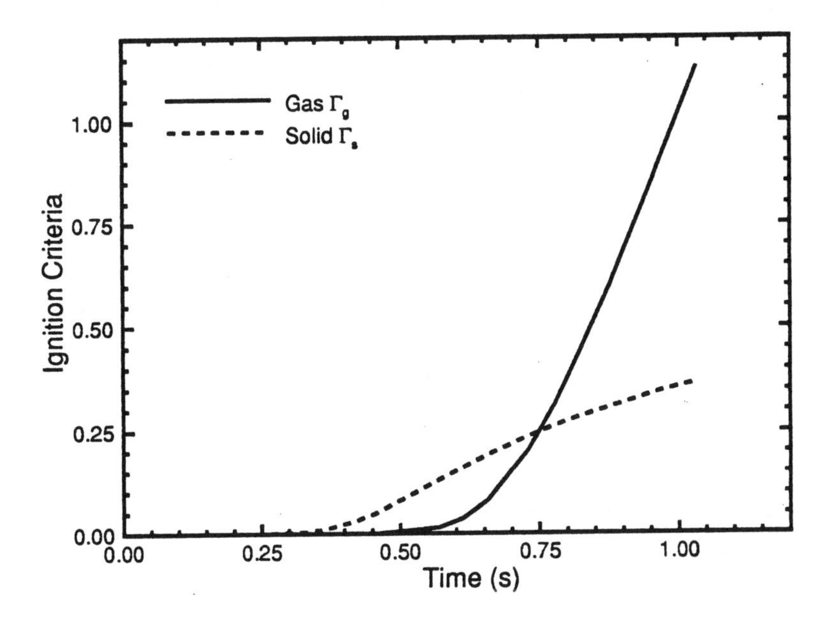

Figure 2: Ignition criteria

3.2 Effect of External Flux Intensity

Prior to the onset of radiation exposure, the condensed fuel and the gas are in thermodynamic equilibrium with the environment. The sustained radiation exposure triggers a host of thermophysical and thermochemical events which eventually lead to ignition. Consequently, the magnitude of the external flux must be instrumental to the ignition process. Figure 3 shows the relationship between the ignition delay time and the external flux intensity. When the radiant flux is high, the relationship appears linear on the log-log scale. Specifically, the asymptote at the high flux range has a slope of -2.25. At the lower flux range, the relationship becomes again linear on the log scale, but with a slope of -1.6. Mathematically these asymptotic relations may be expressed as $t_{ig} \propto I_o^{-n}$, where $n = 1.6$ if I_o is small and $n = 2.25$ if I_o is large. Ignition first occurs in the boundary layer gas mixture when the external flux is large. As the flux decreases, ignition time delay increases, following the asymptotic pattern initially, then more sharply until the slope is almost vertical. Beyond this point, ignition no longer occurs in the gas mixture. Instead, solid phase ignition follows. This so-called ignition phase transition has been observed experimentally [5, 6]. In the present case, the critical flux is between 15.5 and 16.3 W/cm^2.

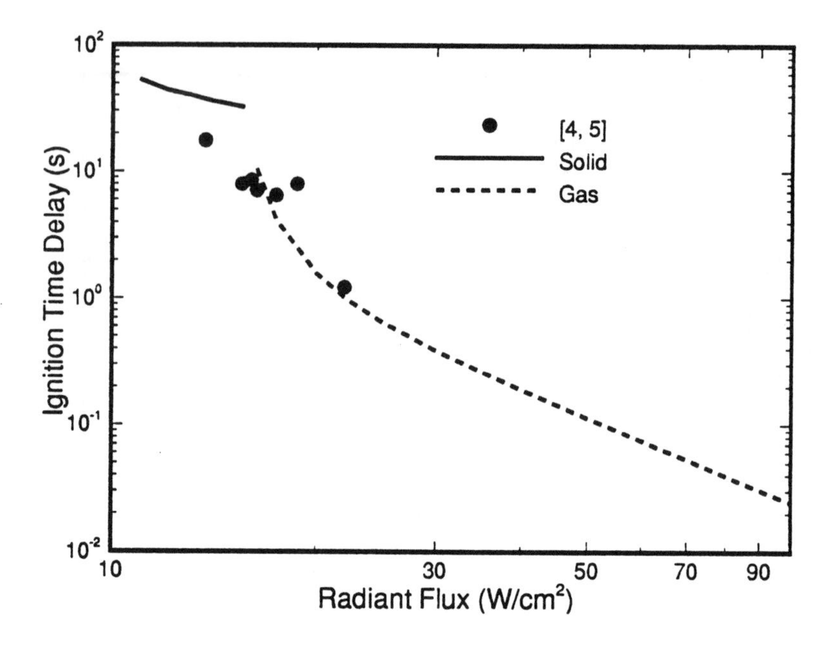

Figure 3: Variation of ignition time delay with external radiation intensity

4 DISCUSSION

The Patankar-Spalding method has been applied to the simultaneous solution of the governing equations of a chemically active multiphase system. The results quantify major radiative ignition phenomena observed experimentally. In particular, the predictions show the ignition phase transition which has practical implications in fire prevention and hazard assessment studies.

Published experimental results on the radiative ignition of PMMA are limited, and are mainly from two sources [4, 6]. The ignition source in these studies is a CO_2 laser beam with emission spectra in the range of 9.3–10.6 μm, and the ignition event is determined by the occurrence of visible flame as detected by a photomultiplier. After adjusting for sample size and orientation effects, these measurements are plotted on Figure 3. The experimental data show a considerable scatter mainly because determining the ignition event in the presence of pyrolysis gases is difficult and imprecise. The overall agreement of the present study is seen to be reasonable. More importantly, the current study predicts a critical radiation flux intensity between 15.5 and 16.3 W/cm^2 compared with the experimental results at approximately 16 W/cm^2.

In addition to the intensity of the external radiation, other factors such as the free stream oxygen content and the spectrum of the external radiation have also been found experimentally to influence the ignition behavior. These factors shall be examined in future studies.

5 CONCLUSION

The ignition phenomenon of a semi-infinite solid fuel under radiative exposure is studied in terms of the interrelations among the heat transfer, mass transfer, and chemical activities occurring both in the boundary layer gas mixture and on the solid surface. A model accounting for the above mechanisms as well as the radiant interactions between the boundary layer gases and the fuel surface is established. The nonlinear governing equations along with the auxiliary conditions were solved by the Patankar-Spalding method. Variable grid spacing, adaptive mesh refinement and adaptive time step sizes were employed to assure numerical stability, minimize computational costs and maintain solution precision. Despite the fact that the thermochemical and absorption data come from a range of different sources and over a period of two decades, the current ignition model predicts the key ignition parameters with excellent accuracy as compared with published experimental results. Further studies on the effect of free stream oxygen content and spectrum of the external radiation are desired.

NOMENCLATURE

A, $\widehat{A}$	original and modified frequency factor
c_p, c_s	specific heat of gas at constant pressure and solid
D	diffusion coefficient
E	activation energy
E^+, E^-	hemispherical total emissive power
h°	enthalpy of formation
I_o	initial external radiant intensity
M	molecular weight
P	pressure
q_r	radiant flux
Q	energy effect of chemical activity
r	surface reflectivity
r_s	stoichiometric oxidizer-fuel mass ratio
T	temperature
u, v	streamwise and transverse velocity component
Y	species mass fraction

Greek symbols

α, κ_p	length- and pressure-based absorption coefficients
ϵ	surface emissivity
θ	dimensionless temperature
λ	thermal conductivity
ω	(mass) rate of reaction

ν	stoichiometric coefficient
Subscripts	
F, O, P, N	fuel, oxidizer, product, and inert species
s	solid
sp	surface pyrolysis
sR	surface reaction
w	wall
∞	free stream

REFERENCES

[1] BROSMER, M.A. and TIEN, C.L. - Radiative Energy Blockage in Large Pool Fires, Combustion Science and Technology, Vol. 51, pp. 21-37, 1987.

[2] OHLEMILLER, T.J. and SUMMERFIELD, M. - Radiative Ignition of Polymeric Materials in Oxygen-Nitrogen Mixtures, Thirteenth Symposium (International) on Combustion, pp. 1087-1094, 1971.

[3] DURBETAKI, P., TINCHER, W.C., CHANG, H., NDUBIZU, C.C. et al. - Effect of Ignition Sources and Fire Retardants on Material Ignition, Sixth Research Report, NSF Grant No. G7-900B, School of Mechanical Engineering, Georgia Institute of Technology, Atlanta, 1978.

[4] KASHIWAGI, T. - Experimental Observations of Radiative Ignition Mechanisms, Combustion and Flame, Vol. 34, pp. 231-244, 1979.

[5] KASHIWAGI, T. - Radiative Ignition Mechanism of Solid Fuels, Fire Safety Journal, Vol. 3, pp.185-200, 1981.

[6] MUTOH, N., HIRANO, T. and AKITA, K. - Experimental Study on Radiative Ignition of Poly Methylmethacrylate, Seventeenth Symposium (International) on Combustion, pp. 1183-1190, 1978.

[7] DURBETAKI, P. and PHUOC, T.X. - Ignition Phase Transition of a Polymer: Convective Exposure, International Journal for Numerical Methods in Engineering, 25, pp. 373–386 (1988).

[8] AMOS, B.T. and FERNANDEZ-PELLO, A.C. - Model of the Ignition and Flame Development on a Vaporizing Combustible Surface in a Stagnation Point, Combustion Science and Technology, Vol. 62, pp. 331-343, 1989.

[9] PARK, S.H., STRETTON, A.J. and TIEN, C.L. - Infrared Radiation Properties of Methyl Methacrylate Vapor, Combustion Science and Technology, Vol. 62, pp. 257-271, 1989.

[10] PARK, S.H. and TIEN, C.L. - Radiation Induced Ignition of Solid Fuels, Int. J. Heat Mass Transfer, 33, pp. 1511-1520 (1990).

[11] PATANKAR, S.V. Numerical Heat Transfer and Fluid Flow, Hemisphere Publishing Corporation, Washington, D.C., 1980.

[12] SPALDING, D. B. GENMIX: A General Computer Program for Two-Dimensional Parabolic Phenomena, Pergamon Press, Oxford, 1977.

[13] LI, X.M. The Effect of Gas-Surface Interactions on Radiative Ignition of PMMA, Ph.D. Dissertation, School of Mechanical Engineering, Georgia Institute of Technology, Atlanta, 1990.

[14] CHAIKEN, R.F., ANDERSEN, W.H., BARSH, M.K., MISHUCK, E. and SCHULTZ,R.D. - Kinetics of the Surface Degradation of Polymethylmethacrylate, The Journal of Chemical Physics, 32, pp. 141-146, 1960.

[15] KAYS, W.M. and CRAWFORD, M.E. Convective Heat and Mass Transfer (second edition), McGraw-Hill, New York, 1980.

[16] ROHM AND HASS CO. Plexiglas Design, and Fabrication Data, PL-1p, 1983.

[17] TIEN, J.S., SINGHAL, S.N., HARROLD, D.P. and PRAHL, J.M. - Combustion and Extinction in the Stagnation-Point Boundary Layer of a Condensed Fuel, Combustion and Flame, Vol. 33, pp. 55-68, 1978.

[18] GRISHIN, A.M. AND ISAKOV, G.N. - Heterogeneous Ignition Modes for PMMA in a Gaseous Oxidizer Stream, Combustion Explosion and Shock Waves, 12, pp. 325-332 (1976).

[19] ISAKOV, G.N. and GRISHIN, A.M. - Experimental Investigation of the Process of Heterogeneous Combustion, Combustion Explosion and Shock Waves, Vol. 10, pp. 166-170, 1974.

[20] LI, X. and DURBETAKI, P. - The Conjugate Formulation of a Radiation Induced Transient Natural Convection Boundary Layer, International Journal for Numerical Methods in Engineering, Vol. 35, pp. 853-870, 1992.

[21] BERGER, M.J. and OLIGER, J. - Adaptive Mesh Refinement for Hyperbolic Partial Differential Equations, Journal of Computational Physics, Vol. 53, pp. 484-512, 1984.

Numerical study of grey-body surface radiation coupled with fluid flow for general geometries using a finite volume multigrid solver

L. Kadinskii and M. Perić
Lehrstuhl für Strömungsmechanik
University of Erlangen–Nürnberg
Cauerstr. 4, D–8520 Erlangen, F. R. G.

The report presents a numerical technique for the simulation of the effects of grey-diffusive surface radiation on the fluid flow using a finite volume procedure for two-dimensional (plane and axi-symmetric) laminar, incompressible or low Mach number flows.

The radiating surface model assumes a non-participating medium, semi-transparent walls and constant elementary surface temperature and radiation fluxes. The calculation of view factors is based on the analytical evaluation for the plane geometry and numerical integration for axisymmetric geometry. A shadowing algorithm was implemented for the calculation of view factors in complex geometries.

The method was tested by comparison with available analytical solutions. Further test calculations were done for the flow and heat transfer in a cavity with a radiating submerged body. As an example of the capabilities of the method, transport processes in metalorganic chemical vapor deposition (MOCVD) reactors were simulated.

1. INTRODUCTION

In many problems of engineering interest the radiation heat transfer plays an important role and it becomes especially important in high-temperature systems. In some applications such as conventional boilers and furnaces, combustion, nuclear fission and fusion, etc., the medium can absorb, emitt and scatter the energy. For a review of the recent advances in this area see Reference [1]. In other types of problems, when the media are not radiatively participating, only radiation exchange among surfaces should be taken into account. This classical problem is relatively simple and well understood. Reference [2,3] discusses radiation phenomena in great detail and presents required theory, formulae and methods for general cases and some applications. References [4,5] considered the radiation heat transfer in the Czochralski crystal growth. In References [6,7] the effects of wall radiation in crystal growth by MOCVD (Metalorganic chemical vapor deposition) method were studied.

The modelling equations in all mentioned publications are solved by single grid methods. Due to excessive computing times, the calculations were limited to relatively coarse grids, and the distinction between the errors due to discretization and those due to the modelling of the physical processes was not possible. In recent years, multigrid finite difference (FD) and finite volume (FV) methods applicable to the solution of coupled, non-linear differential equations have been developed (see [8], and [9] for MOCVD modelling). The application of the multigrid idea results in an approximately linear increase of computing time with grid refinement, allowing much finer grids to be used and, therefore, more accurate solutions to be obtained.

In this paper, a multigrid finite volume solution procedure is employed to predict the non-participating flows and heat transfer in complicated two-dimensional (2D) geometries when the boundary surfaces are considered to be radiating, grey-diffusive and semi-transparent. By systematically refining the grids, it was possible to evaluate the discretization errors and make sure that they were negligibly small on the finest grid. Due to the multigrid solution procedure, computing times were reduced by an order of magnitude or more in comparison with standard methods.

2. MODELLING EQUATIONS

2.1 Flow model equations The gas flows with very low Mach numbers and large temperature differences are considered. These flows can be described by the conservations law for mass, momentum and energy, expressed in a coordinate-free form as follows:

$$\frac{\partial \rho}{\partial t} + \nabla \cdot (\rho \vec{V}) = 0 \quad , \tag{1}$$

$$\frac{\partial (\rho \vec{V})}{\partial t} + \nabla \cdot (\rho \vec{V}\vec{V}) = -\nabla p + \rho \vec{g} + 2\nabla \cdot (\mu \dot{S}) - \frac{2}{3}\nabla(\mu \nabla \cdot \vec{V}) \quad , \tag{2}$$

$$\frac{\partial (\rho c_p T)}{\partial t} + \nabla \cdot (\rho \vec{V} c_p T) = \nabla \cdot (\lambda \nabla T) \quad , \tag{3}$$

where $\vec{V}$ is the velocity vector, T is the temperature, p is the pressure, $\dot{S}$ is the deformation rate tensor, $\vec{g}$ is the gravitational acceleration vector, and c_p is the specific heat at constant pressure. The deformation rate tensor $\dot{S}$ is defined as: $\dot{S} = \frac{1}{2}[\nabla \vec{V} + \nabla \vec{V}^{\,T}]$. The dynamic viscosity, μ , thermal conductivity, λ, and density, ρ, of the gas are functions of temperature:

$$\mu = \mu(T), \;\; \lambda = \lambda(T), \;\; \rho = \rho(T), \;\; P_0 = \rho RT. \tag{4}$$

In the equation of state, P_0 is the constant characteristic pressure and R is the gas constant.

2.2 Boundary conditions The transport equations (1)-(3) have to be supplemented with appropriate boundary conditions on all boundary surfaces of the solution domain, such as walls, symmetry planes or lines, inflow and outflow. The boundary conditions for the momentum equations include a specified velocity profile at the inlet, zero velocities on solid walls (no slip and no penetration), and zero velocity gradients at the outlet ("soft" boundary conditions). For the energy transport equation, a special analysis is required to determine the boundary conditions at walls. At the inlet, the temperature profile is prescribed. At the outlet and symmetry boundaries, zero temperature gradient is assumed.

Enclosures with semi-transparent grey-diffusive radiating walls and non-participating gas inside and ambient outside are considered. In this case the energy flux to the inside wall is composed of two contributions: radiative transfer from surfaces facing the wall and conduction by the gas. Energy is lost from the wall to the ambient by radiation and cooling by the surrounding gas. Thus, the wall boundary conditions take the form:

$$\text{Ambient} - \text{Wall}: \ -\left(\lambda \frac{\partial T}{\partial n}\right)_{wa} = h(T_{wa} - T_a) + q_i^{out} - q_o^{out} \tag{5}$$

$$\text{Gas} - \text{Wall}: \ T_{wg} = T_g, \ \left(\lambda \frac{\partial T}{\partial n}\right)_{wg} = \left(\lambda \frac{\partial T}{\partial n}\right)_g - q_i^{in} + q_o^{in} \tag{6}$$

where the indices "w" and "g" are related to the wall and to the gas, respectively, T_a is the temperature of the ambient, h is the overall heat transfer coefficient given by heat transfer correlation, q^{out} and q^{in} are the radiation fluxes to the wall (index i) and from the wall (index o) on the outer and inner side, respectively, and $\frac{\partial}{\partial n}$ is the normal derivative to the wall. In the case of thin walls, the boundary conditions could be simplified by assuming a constant temperature gradient across the wall. While this approximation makes little difference to the complexity of the wall energy balance, it simplifies the radiation analysis in the case of semi-transparent walls. (If the walls are thick enough, the processes of conduction and radiation transfer in the walls should be taken into account.) Under this assumption and neglecting the temperature difference between both sides of the wall ($T_{wg} \approx T_{wa}$), the energy boundary condition at the wall takes the form:

$$-\left(\lambda \frac{\partial T}{\partial n}\right)_g = h(T_g - T_a) + \underbrace{q_i^{out} - q_o^{out} + q_i^{in} - q_o^{in}}_{q_{net}} \tag{7}$$

To close the model equations, the radiation fluxes have to be specified.

2.3 Radiation model equations These fluxes for the elementary wall surface element k ($k = \overline{1, N}$, where N is the total number of elementary surfaces on the wall) can be calculated using Stefan-Boltzmann equation and definitions of emissivity (E), reflectivity (R), and transmissivity (Tr) as follows [2,3]:

$$q_{i,k}^{in} = \sum_{j=1}^{N} q_{o,j}^{in} F_{kj}, \tag{8}$$

$$q_{o,k}^{in} = \sigma E_k T_k^4 + R_k q_{i,k}^{in} + \sigma Tr_k T_a^4, \tag{9}$$

$$q_{i,k}^{out} = \sigma T_a^4, \tag{10}$$

$$q_{o,k}^{out} = \sigma E_k T_k^4 + R_k q_{i,k}^{out} + Tr_k q_{i,k}^{in}, \tag{11}$$

where σ is the Stefan - Boltzmann constant and F_{kj} are the view factors, cf. Section 3.2. The major assumptions in deriving these equations are grey-diffusive surfaces and uniform temperatures for each radiating surface k. Without introducing uniform temperature surfaces, in equation (8) surface integrals would appear instead of view factors. If radiative properties are wavelength depending, the equations (8)-(11) should be written in differential form. In this case some simplification should be done in order to get a closed form of eqs. (8)-(11), for instance using the band model (see e.g. [3] for the theory and [7] where two-band model was used to approximate optical properties of quartz.)

After summation of fluxes and using relation $E + R + Tr = 1$, q_{net}, the net radiation flux for the kth surface element becomes (cf. eq.(7)):

$$q_{net,k} = E_k\left(q_{i,k}^{in} - 2\sigma T_k^4 + \sigma T_a^4\right). \tag{12}$$

where $q_{i,k}^{in}$ is defined by the system of equations (8)-(9), providing the heat exchange relationship between the radiating boundaries. Thus the heat transfer in the gas and between the walls is coupled through eq. (7), since the temperature of the gas at wall is equal to the wall temperature.

If the jth surface is an opening (inflow or outflow), the radiation flux in eq. (8) is calculated as:

$$q_{j,k}^{in} = \sigma T_a^4 \quad (13)$$

and no special modifications for the system (8)-(9) are required as it was done in [10]. Since the medium is transparent, the gas temperature at the openings is not affected by the radiation.

The system (8)-(9) can be resolved once the view factors are known.

3. NUMERICAL SOLUTION METHOD

The mathematical model described above was implemented in a solution procedure for two-dimensional (plane or axisymmetric) laminar flows described by Demirdžić and Perić [11] and Durst et al [9]. For this reason, only a short description of the solution procedure is given and only the new features are described in detail.

3.1 Discretization method The equations (1)-(3) are discretized on non-orthogonal grids using a finite volume approach with colocated arrangement of variables. Central differences are used to approximate both convection and diffusion fluxes through control volumes (CV) faces. In the momentum equations, the pressure and gravitational forces are treated as source terms. The specific source is evaluated at the center of the CV, and treated as a mean value over the whole CV. The boundary conditions are incorporated in the solution procedure either by specifying the CV face fluxes or calculating them from their discretized expressions using specified boundary values or gradients of the variables.

In eqs. (7)-(9), the boundary CV faces are treated as the elementary surfaces with temperature T_k. The main assumption in this formulation is that each CV face at the boundary is a radiation surface which has constant properties. The discretized boundary conditions (7) couple the discretized energy eq. (3) with the non-linear system (8)-(9), and the temperature of every node on the boundary is coupled with every other node on the boundary. If this matrix system is solved as a coupled system (after some linearization of (8)-(9)) , the computing cost increases significantly, because effective solvers for diagonal matrices can not be applied. An alternative is a decoupled iterative approach, in which the linear system (8)-(9) is solved for given boundary temperatures, and then the resultant heat fluxes are used as specified boundary fluxes [10].

The solution of the resulting coupled set of algebraic equations for the momentum transport is based on the SIMPLE algorithm [12]. The momentum equations are assembled by treating pressure, mass fluxes, and other variables as known (values from previous iterations are used) and the resulting linear systems are solved by applying one iteration of the Strongly Implicit Procedure (SIP) of Stone [13]. With these intermediate velocities, which do not satisfy the continuity equation (1), new mass fluxes are calculated, whose imbalance provides the source term for the pressure correction equation. It is also solved by applying up to 10 iterations of the SIP-solver (inner iterations). The pressure correction is used to correct velocities, pressure and mass fluxes, which then satisfy the continuity requirement. Thereafter, the system (8)-(9) is solved in order to provide boundary fluxes needed for the temperature equation.

The finite-volume approximation of the boundary condition (7) leads to the non-linear equations for the boundary temperature. It was found that simple linearization of eq. (7):

$$-\lambda \frac{\partial T^{new}}{\partial n}\bigg|_k = (T_k^{new} - T_a)\Big[h + 2\sigma E_k (T_k^n + T_a)\Big((T_k^n)^2 + (T_a)^2\Big)\Big] - E_k\left(q_{i,k}^{in} - \sigma T_a^4\right) \quad (14)$$

where T^n is the temperature from the previous iteration and T^{new} is the new approximation for the boundary temperature, in some cases leads to divergence. Instead of that, the equation (7) for every boundary node is iterated until convergence. Thereafter, the temperature equation is assembled and relaxed by applying one SIP iteration. The new density is then calculated from the equation of state (4). Based on these velocity and temperature fields, the coefficients of the difference equations are updated, and a new outer iteration is started. The process is repeated until convergence.

In all calculations, also for the system (8)-(9), underrelaxation is used in ensure the convergence of the procedure. In most of the simulations presented in this paper a relaxation parameter between 0.8 and 1.0 was employed. No special study was done to determine an optimal set of parameters.

The convergence criterion used here is that the sum of absolute residuals in all equation is reduced by five orders of magnitude.

3.2 Multigrid solver In order to speed up the convergence of the outer iterations, the multigrid procedure based on the Full Approximation Scheme (FAS) for non-linear coupled equation systems is used [8]. Details of the described algorithm can be found in Hortmann et al. [8]. In the present implementation, the gain flux q_{net} is calculated on coarse grid by summing the fine grid fluxes and used in the equation (7) without recalculating. At the current finest grid the equation (7) is used together with the system (8)-(9) as it was described for the single grid algorithm. The systematic grid refinement used in the multigrid algorithm to achieve higher convergence rates facilitates also the evaluation of discretization errors. These are proportional to the difference in solutions on two consecutive grids and the order of the discretization scheme [14].

3.3 Calculations of view factors The analysis of radiation interchange among diffusively emitting and reflecting surfaces depends on the knowledge of view factors. For two black bodies the view factor F_{ij} is defined as the fraction of radiant energy leaving the surface i that arrives at the surface j (see. e.g. [2]):

$$F_{ij} = \frac{1}{A_i} \int_{A_i} \int_{A_j} \frac{\cos\theta_i \cos\theta_j}{\pi r_{ij}^2} dA_i dA_j \ , \qquad (15)$$

where A_i and A_j are the areas of surfaces i and j respectively, and θ_i and θ_j are the angles between the normal vectors to surfaces and the line connecting surface centers whose length is r_{ij}

Despite of the flow and radiation problem being considered as 2D, the view factor problem is three-dimensional. Two different strategies for the calculation of the view factors depending on the geometry are presented below.

For the plane geometries the view factor between two elementary surfaces is the factor between two infinite stripes and, therefore, can be evaluated analytically [2].

For axisymmetric geometries there is no general closed form for the view factor evaluation. In this study the view factors are numerically evaluated. The algorithm operates by discretizing azimuthally each surface element into $2N_\theta$ facets as shown in Fig. 1. The view factor between two surface elements can be found as the sum of view factors between a pair of facets using view factor algebra correlations and symmetry:

$$A_i F_{ik} = 4N_\theta \sum_{m=1+n}^{N_\theta+n} \delta_{nm} A_n F_{nm} \ , \qquad (16)$$

where $\delta = 1$, if $m \neq n + N_\theta$; $\delta = 0.5$, if $m = n + N_\theta$, and n is any number between 1 and N_θ.

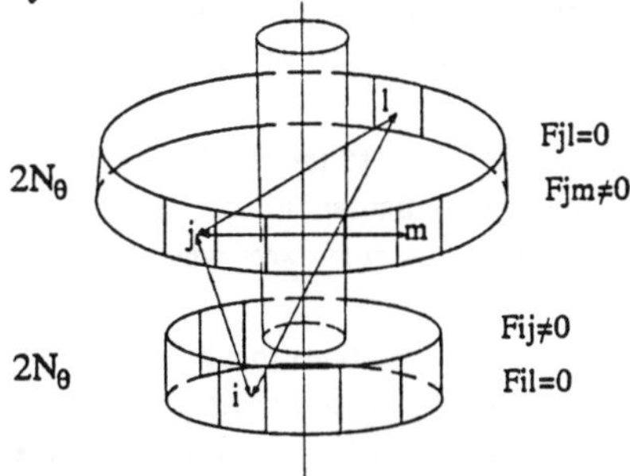

Fig. 1 Numerical evaluation of view factor in axi-symmetric geometries.

Each pair of facets is taken to be either totally visible to each other or to be totally shadowed by the obstacles crossing the line connecting the centroids of these two facets, cf Fig. 1. If the facets can see each other, then the algorithm calculates the view factor between them as:

$$F_{ik} \approx \frac{4N_\theta}{A_i} \sum_{m=1+n}^{N_\theta+n} \delta_{nm} \frac{\cos\theta_n \cos\theta_m}{\pi r_{nm}^2} A_n A_m = \frac{A_k}{\pi N_\theta} \sum_{m=1+n}^{N_\theta+n} \delta_{nm} \frac{\cos\theta_n \cos\theta_m}{\pi r_{nm}^2}. \quad (17)$$

In order to get better approximation for the numerical evaluation of the view factor one modification was done: each elementary surface can be discretized not only azimuthally but also radially or/and axially into N_i and N_k elementary subsurfaces. In this case the *n*th elementary surface consists of N_n elementary subsurfaces and the numerical approximation for the view factor becomes:

$$F_{ik} = \frac{1}{A_i} \sum_{i_m=1}^{N_i} A_{i_m} \sum_{k_l=1}^{N_k} F_{i_m k_l} \quad , \quad (18)$$

where $F_{i_m k_l}$ is given by (17).

After computing the view factors for the finest grid, the view factors for all others coarse grids can be recursively calculated using symmetry property, because every boundary surface of the grid *k-1* consists of two corresponding surfaces on the grid *k*.

The major complication with the view factor calculation is the possibility of partial blocking or shadowing between two surfaces by an intervening body or obstacle or by another boundary surface. The present shadowing algorithm is based on the blockage of line between centroids of separate surfaces. After the testing of the algorithm it was found that the main error occurs in the computation for the neighboring surfaces. In order to improve the results, the following semi-empirical correction technique was suggested. Let F_{in} be the maximal view factor for given index i: $F_{in} = \max_{j=1,N_\bullet} F_{ij}$. The corrected view factor is defined as follows:

$$F_{in}^{corr} = F_{in} + 1. - \sum_{j,j\neq n}^{N_\bullet} F_{jn} \, , \quad (19)$$

if $n > i$, or no correction, if $n < i$. After correction, the change of error for the index n is checked. If the error has not increased, the correction is adopted. This technique is self-consistent, because the change of F_{in} involves the change of F_{ni} due to the reciprocity correlation.

The accuracy of the calculated view factors was examined by comparison with the exact expressions. The error is more sensitive to the axial/radial than to the azimuthal discretization, and inversely proportional to the number of subsurfaces N_{sub}. It also reduces when the azimuthal discretization is refined , but only for $N_\theta < N_\theta^0$, where N_θ^0 is a definite number depending on the configuration. As the main error occurs in the computation for the neighboring surfaces, the contribution of that error is greater for smaller N_θ.

The computational time is asymptotically proportional to the number of all elementary surfaces: $t \sim c \cdot N_\theta \cdot N_{sub}^2$, where c depends on configuration. Another measure of the accuracy is the sum of view factors for each elementary surface, which should be equal to unity.

The results of calculations show that view factor calculations for the grid of 64×64 CV and finer do not need further subdivision of CV faces into subsurfaces, since the errors are then lower than 1%. Because the algorithm is intended very find grids, the applied integration technique is sufficiently accurate.

4. RESULTS

The aim of the present study was to develop a numerical solution procedure which gives accurate predictions of the effects of grey-diffusive surface radiation on the fluid flow. To test the implementation described in the previous sections, several model problems were solved.

4.1 Verification of the code First a simple problem of radiative heat transfer between two infinite parallel plates, with a 1D analytical solution, was chosen. (Due to the non-linear boundary conditions, the 1D analytical solution contains one unknown which can be determined from the non-linear equations). In the calculations the plates were of finite size (100 times the distance between plates). The discrepancies between the numerical and analytical solutions were of the order of 0.1%.

4.2 Flow and heat transfer in a cavity Next the buoyancy-driven flow in a cavity with a radiating submerged body is considered. The temperature of the body is T_0=1000 K and the ambient temperature is T_a=300 K. At all cavity walls the boundary conditions (7) are assumed (the constant *h* was set to zero). All transport properties were chosen for hydrogen at the normal atmospheric pressure: P_0=1 bar. The length of the cavity side wall is H=10 cm. The resulting characteristic dimensionless numbers are

$$Ra = \frac{g\rho_{ref}^2 H^3 (T_0 - T_a)}{\mu_{ref}^2 T_{ref}} Pr = 4.632 \times 10^4 \text{ and } Rad = \frac{\sigma T_0^3 H}{\lambda_{ref}} = 3.126 \times 10^1 \quad (20)$$

Reference values were defined for the reference temperature: $T_{ref} = \frac{1}{2}(T_0 + T_a)$. The finest grid had 128×128 CV and five grid levels. The coarse grids were subsets of the finer grids. Three cases were considered: C1– square cavity with a centered square obstacle whose side length is one quarter of the cavity size, C2– square cavity with shifted obstacle and C3– squeezed cavity C2 (side walls inclined at 45°). The aim of the third test case was to check the effect of grid non-orthogonality. All view factors had to be calculated using shadowing option.

In order to investigate the efficiency of the multigrid algorithm, the calculations for the case C1 were done using both multigrid and single grid methods. These results show that for this moderately fine grid of 128×128 CV the multigrid method is 30–50 times faster then the single grid version. Similar ratios are obtained for flows without radiation.

The results for an asymmetrically placed obstacle in an inclined cavity (C3) are shown in Figs. 2a-c. In this case, the flow is very weak in the bottom left corner; the isotherms there hardly differ from those of the pure heat transfer case. In the upper cavity region, the effects of fluid flow are remarkable. They get more pronounced as the Rayleigh number is increased. The aim of this test case was to investigate the effect of grid non-orthogonality on the performance of the solution method. In this case more terms are treated explicitly (those due to non-ortogonality), and the implementation of boundary conditions becomes more complicated since grid lines are not directed along the normal to wall surfaces. While we have no reference solution to assess the accuracy of the given solution, at least the rate of convergence has not deteriorated compared to the square cavity and the solution looks plausible.

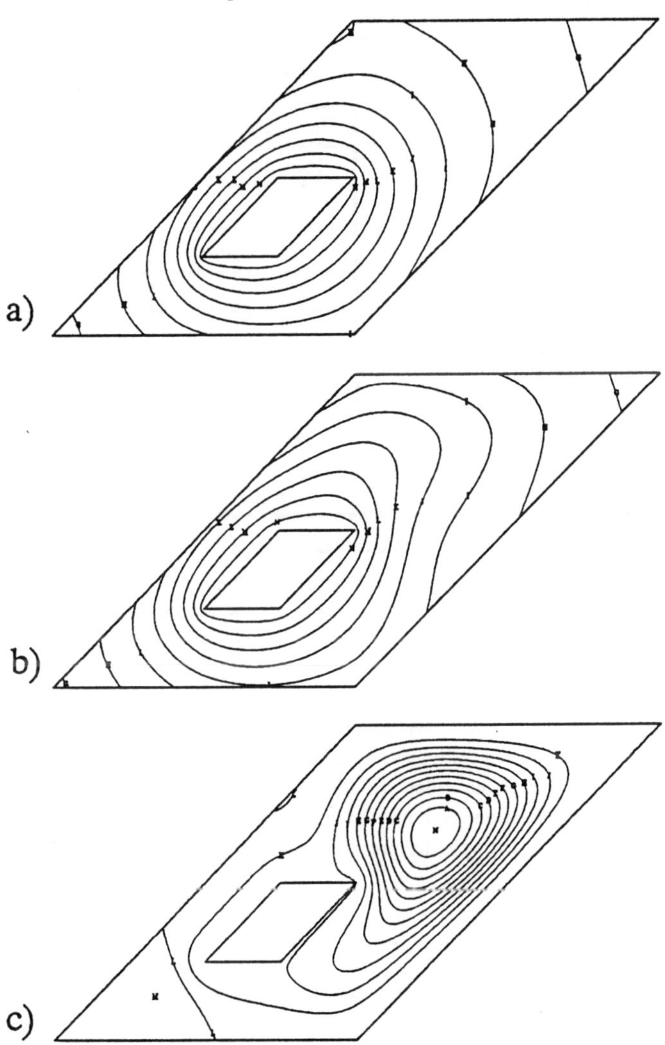

Fig. 2. Isotherms and streamlines for the case C3 : a) Isotherms, pure heat transfer; b) Isotherms, flow and heat transfer (contour levels from 950 K (N) to 600 K (G), in steps of 50 K); c) Streamlines; flow and heat transfer (Ψ_{max}=5.8×10^{-6} (M), Ψ_{min}=−7.4×10^{-5} (N)).

4.3 Flow in MOCVD reactors In order to demonstrate the ability of the present method to analyze complex problems, the flow and heat transfer were studied for horizontal and vertical MOCVD reactor configurations. These problems include blocking surface, various types of obstacles in the solution domain as well as heat sources and buoyancy-driven flow.

The sets of parameters for both cases are collected in Table 1. The view factors for the vertical geometry were calculated with $N_\theta = 80$, $N_{sub} = 2$ and the correction technique was used. Accurate modelling of such flows needs a more realistic description of the radiative properties of the walls and hot zones, as they may differ significantly and may have wave-length depending transmittance as in the case of the quartz glass [7].

Parameters		Horizontal reactor	Vertical reactor
Finest grid (levels)		160×64 CV (5)	288×80 CV (5)
Error of the view factor calculations	e_C	1.29×10^{-2}	2.78×10^{-2}
	e_L	8.52×10^{-5}	1.19×10^{-4}
Re		8.21	14.6
Ra		3.36×10^{4}	5.56×10^{4}
Rad		11.8	23.1
Res_0/Res_n		10^{-6}	10^{-7}
Calculation time (min)		84	175
$Time_{SG}/Time_{MG}$		5.4	46.

Table 1: Parameters and results of calculations for the horizontal and vertical reactors.

Results of calculations for the horizontal reactor are shown in Figs. 3 and 4.. In one case the walls are assumed to be adiabatic, except for the heated section of the inclined lower wall (at the susceptor), which is kept at a constant temperature. In the other case, radiation boundary conditions are applied. Obviously, there is a big difference in flow and temperature distribution, depending on the thermal boundary conditions. In both cases a high numerical accuracy was assured by using fine numerical grids (160×64 CV). Discretization errors were estimated by comparison of solutions on successive grids [8] and were found to be of the order of 0.03 % on the finest grid. The radiation heat transfer causes the recirculation region at the upper wall to extend almost up to the inlet plane (where a uniform block velocity profile was specified in both cases).

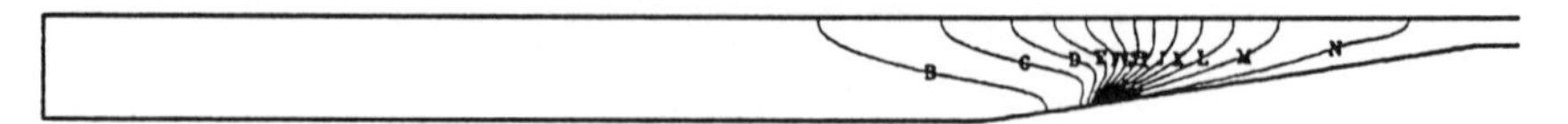

Fig. 3 a.) — Isotherms for the horizontal reactor — adiabatic boundary conditions; contour levels from 950 K (N) to 350 K (B), in steps of 50 K.

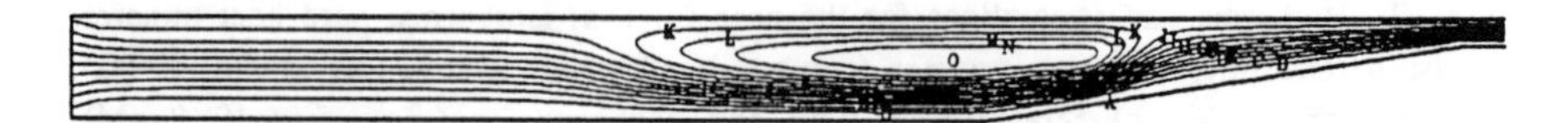

b.) — Streamlines for the horizontal reactor — adiabatic boundary conditions. $\Psi_{max}=1.09\times10^{-4}$ (O), $\Psi_{inflow}=7.39\times10^{-5}$.

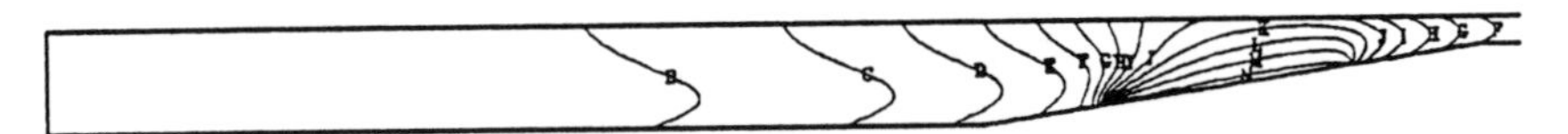

Fig. 4 a.) — Isotherms for the horizontal reactor — radiative boundary conditions; contour levels from 950 K (N) to 350 K (B), in steps of 50 K.

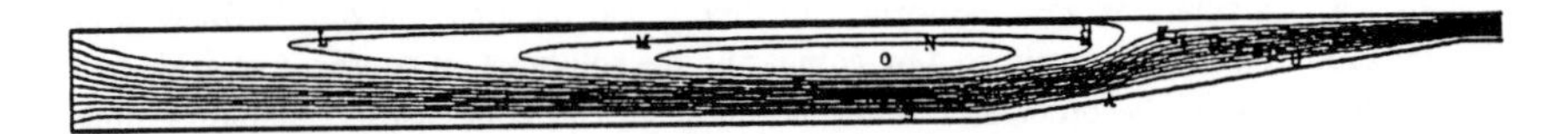

b.) — Streamlines for the horizontal reactor — radiative boundary conditions. $\Psi_{max}=9.51\times10^{-5}$ (O), $\Psi_{inflow}=7.39\times10^{-5}$.

The vertical reactor represents an axi-symmetric geometry with a submerged obstacle, cf. Fig. 5.

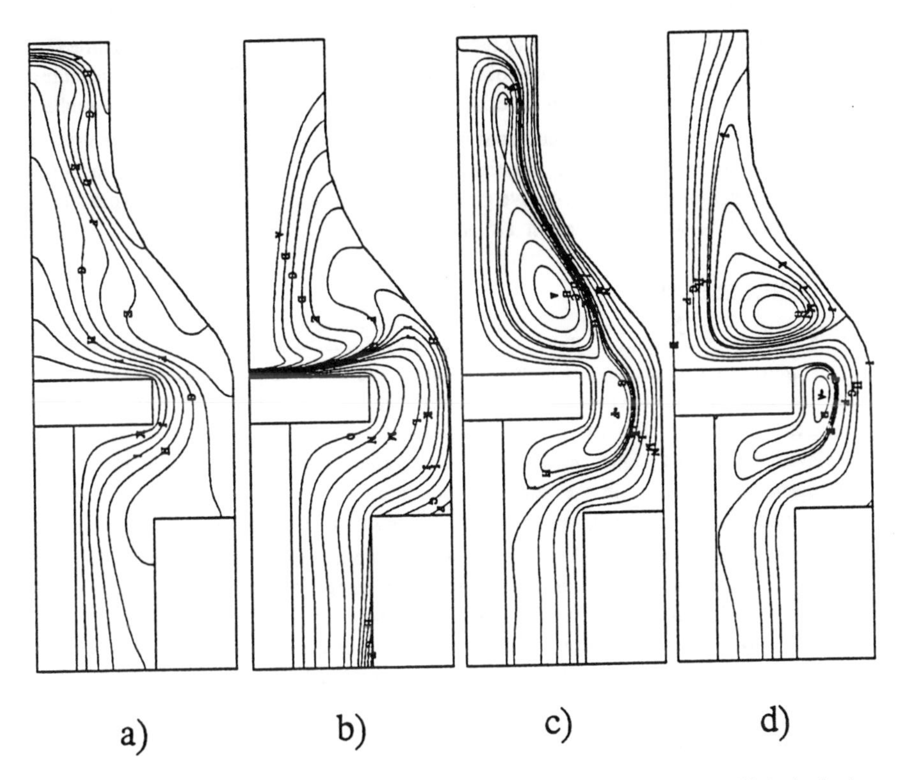

Fig. 5. Isotherms (a,b) and streamlines (c,d) for the vertical reactor with adiabatic (a,c) and radiative boundary conditions (b,d). Temperature contour levels: (a) from 850 K (K) to 350 K (A), in steps of 50 K; (b) (O)–850 K, to (L) 700 K, in steps of 50 K; (K)–675 K, (J)–650 K, to (F)–600 K, in steps of 12.5 K; (E)–575 K, (D)–550 K, to (A)–400 K, in steps of 50 K. $\Psi_{inflow}=6.55\times10^{-6}$, $\Psi_{min}=-9.33\times10^{-6}$ (c), $\Psi_{max}=1.41\times10^{-5}$ (d).

The gas enters the pipe at the top and flows around the disc-shaped obstacle (heated susceptor), leaving through the exhaust at the bottom. Both the disk and the centrally located holder are held at a constant temperature. In one case all other walls were considered adiabatic, in the other radiative boundary conditions were assumed on all boundaries. Calculations were performed on five grids, the finest having 288×80 CV. Discretization errors were estimated by comparing the solutions on consecutive grids and were found to be of the order of 1 % on the finest grid in both cases. Computing time using multigrid solver and comparison with the single grid one are indicated in Table 1.

In case of adiabatic boundary conditions and no radiation, buoyancy forces let gas above the disk flow upwards along symmetry axis like a plume. This upcoming flow is turned back by the incoming gas at the inlet boundary, where a block velocity profile was specified. Thus a large recirculation region is formed, which extends also behind the susceptor. In the case of radiative boundary conditions, a completely different flow and temperature distribution were obtained. The oncoming gas flows along the symmetry axis, impinges onto the disk, flows then radially and separates at the outer disk edge. Another large recirculation region is now attached to the outer wall.

Similar flow patterns and the effect of radiations in vertical reactors were also reported by Fotiadis et al. [6]. The detailed study of transport processes in MOCVD reactors including the complex heat transfer and real material properties is the aim of the forthcoming studies.

The main aim of the present calculations was to demonstrate the efficiency of the present multigrid solution method when calculating flows in complex geometries with radiative boundary conditions. Solutions on five grids (the finest with 288×80 CV) on a workstation SUN SPARC 1+ were obtained in 175 minutes.

5. CONCLUSIONS

In this paper, a 2D mathematical model for grey-diffusive surface radiation combined with the transport equations for mass, momentum and heat was presented. The algorithm for the calculation of view factors provided the needed accuracy for the complex geometries. Through the use of the multigrid algorithm, high numerical efficiency was achieved. By using a sequence of systematically refined grids, it was possible to estimate the discretization errors in the numerical solutions. In examples considered here the discretization error on the finest grids were of the order of or below 1 %.

Concerning the future developments of the method, several new features will be included, like higher-order integration technique using Stokes' theorem and parallel processing using block-structured grids and domain decomposition technique. For MOCVD applications, the complex heat transfer and real material properties as well as chemical reactions and species transport must be considered. This work is under way [9].

6. ACKNOWELEDGMENTS

The authors are grateful to F. Durst and M. Schäfer for valuable discussions and comments. The authors would also like to thank H.-J. Leister for fruitful discussion about the implementation of boundary conditions. The work in the present paper received financial support through the "Bavarian Consortium of High Performance Scientific Computing". This support is thankfully acknowledged.

REFERENCES

1. HOWELL, J.R. — Thermal radiation in participating media: the past, the present and some possible futures, Trans. ASME, Vol. 110, pp. 1220–1229, 1988.
2. SIEGEL, R., HOWELL, J.R. — Thermal radiation heat transfer, McGraw-Hill, New York, 1981
3. SPARROW, E., CESS, R. — Radiation heat transfer, McGraw-Hill, New York, 1978.
4. BORNSIDE, D., KINNEY, T., BROWN., R. — Finite element/Newton method for the analysis of Czochralski crystal growth with diffusive-grey radiative heat transfer, Int. J. Numer. Meth. Eng., Vol. 30, p. 133, 1990.
5. DUPRET, F., NICODEME, P., RYCKMANS, Y., WOUTERS, P., CROCHET, M.J. — Global modelling of heat transfer in crystal growth furnaces, Int. J. Numer. Meth. Fluids, Vol. 33, p. 1849, 1990.
6. FOTIADIS, D.F., KIEDA, S., JENSEN. K.F. — Transport phenomena in vertical reactors for metalorganic vapor phase epitaxy, J. Crystal Growth, Vol. 102, pp. 441–470, 1990.
7. CHINOY, P.B., KAMINSKI, D.A., GHANDHI, S.K. —Effects of thermal radiation on momentum, heat, and mass transfer in a horizontal chemical vapor deposition reactor, Numerical Heat Transfer, Part A, Vol. 19, pp.85–100, 1991.
8. HORTMANN, M., PERIĆ, M., SCHEUERER, G. — Finite volume multigrid prediction of laminar natural convection: bench-mark solutions, Int. J. Numer. methods fluids, Vol. 11, pp. 189–207, 1990.
9. DURST, F., KADINSKII, L., PERIĆ, M., SCHÄFER, M. — Numerical study of transport phenomena in MOCVD reactors using a finite volume multigrid solver, J. Crystal Growth, Vol. 125, pp. 612–626, 1993.
10. ENGELMAN, M., JAMNIA, M.-A. — Grey-body surface radiation coupled with conduction and convection for general geometries, Int. J. Numer. Meth. Fluids, Vol. 13, pp. 1029–1053, 1991.
11. DEMIRDŽIĆ, I., PERIĆ, M. — Finite volume method for prediction of fluid flow in arbitrarily shaped domains with moving boundaries, Vol. 10, pp.771–790, 1990.
12. PATANKAR, S.V., SPALDING, D.B. — A calculation procedure for heat, mass and momentum transfer in three-dimensional parabolic flows, Int. J. Heat Mass Transfer, Vol. 15, pp. 1787–1806, 1972.
13. STONE, H.L. — Iterative solution of implicit approximations of multi-dimensional partial differential equations, SIAM J. Numer. Anal., Vol. 5, pp. 530–558, 1968.
14. DEMIRDŽIĆ, I., LILEK, Ž., PERIĆ, M. — Fluid flow and heat transfer test problems for non-orthogonal grids: bench-mark solutions, Int. J. Numer. methods fluids, Vol. 15, pp. 329–354, 1992.

SMOOTHING OF APPROXIMATE EXCHANGE AREAS

D.A. Lawson

School of Mathematical & Information Sciences, COVENTRY UNIVERSITY, Coventry, CV1 5FB

SUMMARY

When using the zone method to perform radiative heat transfer calculations the exchange areas are key intermediate quantities. The direct exchange areas are defined by a number of complex multi-dimensional integrals. It can be shown that the exchange areas satisfy two simple conditions: the symmetry and conservation conditions. The complexity of the integral definitions means that the direct exchange areas must be calculated numerically. These numerically determined values are approximations and so may not satisfy the symmetry and conservation conditions. It is straightforward to ensure that the calculated values satisfy the symmetry condition but the conservation condition is more complex. A smoothing process takes a set of approximate exchange areas and modifies them slightly to ensure that they satisfy the conservation condition whilst preserving symmetry. In this paper three different smoothing techniques are studied to determine their effect on the exchange areas and also on the resulting zone heat fluxes (which are usually the goal of zone method calculations).

1. THE ZONE METHOD

1.1 Introduction

The zone method, sometimes known as Hottel's zone method [1], is the most widely used method for carrying out engineering calculations of radiative heat transfer in enclosures where radiation is the dominant mode of heat transfer. The basic idea of the method is to divide the bounding surfaces into a number of surface zones and divide the enclosed gas into a number of gas zones and to assume that all radiative properties are uniform in each zone (although they may differ from zone to zone). Then, in order to be able to write down heat balances for each zone it is necessary to define the inter-zone direct exchange areas (DEAs). There are four kinds of DEAs: surface to surface (s_is_j), surface to gas (s_ig_j), gas to gas (g_ig_j) and gas to surface (g_is_j). In the definitions of these quantities given below A_i is the area of surface zone i, V_i is the volume of gas zone i and k_i is the absorption coefficient of the gas in gas zone i.

Surface to surface DEA (s_is_j)
s_is_j/A_i is the fraction of energy emitted by surface zone i which is directly incident upon surface zone j.

Surface to gas DEA (s_ig_j)
s_ig_j/A_i is the fraction of energy emitted by surface zone i which is directly absorbed by gas zone j.

Gas to gas DEA (g_ig_j)
$g_ig_j/4k_iV_i$ is the fraction of energy emitted by gas zone i which is directly absorbed by gas zone j.

Gas to surface DEA (g_is_j)
$g_is_j/4k_iV_i$ is the fraction of energy emitted by gas zone i which is directly incident upon surface zone j.

In these definitions the use of the word "directly" refers to events occurring without the radiation being reflected by any of the surfaces of the enclosure.

1.2 Integral Formulae

It can be shown that the direct exchange areas, as defined above, can be found by evaluating certain multi-dimensional integrals [1]. These integral formulae are:

$$s_is_j = \int_{A_i}\int_{A_j} \frac{\tau(r)\cos\theta_i\cos\theta_j}{\pi r^2} dA_jdA_i \tag{1.1a}$$

$$g_is_j = s_jg_i = \int_{V_i}\int_{A_j} \frac{k_i\tau(r)\cos\theta_j}{\pi r^2} dA_jdV_i \tag{1.1b}$$

$$g_ig_j = \int_{V_i}\int_{V_j} \frac{k_ik_j\tau(r)}{\pi r^2} dV_jdV_i \tag{1.1c}$$

In these formulae r is the length of the beam connecting a general point in one zone with a general point in the other; θ_i is the angle between the beam and the normal to surface zone i; and $\tau(r)$ is the transmittance along the beam defined by

$$\tau(r) = \exp\left[-\int_0^r k(s)\, ds \right] \tag{1.2}$$

where the absorption coefficient k(s) may change along the path of the beam (but is, by the zone method assumptions, constant within each zone).

1.3 Symmetry and Conservation Conditions

The integral formulae for the direct exchange areas (equations (1.1)) clearly show the symmetry of the DEAs. Equations (1.1a) and (1.1c) are symmetric in i and j and so

$$s_i s_j = s_j s_i \tag{1.3a}$$

$$g_i g_j = g_j g_i \tag{1.3b}$$

The symmetry existing between the surface to gas and gas to surface DEAs is inherent in the formula (equation (1.2b) giving

$$s_i g_j = g_j s_i \tag{1.3c}$$

If all the DEAs are assembled into a matrix X as shown below

$$X = \begin{bmatrix} [s_i s_j] & [s_i g_j] \\ [g_i s_j] & [g_i g_j] \end{bmatrix} \tag{1.4}$$

then the symmetry conditions (equations (1.3)) ensure that matrix X is symmetric.

All the energy emitted by a zone must go somewhere: either it is (directly) incident upon a surface zone or it is absorbed by a gas zone before it reaches a surface zone. This means that for any particular zone the sum of all the fractions (to both surface and gas zones) in the definitions of DEAs (in Section 1.1 above) is one. With N surface zones and M gas zones we have, after clearing fractions,

$$\sum_{j=1}^{N} s_i s_j + \sum_{j=1}^{M} s_i g_j = A_i \tag{1.5a}$$

$$\sum_{j=1}^{N} g_i s_j + \sum_{j=1}^{M} g_i g_j = 4k_i V_i \tag{1.5b}$$

Equations (1.5) are known as the conservation condition since they ensure that none of the energy emitted within the enclosure is lost. In terms of the matrix X equations (1.5) prescribe the rows sums. They are summarised as

$$\Sigma x_{ij} = c_i \tag{1.6a}$$

with $\quad c_i = A_i \quad$ for $i \leqslant N$ $\quad$ (1.6b)

and $\quad c_i = 4k_{i-N}V_{i-N} \quad$ for $i > N$ $\quad$ (1.6c)

2. SMOOTHING TECHNIQUES

2.1 Introduction

The complexity of the 4, 5 and 6 dimensional integrals in equations (1.1) is such that they can rarely be evaluated analytically. Numerical methods of integration must be used. A commonly used technique, particularly when general geometries are under consideration is a Monte Carlo method [2]. DEAs calculated by a numerical method are only approximations and so may not satisfy the symmetry and conservation conditions. A set of approximate DEAs can be made symmetric by assembling them into a single matrix and then taking the average of this matrix and its transpose. Ensuring that the DEAs also satsify the conservation condition is a more difficult task. A number of methods have been described in the literature.

2.2 Iterative Smoothing

An iterative method for taking a symmetric set of DEAs (matrix X) and adjusting them to produce a new set (matrix Y) which satisfies the conservation condition (equations (1.6)) has been described in [3]. The elements of the ith row are divided into two sets: S_1 - those elements to be adjusted and S_2 - those elements which are to be unchanged. Initially S_2 contains only those elements of row i which are zero. Then, if there are m elements in the set S_1 the correction to each of these elements is defined by

$$D_i = \frac{c_i - \Sigma x_{ij}}{m} \tag{2.1}$$

Adjusted values of the elements of row i can now be found by

$$y_{ij} = x_{ij} + D_i \qquad \text{for } j=1,\ldots N+M \tag{2.2}$$

If an adjusted value turns out to be negative (which is physically impossible) then this element is moved from S_1 to S_2 and the process is repeated (m will now be smaller making the absolute value of D_i larger). This is repeated until all the adjusted y_{ij} values are positive. At this stage the ith row of matrix Y satisfies the conservation condition (1.6a) but the symmetry of the matrix has been lost. This is retrieved by setting

$$y_{ji} = y_{ij} \qquad \text{for } j=1,\ldots,N+M \tag{2.3}$$

This restores symmetry but destroys the compliance with the conservation condition of previous rows of the matrix. The rows are processed in order, beginnning with the first row. After one sweep through the matrix a new symmetric matrix is obtained which should be closer to satisfying all the conservation constraints. The process is iterated until each row sum is within a specified tolerance of its required value.

In this method all non-zero DEAs in a row are treated equally. Very small values are changed by the same amount as large values (as long as the adjusted values are not negative). It is unlikely that the absolute error of two widely different sized values will be the same and this may lead to over-correction of some values and under-correction of other values.

2.2 Least Squares Smoothing

This method, like the one described in Section 2.1, begins with a set of approximate DEAs (matrix X) which is symmetric and produces from it a set of symmetric DEAs (matrix Y) which also satisfy the conservation condition (equations (1.6)). The full details of the method can be found in [4]. Basically, this method treats the problem as one of constrained minimisation. A function H is defined to measure the size of the disturbance to the original matrix X. This function is defined by

$$H(Y) = \sum_{i=1}^{N+M} \sum_{j=1}^{N+M} \frac{1}{2w_{ij}} (y_{ij}-x_{ij})^2 \tag{2.4}$$

The w_{ij} are weights which allow variable penalties to be assigned to the adjustment of different elements. It is necessary to minimise H subject to the conditions that the rows of matrix Y satsify the conservation condition (1.6) and also (to maintain symmetry) the elements in the ith column must sum to c_i. This constrained minimisation problem can be solved using the method of Lagrange multipliers and the resulting elements of Y are given by

$$y_{ij} = x_{ij} + w_{ij}(\lambda_i+\lambda_j) \tag{2.5}$$

where the Lagrange multipliers λ_i are found by solving the simultaneous linear equations

$$R\underset{\sim}{\lambda} = \underset{\sim}{\delta} \tag{2.6}$$

where the elements of R and $\underset{\sim}{\delta}$ are defined by

$$r_{ij} = w_{ij} \qquad i \neq j \tag{2.7a}$$

$$r_{ii} = w_{ii} + \sum_{j=1}^{N+M} w_{ij} \tag{2.7b}$$

$$\delta_i = c_i - \sum_{j=1}^{N+M} x_{ij} \tag{2.7c}$$

The actual values of y_{ij} obtained depends on the weights w_{ij}. The simplest setting is for them all to be given the value 1. This penalises all changes in the original matrix X equally. This would then be similar to the method described in Section 2.1. However, it does not contain any mechanism to prevent the adjustment of zero elements in the original X matrix. Zero elements arise when two zones cannot "see" each other. For example, a planar zone cannot "see" itself, so the

DEAs from such zones to themselves should always be zero. Numerical integration usually determines zero values correctly and so the smoothing process should not alter these. To prevent zero elements being altered the recommendation in [4] is that the weights w_{ij} be specified by $w_{ij}=x_{ij}^2$. This places an infinite penalty on adjusting zero elements and ensures that all the other elements are adjusted according to their size (and not treated equally as in the iterative smoothing technique). It is this weighting which is used throughout the test problems of the next section.

2.3 Lumped Smoothing

This smoothing technique is based on an unpublished idea of Tucker [5]. To adjust a row of matrix X to ensure that the conservation condition (1.6) is satisfied the shortfall in the row sum is added to the diagonal element. The elements of the adjusted matrix Y are thus given by

$$y_{ij} = x_{ij} \qquad \text{for } i \neq j \qquad (2.8a)$$

$$y_{ii} = x_{ii} + \delta_i \qquad (2.8b)$$

with δ_i specified by equation (2.7c). The symmetry of matrix Y is a consequence of the symmetry of matrix X as the off-diagonal elements are not changed.

The main advantage of this method is its simplicity to implement. There is no need for iteration or for solving simultaneous linear equations. However, as it only changes the diagonal elements it is, in effect, assuming that the numerical integration technique has only made errors in the determination of these values. In fact, the numerically calculated values of the diagonal elements will often be exact. This is because the diagonal elements represent the DEAs from zones to themselves. For planar surface zones these DEA values are zero. This lumped smoothing technique therefore does not preserve any zeros on the diagonal, furthermore it has no safeguards to prevent negative values being introduced.

3. TEST PROBLEMS

3.1 A Unit Cube

The three smoothing techniques were tested on a unit cube enclosure containing a non-participating gas. Each face of the cube is taken as a surface zone. The exact values of the DEAs can be computed analytically for this enclosure zoned in this way [6]. In such a cube, because of the symmetry, there are only two fundamentally different non-zero DEAs: from a zone to an adjacent zone and from a zone to the zone opposite (the DEAs from zones to themselves are all zero). The exact values for these are 0.200044 and 0.199825 respectively.

A set of approximate DEAs were generated using a Monte Carlo technique. These values were made symmetric by averaging them with their transpose. The symmetric DEAs obtained are shown in the table below (in this table opposite zone pairs are (1,6), (2,3) and (4,5)).

0	0.1987	0.2008	0.1970	0.1986	0.2043
0.1987	0	0.1990	0.1996	0.1983	0.1963
0.2008	0.1990	0	0.2015	0.2035	0.2038
0.1970	0.1996	0.2015	0	0.1974	0.2006
0.1986	0.1983	0.2035	0.1974	0	0.2010
0.2043	0.1963	0.2038	0.2006	0.2010	0

Table 1: Approximate DEA values for unit cube.

It can be seen that these DEAs are symmetric, but the row sums are not all one, in other words the conservation condition is not satisfied. These values were used as the matrix X in the three methods described in Section 2 and new DEAs (matrix Y) were obtained. The values obtained are shown in Table 2 below.

ITERATIVE SMOOTHING

0	0.2009	0.1989	0.1981	0.1991	0.2030
0.2009	0	0.1989	0.2026	0.2007	0.1968
0.1989	0.1989	0	0.2004	0.2017	0.2002
0.1981	0.2026	0.2004	0	0.1987	0.2001
0.1991	0.2007	0.2017	0.1987	0	0.1998
0.2030	0.1968	0.2002	0.2001	0.1998	0

LEAST SQUARES SMOOTHING

0	0.2009	0.1989	0.1981	0.1991	0.2030
0.2009	0	0.1990	0.2026	0.2007	0.1968
0.1989	0.1990	0	0.2004	0.2016	0.2001
0.1981	0.2026	0.2004	0	0.1987	0.2002
0.1991	0.2007	0.2016	0.1987	0	0.1999
0.2030	0.1968	0.2001	0.2002	0.1999	0

LUMPED SMOOTHING

0.0007	0.1987	0.2008	0.1970	0.1986	0.2043
0.1987	0.0082	0.1990	0.1996	0.1983	0.1963
0.2008	0.1990	-0.0085	0.2015	0.2035	0.2038
0.1970	0.1996	0.2015	0.0040	0.1974	0.2006
0.1986	0.1983	0.2035	0.1974	0.0014	0.2010
0.2043	0.1963	0.2038	0.2006	0.2010	-0.0059

Table 2: DEA sets obtained from various types of smoothing

The DEAs obtained by each of these smoothing techniques satisfy the conservation condition. It can be seen that lumped smoothing has introduced some negative values (albeit small ones). The iterative and least squares smoothing techniques produce almost the same set of adjusted values. Although iterative smoothing is based on equal absolute changes whilst the least squares technique is based on equal relative changes in this example all non-zero DEAs have almost identical values and so equal absolute and relative changes are equivalent.

The random nature of the Monte Carlo method means that although there are only two different ncn-zero values in the exact DEA matrix the calculated values are all different. Smoothing should help to reduce the differences. Table 3 shows the range of values for each of the two types of DEA (to adjacent and opposite zones).

	Adjacent	Opposite
Averaged	0.1963 - 0.2038	0.1974 - 0.2043
Iterative	0.1981 - 0.2026	0.1987 - 0.2030
Least Squares	0.1981 - 0.2026	0.1987 - 0.2030
Lumped	0.1963 - 0.2038	0.1974 - 0.2043

Table 3: Range of direct exchange area values in different data sets

It can be seen that, in each data set, the range of values spans the exact value. It can also be seen that both iterative and least squares smoothing reduce the sizes of the two ranges and so smoothing has increased the accuracy of the calculated DEA values. The nature of lumped smoothing is such that it does not affect the values of the DEAs to adjacent and opposite zones.

DEAs are not an end in themselves; rather they are a step on the way to determining the heat fluxes out of the zones of the enclosure. In order to determine the heat fluxes it is necessary to specify the surface zone emissivities and temperatures. In this test example all surface zones have emissivity 0.8 and 5 surfaces are given a temperature of 1000K whilst the remaining surface (zone 4) is at 300K. This may be thought of as the hot walls of a furnace radiating to a load on the furnace floor. The table below shows the heat fluxes calculated using each of the four sets of DEAs from tables 1 and 2 and the exact values.

Exact	Original	Iterative	Least Squares	Lumped
8.6528	8.5742	8.5894	8.5882	8.5498
8.6528	8.9755	8.7395	8.7378	8.6380
8.6528	8.3378	8.6638	8.6646	8.7012
-43.2568	-43.0958	-43.2567	-43.2567	-43.1234
8.6455	8.6148	8.6081	8.6081	8.5632
8.6528	8.4187	8.6558	8.6579	8.6713

Table 4: Outward heat fluxes (in kW) for five hot (1000K) and one cold (300K) surfaces

The sum of these fluxes is zero in all cases except using the original values (which do not satisfy the conservation condition). The symmetry of the situation means that the fluxes out of surfaces 1, 2, 3 and 6 are the same. The random nature of the Monte Carlo method means that none of the data sets has these four fluxes equal. However, the difference between the largest and smallest of these values is reduced by a factor of four whichever smoothing technique is used. There is very little to choose between the smoothing techniques. Iterative and least squares are marginally more accurate than lumped, but the errors in the fluxes calculated using the lumped values are less than 2%.

The nature of this test example with so much inherent symmetry and the different non-zero values so close to each other may mean that it is not a fair test of the smoothing techniques. A simple matrix which satsifies the symmetry and conservation conditions in this case is the matrix with 0.2 in all off-diagonal positions and 0 on the diagonal. This matrix is very close to the matrix of exact values and it may be that smoothing is actually moving towards this matrix rather than towards the exact values.

3.2 A Trapezoidal Enclosure

The geometrical symmetry and the uniformity of zone size in the unit cube may work to enhance the effect of smoothing. In this test case an enclosure with a trapezoidal cross-section is considered. The enclosure has height 0.2 at the left end and height 1 at the right. The enclosure has unit width and depth. The surfaces of this enclosure have areas ranging in size from 0.2 to 1.2807 square units. There is a small amount of symmetry (for example, DEAs from the smallest surface to the front and back of the enclosure are the same) but nowhere near as much as for the unit cube. The surface zones of this enclosure are:

Zone 1: Left end, area 0.2
Zone 2: Front, area 0.6
Zone 3: Back, area 0.6
Zone 4: Base, area 1
Zone 5: Roof, area 1.2807
Zone 6: Right end, area 1

The symmetry of the geometry means that DEAs from zone 1 to zones 2 and 3 are the same; similarly those from zone 6 to zones 2 and 3 are the same. The DEAs from zone 4 to zones 1 and 6 can be found analytically (in both cases we have two perpendicular rectangles). Using the formula in [6] we have s_1s_4=0.07744 and s_4s_6=0.200044. These values and the small amount of inherent symmetry allow us to examine the effect of the different smoothing techniques.

Using a Monte carlo method a set of symmetric DEAs for this enclosure was obtained. The values are shown in the table below.

0	0.0286	0.0288	0.0779	0.0269	0.0367
0.0286	0	0.0812	0.1540	0.1860	0.1481
0.0288	0.0812	0	0.1548	0.1854	0.1528
0.0779	0.1540	0.1548	0	0.4121	0.1998
0.0269	0.1860	0.1854	0.4121	0	0.4676
0.0367	0.1481	0.1528	0.1998	0.4676	0

Table 5: Approximate DEAs for trapezoidal enclosure

These DEAs were used as the starting point (matrix X) for each of the smoothing techniques of Section 2. In each case a set of adjusted values was calculated (matrix Y) which satisfies the conservation condition (equation 1.6). These values are shown below.

ITERATIVE SMOOTHING

0	0.0294	0.0283	0.0786	0.0278	0.0358
0.0294	0	0.0811	0.1549	0.1872	0.1474
0.0283	0.0811	0	0.1544	0.1854	0.1508
0.0786	0.1549	0.1544	0	0.4131	0.1989
0.0278	0.1872	0.1854	0.4131	0	0.4670
0.0358	0.1474	0.1508	0.1989	0.4670	0

LEAST SQUARES SMOOTHING

0	0.0287	0.0288	0.0787	0.0270	0.0369
0.0287	0	0.0812	0.1544	0.1878	0.1479
0.0288	0.0812	0	0.1538	0.1851	0.1512
0.0787	0.1544	0.1538	0	0.4150	0.1982
0.0270	0.1878	0.1851	0.4150	0	0.4658
0.0369	0.1479	0.1512	0.1982	0.4658	0

LUMPED SMOOTHING

0.0012	0.0286	0.0288	0.0779	0.0269	0.0367
0.0286	0.0022	0.0812	0.1540	0.1860	0.1481
0.0288	0.0812	-0.0029	0.1548	0.1854	0.1528
0.0779	0.1540	0.1548	0.0014	0.4121	0.1998
0.0269	0.1860	0.1854	0.4121	0.0026	0.4676
0.0367	0.1481	0.1528	0.1998	0.4676	-0.0049

Table 6: DEA sets obtained from different kinds of smoothing

As indicated above the values which we are using as performance indicators are s_1s_4, s_6s_4 which are known exactly, s_1s_2 and s_1s_3 which should be the same and s_6s_2 and s_6s_3 which should also be the same. The nature of lumped smoothing means that it will not adjust any of these values. For ease of comparison these values are extracted from tables 5 and 6 and presented below.

	s_1s_4	s_6s_4	s_1s_2 / s_1s_3	s_6s_2 / s_6s_3
Averaged	0.0779	0.1998	0.0286/0.0288	0.1481/0.1528
Iterative	0.0786	0.1989	0.0294/0.0283	0.1474/0.1508
Least Squares	0.0787	0.1982	0.0287/0.0288	0.1479/0.1512
Lumped	0.0779	0.1998	0.0286/0.0288	0.1481/0.1528

Table 7: Selected direct exchange areas from the four data sets

These values give conflicting evidence about the usefulness of smoothing in increasing the accuracy of specific values. The original values (and hence ones from lumped smoothing) of the s_1s_4 and s_6s_4 DEAs are closer to the exact values than those obtained using both iterative and least squares smoothing. Smoothing has had a detrimental effect on these DEAs. The original values of DEAs s_1s_2 and s_1s_3 were already in good agreement however least squares smoothing does improve this, but iterative smoothing pushes the values further apart. Both techniques improve the poor agreement between the s_6s_2 and s_6s_3 values.

As DEAs are not an end in themselves it is informative to examine the effect of smoothing on the heat fluxes resulting from a full zone method calculation. For this enclosure all surfaces were assumed to have emissivity 0.8, the smallest surface zone (zone 1) was given a temperature of 1200K and all the other zones were cold (300K). (Often in 'real' enclosures the heat source will be the smallest zone.) The only symmetry left by this stage is that the heat flux into surfaces 2 and 3 is the same. The fluxes obtained using the different DEA data sets are given below.

Original	Iterative	Least Squares	Lumped
18.5460	18.5434	18.5431	18.4589
-2.6121	-2.6801	-2.6260	-2.6177
-2.6324	-2.6022	-2.6380	-2.6320
-6.4728	-6.5284	-6.5301	-6.4826
-3.2053	-3.2814	-3.2247	-3.2109
-3.5168	-3.4514	-3.5246	-3.5157

Table 8: Outward heat fluxes (in kW) for one hot (1200K) and five cold (300K) surfaces

The iterative smoothing technique makes the heat fluxes into surfaces 2 and 3 less in agreement than those produced by the original (unconservative) DEAs. This is probably brought about because of the over-correction of values resulting from treating all DEAs equally irrespective of their size. Least squares smoothing brings about a marginal improvement in the agreement showing the benefit of making adjustments proportional (in some way) to the DEA being adjusted.

The sum of the heat fluxes using the original DEAs is 0.1066kW (all the smoothed results sum to zero because of the conservation condition). The simple expedient of lumping the DEA deficit for each row onto the diagonal ensures that conservation is achieved and the difference between the heat fluxes obtained using lumping and the more sophisticated least squares approach is less than 1% for each surface zone.

In this test case a ranking order emerges which shows least squares to be the best technique, followed by lumping with the iterative technique being the worst of the three methods. However, it should be noted that the differences in the zone heat fluxes produced using the three different kinds of smoothing are all less than 2%.

4. CONCLUSIONS

Three techniques for smoothing approximate DEA values which are symmetric but not conservative have been studied. Their effects on DEA values and on the heat fluxes resulting from zone method calculations have been examined. It is found that the least squares smoothing technique which is the most sophisticated mathematically produces the best results. The iterative method is not to be recommended when there are large differences in the size of the zones of the enclosure. The lumped smoothing technique performs surprisingly well given its simplicity and the artificiality it introduces.

Although the different techniques do produce different answers the differences in the final heat fluxes are very small (in fact well below the expected accuracy of the zone method). This is in agreement with the study in [7] which examined the effect of using different weightings in least squares smoothing.

The final conclusion is that least squares smoothing with $w_{ij}=x_{ij}^2$ produces the best results but at the cost of solving a set of simultaneous linear equations. Lumped smoothing produces heat flux values which are little different with virtually no computational overhead.

REFERENCES

1. H.C.Hottel and A.F.Sarofim, Radiative Transfer, McGraw-Hill, 1967.

2. D.K.Edwards, 'Hybrid Monte Carlo Matrix Inversion Formulation of Radiation Heat Transfer with Volume Scattering', 23rd National Heat Transfer Conference, Denver, CO, *HTD-Vol 45*, Symposium on Heat Transfer in Fire and Combustion Systems, pp. 273-278, 1985.

3. J. van Leersum, 'A Method for Determining a Consistent Set of Radiation View Factors from a Set Generated by a Non-exact Method', Int J Heat Fluid Flow, Vol. 10 (1), pp.83-85, 1989.

4. M.E.Larsen and J.R.Howell, 'A Least Squares Smoothing of Direct Exchange Areas in Zonal Analysis', ASME Paper No. 84-HT-40, 1984.

5. R.J.Tucker, Private Communication.

6. R.Siegel and J.R.Howell, Thermal Radiation Heat Transfer, Hemisphere, Washington DC, 1981.

7. C.V.S. Murty and B.S.N.Murty, 'Significance of Exchange Area Adjustment in Zone Modelling', Int J Heat Mass Transfer, Vol 34(2), pp.499-503, 1991.

NOMENCLATURE

A_i	area of surface zone i
c_i	required value for sum of elements in row i of DEA matrix
D_i	size of adjustments to elements in row i of DEA matrix
g_ig_j	direct exchange area from gas zone i to gas zone j
g_is_j	direct exchange area from gas zone i to surface zone j
H	penalty function
k_i	gas absorption coefficient for gas zone i
M	number of gas zones
N	number of surface zones
r	length of beam connecting general points in two zones
s_ig_j	direct exchange area from surface zone i to gas zone j
s_is_j	direct exchange area from surface zone i to surface zone j
V_i	volume of gas zone i
w_{ij}	weight applied to adjustments in ij element of DEA matrix
X	matrix of unadjusted DEA values
x_{ij}	element of matrix X
Y	matrix of adjusted DEA values
y_{ij}	element of matrix Y
δ_i	shortfall in sum of elements of row i of DEA matrix
λ_i	Lagrange multiplier
θ_i	angle between general beam and the normal to surface zone i
$\tau(r)$	transmittance along a beam of length r

IMPROVED BEM SOLUTIONS OF RADIATIVE HEAT TRANSFER PROBLEMS IN PARTICIPATING MEDIA

Ryszard A. Białecki(1)

SUMMARY

An alternative zoning method of solving heat radiation problems in emitting-absorbing media is described. The integral equations of heat radiation are discretized using the Boundary Element Method (BEM). Thus, the existing BEM codes can be, with only minor changes, used to solve radiation problems. Of practical importance is the possibility of using the standard integration error control mechanism of BEM codes. The volume integral arising in the equation of radiation is converted into a surface one. This, and the use of point collocation causes that the entries of the resulting matrix are computed by 2D integration contrary to 4D, 5D and 6D integration required to form the classic zoning matrices. The paper brings also some suggestions concerning an improvement of the Hottel method. It is shown that the latter can be interpreted as a Galerkin solution of the integral equation of heat radiation.

1 INTRODUCTION

Mathematical model of heat radiation comprises the crucial portion of the heat transfer models of high temperature systems. Available numerical methods of handling radiation problems fall into four groups: probabilistic (Monte Carlo) methods, techniques based on integration along the line of sight (flux, discrete ordinates), zoning methods (Hottel's method), hybrid techniques (discrete transfer). An exhaustive discussion of these techniques is available in the literature [1, 2], therefore no attempt is made to repeat this work here.

As the approach discussed in the present paper belongs to the group of zoning methods some questions concerning these techniques should be raised. The disadvantage of the classic Hottel zoning method [4] is the long time needed to form the matrices of the final set of equations. One matrix entry is computed upon evaluating integrals over regions of dimensionality ranging from 4D to 6D. FEM solutions of the heat radiation problems can

[1]Institute of Thermal Technology, Silesian Technical University, Konarskiego 22, 44-101 Gliwice, Poland

be seen as zoning methods. Contrary to other fields of engineering applications, using FEM to solve radiation problems does not become popular. This stems from the fact that this powerful numerical technique is oriented to the solution of differential rather than integral equations.

As the equations discretized in BEM are integral ones, the idea of using BEM in heat radiation context seems to be very promising. The efficiency and relative ease of employing the BEM methodology to handle radiation equations was confirmed in some recent works of the present author [4, 5]. The aim of this paper is to improve and generalize this approach upon adopting some special BEM numerical techniques to deal with heat radiation problems.

The physical situation analyzed in the paper is a nonisothermal, absorbing and emitting medium filling an enclosure formed by diffusively radiating walls. In order to simplify the notation, index λ denoting spectral quantities is skipped. Thus, the equations occurring in the present paper can be interpreted as written either in terms of spectral or total quantities.

2 GOVERNING EQUATIONS OF HEAT RADIATION

The integral equations of heat radiation can be derived in two steps [4]. First, the differential equation of radiative heat transfer is integrated formally along a line of sight L_{rp} linking a current point $\mathbf{r}$ and an observation one $\mathbf{p}$. Then the result is integrated over the entire solid angle for two placements of the observation point. First, the point is located on a solid wall surrounding the participating medium. In this case the integration over solid angle encompasses a hemisphere 2π. Then, the observation point is located within the participating medium. Here, the integration runs over the entire solid angle 4π.

Let S denote the surface bounding a volume V containing a participating medium and T, T^m stand for the temperature distribution of the bounding surface and the participating medium, respectively. Then, the resulting sets of integral equations link the blackbody emissive powers of both the wall $e_b(T)$ and the medium $e_b(T^m)$, the radiative heat flux q^r and the radiative heat source q^r_V. Radiative flux is defined here as a net amount of radiative energy gained by an infinitesimal surface, whereas the radiative heat source is a net amount of radiative energy gained by an infinitesimal volume of the participating medium.

Appropriate integral equations have an appearance

$$q^r(\mathbf{p}) + \epsilon(\mathbf{p})\, e_b[T(\mathbf{p})] = \epsilon(\mathbf{p}) \int_S b(\mathbf{r})\, \tau(\mathbf{r},\mathbf{p})\, K(\mathbf{r},\mathbf{p})\, dS(\mathbf{r}) + \epsilon(\mathbf{p}) \int_V a(\mathbf{r}')\, e_b[T^m(\mathbf{r}')]\, \tau(\mathbf{r}',\mathbf{p}) K_p(\mathbf{r},\mathbf{p})\, dV(\mathbf{r}) \tag{1}$$

$$q^r_V(\mathbf{p}) + 4a(\mathbf{p})\, e_b[T^m(\mathbf{p})] = a(\mathbf{p}) \int_S b(\mathbf{r}) q^r(\mathbf{r})\, \tau(\mathbf{r},\mathbf{p})\, K_r(\mathbf{r},\mathbf{p})\, dS(\mathbf{r}) + a(\mathbf{p}) \int_V a(\mathbf{r}')\, e_b[T^m(\mathbf{r}')]\, \tau(\mathbf{r}',\mathbf{p})\, K_0(\mathbf{r}',\mathbf{p})\, dV(\mathbf{r}') \tag{2}$$

where a is the absorption coefficient of the medium, ϵ is the emissivity of the walls and $\tau(\mathbf{r},\mathbf{p})$ is the transmissivity between points $\mathbf{r}$ and $\mathbf{p}$ computed as a line integral along a line of sight

$$\tau(\mathbf{r},\mathbf{p}) = \exp[-\int_{L_{rp}} a(\mathbf{r}')dL(\mathbf{r}')] \tag{3}$$

and $b(\mathbf{r})$ is the radiosity defined as a sum of radiative energy emitted by an infinitesimal surface and reflected therefrom. Radiosity can be expressed as a linear combination of the emissive power and radiative heat flux as

$$b = e_b + \frac{1-\epsilon}{\epsilon}q^r \tag{4}$$

The kernel functions entering equations (1) and (2) are defined as

$$K(\mathbf{r},\mathbf{p}) = \frac{\cos\phi_r \cos\phi_p}{\pi|\mathbf{r}-\mathbf{p}|^2} \qquad K_p(\mathbf{r}',\mathbf{p}) = \frac{\cos\phi_p}{\pi|\mathbf{r}'-\mathbf{p}|^2}$$
$$K_0(\mathbf{r}',\mathbf{p}) = \frac{1}{\pi|\mathbf{r}'-\mathbf{p}|^2} \qquad K_r(\mathbf{r},\mathbf{p}) = \frac{\cos\phi_r}{\pi|\mathbf{r}-\mathbf{p}|^2} \tag{5}$$

with $|\mathbf{r}-\mathbf{p}|$ denoting the distance between points $\mathbf{r}$ and $\mathbf{p}$ and ϕ_r and ϕ_p being the angles made by the line of sight and normals at points $\mathbf{r}$ and $\mathbf{p}$, respectively.

As both surface and volume integrals arise in the integral equations of heat radiation, the procedure of discretization of radiation transfer equations involves numerical treatment of both types of integrals. The crucial factor here is the volume integral, whose discretization is one order more expensive than the surface one. The amount of CPU can be significantly reduced upon converting the volume integral into a surface one. This is accomplish using an identity valid for arbitrary function $f(\mathbf{r}',\mathbf{p})$ [4]:

$$\int_V \frac{f(\mathbf{r}',\mathbf{p})}{|\mathbf{r}'-\mathbf{p}|^2}\,dV(\mathbf{r}') = \int_S \left\{ \int_{L_{rp}} f(\mathbf{r}',\mathbf{p})\,dL_{rp}(\mathbf{r}') \right\} \frac{\cos\phi_r}{|\mathbf{r}-\mathbf{p}|^2}\,dS(\mathbf{r}) \tag{6}$$

Provided the line integration can be carried out analytically, this reduces the dimensionality of the integral. Applying Eq. (6) to convert the volume integrals in Eqs. (1), (2) into surface ones yields

$$q^r(\mathbf{p}) + \epsilon(\mathbf{p})e_b[T(\mathbf{p})] = \epsilon(\mathbf{p}) \int_S b(\mathbf{r})\tau(\mathbf{r},\mathbf{p})K(\mathbf{r},\mathbf{p})\,dS(\mathbf{r})$$
$$+ \epsilon(\mathbf{p}) \int_S \left\{ \int_{L_{rp}} a(\mathbf{r}')\,e_b[T^m(\mathbf{r}')]\,\tau(\mathbf{r}',\mathbf{p})\,dL_{rp}(\mathbf{r}') \right\} K(\mathbf{r},\mathbf{p})\,dS(\mathbf{r}) \tag{7}$$

$$q_V^r(\mathbf{p}) + 4a(\mathbf{p})e_b[T^m(\mathbf{p})] = a(\mathbf{p}) \int_S b(\mathbf{r})\tau(\mathbf{r},\mathbf{p})K_r(\mathbf{r},\mathbf{p})\,dS(\mathbf{r})$$
$$+ a(\mathbf{p}) \int_S \left\{ \int_{L_{rp}} a(\mathbf{r}')\,e_b[T^m(\mathbf{r}')]\,\tau(\mathbf{r}',\mathbf{p})\,dL_{rp}(\mathbf{r}') \right\} K_r(\mathbf{r},\mathbf{p})\,dS(\mathbf{r}) \tag{8}$$

Formulation (1) and (2) will be referred to hereafter as the standard one whereas Eqs. (7) and (8) will be termed the BEM formulation of heat radiation equations.

3 HOTTEL'S ZONING METHOD AS A GALERKIN SOLUTION

To convert integral Eqs. (1) and (2) into an equivalent sets of algebraic equations one have to discretize surface, volume and line integrals. The latter though not explicitly present in equations enter the definition of the transmissivity τ [c.f. Eq. (3)]. Let the entire volume be subdivided into finite number of volume elements (zones) ΔV_j; $j = 1, 2, \dots L_V$. Similarly, the surface is subdivided into a finite number of surface elements (zones) ΔS_e; $e = 1, 2, \dots E$. It is assumed that the medium and wall temperature distributions as well as those of the radiative heat fluxes and sources are described by step functions. Using the FEM terminology this means that the functions of interest are approximated by shape functions of zeroth order (constant within zone). Substitution of these approximate distributions into the governing equations (1) and (2) produces residuals. Upon using the standard Galerkin approach i.e. making these residuals orthogonal to the approximating functions one arrives at a set of linear equations linking constant zone values of emissive powers and radiative heat fluxes and sources. Using the, somewhat strange, original Hottel's notation the resulting sets of algebraic equations can be written as

$$\left[\frac{q_k^r}{\epsilon_k} + e_b(T_k)\right] \Delta S_k = \sum_{e=1}^{E} \left[e_b(T_e) + \frac{1-\epsilon_e}{\epsilon_e} q_e^r\right] \overline{s_k s_e} + \sum_{j=1}^{L_V} e_b(T_j^m) \overline{s_k g_j} \tag{9}$$

$$[q_{V_i}^r + 4a_i e_b(T_i^m)] \Delta V_i = \sum_{e=1}^{E} \left[e_b(T_e) + \frac{1-\epsilon_e}{\epsilon_e} q_e^r\right] \overline{g_i s_e} + \sum_{j=1}^{L_V} e_b(T_j^m) \overline{g_i g_j} \tag{10}$$

The overbared double letter symbols can be interpreted as entries of matrices and are defined as

$$\begin{aligned}
\overline{s_k s_e} &= \int_{\Delta S_k} \int_{\Delta S_e} K(\mathbf{r}, \mathbf{p}) \tau(\mathbf{r}, \mathbf{p}) \, dS(\mathbf{r}) \, dS(\mathbf{p}) \\
\overline{s_k g_j} &= a_j \int_{\Delta S_k} \int_{\Delta V_j} K_p(\mathbf{r}', \mathbf{p}) \tau(\mathbf{r}', \mathbf{p}) \, dV(\mathbf{r}') \, dS(\mathbf{p}) \\
\overline{g_i s_e} &= a_i \int_{\Delta V_i} \int_{\Delta S_e} K_r(\mathbf{r}, \mathbf{p}') \tau(\mathbf{r}, \mathbf{p}') \, dS(\mathbf{r}) \, dV(\mathbf{p}') \\
\overline{g_i g_j} &= a_i a_j \int_{\Delta V_i} \int_{\Delta V_j} K_0(\mathbf{r}', \mathbf{p}') \tau(\mathbf{r}', \mathbf{p}') \, dV(\mathbf{r}') \, dV(\mathbf{p}')
\end{aligned} \tag{11}$$

These entries are referred to as *surface-surface, gas-surface, surface-gas and gas-gas direct exchange areas*, respectively. Eqs. (9) through (11) are identical with those obtained by Hottel who derived the equations by purely physical reasoning. It is therefore not surprising that the integral equations of transfer (1) and (2) did not arise in his book.

The line integrals (transmissivities) entering the definitions of the direct areas require some additional comments. The transmissivity (3) cannot, at least in its general form, be evaluated analytically. However, when constant temperatures and absorption coefficients within volume elements are assumed, the integration can be carried out analytically. Under such assumptions, the transmissivity can be calculated as

$$J_1 = \tau(\mathbf{r},\mathbf{p}) \approx \tilde{\tau}(1, I_{rp}) = \exp\left(-\sum_{l=1}^{I_{rp}} a_l d_l\right) \tag{12}$$

where a_l is the absorption coefficient within the lth cell intersected by a ray travelling from point $\mathbf{r}$ to $\mathbf{p}$ (along the line of sight L_{rp}), d_l is the length of ray within lth cell. An efficient algorithm for determining d_l is described in the literature [4]. I_{rp} is the number of cells intersected by a ray travelling from point $\mathbf{r}$ to $\mathbf{p}$, and $\tilde{\tau}(1, I_{rp})$ is the approximate value of transmissivity between points $\mathbf{r}$ and $\mathbf{p}$.

As already shown, the classic zoning method can be interpreted as Galerkin's weighted residuals solution of the standard integral equations of heat radiation with weighting and approximating functions being constant in subregions. As the equations of the FEM are typically derived *via* Galerkin technique, the zoning method has some common roots with FEM. Knowing the mathematical interpretation of the zoning method, one can think of improving its accuracy by introducing higher order shape and weighting functions widely used in FEM.

4 BEM DISCRETIZATION

4.1 Discretization

Contrary to standard formulation (1) and (2) volume integrals do not arise in Eqs. (7) and (8). Thus, to convert the BEM equations of heat radiation into sets of algebraic equations it is enough to discretize only the surface and line integrals. Because, similarly as in the standard formulation, the line integration can be carried out analytically, the discretization is limited to surface integrals. The technique of doing this is completely analogous to the standard BEM approach i.e. the surface is subdivided into a number of boundary elements ΔS_e; $e = 1, \ldots E$ with both geometry and distribution of functions approximated using the shape functions. The final form of the algebraic equations is then obtained by nodal collocation being a variant of the weighed residuals technique. It should be stressed that the method requires also a subdivision of the entire volume into isothermal cells ΔV_l; $l = 1, \ldots L_V$ of constant absorption coefficient. However, there is a substantial difference between the volume cell introduced here and the volume finite element present in the Hottel method. While the first are introduced only in order to compute the line integrals analytically, the aim of the second is to discretize the volume integral. From this viewpoint Hottel's technique requires introducing both volume cells and volume finite elements.

Discretization of Eq. (7) yields

$$\mathbf{A}\mathbf{e_b}(T) = \mathbf{B}\mathbf{q}^r + \mathbf{C}\mathbf{e_b}(T^m) \tag{13}$$

where the vectors $\mathbf{e_b}(T)$ and $\mathbf{q}^r$ contain emissive powers and radiative heat fluxes at nodes placed on the boundary while the vector $\mathbf{e_b}(T^m)$ contains emissive powers associated with volume cells. The entries of matrices $\mathbf{A}$ and $\mathbf{B}$ are calculated from relationships

$$A_{kj} = -\sum_e \left\{ \int_{\Delta S_e} K(\mathbf{r},\mathbf{p}_k)\Phi_j^e(\mathbf{r})\,\tau(\mathbf{r},\mathbf{p}_k)\,dS(\mathbf{r}) \right\} + \delta_{kj} \tag{14}$$

$$B_{kj} = \sum_e \left\{ \int_{\Delta S_e} \frac{1-\epsilon(\mathbf{r})}{\epsilon(\mathbf{r})} K(\mathbf{r},\mathbf{p}_k)\Phi_j^e(\mathbf{r})\,\tau(\mathbf{r},\mathbf{p}_k)\,dS(\mathbf{r}) \right\} - \frac{\delta_{kj}}{\epsilon(\mathbf{p}_k)} \tag{15}$$

where Φ_j is the shape function associated with jth node and the summation runs over all boundary elements sharing this node. The transmissivity τ is computed from Eq. (12) as in the Hottel method.

To compute the entries of the matrix $\mathbf{C}$ another(double) line integral should be evaluated. This integral can be, similarly as the J_1 one, computed analytically. Appropriate relationship reads

$$\begin{aligned} J_2 &= \int_{L_{rp}} a(\mathbf{r}')\, e_b[T^m(\mathbf{r}')]\, \tau(\mathbf{r}',\mathbf{p})\, dL_{rp}(\mathbf{r}') \\ &\approx \sum_{l=1}^{I_{rp}} e_b(T_l^m) A_l \exp\left(-\sum_{j=l+1}^{I_{rp}} a_j d_j \right) \end{aligned} \tag{16}$$

where $A_l = 1 - \exp(-a_l d_l)$ is the self absorption of radiation emitted within the lth cell. The summation index j runs over subsequent cells passed by the ray after leaving the lth cell. It should be stressed that all quantities needed to compute the J_2 integral are already determined when evaluating the J_1 one. Finally the entries of the $\mathbf{C}$ matrix can be expressed as

$$C_{kl} = \sum_{j=1}^{N} \sum_e \int_{\Delta S_e} K(\mathbf{r},\mathbf{p}_k)\Phi_j^e(\mathbf{r})\, A_l \exp\left(-\sum_{j=l+1}^{I_{rp}} a_j d_j \right) dS(\mathbf{r}) \tag{17}$$

Discretization of Eq. (8) yields

$$q_{Vi}^r = \sum_{j=1}^{N} L_{ij} q_j^r - \sum_{j=1}^{N} M_{ij} e_b(T_j) + \sum_{l=1}^{L_V} N_{il} e_b(T_l^m) \tag{18}$$

with coefficients L_{ij}, M_{ij}, N_{il} computed similarly as A_{ij}, B_{ij}, C_{kl} in Eq. (13)

5 NUMERICAL INTEGRATION IN HEAT RADIATION

As already mentioned both in standard BEM and in BEM applied to radiation, integral equations are being discretized. Because in both cases these equations are *strongly singular*, the analogy of these two application areas of BEM are much deeper. Strongly singular kernels behave in 3D like $|\mathbf{r} - \mathbf{p}|^{-2}$. The value of such integrals exists only in the Cauchy principle value sense. Therefore, when solving heat radiation problems special numerical integration techniques developed in BEM to deal with such situation should be used. Owing to the same behaviour of the integrand in classic BEM applications and in heat radiation, the integration routines can be copied from BEM codes to radiation ones without any modifications. This obviously saves the time of developing the BEM radiation codes and offers the possibility of taking over the entire experience in handling singular integrals gained by BEM researchers so far.

One row of a BEM matrix contains only one strongly singular integral, the remaining entries are expressed as a sum of regular integrals. Singular integrals correspond to a situation when the collocation point lies within the boundary element over which the integration is carried out. Such kind of integrals is usually handled using an analog of the *rigid body movement condition* stating that for isothermal fields, all heat fluxes must vanish. Thus, after the remaining, regular integrals are computed by quadratures, the singular entry e.g. the singular, diagonal element of the $\mathbf{A}$ matrix occurring in Eq. (13) can be readily obtained from the relationships

$$A_{ii} = -\sum_{j=1, j\neq i}^{N_n} A_{ij} + \sum_{l=1}^{L_V} C_{il} \tag{19}$$

where N_n is the number of nodal points on the boundary.

The cases when the observation point lies very close to the element requires special treatment. In such a case the integrand becomes 'nearly singular' and the standard numerical integration fail to give precise results. This question has been also solved by BEM modellers upon using the concept of adaptive integration. Appropriate procedures can be used in heat radiation context.

One of such approaches developed recently [6] to deal with potential problems ruled by differential equations has been used here in the context of heat radiation. The procedure is based on the Gaussian error bound formula linking the error ε to the relative, minimum distance between the collocation point and the element, and the required number of Gaussian nodes. The equation has a form [7]

$$4(2G_\xi + 1)\left[\frac{D_\xi}{4L_{min}}\right]^{2G_\xi} \leq \varepsilon \tag{20}$$

where L_{min} is the minimum distance between the source point and the boundary element, D_ξ is the maximum length of the element in the direction of local coordinate ξ and G_ξ is the required number of Gaussian nodes in that direction. Similar equation holds for the second local coordinate associated with a boundary element.

Closer inspection of Eq. (20) shows that at small relative distances ($L_{min}/D_{xi} < 0.25$) the integration accuracy cannot be practically increased by taking more quadrature nodes. In this case the domain of integration should be divided into subdomains. This subdivision process can be programmed in such a way that the integration process is self adaptive. Appropriate technique is borrowed from the computer graphics where it is known as a *quadtree algorithm.*

Necessary minimum distance L_{min} is determined solving a curtailed nonlinear programming problem. The accuracy of integration can be checked for a specific case of flat elements. In this case the singular integral vanishes (a flat element cannot irradiate itself) and the error in satisfying the rigid body movement is a measure of the quality of integration. Calculations carried out by the present author shows that for an example with about 700 boundary elements and 100 volume cells the integration accuracy $\varepsilon = 10^{-3}$ produced an error in satisfying the rigid body movement less than 10^{-5}.

Although the integrand arising in heat radiation differs from that occurring in other application fields, the majority of quantities necessary to evaluate the value of the heat radiation integrand are available in standard BEM codes. This concerns the distance between the collocation and integration point and the coordinates of the local normal. The knowledge of these quantities enables one the evaluation of both the kernel functions (5) and the infinitesimal surface element. The new quantities to be computed in heat radiation are restricted to those associated with the line integration. For a frequently occurring case of transparent media the transmissivity is 1 while for isothermal media it is equal to $\exp(-a|\mathbf{r}-\mathbf{p}|)$. Hence, in these cases the amount of changes to be introduced into a standard BEM code to make it capable of dealing with heat radiation is indeed insignificant.

6 TREATMENT OF SYMMETRIES

Symmetry of temperature fields in heat conducting solids is frequently encountered in practical problems. One can neglect the symmetry and use this condition to check the correctness of the obtained solution. However, taking into account symmetry can lead to substantial reduction of both computing and data preparation time.

Standard method of handling these planes in potential problems is to treat them as new boundaries with homogeneous Neumann condition. This approach cannot be used when solving heat radiation problems. Although the radiative heat flux vanishes on the planes of symmetry, but the integral equations of radiation have been derived for diffuse emission and reflection on all bounding surfaces. Therefore, the zero radiative flux condition imposed on the symmetry planes would be equivalent to an erroneous assumption, that the ray reaching these imaginary planes can be reflected therefrom in an arbitrary direction. In fact the plane of symmetry is a specular surface. This can be explained by simple reasoning: consider a domain consisting of two symmetric subdomains and a ray originating at point $\mathbf{r}$ and impinging the plane of symmetry at point $\mathbf{q}$. At this point the ray leaves the first subdomain. Simultaneously, another ray emitted at a point $\mathcal{M}\{\mathbf{r}\}$ being a symmetric image of point $\mathbf{r}$ (and thus belonging to the second subdomain), enters the first subdomain at the same point $\mathbf{q}$. Apart from its direction the ray entering the first subdomain has exactly the same features, as that which leaves it. The direction of the incoming ray is symmetrical with respect to the normal to the plane of symmetry. Hence, this plane should be treated as if it were an ideal mirror.

BEM offers an elegant method of treating symmetries of the sought for fields. An important feature of this approach is, that instead of introducing a new boundary condition on the plane of symmetry, advantage is taken of the symmetry of the kernel functions. Thus, the technique does not introduce the assumption of diffuse reflection on the planes of symmetry and the it can be used to model symmetry in radiation.

Application of this technique to radiation will be described discussing the simplest possible case of a field having one plane of symmetry. The entire domain is divided by this plane into two equal parts. Superscript I and II will be appended to quantities associated with the first and second halves, respectively. Let $\mathcal{M}$ denote the symmetry relation transforming one half of the entire domain into its symmetrical counterpart. Integral BEM equation of heat radiation (the case of transparent medium is assumed for simplicity) can written as

$$\begin{aligned} q^r(\mathbf{p}) &+ \epsilon(\mathbf{p})e_b[T(\mathbf{p})] = \epsilon(\mathbf{p}) \int_{S^I} b(\mathbf{r})K(\mathbf{r},\mathbf{p})\, dS^I(\mathbf{r}) \\ &+ \epsilon(\mathbf{p}) \int_{S^{II}} b(\mathcal{M}\{\mathbf{r}\})K(\mathcal{M}\{\mathbf{r}\},\mathbf{p})\, dS^{II}(\mathcal{M}\{\mathbf{r}\}) \end{aligned} \tag{21}$$

It can be shown using elementary geometric reasoning, that the kernel function K is invariant with respect to symmetry transformation i.e.

$$K(\mathcal{M}\{\mathbf{r}\},\mathbf{p}) = K(\mathbf{r},\mathcal{M}\{\mathbf{p}\}) \tag{22}$$

Making use of the symmetry of kernel function (22) and that of the emissivity and radiosity, and taking into account the obvious invariance of the area of the infinitesimal surface dS, Eq. (21) can be rewritten in a form

$$q^r(\mathbf{p}) + \epsilon(\mathbf{p})e_b[T(\mathbf{p})] = \epsilon(\mathbf{p}) \int_{S^I} b(\mathbf{r})\,[K(\mathbf{r},\mathbf{p}) + K(\mathbf{r},\mathcal{M}\{\mathbf{p}\})]\; dS^I(\mathbf{r}) \tag{23}$$

The analysis of Eq. (23) leads to four conclusions

- only a half of the entire boundary should be discretized. Contrary to the standard treatment of the symmetry, the symmetry planes need not to be discretized,
- symmetry can be accounted for upon replacing the actual kernel function by a sum of two kernel functions. Both functions depend on the same current point $\mathbf{r}$. The second argument is the source point $\mathbf{p}$ and its image $\mathcal{M}\{\mathbf{p}\}$. Thus, this technique originally developed in BEM to be employed in problems ruled by differential equations can be used to handle symmetry in BEM radiation codes practically without any changes,
- imaginary symmetry planes behave like ideal mirrors. Obviously, the method used to handle the symmetry can be used to model specular

reflections from real surfaces without any difficulty. Hence, BEM is capable of handling problems in cavities consisting of diffuse and mirrorlike surfaces,

- as the transmissivity and the kernel function K_r are also invariant with respect to the symmetry transformation, the procedure can be readily generalized to handle nontransparet media. Generalization of the technique to a greater number of symmetry planes and 2D geometry presents no difficulties.

7 COMPARISON OF BEM AND THE HOTTEL METHOD

When comparing the classic Hottel method with BEM following points should be raised.

1. both approaches have the same field of applications, i.e. they can deal with similar physical situations,
2. it is possible in BEM to use higher accuracy approximation of both the involved functions and geometry. This can be accomplished upon using the concept of higher order shape functions.
3. the data sets needed to run a BEM radiation code are practically the same as in other fields of BEM applications enabling one the usage of the available BEM and FEM pre- and postprocessors.
4. reliable routines to control the accuracy of integration are available in BEM
5. treatment of symmetry and specular surfaces can be readily included in BEM codes
6. BEM codes need significantly less CPU times. This comes from two factors
   a) entries of BEM matrices are computed by 2D integration as opposed to 4D, 5D and 6D integration required to evaluate coefficients of zoning matrices.
   b) negligible time of discretization of the integrals describing the radiation generated within the medium.

The latter question demands more explanation. Both formulations of heat radiation equations, the standard one (1), (2) and the BEM (7), (8) contain two integrals. The first is due to surface radiation and has the same shape in both versions. The second integral describes the radiation emitted within the medium. This integral is in the standard formulation a volume one, while in BEM it is a surface integral. The discretization of volume integrals is much more expensive than the same procedure applied to surface integrals. Thus, in Hottel's method the factor controlling the time of matrix generation is the handling of the volume integral. Contrary to this situation, the BEM (surface) analog of this integral is discretized almost 'for free'. The reason for this neat behaviour is that there are two sets of quantities needed to carry out the discretization of the integral in question: the values of the kernel function and the lengths of rays d_l intersected by the limits of cells. All these quantities are computed (also in the zoning method) when discretizing the integral describing the surface radiation. Thus, the discretization of the integral due to medium radiation can be in BEM accomplished parallel to that due to surface radiation. The cost of the numerical treatment of the integral due to participating medium emission is negligible as all necessary quantities are already available.

8 EXAMPLE

An experimental combustion chamber has been considered. The temperatures within the chamber has been measured and the distribution of the absorption coefficient of the combustion gases have been determined upon measuring radiation intensities [8]. 700 boundary constant elements and 100 volume cells have been used to discretize the problem. The resulting distributions of the radiative heat flux on two faces of the chamber are depicted in Fig. 1.

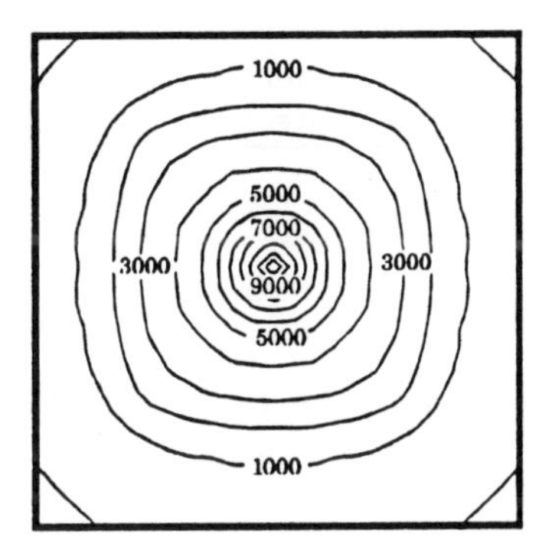

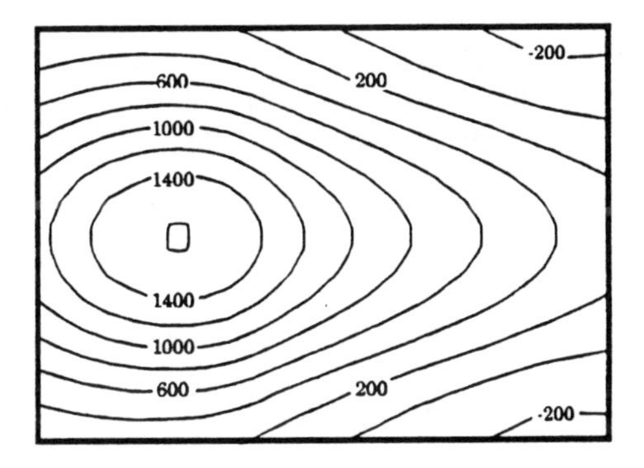

Figure 1: Radiative heat flux distribution over two faces of a combustion chamber. Isolines are marked in W/m^2

ACKNOWLEDGEMENTS
The work has been supported by the Polish Committee of Scientific Research under the grant 3 3113 92 03.

References

[1] HOWELL J.R. Thermal radiation in participating media the past, the present and some futures. Journal of Heat Transfer, Vol. 110, No. 4, pp. 1220–1229, 1988.

[2] VISKANTA R. and MENGÜC M.P. Radiation heat transfer in combustion systems. Progress in Energy Combustion Science, Vol. 13, 97–160, 1987.

[3] HOTTEL, H.C. and SAROFIM, A.F. Radiative Transfer, 2nd ed., McGraw Hill, New York, 1973.

[4] BIALECKI, R.A. Applying the Boundary Element Method to the solution of heat radiation problems in cavities filled by a nongray emitting-absorbing medium, Numerical Heat Transfer, Part A, Vol. 20, pp. 41–64, 1991.

[5] BIALECKI R.A. Solving Heat Radiation Problems Using the Boundary Element Method, Elsevier Applied Science, London 1993, to appear

[6] BIALECKI R.A., DALLNER R. and KUHN G. Minimum distance calculation between a source point and a boundary element, Engineering Analysis with Boundary Elements, submitted for publication.

[7] LACHAT, J.C. and WATSON, J.O. Effective numerical treatment of boundary integral equations: a formulation for three dimensional elastostatics, International Journal for Numerical Methods in Engineering, Vol. 10, 991-1005. 1976

[8] NADZIAKIEWICZ J. Theoretic-experimental model of radiation transport in the gas flame. Zeszyty Naukowe Politechniki Slaskiej, Energetyka, Vol. 105, 1989, (in Polish)

NOMENCLATURE

Symbol	Description
a	absorption coefficient
b	radiosity
$\mathbf{A},\mathbf{B},\mathbf{C}$	BEM radiation matrices
d_l	length of ray within lth cell
D_ξ	length of an edge of an boundary element measured along a local coordinate ξ
e_b	blackbody emissive power
I_{rp}	number of cells intersected by a ray going from $\mathbf{r}$ to $\mathbf{p}$
L_{rp}	line of sight
L_{min}	minimum distance between a source point and a boundary element
K, K_r, K_p, K_0	known kernel functions
q^r, q^r_V	radiative heat flux and source, respectively
$\mathbf{r},\mathbf{p}$	current and observation point, respectively
S	surface bounding the domain
T, T^m	temperature of the surface and the medium, respectively
V	participating medium domain
ϵ	emissivity
ε	relative integration error
τ	transmissivity
ϕ_r, ϕ_p	angle made by the line of sight connecting points
Φ_j	shape function
ξ	local coordinate
$\mathcal{M}$	symmetry relation

SIMULATION OF RADIATION HEAT TRANSFER IN TELEVISION TUBES

E.J.W. ter Maten[1]

[1]Philips Electronics N.V., Philips Research Laboratories, Applied Mathematics Group, Bldng WL-p.1.22, P.O. Box 80.000, 5600 JA Eindhoven, the Netherlands; email: maten@prl.philips.nl.

1 ABSTRACT

Heat transfer inside cathode ray tubes (CRT) is governed dominantly by radiation. Due to this radiation, the temperature of a particular object in the CRT is closely connected with the thermal behaviour of the complete CRT. Because radiation depends on how boundaries of objects see each other, it is necessary to incorporate their viewfactors. This can be modelled effectively by means of an integral operator.
The heat transfer in the interior of the materials is described by the usual heat transfer equation. Hence, the complete problem is governed by a linear partial differential equation that uses a non-linear integral operator in the description of the net fluxes at boundaries.
Finite-element modelling results in a non-linear system of equations that may be solved by some iterative non-linear solver like the Newton-Raphson method. Due to the viewfactors between parts of all components, intermediate linear systems that have to be solved for such a method have a fairly full coefficient matrix. A good initial solution will speed up the convergence of the non-linear solver.
In this paper we describe a construction of such an initial solution for the non-linear solver. Our approach exploits some general geometrical properties, like thinness, of the objects within the cavity. We also study efficient linear solvers that can deal with the intermediate large and full linear systems.

2 GEOMETRY

A CRT forms a closed cavity in which several components are placed (cf. Fig. 1). The inner components are made of thin metal sheet. The outer boundary consists of glass. The inside of the CRT is vacuum. Here, the objects may be isolated, but are thermally coupled through radiation. At all internal boundaries various coatings give rise to a spatially varying emissivity parameter along the boundaries. In addition, the emissivities are temperature-dependent. On specific sides of the inner objects heat is generated by absorbed energy of incoming electrons. At the outer boundary of the cavity heat loss occurs due to radiation to the outside and by exchange of energy between the CRT and the air by means of convection.

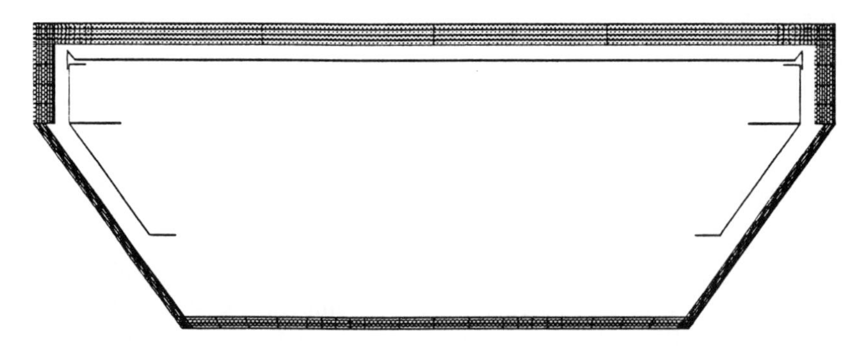

Fig. 1: Simple-configuration cross-section model of a CRT.

3 STEADY-STATE HEAT TRANSFER

We will concentrate on a steady-state model. Our interest lies in the thermal behaviour of the objects inside the CRT, assuming an applied temperature profile at the outer boundary. In our simulations we will be mainly interested in the temperature profile along the horizontal part of the 'mask' (i.e. the thin horizontal object that is just below the top of the cavity). In a thermo-mechanical analysis this temperature profile can be used for studying the doming of the mask in order to limit distortion of the colour display.

In the sequel we denote the complete set of objects by Ω and the union of all boundaries by Γ. The temperature is denoted by T, expressed in

Kelvin. The steady-state heat transfer on Ω is defined by

$$\text{div}(\lambda \nabla T) = 0. \tag{1}$$

Here λ is the thermal conductivity and may depend on the spatial coordinates and on temperature.
We assume that Γ is split into two disjoint components: $\Gamma = \Gamma_{\text{Flux}} \cup \Gamma_{\text{Dir}}$, on which either flux or Dirichlet boundary conditions are specified. In our model Γ_{Dir} is the boundary of the cavity, while Γ_{Flux} consists of all boundaries of the inner components. The net incoming flux Φ_{Heat} exists on Γ_{Flux}. It sums several effects and is defined by

$$\Phi_{\text{Heat}} = \lambda \nabla T \cdot \mathbf{n} = -Q_r + e. \tag{2}$$

The terms at the right-hand side describe different features. The term $Q_r(T)$ describes the net outward flow due to radiation. It is a non-linear function of T. We will consider this term more closely in Section 4. The term e covers additional heat input (due to incoming electrons). It is a given function of the spatial coordinates.
On Γ_{Dir} the temperature T is imposed. It describes the heat input from deflection coils as well as the effects of the slowly time-varying temperature profiles on screen and cone of the CRT.

4 THERMAL RADIATION

The heat flux Q_r, on Γ_{Flux}, is defined as the net effect due to radiation

$$Q_r = W_{out} - W_{in}, \tag{3}$$
$$W_{out} = \varepsilon \sigma T^4 + \rho W_{in}, \tag{4}$$
$$W_{in} = F(W_{out}). \tag{5}$$

The term W_{in} represents the incoming heat flux, or irradiation density. The term W_{out} describes the radiosity density, which is expressed as the emitted radiation plus a portion of W_{in} due to reflection. The scalar function ε denotes the emissivity of the surface and gives the fraction of the total power (σT^4 for black-body radiation) that is emitted. ε incorporates an accumulated effect over all frequencies that occur in the radiation, which is an appropriate assumption for 'gray body radiation'. $\sigma = 5.6703\,10^{-8}$ W/(m^{2}K^4) is the Stefan-Boltzmann constant. The fraction function ρ is a scalar and denotes the reflectivity of the surface. The functions ρ and ε may depend on x, y and on T. They are assumed to obey $\|\rho\|_\infty \leq 1$ and $\|\varepsilon\|_\infty \leq 1$ respectively ($\|.\|_\infty$ being the supremum norm).

In general a portion ρ of W_{in} is reflected, a portion α (absorptivity) is absorbed and a portion τ (transmissivity) is transmitted. We will assume that $\tau = 0$. For surfaces that are classified as 'gray Lambert', one also has $\alpha = \varepsilon$, which results in a simple relation between ρ and ε [8],

$$\rho = 1 - \varepsilon. \tag{6}$$

This supports the introduction of W_{out} in (4) as a linear combination of two effects.

4.1 RADIATION TRANSFER FUNCTION

The radiation heat transfer function F is a bounded linear function from the space $L^2(\Gamma)$ into itself. In a specific boundary point $x \in \Gamma$ the resulting image function sums all fractions of the radiosity from all points of the boundary that are 'seen' by the point x. It is always possible to express F as an integral operator with density function $f(x,s)$:

$$F(v)(x) = \oint_\Gamma f(x,s)v(s)ds, \quad x \in \Gamma. \tag{7}$$

For physical reasons it can be assumed that F and f have the following general properties:

$$F(1) \equiv 1 \text{ and } f \geq 0. \tag{8}$$

In addition, $\|F\| \leq 1$. However, we will sharpen the assumptions with respect to the norms of F and ρ such that $\|F\rho\| < 1$ and $\|\rho F\| < 1$. Notice that ρ is a function on Γ. For bounded ρ the functions ρF and $F\rho$ are defined on $L^2(\Gamma)$ by $(\rho F)(w) = \rho F(w)$ and $(F\rho)(w) = F(\rho w)$ for all $w \in L^2(\Gamma)$, respectively. These assumptions imply that the inverses of $I - \rho F$ and of $I - F\rho$ exist and can be expressed in terms of a Neumann series. One easily verifies that:

$$F(I-\rho F)^{-1} = (I - F\rho)^{-1}F, \tag{9}$$
$$(I-\rho F)^{-1}\rho = \rho(I - F\rho)^{-1}. \tag{10}$$

We can derive the following general explicit expressions for Q_r as a function of T^4.

THEOREM 1

$$Q_r = [(I-\rho F)^{-1} - (I-F\rho)^{-1}F]\varepsilon\sigma T^4 \tag{11}$$
$$= (I-F)(I-\rho F)^{-1}\varepsilon\sigma T^4 \tag{12}$$
$$= (1-\rho)(I-F\rho)^{-1}(I-F)\frac{1}{1-\rho}\varepsilon\sigma T^4 \text{ if } \|\rho\|_\infty < 1. \tag{13}$$

A discrete version of (13) using surfaces with a locally uniform temperature distribution, and assuming (6), can be found in [8,9].

4.2 GEOMETRICAL VIEWFACTORS

The radiation function F is related to the geometrical viewfactors that can be evaluated by several simulation packages [5,9]. For heat radiation between two surfaces A_i and A_j fully diffuse reflection and emission are assumed. Based on Lambert's cosine law [8], one finds that the fraction $\mathcal{F}_{ji}$ of the total radiant energy (with uniform density $W_{out}(s)$, $s \in A_j$) leaving A_j, and arriving directly on A_i, can be determined using the geometry only. The definition implies reversibility (the configuration reciprocity relation)

$$\mathcal{F}_{ji}A_j = \mathcal{F}_{ij}A_i. \tag{14}$$

We can now describe the operator F (cf. the relation (5)) more clearly. We consider N surfaces A_i. Then $\mathcal{F}_{ji}A_jW_{out,j}$ is the resultant energy from A_j that arrives on A_i. Using the principle of conservation of energy on the total energy $A_iW_{in,i}$ that is received on A_i and applying relation (14), we find that $\mathcal{F}$, the discrete equivalent of F, is expressed in terms of $\mathcal{F}_{ij}$

$$W_{in,i} = \sum_j \mathcal{F}_{ij}W_{out,j} = F(W_{out})_i. \tag{15}$$

Therefore, while the 'natural definition' of $\mathcal{F}$ applies to the energy quantities $A_iW_{in,i}$ and $A_jW_{out,j}$, the transposed operator is used in the relation between the densities and will be used as our operator F.
From the principle of conservation of energy there immediately follows

$$\forall j: \ \sum_i \mathcal{F}_{ji} = 1 \Rightarrow \|\mathcal{F}\|_\infty = 1. \tag{16}$$

4.3 NON-COMPACT INTEGRAL OPERATOR

In order to calculate Q_r from a given temperature profile T on Γ, one has to solve an integral equation like

$$(I - K)q = g. \tag{17}$$

For several procedures it is convenient that K is a compact operator [2]. Hence, we are interested in the question whether F is compact or not. As long as x and s are in disjoint components of Γ, we find that $f(x, s)$ is continuous. Assuming that $s \to x$ by means of some parameterisation, one easily deduces that:

- If there is a local open segment $U_x \subset \Gamma$ containing x that is flat, then $\lim_{s\to x} f(x,s) = 0$ (which physically means that flat objects do not radiate to themselves). Note that the result is also valid when $x \in \partial U$ and $s \in U$ for some open and flat U.
- If the segment U_x has a radius of curvature $r < \infty$, the limit exists but is non-zero.

On one-dimensional geometries, the continuity of the kernel function f implies that the integral operator F is compact. In higher dimensions, the situation is more complicated. Then the smoothness of Γ plays an additional role.
If Γ is only piecewise smooth and has two adjacent parts that form a nontrivial angle, the integral operator F is not compact (cf. [1], Theorem 8.2.5 and Lemma 8.2.24, for similar results for the dipole operator K that is closely related to our F; see also [6]).

5 VARIATIONAL FORMULATION

We construct a finite-element discretisation (Ω_h, Γ_h) of the geometry (Ω, Γ) and associated proper sets of test functions v_j and w_j on Ω_h and Γ_h respectively, such that $\mathcal{V}_h = \text{Span}(v_j) \subset H_1(\Omega_h)$ and $\mathcal{W}_h = \text{Span}(w_j) \subset L_2(\Gamma_{h,n})$. Here $\Gamma_{h,n}$ denotes those parts of Γ_h where Φ_{Heat} is specified. For example, the v_j can be piecewise bilinear functions on isoparametrically defined quadrilaterals, while the w_j are piecewise constants on the boundary elements (sides of a quadrilateral). Let $\mathcal{V}_{\Gamma_{h,n}} \subset L_2(\Gamma_{h,n})$ denotes the functions in $\mathcal{V}_h$ restricted to $\Gamma_{h,n}$.
T is approximated by $T = T(x,y) \approx \sum_j T_j v_j(x,y) = T_h(x,y)$. Hence we can identify the approximation T_h of T with $\vec{T} = (\ldots, T_i, \ldots)^T$. In a similar way one can express Q_r, W_{out}, W_{in} and Φ_{Heat}, using the w_j in $L_2(\Gamma_{h,n})$, which results in vectors $\vec{Q}_r$, $\vec{W}_{out}$, $\vec{W}_{in}$ and $\vec{\Phi}_{\text{heat}}$.
In order to determine $\vec{\Phi}_{\text{heat}}$ in the case of radiation, the mappings Π and ¶ are introduced:

$$\Pi : \mathcal{V}_{\Gamma_{h,n}} \longrightarrow \mathcal{W}_h, \tag{18}$$
$$\P : \mathcal{W}_h \longrightarrow \mathcal{V}_{\Gamma_{h,n}}. \tag{19}$$

Using the same functions as trial and weighting functions, the differential equation (1) is approximated by a set of non-linear equations of which the j-th equation is given by

$$0 = -\int_{\Omega_h} \lambda \nabla T_h \cdot \nabla v_j d\Omega_h + \oint_{\Gamma_{h,n}} \P(\Phi_{\text{Heat}}(\Pi(T_h))) v_j d\Gamma_{h,n}. \tag{20}$$

If we further assume a temperature profile that varies only in the tangential direction of the inner objects, (20) can be simplified by neglecting variations in the directions perpendicular to the surface. We obtain

$$0 = -\oint_{\Gamma_{h,n}} \lambda_{\text{Eff}} \nabla T_h \cdot \nabla v_j d\Gamma_{h,n} + \oint_{\Gamma_{h,n}} \P(\Phi_{\text{Heat}}(\Pi(T_h))) v_j d\Gamma_{h,n}. \quad (21)$$

Here λ_{Eff} is the integration of λ over a cross-section of the inner materials. Due to the thinness of the components it will be a small parameter.
In the second integral Φ_{heat} is expressed by (2), in which e is assumed to be specified on $\Gamma_{h,n}$ and $\vec{Q}_r$ is determined by the temperature profile $\Pi(T_h)$ along Γ_h.

5.1 SOLUTION STRATEGY

The determination of geometrical viewfactors usually requires an orientation of the boundary. In practice, one should have a small volume whose sides are covered by 'viewing' elements. Equation (21) actually significantly reduces the number of unknowns by identifying a thin volume with a line element, but still allowing radiation transfer to all sides. The equation (21) is non-linear in T_h. The second integral (the source term) covers the non-linear terms. Hence the equations may be solved efficiently by an iterative method that starts with the solution of a zero flux problem, i.e. the solution of

$$\P(\Phi_{\text{Heat}}(\Pi(T_h))) = 0. \quad (22)$$

This is the approximation of (21) for small λ_{Eff}. It is interesting that this zero flux problem is linear in $\sigma \vec{T}_h^4$. Its solution requires a linear solver that can deal efficiently with large full matrices:

$$\mathcal{C} \ \sigma \vec{T}_h^4 = \vec{e}. \quad (23)$$

This matrix $\mathcal{C}$ will be called the radiation matrix. Its computation also involves the inversion of a full matrix.

5.2 L-MATRIX PROPERTY

The radiation matrix $\mathcal{C}$ posesses the L-matrix property. The result in Theorem 3 is based on the following theorem, which may be proved by applying arguments related to non-negative matrices and the concept of irreducibility [11].

THEOREM 2 *Let the matrix F and the diagonal matrix D have non-negative entries. In addition, assume that*

$$F_{ji} \neq 0 \iff F_{ij} \neq 0 \;\; \forall j,i, \;\; \textit{and} \;\; F_{ii} = 0 \;\; \forall i, \tag{24}$$
$$F\vec{1} = \vec{1}, \;\; \textit{where} \;\; \vec{1} = (1, \ldots, 1)^T, \tag{25}$$
$$\|D\|_\infty < 1. \tag{26}$$

Then $C = (I - FD)^{-1}(I - F)$ is a weakly diagonally dominant matrix, which also has the L-property

$$C_{ji} \leq 0 \;\text{ for }\; j \neq i \;\text{ and }\; C_{jj} > 0.$$

The same properties hold for $E = (I - F)(I - DF)^{-1}$.

In the radiation matrix C in (23), contributions from the applied temperatures can be eliminated. We reduced the size further to that of the final effective matrix. We identified the unknown temperatures in elements on both sides of the thin materials and added the net incoming energies for these elements. The process effectively resembles a 'kneading' of matrix C: within each row contributions from unknowns at two equivalenced nodes are added, while rows for the same resulting unknowns are also 'added' (taking into account the fact that we are dealing with densities rather than with energies).
In (8) it was noted that $F(1) = 1$. Hence, gauging is necessary to solve the flux equation (23). If the geometry is such that each boundary element on the inner objects has a non-trivial viewfactor to a boundary element on which the temperature is applied, the gauging is guaranteed. However, when the geometry is such that internally two parallel strips occur, the reduction to line elements may be necessary in order to be able to solve (23).

THEOREM 3 *In the case of gray Lambert surfaces, the radiation matrix C is a weakly dominant L-matrix only if the meshing along both sides of each thin object is (nearly) the same.*
If the gray Lambert assumption is violated, the diagonal dominancy may be lost. The L-matrix property, however, is maintained.

5.3 LINEAR FLUX EQUATION

The linear flux equation (23) was solved using direct as well as iterative methods. The direct methods are found in the NAG Fortran Library [3]: F04ATF (emphasis on accuracy) and F04ARF (emphasis on memory and cpu time). The iterative methods were: Jacobi iteration, Gauss-Seidel iteration (cf. [11,12]) and the Two-Grid Method (TGM) [10,2,7], that used Gauss-Seidel as a relaxation procedure. The coarse grid, needed by the TGM, was chosen such that it represented the corners of the geometry well, in order to eliminate the non-compactness property in the correction process of the Two-Grid Problem [1] (Section 9.3.1).

5.4 THE NON-LINEAR PROBLEM

Several non-linear methods were used to solve the complete non-linear system (21). We compared iterative non-linear solvers with respect to their ability to solve the problem efficiently. Again intermediate linear or non-linear processes have to deal with full coefficient matrices. In addition, we investigated whether the solution of the flux equation (23) provides a proper initial approximation for the non-linear solvers.
The solvers we applied are: Newton-Raphson (with linear solvers), Non-Linear Gauss-Seidel (including local Newton-Raphson for the pointwise non-linear equation) [4] and a Non-Linear Two-Grid Method (using the Non-Linear Gauss-Seidel as a non-linear smoother) [10].

6 PROBLEM

The simple configuration cross-section model of a CRT showed in Fig. 1 was used to study the heat radiation problem. Built with ANSYS, this model had 717 nodes. ANSYS determined the viewfactors between the several boundary elements. At the boundary of the cavity Dirichlet conditions were taken. This, together with the reduction process mentioned earlier, reduced the number of unknowns to 213. At the top of the outer boundary the temperature was set to 300 K, on the remaining part to 340 K. An additional internal heat source was assumed on the horizontal part of the 'mask' by heat flux generated by incoming electrons. We concentrated on the temperature profile along the mask. A typical result is shown in Fig. 2.
At both ends the effect of including the thermal conductivity is clearly demonstrated. Here the final solution intersects the solution of the flux equation (23). In the other parts the convergence behaviour of the Newton process was monotonic.
It is interesting to note that the solution of the flux equation already shows a qualitative behaviour that is preserved in the non-linear solution. The peaks at both ends are due to the nearby diaphragm that

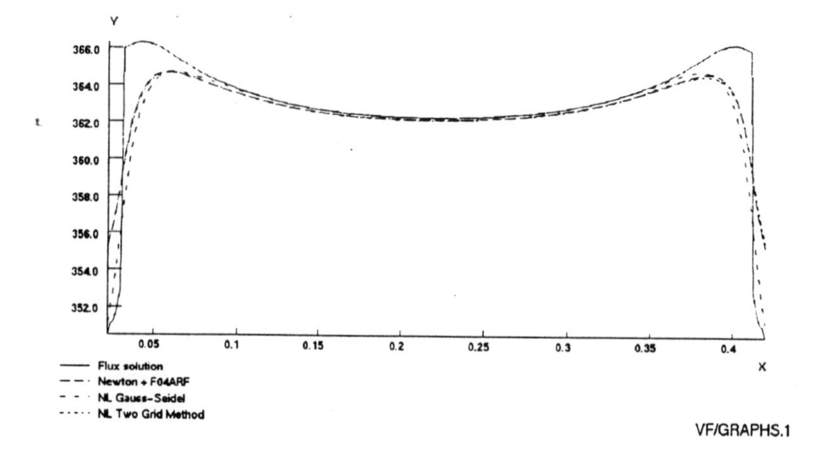

Fig. 2: The flux solution is compared to the solution of the non-linear equation. Note also the low-frequency error in the solution of the last iterand of the Non-Linear Gauss-Seidel. The results of Newton and of the Non-Linear Two-Grid Method are nearly identical.

partly prevents the mask from loosing heat by radiation. Here, essentially, the effect of the viewfactors is demonstrated.

7 CONCLUSIONS

The following conclusions can be drawn:

- The simulation of radiative heat transfer in television tubes requires the solution of a non-linear problem. The size of the discretised problem can be effectively reduced by making use of the thinness of the objects. In addition, the condition of the numerical equations is improved.
- The iterative non-linear solvers need efficient and accurate linear solvers for full matrices.
- The strategy to solve the steady-state non-linear problem is based upon first solving the flux problem (23), which is a linear problem in T^4 with a full coefficient matrix. This solution is a good initial solution for solving the non-linear problem.
- With regard to the linear solvers for the flux equation (23):

- Comparing the cpu time needed for all methods the iterative methods are favoured over the two direct methods.
- For small problems Gauss-Seidel may be preferred.
- For larger problems the Two-Grid Method becomes attractive. Of the four variants no best one can be singled out. An optimal TGM is still to be determined and depends on the stopping criterion for the determination of the coarse grid correction of the TGM.
  The results indicate that a Multigrid method is probably an efficient alternative.

- Newton-Raphson and the Non-Linear Two-Grid Method are the most attractive non-linear solvers. The first is simpler to implement and shows quadratic convergence, the second is more efficient in cpu time, in particular for larger problems. Both methods require for optimal performance the accurate solution of intermediate problems: for Newton-Raphson on the fine grid (when using the Two-Grid Method as the solver: on the coarse grid) and for the Non-Linear Two-Grid Method on the coarse grid. Hence we used a direct linear solver on such a grid. This indicates again that for large problems full Multigrid methods can be useful (linear as well as non-linear).
- The meshing of the problem is governed by requirements of compactness in addition to those imposed by discretisation. This implies that a fine meshing should be governed by accuracy requirements, whereas the coarser grids should still be able to represent the corners in the geometry well.

8 ACKNOWLEDGEMENT

I would like to thank Prof.Dr. P.W. Hemker (CWI-Amsterdam), Dr. A. Reusken (TU Eindhoven) and Dr. E. Egberts (Philips Research Laboratories) for stimulating discussions.

REFERENCES

[1] W. HACKBUSCH – *Integralgleichungen, Theorie und Numerik*, LAMM 68, B.G. Teubner, Stuttgart, 1989.

[2] P.W. HEMKER, H. SCHIPPERS – Multiple grid methods for the solution of Fredholm integral equations of the second kind, Maths. of Comp. 36/153, pp. 215-232,1981.

[3] *The NAG fortran library manual - Mark 15*, Vol. 6, NAG Ltd, Oxford (UK), 1991.

[4] J.M. ORTEGA, W.C. RHEINBOLDT – *Iterative solution of nonlinear equations in several variables*, Academic Press, New York, 1970.

[5] *P/Viewfactor: User Manual*, Release 2.4, PDA Engineering, Costa Mesa (CA, USA), 1989.

[6] H. SCHIPPERS – On the regularity of the principal value of the double-layer potential, J. of Engng. Maths. 16/1, pp. 59-76, 1982.

[7] H.SCHIPPERS – Multigrid methods for boundary integral equations, Numer. Math. 46, pp. 351-363, 1985.

[8] R. SIEGEL, J.R. HOWELL – *Thermal radiation heat transfer*, 3rd Edition, Hemisphere Publ. Corp., 1992.

[9] R.W. SMITH, D.D. KETELAAR – *Radiation matrix generation utility*, ANSYS Rev. 4.3 Tutorial, Swanson Analysis Systems Inc., Houston (PA, USA), 30-03-1987.

[10] K. STÜBEN, U. TROTTENBERG – Multigrid methods: Fundamental algorithms, model problem analysis and applications, *Multigrid methods, proc. conf. Köln-Porz, nov. 23-27, 1981*, Ed. W. Hackbusch, U. Trottenberg, Lect. Notes in Maths. 960, Springer Verlag, Berlin, pp. 1-176, 1982.

[11] R.S. VARGA – *Matrix iterative analysis*, Prentice-Hall Int., London, 1962.

[12] D.M. YOUNG – *Iterative solution of large linear systems*, Academic Press, New York, 1971.